Contemporary Mathematicians

Gian-Carlo Rota
Editor

Hassler Whitney in the 1970s
(Photograph by Sally Whitney)

Hassler Whitney at age 14 in the Swiss Alps

Hassler Whitney
Collected Papers
Volume I

James Eells
Domingo Toledo
Editors

Birkhäuser
Boston · Basel · Berlin
1992

James Eells
Department of Mathematics
University of Warwick
Coventry CV 4 7AL
England

Domingo Toledo
Department of Mathematics
University of Utah
Salt Lake City, Utah 84112
U.S.A.

Library of Congress Cataloging-in-Publication Data

Whitney, Hassler.
The collected papers of Hassler Whitney / edited by
James Eells, Domingo Toledo
p. cm. – (Contemporary mathematicians)
Includes bibliographical references.
ISBN-13: 978-1-4612-7740-8 e-ISBN-13: 978-1-4612-2972-8
DOI: 10.1007/978-1-4612-2972-8
1. Topology. 2. Geometry, Differential. 3. Combinatorial
analysis. I. Eells, James, 1926- II. Toledo, Domingo.
III. Title. IV. Series.
QA611.W4854 1992
514–dc20 91-9885
CIP

Printed on acid-free paper.

Printed and bound by Quinn Woodbine, Woodbine, NJ

9 8 7 6 5 4 3 2 1

The Collected Papers of Hassler Whitney

Contents — Volume 1

(Bracketed numbers are from the Bibliography)

Chapter 2
Differentiable Functions and Singularities 227

Chapter 3
Analytic Spaces 455

Permissions

The Collected Papers of Hassler Whitney

Contents — Volume 2

(Bracketed numbers are from the Bibliography)

Chapter 3
Topology and Algebraic Topology 227

Chapter 4
Geometric Integration Theory 431

Chapter 5
Other Subjects 489

Preface

We present here the mathematical papers of Hassler Whitney. This collection contains all the published papers, with the exception of some short announcements that Whitney did not wish to be included. We also include the introduction to his book *Geometric Integration Theory*, and one previously unpublished manuscript on the four-color problem. The papers are presented under some broad categories: graphs and combinatorics, differentiable functions and singularities, analytic spaces, manifolds, bundles and characteristic classes, topology and algebraic topology, geometric integration theory.

Whitney intended to write an introduction to this collection. Unfortunately he left us no manuscript at the time of his death, May 10, 1989. We had discussed the possibility of using his paper "Moscow 1935 – Topology moving toward America," written for the Centennial of the American Mathematical Society, as part of his introduction to this collection, an idea which he much liked. We therefore include this paper, which contains personal information as well as mathematical reflections, as Whitney's own introduction to these volumes.

Whitney's mathematical style, like his personal style, was that of an explorer and pioneer. One of the pictures included in these volumes shows him as a mountain climber. In mathematics, he preferred to work on undeveloped areas: break new ground and build foundations. During the last twenty years of his life he concentrated his efforts on developing an educational system that builds on the natural tendency in children to be explorers.

We find that his papers are as fresh today as when they were written, and feel that any comments from us on their contents could only detract from their clarity. The influence of his work is too vast to review meaningfully in a few pages, event if we felt competent to do so.

We do find it appropriate to comment on his unpublished manuscript (dating from around 1940) of a book that contained the details of all the results announced in "On the theory of sphere bundles," [42]. It contains a lengthy proof of his duality theorem (the formula for the Stiefel–Whitney classes of the direct sum of two vector bundles) obtained by developing an obstruction theory for direct sums, and computations with cochains. Whitney has referred to that proof as the hardest thing he ever did. In an unguarded moment he at once remarked that he felt that the duality theorem was his most original contribution to mathematics.

J. Eells
D. Toledo

Academic Appointments and Awards

Born March 23, 1907, in New York City
Yale University, Ph. B., 1928; Mus. B., 1929; honorary Sc. D., 1947
Harvard University, Ph. D. 1932
National Research Council Fellow in Mathematics, 1931–33
Instructor to Professor of Mathematics, Harvard University, 1933–52
Professor of Mathematics, Institute for Advanced Study, 1952–77;
Professor Emeritus, 1977–89

Member: American Mathematical Society, Mathematical Association of America, National Council of Teachers of Mathematics, National Academy of Sciences, American Philosophical Society, Swiss Mathematical Society (honorary)

American Mathematical Society:

Colloquium Lecturer, 1946
Vice President, 1948–50
Editor, American Journal of Mathematics, 1944–49
Editor, Mathematical Reviews, 1949–54
Committee on Visiting Lectureship (chairman), 1946–51
Committee for Summer Institutes (chairman), 1953–54

National Science Foundation: Chairman of Panel on Mathematics, 1953–56
Researcher for Applied Mathematics Panel of NDRC of OSRD, 1943–45
Exchange Professor to France from Harvard, 1951–52
Exchange Professor to College de France (Fulbright), 1957
Committee on Support of Research in the Mathematical Sciences of the
National Research Council, 1966–67
L. R. Ford Award for paper, "The Mathematics of Physical Quantities," 1969
National Medal of Science, 1967
Wolf Prize in Mathematics (shared) for 1982
Steele Prize, American Mathematical Society, 1985
Consultant to School Mathematics Study Group, Cambridge Conference in School Mathematics, Education Development

Center, and other groups International Commission on Mathematical Instruction, President, 1979–82

Bibliography of Hassler Whitney

[1] The coloring of graphs, NAS Proc. v. 17, 1931, 122–125 .

[2] Non-separable and planar graphs, NAS Proc. v. 17, 1931, 125–127 .

[3] A theorem on graphs, Annals of Math. (2) v. 32, 1931, 378–390 .

[4] Note on Perron's solution of the Dirichlet problem, NAS Proc., v. 18, 1932, 68–70.

[5] Non-separable and planar graphs, AMS Transac. v. 34, 1932, 339–362.

[6] Congruent graphs and the connectivity of graphs, Am. Jour. Math. v. 54, 1932, 150–168.

[7] Regular families of curves I, NAS Proc., v. 18, 1932, 275–278.

[8] Regular families of curves II, NAS Proc., v. 18, 1932, 340–342.

[9] A logical expansion in mathematics, AMS Bull., v. 38, 1932, 572–579.

[10] The coloring of graphs, Annals of Math., (2) v. 33, 1932, 688–718.

[11] A characterization of the closed 2-cell, AMS Transac., v. 35, 1933, 261–273.

[12] A set of topological invariants for graphs, Am. Jour. Math., v. 55, 1933, 231–235.

[13] On the classification of graphs, Am. Jour. Math., v. 55, 1933, 236–244.

[14] 2-Isomorphic graphs, Am. Jour. Math., v. 55, 1933, 245–254.

[15] Regular families of curves, Annals of Math. (2) v. 34, 1933, 244–270.

[16] Characteristic functions and the algebra of logic, Annals of Math. (2) v. 34, 1933, 405–414.

[17] Planar graphs, Fundamenta Math., v. 21, 1933, 73–84.

[18] Analytic extensions of differentiable functions defined in closed sets, AMS Transac., v. 36, 1934, 63–89.

[19] Derivatives, difference quotients and Taylor's formula, AMS Bull., v. 40, 1934, 89–94.

[20] Differentiable functions defined in closed sets I, AMS Transac., v. 36, 1934, 369–387.

[21] Derivatives, difference quotients and Taylor's formula II, Annals of Math. (2) v. 35, 1934, 476–481.

[22] Functions differentiable on the boundaries of regions, Annals of Math. (2) v. 35, 1934, 482–485.

[23] On the abstract properties of linear dependence, Am. Jour. Math., v. 57, 1935, 509–533.

[24] Differentiable manifolds in Euclidean space, NAS Proc., v. 21, 1935, 462–464, reprinted in Receuil Mathematique. T. 1(43) N. 5 (1936) 783–786.

[25] Sphere-spaces, NAS Proc., v. 21, 1935, 464–468, reprinted in Receuil Mathematique. T. 1(43) N. 5 (1936) 787–791.

[26] A function not constant on a connected set of critical points, Duke Math. J., v. 1, 1935, 514–517.

[27] Differentiable functions defined in arbitrary subsets of Euclidean space, AMS Transac., v. 40, 1936, 309–317.

[28] Differentiable manifolds, Annals of Math. (2) v. 37, 1936, 645–680.

[29] The imbedding of manifolds in families of analytic manifolds, Annals of Math. (2) v. 37, 1936, 865–878.

[30] On regular closed curves in the plane, Compositio Math. 4, 1937, 276–284.

[31] On matrices of integers and combinatorial topology, Duke Math. J., 3, 1937, 35–45.

[32] On the maps of an n-sphere into another n-sphere, Duke Math. J., 3, 1937, 46–50.

[33] The maps of an n-complex into an n-sphere, Duke Math. J., 3, 1937, 51–55.

[34] On products in a complex, NAS Proc., v. 23, 1937, 285–291.

[35] Analytic coordinate systems and arcs in a manifold, Annals of Math. (2) 38, 1937, 809–818.
[36] Topological properties of differentiable manifolds, AMS Bull. 43, 1937, 785–805.
[37] A numerical equivalent of the four color problem, Monatshefte fur Math. und Phys. 3, 1937, 207–213.
[38] Cross sections of curves in 3-space, Duke Math. J., 4, 1938, 222–226.
[39] On products in a complex, Annals of Math. (2) 39, 1938, 397–432.
[40] Tensor products of abelian groups, Duke Math. J., 4, 1938, 495–528.
[41] Some combinatorial properties of complexes, NAS Proc., 26, 43–148.
[42] On the theory of sphere-bundles, NAS Proc., 26, 1940, 148–153.
[43] On regular families of curves, AMS Bull., 47, 1941, 145–147.
[44] On the topology of differentiable manifolds, Lectures in Topology, U. of Michigan Press, 1941, 101–141.
[45] Differentiability of the remainder term in Taylor's formula, Duke Math. J., 10, 1943, 153–158.
[46] Differentiable even functions, Duke Math. J., 10, 1943, 159–160.
[47] The general type of singularity of a set of $2n-1$ smooth functions of n-variables, Duke Math. J., 10, 1943, 161–172.
[48] Topics in the theory of Abelian groups, I. Divisibility of Homomorphisms, AMS Bull., 50, 1944, 129–134.
[49] On the extension of differentiable functions, AMS Bull., 50, 1944, 76–81.
[50] The self-intersections of a smooth n-manifold in $2n$-space, Annals of Math. (2), 45, 1944, 220–246.
[51] The singularities of a smooth n-manifold in ($2n$-1)-space, Annals of Math. (2), 45, 1944, 247–293.
[52] Algebraic topology and integration theory, NAS Proc., v. 33, 1947, 1–6.
[53] Geometric methods in cohomology theory, NAS Proc., v. 33, 1947, 7–9.
[54] Complexes of manifolds, NAS Proc., v. 33, 1947, 10–11.
[55] On ideals of differentiable functions, Am. Jóur. Math. 70, 1948, 635–658.
[56] Relations between the second and third homotopy groups of a simply-connected space, Annals of Math. (2) 50, 1949, 180–202.
[57] Classification of the mappings of a 3-complex into simply-connected spaces, Annals of Math. (2) 50, 1949, 270–284.
[58] An extension theorem for mappings into simply-connected spaces, Annals of Math. (2) 50, 1949, 285–296.
[59] La topologie algebrique et la theorie de l'integration, Colloques Internationaux du CNRS XII, Topologie Algebrique, 1947, 107–113, published by CNRS, Paris, 1949.
[60] (With L. H. Loomis) An inequality related to the isoperimetric inequality, AMS Bull., 55, 1949, 961–962.
[61] On totally differentiable and smooth functions, Pacific J. Math. 1, 1951, 143–159.
[62] r-dimensional integration in n-space, Proc. Int. Cong. Math., 1950, vol. 1, 245–256, Amer. Math. Soc. 1952.
[63] On singularities of mappings of Euclidean spaces, I. Mappings of the plabe into the plane, Annals of Math. (2) 62, 1955, 374–410.
[64] On functions with bounded n-th differences, J. de Maths. Pres et Appl. 36, 1957, 67–95.
[65] Geometric Integration Theory, Princeton University Press, 1957, Princeton, NJ, xv + 397 pages. (Princeton Math. Series 21) Translated into Russian [book].
[66] Elementary structure of real algebraic varieties, Annals of Math. (2) 66, 1957, 545–556.

[67] Singualrities of mappings of Euclidean spaces, Symposium Internacional de Topologia Algebraica, Mexico, 1956, 285–301, Mexico, La Universidad Nacional Autonoma.
[68] (With F. Bruhat) Quelques proprietes fondamentales des ensembles analytiques-reels, Comm. Math. Helv. 33, 1959, 132–160.
[69] (With A. Dold) Classification of oriented sphere bundles over a 4-complex, Annals of Math. (2) 69, 1959, 667–677.
[70] On bounded functions with bounded n-th differences, AMS Proc. 10, 1959, 480–481.
[71] (With A. M. Gleason) The extension of linear functionals defined on H-infinity, Pacific J. Math. 12, 1962, 163–182.
[72] The work of John W. Milnor, Proceedings ICM 1962m Institut Mittag–Leffler Djursholm, Sweden, xlviii-1.
[73] Local properties of analytic varieties, in: differential and combinatorial topology (Symposium in Honor of Marston Morse), Princeton, NJ., Princeton University Press.
[74] Tangents to an analytic variety, Annals of Math (2) 81, 1965, 496–549.
[75] The mathematics of physical quantities. Part I, Mathematical Models for measurement, Am. Math. Monthly 75 (1968), 115–138, Part II, Quantity structures and dimensional analysis, ibid., 237–256.
[76] Logic Fad or tool? Nico 4, 1969, Revue per du Centre Belge de Pedagogie de la Mathematique, 2–14.
[77] On reducibility in the four color problem, unpublished manuscript, 1971.
[78] (With W. T. Tutte) Kempe chains and the four color problem, Utilitas Mathematica 2(1972), 241–281.
[79] Complex Analytic Varieties, Addison–Wesley Pub. Co., Reading, MA 1972, xii +399 pp. [book].
[80] Math Activities, multilithed, Institute for Adv. Study, 1974 [book].
[81] Comment on the division of the plane by lines, Am. Math. Monthly 86(1979), p. 700.
[82] Moscow 1935: Topology moving toward America, in *A Century of Mathematics in America*, Part I, P. L. Duren Ed., 96–117, Amer. Math.

Hassler Whitney was a student of G. D. Birkhoff at Harvard University, where he earned a Ph.D. in 1932. After two years as a National Research Fellow (1931–1933), he served on the Harvard faculty until 1952. He has since been at the Institute for Advanced Study, where he is now Professor Emeritus of Mathematics. A recipient of the National Medal of Science, the Wolf Prize, and the AMS Steele Prize, Professor Whitney is also a member of the National Academy of Sciences. His research has been primarily in the areas of topology and analysis.

Moscow 1935: Topology Moving Toward America

HASSLER WHITNEY

The International Conference in Topology in Moscow, September 4–10, 1935, was notable in several ways. To start, it was the first truly international conference in a specialized part of mathematics, on a broad scale. Next, there were three major breakthroughs toward future methods in topology of great import for the future of the subject. And, more striking yet, in each of these the first presenter turned out not to be alone: At least one other had been working up the same material.

At that time, volume I of P. Alexandroff / H. Hopf, *Topologie*, was about to appear. I refer to this volume as A-H. Its introduction gives a broad view of algebraic topology as then known; and the book itself, a careful treatment of its ramifications in its 636 pages. (It was my bible for some time.) Yet the conference was so explosive in character that the authors soon realized that their volume was already badly out of date; and with the impossibility of doing a very great revision, the last two volumes were abandoned. Yet a paper of Hopf still to come (1942) led to a new explosion, with a great expansion of domains, carried on especially in America.

It is my purpose here to give a general description of the subject from early beginnings to the 1940s, choosing only those basic parts that would lead to later more complete theories, directly in the algebraic treatment of the subject. We can then take a look at some directions of development since the conference, in very brief form, with one or two references for those who wish a direct continuation.

Top Row: 1. E. Čech; 2. H. Whitney; 3. K. Zarankiewicz; 4. A. Tucker; 5. S. Lefschetz; 6. H. Freudenthal; 7. F. Frankl; 8. J. Nielsen; 9. K. Borsuk; 10 ?; 11. J. D. Tamarkin; 12. ?; 13. V. V. Stepanoff; 14. E. R. van Kampen; 15. A. Tychonoff; Bottom Row: 16. C. Kuratowski; 17. J. Schauder; 18. St. Cohn-Vossen; 19. P. Heegaard; 20. J. Różańska; 21. J. W. Alexander; 22. H. Hopf; 23. P. Alexandroff; 24. ?.

I also do not hesitate to draw a few conclusions on our difficulties with new research, with some comments on how research might be improved.

What were early beginnings of "analysis situs"? Certainly a prime example is Euler's discovery and proof that for a polyhedron, topologically a ball, if α_0, α_1 and α_2 denote the numbers of "vertices," "edges" and "faces," then

$$\alpha_0 - \alpha_1 + \alpha_2 = 2. \tag{1}$$

How might one find something like this? Who might think of trying it out? These are questions looking directly for *answers*, rather than at *situations to explore*. For the latter, one might build up a picture:

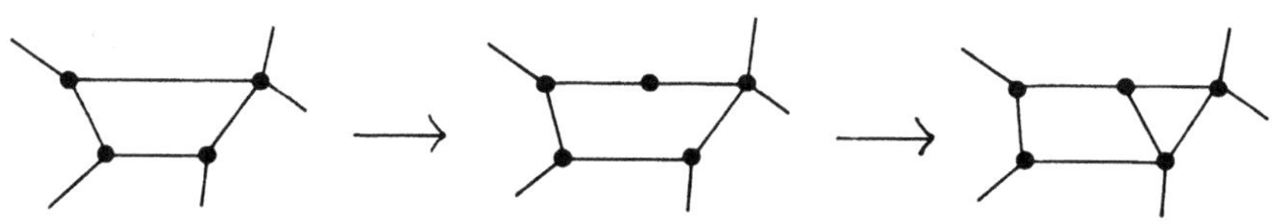

The first step here is to add a vertex, cutting one edge into two: this leaves $\alpha_1 - \alpha_0$ unchanged. The second step is to add an edge joining two vertices; this leaves $\alpha_2 - \alpha_1$ unchanged. Now it needs some playing to see that (1) contains both these facts. We might now say that we *essentially* know the formula (1); just the 2 is missing. That expression, generalizing to $\alpha_0 - \alpha_1 + \alpha_2 - \alpha_3$, etc., is known as the *Euler characteristic*. (Also Descartes discovered it much earlier; see A-H, p. 1.)

Can you be taught how to think? If you are in a particular subject, there may be tricks of the trade for that subject; Polya shows this for some standard parts of mathematical thinking. But trying to learn to carry out research by studying Polya is unlikely to get you far. It is the *situation* you are in which can lead to insights, and any *particular* thinking ways are quite unlikely to apply to different sorts of situations. "Sharpening your wits" on peculiar questions may keep your mind flexible so that new situations can let you think in new directions. Thus Lakatos, "Proofs and refutations" can give you *ideas, samples*, of thoughts; the usefulness is less in *learning* that in *keeping your mind flexible.*

A popular pastime in Königsberg, Germany, was to try to walk over each of its seven bridges once and only once. Euler showed how to organize the situation better and check on the possibility. Can we find a way to *get naturally at this*?

If we started in the island, say at A, and crossed the upper left bridge, why not sit down at C and think it over instead of wandering around aimlessly?

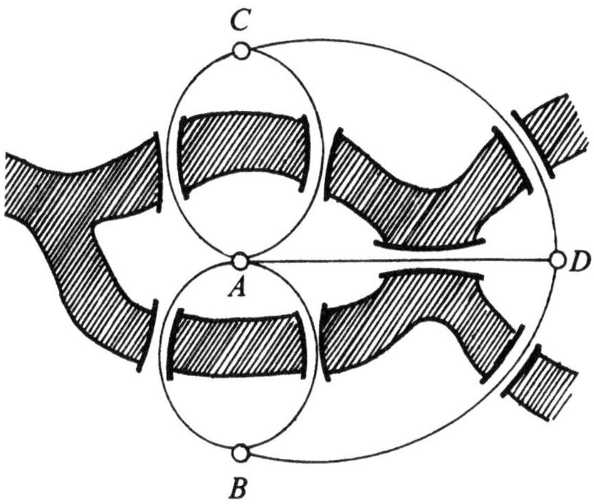

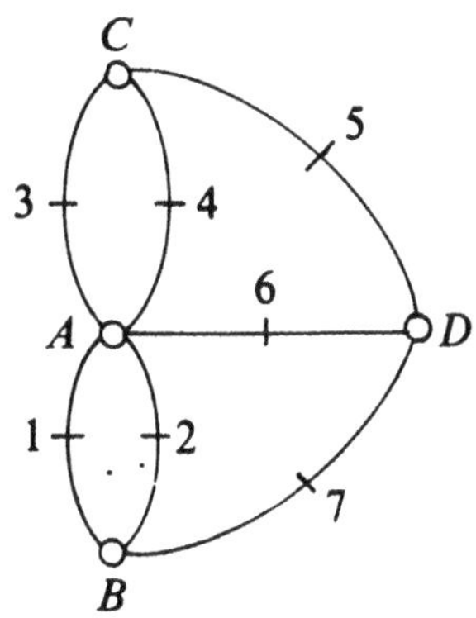

If we *can* find the desired path, we can certainly simplify it by using just the paths shown; and putting a gate in each bridge, we can check on which ones we have crossed. Thus we crossed gate 3, and must next cross either 4 or 5. But then we must cross the other of these gates later, and find ourselves back at C with no way to reach any uncrossed bridges. This is enough for us to start organizing. The final result, applying to any such situation, was given by Euler.

A most famous question is of course the four color problem: Can one color any map on the plane or globe in at most four colors so any adjoining regions are of different colors? A first "proof" was given by Kempe, in 1869, who introduced the important tool of "Kempe chains." The mistake was discovered by Heawood in 1890. A major step in advance was given by G. D. Birkhoff, in a paper in 1913 on "The reducibility of maps." In the early 1930s, when I was at Harvard, exploring the problem among other things, Birkhoff told me that every great mathematician had studied the problem, and thought at some time that he had proved the theorem (I took it that Birkhoff included himself here). In this period I was often asked when I thought the problem would be solved. My normal response became "not in the next half century." The proof by computer (W. Haken and K. Appel) began appearing in non-final form about 1977.

A very important advance in mathematics took place in the mid nineteenth century, with the appearance of Riemann's thesis. Here he made an investigation of "Riemann surfaces," along with basic analytic considerations, in particular, moving from one "sheet" to another by going around branch points. This led to the general question of what a "surface" was, topologically, and the problem of classification. This culminated (Möbius, Jordan, Schäfli, Dyck 1888) in the characterization of closed surfaces (without boundary); they are determined through their being orientable or not, and through their Euler characteristic. Let me note that H. Weyl's book *Die Idee der Riemannsche Fläche*, Leipzig 1913, clarified many basic notions such as neighborhood, manifold, fundamental group, and covering space.

A notable discovery was made by Gauss (who had made deep investigations in differential geometry, with special studies of the earth's surface). This was the expression as a double integral for the "looping coefficient" of two non-intersecting oriented curves C_1 and C_2 in 3-space R^3. Consider all pairs of points P in C_1 and Q in C_2, and

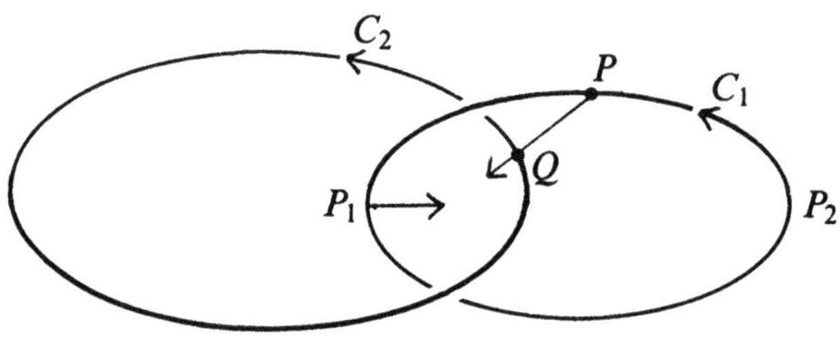

the unit vector from P toward Q:

$$v(P, Q) = \frac{Q - P}{|Q - P|}.$$

With P and Q as in the figure, if we let P' run over a short arc A in C_1 about P and let Q' run similarly over B in C_2, $v(P', Q')$ will clearly run over a little square-like part of the unit 2-sphere S^2 of directions in 3-space R^3.

The whole mapping is a little complex, since we are mapping $C_1 \times C_2$, which is a torus, into S^2. But we can see that it covers S^2 an algebraic number 1 of times, as follows. For each P, look at the image of $v(P, C_2)$. From the figure we see that it is circle-like, down and to the left to start. When P is taken down to P_1, the above circle has moved to the right and up, now going directly around P_1. Continuing down and along C_1 from P_1 to

P_2, $v(P, C_2)$ moves upward, to the left and down again. Thus that part of S^2 directly to the right from P_1 (see the arrow at that point) is swept over just once in the total sweep. We now use general theory (see A-H for instance) that says that S^2 must be covered some integral number of times, hence once (algebraically).

Gauss gave a numerical form to the double integral, in the general case of non-intersecting curves (see A-H, p. 497). If the result is not zero, the looping coefficient is not zero, and being invariant under deformations, one curve cannot be separated to a distance from the other without cutting through the other.

Kronecker considered the common zeros of a set of functions $f_1, \ldots, f_k$. Equivalently, consider the vector field $v(p) = (f_1(p), \ldots, f_k(p))$, and its zeros. This leads to the "Kronecker characteristic," generalizing the Gauss integral to higher dimensions. See A-H for some details.

All this work was growing and expanding at the end of the last century. But I call this the end of the early period, since Poincaré's studies, from 1895 on, gave a better general organization and important new directions of progress. The essentials of the early period were described in the article by Dehn and Heegaard, *Enzyklopädie der Mathematischen Wissenschaften*, III A B 3, 1907, and a very nice exposition of the analytical aspects was given by Hadamard in an appendix to Tannery, *Introduction à la théorie des fonctions d'une variable*, 2nd ed., 1910.

Turning now to the middle period, Poincaré set out to make a deep study of n-dimensional manifolds (locally like a part of n-space); these were basic in his work on dynamical systems. He cut them into "n-cells," each bounded by $(n-1)$-cells; and each of the latter is a face of two n-cells. Each $(n-1)$-cell is bounded by $(n-2)$-cells, and so on. Moreover "r-chains," written $\sum a_i \sigma_i^r$, associating an integer a_i with each r-cell σ_i^r, were defined. Now using ∂ for boundary, each boundary $\partial\sigma_i^r$ can be seen as an $(r-1)$-chain, and for a general r-chain A^r as above, $\partial A^r = \sum a_i \partial\sigma_i^r$.

For any σ_i^r, with a given orientation, an orientation of each of its boundary cells σ_j^{r-1} is induced, and $\partial\sigma_i^r$ is the sum of these with the induced orientations (see below). And since each σ_k^{r-2} in the boundary of σ_i^r is a face of just two σ_j^{r-1}, with opposite orientations induced, we have $\partial\partial\sigma_i^r = 0$, and hence $\partial\partial A^r = 0$ for all r-chains A^r.

A special case is the "simplicial complex," composed of "simplexes." In n-space R^n, an r-simplex is the convex hull $p_0 p_1 \cdots p_r$ of a set of points $p_0, \ldots, p_r$ lying in no $(r-1)$-plane. In barycentric coordinates, the points of $p_0 \cdots p_r$ are given by $\sum a_i p_i$, each $a_i \geq 0$, $\sum a_i = 1$. (This point is the center of mass of a set of masses, in amount a_i at each p_i.)

Any R^r can be oriented in two ways. Choosing an ordered set of r independent vectors $v_1, \ldots, v_r$ determines an orientation. A continuous change to another set $v_1', \ldots, v_r'$ gives the same orientation if independence was maintained.

A simplex $\sigma^r = p_0 \cdots p_r$ has a natural orientation, given by the ordered set $p_1 - p_0, \ldots, p_r - p_0$ of vectors. Note that the ordered set $p_1 - p_0, p_2 - p_1, \ldots, p_r - p_{r-1}$ is equivalent. The induced orientation of the face $p_1 \cdots p_r = \sigma^{r-1}$ of σ^r is defined by choosing $v_1, \ldots, v_r$ to orient σ^r, with $v_2, \ldots, v_r$ in σ^{r-1} (orienting it) and v_1 pointing from $p_0 \cdots p_r$ out of σ^{r-1}, as used just above. This holds true for the second set of vectors chosen above for $p_0 \cdots p_r$; and this shows also that $p_1 \cdots p_r$ has that orientation. In this way we may find the full expression for $\partial (p_0 \cdots p_r)$.

Some instances of this relation are

$$\partial (p_0 p_1) = p_1 - p_0, \qquad \partial (p_0 p_1 p_2) = p_1 p_2 - p_0 p_2 + p_0 p_1.$$

Later we will note that Kolmogoroff and Alexander might have found the correct products in cohomology by 1935 if they had kept such relations in mind, along with the relationship with differential geometry (typified by de Rham's theorem).

In accordance with the influence of Emmy Noether in Göttingen in the mid twenties, we shift now to group concepts to simplify the work. If an r-chain A^r has no boundary, $\partial A^r = 0$, we call it a *cycle*. Under addition, the cycles form a group Z^r. Similarly we have the group of r-boundaries, B^r, which is a subgroup of Z^r since $\partial\partial A^{r+1} = 0$ always. The factor, or difference, group, $H^r = Z^r \bmod B^r$, is the rth *homology group* of the complex. Any finite part of H^r (its elements of finite order) is the "torsion" T^r.

For an example of the above ideas we look at the real projective plane P^2. It can be described topologically as a closed disk σ^2, with opposite points

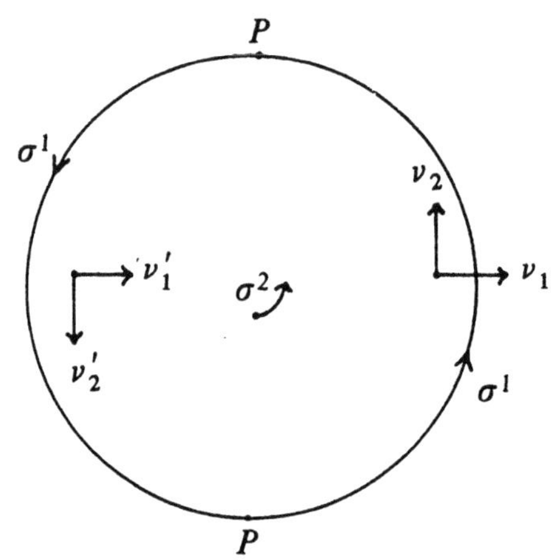

6

on the boundary ("points at infinity") identified. The simplest possible cutting into cells is shown. The boundary relations are

$$\partial\sigma^1 = p - p = 0, \qquad \partial\sigma^2 = 2\sigma^1.$$

Thus the one-dimensional homology group has a single non-zero element (i.e., $H^1 = T^1$), with σ^1 as the representative cycle. If we carry a pair of independent vectors (v_1, v_2) from σ^2 across σ^1, leading back into σ^2 on the other side, we see that the pair has shifted orientation: P^2 is non-orientable.

When a manifold is cut into cells of a reasonably simple nature, we may form the "dual subdivision" as follows. Put a new vertex in each n-cell. Let a new 1-cell cross each $(n-1)$-cell, joining two new vertices. Let a new (piecewise linear) 2-cell cross each $(n-2)$-cell, finding a boundary waiting for it, and so on. The figure shows a portion of the construction for $n = 2$.

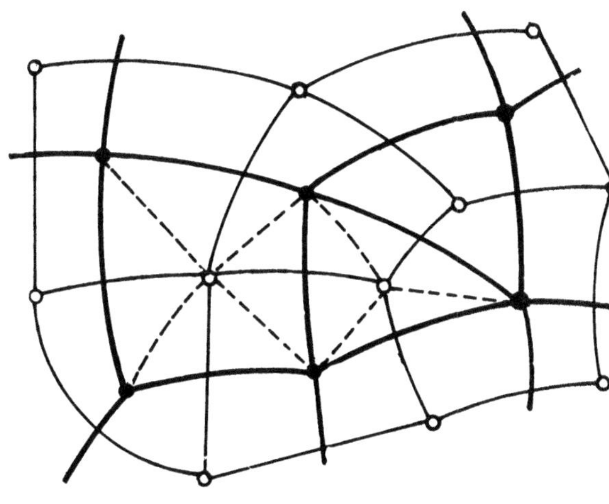

There is a one-one correspondence between r-cells of the original complex K (shown in heavy lines) and $(n-r)$-cells of the dual K_D (shown in lighter lines), and incidence between cells of neighboring dimensions is preserved. If the manifold is orientable, this shows that homology in K is the same as cohomology in K_D (except in extreme dimensions). From this we can see that the Betti numbers (ranks of the H^r) coincide in dimensions r and $n-r$, and the torsion numbers in dimensions r and $n-r-1$. This is the "Poincaré duality" in a complex formed from an orientable manifold.

Note also (see the dashed lines in the figure) that K and K_D have a common simplicial subdivision, the "barycentric subdivision" K^* of K. Also invariance of the homology groups under subdivisions is not hard to show; Alexander proved topological invariance of the ranks of the H^r in 1915. If we examine a cycle A^r of K and a cycle B^s of K_D, with $r+s \geq n$, the intersection is seen to be expressible as a cycle C^{r+s-n} of K^*. This is a generalization of the intersection of submanifolds of M^n, of great importance in algebraic geometry for instance (Lefschetz, Hodge). It is quite clear (that is, until 1935) that there is nothing of this sort in general complexes.

Poincaré applied considerations like these to his work in dynamical theory (for instance, the three body problem). But he could not prove a simply stated fact needed about area preserving transformations of a ring shaped surface. However, G. D. Birkhoff succeeded in proving this theorem in 1913.

The fundamental group and covering spaces were also studied in detail by Poincaré. In a space K, with a chosen point P, a curve C starting and ending at P defines an element of the fundamental group; any deformation of C, keeping the ends at P, defines the same element. One such curve followed by another gives the product of the two elements. The identity is defined by any curve which can be "shrunk to a point" (hence to P). The fundamental group is in general noncommutative. A space with vanishing fundamental group is called "simply connected."

Great efforts were expended by Poincaré to understand 3-dimensional manifolds. In particular, he conjectured that the 3-sphere was the only simply connected 3-manifold. This is as yet unproved.

Alexander proved an entirely new kind of "duality theorem" in 1922: Given a complex K imbedded homeomorphically in an n-sphere S^n, there is a strict relation between the homology groups of K and of $S^n - K$.

Alexander also gave in 1924 a remarkable example (using ideas of Antoine) of a simply connected surface S^* (homeomorphic image of S^2) in S^3, cutting S^3 into two regions, one of which is not simply connected. We begin with the surface of a cylinder, stretched and bent around to have its two ends facing each other; the figure shows these facing ends, the gap G between them partly filled.

We pull out, from each side of the gap, a piece, pulled into a cylindrical piece with a gap (like the original cylinder), these two pieces looped together, as shown; there are now two much smaller gaps, G_1 and G_2. We next act in the same manner with each of these gaps, giving G_{11} and G_{12} in G_1 and G_{21} and G_{22} in G_2, and continue. The limiting surface S^* has the stated properties. In fact, a loop going around each gap $G_{k_1 \cdots k_s}$ gives an infinite set of independent generators in the fundamental group of the outside of S^* in S^3, as we see easily. (The inside of S^* is simply connected.)

Going back to the early 1910s, Lebesgue discovered (1911) that a region of R^n, if cut into sufficiently small closed pieces, must contain at least $n+1$

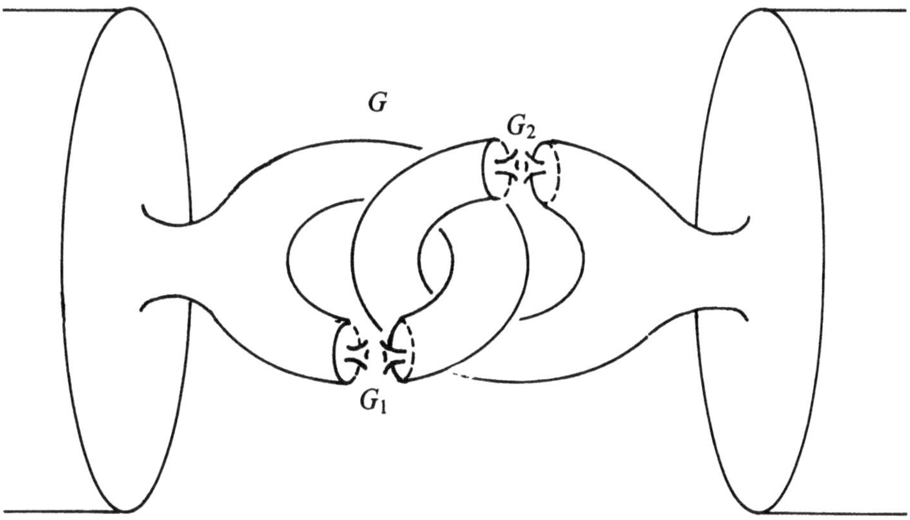

of these pieces with a common point. This (when proved) gives a topological definition of that number n for R^n.

L. E. J. Brouwer proved this in 1913. He was very active, with a general proof of invariance of dimension (a general definition of dimension was given by Menger and Urysohn), mappings of complexes and manifolds, studied through simplicial approximations, the Jordan separation theorem in n-space, coverings and fixed points of mappings, and other things. Alexandroff and Hopf were so inspired by all this that they dedicated their volume A-H to him. (If f is a mapping of a simplicial complex K into R^n, and $f'(p_i)$ is near $f(p_i)$ for each vertex p_i of K, the corresponding simplicial approximation is defined by letting f' be linear over simplexes: $f'(\sum a_i p_i) = \sum a_i f'(p_i)$. By subdividing into smaller simplexes, the approximation f' can be made closer to f.)

In the 1920s there was considerable rivalry between S. Lefschetz and W. V. D. Hodge in the applications of topology to algebraic geometry. In a Riemannian manifold M^n, a principal question was to find the "periods" of a differential form ω, that is, the integrals $\int_{C^r} \omega$ over cycles C^r which would form a base for the homology in dimension r. In later work, the forms would be required to be harmonic.

At one time I was visiting Hodge in Cambridge. In our taking a walk together, he said "Lefschetz claimed to have proved that theorem before I did; but I really did prove it first; besides which the theorem was false!" He liked intriguing questions, so I asked him one that was recently going around Princeton: A man walked south five miles, then east five miles, then north five miles, and ended up where he had started. What could you say about where he had started? (Or more popularly, what color was the bear?) He insisted it must be the north pole, and proceeded to give a careful proof; but I got the sense he did not really believe his proof was correct. (Try Antarctica.)

A contrasting situation was less happy. With both Alexander and Lefschetz in Princeton, they naturally had many discussions on topology. But Alexander became increasingly wary of this; for Lefschetz would come out with results, not realizing they had come from Alexander. Alexander was a strict and careful worker, while Lefschetz's mind was always full of ideas swimming together, generating new ideas, of origin unknown. I saw this well in my year, 1931–1932, as a National Research Fellow in Princeton. I believe that Lefschetz never felt good about Veblen choosing Alexander, not him, as one of the first professors at the new Institute for Advanced Study. Let me mention here the famous Lefschetz formula for the algebraic number of fixed points of a self-mapping of a space, an example of Lefschetz's great power.

The basic work on integration of differential forms in manifolds was given by G. de Rham in his thesis (1931, under E. Cartan). A complete identity was shown in the homology structure of Riemannian manifolds, as seen through the algebraic structure of a subdivision or through integrating differential forms; moreover, the intersections of submanifolds were related in the natural manner to the products of differential forms.

There are three more recent books with fine accounts of this theory in extended form: W. V. D. Hodge, *The theory and applications of harmonic integrals*, Cambridge University Press, 1941; G. de Rham, *Variétées Differentiables*, Hermann, Paris, 1955; and H. Flanders, *Differential Forms*, Academic Press, 1963. The third is especially helpful to the untutored reader. I myself was greatly intrigued by de Rham's work, and studied his thesis assiduously when it appeared. Of course I looked forward to meeting him; I did not suspect the happy occasion in which this would take place.

In the late twenties, Alexandroff and Hopf spent considerable time in Göttingen, especially influenced by E. Noether (as mentioned above). I was there for three weeks in early summer, 1928, after graduating from Yale, to get the sense of a great physics and mathematics center. I had physics notes to review, which I thought would go quickly; instead I found that I had forgotten most of it, in spite of much recent physics study. Seeing Hilbert-Ackermann, *Grundzüge der Theoretischen Logik*, in a bookstore, I got it and started working on it, along with George Sauté, a math student from Harvard. So I soon decided that since physics required learning and remembering facts, which I could not do, I would move into mathematics. I have always regretted my quandary, but never regretted my decision.

Those weeks I was staying in the house of Dr. Cairo, along with some physicists, Paul Dirac in particular. We became quite friendly, and discussed many things together. One was the problem of expressing all possible natural numbers with at most four 2's, and common signs. For example we can write $7 = \sqrt{[(2/.2 - .2)/.2]}$. We finally discovered a simple formula, which uses a transcendental function taught in high school. (I'll let you look for it; it starts with a minus sign.)

The authors of A-H speak also of a fruitful winter of 1931 in Princeton, influenced by Veblen, Alexander, and especially Lefschetz. The next autumn I found this also. At one time there were seven separate seminars going on together; one of them was devoted to my proof (just discovered) of a characterization of the closed 2-cell. One of my talks was to be on my cutting up process. But a few days before, I was horrified to find that there was a bad mistake in the proof. I worked desperately hard the next two days, and found a valid proof. Later, at the Moscow conference, Kuratowski told me that he especially liked that proof, for he had tried very hard to carry out such a process, but could not. Conversely, I had greatly appreciated his characterization of planar graphs through their containning neither of two graph types: five vertices, each pair joined, and two triples of vertices, each pair from opposite triples being joined. I did, however, find how to use my characterization of planar graphs through dual graphs to give his theorem.

By the time of the conference, Heinz Hopf had become my favorite writer (and I later became a personal friend). I found his papers always very carefully written, with fine introductory sections, describing purposes and tools (and he made some similar comments on my writings; he told me he "learned cohomology" from my 1938 paper). I still want to speak of two of Hopf's theorems published before the conference. One was the classification of mappings of an n-complex K^n into the n-sphere S^n; it required working separately with the Betti numbers and the torsion numbers. The other described a simple analytic mapping of S^3 onto S^2 which could not be shrunk to a point; yet homology could not suggest its existence. The latter theorem was a basic step forward in studying the homotopy groups, to be presented at the conference. Also, it showed that formally the above-mentioned classification theorem could not easily be extended to higher dimensions K^m, $m > n$.

How did people learn topology at that time? For point set theory, Hausdorff's *Mengenlehre* was the bible. Menger's *Dimensionstheorie* was a help (superseded later by the Hurewicz-Wallman book). For "combinatorial" topology, Veblen's book *Analysis situs* was a very useful book in the 1920s. Kerekjarto's *Topologie* was a help (he disliked Bessel-Hagen; look up the reference to the latter in his index). Lefschetz's *Topology* (1930) became at one a basic reference; but it was very difficult to read. I failed completely to understand some broad sections. But soon Seifert-Threlfall *Lehrbuch der Topologie* appeared, a very fine book; it was admirable for students, and its chapters on the fundamental group and covering spaces remain a good source for these topics.

Finally, the foreword to A-H was written soon after the Moscow conference. But, as mentioned earlier, one tragic result of the conference was the abandonment of later volumes.

It is high time that we turned to the conference itself. Who was there? Most of the world leaders, that is, in the combinatorial direction. There was Heegaard, representing the old-timers. (Replying to his invitation, he wrote, "I could not resist coming and meeting the greats of present day topology.") Representing the great Polish school of point set theory were W. Sierpinski (but he could not come, I believe) and K. Kuratowski.

Two great figures who could have added immeasurably to the conference had they been there, were Marston Morse (analysis in the large) and S. S. Chern (differential geometry, in the complex domain in particular). Apart from these (and Veblen, no longer active in this direction) there were, from America, Alexander, Lefschetz, J. von Neumann, M. H. Stone, and P. A. Smith; also W. Hurewicz and A. Weil (later to be U.S. residents). There were Hopf and de Rham from Switzerland, J. Nielson from Copenhagen, E. Čech from Czechoslovakia, and Alexandroff, Kolmogoroff (not usually thought of as a topologist) and Pontrjagin from U.S.S.R. Then there were younger people: Garrett Birkhoff, A. W. Tucker and myself from America; Borsuk, Cohn-Vossen, D. van Danzig, E. R. van Kampen (becoming a U.S. resident), G. Nöbeling, J. Schauder, and others.

The Proceedings of the conference came out as No. 5 of vol. 1 (43), of *Recueil Math* or *Matematischiskii Sbornik*, 1936. All papers were either published or listed here. There were about 40 members in all; a number of them missed being in the official photograph (see page 88).

For many of us, coming to the conference was a very special event. And since I was one of three from America that met in Chamonix to climb together beforehand, I tell something about this. But to start, how did Alexander and de Rham first meet? Alexander told me (when he and I were at the Charpoua hut above Chamonix in 1933) how he and his guide Armand Charlet (the two already forming a famous team) were crossing the enormous rock tower, the Dru, from this same hut a few years before. They and another party crossed paths near the top; so since each had left a pair of ice axes at the glacier, they decided to pick up the other party's axes when they reached the glacier again. With all back at the hut, two of them discovered that they knew each other by name very well: Alexander and de Rham.

I had had the great fortune to spend two years in school in Switzerland, in 1921–1923, including three summers. Besides learning French one year and German the next, I had essentially one subject of study: the Alps. The first of these years my next elder brother, Roger, was with me. We were very lucky in having an older boy, Boris Piccioni, quite experienced in climbing, in school with us; and in a neighboring school teacher, M. le Coultre, who was a professional guide also, inviting us all on three climbing trips, which included training in high alpine climbing. As a further consequence, nearly all my climbing has been without guides.

In 1933 Alexander and I met for several fine climbs at Chamonix, then went on to Saas Fee for more climbing. We next went up to the Weisshaon hut, below the east side of the great Weisshorn, with the idea of trying an apparently unclimbed route, the E. ridge of the Schallihorn, a smaller peak just south of the Weisshorn. At the hut, there was Georges de Rham, with a friend Nicolet! They had just climbed the Weisshorn by the N. ridge and descended the E. ridge; tomorrow they would climb the E. ridge again, to descend the much more difficult S. ridge, the "Schalligrat." So we were all off early the next morning. Alexander and I found our ridge easier than expected, and never put on the rope during the ascent. (Near the top we found a bottle; it was apparently from a party traversing to the top part of the ridge in 1895.) The descent (now we were roped) was over the N. ridge and down to the Schallijoch (where we heard calls of greeting from the other party). The others watched our route going down the glacier, aiding their own descent, which was partly after dark. From this time on, de Rham and I often met during the summers, and did much fine climbing together. It seems that he was renowned in Switzerland as much for his climbing as for his mathematics. In the summer of 1939, my finest alpine climbing season, he and Daniel Bach and I crossed the Schallihorn by "our ridge" (now its third ascent), and went on to climb the "Rothorngrat" and Ober Gabelhorn (we having first climbed the Matterhorn). Georges' new "vibram"-soled boots were giving him trouble, so he stopped now, while Daniel and I returned to the Weisshorn hut and made a one-day traverse of the Weisshorn by the Schalligrat and N. ridge, closing the season. And imagine my surprise when, some years later, I bought a wonderful picture book "La Haute Route" of the high peaks, by Georges' friend André Roch, and saw the first picture in it: Daniel and I on the Schallihorn (taken by Georges)!

To return to 1935: Alexander, Paul Smith and I met at Chamonix, climbed the Aiguille de Peigne together, then went on to further climbing; but the weather was turning bad, and we soon had to go on toward Moscow. (de Rham was already in Warsaw.) Alexander drove me to Berlin, and we took the night train from there.

What was the main import of the conference? As I see it, it was threefold:

1. It marked the true birth of cohomology theory, along with the products among cocycles and cycles.

2. The pair of seemingly diverse fields, homology and homotopy, took root and flourished together from then on.

3. An item of application, vector fields on manifolds, was replaced by an expansive theory, of vector bundles.

Yet seven years later, a single paper of Hopf would cause a renewed bursting open of the subject in a still more general fashion.

We now look at the remarkable way in which these matters developed at the conference.

The first major surprise was from Kolmogoroff, an unlikely person at the conference, who presented a multiplication theory in a complex, applying it also to more general spaces. The essence of the definition lies in the expression

$$(p_0 \cdots p_r) \times (q_0 \cdots q_s) = (p_0 \cdots p_r q_0 \cdots q_s),$$

provided that the right-hand side is a simplex; besides, an averaging over permutations is taken. (One obvious problem is that the product seems to be of dimension $r+s+1$, one more than it should be.)

When he had finished, Alexander announced that he, also, had essentially the same definition and results. (Both had papers in press.) From the reputations of these mathematicians, there must be something real going on; but it was hard to see what it might be. I digress for a moment to say what happened to this product. Within a few months, E. Čech and I both saw a way to rectify the definition. We each used a fixed ordering of the vertices of a simplicial complex K, and defined everything in terms of this ordering. The basic definition was simply (with the vertices in proper order)

$$(p_0 \cdots p_r) \smile (p_r \cdots p_{r+s}) = (p_0 \cdots p_r \cdots p_{r+s}),$$

whenever the latter is a simplex of K. Alexander at once saw the advantage of this, and rewrote his paper from this point of view (*Annals of Math.*, 1936).

Another event at the conference was the defining of the homotopy groups in different dimensions of a space, with several simple but important applications, by Hurewicz. Alexander responded by saying he had considered that definition many years (twenty?) earlier, but had rejected it since it was too simple in character and hence could not lead to deep results. Perhaps one lesson is that even simple things may have some value, especially if pushed long distances.

Both E. Čech and D. van Danzig also said that they had considered or actually used the definition of Hurewicz. Thus at the time of the conference, the homotopy groups were very much "in the air."

I now turn to the paper that had the most intense personal interest for me. Hopf presented the results of E. Stiefel (written under Hopf's direction), "Richtungsfelder und Fernparallelismus in n-dimensionalen Mannigfaltigkeiten." It was concerned with the existence of several independent vector fields in a manifold. Both in generality, and (largely) in detail, this was just what I had come to Moscow in order to present myself! Stiefel had more complete results; in particular, that all orientable 3-dimensional manifolds were "parallelizable." On the other hand, I had given a much more general definition; for example, for submanifolds of Euclidean space (or of another manifold), I considered normal vector fields also. Moreover, I considered sphere (or vector, or fiber) bundles over a complex as base space, and found that results were best expressed in terms of *cohomology*, not homology, in the complex (for manifolds it did not matter).

I spoke briefly of these things right after Hopf's talk; but still had to decide afresh how to talk about my own work. Moreover, on my way to the conference I had already become uncertain on how to talk; for I had realized that Hopf's classification of the mappings of K^n into S^n could be presented much more simply in terms of cohomology than of homology. In fact, it seemed to me highly worthwhile to show this in detail, as the possibly first true use of cohomology, and the simplest possible example of its usefulness.

I therefore gave two shorter talks, one giving a fuller account of my work on sphere bundles, and the other, a pretty complete proof of the Hopf theorem with cohomology.

I want to speak briefly of two further presentations. Tucker spoke on "cell spaces," a thesis written under Lefschetz's direction, which gave certain specifications about what can usefully be considered a "complex." This cleared up some important matters which played a real role in both Čech's and my exposition of cohomology and products in our coming papers in the *Annals of Mathematics.* The other was Nöbeling's presentation, which occupied the full last morning of the conference. (I was not there; I had left early for Leningrad, hoping to meet the composer Shostakovich (which did not happen), and to make the five-day boat trip from Leningrad through the Kiel Canal to London, which was quite interesting.)

Nöbeling's talk was to present, in outline, the proof that all topological manifolds can be triangulated. von Neumann reported on this conference as follows: Nöbeling demonstrated amply that he had answers to every possible question that one might think of. (Within the year, van Kampen found the error in the proof. Disproving the theorem took much longer.)

I give a brief description of Hopf's mapping theorem (about $K^n \to S^n$) through cohomology; take $n = 2$ for ease of expression. (See my papers in the *Duke Math Journal*, 1937.) First, "coboundary" is dual to "boundary": If $\partial\sigma = \tau + \cdots$, then $\delta\tau = \sigma + \cdots$. With the language of scalar products, or, better, considering a cochain as being a linear function of chains, we can write

$$\delta\tau \cdot \sigma = \tau \cdot \partial\sigma, \qquad \delta X^r \cdot A^{r+1} = X^r \cdot \partial A^{r+1}.$$

Whereas the "boundary" of a cell makes good geometric sense, the "coboundary" does not. In the figure, $\delta\tau$ stretches into three pieces; but why stretch so far?

In the theorem, our first step is to deform any mapping f into a "normal" mapping. To this end, choose a definite point P of S^2 (the south pole). For each vertex p_i of K, we may deform f into f' so that $f'(p_i) = P$. Of course

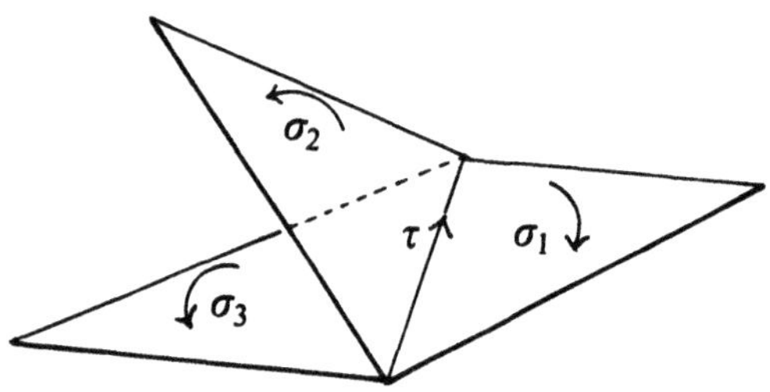

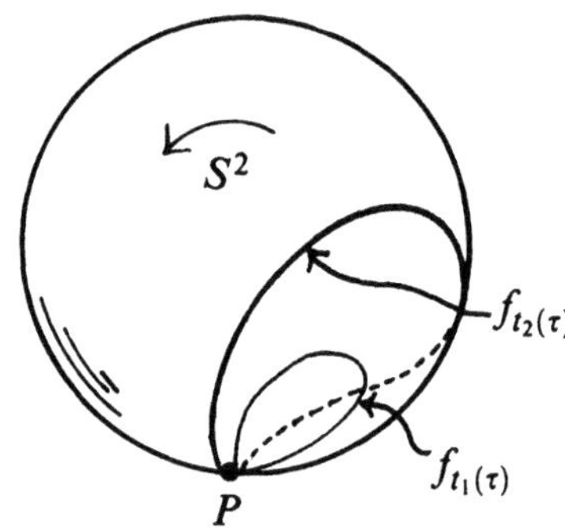

all cells of K with p_i on the boundary must be pulled along some also. Do this for all vertices, so they are now all at P. Now each 1-cell τ_j of K has its ends mapped into P; we may pull τ_j; along S^2 down to P, keeping its ends at P, extending the deformation in any manner through the rest of K. This gives a *normal* mapping f_0, in which the 1-dimensional part K^1 of K lies at P.

Now take any 2-cell σ_i^2 of K. It is a standard theorem (first proved by Brouwer) that (since its boundary is at P) it lies over S^2 with the *degree* $d_i = d(\sigma_i^2)$, this being an integer. (If it only partly covers S^2, each piece of S^2 is covered an algebraic number 0 of times; we may shrink the mapping to P, keeping $\partial\sigma_i^2$ at P.)

We remark in passing that when τ is pulled down to P, how far we choose to extend the deformation into 2-cells depends on how far those 2-cells reach beyong τ; thus $\delta\tau$ plays a role in the proceedings.

Let us write $X(\sigma_i^2) = d_i$ for each σ_i^2. (Or we could write $X = \sum d_i\sigma_i^2$.) This is the cochain X defined by the deformed mapping.

We could deform a normal mapping f_0 into a different normal mapping f_1 as follows. Choose a 1-cell τ, and sweep it up and over S^2, and down the other side back to P, keeping its ends always at P. (In the figure, we show two stages, f_{t_1} and f_{t_2}, $0 < t_1 < t_2 < 1$, of this deformation of τ.) Extend the deformation over the neighboring 2-cells of K. For each σ_i^2 of K which has τ on its boundary (positively), d_i^2 is increased by 1; thus the change in the corresponding cochain is simply to add $\delta\tau$. In this manner we may add δY to the cochain X, for any 1-dimensional cochain Y; but the cohomology class of X remains unchanged.

Since there is no 3-dimensional part in K, all 2-cochains are cocycles; thus there is a definite 2-dimensional cohomology class associated with the original mapping.

We have one thing still to prove: Given *any* deformation of a normal mapping f_0 into another one, f_1, the same cohomology class is defined. We use a standard technique to do this. Let $f_t (0 \leq t \leq 1)$ denote the deformation. Set $F(t, p) = f_t(p)$. Now F is a mapping of the product space $I \times K$ into S^2, where I is the unit interval $(0, 1)$.

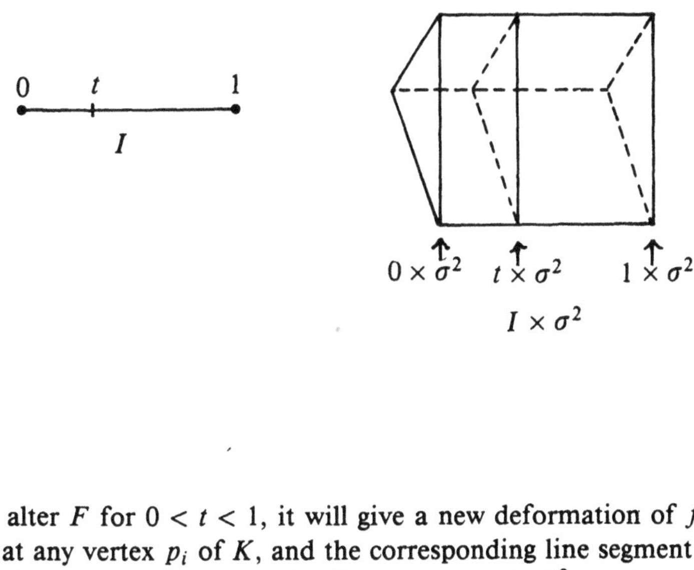

If we alter F for $0 < t < 1$, it will give a new deformation of f_0 into f_1. So look at any vertex p_i of K, and the corresponding line segment $I \times p_i$ of $I \times K$. This segment is mapped by F into a curve in S^2 starting and ending at P. We may alter $F(t, p_i)$ by pulling this curve along S^2 down to P, and extending the mapping to the rest of $I \times K$. Doing this for each vertex of K, we have defined a new deformation of f_0 into f_1, in which each vertex of K remains at P. We have already seen that this implies that the cohomology classes of f_0 and f_1 differs by a coboundary, and the proof is complete.

I have spoken of my great interest in the papers of Hopf and de Rham. In the mid thirties I saw my main job of coming years to be the extending of the general subject of sphere bundles. For this purpose, I was very anxious to have the basic foundations in as nice a state as possible; I could not work with concepts that were at all vague in my mind. As part of this, I wanted to work both with algebraic and with differential methods; hence I needed, as far as possible, a common foundation of both (and Hopf and de Rham were my best models). But these subjects had been quite separate, and hence the notations used were very different in character. So I tried to devise notations that could allow the fields to work more closely together.

The use of "contravariant" and "covariant" vectors raised a quandary. A covariant tensor was one whose components (depending on the coordinate system) transform "like" or "with" the partial derivatives of a function. But for me, the basic object was a vector space, and its elements, vectors, should be the base of operations: *They* should be called "covariant" (if anything), not contravariant. Also homology dealt directly with geometric things, and should have the prefix (if any) "co" not "contra." In any case, I would not use prefixes differently in homology and in differential geometry. So I started publishing, using the term "cohomology" for the new topic, omitting any use at all of "contra," and disregarding the (for me) wrong use of "co." This was picked up quickly by others, and the inherent reverse of "co" and "contra" remains.

There was a further block to my progress. I had to handle tensors; but how could I when I was not permitted to *see* them, being only allowed to *learn* about their changing costumes under changes of coordinates? I had somehow to grab the rascals, and look straight at them. I could *look at* a pair of vectors, "multiplied": $u \vee v$. And here, I needed $u \vee v = -v \vee u$. So I managed to construct the rest of the beasts, in "tensor products of abelian groups." (*Duke Math Journal*, 1938). Before long I noticed that neat form, using less space, was the sine qua non of mathematical writing: the CORRECT definition of the tensor product of two vector spaces must use the linear functionals over the linear functionals over one of them. So this is the way in which later generations learned them.

Only in 1988 did I make a further discovery (or rediscovery?): A typical "differential 2-form" is $u \vee v$; and this is already a product! Any simplex, say $p_0p_1p_2$, is a bit of linear space, and writing $v_{ij} = p_j - p_i$, a natural associated 2-form is $v_{01} \vee v_{12}$. Another such product in $p_0p_1p_2p_3$ is, for instance,

$$v_{01} \vee (v_{12} \vee v_{23}).$$

Then why not write

$$(p_0p_1) \smile (p_1p_2p_3) = p_0p_1p_2p_3,$$

whenever the right hand side is a simplex? From the basic definition only of differential forms associated with simplexes (or a simplicial complex), nothing could be more natural. (This was soon called the "cup" product.)

Why did not Kolmogoroff and Alexander (and lots of others) think of it? I think this is a real lesson to be learned: Keep wide contexts and broad relations in mind; new connections and extended methods may show up.

I mention still the "cap" product, typified by $p_1p_2p_3 \frown p_0p_1p_2p_3 = p_0p_1$, and for cochains X and Y and chains A,

$$X \frown (Y \frown A) = (X \smile Y) \frown A.$$

What was the aftermath of the conference? To a large extent, the younger generation took over. In the U.S.S.R., L. Pontrjagin was coming into full flower (in particular, with topological groups and duality). J. Leray (France) and J. Schauder (Poland), collaborating in large part, and S. S. Chern (China, U.S.A. and elsewhere) were bringing powerful tools into play. New domains such as sheaves and spectral sequences were playing a big role.

In the U.S.A., N. E. Steenrod, S. Eilenberg, and S. Mac Lane were playing an increasing role, especially in building edifices from extremely general principles (with categories and functors for the foundation).

We are getting into the 1940s, with an astounding pair of papers by H. Hopf about to arrive on the scene. The first of these papers, "Fundamentalgruppe und zweite Bettische Gruppe," was communicated to the *Commentarii Mathematicii Helvetici* on September 12, 1941. In this paper Hopf gives an algebraic construction of a certain group G_1^* from any given group G. Now let K be a complex, with second homology (Betti) group B^2 and with fundamental group G. Let S^2 be the subgroup of B^2 formed from the "spherical cycles," continuous images of the 2-sphere. Then Hopf proves that the factor group B^2/S^2 is isomorphic with G_1^*. Thus the fundamental group of a complex has a strong influence on the second homology group. For example, if G is a free abelian group of rank p, then G_1^* is a free abelian group of rank $p(p-1)/2$, so this is a lower bound for the rank of the second homology group.

The construction of G_1^* is as follows. Represent the fundamental group G as a factor group F/R, where F is a free group and R is a subgroup (of relations in F). Define the commutator subgroups $[F, R]$ (generated by all commutators $frf^{-1}r^{-1}$ for $f \in F$ and $r \in R$) and $[F, F]$ (using $f_1f_2f_1^{-1}f_2^{-1}$). Then

$$G_1^* = (R \frown [F, F])/[F, R].$$

(Actually, this group G_1^* had been defined much earlier, by I. Schur, in Berlin.)

At this point I turn over a description of happenings to S. Mac Lane, "Origins of the cohomology of groups", *L'Enseignement Mathématique*, vol. 24, fasc. 1–2, 1978. This is an admirable paper, full of descriptions of modern fields of work and their interrelations, all showing the enormous influence of the above paper of Hopf (and another to follow a few years later). Here is one quote from Mac Lane.

> Hopf's 1942 paper was the starting point for the cohomology and homology of groups; indeed this Hopf group G_1^* is simply our present second homology group $H_2(G, Z)$. This idea and this paper were indirectly the starting point for several other developments: Invariants of group presentations; cohomology of other algebraic systems; functors and duality; transfer and Galois cohomology; spectral sequences; resolutions; Eilenberg-Mac Lane spaces; derived functors and homological algebra; and other ideas as we will indicate below.

I had the great pleasure of reviewing Hopf's paper for *Math Reviews* (the review appeared already in November 1942). I also, many years later, wrote an informal paper whose purpose was to bring out the essential reasoning (commonly geometric in character) of various basic theorems. This paper ended with the quoted formula of Hopf. It was published as "Letting research come naturally," *Math. Chronicle* 14 (1985) (Auckland, New Zealand) 1–19. I mention just one crucial idea of Hopf: that of "homotopy boundary," λ, from which everything flowed: From a fixed point P of K, choose a (simplicial) path to any 2-simplex σ^2, go around its boundary, and back to P:

$$\lambda = \gamma\partial\sigma^2\gamma^{-1}.$$

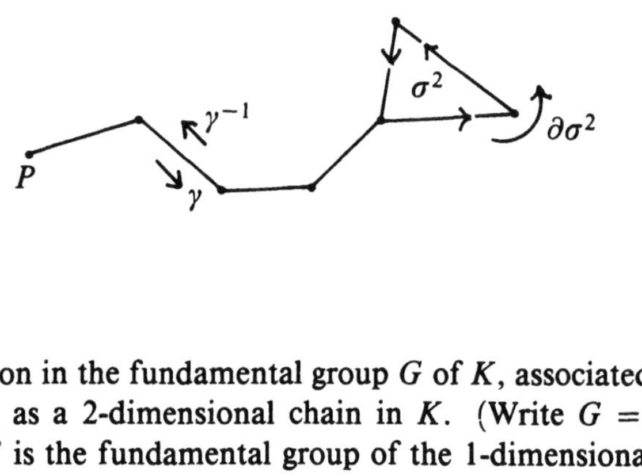

Now λ (in R) is a relation in the fundamental group G of K, associated with the cell σ^2, considered as a 2-dimensional chain in K. (Write $G = F/R$, where the free group F is the fundamental group of the 1-dimensional part K^1 of K. Now $\lambda \in R \subset F$.)

I will say a few words still about my own work in the direction of topology. My one fairly full account of researches in sphere bundles appeared in *Lectures in Topology*, University of Michigan Press, 1941, under the title "On the topology of differentiable manifolds." This paper is certainly the basis of some awards I have received in later years. Largely to help prepare good foundations for my planned book on sphere bundles, I wrote the book *Geometric integration theory*, Princeton University Press, 1957. This volume received quite unexpected acclaim over many years. On the other hand, in this period, Warren Ambrose once said to me "I calculate that the publication date of your book on sphere bundles is receding at the rate of two years per year." He was quite right. I was having increasing difficulty in finding good geometric foundations for the topological aspects. Then I had an unexpected piece of good fortune; rather, two. Steenrod's book on fiber bundles appeared, doing a far better job than I could possibly have done, and I was invited to join the faculty at the Institute for Advanced Study.

I also had new sources of inspiration: Henri Cartan (we played music together at times, piano and violin) was combining topology with analytic studies (several complex variables), which I studied with determination, getting preparation for my final book on complex analytic varieties (whose unfulfilled purpose was to help in the foundations of singularity theory). And R. Thom and I had started working (independently) on singularities of mappings, my last major field of work. He had also given a very simple and general proof of my "duality theorem," really the formula for the characteristic classes of the product of two sphere bundles over the same base space, which I had carried out through a full examination of the geometric definitions. (And later, under the general definitions of Eilenberg-Steenrod, the theorem became part of the definition of sphere bundles.)

I did still write some further papers, with which I was amply satisfied, about ideals of differentiable functions (*Am. J. of Math*, 1943), On totally differentiable and smooth functions (*Pacific J. Math.*, 1951), On functions with bounded *n*th differences (*J. de Math. Pures at appliquées*, 1957), somewhat outside my normal fields of work. But younger workers, in America in particular, were taking over in a strong way; I mention John Mather especially, in singularities of mappings. So, to return to the title of this paper, I have seen general topological and algebraic methods flourishing all over the world increasingly, as the "center of mass" of such studies moves still nearer to the U.S.A. shores.

Chapter 1

Graphs and Combinatorics

A THEOREM ON GRAPHS.[1]

BY HASSLER WHITNEY.

I. Results of this paper.

1. Let a finite number of curves, or edges, whose end-points we call vertices, intersect at no other points than these vertices. Let the system be connected, that is, any two vertices are joined by a succession of edges, each two successive edges having a vertex in common. This forms a graph. A graph is planar if it can be mapped in a 1–1 continuous manner on a plane (or a sphere). If the vertices a, b are joined by an edge, we shall call the edge joining them ab, and shall say a touches b for short. A set of distinct vertices, $a, b, c, \cdots, e, f$, together with a set of distinct edges joining them in cyclic order, $ab, bc, \cdots, ef, fa$, we shall call a circuit.

A planar graph lying on the surface of a sphere divides this surface into a number of simply connected regions. The boundary of each of these regions may be a circuit. If so, we shall call these circuits elementary polygons. If all these polygons are n-gons, n fixed, we say the graph is composed of elementary n-gons.

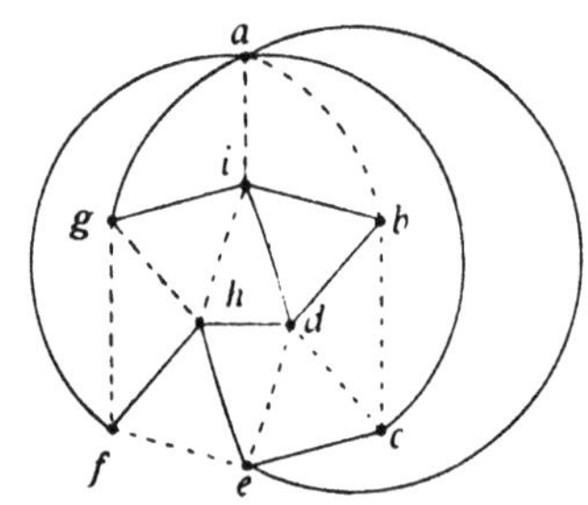

Fig. 1.

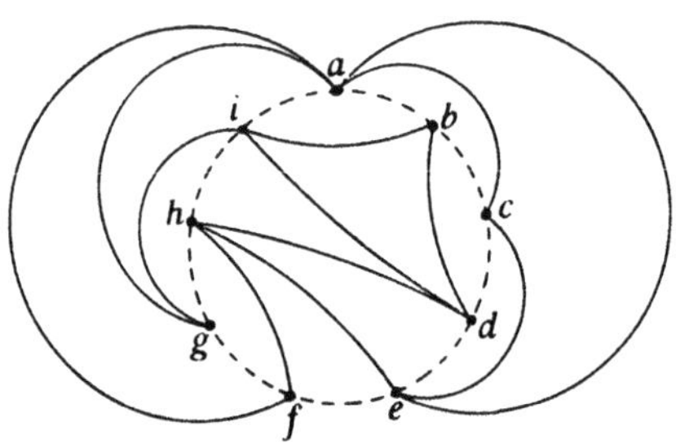

Fig. 2.

2. The fundamental theorem of this paper is the following:

THEOREM I. *Given a planar graph composed of elementary triangles, in which there are no circuits of 1, 2, or 3 edges other than these elementary triangles, there exists a circuit which passes through every vertex of the graph.*

The problem of finding graphs for which this is so has been studied by several people.[2] This seems to be the first case when a large class of planar graphs has been shown to have this property.

3. This theorem gives immediately the following:

NORMAL FORM. *Given any graph as described in Theorem* I, *containing*

[1] Received April 7, and July 14, 1930.—Presented to the American Mathematical Society, Febr. 22, 1930.

[2] See St. Laguë, A., Les Réseaux, Mémorial des Sciences Math., fasc. 18, Paris (1926).

n vertices, we can construct a graph homeomorphic with it as follows: Draw a regular polygon of n sides, and draw diagonals, no two of which cross, dividing the inside of the polygon into triangles. Similarly draw circular arcs, no two of which cross, dividing the outside of the polygon into circular triangles.

We have merely to find the circuit given by Theorem I, and distort it into the polygon.

4. A theorem on maps deducible immediately from Theorem I is the following, as we shall see later:

THEOREM II. *Given a map on the surface of a sphere containing at least three regions in which:*

(A_1) *The boundary of each region is a single closed curve without multiple point,*

(B) *Exactly three boundary lines meet at each vertex,*

(A_2) *No pair of regions taken together with any boundary lines separating them form a multiply connected region,*

(A_3) *No three regions taken together with any boundary lines separating them form a multiply connected region,*

we may draw a closed curve which passes through each region of the map once and only once, and touches no vertex.

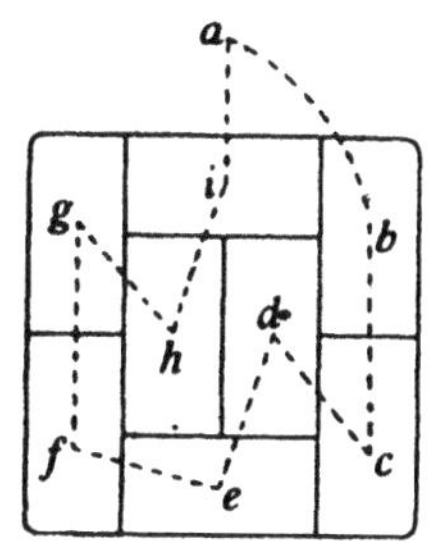

Fig. 3.

5. By means of Theorem II and a lemma to be proved, we have a solution of a conundrum, which we leave to the end of the paper.

6. Finally, Theorem I gives us a new statement of the four color map problem. Given any map on the surface of a sphere, we "color" it by assigning to each region a color in such a way that no two regions with a common boundary are of the same color. Given any polygonal configuration as described in 3., we "color" it by assigning to each vertex of the polygon a color in such a way that no two vertices which are joined by a line, either a side of the polygon or a diagonal, are of the same color.

EQUIVALENT STATEMENT OF THE FOUR COLOR MAP PROBLEM. *If every polygonal configuration as described in* 3. *can be colored in four colors, then every map on the surface of a sphere can be colored in four colors, and conversely.*

2. Proof of Theorem I.

We consider only the graphs defined in § 1, 2. As the graph is composed of elementary triangles, there are at least three vertices present. If there are only three, the theorem is obvious. We shall assume from here on that there are at least four vertices present.

As there are no circuits of one or two edges, no vertex touches itself, and any two vertices are joined by at most a single edge.

There is no vertex touching but a single other vertex. For then the boundary of the region surrounding this other vertex would not be a circuit, and therefor not an elementary triangle.

There is no vertex touching only two others. For suppose a touched b and c alone. Then the edges ab and ac would each be sides of two triangles, whose third sides are both edges bc. But there is only one edge bc, as two vertices are joined by at most a single edge. The two triangles thus cover the whole surface of the sphere, and there are thus only three vertices in the graph, contrary to hypothesis.

Consider a vertex a touching other vertices $b, c, \cdots, f$. We read the edges emenating from a in a counter-clockwise sense, and say, a touches $b, c, \cdots, f$ in cyclic order; or, a touches b, next $c, \cdots$, next f, next b.

Remembering now that the graph is composed of elementary triangles, we have the three properties:

(α) *Each vertex touches at least three other vertices in cyclic order, distinct from each other and distinct from the first,*

(β) *If* a *touches* b *and next* c, *then* b *touches* c *and next* a,

(γ) *There are no triangles other than elementary triangles.*

These properties, together with the fact that the graph lies on a sphere, is all we need to prove the following lemma, from which the theorem is deduced.

LEMMA. *Consider a circuit* R *in a graph of the type considered in Theorem* I, *together with the vertices and edges on one side, which we shall call the inside. Let* A *and* B *be two distinct vertices of* R, *dividing* R *into the two parts* R_1 *and* R_2, *in each of which we include both* A *and* B. *Suppose*

(1) *No pair of vertices of* R_1 *touch each other inside* R *(are joined by an edge which lies inside* R*), and*

(2) *Either no pair of vertices of* R_2 *touch each other inside* R, *or else there is a vertex* C *in* R_2 *distinct from* A *and* B, *dividing* R_2 *into the two parts* R_3 *and* R_4, *in each of which we include* C, *such that no pair of vertices of* R_3 *and no pair of vertices of* R_4 *touch each other inside* R.

Then we can draw a line from A *to* B, *passing only along edges of and inside* R, *and passing through each vertex of and inside* R *once and only once.*

In brief, if we can divide the circuit R into either two or three parts, such that in any part, including end vertices, no pair of vertices touch each other inside R, we can then draw the required curve from any one end vertex to any other end vertex of these parts.

The theorem is an immediate consequence of the lemma. For consider any elementary triangle of the graph, containing the vertices A, B, C, which we call the circuit R. The rest of the graph we call the inside of the circuit. As each pair of vertices of R touch as a part of the circuit, and any two vertices are joined by at most one edge, it follows that no pair of them touch inside R. Thus the conditions of the lemma are fulfilled, and we can pass from A to B through every vertex of R and every vertex inside R, that is, through every vertex of the graph. We now pass from B directly to A, forming a closed curve. The edges passed over by the curve form the desired circuit.

Proof of the lemma. Assume the lemma is true for all circuits which, with the vertices inside, contain m vertices, $m = 3, 4, \cdots, n-1$. It is obviously true for the case where $m = 3$. We will prove it for all circuits which, with the vertices inside, contain n vertices. Then, by mathematical induction, it is true in general.

Take any circuit R therefore, which, with the vertices inside, contains n vertices. Let the vertices of the circuit be $A, a_1, a_2, \cdots, a_\alpha, B, b_1, b_2, \cdots, b_\beta, C, c_1, c_2, \cdots, c_\gamma, A$, (reading in a clockwise sense). We assume that no pair of the vertices $A, a_1, \cdots, a_\alpha, B$, no pair of the vertices $B, b_1, \cdots, b_\beta, C$, (or with C replaced by A, if there is no C), and no pair of the vertices $C, c_1, \cdots, c_\gamma, A$ touch inside the circuit. The vertices $C, c_1, \cdots, c_\gamma$ may be missing from the circuit, as may also the vertices $a_1, \cdots, a_\alpha$ or $b_1, \cdots, b_\beta$. We wish to draw the required curve from A to B.

We will divide the proof into four parts, according to what pairs of vertices of the circuit touch inside the circuit:

Case (1). Some vertex a_q touches a vertex b_r, C, or c_s inside R.

Case (2). There are no edges of the above form, but either B touches a vertex c_s or A touches a vertex b_r inside R.

Case (3). No pairs of vertices of the circuit touch inside the circuit.

Case (4). Some vertex b_r touches a vertex c_s inside R, but there are no edges of other forms between vertices of the circuit inside the circuit.

Case (1). Assume there is an edge of one of the forms $a_q C$, $a_q c_s$. The case where there is no edge as above, but there is an edge of the form $a_q b_r$, is reduced to this case by interchanging the rôles of A and B and of c_s and b_r. Suppose the edge nearest A is $a_i c_k$. If it is $a_i C$, we call C, c_k. The meaning of "nearest A" is obvious. Now either, Case (1a), c_k touches none of the vertices $a_{i+1}, \cdots, a_\alpha, B$, or, Case (1b), a_i touches none of the vertices $C, c_1, \cdots, c_{k-1}$ inside the circuit $c_k, a_i, \cdots, a_\alpha, B, b_1, \cdots, b_\beta, C, c_1, \cdots, c_k$. If c_k is C, the latter condition is satisfied automatically.

Consider Case (1a). We shall draw the required curve in two steps: first from A to c_k, then from c_k to B.

If first, Case (1a$_1$), a_i is not a_1, a_{i-1} exists, and does not touch c_k inside the circuit, as the edge $a_i c_k$ was the edge of this form nearest A. Therefore a_i must touch some vertex in between a_{i-1} and c_k. For if a_i touched a_{i-1} and next c_k, a_{i-1} would touch c_k and next a_i, by (β). Thus a_{i-1} would touch c_k between a_{i-2} (or A) and a_i, and the edge $a_{i-1}\,c_k$ would therefor be inside the circuit, which it cannot be, again as the edge $a_i c_k$ was the edge of this form nearest A. As a_i touches no vertices of the set $c_{k+1}, \cdots, c_\gamma$, $A, a_1, \cdots, a_{i-1}$ inside the circuit, any vertices it touches between a_{i-1} and c_k must be vertices inside the circuit R. Call them in order $p_1, p_2, \cdots, p_\varphi$. Then, by ($\beta$), a_{i-1} touches p_1, p_1 touches $p_2, \cdots$, and p_φ touches c_k. We have thus formed a circuit $A, a_1, \cdots, a_{i-1}, p_1, \cdots, p_\varphi, c_k, \cdots, c_\gamma, A$. No pair of the vertices $A, a_1, \cdots, a_{i-1}$ touch inside this circuit, as none of the set $A, a_1, \cdots, a_\alpha, B$ touched inside the circuit R. Similarly no pair of the set $c_k, \cdots, c_\gamma, A$ touch inside the circuit. Finally, no pair of the set $a_{i-1}, p_1, \cdots, p_\varphi, c_k$ touch inside the circuit. For suppose for instance p_g touched p_h inside, $h > g$. a_i does not touch p_g and next p_h, as p_g and p_h would then touch as a part of the circuit, and therefor not inside the circuit. Therefor a touches a vertex p_s in between. But then a_i, p_g and p_h form a triangle, with p_s on one side, and other vertices, as A, on the other side, which is therefor not an elementary triangle, in contradiction to (γ). Thus all the conditions of the lemma are satisfied for this circuit, and there are fewer than n vertices in and within the circuit. We can therefor draw a line from A to c_k passing through every vertex of and inside the circuit.

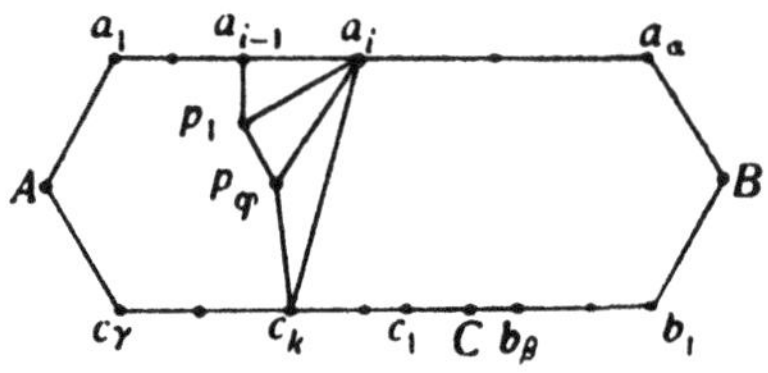

Fig. 4.

If next a_i is a_1, suppose, Case (1a$_2$), c_k is not c_γ. (If the edge nearest A is $a_1\,C$, suppose there is a vertex c_1 in R.) By hypothesis, c_k does not touch A inside the circuit. Therefor a_1 touches vertices between A and c_k. For otherwise, a_1 would touch A and next c_k, and therefor A would touch c_k and next a_1, by (β). But as c_k is not c_γ, A would touch c_k between c_γ and a_1, and the edge $A c_k$ would be inside the circuit. As a_1 does not touch $c_{k+1}, \cdots, c_\gamma$ inside the circuit, the vertices it touches between A and c_k must be vertices not in R. Call these vertices in order $p_1, \cdots, p_\varphi$. We get thus a circuit $A, p_1, \cdots, p_\varphi, c_k, \cdots, c_\gamma, A$. No pair of the set of vertices $c_k, \cdots, c_\gamma, A$ touch inside the circuit, nor do any of the set A, $p_1, \cdots, p_\varphi, c_k$, using exactly the same reasoning as in Case (1a$_1$). Thus the lemma applies to this circuit, and we pass from A to c_k, passing through every vertex of and inside the circuit.

In each of the two Cases (1a$_1$) and (1a$_2$) we have now passed through every vertex of and inside R which is on A's side of the edge $a_i c_k$. For consider the circuit $a_i, c_k, p_\varphi \cdots, p_1, a_{i-1}, a_i$, (or with a_{i-1} replaced by A, if a_i is a_1). As a_i touches every other vertex of the circuit, there can be no vertices inside the circuit. For if there were a vertex d inside the circuit, it must then lie inside one of the triangles a_i, p_1, a_{i-1} (or A), a_i, or a_i, p_2, p_1, a_i, or $\cdots$, or a_i, c_k, p_φ, a_i. In any case, (γ) would be violated. We have thus only to pass from c_k to B on B's side of the edge $a_i c_k$, that is, through the circuit $c_k, a_i, \cdots, a_\alpha, B, b_1, \cdots, b_\beta, C, c_1, \cdots, c_k$.

We have still to consider in Case (1a) the Case ($1a_3$), where the edge nearest A was the edge $a_1 c_\gamma$ (or $a_1 C$, when there is no c_γ). Draw a line directly from A to c_γ (or C). As there are no vertices inside the circuit A, a_1, c_γ, A (or A, a_1, C, A) by (γ), we have left to pass through only vertices of and inside the same circuit as in Cases ($1a_1$) and ($1a_2$).

But we can do this, by the lemma. For, no pair of the set $a_i, \cdots, a_\alpha, B$ touch inside the circuit. Also, c_k touches none of these vertices inside the circuit, by the hypothesis of Case (1a). Therefor none of the vertices $c_k, a_1, \cdots, a_\alpha, B$ touch inside the circuit. Nor do any of the set $B, b_1, \cdots, b_\beta, C$, or any of the set $C, c_1, \cdots, c_k$, (if these are present), by the original hypotheses. The circuit is thus divided into two or three parts, depending on whether c_k is C or not, and the lemma applies in either case. We thus pass from c_k to B, completing the required curve from A to B. This disposes of Case (1a).

Consider Case (1b), where a_i touches none of the vertices $C, c_1, \cdots, c_{k-1}$, inside the circuit (if any are present). In this case, instead of passing from A to c_k through all the vertices of and inside the circuit $A, a_1, \cdots, a_i, c_k, \cdots, c_\gamma, A$, except a_1, the same steps show we can pass from A to a_i through every vertex of and inside this circuit except c_k. We now apply the lemma to pass from a_i to B. For, no pair of vertices of the set $C, c_1, \cdots, c_k$ touch inside the circuit $a_i, \cdots, a_\alpha, B, b_1, \cdots, b_\beta, C, c_1, \cdots, c_k, a_i$, and a_i touches none of these vertices inside the circuit; therefor none of the set $C, c_1, \cdots, c_k, a_i$ touch inside the circuit. Also, no pair of the set $a_i, \cdots, a_\alpha, B$, and no pair of the set $B, b_1, \cdots, b_\beta, C$ touch inside the circuit. The proof for Case (1) is now complete.

Case (2). Suppose B touches a vertex c_s inside the circuit. Of all such vertices, let the one nearest A be c_k. Exactly as we before passed from A to c_k, going through all the vertices on A's side of the edge $a_i c_k$, we now pass from A to c_k, going through all the vertices on A's side of the edge $B c_k$. We have now only to pass from c_k to B, going through all the vertices on the other side of the edge $B c_k$. But we can do this, by the lemma. For the vertices c_k, B do not touch inside the circuit $c_k, B, b_1, \cdots, b_\beta, C, c_1, \cdots, c_k$. Also, no vertices of the set $B, b_1, \cdots, b_\beta, C$, and no vertices of the set $C, c_1, \cdots, c_k$ touch inside the circuit.

The proof is the same if A touches some vertex b_r inside the circuit.

Case (3). No vertices of the circuit touch inside the circuit. As any circuit contains at least three vertices, there is at least one other vertex besides A and B in the circuit. Thus if we call the vertices of the circuit $A, a_1, \cdots, a_\alpha, B, b_1, \cdots, b_\beta, A$, either a_1 or b_β, say b_β, is present. Draw a line from A to b_β. We have still to pass from b_β to B.

Suppose, Case (3a), a_1 is also present in the circuit. As a_1 and b_β do not touch inside the circuit, A does not touch b_β and next a_1, and A touches therefor other vertices in between. Calling these in order $p_1, \cdots, p_\varphi$, we have a circuit $b_\beta, p_1, \cdots, p_\varphi, a_1, \cdots, a_\alpha, B, b_1, \cdots, b_\beta$, where at least b_β, p_1, a_1 and B are present. The lemma applies to this circuit. For, no pair of the vertices $b_\beta, p_1, \cdots, p_\varphi, a_1$, no pair of the vertices $a_1, \cdots, a_\alpha, B$, and no pair of the vertices $B, b_1, \cdots, b_\beta$ touch inside the circuit. There are no vertices inside the circuit $b_\beta, A, a_1, p_\varphi, \cdots, p_1, b_\beta$, as A touches all the other vertices of this circuit.

Suppose now, Case (3b), a_1 is not present in the circuit, but $b_1 \neq b_\beta$ is. Then, as B does not touch b_β inside the circuit, A touches vertices between b_β and B, and we obtain the circuit $b_\beta, p_1, \cdots, p_\varphi, B, b_1, \cdots, b_\beta$, to which the lemma applies. For, no pair of the vertices $b_\beta, p_1, \cdots, p_\varphi, B$, and no pair of the vertices $B, b_1, \cdots, b_\beta$ touch inside the circuit.

Consider now Case (3c), where the circuit R consists only of the vertices $A, B, b_1 = b_\beta, A$. If there are no vertices inside the circuit, we pass directly from b_β to B. If there are vertices inside the circuit, A touches vertices between b_β and B, and we obtain the circuit $b_\beta, p_1, \cdots, p_\varphi, B, b_\beta$, to which the lemma applies, as in Case (3b).

Case (4). No pair of vertices of the circuit R touch inside except for edges of the form $b_r c_s$. Of all such edges, let the one furthest from the vertex C be the edge $b_j c_k$. We will carry through the proof for this case in three steps:

(1) A chain of vertices $p_1, \cdots, p_\varphi$ with the edges joining them stretching from b_j to A and to a_1 (or to B, if there is no a_1), will be found.

(2) A subset of these vertices with the edges joining them will form another chain, $q_1, \cdots, q_\theta$.

(3) The required curve will be drawn from A to b_j on A's side of this latter chain, and from b_j to B on the other side of the chain.

(1) *The chain of p's.* As b_{j-1} (or B, if there is no b_{j-1}) does not touch c_k, the edge $b_j c_k$ being the one furthest from C, b_j touches a vertex in between, which is inside the circuit R. Call p_1 the vertex b_j touches just before c_k. Then p_1 touches c_k and forms the first vertex of the chain. If p_1 touches A, the first part of the chain is finished. If not, let c_{p_1} be the vertex of the set $c_k, \cdots, c_\gamma$ nearest A which it touches.

Suppose we have constructed the chain as far as the vertex p_i, which does not touch A, and c_{p_i} is the vertex nearest A which p_i touches. Assume the following properties hold:

(a) All the p's are distinct.

(b) Each p_s, $s < i$, touches the vertex p_{s+1}, and each touches a vertex c_{p_s}.

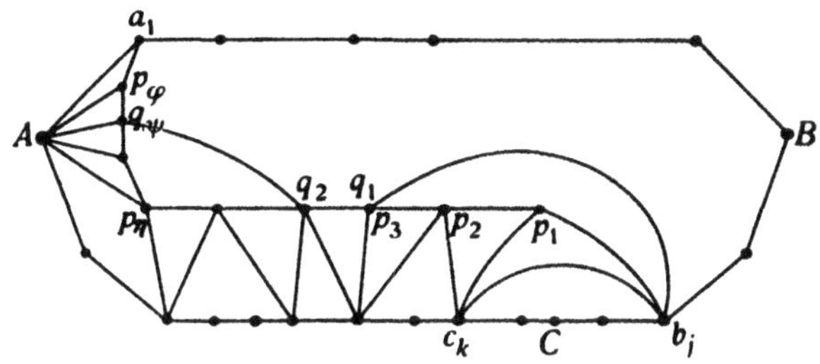

Fig. 5.

(c) No p_s touches any of the vertices $c_{p_i}, \cdots, c_\gamma$ inside the circuit $A, a_1, \cdots, a_\alpha, B, b_1, \cdots, b_j, p_1, \cdots, p_i, c_{p_i}, \cdots, c_\gamma, A$.

These properties are seen to hold when we have found the first vertex of the chain, p_1. Having found p_i, we find the next vertex, p_{i+1}, as follows. As p_i does not touch c_{p_i+1}, (or A, if c_{p_i} is c_γ), inside the circuit, c_{p_i} touches a vertex in between. Any such vertex is not a vertex of the circuit R, nor is it any of the vertices $p_1, \cdots, p_i$, by the above assumptions. Call p_{i+1} the vertex c_{p_i} touches next after p_i. If p_{i+1} touches A, the first part of the chain is finished. Otherwise, let $c_{p_{i+1}}$ be the vertex nearest A that p_{i+1} touches (which may be c_{p_i}). Now p_{i+1} is distinct from all former p's, p_i touches p_{i+1}, p_{i+1} touches $c_{p_{i+1}}$, and no vertex $p_1, \cdots, p_{i+1}$ touches $c_{p_{i+1}}$ or any vertex nearer A inside the new circuit. Thus the same properties still hold, and we continue finding vertices of the chain.

We note that, although p_{i+1} touches c_{p_i}, it touches no vertex c_s nearer C than c_{p_i}. Thus if p_i touches c_s, p_j touches c_t, and $j > i$, then $t \geqq s$.

We must eventually reach A. For each time a vertex p_i does not touch A, we find a new vertex p_{i+1}, all the vertices p_s are distinct, and there are only a finite number of vertices inside the circuit.

Call the last vertex of this chain p_η. If p_η touches a_1 (or B, if there is no a_1), call it also p_φ. Otherwise, A touches vertices in between, none of which are vertices of the circuit R or of the chain $p_1, \cdots, p_\eta$. Call these in order $p_{\eta+1}, \cdots, p_\varphi$. We now have a chain of vertices $p_1, \cdots, p_\varphi$, stretching from b_j to a_1(or B), each of which touches a vertex c_s or A.

(2) *The chain of q's.* Mark in now any edges there may be joining the vertices $b_j, p_1, \cdots, p_\varphi, a_1 (B)$ inside the circuit we now have, which includes the p's and B. Call q_1 the vertex of the set $p_1, \cdots, p_\varphi$ nearest $a_1 (B)$ which b_j touches (which may be p_1). Thus q_1 exists. Having found q_i, if it touches a_1 (B), we cass call it q_θ. Otherwise, we take as q_{i+1} the vertex of the set $p_1, \cdots, p_\varphi$ nearest a_1 (B) which q_i touches. Continue

in this manner till we reach $a_1(B)$. Now every vertex q_i touches a vertex c_s or A. Also, no vertices of the set $b_j, q_1, \cdots, q_\theta, a_1(B)$ touch inside the circuit $b_j, q_1, \cdots, q_\theta, a_1, \cdots, a_\alpha, B, b_1, \cdots, b_j$ (where the a's may be missing), on account of the construction of the chain. As, also, no pair of the vertices $a_1, \cdots, a_\alpha, B$, and no pair of the vertices $B, b_1, \cdots, b_j$ touch inside the circuit, we can apply the lemma and draw a line from b_j to B, passing through every vertex of and inside this circuit.

(3.) *The curve.* If there are no vertices q_s touching A, call $a_1(B)$, q_ψ. Otherwise, call the first vertex q_s which touches A, q_ψ. To finish the proof of the lemma, we have only to pass from A to b_j through every vertex on c_k's side of, but not in, the chain $b_j, q_1, \cdots, q_\psi, A$. For if q_ψ is $a_1(B)$, the chains $b_j, q_1, \cdots, q_\psi$ and $b_j, q_1, \cdots, q_\theta, a_1(B)$ are identical, and we have passed through every vertex of and on B's side of the chain in passing from b_j to B. If q_ψ is not $a_1(B)$, consider the circuit $A, a_1(B), q_\theta, \cdots, q_\psi, A$, (where q_ψ may be q_θ). As A touches each of these vertices, there can be no vertices inside the circuit, by (γ). Thus all the vertices we have not passed through on c_k's side of the chain $b_j, q_1, \cdots, q_\theta, a_1(B), A$, are also on c_k's side of the chain $b_j, q_1, \cdots, q_\psi, A$.

We will pass from A to b_j in two steps: first from A to c_k, on A's side of the edge $b_j c_k$, then from c_k to b_j, on C's side of the same edge.

Mark in all edges between the q's and the c's. Remembering that each vertex q_i, $i < \psi$, touches a vertex c_s, and that if q_i touches c_s, q_j touches c_t, and $j > i$, then $t \geqq s$, we see that these edges divide the section of the graph we must pass through into a number of sections, each of which we will pass through in turn.

Suppose q_ψ touches a vertex of the set $c_k, \cdots, c_\gamma$. Call the one nearest A that q_ψ touches c_g. If c_g is c_γ, there are no vertices inside the circuit A, q_ψ, c_γ, A, and we pass directly from A to c_γ. Otherwise, c_g does not touch A inside the circuit, and therefor q_ψ touches other vertices in between. Call these vertices in order $r_1, \cdots, r_\mu$. There are no vertices inside the circuit $A, q_\psi, c_g, r_\mu, \cdots, r_1, A$. Thus we need only pass from A to c_g through all the vertices of and inside the circuit $A, r_1, \cdots, r_\mu, c_g, \cdots, c_\gamma, A$. But we can do this, by the lemma. For, no pair of the vertices $A, r_1, \cdots, r_\mu, c_g$, and no pair of the vertices $c_g, \cdots, c_\gamma, A$ touch inside the circuit.

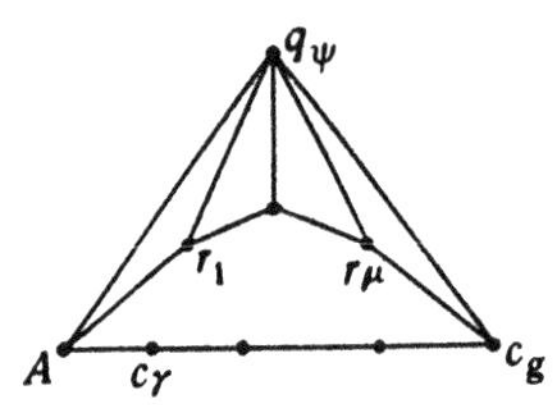

Fig. 6.

If q_ψ touches any more vertices of the set $c_k, \cdots, c_\gamma$, we pass through each of the sections thus formed in turn in exactly the same manner, till we reach the last c that q_ψ touches, c_h.

If the vertex nearest A of the c's that $q_{\psi-1}$ touches is c_i, we must now pass through the section bounded by $c_h, q_\psi, q_{\psi-1}, c_i, \cdots, c_h$.

If q_ψ did not touch any vertex c_s, we would have this section to pass through in the first place, c_h being replaced by A.

If c_i is c_h, this section is a triangle which contains no vertices inside, and we consider the next section. Suppose therefor c_i is not c_h. As then $q_{\psi-1}$ does not touch c_h, q_ψ touches vertices in between, none of which are any of the set $c_i, \cdots, c_h$. We obtain thus a chain of vertices stretching from c_h to $q_{\psi-1}$, of which the last is say d. Similarly, we obtain a chain of vertices stretching from q_ψ to c_i, of which the first is d. As there are no vertices inside the circuit $c_h, q_\psi, q_{\psi-1}, c_i, \cdots, d, \cdots, c_h$, we have only to pass from c_h to c_i through the circuit $c_h, \cdots, d, \cdots, c_i, \cdots, c_h$. We can do this, by the lemma. For, no vertices of the set $c_h, \cdots, d$, none of the set $d, \cdots, c_i$, and none of the set $c_i, \cdots, c_h$ touch inside the circuit.

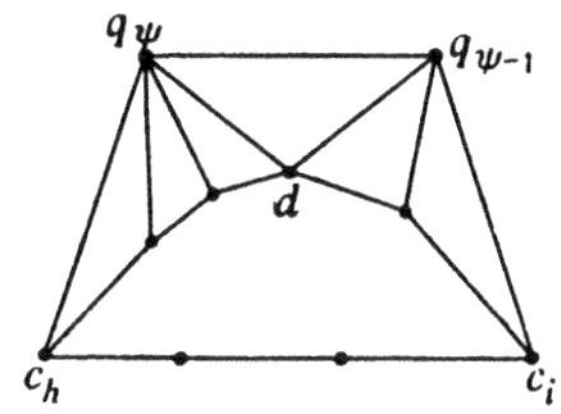

Fig. 7.

We pass in this manner through each section in turn, till we reach c_k. The last section, in particular, is bounded by the vertices $c_f, q_1, b_j, c_k, \cdots, c_f$, where c_f is either c_k or the vertex nearest c_k of the c's that q_1 touches. Thus here, b_j takes the place of what would otherwise be the next q.

We have now but to pass from c_k to b_j on C's side of the edge $b_j c_k$. We can do this, by the lemma. For, the vertices c_k, b_j, no pair of the set $b_j, \cdots, b_\beta, C$, and no pair of the set $C, c_1, \cdots, c_k$ touch inside the circuit thus described.

The proof of the lemma, and therefor of Theorem I, is now complete.

3. Proofs of the theorems on maps.

The dual representation. Given a map on the surface of a sphere, we find the dual representation in the form of a graph as follows. Mark in each region of the map a point, which will be a vertex of the graph, and which we shall call by the same name as the region of the map in which it lies. Across each boundary line of the map draw a line connecting the vertices in the two regions the boundary separates, forming an edge of the graph.

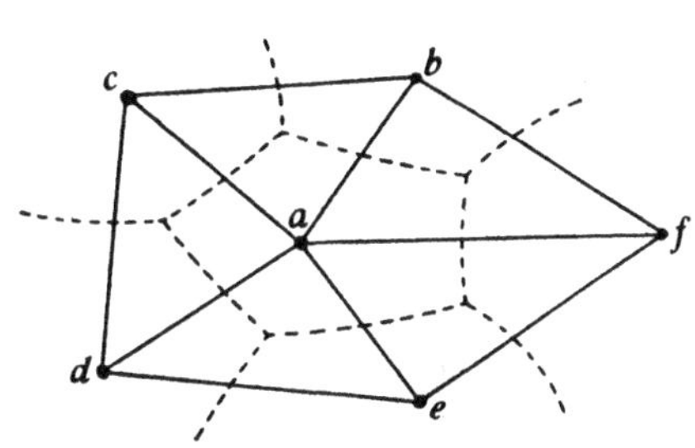

Fig. 8.

Now surrounding each vertex of the map there is a region of the graph bounded by a set of edges.

Proof of Theorem II. We will show first that in any map of the type considered in Theorem II, the dual graph holds to the properties (α), (β) and (γ) of § 2.

Each region of the map is simply connected, on account of (A_1). Each boundary is a boundary between two distinct regions. For suppose there were a boundary line QR running through a single region a. We could then, starting from a point P of QR, move into a on one side of QR, run along a path remaining always in a, and get back to P on the other side of QR. Let us now run around the boundary of a. At some time we pass along the boundary line QR. We are now inside the path we have drawn through a, and as the boundary of a is a closed curve, we must get out again. But we can only get out by passing through P, which contradicts (A_1).

Suppose we run around the boundary of a region a in a counter-clockwise sense. We are on successively sections of the boundary separating a from other regions $b, c, \cdots, f$, in cyclic order. Thus in the dual graph, a touches $b, c, \cdots, f$, in cyclic order, and these vertices are distinct from a.

Suppose a touches b and next c. Then if we pass around the boundary of the region a in a counter-clockwise sense, two successive sections of this boundary will be C, separating a and b, and B, separating a and c. C and B will meet at the vertex V. By (B), only one other boundary line abutts at V. Call it A. It must thus separate the regions b and c. Run now around the boundary of b in a counter-clockwise sense. Two successive sections of this boundary will be A and C. Thus we see that the vertex b touches c and next a, proving property (β).

Suppose now a touches in order $b, c, d, \cdots, f$. These vertices are then all distinct. For consider any two of the vertices a touches, say b and d. If a touches b and next d, or d and next b, then b touches d, and therefor b and d are distinct. Suppose now a touches a vertex c after b and before d, and a vertex f after d and before b. Here again b and d must be distinct, for otherwise the regions a and b would form a multiply connected region, separating c and f, contrary to (A_2).

Except in a map of three regions, for which Theorem II is obvious, each region of the map touches at least three others. For if there were a region touching only one or two others, that region or pair of regions would form a multiply connected region, contrary to (A_1) or (A_2). Thus each vertex of the dual graph touches at least three others. This finishes the proof of property (α).

Finally, there are no triangles in the graph other than elementary triangles. For if there were such a triangle, the regions of the map surrounding it would form a multiply connected region, contrary to (A_3). The properties (α), (β) and (γ) are now proved.

Now, applying Theorem I to the dual graph, we find a circuit passing through every vertex of the graph. This circuit is the desired closed curve passing through every region of the map.

Proof of the equivalent statement of the four color map problem. Elementary considerations in the four color map problem show that if any map of the type considered in Theorem II can be colored in four colors, then any map on the surface of a sphere can be colored in four colors. We need therefor consider only maps of the above type.

Put the dual graph of such a map in the normal form. Suppose we can color this polygonal configuration in four colors. We then color each region of the map with the same color as the corresponding vertex of the dual graph. Any two regions with a common boundary correspond to two vertices of the graph which are joined by an edge, and are therefor of different colors.

The converse is obvious, as every polygonal configuration is the dual of a map.

Conundrum. Suppose a man, living in a certain country (state), wishes to visit all the countries about him, but does not wish to pass through any country more than once on his voyage. Can he do it? If the region he wishes to visit covers the entire globe, he can do it if the countries make up a map of the type considered in Theorem II. Suppose now the region covers but a portion of the globe. If, upon replacing the rest of the globe by a single country, we obtain a map of the type considered, he can do it also. We have but to apply the lemma to the ring of countries about the added country. By (A_3), no pair of the countries of this ring touch inside the ring. Therefor, picking out any two adjacent countries of the ring, A, B, we draw a line from one to the other, passing through every country the man wishes to visit. We now join the two ends of this line, completing the man's path.

More generally, whenever the conditions of the lemma are satisfied by the ring, calling some two adjacent countries A and B, we obtain the desired path.

4. Further remarks.

Necessity of (A_3). Theorem I would not be true if the assumption that there are no circuits of three edges other than the elementary triangles were omitted. That is, Theorem II would not be true if the assumption (A_3) were omitted. The following example shows this.[3]

[3] This example of such a map containing the least number of regions was communicated to me by C. N. Reynolds.

The number P_n. In constructing the normal form for a graph, we divide an n-sided polygon into triangles by diagonals. It is interesting to know in how many ways we can do this. The formula for this number was found by Euler. A simple proof was first given by Lamé:[4]

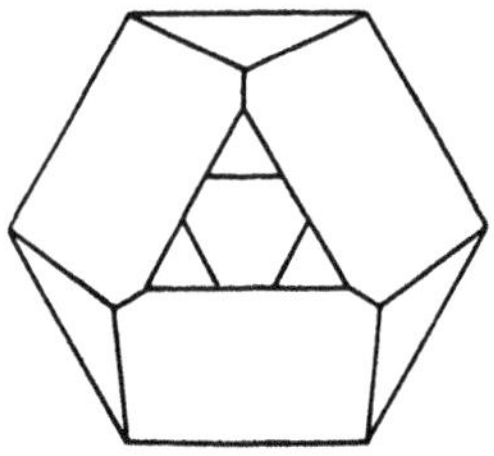

Fig. 9.

$$P_n = 2^{n-3} \frac{3 \cdot 5 \cdot 7 \cdots (2n-5)}{3 \cdot 4 \cdot 5 \cdots (n-1)}.$$

As we divide both the inside and outside of the polygon into triangles, we can construct in this manner P_n^2 different figures. Of course these are not all graphs of the type considered, and many of them give the same graph. For instance, there are 96 different circuits in the graph, Fig. 1.

[4] J. Math. Pures Appl. (1), 3 (1838), pp. 505–507.

HARVARD UNIVERSITY.

NON-SEPARABLE AND PLANAR GRAPHS*

BY

HASSLER WHITNEY

Introduction. In this paper the structure of graphs is studied by purely combinatorial methods. The concepts of rank and nullity are fundamental. The first part is devoted to a general study of non-separable graphs. Conditions that a graph be non-separable are given; the decomposition of a separable graph into its non-separable parts is studied; by means of theorems on circuits of graphs, a method for the construction of non-separable graphs is found, which is useful in proving theorems on such graphs by mathematical induction. In the second part, a dual of a graph is defined by combinatorial means, and the paper ends with the theorem that a necessary and sufficient condition that a graph be planar is that it have a dual.

The results of this paper are fundamental in papers by the author on *Congruent graphs and the connectivity of graphs*† and on *The coloring of graphs*.‡

I. Non-separable graphs

1. **Definitions.**§ A *graph* G consists of two sets of symbols, finite in number: *vertices*, $a, b, c, \cdots, f$, and *arcs*, $\alpha(ab), \beta(ac), \cdots, \delta(cf)$. If an arc $\alpha(ab)$ is present in a graph, its *end vertices* a, b are also present. We may write an arc $\alpha(ab)$ or $\alpha(ba)$ at will; we may write it also ab or ba if no confusion arises,—if there is but a single arc joining a and b in G. We say the vertices a and b are *on* the arc $\alpha(ab)$, and the arc $\alpha(ab)$ is *on* the vertices a and b. The *null graph* is the graph containing no arcs or vertices.

The obvious geometrical interpretation of such a graph, or *abstract graph*, is a *topological graph*, let us say. Corresponding to each vertex of the abstract graph, we select a point in 3-space, a vertex of the topological graph. Corresponding to each arc $\alpha(ab)$ of the abstract graph, we select an arc joining the corresponding vertices of the topological graph. An arc is here a set of points in (1, 1) correspondence with the unit interval, its end vertices corresponding

* Presented to the Society, October 25, 1930; received by the editors February 2, 1931. An outline of this paper will be found in the Proceedings of the National Academy of Sciences, vol. 17 (1931), pp. 125–127.

† American Journal of Mathematics, vol. 54 (1932), pp. 150–168.

‡ An outline will be found in the Proceedings of the National Academy of Sciences, vol. 17 (1931), pp. 122–125.

§ Compare Ste. Laguë, *Les Réseaux*, Mémorial des Sciences Mathématiques, fascicule 18, Paris, 1926.

with the ends of the interval. Moreover, we let no arc pass through other vertices or intersect other arcs. We shall consider topological graphs no further till we come to the section on planar graphs.

An *isolated vertex* is a vertex which is not on any arc. A *chain* is a set of one or more distinct arcs which can be ordered thus: ab, bc, cd, $\cdots$, ef, where vertices in different positions are distinct, i.e. the chain may not intersect itself. A *suspended chain* is a chain containing two or more arcs such that no vertex of the chain other than the first and last is on other arcs, and these two vertices are each on at least two other arcs. A *circuit* is a set of one or more distinct arcs which can be put in cyclic order, ab, bc, $\cdots$, ef, fa, vertices being distinct as in the case of the chain. A *k-circuit* is a circuit containing k arcs. Thus, the arc $\alpha(aa)$ is a 1-circuit; the two arcs $\alpha(ab)$, $\beta(ab)$ form a 2-circuit.

A graph is *connected* if any two of its vertices are joined by a chain. Obviously, if a and b are joined by a chain, and b and c are joined by a chain, then a and c are joined by a chain. Any graph consists of a certain number of *connected pieces* (one, if the graph is connected). In particular, an isolated vertex is one of the connected pieces of a graph. A graph is called *cyclicly connected* if any two of its vertices are contained in a circuit. If G_1, G_2, $\cdots$, G_m are a set of graphs, no two of which have a common vertex (or arc, therefore), we say the graph G, formed of the arcs and vertices of all these graphs, is the *sum* of these graphs. Thus, a graph is the sum of its connected pieces. A *forest* is a graph containing no circuit. A *tree* is a connected forest. A *subgraph* H of G is a graph containing a subset (in particular, all or none), of the arcs of G, and those vertices of G which are on these arcs.

2. **Rank and nullity.*** Given a graph G which contains V vertices, E arcs, and P connected pieces, we define its *rank* R, and its *nullity* (or *cyclomatic number* or first Betti number) N, by the equations

$$R = V - P,$$
$$N = E - R = E - V + P.$$

If G contains the single arc ab, it is of rank 1, nullity 0, while if it contains the single arc aa, it is of rank 0, nullity 1.

The first two theorems follow immediately from the definitions of rank and nullity:

THEOREM 1. *If isolated vertices be added to or subtracted from a graph, the rank and nullity remain unchanged.*

* These are just the rank and nullity of the matrix H_1 of Poincaré. See Veblen's Colloquium Lectures, *Analysis Situs*.

THEOREM 2. *Let the graph G' be formed from the graph G by adding the arc ab. Then*

(1) *if a and b are in the same connected piece in G, then*

$$R' = R, \quad N' = N + 1;$$

(2) *if a and b are in different connected pieces in G, then*

$$R' = R + 1, \quad N' = N.$$

THEOREM 3. *In any graph G,*

$$R \geqq 0, \quad N \geqq 0.$$

For let G_1 be the graph containing the vertices of G but no arcs. Then if R_1 and N_1 are its rank and nullity,

$$R_1 = N_1 = 0.$$

We build up G from G_1 by adding the arcs one at a time. The theorem now follows from Theorem 2.

THEOREM 4. *A forest G is a graph of nullity* 0, *and conversely.*

Suppose first G contained a circuit P. We shall show that the nullity of G is >0. We build up G arc by arc, adding first the arcs of the circuit P. In adding the last arc of the circuit, the nullity is increased by 1, as this arc joins two vertices already connected. (This argument holds even if the circuit is a 1-circuit.) But in adding the rest of the arcs, the nullity is never decreased, by Theorem 2. Thus the nullity of G is >0.

Now suppose G is a forest, and therefore contains no circuit. Build up G arc by arc. Each arc we add joins two vertices formerly not connected. For otherwise, this arc, together with the arcs of a chain connecting the two vertices, would form a circuit. Therefore, by Theorem 2, the nullity remains always the same, and is thus 0.

3. **Theorems on non-separable graphs.** We introduce the following

Definitions. Let H_1, which contains the vertex a_1, and H_2, which contains the vertex a_2, be two graphs without common vertices. Let us rename a_1 a, and rename the arcs of H_1 on a_1 accordingly; that is, if a_1b is an arc on a_1, we rename it ab. Rename also a_2 a, and rename the arcs of H_2 accordingly. H_1 and H_2 have now the vertex a in common; they form the graph G, say. We say G is formed by letting the vertex a_1 of H_1 *coalesce* with the vertex a_2 of H_2, or, by joining H_1 and H_2 at a vertex. Geometrically, we pull the vertices a_1 and a_2 together to form the single vertex a.

Let G be a connected graph such that there exist no two graphs H_1 and

H_2, each containing at least one arc, which form G if they are joined at a vertex. Then G is called *non-separable*. Geometrically, a connected graph is non-separable if we cannot break it at a single vertex into two graphs, each containing an arc. For example, the graph consisting of the two arcs ab, bc is separable, as is the graph consisting of the two arcs $\alpha(aa)$, $\beta(aa)$. A graph containing but a single arc is non-separable, as is the graph containing only the arcs $\alpha(ab)$, $\beta(ab)$.

If G is not non-separable, we say G is *separable*. Thus, a graph that is not connected is separable. Suppose some connected piece G_1 of G is separable. If H_1 and H_2 joined at the vertex a form G_1, we say a is a *cut vertex* of G. We have consequently

THEOREM 5. *A necessary and sufficient condition that a connected graph be non-separable is that it have no cut vertex.*

THEOREM 6. *Let G be a connected graph containing no 1-circuit. A necessary and sufficient condition that the vertex a be a cut vertex of G is that there exist two vertices b, c in G, each distinct from a, such that every chain from b to c passes through a.*

First suppose a is a cut vertex of G. Then, by definition, H_1 and H_2, each containing at least one arc which is not a 1-circuit, form G if they are joined at a. Let b be a vertex of H_1 and c a vertex of H_2, each distinct from a. As a is the only vertex in both H_1 and H_2, every chain from b to c in G passes through a.

Suppose now every chain from b to c in G passes through a. Remove the vertex a and all the arcs on a. The resulting graph G' is not connected, b and c being in different connected pieces. Let H_1' be that connected piece of G' containing b, and let H_2' be the rest of G'. Replace a by the two vertices a_1 and a_2. Now put back the arcs we removed, letting them touch a_1 if their other end vertices are in H_1', and letting them touch a_2 otherwise. Let H_1 and H_2 be the resulting graphs. Then H_1 and H_2 each contain at least one arc, and they form G if the two vertices a_1, a_2 are made to coalesce. Hence, by definition, a is a cut vertex of G.

THEOREM 7. *Let G be a graph containing no 1-circuit and containing at least two arcs. A necessary and sufficient condition that G be non-separable is that it be cyclicly connected.**

If G is not connected, the theorem is obvious. Assume therefore G is connected.

---

* A similar theorem has been proved for more general continuous curves by G. T. Whyburn, Bulletin of the American Mathematical Society, vol. 37 (1931), pp. 429–433.

Suppose first G is separable. Then, by Theorem 5, G has a cut vertex a, and by Theorem 6, there are two vertices b, c in G such that every chain from b to c passes through a. Hence there is no circuit in G containing b and c.

Suppose now there exist two vertices b, c in G which are contained in no circuit. Let $bd, de, \cdots, gc$ be some chain from b to c.

Case 1. There exists a circuit containing b and d. In this case, let a be the last vertex of the chain which is contained in a circuit passing also through b. Let f be the next vertex of the chain. Then every chain from f to b passes through a. For suppose the contrary. Let C be a chain from f to b not passing through a. Let P be a circuit containing b and a. Follow C from f till we first reach a vertex of P. Follow the circuit P now as far as b if b was not the vertex we reached, and continue along P till we reach a. Passing from a to f along the arc af completes a circuit containing both b and f, contrary to hypothesis. Hence, by Theorem 6, a is a cut vertex of G, and therefore G is separable.

Case 2. There exists no circuit containing b and d. Then there is but a single arc joining b and d, and they are joined by no other chain. As G is connected and contains at least two arcs, there is either another arc on b or another arc on d, say the first. The other case is exactly similar. If we add a vertex b' and replace the arc bd by the arc $b'd$, b and d are no longer joined by a chain, and hence the resulting graph G' is not connected. Let H_1 be that part of G' containing the arc $b'd$, and let H_2 be the rest of G'. As there is still an arc on b, H_2 contains at least one arc. Letting the vertices b and b' coalesce forms G, and hence G is separable. The proof is now complete.

THEOREM 8. *A non-separable graph G containing at least two arcs contains no* 1-*circuit and is of nullity* >0. *Each vertex is on at least two arcs.*

Suppose G contained a 1-circuit. Call it H_1. Let H_2 be the rest of the graph. Then H_1 and H_2 have but a single vertex in common, and thus G is separable.

Next, by Theorem 7, G is cyclicly connected. As G contains no 1-circuit, G contains at least two vertices. Containing these there is a circuit. Therefore, by Theorem 4, the nullity of G is >0.

Finally, if there were a vertex on no arcs, G would not be connected. If there were a vertex a on the single arc ab, b would be a cut vertex of G.

THEOREM 9. *Let G be a graph of nullity* 1 *containing no isolated vertices, such that the removal of any arc reduces the nullity to* 0. *Then G is a circuit.*

By Theorem 4, G contains a circuit. Suppose G contained other arcs besides. Removing one of these, the nullity remains 1, as the circuit is still present, contrary to hypothesis. There are no other vertices in G, as G contains no isolated vertices. Hence G is just this circuit.

THEOREM 10. *A non-separable graph G of nullity 1 is a circuit.*

If G contains but a single arc, it is a 1-circuit, being of nullity 1. Suppose G contains at least two arcs. By Theorem 8, it contains no 1-circuit. By Theorem 7, it is cyclicly connected. Remove any arc ab from G; a and b are still connected, and therefore, by Theorem 2, the nullity of G is reduced to 0. Hence, by Theorem 9, G is a circuit.

The converses of the last two theorems are obviously true.

4. **Decomposition of separable graphs.** If the graph G contains a connected piece which is separable, we may separate that piece into two graphs, these graphs having formerly but a single vertex in common. We may continue in this manner until every resulting piece of G is non-separable. We say G is separated into its *components.*

LEMMA. *Let the connected separable graph G be decomposed into the two pieces H_1 and H_2 which had only the vertex a in common in G. Then every non-separable subgraph of G is contained wholly in either H_1 or H_2.*

Suppose the contrary. Then some non-separable subgraph I of G is not contained wholly in either H_1 or H_2. Let I_1 be that part of I in H_1, and I_2 that part in H_2; I_1 and I_2 have at most the vertex a in common. I_1 and I_2 each contain at least one arc. For otherwise, if I_1, say, contained no arc, as it contains a vertex distinct from a, it would not be connected. Thus I is separable into the pieces I_1 and I_2, a contradiction again.

THEOREM 11. *Every non-separable subgraph of G is contained wholly in one of the components of G.*

This follows upon repeated application of the above lemma.

THEOREM 12. *A graph G may be decomposed into its components in a unique manner.*

Suppose we could decompose G into the components $H_1, H_2, \cdots, H_m$, and also into the components $H_1', H_2', \cdots, H_n'$. We shall show that these sets are identical. Take any H_i. It is a non-separable subgraph of G, and thus is contained in some component H_j', by Theorem 11. Similarly, H_j' is contained in some component H_k. Thus H_i is contained in H_k, and they are therefore identical. Hence H_i and H_j' are identical. In this manner we show that each H_k is identical with some H_l', and each H_l' is identical with some H_k, proving the theorem.

THEOREM 13. *Let $H_1, H_2, \cdots, H_m$ be the components of G. Let $R_1, R_2, \cdots, R_m$, and $N_1, N_2, \cdots, N_m$ be their ranks and nullities. Then*

$$R = R_1 + R_2 + \cdots + R_m,$$

$$N = N_1 + N_2 + \cdots + N_m.$$

Let G' be G separated into its components, and let R' be the rank of G'. G is formed from G' by letting vertices of different components coalesce. Each time we join two pieces, the number of vertices and the number of connected pieces are each reduced by 1, so that the rank remains the same. Thus

$$R = R'.$$

Now

$$V' = V_1 + V_2 + \cdots + V_m,$$

$$P' = P_1 + P_2 + \cdots + P_m$$

(where each $P_i = 1$). Subtracting,

$$R = R' = R_1 + R_2 + \cdots + R_m.$$

As also

$$E = E_1 + E_2 + \cdots + E_m,$$

it follows that

$$N = N_1 + N_2 + \cdots + N_m.$$

For a converse of this theorem, see Theorem 17.

THEOREM 14. *Divide the arcs of the non-separable graph G into two groups, each containing at least one arc, forming the subgraphs H_1 and H_2, of ranks R_1 and R_2. Then*

$$R_1 + R_2 > R.$$

Let the connected pieces of H_1 be $H_{11}, \cdots, H_{1m}$ (there may be but one piece, H_{11}), and let those of H_2 be $H_{21}, \cdots, H_{2n}$. Then obviously

$$R_1 = R_{11} + \cdots + R_{1m},$$

$$R_2 = R_{21} + \cdots + R_{2n},$$

whence

$$R_1 + R_2 = R_{11} + \cdots + R_{1m} + R_{21} + \cdots + R_{2n}.$$

Let G' be the sum of the graphs $H_{11}, \cdots, H_{2n}$. Then G' is of rank $R_{11} + \cdots + R_{2n}$. We form G from G' by letting vertices of the graphs $H_{11}, \cdots, H_{2n}$ coalesce. Each time we let vertices of different connected pieces coalesce, the rank is unaltered. Each time we let vertices in the same connected piece coalesce, the rank is reduced by 1. This latter operation happens at least once. For otherwise, let a_1 and a_2 be the last two vertices we let coalesce. Then a_1 and a_2 were formerly in two different pieces, I_1 and I_2. Thus

I_1 and I_2 joined at a vertex form G, and G is separable, contrary to hypothesis. Thus the rank of G is less than the rank of G', that is,

$$R < R_{11} + \cdots + R_{2n}.$$

Hence

$$R_1 + R_2 > R.^*$$

Theorems 13 and 14 give

THEOREM 15. *A necessary and sufficient condition that a graph be non-separable is that there exist no division of its arcs into two groups H_1 and H_2, each containing at least one arc, so that*

$$R = R_1 + R_2.$$

5. **Circuits of graphs.** We shall say two non-separable graphs, each containing at least one arc, form a *circuit of graphs*, if they have at least two common vertices. (They may also have common arcs.) Thus the two graphs G_1: $\alpha(ab)$ and G_2: $\alpha(ab)$ (which are the same graph) form a circuit of graphs. However, the two graphs G_1: $\alpha(aa)$ and G_2: $\beta(aa)$, having but one common vertex, do not form a circuit of graphs. We shall say three or more non-separable graphs form a *circuit of graphs* if we can name them $G_1, G_2, \cdots, G_m$ in such a way that G_1 and G_2 have just the vertex a_1 in common, G_2 and G_3 have just the vertex a_2 in common, $\cdots$, G_m and G_1 have just the vertex a_m in common, these vertices are all distinct, and no other two of these graphs have a common vertex. Thus the three graphs G_1:ab, G_2:bc, G_3:ca form a circuit of graphs.

We note that there can be no 1-circuit in a circuit of graphs; also, no subset of the graphs in a circuit of graphs form a circuit of graphs. We may think of a circuit of graphs as forming a single graph.

THEOREM 16. *A circuit of graphs G is a non-separable graph.*

First suppose there are but two graphs, G_1 and G_2, present. Suppose G were separable. Then it is separable into at least two components $H_1, H_2, \cdots, H_k$. By Theorem 11, G_1 and G_2 are each contained wholly in one of these components. As G_1 and G_2 together form G, there are just two components, and they are G_1 and G_2. These, when joined at a vertex, form G. But this is contrary to the hypothesis that G_1 and G_2 have at least two vertices in common.

Next suppose there are more than two graphs present. Let C_1 be a chain in G_1 joining a_m and a_1, let C_2 be a chain in G_2 joining a_1 and $a_2, \cdots$, let C_m be a chain in G_m joining a_{m-1} and a_m. These chains taken together form a cir-

* This theorem may also be proved easily from Theorem 17.

cuit P passing through all the graphs. Now separate G into its components. By Theorem 11 (see the converse of Theorem 10), P is contained in one of these components. The same is true of each of the graphs $G_1, G_2, \cdots, G_m$, and hence these graphs are all contained in the same component. Thus G is itself this component, that is, G is non-separable.

THEOREM 17. *Let $G_1, \cdots, G_m$ be a set of non-separable graphs, each containing at least one arc, and let G be formed by letting vertices and arcs of different graphs coalesce. Then the following four statements are all equivalent:*

(1) *$G_1, \cdots, G_m$ are the components of G.*

(2) *No two of the graphs $G_1, \cdots, G_m$ have an arc in common, and there is no circuit in G containing arcs of more than one of these graphs.*

(3) *No subset of these graphs form a circuit of graphs.*

(4) *If $R, R_1, \cdots, R_m$ are the ranks of $G, G_1, \cdots, G_m$ respectively, then*

$$R = R_1 + \cdots + R_m.$$

We note that we cannot replace the word rank by the word nullity in (4). For let G be the graph containing the arcs $\alpha(ab), \beta(ab), \gamma(ab)$. Let G_1 contain α and β, and G_2, β and γ. Then the nullity of G is the sum of the nullities of G_1 and G_2, but G_1 and G_2 are not the components of G. We shall prove

(a) if (1) holds, (2) holds,

(b) if (2) holds, (3) holds,

(c) if (3) holds, (1) holds, establishing the equivalence of (1), (2) and (3);

(d) if (1) holds, (4) holds, and finally

(e) if (4) holds, (3) holds, establishing the equivalence of (4) and the other statements.

(a) If (1) holds, (2) holds. For first, in forming G from its components $G_1, \cdots, G_m$, we let vertices alone coalesce, and thus no two of the graphs have an arc in common. Also, there is no circuit in G containing arcs of more than one of the graphs; for each circuit, being a non-separable graph, is contained entirely in one of the components of G, by Theorem 11.

(b) If (2) holds, (3) holds. For suppose the contrary. If, first, some two graphs, say G_1 and G_2, form a circuit of graphs, they have at least two vertices in common, say a and b. Join a and b by a chain C in G_1 and by a chain D in G_2. By hypothesis, G_1 and G_2 have no arcs in common, and thus the arcs of C and D are distinct. From a follow along C till we first reach a vertex d of D. From d follow along D till we get back to a. We have formed thus a circuit containing arcs of both G_1 and G_2, contrary to hypothesis.

Now suppose the graphs $G_1, \cdots, G_k$, $k>2$, formed a circuit of graphs. In the proof of Theorem 16 we found a circuit passing through all the graphs of such a circuit of graphs, again contrary to hypothesis.

(c) If (3) holds, (1) holds. Assuming that no subset of the graphs $G_1, \cdots, G_m$ forms a circuit of graphs, we will show first that some one of these graphs has at most a single vertex in common with other of the graphs. For suppose each graph had at least two vertices in common with other graphs. Then G_1 has a vertex a_1 in common with some graph, say G_2. As G_2 has at least two vertices in common with other graphs, it has a vertex a_2, distinct from a_1, in common with another graph, say G_3. If we continue in this manner, we must at some point get back to a graph we have already considered.

Now starting with G_1, consider the graphs in order, and let G_i be the first one which has a vertex in common with one of the preceding graphs other than the vertex a_{i-1}, which we know already it has in common with G_{i-1}. Now of the graphs $G_{i-1}, G_{i-2}, \cdots, G_1$, let G_j be the first with which G_i has a common vertex, other than the vertex a_{i-1}. First suppose G_j is G_{i-1}. Then G_i and G_{i-1} have at least two vertices in common, and they form therefore a circuit of graphs, contrary to hypothesis. Next suppose G_j is not G_{i-1}. Then on account of the choice of G_i and G_j, G_j and G_{j+1} have just one common vertex a_j, G_{j+1} and G_{j+2} have just one common vertex $a_{j+1}, \cdots, G_i$ and G_j have just one common vertex a_i (for otherwise G_i and G_j would form a circuit of graphs), and no other two of these graphs have a vertex in common. These vertices $a_j, a_{j+1}, \cdots, a_i$ are all distinct. For, on account of the construction of the chain of graphs, two succeeding vertices a_k and a_{k+1} are distinct. a_i and a_j are distinct, for otherwise G_i and G_{j+1} would have a common vertex, etc. These graphs $G_j, G_{j+1}, \cdots, G_i$ form therefore a circuit of graphs, contrary to hypothesis.

Some graph therefore, say G_1, has at most a single vertex in common with the other graphs. Thus either it is separated from them, or we can separate it at a single vertex. Now among the graphs $G_2, \cdots, G_m$, there is also no circuit of graphs, so again we can separate one of them, say G_2. Continuing, we have finally separated G into its components $G_1, G_2, \cdots, G_m$.

(d) If (1) holds, (4) holds. This is just Theorem 13.

(e) If (4) holds, (3) holds. Let G' be the sum of the graphs $G_1, \cdots, G_m$. We form G from G' by letting vertices and arcs of different graphs coalesce. Each time we let two vertices coalesce, either (α) the two vertices were formerly in different connected pieces, in which case the rank is unchanged, or (β) the two vertices were in the same connected piece, in which case the rank is reduced by 1. Letting arcs alone coalesce (their end vertices having already coalesced) does not alter the rank. Thus in any case, the rank is never increased. To begin with, the rank of G' is $G_1 + \cdots + G_m$, and by hypothesis, the rank of G is $G_1 + \cdots + G_m$. Thus the rank is never altered, and (β) never occurs. Hence, obviously, no circuit of graphs is formed in forming G from G'. This completes the proof of the theorem.

6. **Construction of non-separable graphs.** We prove the following theorem:

THEOREM 18. *If G is a non-separable graph of nullity $N>1$, we can remove an arc or suspended chain from G, leaving a non-separable graph G' of nullity $N-1$.*

Assume the theorem is true for all graphs of nullity $2, 3, \cdots, N-1$. We shall prove it for any graph of nullity N (including the case where $N=2$). This will establish the theorem in general.

Take any non-separable graph G of nullity $N>1$. It contains at least two arcs, and therefore, by Theorem 8, it contains no 1-circuit. Remove from G any arc ab, forming the graph G_1. If G_1 is non-separable, we are through. Suppose therefore G_1 is separable, and let its components be $H_1, H_2, \cdots, H_{m-1}$. G_1 is connected, for between any two vertices c, d there exists a circuit in G by Theorem 7, and therefore there is a chain joining them in G_1.

Let H_m consist of the arc ab. By Theorem 17, no subset of the graphs $H_1, \cdots, H_{m-1}$ form a circuit of graphs, while some subset of the graphs $H_1, \cdots, H_m$ form a circuit of graphs. We shall show that the whole set of graphs $H_1, \cdots, H_m$ form a circuit of graphs. Otherwise, some proper subset, which includes H_m, form a circuit of graphs.

Let H be the graph formed from this circuit of graphs by dropping out H_m. By Theorem 16, the circuit of graphs is a non-separable graph; hence H is connected. All the arcs in G_1 not in the circuit of graphs, form a graph I. Let I_1 be a connected piece of I. Then I_1 has at most a single vertex in common with the rest of G. For suppose I_1 had the two vertices c and d in common with H. From c follow along some chain towards d in H till we first reach a vertex e in I_1. From e follow back along some chain in I_1 to c. We have formed thus a circuit containing arcs of both H and I_1. But as H consists of a certain subset of the components of G_1, this circuit contains arcs of at least two components of G_1, contrary to Theorem 17. Thus I_1 has at most a single vertex in common with the rest of G, and hence G is separable, contrary to hypothesis. Thus $H_1, \cdots, H_m$ form a circuit of graphs, that is, G is formed of a circuit of graphs.

As we assumed G_1 was separable, $m \geqq 3$. Therefore we can order the graphs so that H_1 and H_2 have just the vertex a_1 in common, $\cdots$, H_{m-1} and H_m have just the vertex $a_{m-1}=b$ in common, and H_m and H_1 have just the vertex $a_m=a$ in common. Moreover, these vertices are all distinct, and no other two of the graphs $H_1, \cdots, H_m$ have a common vertex.

As the nullity of G was >1, the nullity of G_1 is >0. By Theorem 13, this is the sum of the nullities of $H_1, \cdots, H_{m-1}$. Therefore the nullity of some one of these graphs, say H_i, is >0.

Suppose first the nullity of H_i is 1. Then, by Theorem 10, H_i is a circuit, consisting of two chains joining a_{i-1} and a_i. Remove one of these chains from G. This leaves a graph G', which again is a circuit of graphs. For the graph H_i we replace by an ordered set of non-separable graphs, each consisting of one of the arcs of the chain we have left in H_i.

Suppose next the nullity of H_i is >1. It is less than N, as H_i is contained in G_1, whose nullity is $N-1$. Therefore, by induction, we can remove an arc or a suspended chain, leaving a non-separable graph H_i' of nullity one less. If neither a_{i-1} nor a_i has thus been removed, we again have a circuit of graphs. Suppose a_i but not a_{i-1} was removed. Replace that part of the chain we removed joining a_i and a vertex of H_i distinct from a_{i-1}. Here again we have a circuit of graphs, H_i being replaced by H_i' and a set of arcs. The case is the same if a_{i-1} but not a_i was removed. If finally, both a_i and a_{i-1} were in the chain we removed, we put back all of the chain but that part between these two vertices. Here again, the resulting graph G' is a circuit of graphs.

Thus in all cases we can drop out from G an arc or suspended chain, leaving a circuit of graphs. By Theorem 16, the resulting graph G' is non-separable. As also the nullity of G' is one less than the nullity of G, the theorem is now proved.

As a consequence of this theorem, Theorem 8, and Theorem 10, we have

THEOREM 19. *We can build up any non-separable graph containing at at least two arcs by taking first a circuit, then adding successively arcs or suspended chains, so that at any stage of the construction we have a non-separable graph.*

It is easily seen that, conversely, any graph built up in this manner is non-separable. For each time we add an arc or suspended chain, these arcs, each considered as a graph, together with the non-separable graph already present, form a circuit of graphs.

II. DUALS, PLANAR GRAPHS

7. Congruent graphs. We introduce the following

Definitions. Given two graphs G and G', if we can rename the vertices and arcs of one, giving distinct vertices and distinct arcs different names, so that it becomes identical with the other, we say the two graphs are *congruent.** (We used formerly the word "homeomorphic.")

* See the author's American Journal paper, cited in the introduction.

The geometrical interpretation is that we can bring the two graphs into complete coincidence by a (1, 1) continuous transformation.

Two graphs are called *equivalent* if, upon being decomposed into their components, they become congruent, except possibly for isolated vertices.

8. **Duals.** Given a graph G, if H_1 is a subgraph of G, and H_2 is that subgraph of G containing those arcs not in H_1, we say H_2 is the *complement* of H_1 in G.

Throughout this section, R, R', r, r', etc., will stand for the ranks of G, G', H, H', etc., respectively, with similar definitions for V, E, P, N.

Definition. Suppose there is a (1, 1) correspondence between the arcs of the graphs G and G', such that if H is any subgraph of G and H' is the complement of the corresponding subgraph of G', then

$$r' = R' - n.$$

We say then that G' is a *dual* of G.*

Thus, if the nullity of H is n, then H' (including all the vertices of G') is in n more connected pieces than G'.

THEOREM 20. *Let G' be a dual of G. Then*

$$R' = N,$$
$$N' = R.$$

For let H be that subgraph of G consisting of G itself. Then

$$n = N.$$

If H' is the complement of the corresponding subgraph of G', H' contains no arcs, and is the null graph. Thus

$$r' = 0.$$

But as G' is a dual of G,

$$r' = R' - n.$$

These equations give

$$R' = N.$$

The other equation follows when we note that $E' = E$.

THEOREM 21. *If G' is a dual of G, then G is a dual of G'.*

Let H' be any subgraph of G', and let H be the complement of the corresponding subgraph of G. Then, as G' is a dual of G,

* While this definition agrees with the ordinary one for graphs lying on a plane or sphere, a graph on a surface of higher connectivity, such as the torus, has in general no dual. (See Theorems 29 and 30.)

$$r' = R' - n.$$

By Theorem 20,

$$R' = N.$$

We note also,

$$e + e' = E.$$

These equations give

$$\begin{aligned} r = e - n = e - (R' - r') &= e - N + (e' - n') \\ &= E - N - n' = R - n'. \end{aligned}$$

Thus G is a dual of G'.

Whenever we have shown that one graph is a dual of another graph, we may now call the graphs "dual graphs."

LEMMA. *If a graph G is decomposed into its components, the rank and nullity of any subgraph H is left unchanged.*

For each time we separate G at a vertex, H is either unchanged or is separated at a vertex. Hence neither its rank nor its nullity is altered. (See the proof of Theorem 13.)

THEOREM 22. *If G' and G'' are equivalent and G' is a dual of G, then G'' is a dual of G.*

Let H be any subgraph of G, and let H' be the complement of the corresponding subgraph of G'. Let G_1' and G_1'' be G' and G'' decomposed into their components. Then G_1' and G_1'' are congruent. H' turns into a subgraph H_1' of G'. Let H_1'' be the corresponding subgraph of G_1'', and H'' the same subgraph in G''. Then

$$r_1' = r_1''.$$

But by the above lemma,

$$r' = r_1', \qquad r'' = r_1''.$$

Hence

$$r' = r''.$$

As a special case of this equation, letting H' be the whole of G', we have

$$R' = R''.$$

As G' is a dual of G,

$$r' = R' - n.$$

Therefore

$$r'' = R'' - n,$$

and G'' is a dual of G.

The converse of this theorem is not true. For define the three graphs G: $\alpha(ab)$, $\beta(ab)$, $\gamma(ac)$, $\delta(cb)$, $\epsilon(ad)$, $\zeta(db)$;

$G': \alpha'(a'b'), \beta'(c'd'), \gamma'(a'd'), \delta'(a'd'), \epsilon'(b'c'), \zeta'(b'c');$
$G'': \alpha''(a''b''), \beta''(b''c''), \gamma''(a''d''), \delta''(a''d''), \epsilon''(c''d''), \zeta''(c''d'').$
G' and G'' are both duals of G, but they are not congruent.*

THEOREM 23. *Let $G_1, \cdots, G_m$ and $G_1', \cdots, G_m'$ be the components of G and G' respectively, and let G_i' be a dual of G_i, $i=1, \cdots, m$. Then G' is a dual of G.*

Let H be any subgraph of G, and let the parts of H in $G_1, \cdots, G_m$ be $H_1, \cdots, H_m$. Let H_i' be the complement of the subgraph corresponding to H_i in G_i', $i=1, \cdots, m$, and let H' be the union of $H_1', \cdots, H_m'$ in G'. Then H' is the complement of the subgraph in G' corresponding to H in G. Using the proof of Theorem 13, we find that

$$r' = r_1' + \cdots + r_m',$$

and

$$n = n_1 + \cdots + n_m.$$

As also

$$R' = R_1' + \cdots + R_m'$$

and

$$r_i' = R_i' - n_i \qquad (i = 1, \cdots, m),$$

adding these last equations gives

$$r' = R' - n,$$

and hence G' is a dual of G.

THEOREM 24. *Let $G_1, \cdots, G_m$ and $G_1', \cdots, G_m'$ be the components of the dual graphs G and G', and let the correspondence between these two graphs be such that arcs in G_i correspond to arcs in G_i', $i=1, \cdots, m$. Then G_i and G_i' are duals, $i=1, \cdots, m$.*

Let H_1 be any subgraph of G_1, let H' be the complement of the corresponding subgraph in G', and let H_1' be the complement in G'. Then $H_1', G_2', \cdots, G_m'$ form H'. By Theorem 13, we find

$$R' = R_1' + R_2' + \cdots + R_m'$$

and

$$r' = r_1' + R_2' + \cdots + R_m'.$$

Now

$$r' = R' - n_1,$$

hence

$$r_1' = R_1' - n_1,$$

and G_1' is a dual of G_1. Similarly for $G_2', \cdots, G_m'$.

* See the author's American Journal paper, however.

THEOREM 25. *Let G and G' be dual graphs, and let $H_1, \cdots, H_m$ be the components of G. Let $H_1', \cdots, H_m'$ be the corresponding subgraphs of G'. Then $H_1', \cdots, H_m'$ are the components of G', and H_i' is a dual of H_i, $i=1, \cdots, m$.*

H_1 is the subgraph of G corresponding to H_1' in G'. Its complement is I_1, the graph formed of the arcs of $H_2, \cdots, H_m$. Obviously $H_2, \cdots, H_m$ are the components of I_1. Hence, by Theorem 13, the nullity of I_1 is $n_2+n_3+\cdots+n_m$. Thus, as G' is a dual of G,

$$r_1' = R' - (n_2 + n_3 + \cdots + n_m).*$$

Similarly,

$$r_2' = R' - (n_1 + n_3 + \cdots + n_m),$$

$$\cdots\cdots\cdots\cdots\cdots\cdots$$

$$r_m' = R' - (n_1 + n_2 + \cdots + n_{m-1}).$$

Adding these equations gives

$$r_1' + r_2' + \cdots + r_m' = mR' - (m-1)(n_1 + n_2 + \cdots + n_m).$$

As $H_1, H_2, \cdots, H_m$ are the components of G,

$$N = n_1 + n_2 + \cdots + n_m.$$

Also, as G and G' are duals, by Theorem 20,

$$R' = N.$$

Hence

$$\begin{aligned} r_1' + r_2' + \cdots + r_m' &= mR' - (m-1)R' \\ &= R'. \end{aligned}$$

Let now $H_{11}', \cdots, H_{1k_1}'$ be the components of H_1' (there may be but one) and similarly for $H_2', \cdots, H_m'$. Then, by Theorem 13,

$$r_1' = r_{11}' + \cdots + r_{1k_1}',$$

$$\cdots\cdots\cdots\cdots$$

$$r_m' = r_{m1}' + \cdots + r_{mk_m}'.$$

Adding these equations gives

$$\sum_{i,j} r_{ij}' = r_1' + \cdots + r_m' = R'.$$

As the graphs $H_{11}', \cdots, H_{mk_m}'$ are non-separable, Theorem 17 tells us that they are the components of G'. Hence G' has at least as many components as

* Which equals n_1.

G. Similarly, G has at least as many components as G'. They have therefore the same number, m, of components.

There are therefore m graphs in the set $H_{11}', \cdots, H_{mk_m}'$. But there is at least one such graph in each graph $H_1', \cdots, H_m'$, and there is therefore exactly one in each. Hence each graph H_{i1}' fills out the graph H_i', and the two sets of graphs $H_{11}', \cdots, H_{mk_m}'$ and $H_1', \cdots, H_m'$ are identical, that is, $H_1', \cdots, H_m'$ are the components of G'.

The rest of the theorem follows from Theorem 24.

As a special case of this theorem, we have

THEOREM 26. *A dual of a non-separable graph is non-separable.*

9. **Planar graphs.** Up till now, we have been considering abstract graphs alone. However, the definition of a planar graph is topological in character. This section may be considered as an application of the theory of abstract graphs to the theory of topological graphs.

Definitions. A topological graph is called *planar* if it can be mapped in a (1, 1) continuous manner on a sphere (or a plane). For the present, we shall say that an abstract graph is *planar* if the corresponding topological graph is planar. Having proved Theorem 29, we shall be justified in using the following purely combinatorial definition: *A graph is planar if it has a dual.*

We shall henceforth talk about "graphs" simply, the terms applying equally well to either abstract or topological graphs.

LEMMA. *If a graph can be mapped on a sphere, it can be mapped on a plane, and conversely.*

Suppose we have a graph mapped on a sphere. We let the sphere lie on the plane, and rotate it so that the new north pole is not a point of the graph. By stereographic projection from this pole, the graph is mapped on the plane. The inverse of this projection maps any graph on the plane onto the sphere.

By the *regions* of a graph lying on a sphere or in a plane is meant the regions into which the sphere or plane is thereby divided. A given region of the graph is characterized by those arcs of the graph which form its boundary. If the graph is in a plane, the outside region is the unbounded region.

LEMMA. *A planar graph may be mapped on a plane so that any desired region is the outside region.*

We map the graph on a sphere, and rotate it so that the north pole lies inside the given region. By stereographic projection, the graph is mapped onto the plane so that the given region is the outside region.

We return now to the work in hand.

THEOREM 27. *If the components of a graph G are planar, G is planar.*

Suppose the graphs G_1 and G_2 are planar, and G' is formed by letting the vertices a_1 and a_2 of G_1 and G_2 coalesce. We shall show that G' is planar. Map G_1 on a sphere, and map G_2 on a plane so that one of the regions adjacent to the vertex a_2 is the outside region. Shrink the portion of the plane containing G_2 so it will fit into one of the regions of G_1 adjacent to a_1. Drawing a_1 and a_2 together, we have mapped G' on the sphere.* The theorem follows as a repeated application of this process.

THEOREM 28. *Let G and G' be dual graphs, and let* $\alpha(ab)$, $\alpha'(a'b')$ *be two corresponding arcs. Form* G_1 *from G by dropping out the arc* $\alpha(ab)$, *and form* G_1' *from G' by dropping out the arc* $\alpha'(a'b')$, *and letting the vertices a' and b' coalesce if they are not already the same vertex. Then* G_1 *and* G_1' *are duals, preserving the correspondence between their arcs.*

Let H_1 be any subgraph of G_1 and let H_1' be the complement of the corresponding subgraph of G_1'.

Case 1. Suppose the vertices a' and b' were distinct in G'. Let H be the subgraph of G identical with H_1. Then

$$n = n_1.$$

Let H' be the complement in G' of the subgraph corresponding to H. Then

$$r' = R' - n.$$

Now H' is the subgraph in G' corresponding to H_1' in G_1', except that H' contains the arc $\alpha'(a'b')$, which is not in H_1'. Thus if we drop out $\alpha'(a'b')$ from H' and let a' and b' coalesce, we form H_1'. In this operation, the number of connected pieces is unchanged, while the number of vertices is decreased by 1. Hence

$$r_1' = r' - 1.$$

As a special case of this equation, if H' contains all the arcs of G', we find

$$R_1' = R' - 1.$$

These equations give

$$r_1' = R_1' - n_1.$$

Thus G_1' is a dual of G_1.

Case 2. Suppose a' and b' are the same vertex in G'. In this case, defining H and H' as before, we form H_1' from H' by dropping out the arc $\alpha'(a'a')$. This leaves the number of vertices and the number of connected pieces un-

* Here and in a few other places we are using point-set theorems which, however, are geometrically evident.

changed. Thus two of the equations in Case 1 are replaced by the equations

$$r_1' = r_1, \quad R_1' = R_1.$$

The other equations are as before, so we find again that G_1' is a dual of G_1. The theorem is now proved.

THEOREM 29. *A necessary and sufficient condition that a graph be planar is that it have a dual.*

We shall prove first the necessity of the condition. Given any planar graph G, we map it onto the surface of a sphere. If the nullity of G is N, it divides the sphere into $N+1$ regions. For let us construct G arc by arc. Each time we add an arc joining two separate pieces, the nullity and the number of regions remain the same. Each time we add an arc joining two vertices in the same connected pieces, the nullity and the number of regions are each increased by 1. To begin with, the nullity was 0 and the number of regions was 1. Therefore, at the end, the number of regions is $N+1$.

We construct G' as follows: In each region of the graph G we place a point, a vertex of G'. Therefore G' contains $V'=N+1$ vertices. Crossing each arc of G we place an arc, joining the vertices of G' lying in the two regions the arc of G separates (which may in particular be the same region, in which case this arc of G' is a 1-circuit). The arcs of G and G' are now in (1, 1) correspondence.

G' is the dual of G in the ordinary sense of the word. We must show it is the dual as we have defined the term.

Let us build up G arc by arc, removing the corresponding arc of G' each time we add an arc to G. To begin with, G contains no arcs and G' contains all its arcs, and at the end of the process, G contains all its arcs and G' contains no arcs. We shall show

(1) each time the nullity of G is increased by 1 upon adding an arc, the number of connected pieces in G' is reduced by 1 in removing the corresponding arc, and

(2) each time the nullity of G remains the same, the number of connected pieces in G' remains the same.

To prove (1) we note that the nullity of G is increased by 1 only when the arc we add joins two vertices in the same connected piece. Let ab be such an arc. As a and b were already connected by a chain, this chain together with ab forms a circuit P. Let $a'b'$ be the arc of G' corresponding to ab. Before we removed it, a' and b' were connected. Removing it, however, disconnects them. For suppose there were still a chain C' joining them. As a' and b' are on opposite sides of the circuit P, C' must cross P, by the Jordan Theorem, that is, an arc of C' must cross an arc of P. But we removed this arc of C' when we put in the arc of P it crosses. (1) is now proved.

The total increase in the nullity of G during the process is of course just N. Therefore the increase in the number of connected pieces in G' must be at least N. But G' was originally in at least one connected piece, and is at the end of the process in $V=N+1$ connected pieces. Thus the increase in the number of connected pieces in G' is just N (hence, in particular, G' itself is connected) and therefore this number increases only when the nullity of G increases, which proves (2).

Let now H be any subgraph of G, let H' be the complement of the corresponding subgraph of G', and let H' include all the vertices of G'. We build up H arc by arc, at the same time removing the corresponding arcs of G'. Thus when H is formed, H' also is formed. By (1) and (2), the increase in the number of connected pieces in forming H' from G' equals the nullity of H, that is,

$$p' - P' = n.$$

But

$$r' = V' - p', \quad R' = V' - P',$$

as G' and H' contain the same vertices. Therefore

$$r' = R' - n,$$

that is, G' is a dual of G.

To prove the sufficiency of the condition, we must show that if a graph has a dual, it is planar. It is enough to show this for non-separable graphs. For if the separable graph G has a dual, its components have duals, by Theorem 25, hence its components are planar, and hence G is planar, by Theorem 27. This part of the theorem is therefore a consequence of the following theorem:

THEOREM 30. *Let the non-separable graph G have a dual G'. Then we can map G and G' together on the surface of a sphere so that*

(1) *corresponding arcs in G and G' cross each other, and no other pair of arcs cross each other, and*

(2) *inside each region of one graph there is just one vertex of the other graph.*

The theorem is obviously true if G contains a single arc. (The dual of an arc ab is an arc $a'a'$, and the dual of an arc aa is an arc $a'b'$.) We shall assume it to be true if G contains fewer than E arcs, and shall prove it for any graph G containing E arcs. By Theorem 8, each vertex of G is on at least two arcs.

Case 1. G contains a vertex b on but two arcs, ab and bc. As G is non-separable, there is a circuit containing these arcs. Thus dropping out one of them will not alter the rank, while dropping out both reduces the rank by 1. As G' is a dual of G, the arcs corresponding to these two arcs are each of nullity 0, while the two arcs taken together are of nullity 1. They are thus of the form $\alpha'(a'b')$, $\beta'(a'b')$, the first corresponding to ab, and the second, to bc.

Form G_1 from G by dropping out the arc bc and letting the vertices b and c coalesce, and form G_1' from G' by dropping out the arc $\beta'(a'b')$. By Theorem 28, G_1 and G_1' are duals, preserving the correspondence between the arcs. As these graphs contain fewer than E arcs,* we can, by hypothesis, map them together on a sphere so that (1) and (2) hold; in particular, $\alpha'(a'b')$ crosses ac. Mark a point on the arc ac of G_1 lying between the vertex c and the point where the arc $\alpha'(a'b')$ of G' crosses it. Let this be the vertex b, dividing the arc ac into the two arcs ab and bc. Draw the arc $\beta'(a'b')$ crossing the arc bc. We have now reconstructed G and G', and they are mapped on a sphere so that (1) and (2) hold.

Case 2. Each vertex of G is on at least three arcs. As then G contains no suspended chain, and G is not a circuit and therefore is of nullity $N>1$, we can, by Theorem 18, drop out an arc ab so that the resulting graph G_1 is non-separable. G' is non-separable, by Theorem 26, and hence the arc $a'b'$ corresponding to ab in G is not a 1-circuit. Drop it out and let the vertices a', b' coalesce into the vertex a_1', forming the graph G_1'. By Theorem 28, G_1 and G_1' are duals, and thus G_1' also is non-separable.

Consider the arcs of G' on a'. If we drop them out, the resulting graph G'' has a rank one less than that of G'. For if its rank were still less, G'' would be in at least three connected pieces, one of them being the vertex a'. Let c and d be vertices in two other connected pieces of G''. They are joined by no chain in G'', and hence every chain joining them in G' must pass through a', which contradicts Theorem 6. If we put back any arc, the rank is brought back to its original value, as a' is then joined to the rest of the graph. Hence, G' being a dual of G, the arcs of G corresponding to these arcs are together of nullity 1, while dropping out one of them reduces the nullity to 0. Therefore, by Theorem 9, these arcs form a circuit P. One of these arcs is the arc ab. The remaining arcs form a chain C. Similarly, the arcs of G corresponding to the arcs of G' on b' form a circuit Q, and this circuit minus the arc ab forms a chain D. C and D have the vertices a and b as end vertices. Also, the arcs of G_1 corresponding to the arcs of G_1' on a_1' form a circuit R. These arcs of G_1' are the arcs of G' on either a' or b', except for the arc $a'b'$ we dropped out. Thus the arcs of G_1 forming the circuit R are the arcs of the chains C and D.

As G_1 and G_1' contain fewer than E arcs, we can map them together on a

* Obviously G_1 is non-separable.

sphere so that properties (1) and (2) hold. a_1' lies on one side of the circuit R, which we call the inside. Each arc of R is crossed by an arc on a_1', and thus there are no other arcs of G_1' crossing R. There is no part of G_1' lying inside R other than a_1', for it could have only this vertex in common with the rest of G_1', and G_1' would be separable. Also, there is no part of G_1 lying inside R, for any arc would have to be crossed by an arc of G_1', and any vertex would have to be joined to the rest of G_1 by an arc, as G_1 is non-separable.

Let us now replace a_1' by the two vertices a' and b', and let those arcs abutting on a_1' that were formerly on a' be now on a', and those formerly on b', now on b'. As the first set of arcs all cross the chain C, and the second set all cross the chain D, we can do this in such a way that no two of the arcs cross each other. We may now join a and b by the arc ab, crossing none of these arcs. This divides the inside of R into two parts, in one of which a' lies, and in the other of which b' lies. We may therefore join a' and b' by the arc $a'b'$, crossing the arc ab. G and G' are now reconstructed, and are mapped on the sphere as required. This completes the proof of the theorem, and therefore of Theorem 29.

THEOREM 31. *A necessary and sufficient condition that a graph be planar is that it contain neither of the two following graphs as subgraphs:*

G_1. *This graph is formed by taking five vertices a, b, c, d, e, and joining each pair by an arc or suspended chain.*

G_2. *This graph is formed by taking two sets of three vertices, a, b, c, and d, e, f, and joining each vertex in one set to each vertex in the other set by an arc or suspended chain.*

This theorem has been proved by Kuratowski.* It would be of interest to show the equivalence of the conditions of the theorem and Theorem 29 directly, by combinatorial methods. We shall do part of this here, in the following theorem:†

THEOREM 32. *Neither of the graphs G_1 and G_2 has a dual.*

Suppose the graph G_1 had a dual. By Theorem 28, if G_1 contains a suspended chain, we can drop out one of its arcs and let the two end vertices coalesce, and the resulting graph will have a dual. Continuing, we see that the graph G_3, in which each pair of vertices of the set a, b, c, d, e are joined by an arc, must have a dual. Similarly, if G_2 has a dual, then the graph G_4, in which each vertex of the set a, b, c is joined to each vertex of the set d, e, f by an arc, must have a dual. Both of these are impossible.

* Fundamenta Mathematicae, vol. 15 (1930), pp. 271–283.

† The other half has recently been proved by the author. See Bulletin of the American Mathematical Society, abstract (38-1-39). (Note added in proof.)

(a) *The graph* G_3. To avoid subscripts, let us call it G. Suppose it had a dual, G'. Then

$$R = N' = 4,$$
$$N = R' = 6,$$
$$E = E' = 10.$$

If G' has isolated vertices, we drop them out, which does not alter its relation to G.

(1) There are no 1-circuits, 2-circuits or triangles in G'. For if there were, dropping out the corresponding arcs of G would have to reduce the rank of G. But we cannot reduce its rank without dropping out at least four arcs.

(2) G' contains at least five quadrilaterals. For if we drop out the four arcs on any vertex of G, the rank is reduced by 1, and if we put back any of these arcs, the rank is brought back to its original value; Theorem 9 now applies.

(3) At least two of these quadrilaterals have an arc in common, as there are but ten arcs in G'.

There are just two ways of forming two quadrilaterals out of fewer than eight arcs without forming any 2-circuits or triangles. One of these graphs, I_1', contains the arcs $a'b'$, $b'e'$, $a'c'$, $c'e'$, $a'd'$, $d'e'$. The other, I_2', contains the arcs $a'e'$, $e'f'$, $f'b'$, $b'a'$, $e'c'$, $c'd'$, $d'f'$. But there is no subgraph of the type I_1' in G', for this subgraph is of rank 4 and nullity 2, and there would have to be a subgraph of G of rank 2 and nullity 2, and such a graph contains a 1- or a 2-circuit, of which there are none in G. Hence G' contains a subgraph I_2'.

(4) Each vertex of G' is on at least three arcs, as there are no 1- or 2-circuits in G.

Each of the vertices a', b', c', d' of I_2' is on but two arcs. Hence there must be another arc on each of these vertices. As I_2' contains seven arcs, and G' contains but ten, one of the three arcs left must join two of these vertices. But if we add an arc $a'b'$ or $c'd'$, we would form a 2-circuit; if we add an arc $a'c'$ or $b'd'$, we would form a triangle; if we add an arc $a'd'$ or $b'c'$, we would form a graph of the type I_1'. As G' contains none of these graphs, we have a contradiction.

(b) *The graph* G_4. Let us call it G. If it has a dual G', then

$$R = N' = 5,$$
$$N = R' = 4,$$
$$E = E' = 9.$$

We proceed exactly as for the graph G_3. In outline:

(1) G' contains no 1- or 2-circuits.

CONGRUENT GRAPHS AND THE CONNECTIVITY OF GRAPHS.*

By HASSLER WHITNEY.

Introduction. We give here conditions that two graphs be congruent and some theorems on the connectivity of graphs, and we conclude with some applications to dual graphs. These last theorems might also be proved by topological methods. The definitions and results of a paper by the author on "Non-separable and planar graphs," † will be made use of constantly. We shall refer to this paper as N. For convenience, we shall say two arcs touch if they have a common vertex.

1. *Congruent graphs.* For a definition, see N, 7.‡ In this section we consider no graphs containing 1- or 2-circuits; that is, each arc joins two distinct vertices, and any two vertices are joined by at most a single arc.

THEOREM 1. *Let G and G' be two connected graphs, neither of which consists of three arcs of the form ab, ac, ad. Let there be a 1-1 correspondence between their arcs so that to any two arcs having a common vertex in one graph correspond two arcs having a common vertex in the other. Then G and G' are congruent.*

Case A. *At least one of the graphs, say G, is of one of the following forms:*

$$G_a : \alpha(ab), \beta(ac), \gamma(bc), \delta(bd),$$
$$G_b : \alpha(ab), \beta(ac), \gamma(bc), \delta(bd), \epsilon(cd),$$
$$G_c : \alpha(ab), \beta(ac), \gamma(bc), \delta(bd), \epsilon(cd), \zeta(ad).$$

Consider first the graph G_a. Let G'_a be the corresponding graph, and let α', β', γ', δ' be the corresponding arcs of G'_a. As α and β touch, α' and β' touch. Hence they are of the form $\alpha'(a'b')$, $\beta'(a'c')$. δ touches α but not β; hence δ' touches α' but not β'; it has therefore b', but neither a' nor c', as end vertex, and is of the form $\delta'(b'd')$. Now γ touches α, β and δ. Hence γ' touches α', β' and δ'. γ' is not of the form $\gamma'(a'b')$, for then $\alpha'(a'b')$ and

* Presented to the American Mathematical Society, February 28, 1931.

† An outline of this paper will be found in the *Proceedings of the National Academy*, Vol. 17 (1931), pp. 125-127. We refer to this outline as P. The full paper has not yet appeared.

‡ P, 3. We formerly used the term *homeomorphic*.

$\gamma'(a'b')$ would form a 2-circuit. It is thus either $\gamma'(b'c')$ or $\gamma'(a'd')$. In the first case, upon dropping primes, G'_a becomes identical with G_a. In the second case, we interchange the names a' and b', c' and d', β' and δ'. Upon dropping primes, G'_a becomes identical with G_a. Thus in either case, G_a and G'_a are congruent.

Consider next the graph G_b. We form G_b from G_a by adding the arc $\epsilon(cd)$. ϵ touches all the preceding arcs but α. The same must be true of ϵ'. and hence ϵ' is of the form $\epsilon'(c'd')$. Just as above, G_b and G'_b are congruent.

Consider finally the graph G_c. We form it from G_b by adding the arc $\zeta(ad)$. ζ touches all the preceding arcs but γ. The same must be true of ζ'. Hence, if γ' is of the form $\gamma'(b'c')$, ζ' is of the form $\zeta'(a'd')$; if γ' is of the form $\gamma'(a'd')$, ζ' is of the form $\zeta'(b'c')$. In either case, just as before, G_c and G'_c are congruent.

Before proceeding, we shall prove the following

LEMMA. *Let G and G' satisfy the conditions of the theorem. Suppose G contains G_a as a subgraph, and contains also an arc which has just one end vertex in G_a. Let G'_a be the subgraph of G' containing those arcs corresponding to the arcs of G_a. Then G_a and G'_a are congruent, preserving the correspondence between their arcs.*

That is, we can rename the vertices and arcs of G'_a so that arcs that formerly corresponded become identical. By the proof of case A,* G_a and G'_a are congruent, and the correspondence between their arcs is preserved if γ' is of the form $\gamma'(b'c')$. Suppose not. Then γ' is of the form $\gamma'(a'd')$. Let η be an arc of G with one end vertex in G_a. This vertex is either, a, b, c, or d, and η touches α and β; α, γ and δ; β and γ; or δ; but not other arcs of G_a. There is no way of adding an arc to G' (without forming a 2-circuit) so that the similar conditions hold, and we have a contradiction, proving the lemma.

We return now to the theorem.

Case B. *Neither G nor G' is any of the graphs G_a, G_b, G_c.* In this case, we shall prove the stronger theorem, that G and G' are congruent, preserving the correspondence between their arcs. We shall consider first non-separable, then separable, graphs.

Case B1. *At least one of the graphs, say G, is non-separable.*

There are three cases to consider here.

* We see, as in Case B1c, that G'_a is connected.

Case B1a. *G is of nullity* 0. By N, theorem 8, G contains but a single arc, ab. Hence G' contains the single arc $a'b'$, and G and G' are congruent.

Case B1b. *G is of nullity* 1. Then, by N, theorem 10,* G is a circuit, and contains therefore at least three arcs. Suppose first G contains just three arcs. They are then of the form $\alpha(ab)$, $\beta(ac)$, $\gamma(bc)$. α' and β' of G' must touch, and they are of the form $\alpha'(a'b')$, $\beta'(a'c')$. γ' must touch both α' β', and is either of the form $\gamma'(b'c')$ or $\gamma'(a'd')$. But by the hypothesis of the theorem, G' is not of the form $a'b'$, $a'c'$, $a'd'$; it is therefore of the form $a'b'$, $a'c'$, $b'c'$. G and G' are thus congruent, preserving the correspondence between their arcs.

Suppose next G contains at least four arcs, $\alpha(ab)$, $\beta(bc)$, $\gamma(cd)$, $\delta(de)$, $\cdots$, $\zeta(fa)$. Let α', β', γ', δ', $\cdots$, ζ' be the corresponding arcs of G'. α' and β' must touch, and they are therefor $\alpha'(a'b')$, $\beta'(b'c')$. γ touches β but not α; the same must be true of γ', and it is therefore $\gamma'(c'd')$. δ touches γ but neither α nor β; hence δ' is $\delta'(d'e')$. We continue in this manner. Finally, ζ touches the preceding arc and α, but none of the other arcs. The same must be true of ζ', and hence it is of the form $\zeta'(f'a')$. Thus G and G' are congruent as required.

Case B1c. *G is of nullity $N > 1$.* We will assume the theorem is true if G is of nullity $< N$, and will prove it for the case that G is of nullity N, (including the case where $N = 2$). This will establish the proof for this case.

By N, theorem 18,† we can remove an arc or suspended chain C from G, leaving a non-separable graph G_1 of nullity $N - 1$. Let C' be the corresponding arc or arcs of G'. Remove them and any isolated vertices there may then be, forming the graph G'_1. As G_1 is connected, G'_1 is connected. For order the arcs of G_1 so that each arc other than the first touches one of the preceding arcs. Ordering the arcs of G'_1 in the same manner as the corresponding arcs of G_1, each arc other than the first touches one of the preceding arcs; thus G'_1 is connected. C' is a chain. For if C is a single arc, C' is a single arc. Otherwise, the proof is exactly like the proof in the last part of case B1b.

We shall divide case B1c into two cases.

Case B1c$_1$. *None of the vertices of C' other than its two end vertices are in G'_1.* We shall show first that G_1 and G'_1 are congruent, preserving the correspondence between their arcs. G_1 is not of the form ab, ac, ad, as it is non-separable. Suppose G'_1 were of this form. Then each pair of arcs

* P, theorem 6.

† P, theorem 7.

of G_1 have a common vertex, and G_1 is of the form ab, ac, bc. As G contains no 2-circuit, C contains at least two arcs. The first of these, δ say, has just one vertex in G_1, and touches therefor just two of the arcs of G_1. The corresponding arc δ' of C' must touch just two of the arcs $a'b'$, $a'c'$, $a'd'$, and it is therefor of one of the forms $b'c'$, $b'd'$, $c'd'$. But by the hypothesis of this case, δ' has but a single end vertex in G'. This is a contradiction, so G'_1 is not of the form described.

Suppose neither G_1 nor G'_1 are one of the graphs G_a, G_b or G_c of case A. Then the hypotheses of case B1 are fulfilled, and G_1 is of nullity $N-1$, and hence G_1 and G'_1 are congruent as required.

Suppose next one of the graphs G_1, G'_1 was one of the graphs G_a, G_b, G_c. If G_1 is the graph G_c, C contains an arc with just one end vertex in G_c, as G contains no 2-circuit. But G_c contains G_a as a subgraph. Let G'_a be the arcs of G' corresponding to G_a. Then, by the lemma, G_a and G'_a are congruent, preserving the correspondence between their arcs. Hence, following the proof of case A, we see that G_1 and G'_1 are congruent as required. If, next, G_1 is the graph G_b, C contains at least two arcs. For if C contained but one arc, this arc would be $\zeta(ad)$, and G would be the graph G_c. This is ruled out by the hypothesis of case B. Again, G contains G_a as a subgraph, and there is an arc with just one end vertex in G_a. Hence G_a and G'_a, and therefor G_1 and G'_1, are congruent as required. Finally, G_1 is not G_a, for G_a is separable. If now G'_1 were one of the graphs G_a, G_b, G_c, G_1 would be also, by case A, and we are back to the case we have considered. Thus in all cases G_1 and G'_1 are congruent as required.

Rename the vertices and arcs of G'_1 so that upon dropping primes, G'_1 and G_1 become identical. We can do this, preserving the correspondence between the arcs. Suppose the chain C had a and b as end vertices in G_1. We shall show that the end vertices of C' in G'_1 are a' and b'. Let A be the arcs of G_1 on a, and B, those on b. By N, theorem 8, there are at least two arcs in A, as G_1 is of nullity $N-1 \geqq 1$; thus a is the only vertex on them all. Similarly, b is the only vertex on all the arcs B. Let A', B' be the corresponding arcs of G'. They are then on a' and b', respectively.

Suppose first C contained but a single arc, $\alpha(ab)$. As it touches all the arcs of both A and B, the corresponding arc α' of C' touches all the arcs of both A' and B'. Suppose α' were not $\alpha'(a'b')$. Say α' is not on a'. As it must touch all the arcs of A', there are just two such arcs, $a'c'$ and $a'd'$, and α' is $\alpha'(c'd')$. Neither c' nor d' is b'. For then G'_1 would contain the arc $a'b'$, hence G_1 would contain the arc ab, and this arc with the arc α would form a 2-circuit. As $\alpha'(c'd')$ touches all the arcs of B, there are just two such arcs, $b'c'$ and $b'd'$. There are no more arcs in G'_1. For if there were, as there are no more on a' or b', there must be arcs on c' or d'. Hence there are arcs on c or d in G_1. But then α' would touch these arcs in G'_1, and α would not touch the corresponding arcs in G_1. Hence G' is just the graph G_b. But this is ruled out by the hypotheses of case B. Therefore α' is $\alpha'(a'b')$, and G and G' are congruent as required.

Suppose next C contains at least two arcs. The first arc, α, has a in common with the arcs A. As C' has only its end vertices in G'_1, by hypothesis, the arc α' corresponding to α has just one vertex in G_1, and this vertex must be on all the arcs of A'. It is therefor the vertex a'. Similarly, the other end vertex of C' is b', and thus we can rename the vertices and arcs of C' with primed letters so that, upon dropping primes, it becomes identical with C. Therefore G and G' are congruent, as required. This disposes of case B1c$_1$.

Case B1c$_2$. *C' has a vertex besides its two end vertices in G'_1.* Then C' contains just two arcs. For suppose C' contained more than two arcs. Then C would contain more than two arcs. None of these but the first and last touch arcs of G_1. Hence none of the arcs of C'_1 but the first and last have vertices in G'_1, and thus there would be no vertices of C' but the first and last in G'_1.

Let $\alpha(ac)$, $\beta(cb)$ be the two arcs of C, and let $\alpha'(a'c')$, $\beta'(c'b')$ be those of C'. Then a', b' and c' are all in G'_1. There is an arc $\gamma'(c'd')$ in G'_1, where d' is neither a' nor b', as G' contains no 2-circuit. This arc touches both α' and β'. Hence the corresponding arc γ of G touches both α and β, and is therefore $\gamma(ab)$. As G_1 is of nullity $N-1 \geqq 1$, G contains more arcs. As G_1 is non-separable, there must be more arcs on both a and b, for otherwise one of these vertices would be a cut vertex. Let $\delta(ad)$ be an arc on a. It touches α and γ but not β. Hence the corresponding are δ' of G' is $\delta'(a'd')$. Similarly, if $\epsilon(be)$ is an arc on b in G_1, the corresponding arc in G'_1 is $\epsilon'(b'd')$. As now δ' and ϵ' touch, δ and ϵ touch, and thus e is d. G contains no more arcs. For if it did, it would contain an arc on one of the vertices a, b, d. But we cannot fit a similar arc into G'. G is therefore the graph G_b. But this was ruled out, so case B1c$_2$ does not exist. Case B1c, and hence case B1, is now disposed of.

Case B2. *Both G and G' are separable.* Let $H_1, \cdots, H_m$ be the components of G, and let $H'_1, \cdots, H'_n$ be those of G'. Let I'_i be the subgraph of G' containing those arcs corresponding to the arcs of H_i, $i=1, \cdots, m$, and let I_j be the subgraph of G containing those arcs corresponding to the arcs of H'_j, $j=1, \cdots, n$.

Case B2a. One of the graphs $I_1, \cdots, I_n, I'_1, \cdots, I'_m$, say I'_1, is of the form $\alpha'(a'b')$, $\beta'(a'c')$, $\gamma'(a'd')$. Then H_1, being non-separable, is of the form $\alpha(ab)$, $\beta(ac)$, $\gamma(bc)$. There are other arcs in G, and hence there is an arc on one of the vertices a, b, c. Let such an arc be $\delta(ad)$. The corresponding arc δ' in G' is $\delta'(b'c')$. There are no more arcs in G'. For suppose there were an arc with both end vertices in I'_1. It is then either $\epsilon'(b'd')$ or $\zeta'(c'd')$, say the first. The corresponding arc in G must be $\epsilon(bd)$. But then H_1 is not a component of G, for $\alpha(ab)$, $\epsilon(bd)$, $\delta(ad)$ is a circuit with arcs in H_1 and arcs not in H_1, contrary to N, theorem 17.* There cannot be an arc with just one end vertex in I'_1, for the lemma would be contradicted. Thus case B2a does not exist.

Case B2b. *No one of the graphs $I_1, \cdots, I_n, I'_1, \cdots, I'_m$ is of the form ab, ac, ad.* Consider first any one of the graphs $H_1, \cdots, H_m$, say H_1, which is one of the graphs G_b, G_c. (No one is G_a, as G_a is separable.) Consider that part of H_1, G_a. As G is connected, there is an arc with just one end vertex in G_a (i. e. in H_1). Hence, by the lemma, G_a and the corresponding arcs of I'_1, G'_a, are congruent, preserving the correspondence between their arcs. Hence, as is seen from the proof of case A, H_1 and I'_1 are congruent similarly. Consider next any component, say H_2, which is not either G_b or G_c. Then the corresponding graph, I'_2, is neither of these graphs, for otherwise, case A would be contradicted. As H_2 is connected, I'_2 is connected. Thus, by case B1, H_2 and I'_2 are congruent, preserving the correspondence between their arcs. Thus H_i and I'_i are congruent as required, $i = 1, \cdots, m$. Similarly, H'_j and I_j are congruent as required, $j = 1, \cdots, n$.

Now I'_1 is a non-separable subgraph of G', and is therefor contained in one of the components $H'_1, \cdots, H'_n$, say H'_1, of G', by N, theorem 11. Similarly, I_1 is contained in one of the components $H_1, \cdots, H_m$, say H_j, of G. Hence H_1 is contained in H_j, and thus they are identical. Thus the arcs of H_1 and H'_1 correspond, and they are congruent, preserving this correspondence. Similarly for the other components. Rename the vertices and arcs in each H'_i with the subscript i and primes so H'_i becomes identical with H_i when these subscripts and primes are dropped.

Let H_1 and H_2 be two components of G with a common vertex, a, (they have but one common vertex, as otherwise they would form a circuit of graphs, contrary to N, theorem 17 *), and let H_{12} be the graph containing the arcs of both. Let H'_1, H'_2, H'_{12} be the corresponding subgraphs of G'. We shall show that H_{12} and H'_{12} are congruent as required, supposing only that

* P, theorem 5.

H_1, H_2, H'_1, H'_2 are connected. We next show that H_{123}, composed of H_{12} and H_3 together, is congruent as required with the corresponding subgraph H'_{123} of G'. Continuing, we find finally that G and G' are congruent as required.

Let A_1 and A_2 be the arcs of H_1 and H_2 on a. Let A'_1 and A'_2 be the corresponding arcs of H'_1 and H'_2. Then the arcs A'_1 are all on a'_1, and the arcs A'_2 are all on a'_2. We must show that a'_1 and a'_2 are the same vertex in G'. Each arc in A_2 touches an arc in A_1, and hence each arc in A'_2 touches an arc in A'_1. Therefor H'_1 and H'_2 have a common vertex. Call it x'. Moreover, any arc in A'_2 has just one vertex in H'_1, as H'_1 and H'_2 have at most a single vertex in common, being two components of G'.

Suppose first there is but a single arc in H'_1, $a'_1b'_1$. Then if a'_1 is not x', b'_1 is x', and we interchange a'_1 and b'_1 so that a'_1 becomes x'. Suppose next H'_1 contains at least two arcs. If there are two arcs in A'_1, they have only the vertex a'_1 in common, and hence any arc in A'_2 has this vertex as an end vertex; that is, x' is a'_1. If A'_1 contains but a single arc $a'_1b'_1$, there are arcs on b'_1, as H'_1 is connected. ab being the corresponding arc of H_1, there are corresponding arcs on b. The arcs of A_2 touch ab but not these other arcs on b. Hence the arcs of A'_2 touch $a'_1b'_1$ but not these other arcs on b'. They are therefore on the vertex a'_1; that is, x' is a'_1.

Thus in all cases, if x' is not a'_1, we can make it so, preserving the correspondence between the arcs of G and G'. Similarly, if x' is not a'_2, we can make it so. Hence, if a'_1 is not a'_2, we can make it so. As finally, H'_1 and H'_2 have but one common vertex, H_{12} and H'_{12} are congruent as required.

The only fact we need notice to complete the proof is that if I_1 and I_2 are two connected groups of components of G, they have at most a single vertex in common, and the same is true for G'. For otherwise, we could find a circuit entering both of these groups of components, contrary to N, theorem 17.* The proof for case B2b, therefore for case B2, therefore for case B, and therefore of the theorem, is now complete.

THEOREM 2. *Let G and G' be two triply connected graphs, and let there be a 1-1 correspondence between their arcs so that to any set of arcs forming a circuit in one graph, corresponds a set of arcs forming a circuit in the other. Then G and G' are congruent.*

For a definition of "triply connected," see the next section. We shall show that to any two arcs that touch in one graph, there correspond two arcs that touch in the other. Theorem 1 then applies.

* P, theorem 5.

Let $\alpha(ab)$, $\beta(bc)$ be any two arcs in G with the common vertex b. We shall show that the corresponding arcs α', β' in G' have a common vertex. As G is triply connected, we cannot disconnect it without dropping out at least three vertices. Therefore if we drop out the vertex b and the arcs on it, forming the graph G_1, we cannot disconnect this graph without dropping out at least two vertices; that is, G_1 has no cut vertex. Therefore, by N, theorems 5 and 7,* the vertices a and c are contained in a circuit P in G_1. P consists of two chains B and C, joining a and c. Now put back b and the arcs on b. The two arcs ab, bc form a third chain, A, joining a and c. The two chains A and C form a circuit Q, and the two chains A and B form a circuit R. Moreover, no other subset of the arcs of these three chains form a circuit.

The arcs in G' corresponding to these three circuits form three circuits, P', Q', R'. Suppose now the arcs α' and β' had no common vertex. These arcs we shall call A'. Let B' be those arcs corresponding to B, and C', those corresponding to C. Let α', β' be the arcs $\alpha'(a'b')$, $\beta'(c'd')$.

Now as A' and B' together form a circuit, B' consists of two chains, one of them joining a' and either c' or d', say d', and the other joining b' and c'. Similarly, C' is in two chains; either one joins a' and d' and the other joins b' and c', or one joins a' and c' and the other joins b' and d'. The first case cannot be, as then there would be a chain in both B' and C' joining a' and d', and these two chains, or parts of them, form a circuit; but the whole of B' and C' forms a circuit, and therefore no part of them does. Therefore the second case occurs. Consider the following three chains: One part of A', $\alpha'(a'b')$, one part of B', that part joining b' and c', and one part of C', that part joining c' and a'. As B' and C' together form a circuit, they have no vertices in common other than the end vertices of the chains forming them. Hence the three chains described form a circuit, containing arcs of A', B' and C'. The corresponding arcs of G must form a circuit. But this is not the case, as we have seen, and we have a contradiction. Hence α' and β' have a vertex in common, and the theorem is proved.

This theorem is not true for all non-separable graphs. An example is given by the two graphs G', G'', given in N, after the proof of theorem 22.

THEOREM 3. *Let there be a 1-1 correspondence between the arcs of the triply connected graphs G and G' so that to any set of arcs of one graph forming a subgraph of nullity 1 there corresponds a set of arcs of the other graph forming a subgraph of nullity 1. Then G and G' are congruent.*

We shall prove that circuits correspond to circuits. Theorem 2 then

* P, theorem 1.

applies. Consider any circuit P of G. It is of nullity 1. Hence the corresponding arcs P' of G' are of nullity 1. Now drop out any arc from P. The nullity is reduced to 0. Hence the corresponding arcs of G' cannot be of nullity 1, and are therefore of nullity 0. Therefore, by N, theorem 9, P' is a circuit. Similarly, if P' is a circuit in G', the corresponding arcs P form a circuit in G.

2. *The connectivity of graphs.* In this section, we allow the graphs to contain 2-circuits, but no 1-circuits.

Definitions. Let G be a graph containing at least $n+1$ vertices, such that it is impossible to drop out $n-1$ or fewer vertices and the arcs on them in such a manner that the resulting graph is not connected. We shall say then that G is n-tuply connected. (We consider only numbers $n \geqq 1$). If G is n-tuply connected but not $(n+1)$-tuply connected, we say its connectivity is n.

For $n=1$, we have: A graph G is simply connected if it contains at least two vertices, and it cannot be disconnected by dropping out no vertices; i. e., G itself is connected. For $n=2$, N, theorems 5 and 6 give: a non-separable graph containing at least three vertices is doubly connected, and conversely.

THEOREM 4. *A necessary and sufficient condition that a graph G be n-tuply connected is that G contain at least $n+1$ vertices, and there exist no two graphs H' and H'' such that for some number $k \leqq n-1$, H' contains the vertices $a'_1, \cdots, a'_k$, b', H'' contains the vertices $a''_1, \cdots, a''_k$, b'', and if we let a'_i and a''_i coalesce, forming the vertex a_i, $i=1, \cdots, k$, we thereby form G.*

For $n=1$, the theorem is: A necessary and sufficient condition that G be simply connected is that G contain at least two vertices, and is not formed of two connected parts H', H''; i. e. G is connected. (For we let $k=n-1$ $=0$ vertices of H' and H'' coalesce to form G.) The theorem is trivial for this case.

Consider now numbers $n>1$. The theorem is trivial if G contains fewer than $n+1$ vertices; assume therefore G contains at least $n+1$ vertices. Suppose first G is formed from two graphs, H', H'' as described. Drop out the vertices $a_1, \cdots, a_k$, and the arcs on them. As H' and H'' had only these vertices in common, b' and b'' are now in different connected parts. Hence G is not n-tuply connected.

Suppose now G is not n-tuply connected. Then we can drop out $k \leqq n-1$ vertices $a_1, \cdots, a_k$, so that there remain two vertices b', b'', which are in different connected parts in the resulting graph G_1. Let H'_1 be that part of G_1 containing b', and let H''_1 be the rest of G_1. Replace each vertex a_i of G by the two vertices a'_i, a''_i $(i=1,\cdots,k)$ and replace the arcs we removed in the following manner: An arc a_ic, where c is in H'_1, we replace by the arc a'_ic. An arc a_id, where d is in H''_1, we replace by the arc a''_id. An arc a_ia_j we replace by the arc $a'_ia'_j$. In this manner we have added the vertices $a'_1,\cdots,a'_k$ to H'_1, forming the graph H', and we have added the vertices $a''_1,\cdots,a''_k$ to H''_1, forming the graph H''. Now H' contains the vertex b' distinct from $a'_1,\cdots,a'_k$, H'' contains the vertex b'' distinct from $a''_1,\cdots,a''_k$, and letting a'_i and a''_i coalesce $(i=1,\cdots,k)$ forms G. This completes the proof.

THEOREM 5. *If the graph G containing at least two vertices can be disconnected by dropping out $n-1$ or fewer arcs, it is not n-tuply connected.*

The theorem is trivial if $n=1$. Assume $n>1$. Drop out only just enough arcs to disconnect G. The resulting graph is then in two parts, H and H', and each arc we dropped out joins a vertex in H to a vertex in H'. Let these arcs be a_1b_1, and $a_2b_2,\cdots,a_k b_k$, where $a_1,\cdots,a_k$ are in H and $b_1,\cdots,b_k$ are in H'. Then $k<n$. The vertices $a_1,\cdots,a_k$, and also the vertices $b_1,\cdots,b_k$, may not all be distinct.

Case 1. There are two vertices a_i and b_j which are joined by no arc. Consider first the arc a_1b_1. As it does not join a_i and b_j, either a_1 is distinct from a_i or b_1 is distinct from b_j, say the first. Drop out a_1 and the arcs on it. Consider next the arc a_2b_2. One of the vertices a_2, b_2, is distinct from both a_i and b_j. Drop out this vertex, if it is not already dropped out, and its arcs. Continue in this manner. At the end of the process, we have dropped out fewer than n vertices, and we have dropped out all the arcs $a_1b_1,\cdots,a_k b_k$. The vertices a_i and b_j are still in the graph, but are joined by no chain. We have thus dropped out less than n vertices and their arcs, disconnecting G, and hence G is not n-tuply connected.

Case 2. Each vertex a_i is joined to each vertex b_j by an arc. Say there are p distinct vertices in the set $a_1,\cdots,a_k$, and q distinct vertices in the set $b_1,\cdots,b_k$. There are then pq arcs in the set $a_1b_1,\cdots,a_kb_k$, or more, if G contains 2-circuits. Now as $k<n$,

$$pq<n.$$

As

$$pq+1-(p+q)=(p-1)(q-1)\geqq 0, p+q\leqq pq+1<n+1.$$

If G contains n or fewer vertices, it is not n-tuply connected. Assume it contains at least $n+1$ vertices. Then, as there are but $p+q$ vertices in the set $a_1, \cdots, a_k, b_1, \cdots, b_k$, there is a vertex c in G not in this set.

Dropping out the arcs $a_1b_1, \cdots, a_kb_k$ from G leaves the two connected parts H and H', one of which, say H, contains the vertex c. Drop out then the vertices $a_1, \cdots, a_k$ and their arcs from G. We have dropped out fewer than n vertices and their arcs, disconnecting G, as b_1 and c are joined by no chain, and thus G is not n-tuply connected.

THEOREM 6. *If a vertex and its arcs be dropped out of the n-tuply connected graph G, the resulting graph G′ is (n — 1)-tuply connected.*

For G' contains at least n vertices, and we must drop out at least $n-1$ vertices to disconnect it.

THEOREM 7. *A necessary and sufficient condition that a graph containing no 2-circuit be n-tuply connected is that any two of its vertices be joined by n distinct chains.**

For $n=1$, the theorem is obvious. Suppose $n=2$. If the graph contains but two vertices, the theorem is obvious. If the graph contains at least three vertices, the theorem becomes theorem 7 of N.† Assume the theorem is true for all numbers $1, 2, \cdots, n-1$; we shall prove it for $n=n$. This will establish it in general.

We shall prove first the sufficiency of the condition. To show that G contains at least $n+1$ vertices, let a and b be any two of its vertices; they are joined by n distinct chains. As there is at most a single arc ab, there must be $n-1$ distinct chains joining a and b, each of which contains at least two arcs. Each chain contains at least one vertex in its interior, so there are at least $n-1+2=n+1$ vertices in G. Now suppose we could disconnect G by dropping out $k<n$ vertices. Some two remaining vertices, a, b, are now joined by no chain. Hence in the original graph each chain joining them must pass through one of the k vertices we dropped out, and there are not n distinct chains therefore, a contradiction.

To prove the necessity of the condition, we must show that for $n>2$, if G is an n-tuply connected graph and a, f are any two of its vertices, then a and f are joined by n distinct chains. As $n>2$, G is non-separable, and

* i. e. chains which have only the two given vertices in common. Similar theorems have been proved by K. Menger, "Zur allgemeinen Kurventheorie," *Fundamenta Mathematicae*, Vol. 10 (1926), pp. 96-115, Satz β; N. E. Rutt, "Concerning the Cut-Points of a Continuous Curve When the Arc Curve *ab* Contains Exactly *n* Independent Arcs," *American Journal of Mathematics*, Vol. 51 (1929), pp. 217-246.

† P, theorem 1.

hence, by N, theorem 7,* there is a circuit passing through a and f. One of the two chains joining them, say the chain $D = ab, bc, cd, \cdots, ef$, contains at least two arcs. Form the graph G_2 from G by dropping out the vertex b and the arcs on it. By theorem 6, G_2 is $(n-1)$-tuply connected, and hence there are $n-1$ distinct chains joining a and c. Now replace the vertex b and the arcs on b. The arcs ab, bc form another chain from a to c, so that there are in G n distinct chains joining a and c. Now if f is not joined to a by n distinct chains, there is a first vertex of the chain D (coming after c), say d, which is not, while the vertex preceding it, say c, is. We shall show this leads to a contradiction.

Let $A_1, A_2, \cdots, A_n$ be n distinct chains from a to c. Form the graph G_1 from G by dropping out the vertex c and the arcs on c. There are in G_1 $n-1$ distinct chains $B_1, B_2, \cdots, B_{n-1}$ from d to a. We shall find in G n distinct chains $C_1, C_2, \cdots, C_n$ from d to a.

Case 1. d is not on any of the chains $A_1, \cdots, A_n$. The method is to find first a set of chains $C'_1, \cdots, C'_{n-1}$, with the following properties:

(1) For some number l, $0 \leqq l \leqq n-1$, l of the chains, say C'_{n-l}, $C'_{n-l+1}, \cdots, C'_{n-1}$ are identical with the chains $B_{n-l}, B_{n-l+1}, \cdots, B_{n-1}$.

(2) Each other chain C'_i, (of which there are $n-1-l$), is identical with a part of B_i, that part stretching from d to a certain vertex, say b_i, of one of the chains $A_1, \cdots, A_n$, say A_i, and these vertices lie on distinct chains.

(3) No one of the chains C'_i has a vertex in common with either a vertex of that part of any chain A_j lying between a and b_j, $j = 1, 2, \cdots, n-1-l$, or with any vertex of any of the chains $A_{n-l}, A_{n-l+1}, \cdots, A_n$, other than the vertex a. (Of course no chain B_i passes through the vertex c).

We can then construct the chains $C_1, \cdots, C_n$ as follows: For $i = 1, 2, \cdots, n-1-l$, C_i consists of C'_i plus that part of A_i lying between a and b_i; for $j = n-l, n-l+1, \cdots, n-1$, C_j is exactly C'_j; and C_n consists of the arc cd plus the chain A_n. These are n distinct chains from d to a, as required.

We shall find the chains $C'_1, \cdots, C'_{n-1}$ in the following manner: First follow B_1 from d towards a till we reach a vertex of one of the chains $A_1, \cdots, A_n$, or the vertex a itself. This much of B_1 might serve as C'_1. Next follow B_2 towards a, and alter C'_1 if necessary, so we have two chains that might serve as C'_1, C'_2. Next follow B_3 towards a, and alter the other two chains if necessary till we have three distinct chains which might serve as C'_1, C'_2, C'_3. Continue in this manner till we have found the $n-1$ re-

* P, theorem 1.

11

quired chains. At each stage therefore, the properties (1), (2), (3) hold, with $n-1$ replaced by some number $k \leqq n-1$.

To turn to the actual construction, we find C'_1 as described. Next follow B_2 from d towards a. Suppose first B_1 was entirely distinct from the chains $A_1, \cdots, A_n$, so that C'_1 is B_1. If we reach a before touching any of the chains $A_1, \cdots, A_n$, B_2 forms C'_2. If we reach first a vertex b_2 of the chain A_2, say, this much of B_2 forms C'_2. In either case, (1), (2) and (3) hold. Suppose next B_1 had the vertex b_1 in common with A_1 say. If we reach either the vertex a or a vertex b_2 of one of the chains $A_2, \cdots, A_n$, say A_2, before reaching any other vertex of the chains $A_2, \cdots, A_n$, or any vertex of $A_1(ab_1)$, (that is, that part of A_1 lying between a and b_1), this much of B_2, that is, B_2 itself or $B_2(db_2)$, forms C'_2, and again (1), (2) and (3) hold. We are thus in difficulty only if we reach a vertex b'_2 of $A_1(ab_1)$ before reaching either a or a vertex of one of the chains $A_2, \cdots, A_n$.

In this case, let C'_2 be $B_2(db'_2)$. Let us follow B_1 further from b_1 towards a. If we reach either a or a vertex b'_1 of one of the chains $A_2, \cdots, A_n$ before reaching a vertex of $A_1(ab'_2)$, this much of B_1 added to the original C'_1 forms the new C'_1, and again (1), (2) and (3) hold. We are in difficulty only if we reach first a vertex, say b''_1, of $A_1(ab'_2)$. In this case, let $B_1(db''_1)$ be the new C'_1. Follow now B_2 from b'_2 towards a, forming more of the new C'_2. Again, we are in difficulty only if we reach a vertex of $A_1(ab''_1)$ before reaching either a or a vertex of one of the chains $A_2, \cdots, A_n$. Continuing, the process must at some time come to an end, as there are only a finite number of vertices in the chain A_1. We now have two chains C'_1, C'_2, as required.

We shall assume now we have found $k-1$ chains $C'_1, \cdots, C'_{k-1}$, as required, and we shall find the k'th chain C'_k. We can rename chains and vertices so that for some number $l \leqq k-1$, l of the chains, C'_{k-l}, C'_{k-l+1}, $\cdots$, C'_{k-1} are identical with $B'_{k-l}, B'_{k-l+1}, \cdots, B'_{k-1}$, respectively, and each other chain C'_i is identical with $B_i(db_i)$, $i=1, 2, \cdots, k-1-l$, where b_i lies on the chain A_i.

Follow B_k from d towards a. If we reach either the vertex a or a vertex b_{k-l} of one of the chains $A_{k-l}, A_{k-l+1}, \cdots, A_n$, say A_{k-l}, before reaching any other vertex of these chains or any vertex of $A_i(ab_i)$, $i=1, 2, \cdots, k-1-l$, this much of B_k forms C'_k, and (1), (2) and (3) hold. We are in difficulty only if we reach first a vertex b' of $A_i(ab_i)$, $1 \leqq i \leqq k-1-l$. Let this much of B_k be C'_k. We alter C'_i as follows: Follow B_i from b_i towards a. If we reach a or a vertex of one of the chains $A_{k-l}, \cdots, A_n$ first, this much of B_i together with $B_i(db_i)$ forms the new C'_i, and again (1), (2) and (3) hold. Suppose we reached first the vertex b'' of $A_j(ab_j)$, where $1 \leqq j \leqq$

$k-1-l$, $j \neq i$, or of $A_i(ab')$. We then follow B_j or B_k, as the case may be, further towards a, forming more of the new C'_j or C'_k. Again, we are in difficulty only if we reach a vertex of A_s, $1 \leqq s \leqq k-1-l$, lying between a and the vertex b_s (or b' or b''). In this case, we follow B_s (or B_k or B_i) further towards a. The process must eventually come to an end, as there are but a finite number of vertices in the graph. We now have the required chains $C'_1, \cdots, C'_k$.

Putting $k = n-1$, we have the required $n-1$ chains $C'_1, \cdots, C'_{n-1}$, and the proof for case 1 is complete.

Case 2. d is on one of the chains $A_1, \cdots, A_n$, say A_n. Let A_n now stand for only that part of the former A_n lying between a and d. Exactly as we before found $n-1$ distinct chains from d to a which did not touch one of the chains $A_1, \cdots, A_n$ (which chain we called A_n), we now find a similar set of chains. If, first, A_n is a chain we have not touched, it forms the required n-th chain from d to a. If A_i, $i \neq n$, was a chain we have not touched, this chain plus the arc cd forms the required n-th chain. The proof for case 2 is now complete also.

3. *Applications to dual graphs.* In this section we remove the restriction on the graphs we consider, that they contain no 1- or 2-circuits, for during the proofs of the following theorems we may run into graphs with such circuits, and we do not wish to have to avoid them.

A planar graph cannot be more than quintuply connected. For suppose there were a sextuply connected graph. Then there is such a graph containing no 2-circuits. By theorem 5, each vertex is on at least six arcs. But this cannot be, as is easily seen from Euler's formula.* However, there are quintuply connected planar graphs, for instance, the dual graph of the dodecahedron (containing twelve vertices, each of which is on five arcs).

Dual graphs may not have the same connectivity. (The dual of the graph just mentioned is but triply connected, each vertex being on but three arcs.) However, we might define a certain notion, the connectivity of a graph *in the large.* Then to a certain extent, we have the theorem that *dual graphs have the same connectivity in the large.* This will be understood on referring to theorem 9 below. We will not go further into the matter here.

THEOREM 8. *Let G and G' be connected† dual graphs. Let a be a*

* This is the equivalent of the theorem that a map on a sphere contains at least one region with less than six sides.

† The hypothesis that the graphs be connected is obviously unnecessary. We include it merely for convenience.

vertex of G, and let A be the arcs on a. Let A', the corresponding arcs of G', form a circuit. Let b' and c' be two distinct vertices of this circuit, dividing it into the two chains B' and C'. Let B and C be the corresponding arcs of G. Form G'_1 from G' by letting b' and c' coalesce into the vertex a'. Form G_1 from G by replacing a by the two vertices b and c, and letting the arcs B end on b, and the arcs C, on c. Then G_1 and G'_1 are duals, preserving the correspondence between their arcs.

Let R, R', r, r', etc., stand for the ranks of G, G', H, H', etc., respectively. Letting b' and c' coalesce in G' reduced the number of vertices by one, and hence

$$R'_1 = R' - 1.$$

Let H_1 be any subgraph of G_1, and let H'_1 be the complement of the corresponding subgraph of G'_1. We must show that

$$r'_1 = R'_1 - n_1.$$

Let H and H' be the subgraphs of G and G' containing the same arcs as H_1 and H'_1, respectively. Then, as G and G' are duals,

$$r' = R' - n.$$

Case 1. b' and c' are in different connected pieces in H'. Then b and c are in the same connected piece in H_1. For add to H' every arc of G' we can without connecting b' and c', forming the graph I'. Let D' be the remaining arcs of G', and let D be the corresponding arcs in G. Then D is the complement in G of the arcs corresponding to I' in G'. I' is in two connected pieces, while adding any arc of D' renders the graph connected. Therefore, as G' is a dual of G, the arcs D are of nullity 1, while dropping out any one of them reduces the nullity to 0. Hence, by N, theorem 9, D is a circuit. As b' and c' are not connected in I', at least one of the arcs of B' and at least one of the arcs of C' are missing from I', and therefore at least one of the arcs of B and at least one of the arcs of C are present in D. In forming G_1 from G, the circuit D is broken at the vertex a, and becomes a chain D_1 in G_1. As D_1 contains still an arc of B and an arc of C, D_1 joins b and c. As I' contains every arc of H', D_1 is contained in H_1. Hence b and c are connected in H_1.

In forming H'_1 from H' by letting b' and c' coalesce, the numbers of vertices and of connected pieces are each reduced by 1, and hence

$$r'_1 = r'.$$

In forming H from H_1 by letting b and c coalesce, the the number of vertices is reduced by 1, while the number of connected pieces and of arcs remains the same. Hence

$$n_1 = n - 1.$$

These equations with the equations above give

$$r'_1 = R'_1 - n_1,$$

as required.

Case 2. b' and c' are connected in H'. Then b and c are not connected in H_1. For suppose there were a chain D_1 joining them. The arcs of this chain form a circuit D in G, containing one arc of B and one arc of C. Let D' be the corresponding arcs of G', and let I' be the complement of D'. Then as G' is a dual of G, I' is in two connected pieces, while adding any arc of D' connects the graph. Let $d'e'$ be the arc of B' contained in D', where d' is nearest b', e' is nearest c'. (d' may be b', and e' may be c'). As D' contains no other arc of B', b' and d' are connected in I', as are c' and e'. Adding the arc $d'e'$ to I' connects I'. Therefore d' and e', and therefore also b' and c', were formerly not connected. As I' contains every arc of H', b' and c' are not connected in H' either, a contradiction.

As b' and c' are connected in H', letting them coalesce reduces the rank of H', that is,

$$r'_1 = r' - 1.$$

As b and c are not connected in H', letting them coalesce does not alter the rank or nullity of H_1, that is,

$$n_1 = n.$$

These equations with the first equations give

$$r'_1 = R'_1 - n_1$$

again, as required.

THEOREM 9. *Let G be a non-separable planar graph with the following properties. If the vertices $a_1, a_2, \cdots, a_n$ and their arcs are dropped out, the resulting graph is in two connected pieces, H_1 and H_2, while no proper subset of these vertices, if dropped out, disconnect the graph. Either H_1 is of nullity > 0 or one of the vertices a_i is joined to H_1 by at least two arcs, and the same is true of H_2. Then if G' is a dual of G, G' can be disconnected by dropping out a corresponding set of n vertices and their arcs.*

A definition of a "corresponding set of vertices" is given during the proof.

Let A_i be the arcs on a_i, let $A_i(H_1)$ be those arcs joining a_i to vertices in H_1, let $A_i(H_2)$ be those joining a_i to vertices in H_2, and let $A_i(A)$ be those joining a_i to other vertices of the set $a_1, \cdots, a_n$, $i = 1, \cdots, n$. There are arcs in $A_i(H_1)$ and $A_i(H_2)$ for each i. For suppose for instance there were no arcs in $A_1(H_1)$. Then if we drop out the vertices $a_2, \cdots, a_n$ and their arcs from G, a_1 is now connected only to H_2, and hence H_1 and H_2 are not connected, contrary to hypothesis. Let A'_i, $A'_i(H_1)$, $A'_i(H_2)$, $A'_i(A)$ be the arcs of G' corresponding to A_i, $A_i(H_1)$, $A_i(H_2)$, $A_i(A)$ respectively, $i = 1, \cdots, n$.

The arcs A_i, if dropped out, disconnect G.† Hence the arcs A'_i form a circuit, $i = 1, \cdots, n$. Let us form G^*_1 from G by dropping out the arcs $P = A_1(H_1), A_2(H_1), \cdots, A_n(H_1)$. G^*_1 is in two connected pieces, one of them being H_1, the other being H_2 together with the vertices $a_1, \cdots, a_n$. Putting back any of these arcs connects the two pieces. Hence the corresponding arcs of G' form a circuit P'.

The arcs $A'_i(H_1)$ for any i form a chain. For suppose not. Then for some i, $A'_i(H_1)$ is composed of k chains $C'_1, \cdots, C'_k$. The rest of the circuit A'_i is composed of k chains $D'_1, \cdots, D'_k$, and each chain D'_j joins some two chains of the set $C'_1, \cdots, C'_k$. We formed G^*_1 from G by dropping out the arcs P. Let us form G^*_2 by dropping out also the arcs corresponding to the arcs of D'_1. As D'_1 has its end vertices in P', P' and D'_1 are together of nullity 2. Hence G^*_2 must be in three connected pieces. But this is not so. For the arcs of $A_i(H_2)$ and $A_i(A)$ all join a_i to that part of G^*_1 containing H_2, and D_1 contains but a part of these arcs. Hence a_i is still joined to the rest of this part of G^*_1, and G^*_2 contains the same number, 2, of connected pieces as G^*_1, a contradiction. This proves the statement.

Let us interchange the names of the chains $A'_1(H_1), A'_2(H_1), \cdots, A'_n(H_1)$ so that they occur in that order in P', and rename the vertices $a_1, a_2, \cdots, a_n$ and their arcs accordingly. Let a'_1 be the vertex joining $A'_1(H_1)$ and $A'_2(H_1), \cdots$, let a'_{n-1} be the vertex joining $A'_{n-1}(H_1)$ and $A'_n(H_1)$, and let a'_n be the vertex joining $A'_n(H_1)$ and $A'_1(H_1)$. These vertices form a *corresponding set of vertices* to the vertices $a_1, \cdots, a_n$. If we map the graphs together on a sphere ‡ as described in N, theorem 30, we can draw a closed curve which passes successively through the vertices $a_1, a'_1, a_1, a'_2, \cdots, a_n, a'_n$, and touches no other vertices or arcs. We shall show that $a'_1, \cdots, a'_n$, if dropped out, disconnect G'.

Form G'_{n-1} from G' by letting a'_n and a'_{n-1} coalesce. Form G_{n-1} from G

† The resulting graph is in just two pieces as a_i is not a cut vertex of G, G being non-separable.

‡ So that each arc of one graph cuts the corresponding arc of the other, and otherwise the graphs do not intersect.

by replacing the vertex a_n by the two vertices $a_n(H_1)$ and $a_n(H_2)$, letting the arcs $A_n(H_1)$ end on $a_n(H_1)$, and letting the arcs $A_n(H_2)$ and $A_n(A)$ end on $a_n(H_2)$. By theorem 8, G_{n-1} and G'_{n-1} are duals. Form G'_{n-2} from G'_{n-1} by letting a'_{n-1} and a'_{n-2} coalesce. Form G_{n-2} from G_{n-1} by replacing the vertex a_{n-1} by the two vertices $a_{n-1}(H_1)$ and $a_{n-1}(H_2)$, and replacing the arcs on a_{n-1} as above.* Then G_{n-2} and G'_{n-2} are duals. Continuing, we have finally the dual graphs G_1 and G'_1, where $a'_1, a'_2, \cdots, a'_n$ have all coalesced to form a single vertex in G'_1, and a_i has been replaced by the two vertices $a_i(H_1)$, $a_i(H_2)$, $i = 2, 3, \cdots, n$, in G_1.

Let I_1 be H_1 together with the vertices a_1, $a_2(H_1), \cdots, a_n(H_1)$, and the arcs $A_1(H_1), \cdots, A_n(H_1)$. Define I_2 similarly. Then I_1 and I_2 have but the vertex a_1 in common, and together they form G_1. Let $I_{11}, \cdots, I_{1k}$ be the components of I_1, (there may be but one), and let $I_{21}, \cdots, I_{2l}$ be those of I_2. I_1 is of nullity > 0. For if H_1 is of nullity > 0, I_1 is, as it contains H_1. Otherwise, by hypothesis, there are at least two arcs in one of the sets of arcs $A_i(H_1)$. Let these be the arcs a_ib and a_ic in G. b and c are joined by a chain in H_1, as H_1 is connected; this chain together with the arcs a_ib and a_ic forms a circuit in I_1. The statement now follows from N, theorem 4. Similarly, I_2 is of nullity > 0. Thus one of the components I_{1i} of I_1, and one of the components I_{2j} of I_2, is of nullity > 0 by N, theorem 13.†

Let $I'_{11}, \cdots, I'_{2l}$ be the subgraphs of G'_1 whose arcs correspond to the arcs of $I_{11}, \cdots, I_{2l}$. By N, theorem 25,‡ $I'_{11}, \cdots, I'_{2l}$ are the components of G'_1, and I'_{11} is a dual of $I_{11}, \cdots, I'_{2l}$ is a dual of I_{2l}. As I_{1i} and I_{2j} are of nullity > 0, I'_{1i} and I'_{2j} are of rank > 0, and they contain therefore no 1-circuits. Drop out any 1-circuits there may be in G'_1. The resulting graph J' is separable, as it contains both I'_{1i} and I'_{2j}. As it is connected, it has a cut vertex x'. J' is formed therefore of two graphs J'_1 and J'_2, each containing an arc and hence at least two vertices, which have only the vertex x' in common. If x' is dropped out of J', J' is disconnected. Hence if x' is dropped out of G'_1, forming the graph K'_1, K'_1 is not connected.

x' is the vertex a'_1. For suppose it were not. Form K' from G' by dropping out x' and its arcs. As G is doubly connected, we see easily that G' is also, by N, theorem 26,§ and hence K' is connected. K'_1 is formed from K' by letting the vertices $a'_1, \cdots, a'_n$ coalesce, and hence K'_1 also is con-

* It is easily seen that a_{n-1} is not a cut vertex of G_{n-1}, and hence A'_{n-1} is still a circuit in G'_{n-1}.

† P, theorem 3.

‡ P, theorem 13.

§ P, theorem 14.

nected, a contradiction. Now drop out the vertices $a'_1, \cdots, a'_n$ and their arcs from G'. We are left with the graph K'_1 which is not connected, and the theorem is proved.

THEOREM 10. *A dual G' of a triply connected graph G containing no 1- or 2-circuits is a similar graph.*

G' contains no 1- or 2-circuits; for if it did, dropping out the corresponding one or two arcs of G would disconnect this graph, contrary to theorem 5.

G' contains at least four vertices. For if it contained less than four vertices, it would contain at most three arcs, and would be of nullity $\leqq 1$, and G would be of rank $\leqq 1$, and would contain but two vertices.

Finally, G' cannot be disconnected by dropping out but two vertices. For suppose it could. If dropping out a'_1 and a'_2 and their arcs leaves the two parts H'_1 and H'_2, then, as G cannot be similarly disconnected, either a'_1 and a'_2 are each joined to H'_1 by but a single arc, or they are each joined to H'_2 by but a single arc, say the first, by theorem 9. The two arcs joining a'_1 and a'_2 to H'_1, if dropped out, disconnect G'. Hence the corresponding two arcs of G form a circuit. But this is contrary to the hypothesis that G contains no 2-circuit. Thus G' is triply connected.

THEOREM 11. *A triply connected planar graph containing no 1- or 2-circuit has a unique dual.*

For suppose G' and G'' were both duals of the triply connected graph G. Let H' be any subgraph of G' of nullity 1. Let H'' be the corresponding subgraph of G'' (the correspondence being given through the graph G), and let H be the complement of the corresponding subgraph of G. Then

$$r = R - n',$$
$$r = R - n'',$$

and hence

$$n'' = n',$$

and H'' is of nullity 1. Similarly, if H'' is any subgraph of G'' of nullity 1, the corresponding subgraph H' of G' is of nullity 1. Moreover, by theorem 10, G' and G'' are both triply connected, and neither contains any 1- or 2-circuits. Therefore, by theorem 3, G' and G'' are congruent; that is, the dual of G is unique.

HARVARD UNIVERSITY.

THE COLORING OF GRAPHS.[1]

By Hassler Whitney.[2]

Introduction. In another paper, L,[3] the author has given a proof of a formula for $M(\lambda)$, the number of ways of coloring a graph in λ colors, due to Birkhoff. The numbers m_{ij}, in terms of which $M(\lambda)$ is expressed, are here studied in detail; a method of calculating them is given.

In Part I elementary properties of the m_{ij} and the m_i and of $M(\lambda)$ are given. The expression for the m_{ij} in terms of the broken circuits of the graph gives rise to sets of numbers α_i and β_i, in terms of which $M(\lambda)$ may also be expressed. In Part II it is shown that m_{ij} may be expressed as a polynomial in numbers $N_1, N_2, \cdots$, numbers of *non-separable* subgraphs, which are easier to count in a given graph than are the m_{ij}. Part III is devoted to the theory of an algebraic transformation. This is made use of in Part IV to show that if we know the linear terms f_{ij} of the above polynomial, the whole polynomial may be calculated. Finally, the polynomials for the f_{ij}, m_{ij} and m_i are calculated for a few values of the subscripts.

---

[1] Received February 16, 1932.—Presented to the American Mathematical Society, Oct. 25, 1930. This is, in revised form, the author's Harvard thesis. The section on recursion formulas has been left out; Part V has been added. An outline of the paper will be found in the Proceedings of the National Academy of Sciences, 17 (1931), pp. 122–125.

[2] National Research Fellow.

[3] References will be made to the following papers by the author.

I. A logical expansion in mathematics (L); Bulletin of the American Mathematical Society, 38 (1932), pp. 572–579.

II. Non-separable and planar graphs (N); Transactions of the American Mathematical Society, 34 (1932), pp. 339–362.

III. Characteristic functions and the algebra of logic (C. F.); to appear in the Ann. of Math.

IV. Congruent graphs and the connectivity of graphs (C. G.), American Journal of Mathematics, 54 (1932), pp. 150–168.

The notations in N will be made use of. We recall the following. The *rank* R and *nullity* N of a graph G containing V vertices and E arcs which is in P connected pieces are given by the equations $R = V - P$, $N = E - R = E - V + P$. A *subgraph* H of G is determined by naming a subset of the arcs of G. A graph is *non-separable* if it is connected and cannot be broken at a single vertex into two pieces. The *components* of a graph are the non-separable parts of the graph. The *sum* of two graphs G_1 and G_2 is the graph $G_1 + G_2$ containing the arcs and vertices of both graphs, provided they had no vertices in common. If they have common vertices, we consider these vertices as distinct in $G_1 + G_2$.

A graph is *colored* by assigning to each vertex a color in such a way that no two vertices which are joined by an arc are of the same color.

With reference to the four color map problem, we mention the following theorem. *If G is a planar graph and G' is a dual of G, and m_{ij} and m'_{ij} are their numbers, then*

$$m'_{ij} = m_{R-j,N-i}.$$[4]

This follows immediately from the definition of dual graphs in N. Consider the class of graphs G with numbers m_{ij} for which the above numbers m'_{ij} are also the numbers for a graph G'; this class then includes all planar graphs (see N, Theorem 29). Hence a proposition which implies the four color map theorem is the following: *For any graph G of the above class,* $\sum_{i,j}(-1)^{i+j} m_{ij} 4^{V-i} > 0$. (We must, however, exclude graphs for which $m_{01} \neq 0$; see § 3.) This proposition is stronger than the four color map theorem; for there are graphs in the above class which are not planar.[5]

I. The Polynomial $M(\lambda)$; the Numbers α_i and β_i.

1. Some properties of the m_{ij}. We showed in L that if there were m_{ij} subgraphs of G of rank i, nullity j, then the number $M(\lambda)$ of ways of coloring G is[6]

$$M(\lambda) = \sum_{i,j}(-1)^{i+j} m_{ij} \lambda^{V-i} = \sum_i m_i \lambda^{V-i}, \tag{1.1}$$

$$m_i = \sum_j (-1)^{i+j} m_{ij}. \tag{1.2}$$

If G_1 and G_2 are the two graphs mentioned at the end of L (T_{21} and T_{32} in Fig. 2), whose arcs are $\alpha(ab)$, $\beta(ac)$, $\gamma(bc)$ and $\alpha(ab)$, $\beta(ac)$, $\gamma(bc)$, $\delta(bd)$, $\varepsilon(cd)$ respectively, we find for their numbers m_{ij} those given in the following tables.

G_1 (rows: rank; columns: nullity)

rank \ nullity	0	1
0	1	0
1	3	0
2	3	1

G_2

	0	1	2
0	1	0	0
1	5	0	0
2	10	2	0
3	8	5	1

From these tables we immediately find their numbers m_i as given at the end of L. For instance, in G_1, m_2, the coefficient of λ, is $m_{20} - m_{21} = 3 - 1 = 2$.

[4] The formula of Theorem 8 of the Proceedings paper should be corrected.

[5] An example was recently found (May 13) by R. M. Foster: G contains the arcs ab, ac, bc, ad, de, ec, bf, fg, gc, eg, dg, fe. The corresponding graph G' is this same graph.

[6] Throughout this paper, sums are carried out over all non-zero terms unless otherwise stated. We put always $\binom{p}{q} = 0$ if $q < 0$ or $q > p$.

m_{00} is always 1, for there is always exactly one subgraph of any graph (even of the null graph, containing no arcs or vertices), of rank 0, nullity 0, namely, that containing no arcs. If the graph contains no 1-circuits, each arc is of rank 1, nullity 0, and each subgraph of rank 1, nullity 0 contains just one arc. Thus $m_{10} = E$. We shall always assume in the future that there are no 1-circuits in the graph, unless expressly mentioned. Then also of course $m_{01} = 0$. In any case, $m_{10} + m_{01} = E$. There is just one subgraph of G of rank R, nullity N, namely G itself, and $m_{RN} = 1$. Thus

(1.3) $$m_{00} = m_{RN} = 1, \qquad m_{10} = E \qquad (\text{or } m_{10} + m_{01} = E).$$

Of course

(1.4) $$m_{ij} = 0 \text{ if } i < 0, \ j < 0, \ i > R \text{ or } j > N.$$

There is a fundamental set of relations between the m_{ij}:

(1.5) $$m_{i0} + m_{i-1,1} + \cdots + m_{0i} = \binom{E}{i};$$

for the sum on the left is the total number of subgraphs of G of i arcs. We shall see in Part IV that there is no other linear, or even polynomial, equality connecting the m_{ij}.

2. Some properties of the m_i. We can derive immediately a number of properties of the coefficients m_i of $M(\lambda)$ from the interpretation of $(-1)^i m_i$ in terms of the broken circuits of G given in L. Let us show that $(-1)^R m_R > 0$. We drop out all the arcs of G we can without disconnecting any vertices that were formerly connected. The resulting graph H is a forest[7] of rank R. Order the arcs of G in a definite manner, choosing the arcs not in H first. Every circuit in G contains an arc not in H, and therefore every broken circuit contains an arc not in H. Thus H is a subgraph of G of R arcs not containing all the arcs of any broken circuit, and hence $(-1)^R m_R > 0$. Taking any subgraph of H, we see that

(2.1) $$(-1)^i m_i > 0, \qquad i = 0, 1, \cdots, R.$$

Also

(2.2) $$m_i = 0, \qquad i < 0 \ \text{ or } \ i > R$$

from eq. (1.4). Hence $M(\lambda)$ has no constant term, as it would be m_V; but $V > R$ always. If G is connected, $R = V - 1$, and the coefficient of λ in $M(\lambda)$ is $m_R \neq 0$.

3. Some properties of $M(\lambda)$. As $M(\lambda)$ has no constant term, $M(0) = 0$. This is obvious, as no graph can be colored in no colors.

Let us find $M(1)$.

$$M(1) = \sum_i \sum_j (-1)^{i+j} m_{ij},$$

[7] A graph of nullity 0.

or, if we sum diagonally,

$$M(1) = \sum_i (-1)^i \sum_k m_{i-k,k} = \sum_i (-1)^i \binom{E}{i}$$

by eq. (1.5). Hence

$$M(1) = (1-1)^E = 0$$

if $E>0$. If $E = 0$, $M(1) = m_{00} = 1$. Thus a graph containing an arc cannot be colored in one color; if it contains no arc, there is one way of coloring it in one color.

Let us find $M(\lambda)$ for a circuit of E arcs. Every subgraph of the circuit is of nullity 0 except the whole graph, which is of nullity 1. Hence

$$m_{i0} = \binom{E}{i}, \qquad 0 \leqq i \leqq E-1, \qquad m_{E-1,1} = 1.$$

Therefore

$$\begin{aligned} M(\lambda) &= \sum_{i=0}^{E} (-1)^i \binom{E}{i} \lambda^{V-i} - (-1)^E \lambda^{V-E} + (-1)^E \lambda^{V-(E-1)} \\ &= (\lambda-1)^E + (-1)^E (\lambda-1), \end{aligned} \tag{3.1}$$

remembering that $V = E$. We can also prove this by noting that there is just one broken circuit, which contains $E-1$ arcs. In particular, $M(2) = 1+(-1)^E$, which is 2 if E is even and 0 if E is odd.

Suppose the graph G contained a 1-circuit aa. Then in any coloring of the graph, a must be colored in a different color from itself, which is not possible. Consequently $M(0) = M(1) = \cdots = 0$, and thus $M(\lambda) \equiv 0$. We can prove this from L, eq. (11), by pairing off terms, so that the two terms of a pair are the same except that one contains the factor A_{aa} while the other does not. The two terms of each pair cancel.

4. **The numbers α_i and β_i.** We shall illustrate these numbers by means of an example, using the graph G_2 of § 1. Its circuits are $\alpha\beta\gamma$, $\gamma\delta\varepsilon$, $\alpha\beta\delta\varepsilon$; its broken circuits are $\alpha\beta$, $\gamma\delta$, $\alpha\beta\delta$. $(-1)^i m_i$ is the number of subgraphs of G_2 of i arcs not containing both α and β and not containing both γ and δ. For under these conditions, it does not contain α, β and δ either. Let R_i be the set of subgraphs of G_2 of i arcs, let $n_i(\alpha)$ be the number which contain the arc α, etc. Then

$$(-1)^i m_i = n_i(\overline{\alpha \cdot \beta} \cdot \overline{\gamma \cdot \delta}).$$

Expanding this as in C. F. into the second normal form, we have

$$\begin{aligned} (-1)^i m_i &= n_i(1-\alpha\beta-\gamma\delta+\alpha\beta\gamma\delta) \\ &= n_i(1)-n_i(\alpha\beta)-n_i(\gamma\delta)+n_i(\alpha\beta\gamma\delta). \end{aligned} \tag{4.1}$$

Consider a typical term, such as $n_i(\alpha\beta)$. This is the number of subgraphs of i arcs which contain both α and β. To form such a subgraph, we must pick out the two arcs α and β; we have left then $i-2$ arcs which we can pick out from the remaining $5-2=3$ arcs of G_2 in $\binom{5-2}{i-2}$ ways. This then is the value of the term. The sign of the term is $(-1)^l$, where l is the number of broken circuits combined in forming the term; in the above case, 1. In general, if the term is formed from l broken circuits, and if it contains k distinct arcs, then the value of the term is $(-1)^l\binom{E-k}{i-k}$.

Definition. Let α_{kl} be the number of sets of l broken circuits which together contain k arcs. Put

$$\alpha_k = \sum_l (-1)^l \alpha_{kl}. \tag{4.2}$$

Then

$$(-1)^i m_i = \sum_{k,l} (-1)^l \binom{E-k}{i-k} \alpha_{kl} = \sum_k \binom{E-k}{i-k} \alpha_k.$$

We note that α_k is just the sum of all the coefficients in eq. (4.1) which contain k arcs, and thus α_k is the α_k defined in C. F. at the end of § 10.

If we expand eq. (4.1) into the third normal form, we find

$$\begin{aligned}(-1)^i m_i = n_i[\bar\alpha\bar\gamma+\bar\alpha\bar\delta+\bar\beta\bar\gamma+\bar\beta\bar\delta-(\bar\alpha\bar\gamma\bar\delta+\bar\alpha\bar\beta\bar\gamma+\bar\alpha\bar\beta\bar\delta+\bar\beta\bar\gamma\bar\delta)\\ +\bar\alpha\bar\beta\bar\gamma\bar\delta].\end{aligned} \tag{4.3}$$

We define β_k as being the sum of the coefficients of all terms in this equation containing k arcs, again as in C. F.

With the help of C. F., eq.'s (22) and (23), we can express any of the sets m_i, α_i, β_i in terms of any other of the sets. The relations are[8]

[8] The formulas not already proved may be proved from the others by use of the relation

$$\sum_s (-1)^s \binom{p-s}{m}\binom{q}{s} = \binom{p-q}{m-q}, \quad \text{if } p \geqq q;$$

see Netto, Lehrbuch der Combinatorik, Leipzig, 1927, § 158, (27). For instance, to prove the formula for α_i in terms of m_k, put

$$\begin{aligned}(-1)^i a_i &= \sum_k \binom{E-k}{i-k} m_k = \sum_k \binom{E-k}{i-k}(-1)^k \sum_l \binom{E-l}{k-l}\alpha_l\\ &= \sum_l \alpha_l \sum_k (-1)^k \binom{E-k}{E-i}\binom{E-l}{k-l}.\end{aligned}$$

Replacing k by $k+l$, we find

$$\begin{aligned}(-1)^i a_i &= \sum_l (-1)^l \alpha_l \sum_k (-1)^k \binom{E-l-k}{E-i}\binom{E-l}{k}\\ &= \sum_l (-1)^l \alpha_l \binom{0}{l-i} = (-1)^i \alpha_i.\end{aligned}$$

Thus $a_i = \alpha_i$, as required.

$$
\begin{aligned}
(-1)^i m_i &= \sum_k \binom{E-k}{i-k} \alpha_k &&= \sum_k \binom{E-k}{i} \beta_k, \\
(-1)^i \alpha_i &= \sum_k \binom{E-k}{i-k} m_k &&= \sum_k \binom{k}{i} \beta_k, \\
(-1)^i \beta_i &= (-1)^E \sum_k \binom{k}{E-i} m_k &&= \sum_k \binom{k}{i} \alpha_k.
\end{aligned} \tag{4.4}
$$

The numbers for G_2 are

$$
\begin{array}{lllll}
\alpha_0 = 1, & & \alpha_2 = -2, & & \alpha_4 = 1; \\
 & & \beta_2 = 4, & \beta_3 = -4, & \beta_4 = 1; \\
m_0 = 1, & m_1 = -5, & m_2 = 8, & m_3 = -4, & m_4 = 0.
\end{array}
$$

The first non-zero β_i is $\beta_{E-R} = \beta_N$. Remembering that $m_k = 0$, $k > R$, we find from the last equation above

$$
\beta_N = (-1)^R m_R, \tag{4.5}
$$

which is the last non-zero coefficient in $M(\lambda)$; the coefficient of λ, if the graph is connected.

In terms of the α_i, we have

$$
M(\lambda) = \lambda^{V-E} \sum_i (-1)^i \alpha_i (\lambda - 1)^{E-i}. \tag{4.6}
$$

For, calling this expression $\varphi(\lambda)$,

$$
\begin{aligned}
\varphi(\lambda) &= \lambda^{V-E} \sum_i (-1)^E \alpha_i (1-\lambda)^{E-i} \\
&= \lambda^{V-E} \sum_i (-1)^E \alpha_i \sum_k (-1)^k \binom{E-i}{k} \lambda^k
\end{aligned}
$$

by the binomial formula,

$$
= \sum_k \lambda^{V-E+k} (-1)^{E-k} \sum_i \binom{E-i}{k} \alpha_i,
$$

or, replacing k by $E-k$,

$$
\begin{aligned}
\varphi(\lambda) &= \sum_k \lambda^{V-k} (-1)^k \sum_i \binom{E-i}{E-k} \alpha_i \\
&= \sum_k m_k \lambda^{V-k} = M(\lambda).
\end{aligned}
$$

We shall see by examples in Part V that the α_i are given by simpler formulas in terms of the numbers of non-separable subgraphs of G than are the m_i.

II. m_{ij} in Terms of Non-Separable Subgraphs.

5. The numbers N_1, N_2, $\cdots$. To find m_{ij}, we count all the subgraphs of G of rank i, nullity j. In this part we shall show that it is not necessary to count all the subgraphs; we need count only the non-separable subgraphs. We say a subgraph is non-separable if it becomes so when any isolated vertices there may be are dropped out. This operation alters neither the rank nor the nullity of the subgraph.

We note that Theorem 7A holds even if we allow the graphs to contain 1-circuits. If G' is G except for its m_{01} 1-circuits, then obviously

$$m_{ij} = \sum_k \binom{m_{01}}{k} m'_{i,j-k}. \tag{5.1}$$

Let us say two graphs are of the same type if they are congruent. Make a list of all types of non-separable graphs; call these types T_1, T_2, T_3, $\cdots$. There are of course but a finite number of types of graphs with a given number of arcs. Let N_i be the number of subgraphs of G of type T_i, $i = 1, 2, \cdots$. The fundamental theorem of this part is that m_{ij} for any graph G is given by a unique polynomial in the numbers N_1, N_2, $\cdots$ for G. When we speak of a polynomial, we mean always a polynomial in a finite number of variables. Thus we do not allow expressions such as $N_1 + N_2 + N_3 + \cdots$, even if, for any given graph, there are but a finite number of non-vanishing terms.

THEOREM 5A. *Any polynomial in the numbers N_1, N_2, $\cdots$, which equals* 0 *for all graphs is identically* 0.

Suppose the theorem false. Then there is a polynomial which is 0 for all graphs but not $\equiv 0$. Let P be such a polynomial containing the least possible number, say q, of variables. Let us call these variables N_1, N_2, $\cdots$, N_q, where T_1, corresponding to N_1, is one of the types of graphs concerned containing the least number of arcs.

Arrange the polynomial according to descending powers of N_1, and say it is of degree α in this variable:

$$P(N_1, N_2, \cdots, N_q) = P_0(N_2, \cdots, N_q) N_1^\alpha + P_1(N_2, \cdots, N_q) N_1^{\alpha-1} + \cdots + P_\alpha(N_2, \cdots, N_q), \quad P_0 \not\equiv 0.$$

Take now any graph G, and form from it graphs $G_1, G_2, \cdots, G_\alpha$, each being formed from the last by adding a graph of type T_1, letting it have no vertices in common with the preceding graphs. As a graph of type T_1 contains no subgraph of any of the types T_2, $\cdots$, T_q, we have in this manner added no subgraphs of any of these types. Hence the number N_1 for each graph G_i is one greater than that for the preceding graph, G_{i-1}, while the numbers N_2, $\cdots$, N_q are the same. P vanishes for these distinct values of N_1 and the coefficients P_0, P_1, $\cdots$, P_α are constant for these graphs, and hence these coefficients must vanish for these graphs. They vanish, in particular, for the graph G. But G was any graph, so they vanish for all graphs. But this is a contradiction. For, if q was > 1,

P_0 is a polynomial in a smaller number of variables, vanishing for all graphs without being identically 0, and if q was 1, P_0 is a constant $= 0$, and P was not of degree α in N_1.

THEOREM 5B. *Theorem* 5A *holds if the words "for all graphs" are replaced by the words "for all non-separable graphs".*

Using the same notation as before, let G be any non-separable graph. The only change necessary is to make $G_1, \cdots, G_\alpha$ non-separable also. Suppose first that the numbers $N_1, \cdots, N_q$ do not include E, the number of arcs in G. Let s be the greatest number of arcs in a graph of any of the types $T_1, \cdots, T_q$. We add a graph of type T_1 to G, and then join two of its vertices to two vertices of G by distinct chains, each containing at least $s/2$ arcs. The resulting graph G_1 is non-separable. Moreover, in adding the two chains we have added no subgraphs of any of the types $T_2, \cdots, T_q$. $G_2, \cdots, G_\alpha$ are formed similarly. If, however, the polynomial contains the variable E, we merely add a suspended chain of s arcs each time, increasing E by s, and leaving the other N_i unchanged.

6. m_{ij} **in terms of** $N_1, N_2, \cdots$. Given a graph G, consider the set S of all its subgraphs of rank i, nullity j. Drop out any isolated vertices there may be from each subgraph. Let us list all possible graphs of rank i, nullity j which contain no isolated vertices (many may be separable), and divide them into types $U_1, U_2, \cdots, U_p$. We say here two graphs are of the same type if they are equivalent.[8a] Each subgraph of S is of one of these types; say S_k, containing s_k members, is the set of subgraphs of type U_k (any or all of the s_k may be 0). Then

$$m_{ij} = s_1 + s_2 + \cdots + s_p. \tag{6.1}$$

We shall show how to evaluate each s_k.

Let H be a graph of type U_k, let $H_1, H_2, \cdots, H_q$ be its components, named in a definite order, (there will be but one component if H is non-separable), and say H_i is of type T_i, $i = 1, \cdots, q$. (Some of these types may be the same.) We now define a set R_k as follows: Any member of R_k consists of an ordered set of non-separable subgraphs of G. We name first in G a subgraph H_1' of type T_1, next a subgraph H_2' of type T_2 (which may have parts in common with H_1'), $\cdots$, and finally a subgraph H_q' of type T_q (which may have parts in common with any of the preceding graphs). Let n_k be the number of sets of graphs in R_k. If, for instance, H consists of two arcs, its components H_1 and H_2 each being an arc, R_k is the set of ordered pairs of arcs in G, the two arcs being distinct or not, and $n_k = E^2$.

[8a] Two graphs are equivalent if they have the same components.

Consider any member of R_k. It may happen that some of the subgraphs picked out of G to form this member, say the first three, form a circuit of graphs in G. If so, we shall say that this member of R_k has the property

$$H_1 \circ H_2 \circ H_3.$$

We are interested in those members of R_k in which no subset of the subgraphs form a circuit of graphs in G. These are

$$\begin{aligned} &R_k(\overline{H_1 \circ H_2} \cdot \overline{H_1 \circ H_3} \cdots \overline{H_{q-1} \circ H_q} \cdot \overline{H_1 \circ H_2 \circ H_3} \cdots \overline{H_1 \circ H_2 \circ \cdots \circ H_q}) \\ &= R_k(\bar{A}_{12} \cdot \bar{A}_{13} \cdots \bar{A}_{q-1,q} \cdot \bar{A}_{123} \cdots \bar{A}_{12\cdots q}) = R_k(A). \end{aligned}$$

Remembering that the types $T_1, \cdots, T_q$ may not be distinct, let us pick out the distinct types $T_1', T_2', \cdots, T_r'$ from these. Say the first τ_1 types $T_1, \cdots, T_{\tau_1}$ are the same type as T_1', etc. Of course

$$\tau_1 + \tau_2 + \cdots + \tau_r = q. \tag{6.2}$$

We shall now show that

$$s_k = \frac{1}{\tau_1!\, \tau_2! \cdots \tau_r!} n_k(A). \tag{6.3}$$

To this end we establish a correspondence between the members of S_k and of $R_k(A)$. Consider first any member of $R_k(A)$; say $H_1', \cdots, H_q'$ are the subgraphs of G named, where H_i' is of type T_i. Let H' be the subgraph of G containing the vertices and arcs of all these graphs. No subset of these graphs form a circuit of graphs, and hence these graphs are the components of H', by N, Theorem 17. Therefore H' is of the type U_k, and is a member of S_k. Thus to each member of $R_k(A)$ corresponds just one member of S_k.

Consider now any member H' of S_k in G, whose components are $H_1', \cdots, H_q'$, H_i' being of type T_i. There are exactly $\tau_1!\, \tau_2! \cdots \tau_r!$ corresponding members of $R_k(A)$, obtained as follows. $H_1', \cdots, H_{\tau_1}'$ all being of the same type T_1', and these being distinct in G, we name first any one of these, then any other, etc., which we can do in $\tau_1!$ ways, forming the first part of an ordered set of graphs of $R_k(A)$. The next τ_2 subgraphs we can name in $\tau_2!$ ways, etc. The equation now follows. This is true in particular if all the types $T_1, \cdots, T_q$ are different or all the same; also if H' is non-separable, in which case, there is but one type $T_1 = T_1'$, and $\tau_1 = 1$.

Summing over k, we have

$$m_{ij} = \sum_k \frac{1}{\tau_1!\, \tau_2! \cdots \tau_r!} n_k(A). \tag{6.4}$$

This expression is evaluated by expanding A by the logical expansion:

$$m_{ij} = \sum_k \frac{1}{\tau_1!\,\tau_2!\cdots\tau_r!}\, n_k[1-(A_{12}+A_{13}+\cdots+A_{q-1,q} + A_{123}+\cdots+A_{12\cdots q})+(A_{12}A_{13}+\cdots)-\cdots]. \tag{6.5}$$

7. Evaluation of eq. (6.5). We shall show that a typical term of the above expression is a polynomial in the numbers $N_1, N_2, \cdots$. This will prove that m_{ij} is a polynomial in these numbers.

Consider the term

$$n_k(A_{i_1 i_2\cdots i_\alpha} A_{j_1 j_2\cdots j_\beta}\cdots A_{l_1 l_2\cdots l_\delta}) = n_k(B_1 B_2\cdots B_u) = n_k(B), \tag{7.1}$$

each B_i stating that a certain subset of the graphs $H_1', \cdots, H_q'$ form a circuit of graphs in G. This term is the number of ways of naming in G first a subgraph H_1' of type $T_1, \cdots$, finally a subgraph H_q' of type T_q, so that the relations $B_1, \cdots, B_u$ all hold.

Divide the relations $B_1, \cdots, B_u$ into sets as follows. Suppose two of these included the same graph; e. g. if

$$B_1 = H_1 \circ H_2, \qquad B_2 = H_2 \circ H_4 \circ H_5,$$

then B_1 and B_2 both include the graph H_2. Say then that B_1 and B_2 are connected. If B_i and B_j, also B_j and B_k, are connected, say that B_i and B_k are connected. We put two relations into the same set if they are connected, otherwise, into different sets. Let C_1 stand for all the relations in the first set, C_2, for all those in the second set, $\cdots$, $C_{v'}$ for all those in the last set. Then

$$B = C_1 C_2 \cdots C_{v'}. \tag{7.2}$$

Divide now the graphs $H_1, H_2, \cdots, H_q$ into sets as follows: All the graphs appearing in any relation B_i of C_1 we put in the first set Q_1, all appearing in a relation of C_2 we put in the second set $Q_2, \cdots$, all appearing in the last set $C_{v'}$ we put in the set $Q_{v'}$. There may be some graphs left over (if we are considering the first term of eq. (6.5), there are no B's, and none of the graphs $H_1, \cdots, H_q$ have been placed in sets); if so, put each of these in a set by itself, calling these sets $Q_{v'+1}, \cdots, Q_v$. If we define the characteristic functions

$$C_i = 1, \qquad i = v'+1, \cdots, v, \tag{7.3}$$

then we may write also

$$B = C_1 C_2 \cdots C_v, \tag{7.4}$$

46

where $v = v'$ if there were no graphs left over. Now each graph H_i falls into exactly one of the sets $Q_1, \cdots, Q_v$.

In determining $n_k(B)$, we picked out the graphs $H_1', \cdots, H_q'$ from G in a definite order. Let us fix this order so that the first graphs fall into Q_1, the next fall into Q_2, etc. We can now say that $n_k(B)$ equals the number of ways of picking out from G first the graphs of Q_1 in a definite order so that C_1 holds, next the graphs of Q_2 in a definite order so that C_2 holds, $\cdots$, finally the graphs of Q_v in a definite order so that C_v holds. (For values of $i > v'$, the words "so that C_i holds" may of course be left out.) Hence if $n_k'(C_i)$ is the number of ways of picking out from G graphs of the set Q_i in a definite order so that C_i holds, then

$$n_k(B) = n_k'(C_1)\, n_k'(C_2) \cdots n_k'(C_v). \tag{7.5}$$

We shall show that each term of this product, such as $n_k'(C_j)$, is a homogeneous linear polynomial in the numbers $N_1, N_2, \cdots$, showing that $n_k(B)$ is a homogeneous polynomial of degree v in these numbers.

Let us say $H_1, H_2, \cdots, H_z$ are the graphs in Q_j. Name graphs of these types in G so that C_j holds. These subgraphs of G together form a subgraph I of G. I is non-separable. For if first C_j contains no B's, i. e. $C_j = 1$, then there is but a single graph H_1' in Q_j, and I is H_1', a non-separable graph. Otherwise, I consists of a number of these graphs, various ones of which form circuits of graphs. Suppose

$$C_j = B_{i_1} B_{i_2} \cdots B_{i_l}.$$

Consider that part of I, I_1, formed of the graphs forming the circuit of graphs corresponding to B_{i_1}. By N, Theorem 16, I_1 is non-separable. Similarly, $I_2, \cdots, I_l$ are non-separable. Now the relations $B_{i_1}, \cdots, B_{i_l}$ are connected. Hence there is a graph H_s' (containing at least two vertices) in both B_{i_1} and some other relation, say B_{i_2}, i. e. in both I_1 and I_2. Thus these two graphs form a circuit of graphs, and the two together form a non-separable graph. Continuing in this manner, we see finally that I is non-separable.

Fit the graphs $H_1, \cdots, H_z$ of Q_j together in all possible ways so that the relation C_j holds. We get various non-separable graphs, of types $T_1'', T_2'', \cdots, T_w''$ say. Suppose I is of type T_i''. Let η_i equal the number of ways of fitting first H_1, next $H_2, \cdots$, finally H_z into I so that C_j holds and I is completely filled out. Then if there are N_i subgraphs in G of type T_i'', $i = 1, \cdots, w$,

$$n_k'(C_j) = \eta_1 N_1 + \eta_2 N_2 + \cdots + \eta_w N_w. \tag{7.6}$$

To prove this, consider the correspondence between the sets of graphs counted in $n'_k(C_j)$ and the subgraphs of G of types $T_1'', \cdots, T_w''$. Corresponding to each way of picking out a set of graphs from G in counting $n'_k(C_j)$ is a non-separable subgraph I of G of one of these types. Consider any such subgraph I of type T_i'' of G. There are exactly η_i ways of naming subgraphs in G properly so that these graphs together form I in G. Hence each subgraph of this type contributes an amount η_i to $n'_k(C_j)$; as there are N_i such subgraphs in G, they all contribute an amount $\eta_i N_i$, and the result is established.

As the numbers $\tau_1, \tau_2, \cdots, \tau_r$, and $\eta_1, \eta_2, \cdots, \eta_w$ depend only on the types of graphs considered and on i and j, not on the graph G, we have proved all but the statement of uniqueness in the

THEOREM 7 A. *m_{ij} for any graph G is given by a unique polynomial in the numbers $N_1, N_2, \cdots$, containing no constant term, the coefficients being independent of the graph G.*

To prove the uniqueness of the polynomial, suppose there were two such polynomials. Then their difference is a polynomial which is 0 for all graphs, and hence is identically 0, by Theorem 5 A; the two polynomials are therefore identically equal.

Examples. We consider here only graphs containing no 1- or 2-circuits.

m_{20}: Each subgraph H of G of rank 2, nullity 0, consists of two distinct arcs, H_1 and H_2. As H_1 and H_2 are of the same type, $\tau_1 = 2$. Hence, by eq. (6.5),

$$m_{20} = \frac{1}{2!} n(\overline{H_1 \circ H_2}) = \frac{1}{2} n(1 - H_1 \circ H_2).$$

$n(1)$ is the number of ways of picking out first any arc, then again any arc from G. This is E^2. $n(H_1 \circ H_2)$ is the number of ways of picking out first an arc H_1, then an arc H_2, such that the two form a circuit of graphs in G. As there are no 2-circuits in G, these arcs must be identical. This number is thus E. This gives the obvious result

$$m_{20} = \frac{1}{2}(E^2 - E) = \binom{E}{2}.$$

m_{31}: A subgraph H of G of rank 3, nullity 1 consists either of a quadrilateral, or of a triangle H_1 and an arc H_2. The number of subgraphs of the first type is N_{31} say. The number in the second set is, as $\tau_1 = \tau_2 = 1$, $n(1 - H_1 \circ H_2)$. Now $n(1) = E N_{21}$ if there are N_{21} triangles in G. If H_1 and H_2 are to form a circuit of graphs, the arc H_2 must be one of the arcs of the triangle H_1. Together they form a triangle I. H_1 can be fitted into I in one way; the arc can be fitted in in three ways. Hence $\eta_1 = 3$, and $n(H_1 \circ H_2) = 3 N_{21}$. Adding,

46*

$$m_{31} = N_{31} + (E-3) N_{21},$$

a result which can be easily verified.

8. The types of non-separable subgraphs contributing to a given m_{ij}. Given two numbers i, j, we wish to know what numbers $N_1, N_2, \cdots$, appear in the polynomial for m_{ij}. We are interested especially in the linear terms of the polynomial; c. f. Part IV. If the number N_s corresponding to the type T_s of graph appears in some term of the expansion (6.5) for m_{ij}, we shall say a graph of this type *contributes* to m_{ij}. However, it might possibly happen that the terms cancel out so that N_s does not actually appear in the polynomial for m_{ij}.

As we saw in § 7, a term such as $n_k(B)$ in eq. (6.5) is linear if all the graphs $H_1, H_2, \cdots, H_q$ fall into a single set Q_1, i. e. if $v = 1$. This happens when the relations $B_1, \cdots, B_u$ are connected and each graph H_i falls into one of these relations, or if H is non-separable.

From the considerations of § 6 and § 7 we have the

THEOREM 8A. *The necessary and sufficient condition that a non-separable graph I contribute to the linear terms of m_{ij} is that there exists a graph H of rank i, nullity j, with the components $H_1, H_2, \cdots, H_q$ $(q \geqq 1)$, such that if these graphs are allowed to coalesce in the proper manner, I is formed.*

As an immediate application, an arc contributes to m_{i0} only. It is seen that a non-separable graph can only contribute to m_{ij} if it contributes to the linear terms of some m_{kl}, $k \leqq i$, $l \leqq j$. (This follows also directly from the facts in Part IV.)

LEMMA A. *A non-separable graph of rank i, nullity j, contributes to m_{ij} with the coefficient* 1.

These are the only non-separable graphs that are counted directly in eq. (6.5).

LEMMA B. *A non-separable graph of rank $i-k$, nullity $j+k$, contributes to the linear terms of m_{ij} if $k>0$.*

Let I be such a graph. As it is of nullity >0, it contains at least two arcs. Pull out one of its arcs, and let I' be the new graph, composed of this arc, I_1', and the remaining part of I, I_2'. I' is of rank $i-k+1$, nullity $j+k-1$. Separate I_2' into its components if it is separable, which does not alter its rank or nullity. If I' is still of nullity $>j$, we take one of its components of nullity >0, and pull out an arc as before. Continuing, we obtain finally a set of non-separable graphs, the whole being of rank i, nullity j, which, if pieced together properly, form I. The lemma now follows from Theorem 8A.

LEMMA C. *A non-separable graph of rank $i-k-l$, nullity $j+k$, contributes to the linear terms of m_{ij} if $k \geqq 0$, $l>0$.*

Let I be such a graph. Add to it l arcs, forming the graph I' in $l+1$ connected pieces. I' is of rank $i-k$, nullity $j+k$. Pull out arcs from this graph as in Lemma B until we obtain a graph of rank i, nullity j. I is formed from this graph by reconstructing I', then letting the arcs coalesce with arcs of the resulting piece.

Various graphs of rank $i-k$, nullity $j-l$, may also contribute. If so, similar graphs of rank $i-k-s$, nullity $j-l$, contribute.

LEMMA D. *No non-separable graph containing more than* $i+j$ *arcs contributes.*

Obviously, from Theorem 8 A.

LEMMA E. *No non-separable graph of rank* $i+l$, *nullity* $j-k$, *contributes to* m_{ij} *if* $l \geqq 0$, $k>0$.

For suppose I were such a graph which contributes to m_{ij}. Then, by Theorem 8 A, I is formed by letting the components $H_1, H_2, \cdots, H_q$, of which there are at least two, of a graph H of rank i, nullity j, coalesce. But then, by N, Theorem 14, I is of rank $<i$, a contradiction.

LEMMA F. *If* m_{ij}, $j>0$, *is* >0 *for some graph* G, *then there exists a non-separable graph* I *of rank* i, *nullity* j *(which contributes).*

G contains a subgraph H of rank i, nullity j. If it is separable, let $H_1, \cdots, H_q$ be its components. Let an arc of H_1 coalesce with an arc of H_2; this reduces the rank of the whole by one, but leaves the nullity unchanged. Continue in this manner until we have a non-separable graph H' of rank $i-q+1$, nullity j. Let ab be an arc of H'; replace it by two arcs, ac and cb, c being a new vertex. This increases the rank by one but leaves the nullity unchanged. Continuing, we have finally the non-separable graph I as required.

These results are collected in Fig. 1.

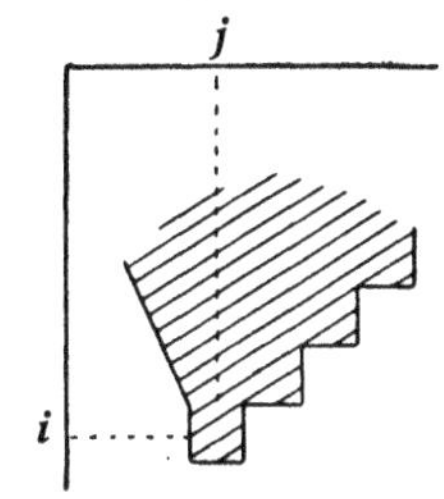

Fig. 1.
Showing those types of non-separable graphs contributing to the linear terms of a given m_{ij}

III. THE TRANSFORMATION T.

9. **Definition and elementary properties of T.** Let there be a set of elements $a, b, c, \cdots$, such that for any two elements a, b, there is a "sum element" $a+b=b+a$ also in the set. (We shall later interpret these elements as graphs.) Let $f_i(a)$, $i=1, 2, \cdots$, be a set of numbers defined for each element a, with the property that

$$f_i(a+b)=f_i(a)+f_i(b) \tag{9.1}$$

for any two elements a, b. We wish to find a transformation T from these sets of numbers to sets of numbers $m_i(a)$, $i=0, 1, 2, \cdots$, with the properties that $m_0 \equiv 1$ and

$$
(9.2)\qquad \begin{aligned} m_i(a+b) &= m_0(a)\,m_i(b)+m_1(a)\,m_{i-1}(b)+\cdots+m_i(a)\,m_0(b) \\ &= \sum_k m_k(a)\,m_{i-k}(b).^9 \end{aligned}
$$

The following symbol will be found convenient: $R_i^k(f)$ is the sum of all terms formed by multiplying together k numbers $f_{j_1}, f_{j_2}, \cdots, f_{j_k}$, whose subscripts (which need not be distinct) add up to i:

$$
(9.3)\qquad j_1+j_2+\cdots+j_k = i,
$$

and two terms are called different if their factors are arranged in different orders. Put also

$$
(9.4)\qquad R_0^0(f) = 1, \quad R_i^0(f) = 0, \quad i>0, \quad R_i^k(f) = 0, \quad i<k.
$$

Thus for example

$$
(9.5)\qquad \begin{aligned} R_4^1(f) &= f_4, \\ R_4^2(f) &= f_1f_3+f_2f_2+f_3f_1 = 2f_1f_3+f_2^2, \\ R_4^3(f) &= f_1f_1f_2+f_1f_2f_1+f_2f_1f_1 = 3f_1^2f_2, \\ R_4^4(f) &= f_1f_1f_1f_1 = f_1^4. \end{aligned}
$$

Now for any element a, let $f_1(a)$, $f_2(a)$, $\cdots$, be the coefficients of a power series in x:

$$
F_a(x) = f_1(a)\,x+f_2(a)\,x^2+\cdots.
$$

We shall define $m_i(a)$ as being the coefficient of x^i in the power series expansion of $e^{F_a(x)}$:

$$
\begin{aligned} e^{F_a(x)} &= 1+F_a(x)+\frac{1}{2!}F_a^2(x)+\cdots \\ &= m_0(a)+m_1(a)x+m_2(a)\,x^2+\cdots = M_a(x). \end{aligned}
$$

Equating coefficients determines the m_i. The first few are

$$
(9.6)\qquad \begin{aligned} m_0 &= 1, \quad m_1 = f_1, \quad m_2 = f_2+\frac{1}{2}f_1^2, \\ m_3 &= f_3+\frac{1}{2!}(f_1f_2+f_2f_1)+\frac{1}{3!}f_1^3 = f_3+f_1f_2+\frac{1}{6}f_1^3, \\ m_4 &= f_4+f_1f_3+\frac{1}{2}f_2^2+\frac{1}{2}f_1^2f_2+\frac{1}{24}f_1^4, \end{aligned}
$$

and the general term is found to be

$$
m_i = \sum_k \frac{1}{k!}R_i^k(f).
$$

[9] See footnote [6] in § 1, Part. I.

m_i is a polynomial in $f_1, f_2, \cdots, f_i$, for $i > 0$. If we collect terms, we find the following expression for the transformation T:

$$(9.7) \qquad m_i = \sum_k \frac{1}{k!} R_i^k(f) = \sum_{\Sigma k\alpha_k = i} \frac{1}{\alpha_1!\, \alpha_2! \cdots \alpha_i!} f_1^{\alpha_1} f_2^{\alpha_2} \cdots f_i^{\alpha_i},$$

where in the second sum all terms with $\alpha_1 + 2\alpha_2 + \cdots + i\alpha_i = i$ appear.

To find the inverse of T, we compare coefficients of x^i in the equation

$$F_a(x) = \log M_a(x).$$

We find

$$(9.8) \qquad f_i = \sum_k \frac{(-1)^{k-1}}{k} R_i^k(m) = \sum_{\Sigma k\alpha_k = i} \frac{(-1)^s s!}{\alpha_1!\, \alpha_2! \cdots \alpha_i!} m_1^{\alpha_1} m_2^{\alpha_2} \cdots m_i^{\alpha_i},$$

where

$$s = \alpha_1 + \alpha_2 + \cdots + \alpha_i - 1.$$

The first few f_i are

$$(9.9) \quad f_1 = m_1, \qquad f_2 = m_2 - \frac{1}{2} m_1^2, \qquad f_3 = m_3 - m_1 m_2 + \frac{1}{3} m_1^3.$$

If we write T in the form

$$(9.10) \qquad m_i = f_i + \sum_{k \geq 2} \frac{1}{k!} R_i^k(f),$$

each term of the sum contains $k \geqq 2$ factors. Similarly for T^{-1}. We have therefore the

THEOREM 9A. *$m_i[f_i]$, $i > 0$, equals $f_i[m_i]$ plus a polynomial in $f_1, f_2, \cdots, f_{i-1}[m_1, m_2, \cdots, m_{i-1}]$, containing no constant or linear terms.*

We come now to the fundamental theorem.

THEOREM 9B. *If*

$$f_i(a+b) = f_i(a) + f_i(b),$$

then

$$m_i(a+b) = \sum_k m_k(a)\, m_{i-k}(b),$$

and conversely.

Define the power series $F_a(x)$, $F_b(x)$, $F_{a+b}(x)$, $M_a(x)$, $M_b(x)$, $M_{a+b}(x)$ as we formerly defined $F_a(x)$, $M_a(x)$. If now $f_i(a+b) = f_i(a) + f_i(b)$, then

$$F_{a+b}(x) = F_a(x) + F_b(x).$$

Hence

$$M_{a+b}(x) = e^{F_{a+b}(x)} = e^{F_a(x) + F_b(x)} = e^{F_a(x)} e^{F_b(x)} = M_a(x)\, M_b(x).$$

Equating coefficients gives the first half of the theorem. The second half is proved similarly.

10. Uniqueness of T and T^{-1}. We shall show that T and T^{-1} are the only transformations obeying Theorems 9A and 9B. Consider first T.

THEOREM 10A. *Let U be a transformation from numbers $f_1, f_2, \cdots$, to numbers $m_0, m_1, m_2, \cdots$, with the following properties.*

(1) $m_0 \equiv 1$.

(2) *Each m_i, $i>0$, equals f_i plus a polynomial in $f_1, f_2, \cdots, f_{i-1}$, containing no constant or linear terms.*

(3) *If $f_i(a)$ and $f_i(b)$ are arbitrary sets of numbers, and $f_i(a+b)$ is defined by the equation* (9.1), *then the corresponding sets $m_i(a)$, $m_i(b)$, $m_i(a+b)$ are related by the equation* (9.2).

Then U is exactly T.

If we take $i=1$, we find m_1 must equal f_1. Assuming the expressions for $m_1, m_2, \cdots, m_{i-1}$ are uniquely determined by the above conditions, we shall show that that for m_i is also.

By hypothesis, we can write

$$m_i = f_i + \sum A(\alpha_1, \alpha_2, \cdots, \alpha_{i-1}) f_1^{\alpha_1} f_2^{\alpha_2} \cdots f_{i-1}^{\alpha_{i-1}}, \tag{10.1}$$

the sum containing no constant or linear terms; that is, in each term of Σ,

$$\alpha_1 + \alpha_2 + \cdots + \alpha_{i-1} \geqq 2. \tag{10.2}$$

Let now $f_i(a)$, $f_i(b)$ be arbitrary sets of numbers, and define $f_i(a+b)$ as in condition (3). If we put

$$M = m_i(a+b) - m_i(a) - m_i(b), \tag{10.3}$$

then by conditions (1) and (3),

$$M = m_1(a)\, m_{i-1}(b) + m_2(a)\, m_{i-2}(b) + \cdots + m_{i-1}(a)\, m_1(b), \tag{10.4}$$

which contains neither $m_i(a)$ nor $m_i(b)$. But also

$$\begin{aligned}
M &= f_i(a+b) + \sum\nolimits_3 A(\gamma_1, \cdots, \gamma_{i-1}) f_1^{\gamma_1}(a+b) \cdots f_{i-1}^{\gamma_{i-1}}(a+b) \\
&\quad - f_i(a) - \sum\nolimits_1 A(\alpha_1, \cdots, \alpha_{i-1}) f_1^{\alpha_1}(a) \cdots f_{i-1}^{\alpha_{i-1}}(a) \\
&\quad - f_i(b) - \sum\nolimits_2 A(\beta_1, \cdots, \beta_{i-1}) f_1^{\beta_1}(b) \cdots f_{i-1}^{\beta_{i-1}}(b) \\
&= \sum\nolimits_3 A(\gamma_1, \cdots, \gamma_{i-1}) [f_1(a)+f_1(b)]^{\gamma_1} \cdots [f_{i-1}(a)+f_{i-1}(b)]^{\gamma_{i-1}} \\
&\quad - \sum\nolimits_1 - \sum\nolimits_2 \\
&= \sum\nolimits_3 A(\gamma_1, \cdots, \gamma_{i-1}) \left[\sum_{\alpha_1} \binom{\gamma_1}{\alpha_1} f_1^{\alpha_1}(a) f_1^{\gamma_1-\alpha_1}(b) \right] \cdots \qquad (10.5) \\
&\quad \times \left[\sum_{\alpha_{i-1}} \binom{\gamma_{i-1}}{\alpha_{i-1}} f_{i-1}^{\alpha_{i-1}}(a) f_{i-1}^{\gamma_{i-1}-\alpha_{i-1}}(b) \right] - \sum\nolimits_1 - \sum\nolimits_2 \\
&= \sum_{\gamma_1 \cdots \gamma_{i-1}}{}_3\, A(\gamma_1, \cdots, \gamma_{i-1}) \sum_{\alpha_1 \cdots \alpha_{i-1}}{}' \binom{\gamma_1}{\alpha_1} \cdots \binom{\gamma_{i-1}}{\alpha_{i-1}} \\
&\quad \times f_1^{\alpha_1}(a) \cdots f_{i-1}^{\alpha_{i-1}}(a) f_1^{\gamma_1-\alpha_1}(b) \cdots f_{i-1}^{\gamma_{i-1}-\alpha_{i-1}}(b),
\end{aligned}$$

where in $\sum'$ all terms in which

$$\alpha_1 = \alpha_2 = \cdots = \alpha_{i-1} = 0 \quad \text{or} \quad \alpha_1 = \gamma_1,\ \alpha_2 = \gamma_2, \cdots, \alpha_{i-1} = \gamma_{i-1},$$

i. e. in which b alone or a alone appears, are left out, as these are cancelled out by terms of $\sum_1$ and $\sum_2$.

If in eq. (10.4) we express each m_k in terms of the f_l, which we can do in a unique manner, by hypothesis, the resulting expression must be identically equal to the right hand side of eq. (10.5) for all values of $f_l(a)$ and $f_l(b)$, as these values are arbitrary.

This fact is sufficient to determine $A(\gamma_1, \cdots, \gamma_{i-1})$ for all sets of γ's obeying eq. (10.2) (with α_i replaced by γ_i). For choose any such set. This is a factor of the coefficient of at least one term in (10.5). For instance, if $\gamma_p \neq 0$, it is a factor of the coefficient of

$$f_p(a) f_1^{\gamma_1}(b) \cdots f_{p-1}^{\gamma_{p-1}}(b) f_p^{\gamma_p - 1}(b) f_{p+1}^{\gamma_{p+1}}(b) \cdots f_{i-1}^{\gamma_{i-1}}(b).$$

For, if $\gamma_p > 1$, $f_p(a)$ and $f_p^{\gamma_p - 1}(b)$ are both present, and if $\gamma_p = 1$, some $\gamma_q > 0$, and $f_p(a)$ and $f_q^{\gamma_q}(b)$ are both present. The whole coefficient of this term is

$$A(\gamma_1, \cdots, \gamma_{i-1}) \binom{\gamma_1}{0} \cdots \binom{\gamma_p}{1} \cdots \binom{\gamma_{i-1}}{0} = \gamma_p A(\gamma_1, \cdots, \gamma_{i-1}).$$

As this must equal the corresponding coefficient in (10.4), $A(\gamma_1, \cdots, \gamma_{i-1})$ is determined. There is thus at most a single transformation with the given properties. But T is such a transformation; hence U is T.

To prove that T^{-1} is unique, we state a theorem corresponding to the above theorem, with the m_i and f_i interchanged. We assume of course that $m_0 \equiv 1$. Now the transformation U' given in this theorem obviously has an inverse. Moreover, this inverse is seen to obey the conditions of Theorem 10A, and it is thus T. Hence U' is T^{-1}.

11. **Extension of T to the case of two subscripts.** We now consider sets of numbers f_{ij}, $(i, j) \neq (0, 0)$, and m_{ij}, where $m_{00} = 1$. Let $R_{ij}^k(f)$ be defined just as we defined R_i^k, with the condition that the first subscripts add up to i, and the second ones add up to j. Thus for example

$$\begin{aligned} R_{22}^1(f) &= f_{22}, \quad R_{22}^2(f) = 2(f_{10} f_{12} + f_{20} f_{02} + f_{01} f_{21}) + f_{11}^2, \\ R_{22}^3(f) &= 3 f_{10}^2 f_{02} + 6 f_{10} f_{01} f_{11} + 3 f_{01}^2 f_{20}, \quad R_{22}^4(f) = 6 f_{10}^2 f_{01}^2. \end{aligned} \tag{11.1}$$

Letting $f_{ij}(a)$ be the coefficient of $x^i y^j$ in a power series $F_a(x, y)$, we define $m_{ij}(a)$ as being the coefficient of $x^i y^j$ in $e^{F_a(x,y)}$, etc. Just as before, we arrive at the transformations T and T^{-1}:

$$m_{ij} = \sum_k \frac{1}{k!} R_{ij}^k(f) = \sum_{\substack{\sum_p p\alpha_{pq}=i \\ \sum_q q\alpha_{pq}=j}} \frac{1}{\prod_{p,q} \alpha_{pq}!} \prod_{p,q} f_{pq}^{\alpha_{pq}}, \tag{11.2}$$

$$f_{ij} = \sum_k \frac{(-1)^{k-1}}{k} R_{ij}^k(m) = \sum_{\substack{\sum_p p\alpha_{pq}=i \\ \sum_q q\alpha_{pq}=j}} \frac{(-1)^s s!}{\prod_{p,q} \alpha_{pq}!} \prod_{p,q} m_{pq}^{\alpha_{pq}}, \tag{11.3}$$

where

$$s = \sum_{p,q} \alpha_{pq} - 1.$$

Thus, for example,

$$\begin{aligned} m_{22} &= f_{22}+f_{10}f_{12}+f_{20}f_{02}+f_{01}f_{21}+\frac{1}{2}f_{11}^2 \\ &\quad +\frac{1}{2}f_{10}^2f_{02}+\frac{1}{2}f_{01}^2f_{20}+f_{10}f_{01}f_{11}+\frac{1}{4}f_{10}^2f_{01}^2, \\ f_{22} &= m_{22}-m_{10}m_{12}-m_{20}m_{02}-m_{01}m_{21}-\frac{1}{2}m_{11}^2 \\ &\quad + m_{10}^2m_{02}+m_{01}^2m_{20}+2m_{10}m_{01}m_{11}-1\tfrac{1}{2}m_{10}^2m_{01}^2. \end{aligned} \tag{11.4}$$

As before, we prove the theorems

THEOREM 11A. *$m_{ij}[f_{ij}]$, $(i, j) \neq (0, 0)$, equals $f_{ij}[m_{ij}]$ plus a polynomial in the numbers $f_{pq}[m_{pq}]$, $p \leqq i$, $q \leqq j$, $(p, q) \neq (i, j)$, $(p, q) \neq (0, 0)$, which contains no constant or linear terms.*

THEOREM 11B. *If*

$$f_{ij}(a+b) = f_{ij}(a)+f_{ij}(b), \tag{11.5}$$

then

$$m_{ij}(a+b) = \sum_{p,q} m_{pq}(a)\, m_{i-p,j-q}(b), \tag{11.6}$$

and conversely.

THEOREM 11C. *The transformations T and T^{-1} in two subscripts are unique in the same sense that the transformations in one subscript were.*

The transformation in one subscript can be obtained from that in two by making the second subscript always zero. The theory of the transformation holds of course with any number of subscripts. It is the case of two subscripts we shall use.

IV. THE NUMBERS f_{ij}.

12. Definition of the f_{ij}. In this part we let the elements $a, b, c, \cdots$, of Part III be graphs $G_1, G_2, G_3, \cdots$. If G_1 and G_2 are any two graphs, let G_1+G_2 be their sum.

THEOREM 12A. *Given any two graphs G_1 and G_2,*

$$m_{ij}(G_1+G_2) = \sum_{p,q} m_{pq}(G_1)\, m_{i-p,j-q}(G_2). \tag{12.1}$$

For let H_1 be any subgraph of G_1 of rank p, nullity q, and let H_2 be any subgraph of G_2 of rank $i-p$, nullity $j-q$. Then H_1+H_2 is a subgraph of G_1+G_2 of rank $p+(i-p)=i$, nullity $q+(j-q)=j$. Conversely, any subgraph of G_1+G_2 of rank i, nullity j is made up of a subgraph H_1 of G_1 and a subgraph H_2 of G_2 (either of which may be the null graph), whose ranks and nullities add up to i and j respectively. From this the theorem follows.

Definition. Given any graph G, we define the numbers $f_{ij}(G)$, $(i,j) \neq (0,0)$ by the transformation T^{-1}:

$$f_{ij} = \sum_k \frac{(-1)^{k-1}}{k} R_{ij}^k(m). \tag{12.2}$$

As $m_{00}=1$ always, Theorem 11B, Part III gives the

THEOREM 12B. *Given any two graphs G_1 and G_2,*

$$f_{ij}(G_1+G_2) = f_{ij}(G_1)+f_{ij}(G_2). \tag{12.3}$$

THEOREM 12C. *$f_{ij}(G)$ for any graph G is given by a unique polynomial in the numbers N_1, N_2, $\cdots$, containing no constant term.*

This follows immediately from Theorem 7A, Part II, as f_{ij} is a polynomial in the m_{ij} containing no constant term.

13. **Generalized graphs.** With any graph, or *real graph*, is associated a finite set of integers, positive or zero, N_1, N_2, $\cdots$, N. Consider now any finite set of integers, positive, negative or zero, N_1, N_2, $\cdots$, N_q, each N_i corresponding to a definite type T_i of non-separable graph. This collection of numbers we call a *generalized graph* G; we shall say that G contains N_i subgraphs of type T_i, $i=1, 2, \cdots, q$. We may of course add to or remove from a generalized graph any number of numbers which are zero without altering it. We may consider a real graph as a special case of a generalized graph.

Consider now two generalized graphs G and G', whose numbers are N_i and N_i'. We let $G+G'$, the *sum* of these graphs, be the set of numbers

$$N_1+N_1',\ N_2+N_2',\ \cdots,\ N_q+N_q', \tag{13.1}$$

where q is large enough so that all the numbers of both G and G' are included. This holds of course if G and G' are real graphs. Similarly for the difference of two generalized graphs, or any other linear combination.

Various generalized graphs may be expressible in terms of real graphs. This is, in fact, true for all generalized graphs, as the following theorems show.

THEOREM 13A. *Given any type of non-separable graph T_i, the generalized graph G containing one subgraph of this type, and none of any other type, ($N_i = 1$, $N_j = 0$, $j \neq i$), can be expressed as a linear combination of real graphs.*

Let $T_1, T_2, \cdots, T_k$ be all types other than T_i of non-separable graphs which are subgraphs of a graph of type T_i. We arrange them in order, choosing first those with a single arc, next those with two arcs, etc. Types with the same number of arcs we arrange in any order. Then no graph G' of type T_p contains any graph G'' of type T_q as a subgraph, $q > p$.

Let G_1 be a graph of type T_i. Suppose G_1 contains c_2 subgraphs of type T_k. We then subtract from G_1 c_2 graphs of type T_k, forming the generalized graph G_2. In doing this, we have not altered the number of subgraphs of type T_i. G_2 contains therefore one subgraph of type T_i, and zero subgraphs of type T_k. Suppose G_2 contains c_3 subgraphs of type T_{k-1}. Subtract then c_3 graphs of type T_{k-1}, etc. This process gives us the required graph.

Example. Let G_{32} be the graph whose arcs are ab, ac, bc, bd, cd, and say G_{32} is of type T_{32}. G_{32} contains a quadrilateral. Let G_{31} be a quadrilateral; then $G_{32} - G_{31}$ is a generalized graph containing no quadrilateral. This graph contains two triangles. Let G_{21} be a triangle; then $G_{32} - G_{31} - 2\,G_{21}$ contains no triangles. As G_{32} contains five arcs, G_{31} contains four, and G_{21} contains three, $G_{32} - G_{31} - 2\,G_{21}$ contains $5 - 4 - 2 \cdot 3 = -5$ arcs. Let G_{10} be an arc. Then

$$G = G_{32} - G_{31} - 2\,G_{21} + 5\,G_{10}$$

is a generalized graph containing one subgraph of type T_{32}, but none of any other type.

THEOREM 13B. *Any generalized graph, whose numbers are $N_1, N_2, \cdots, N_q$, may be expressed as a linear combination of real graphs.*

Let G_i be a generalized graph containing one subgraph of type T_i and none of any other type. We need merely express each G_i in terms of real graphs, and put

$$G = N_1 G_1 + N_2 G_2 + \cdots + N_q G_q. \tag{13.2}$$

14. The numbers f_{ij} for generalized graphs. Given the generalized graph $G = (N_1, N_2, \cdots, N_q)$, we define $f_{ij}(G)$ by the polynomial of Theorem 12C. $m_{ij}(G)$ may of course be defined by the polynomial of Theorem 7A. It is to be noted that, for generalized graphs, there will be in general an infinite number of non-zero f_{ij} and m_{ij}.

THEOREM 14A. *If G is a generalized graph, made up of the real graphs $G_1, G_2, \cdots, G_l$:*

$$G = G_1 + \cdots + G_k - G_{k+1} - \cdots - G_l, \tag{14.1}$$

then

$$f_{ij}(G) = f_{ij}(G_1) + \cdots + f_{ij}(G_k) - f_{ij}(G_{k+1}) - \cdots - f_{ij}(G_l). \tag{14.2}$$

By Theorem 12B, this is true if there are no minus signs present, for then G is a real graph. Assume it is true if there are fewer than $l-k$ minus signs present; we shall prove it for the case of $l-k$ minus signs.

Put

$$G' = G + G_l = G_1 + \cdots + G_k - G_{k+1} - \cdots - G_{l-1}.$$

As G' contains but $l-k-1$ minus signs,

$$f_{ij}(G') = f_{ij}(G_1) + \cdots + f_{ij}(G_k) - f_{ij}(G_{k+1}) - \cdots - f_{ij}(G_{l-1}). \tag{a}$$

Let us add to G' s graphs congruent with G_l:

$$G^{(s)} = G' + G_l + \cdots + G_l = G' + sG_l.$$

As only $l-k-1$ minus signs are used in expressing this graph in terms of real graphs, we find for $s \geqq 0$

$$f_{ij}(G' + sG_l) = f_{ij}(G') + sf_{ij}(G_l).$$

Let $N_1', N_2', \cdots$, be the numbers for G', and let $N_1'', N_2'', \cdots$, be those for G_l. Then the numbers for $G' + sG_l$ are

$$N_1' + sN_1'', \ N_2' + sN_2'', \ \cdots,$$

whether s is positive or negative. Considering $N_1', N_2', \cdots, N_1'', N_2'', \cdots$, as constants, $f_{ij}(G' + sG_l)$ is a polynomial in s alone:

$$f_{ij}(G' + sG_l) = P(s),$$

with the property that

$$P(s) = f_{ij}(G') + sf_{ij}(G_l) = A + Bs$$

if s is any positive integer or zero. Therefore

$$P(s) \equiv A + Bs.$$

Now $f_{ij}(G'+sG_l) = P(s)$ whether s is positive or negative. Hence, putting $s = -1$, we have

$$f_{ij}(G) = A-B = f_{ij}(G')-f_{ij}(G_l)$$

which, with eq. (a), gives the desired result.

THEOREM 14B. *Let G and G' be any two generalized graphs. Then*

$$f_{ij}(G+G') = f_{ij}(G)+f_{ij}(G').$$

Express G and G' in terms of real graphs as in Theorem 14A. $G+G'$ in terms of real graphs is the sum of these expressions. $f_{ij}(G)$ is as given in Theorem 14A, and similarly for $f_{ij}(G')$, $f_{ij}(G+G')$. Comparing these three expressions gives the above equation.

The theorem obviously holds for any linear combination of generalized graphs.

THEOREM 14C. *$f_{ij}(G)$, a polynomial in the numbers $N_1, \cdots, N_q$, is linear and homogeneous.*

Let us put

$$f_{ij}(G) = P(N_1, N_2, \cdots, N_q).$$

Let G_k be a generalized graph containing one subgraph of type T_k and none of any other types, $k = 1, 2, \cdots, q$. Put

$$a_k = f_{ij}(G_k) = P(0, \cdots, 1, \cdots, 0), \tag{14.4}$$

the 1 appearing in the k'th place. Define the graph G by the equation

$$G = N_1 G_1 + N_2 G_2 + \cdots + N_q G_q,$$

the numbers $N_1, \cdots, N_q$ being any set of integers, positive, negative or zero. Then

$$f_{ij}(G) = N_1 f_{ij}(G_1) + N_2 f_{ij}(G_2) + \cdots + N_q f_{ij}(G_q)$$

by the last theorem, and thus

$$P(N_1, N_2, \cdots, N_q) = a_1 N_1 + a_2 N_2 + \cdots + a_q N_q.$$

As the polynomials on the two sides of this equation are equal for all integral values of the variables $N_1, \cdots, N_q$, they are identically equal:

$$f_{ij}(G) \equiv a_1 N_1 + a_2 N_2 + \cdots + a_q N_q, \tag{14.5}$$

and the theorem is proved.

We can, if we wish, express the polynomial $f_{ij}(G)$ in the form

$$f_{ij}(G) = a_1 N_1 + a_2 N_2 + \cdots, \tag{14.6}$$

writing down a term corresponding to every type of non-separable graph. All but a finite number of the coefficients will be zero. We can find any coefficient by the

THEOREM 14 D. *The coefficient a_k of N_k in $f_{ij}(G)$ is given by eq.* (14.4).

This is true if G in terms of real graphs contains the term G_k, as we have just seen. If it does not, we must show that $a_k = 0$. Take the expression for G and add the term $0 \cdot G_k$. The proof of Theorem 14 C still holds, and the right hand side of eq. (14.5) now has the added term $a_k N_k$. This can only be true if $a_k = 0$.

15. Relation between the m_{ij} and the f_{ij}.

THEOREM 15 A. *Consider the expression for $m_{ij}(G)$ for any graph G as a polynomial in the numbers $N_1, \cdots, N_q$. $f_{ij}(G)$ is exactly the linear terms of this polynomial.*

By definition,

$$f_{ij}(G) = m_{ij}(G) + \sum_{k \geq 2} \frac{(-1)^{k-1}}{k} R_{ij}^k [m(G)]. \tag{15.1}$$

If we express the right hand side in terms of the numbers $N_1, N_2, \cdots$, we obtain the polynomial for $f_{ij}(G)$, as this expression in unique, by Theorem 12 C. Hence, by Theorem 14 C, all terms vanish but the linear terms. The sum on the right hand side is a polynomial in $m_{pq}(G)$ containing no constant or linear terms, and each $m_{pq}(G)$ is a polynomial in $N_1, N_2, \cdots$, containing no constant terms. Hence the sum contains no linear terms in $N_1, N_2, \cdots$, and thus the only linear terms are those in $m_{ij}(G)$, which proves the theorem.

This fact is important for the following reason.

If we know the linear terms f_{pq} of m_{pq} for all numbers $p \leqq i$, $q \leqq j$, then we can calculate m_{ij}, using the transformation T.

We shall make use of this fact in the next part.

Example. Let us find f_{30} for graphs containing no 1- or 2-circuits by eq. (14.5). Let G be a graph containing a single arc, let G' be a triangle, and let G'' be a generalized graph containing one triangle and no other subgraphs. Then

$$G'' = G' - 3G.$$

As a single arc and a triangle are the only types of graphs that can contribute to f_{30} (cf. Part II),

$$\begin{aligned} f_{30} &= f_{30}(G)E + f_{30}(G'')N_{21} \\ &= f_{30}(G)E + [f_{30}(G') - 3f_{30}(G)]N_{21}, \end{aligned}$$

if N_{21} is the number of triangles in G, by eq. (14.5). Now

$$f_{30} = m_{30} - m_{10}\, m_{20} + \frac{1}{3} m_{10}^3,$$

and

$$\begin{aligned} m_{10}(G) &= 1, & m_{20}(G) &= 0, & m_{30}(G) &= 0, \\ m_{10}(G') &= 3, & m_{20}(G') &= 3, & m_{30}(G') &= 0. \end{aligned}$$

Hence

$$f_{30}(G) = \frac{1}{3}, \quad f_{30}(G') = -9+9 = 0,$$

and thus

$$f_{30} = \frac{1}{3} E - N_{21}.$$

Thence, as $f_{10} = E$ and $f_{20} = -\frac{1}{2} E$,

$$m_{30} = f_{30} + f_{10} f_{20} + \frac{1}{6} f_{10}^3 = \binom{E}{3} - N_{21}.$$

Let us find a relation between the f_{ij} analogous to eq. (1.5), Part I. To do this, we replace each term in this equation by its expression in the numbers $N_1, N_2, \cdots$ (the right hand side is already in that form). By Theorem 5A, Part II, this equation, being true for any graph, is an identity. Therefore in particular, the linear terms on the left hand side equal the linear terms on the right hand side; that is,

$$f_{i0} + f_{i-1,1} + \cdots + f_{0i} = \frac{(-1)^{i-1}}{i} E. \tag{15.2}$$

16. Independence of the m_{ij} and f_{ij}. We give here some theorems on the m_{ij} and f_{ij} similar to Theorems 5A and 5B, Part II, on the N_i.

THEOREM 16A. *If we except the numbers m_{i0}, $i \neq 1$, there is no polynomial in the remaining m_{ij} which equals 0 for all graphs and yet is not identically 0.*

If we consider only graphs with no 1- or 2-circuits, we consider only m_{ij} which are not 0 for such graphs. (Thus for instance, we do not consider m_{11}.)

Suppose there were such a polynomial. Consider then such a polynomial P in the least number of variables. Of all the numbers m_{kl} appearing, let m_{ij} be that variable, chosen from the set for which the first subscript is as great as possible, for which the second subscript is the greatest. Let us write the polynomial in the form

$$P = Q_0\, m_{ij}^{\alpha} + Q_1\, m_{ij}^{\alpha-1} + \cdots + Q_{\alpha-1}\, m_{ij} + Q_{\alpha},$$

where the Q's are polynomials in the remaining m_{kl}, and Q_0 is not zero for all graphs.

Express now each of the m_{kl} in terms of the numbers $N_1, N_2, \cdots$. By Lemma F of § 8, Part II, there is a non-separable graph of rank i, nullity j, which contributes to m_{ij} with the coefficient 1; it contributes to none of the other m_{kl}. Let us say there are N_{ij} subgraphs of this type in a given graph G. Then if N_p is the set of numbers excepting N_{ij},

$$m_{ij} = N_{ij}+R(N_p).$$

Let us put

$$Q_s(m_{kl}) = Q'_s(N_p),$$

where Q'_s does not contain N_{ij}. Now

$$\begin{aligned} P &= Q'_0[N_{ij}+R]^{\alpha}+Q'_1[N_{ij}+R]^{\alpha-1}+\cdots+Q'_{\alpha} \\ &= Q'_0 N_{ij}^{\alpha}+(\alpha Q'_0 R+Q'_1)N_{ij}^{\alpha-1}+\cdots, \end{aligned}$$

and N_{ij} appears only where shown. By Theorem 5A, Part II, this polynomial, being 0 for all graphs, must vanish identically. Hence the coefficients of the powers of N_{ij}, in particular, Q'_0, must vanish identically. Hence $Q_0 = Q'_0$ equals zero for all graphs. We have here a contradiction, and the theorem is proved.

THEOREM 16B. *Theorem* 16A *holds if the words "for all graphs" are replaced by the words "for all non-separable graphs".*

We merely use Theorem 5B instead of Theorem 5A.

THEOREM 16C. *Theorems* 16A *and* 16B *hold if "m_{ij}" is replaced by "f_{ij}".*

We replace "m_{ij}" by "f_{ij}" throughout the proofs.

We of course cannot include m_{i0} or f_{i0}, $i \neq 1$, in these theorems, as they are determined from the other m_{kl} or f_{kl} by eq. (1.5) of Part I or eq. (15.2) of this part.

17. **On inequalities.** The theorems of the last article tell us nothing about the existence or non-existence of inequalities on the m_{ij} or f_{ij}. There are many such inequalities, as we now show.

A fundamental inequality is the obvious one:

$$m_{ij} \geqq 0. \tag{17.1}$$

Another set of inequalities is given by eq. (2.1), Part I. Again, the number of ways of coloring any graph is $\geqq 0$:

$$\sum_i \lambda^{V-i} \sum_j (-1)^{i+j} m_{ij} \geqq 0 \tag{17.2}$$

if λ is an integer $\geqq 0$.

Inequalities on the m_i may be deduced from the interpretation of the number $(-1)^i m_i$ in terms of the broken circuits of the graph. We give one example:

$$|m_{i-1}| \geqq \frac{i}{E-i+1}|m_i|. \tag{17.3}$$

47

To prove this, take each of the $|m_i|$ subgraphs of G of i arcs containing no broken circuit, and drop out any arc. We form thus subgraphs of $i-1$ arcs containing no broken circuit in $i|m_i|$ ways. Consider any such subgraph. It was formed by dropping out one of the $E-(i-1)$ arcs in the rest of G, and was thus formed at most $E-i+1$ times. From this the inequality follows.

Further, some polynomials in the f_{ij} (or m_{ij}) equal a number N_{kl}, which is of course $\geqq 0$. For instance, for graphs containing no 1- or 2-circuits,

$$f_{31} = -3N_{21}+N_{31}, \qquad f_{21} = N_{21}.$$

Therefore

(17.4) $$N_{31} = f_{31}+3f_{21} \geqq 0.$$

Of course any inequality on the f_{ij} gives rise to an inequality on the m_{ij}, and conversely.

A fundamental problem, and one of extreme difficulty, is that of discovering all the inequalities.

V. Calculation of the f_{ij} and m_{ij}.

18. **Graphs containing no 1- or 2-circuits.** In such graphs, $m_{01} = m_{11} = 0$, or, $f_{01} = f_{11} = 0$. If there is but a single type of non-separable graph of rank i, nullity j, we shall call this type T_{ij}; if there are several, we shall call them ${}^1T_{ij}$, ${}^2T_{ij}$, $\cdots$, ${}^qT_{ij}$. Define N_{ij}, or ${}^1N_{ij}$, $\cdots$, ${}^qN_{ij}$ correspondingly. If there is more than one T_{ij}, put $N_{ij} = \sum_k {}^kN_{ij}$. Thus N_{ij} is the number of non-separable subgraphs of rank i, nullity j. The non-separable graphs of ranks one to four are given in the figure.

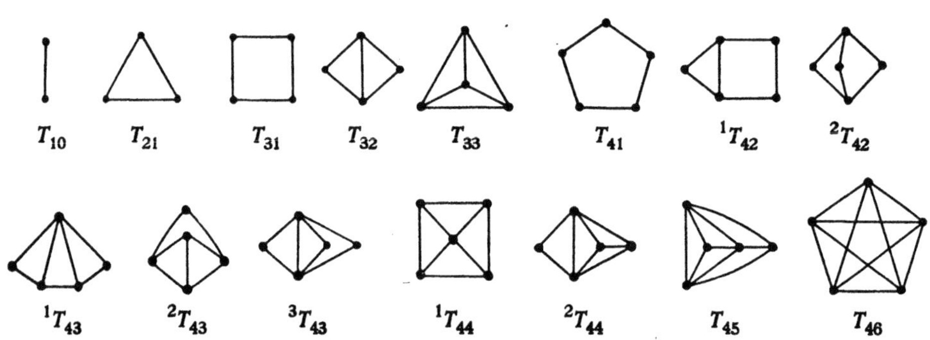

Fig. 2.

We shall calculate the f_{ij} in terms of the ${}^kN_{pq}$ in the following order: $f_{10}, f_{20}, f_{21}, f_{30}, f_{31}, \cdots, f_{40}, \cdots$. To find an f_{ij}, having found the previous f_{kl}, we first write down m_{ij} in terms of the f_{kl} by eq. (11.2), Part III. We know all the f_{kl} occurring in this expression except f_{ij}. Now, by eq. (14.5), Part IV,

$$f_{ij} = a_{i_1} N_{i_1} + a_{i_2} N_{i_2} + \cdots + a_{i_p} N_{i_p},$$

where the corresponding types $T_{i_1}, \cdots, T_{i_p}$ are restricted as seen in § 8, Part II. Let $G_{i_1}, \cdots, G_{i_p}$ be non-separable graphs of these types, arranged in such an order that no one contains any one following as a subgraph. Consider first G_{i_1}, and let its numbers be $N_1(G_{i_1})$, $N_2(G_{i_1})$, $\cdots$, $N_q(G_{i_1})$. Substitute these values in the expression for m_{ij}. $m_{ij}(G_{i_1})$ will be 0 if G_{i_1} is not of rank i, nullity j, and $N_{i_2}, \cdots, N_{i_p}$ are all 0 for G_{i_1}. Hence in the resulting expression a_{i_1} is the only unknown, and we can solve for it. We can now find a_{i_2}, using G_{i_2}, etc.

Several facts will aid us. If G_{i_k} is of rank i, nullity j, then $a_{i_k} = 1$. If $j > 0$, then a_{10}, the coefficient of $E = N_{10}$, is 0 (see Theorem 8A, Part II). f_{i0}, $i > 1$, is always most simply determined by eq. (15.2), Part IV. We turn now to the calculations.

[1,0] $$m_{10} = E, \qquad f_{10} = E.$$

[2,0] $$f_{20} + f_{11} + f_{02} = f_{20} = -\frac{1}{2} E.$$

[2,1] $$m_{21} = N_{21}, \qquad f_{21} = N_{21}.$$

[3,0] $$f_{30} + f_{21} = \frac{1}{3} E, \qquad \therefore f_{30} = \frac{1}{3} E - N_{21}.$$

[3,1] $$m_{31} = f_{10} f_{21} + f_{31} = E N_{21} + (c N_{21} + N_{31}).$$

For a graph G_{21} of type T_{21},

$$m_{31} = 0, \qquad E = 3, \qquad N_{21} = 1, \qquad N_{31} = 0.$$
$$\therefore \quad 0 = 3 \cdot 1 + c \cdot 1, \qquad c = -3.$$
$$\therefore \quad f_{31} = -3 N_{21} + N_{31}.$$

[3,2], [3,3] $$f_{32} = N_{32}, \qquad f_{33} = N_{33}.$$

[4,0] $$f_{40} + f_{31} = -\frac{1}{4} E, \qquad \therefore f_{40} = -\frac{1}{4} E + 3 N_{21} - N_{31}.$$

$$m_{41} = \left(\frac{1}{2} f_{10}^2 f_{21} + f_{10} f_{31} + f_{20} f_{21}\right) + f_{41}$$

[4,1] $$= \left(\frac{1}{2} E^2 N_{21} - 3 \frac{1}{2} E N_{21} + E N_{31}\right)$$
$$+ (c N_{21} + d N_{31} + e N_{32} + N_{41}).$$

47*

$$G_{21}\colon E = 3, \quad N_{21} = 1.$$
$$\therefore 0 = \frac{1}{2}\cdot 3^2\cdot 1 - 3\frac{1}{2}\cdot 3\cdot 1 + c, \qquad c = 6.$$
$$G_{31}\colon E = 4, \ N_{31} = 1.$$
$$\therefore 0 = 4 + d, \qquad d = -4.$$
$$G_{32}\colon E = 5, \quad N_{21} = 2, \quad N_{31} = 1, \quad N_{32} = 1.$$
$$\therefore 0 = 25 - 35 + 5 + 12 - 4 + e, \qquad e = -3.$$
$$\therefore f_{41} = 6N_{21} - 4N_{31} - 3N_{32} + N_{41}.$$

Similarly,

$$\begin{aligned}
f_{42} &= -\frac{1}{2}N_{21} - 6N_{32} - 6N_{33} + N_{42},\\
f_{43} &= -6N_{33} + N_{43}, \quad f_{4j} = N_{4j}, \qquad j = 4, 5, 6;\\
f_{50} &= \frac{1}{5}E - 6N_{21} + 4N_{31} + 2N_{32} - N_{41},\\
f_{51} &= -9N_{21} + 10N_{31} + 17N_{32} + 11N_{33} - 5N_{41} - 3\,{}^1N_{42}\\
&\qquad - 3\,{}^2N_{42} + N_{51},\\
f_{52} &= 3N_{21} + 19N_{32} + 36N_{33} - 7\,{}^1N_{42} - 6\,{}^2N_{42} - 5\,{}^1N_{43}\\
&\qquad - 6\,{}^2N_{43} - 7\,{}^3N_{43} + N_{52},\\
f_{53} &= -2N_{32} + 9N_{33} - 9\,{}^1N_{43} - 7\,{}^2N_{43} - 10\,{}^3N_{43} - 8\,{}^1N_{44}\\
&\qquad - 8\,{}^2N_{44} + N_{53},\\
f_{54} &= -4N_{33} - 8\,{}^1N_{44} - 9\,{}^2N_{44} - 9N_{45} + N_{54},\\
f_{55} &= -9N_{45} + N_{55},\\
f_{56} &= -10N_{46} + N_{56}, \quad f_{5j} = N_{5j}, \qquad j = 7, 8, 9, 10.
\end{aligned} \tag{18.1}$$

From these formulas we can write down the m_{ij}. Thence we find the m_i, the coefficients of $M(\lambda)$. The first few are

$$\begin{aligned}
m_0 &= 1, \quad m_1 = -E, \quad m_2 = \binom{E}{2} - N_{21},\\
m_3 &= -\binom{E}{3} + (E-2)N_{21} + N_{31} - N_{32} + N_{33}.
\end{aligned} \tag{18.2}$$

We can now find the α_i from eq. (4.4), Part I. The first six are

$$\begin{aligned}
&\alpha_0 = 1, \quad \alpha_1 = 0, \quad \alpha_2 = -N_{21}, \quad \alpha_3 = -N_{31} + N_{32} - N_{33},\\
&\alpha_4 = \binom{N_{21}}{2} - 3N_{33} + \sum_{j=1}^{6}(-1)^j N_{4j},\\
&\alpha_5 = N_{21}(N_{31} - N_{32} + N_{33}) + 6N_{33} + {}^2N_{42} - 3\,{}^2N_{43} - {}^3N_{43}\\
&\qquad + 4\,{}^1N_{44} + 3\,{}^2N_{44} - 4N_{45} + 4N_{46} - \sum_{j=1}^{10}(-1)^j N_{5j}.
\end{aligned} \tag{18.3}$$

From these formulas and eq. (4.4) we can calculate the expressions for m_4 and m_5, which are rather lengthy.

19. **Regular maps.** It is well known that in studying the four color map problem, it is sufficient to restrict oneself to certain special maps; in particular, maps which we shall call regular, obeying the following conditions:[10]

(1) All the vertices are triple.

(2) There are no rings of one, two, three or four regions; that is, no set of four or fewer regions, together with all boundary lines separating them, form a multiply connected region.

(3) There are no rings of five regions except about a single region.

The numbers m_{ij} have been defined in terms of the dual graph G of the map. G', the graph consisting of the boundary lines and vertices of the map, is the dual of G in the sense defined in N. We can state the above conditions in terms of G and G' as follows:

(1′) Each vertex of G' is on exactly three arcs.

(2′) G is quintuply connected.[11] It follows that G contains no 2-circuit. For otherwise, dropping out the two corresponding arcs of G' would separate it into the pieces H_1' and H_2', each of which is seen to be of nullity >0. But from this it could be shown that dropping out the two vertices of the 2-circuit in G disconnects G.[12] Consequently G and G' determine each other uniquely.[13]

(3′) If dropping out five vertices from G disconnects G, one of the resulting pieces contains just one vertex.

The numbers $N_1, N_2, \cdots$, involved in $m_1, \cdots, m_5$, can be expressed in terms of the following quantities for all regular maps:

V = no. of regions in the map (no. of vertices in the dual graph).

$E = 3(V-2)$ = no. of boundary lines (no. of arcs).

$T = 2(V-2)$ = no. of vertices ($= N_{21}$).

Q_5 = no. of 5-sided regions (no. of vertices on five arcs).

R_6 = no. of rings of six regions, no two having a common boundary unless they are adjacent in the ring (no. of 6-circuits in which only vertices adjacent in the circuit are joined by arcs in the graph).

We derive the following formulas.

[10] See G. D. Birkhoff, The reducibility of maps, American Journal of Mathematics, 35 (1912), pp. 115–128.

[11] See C. G., § 2 (p. 158).

[12] Loc. cit., Theorem 9.

[13] Loc. cit., Theorems 10 and 11.

$$f_{10} = E, \qquad f_{20} = -\frac{1}{2}E, \qquad f_{21} = \frac{2}{3}E,$$
$$f_{30} = -\frac{1}{3}E, \qquad f_{31} = -E, \qquad f_{32} = E,$$
(19.1) $$f_{40} = \frac{3}{4}E, \quad f_{41} = -E+Q_5, \quad f_{42} = -2\frac{1}{3}E, \quad f_{43} = 2E,$$
$$f_{50} = \frac{1}{5}E - Q_5, \quad f_{51} = 3\frac{2}{3}E + R_6, \quad f_{52} = -3E+15Q_5,$$
$$f_{53} = -6E+10Q_5, \quad f_{54} = 4\frac{2}{3}E+5Q_5, \; f_{55} = Q_5.$$

(19.2)
$$m_0 = 1, \quad m_1 = -E, \quad m_2 = \binom{E}{2} - T,$$
$$m_3 = -\binom{E}{3} + (E-2)T,$$
$$m_4 = \binom{E}{4} - \binom{E-2}{2}T + \binom{T}{2} - Q_5,$$
$$m_5 = -\binom{E}{5} + \binom{E-2}{3}T - (E-4)\binom{T}{2} + (E-8)Q_5 + R_6;$$

(19.3)
$$\alpha_0 = 1, \quad \alpha_1 = 0, \quad \alpha_2 = -T, \quad \alpha_3 = 0,$$
$$\alpha_4 = \binom{T}{2} - Q_5, \qquad \alpha_5 = 4Q_5 - R_6.$$

The regular map of the fewest number of regions is the dodecahedron, containing twelve regions, each touching five others. For this map

$$V = 12, \quad E = 30, \quad T = 20, \quad Q_5 = 12, \quad R_6 = 40.$$

We can thus calculate the first six coefficients. The whole polynomial, which was calculated by direct means, is

$$M(\lambda) = \lambda(\lambda-1)(\lambda-2)(\lambda-3)[\lambda^8 - 24\lambda^7 + 260\lambda^6 - 1670\lambda^5 + 6999\lambda^4 - 19698\lambda^3 + 36408\lambda^2 - 40240\lambda + 20170].$$

NOTE ADDED IN PROOF. R. M. Forster has calculated the m_{ij} for a large number of graphs, using the following formula. If G contains the arc ab, $G' = G - ab$, and G'' is formed from G' by letting a and b coalesce, then

(19.4) $$m_{ij}(G) = m_{ij}(G') + m_{i-1,j}(G''),$$

as is easily proved.

PRINCETON UNIVERSITY.

A SET OF TOPOLOGICAL INVARIANTS FOR GRAPHS.†

By HASSLER WHITNEY.‡

1. *Introduction.* A linear graph, or let us say, a *topological graph,* $\mathfrak{Y}$, is a point set consisting of a finite number of points, or vertices, and a finite number of open arcs (topological images of an open segment) which do not intersect, joining pairs of these points. If we consider the vertices and arcs as abstract elements instead of as point sets, and name the two vertices which each arc joins, we obtain the corresponding *abstract graph* G. Corresponding to each abstract graph G we can form a topological graph $\mathfrak{Y}$.

Now we may take an arc of $\mathfrak{Y}$ and consider it as the sum of three point sets, a vertex (an inner point of the arc) and two arcs. If ab is the corresponding arc of G joining the vertices a and b, the above subdivision of the arc of $\mathfrak{Y}$ corresponds to replacing the arc ab by the two arcs ac and cb, c being a new vertex. The resulting graph G' represents the same point set $\mathfrak{Y}$.

If two abstract graphs G and G' can be reduced to the same abstract graph G^* by a process of subdivision, then they can be made to represent the same topological graph $\mathfrak{Y}$. Conversely, if they both represent a topological graph $\mathfrak{Y}$, then they can be reduced to the same abstract graph G^* by subdivision.§ It is therefore natural to define two such abstract graphs G and G' as being topologically equivalent or homeomorphic. Any number defined for all graphs which is unaltered when an arc is subdivided is called an *invariant under subdivision*; by the above remark, we may also call it a *topological invariant.*

A very simple topological invariant for a graph is the number of con-

---

† Presented to the American Mathematical Society, December 28, 1931. References:

I. "Non-separable and planar graphs," *Transactions of the American Mathematical Society*, Vol. 34 (1932), pp. 339-362.

II. "A logical expansion in mathematics," *Bulletin of the American Mathematical Society*, Vol. 38 (1932), pp. 572-579.

III. "Congruent graphs and the connectivity of graphs," *American Journal of Mathematics*, Vol. 54 (1932), pp. 150-168.

IV. "The coloring of graphs," *Annals of Mathematics*, Vol. 33 (1932), pp. 688-718.

‡ National Research Fellow.

§ See O. Veblen's "Colloquium Lectures," *Analysis Situs*, Ch. I, § 12.

nected pieces; another is the nullity † (cyclomatic number). In this paper we give a set of topological invariants which come from a set of numbers m_{ij} defined by the author.‡

2. *The invariants.* Given the table of the m_{ij} for a graph G, if we sum over the elements in each row with alternating signs, we get the m_i, the coefficients of the polynomial $M(\lambda)$ for the number of ways of coloring G in λ colors. Suppose, instead, we sum over the columns; we get a set of numbers p_i, which we shall show are topological invariants of the graph. The numbers are, if G is of rank R, nullity N,

$$(1) \qquad p_i = \sum_j (-1)^{i+j} m_{R-j,N-i}.$$

The number of non-zero numbers p_i (if there are any) equals one plus the nullity N of G. For a graph with no arcs, $p_0 = 1$, and $p_i = 0$, $i \neq 0$.

Suppose G is planar, and has a dual G'.§ From the definition of dual graphs it follows immediately that if m'_{ij} are the numbers for G', then

$$(2) \qquad m'_{ij} = m_{R-j,N-i}.$$

Hence

$$(3) \qquad m'_i = \sum_j (-1)^{i+j} m'_{ij} = \sum_j (-1)^{i+j} m_{R-j,N-i} = p_i,$$

that is, *if G has a dual G', then the numbers p_i are the coefficients m'_i of $M'(\lambda)$.*

From this it follows immediately that the p_i are topological invariants for G, provided G is planar. For take any arc ab of G and replace it by the two arcs $ac + cb$, c being a new vertex, forming the graph G^*. If G' is a dual of G and $a'b'$ is the arc corresponding to ab, then we can form a dual G'' of G^* by adding to G' another arc $a'b'$, as is easily proved. Now any coloring of G' is a coloring of G'' and conversely, and hence $m'_i = m_i''$. As $p_i = m'_i$ and $p^*_i = m_i''$, it follows that $p_i = p^*_i$.

We give now a direct and general proof that $p_i = p^*_i$. Divide the subgraphs of nullity $N - i = N^* - i$ of G^* into two groups: (1) those containing the arc ac, and (2) those not containing ac. Consider first the subgraphs in (2); we pair them off, letting correspond to each subgraph H_1 not containing the arc cb, the subgraph $H_2 = H_1 + cb$. Say H_1 is of rank

† See I, § 2. If a graph G containing E arcs and V vertices is in P connected pieces, then its rank R and nullity N are defined by the equations $R = V - P$, $N = E - R = E - V + P$.

‡ See II, § 6. A subgraph H of G is determined by naming a subset of the arcs of G. m_{ij} is the number of subgraphs of G of rank i, nullity j.

The author was mistaken in supposing that the m_{ij} are the same as Birkhoff's (i, k).

§ See I, § 8.

$R^* - j$; then H_2 is of rank $R^* - j + 1$. H_1 contributes to $m^*_{R^*-j,N^*-i}$, and thus to p^*_i with the sign $(-1)^{i+j}$; H_2 contributes to $m^*_{R^*-j+1,N^*-i}$, and thus to p^*_i with the opposite sign. The two contributions cancel; thus the subgraphs in (2) together contribute nothing. We have left only the subgraphs of the first group to consider.

To each subgraph H^* of the first group we let correspond that subgraph H of G formed by dropping out the arc ac and letting the vertices a and c coalesce. (Thus the arc cb, if present, is replaced by the arc ab.) This is a $1-1$ correspondence between all the subgraphs of G of nullity $N-i$ and the subgraphs of G^* in the first group. If H^* is of rank R^*-j, then H is of rank $R^*-j-1=R-j$. Hence the contribution of H^* to p^*_i is the same as the contribution of H to p_i. It follows that $p^*_i = p_i$, as required.

3. *Broken cut sets of arcs.* We give here an interpretation of the p_i dual to the interpretation of the m_i in terms of broken circuits (see II, § 7). Suppose that dropping out a set of arcs $\alpha, \beta, \cdots, \delta$ from a graph G increases the number of connected pieces in G, while dropping out no proper subset of them does. We then say these arcs form a *cut set of arcs.* List the arcs of G in a definite order. From each cut set of arcs we drop out the last arc, forming the corresponding *broken cut set of arcs.* If G contains a cut arc, then we can consider the null subgraph of G as the corresponding broken cut set.

If G is planar and G' is a dual of G, then cut sets of arcs in G correspond to circuits in G' and broken cut sets, to broken circuits (see I, Theorem 9). As an example, let G^* be the graph with vertices a^*, b^*, c^*, and arcs $\alpha^*(a^*b^*)$, $\beta^*(a^*b^*)$, $\gamma^*(b^*c^*)$, $\delta^*(a^*c^*)$, $\epsilon^*(a^*c^*)$; then G^* has as dual the graph G given in the beginning of II, § 7. The cut sets of G^* are α^*, β^*, γ^*, and γ^*, δ^*, ϵ^*, and α^*, β^*, δ^*, ϵ^*, and the broken cut sets are α^*, β^*, and γ^*, δ^*, and α^*, β^*, δ^*. For G^*, $p^*_0 = 1$, $p^*_1 = -5$, $p^*_2 = 8$, $p^*_3 = -4$.

We now prove that $(-1)^i p_i$ *is the number of subgraphs of i arcs of G which do not contain all the arcs of any broken cut set of arcs.*

The proof follows the proof of the corresponding theorem in II. Arrange the broken cut sets $P_1, P_2, \cdots, P_\sigma$ in order so that, for any k, naming the arcs of G one by one in the given order, all the arcs of $P_1, \cdots, P_{k-1}$ have been named by the time that all the arcs of P_k have been. Arrange the subgraphs of G into sets $S_1, S_2, \cdots, S_\sigma, S_{\sigma+1}$ (some of which may be empty), putting into S_1 all those subgraphs containing no arcs of P_1, into S_k, $1 < k \leqq \sigma$, all those containing at least one arc from each of the broken cut sets $P_1, \cdots, P_{k-1}$, but containing no arc of P_k, and into $S_{\sigma+1}$, all remaining subgraphs.

6

Consider the subgraphs in any S_k, $1 \leqq k \leqq \sigma$. Let ab be the arc of the cut set which is dropped out in forming the broken cut set P_k. To each subgraph H_1 of S_k containing ab corresponds a subgraph H_2 of S_k not containing ab, and conversely, as ab is in none of the broken cut sets $P_1, \cdots, P_k$. Say H_1 is of rank $R - j$, nullity $N - i$. As the arcs of P_k together with the arc ab form a cut set of arcs and H_1 contains no arcs of P_k, dropping out ab from H_1 disconnects a and b, and thus H_2 is of rank $R - j - 1$, nullity $N - i$. H_1 contributes to $m_{R-j,N-i}$, and thus to p_i with the sign $(-1)^{i+j}$, and H_2 contributes to p_i with the opposite sign; the contributions of H_1 and H_2 to p_i thus cancel. The subgraphs of $S_1, \cdots, S_\sigma$ contribute therefore nothing.

We have left the subgraphs in $S_{\sigma+1}$, that is, the subgraphs which contain at least one arc from each broken cut set $P_1, \cdots, P_\sigma$. Consider any such subgraph of nullity $N - i$; it contains an arc of each cut set, and it has therefore the same rank R as G;† hence it contains $R + (N - i) = E - i$ arcs, if G contains E arcs. Say there are l_i such subgraphs in $S_{\sigma+1}$; they contribute an amount $(-1)^i l_i$ to p_i. Now the subgraphs of $S_{\sigma+1}$ are exactly the complements of the subgraphs of i arcs of G which do not contain all the arcs of any broken cut set, and the theorem is proved.

From this interpretation of the numbers p_i, it is again easily seen that they are topological invariants.

Note that *for any graph containing a cut arc, and only for these, every* $p_i = 0$. This corresponds to the fact that for any graph containing a 1-circuit, every $m_i = 0$.

4. *Separable graphs.* Suppose that G is the union of two graphs G' and G'' which have at most a single vertex in common. Then

$$p_i = \sum_k p'_k p''_{i-k}. \ddagger \tag{4}$$

As a result, if $G_1, G_2, \cdots, G_n$ are the components of G and we know all the $p_i(G_s)$, we can calculate the $p_i(G)$. Components which are isolated vertices may of course be forgotten altogether.

To prove (4), arrange the arcs of G in a fixed order. Now the arcs of any cut set in G, hence also of any broken cut set in G, lie wholly in G' or G''. For suppose there were a cut set containing an arc ab in G' and an arc cd in G''. If the arcs of the cut set are dropped out of G, a and b are dis-

† We can reduce the rank of a graph only by dropping out all the arcs of some cut set.

‡ The same formula holds for the m_i.

connected. If we put back the arc cd, there is then a chain joining a and b. But any chain from a to b must lie wholly in G', and thus does not contain cd, a contradiction, proving the statement. Let $P_1, \cdots, P_\sigma$ be the broken cut sets in G', and $P_{\sigma+1}, \cdots, P_\tau$, those in G''; then $P_1, \cdots, P_\tau$ are those in G.

Let H' be any subgraph of G' of k arcs not containing any broken cut set, and let H'' be a similar subgraph of G'' of $i - k$ arcs. Then $H = H' \dot{+} H''$ is a subgraph of G of i arcs not containing any broken cut set. Conversely, any such subgraph H consists of two such subgraphs H' and H''. The number of such subgraphs H equals the number of such pairs $H', H'', = \sum_k (-1)^k p'_k \cdot (-1)^{i-k} p''_{i-k}$, as required.

5. *Completeness of the invariants.* It is easily seen that if two graphs G and G' are 2-homeomorphic (see the following paper), then $p_i = p'_i$. The question arises, if $p_i = p'_i$, then are G and G' 2-homeomorphic? (If they are triply connected, they would then be isomorphic.) This is not true, as is shown by the two following graphs.†

G: ab, bc, cd, dd_1, d_1e, ee_1, e_1f, fa, ag, bg, cg, dg, eg, fg, ac, d_1e_1;
G': ab, bc, cc_1, c_1d, dd_1, d_1e, ef, fa, ag, bg, cg, dg, eg, fg, ac, c_1d_1.

G and G' are evidently not 2-homeomorphic. That $p_i = p'_i$ may be seen as follows. First, each graph is a dual of itself; hence $p_i = m_i$, $p'_i = m'_i$. Form G^* from G by dropping out the vertex b and the arcs on it, and form G'' from G' by dropping out the same vertex and arcs. Then $M(\lambda) = (\lambda - 3) M^*(\lambda)$, $M'(\lambda) = (\lambda - 3) M''(\lambda)$. But G^* and G'' are isomorphic, hence $M^*(\lambda) = M''(\lambda)$, therefore $M(\lambda) = M'(\lambda)$, therefore $m_i = m'_i$, and thus $p_i = p'_i$.

The following question is as yet unanswered. If two graphs have the same m_{ij}, are they 2-isomorphic? ‡ In the above example, $m_{32} = 10$ and $m'_{32} = 9$.

PRINCETON UNIVERSITY.

† This was discovered by R. M. Foster.

‡ The hypothesis that $m_{ij} = m'_{ij}$ is a greatly weakened form of the hypothesis of the theorem in the paper "2-isomorphic graphs," *American Journal of Mathematics*, Vol. 55 (1933), pp. 245-254.

ON THE CLASSIFICATION OF GRAPHS.†

By Hassler Whitney.‡

1. *Introduction.* R. M. Foster § has given an enumeration of graphs, for use in electrical theory. He uses two distinct methods, classifying the graphs according to their nullity, and according to their rank. In either case, only a certain class of graphs is listed; the remaining graphs are easily constructed from these. In the present paper we give theorems sufficient to put the first method of classification on a firm foundation.

In this method (see §§ 8 and 9), only the elementary graphs (see § 4), or graphs whose connected pieces are elementary, are listed. These graphs are most easily formed from the basic graphs (see § 5), and these, from the basic graphs of nullity one less. This manner of constructing the graphs, and in particular, the important notion of basic graphs, is due to Foster. The definition of elementary graphs and the proofs are, in general, due to the author. We assume here a knowledge of the first half of the paper I.¶

2. Following the terminology of electrical theory, we shall say that two arcs ab, bc, are *in series,* if the vertex b is on no other arc. The vertices a and c need not be distinct, but they must be distinct from b. Two arcs ab, ab, joining the same two distinct vertices, we shall say are *in parallel.*

We shall consider operations on graphs of the following types.

(1a) Replace an arc ab by two arcs ac and cb in series (c being a new vertex).

(1b) Replace two arcs ac and cb in series by a single arc ab, dropping out the vertex c.

In these operations, a and b need not be distinct.

(2) Break the graph at a single vertex into two connected pieces, or join two connected pieces at a single vertex.

---

† Presented to the American Mathematical Society, December 28, 1931, under the title "Basic graphs."

‡ National Research Fellow.

§ Ronald M. Foster, "Geometrical circuits of electrical networks," *Bell Telephone System Technical Publications,* Monograph B-653; also in the *Transactions of the American Institute of Electrical Engineers,* Vol. 51 (1932), pp. 309-317. See in this connection a paper by the author, "2-isomorphic graphs," *American Journal of Mathematics,* Vol. 55 (1933), pp. 245-254.

¶ For references, see the preceding paper.

By operations of this sort we can make one graph isomorphic with another if its components are respectively isomorphic with the components of the other.

(3) Suppose $G = H_1 \dotplus H_2$,† where H_1 and H_2 have the vertices a and b and no others in common, and a and b are connected in both H_1 and H_2. If $ac_1, ac_2, \cdots, ac_m$, and $bd_1, bd_2, \cdots, bd_n$ are the arcs of H_1 on a and b respectively (there is at least one arc in each set), replace these by the arcs $bc_1, bc_2, \cdots, bc_m, ad_1, ad_2, \cdots, ad_n$. We shall say simply, turn H_1 around at the vertices a and b.

Later on we shall also have to consider operations of the following type.

(4) If ab is an arc of G, drop out this arc and let the vertices a and b coalesce.

We note that none of these operations alter the nullity of the graph.

3. We define now certain relations between two graphs. If one graph is formed from another by employing at most operations designated in the first column, then we say the graphs are related as shown in the second column:

no operations	isomorphic ‡
(2)	1-isomorphic §
(2) and (3)	2-isomorphic ¶
(1a) and (1b)	homeomorphic
(1a), (1b) and (2)	1-homeomorphic
(1a), (1b), (2) and (3)	2-homeomorphic.

These relations are all reflexive, symmetric and transitive.

If two topological graphs † are homeomorphic in the topological sense, then the corresponding abstract graphs are homeomorphic in the above sense, and conversely.‖

THEOREM 1. *Any graph 2-homeomorphic with a non-separable graph of nullity* > 0 *is non-separable (and of nullity* > 0*).*

Let G be a non-separable graph of nullity > 0. Suppose first G' is

† $H_1 \dotplus H_2$ is the graph containing the arcs and vertices of both H_1 and H_2. If H_1 and H_2 have no common vertices, then $H_1 \dotplus H_2 = H_1 + H_2$. $H_1 \cdot H_2$ is that graph whose arcs and vertices are in both H_1 and H_2.

‡ See I, § 7; we formerly used the term "congruent." The operation of changing names of vertices and arcs we shall consider as trivial, and shall allow it at any time without mention.

§ The term equivalent was used in I.

¶ Equivalent in the sense of R. M. Foster.

‖ See the preceding paper.

formed from G by an operation of type (1a). If G' is separable, then $G' = I'_1 + I'_2$, $I'_1 \cdot I'_2 =$ a single vertex a, and I'_1 and I'_2 each contain an arc. G is formed from G' by replacing two arcs in series $bd + dc$ by the single arc bc. Then bd and dc both lie in I'_1 or in I'_2. For if not, then $d = a$, and thus any chain from b to c in G' passes through d. Hence b and c are joined in G only through the arc bc, and G is not cyclicly connected. But as G is of nullity > 0 and b and d are distinct, G contains at least two arcs, contradicting I, Theorem 7. Say bd and dc are both in I'_1. Replacing these by the arc bc, I'_1 goes into a graph I_1; put $I_2 = I'_2$. $G = I_1 \dotplus I_2$ is seen to be separable, a contradiction.

The case that G' is formed from G by an operation of type (1b) is similar. Operations of type (3) obviously leave a non-separable graph non-separable. As no operations of type (2) are possible in a non-separable graph, the theorem is proved.

4. *Definitions.* A graph is called *elementary* if it is non-separable and is not 2-homeomorphic with any graph with fewer arcs. Suppose an arc α in a graph G, if dropped out, disconnects G. We then call α a *cut arc* of G. If the two arcs α and β disconnect G if dropped out, while neither is a cut arc, then we say they form a *cut pair of arcs* of G.

THEOREM 2. *If G is 2-homeomorphic with the elementary graph G', then G can be formed from G' by operations of types* (1a) *and* (3) *alone.*

If not, then form G from G', using the fewest possible number of operations of type (1b), and say $G' = G_0, G_1, G_2, \cdots, G_n = G$, are the successive graphs formed. (By the last theorem, operations of type (2) cannot occur.) We suppose G' is of nullity $N > 1$; the theorem is evident otherwise.

Say the first time an operation of type (1b) is employed is in forming G_i from G_{i-1}; the arcs in series $ac + cb$ are replaced by the arc ab. If ab is dropped out of G_i, a graph G^* is formed; let $G_i(1), \cdots, G_i(m-1)$ be its components of nullity > 0 (of which there is at least one), and $H_i(1), \cdots, H_i(p_i - 1)$, its components consisting of a single arc, if there are any (see I, Theorem 8). If we put $H_i(p_i) = ab$, then the graphs $G_i(1), \cdots, G_i(m-1)$, $H_i(1), \cdots, H_i(p_i)$ form a circuit of graphs, as is seen from the first part of the proof of I, Theorem 18.

We shall now show that for each number k, $0 \leqq k \leqq n$, we can put $G_k = G_k(1) \dotplus \cdots \dotplus G_k(m)$, and if $H_k(1), \cdots, H_k(p_k)$ are the arcs of $G_k(m)$, then

(a) Each graph $G_k(1), \cdots, G_k(m-1)$ is non-separable and of nullity > 0,

(b) The graphs $G_k(1),\cdots,G_k(m-1)$, $H_k(1),\cdots,H_k(p_k)$ form a circuit of graphs, and

(c) When G_k is formed from $G_{k\pm1}$, each graph $G_{k\pm1}(s)$ goes thereby into $G_k(s)$, $(s=1,2,\cdots,m)$.

This is true when G_i is formed from G_{i-1} and conversely; we shall show that it is true for G_k when it is formed from $G_{k\pm1}$. If G_k is formed from $G_{k\pm1}$ by an operation of type (1a), this is obvious. An arc of some graph $G_{k\pm1}(s)$ is replaced by a pair of arcs in series; $G_{k\pm1}(s)$ goes thus into $G_k(s)$, which is non-separable if $s<m$ (Theorem 1), and is a set of arcs if $s=m$. (b) obviously holds. Suppose an operation of type (1b) was employed; then two arcs α and β in series are replaced by a single arc γ. As each vertex of $G_{k\pm1}(j)$ is on at least two arcs of $G_{k\pm1}(j)$, $(j=1,\cdots,m-1)$ (I, Theorem 8), α and β lie in the same graph $G_{k\pm1}(s)$; we let the rest of $G_{k\pm1}(s)$ together with γ form $G_k(s)$. The other properties above are easily verified.

Suppose now G_k was formed from $G_{k\pm1}$ by an operation of type (3). Then $G_{k\pm1}=I_1\dot{+}I_2$, and when I_1 is turned around at the vertices a and b, G_k is formed. If either I_1 or I_2 is contained wholly in one of the graphs $G_{k\pm1}(1),\cdots,G_{k\pm1}(m-1)$, the properties are quickly verified. Suppose not; then each graph $G_{k\pm1}(1),\cdots,G_{k\pm1}(m-1)$ is contained wholly in one of the graphs I_1, I_2. For otherwise some graph $G_{k\pm1}(s)$, $s<m$, contains arcs of both I_1 and I_2, and I_1 and I_2 each contain arcs in graphs $G_{k\pm1}(j)$, $(j\neq s)$. Following around the circuit of graphs from an arc of I_1 to an arc of I_2, keeping away from $G_{k\pm1}(s)$, a vertex, say b, is found common to I_1 and I_2 and not lying in $G_{k\pm1}(s)$. Thus I_1 and I_2 have but a single vertex a in common in $G_{k\pm1}(s)$, and this graph is separable, contrary to hypothesis. Now as each graph $G_{k\pm1}(1),\cdots,G_{k\pm1}(m-1)$, $H_{k\pm1}(1),\cdots,H_{k\pm1}(p_{k\pm1})$ lies wholly in I_1 or in I_2, the effect of the operation is merely to alter the arrangement of these graphs in the circuit of graphs, and the properties are again verified.

We can thus divide the operations forming G from G' into two groups: those altering one of the graphs $G_k(1),\cdots,G_k(m-1)$, and those changing the number of arcs in $G_k(m)$ or altering the arrangement of the graphs in the circuit of graphs. We can evidently form G from G' by first performing all the operations in the first group, and then performing all those in the second group. An operation of type (1b) occurs in the second group, in forming G_i from G_{i-1}. Now $G'=G_0$ is elementary, and thus contains no cut pair of arcs (see Theorem 6); hence $G_0(m)$ contains but a single arc: $p_0=1$. We can replace the operations in the second group by the following: replace $H_0(1)$ by p_n arcs in series; then, by operations of type (3), arrange the graphs (including these arcs) properly in the circuit of graphs. We have thus formed G using fewer operations of type (1b), a contradiction, proving the theorem.

We can strengthen this theorem in the following one.

THEOREM 3. *Under the same conditions as in the last theorem, G can be formed from G' by employing first operations of type* (1a) *alone, then operations of type* (3) *alone.*

Take each arc of G which will be replaced by other arcs, and replace it at once by as many arcs in series as it will turn into. We then perform the operations of type (3), being careful merely to break the graph at the proper point each time.

An immediate consequence of Theorem 2 is

THEOREM 4. *Any two 2-homeomorphic elementary graphs are 2-isomorphic.*

The following theorem will be useful in later work.

THEOREM 5. *If a non-separable graph has a cut pair of arcs, and the arcs are not in series, then the four end vertices of these arcs are all distinct.*

For if ab and ac were a cut pair of arcs and there were an arc ad in the graph, a would be a cut vertex.

THEOREM 6. *A necessary and sufficient condition that a non-separable graph G of nullity > 0 be elementary is that it contain no cut pair of arcs.*

We prove first the necessity of the condition. Assuming that G is elementary, we shall show that it has no cut pair of arcs. G contains no two arcs in series, as otherwise, replacing them by a single arc gives a 2-homeomorphic graph with fewer arcs, a contradiction. Suppose G contained a cut pair of arcs ab, cd, not in series. Then these four vertices are distinct, by the last theorem. Dropping out these two arcs leaves two connected graphs H'_1 and H'_2, one containing the vertices a and c say, and the other, the vertices b and d. Put $H_1 = H'_1 \dot{+} ab$, $H_2 = H'_2 \dot{+} cd$. Turning H_2 around at the vertices b and c gives a graph G' 2-homeomorphic with G, with two arcs ab and bd in series; again, we find a 2-homeomorphic graph with fewer arcs.

To prove the sufficiency of the condition, suppose G is not elementary; then it is 2-homeomorphic with an elementary graph G'. By Theorem 2, G can be formed from G' by operations of types (1a) and (3) alone. Say the last operation of type (1a) was to replace an arc γ by two arcs in series

α and β; α and β are a cut pair of arcs in the resulting graph. But operations of type (3) leave these arcs a cut pair, and thus G contains a cut pair of arcs.

5. *Definitions.* A graph is called *cubic* if each vertex is on exactly three arcs. Any cubic elementary graph, also a 1-circuit, is called a *basic graph.* The basic graphs of nullities one, two, and three are: *aa*; *ab, ab, ab*; *ab, ac, ad, bc, bd, cd.* There are two, four, and fourteen basic graphs of nullities four, five, and six respectively (see Foster's paper).

THEOREM 7. *A basic graph G of nullity > 2 contains no* (1- *or*) *2-circuits.*

For suppose G had a 2-circuit *ab, ab.* There is only one other arc α on a, and one other arc β on b, and neither of these is an arc *ab.* α and β are a cut pair of arcs, contradicting Theorem 6.

THEOREM 8. *A basic graph G of nullity > 2 is triply connected.*†

Obviously G contains at least four vertices. Suppose G could be disconnected into the two parts H_1, H_2, by dropping out the two vertices a, b. If, first, there is an arc ab in G, then there is but a single arc joining a to H_1, and a single arc joining b to H_1. These arcs form a cut pair in G, contradicting Theorem 6. If there is no arc ab, one of the graphs H_1, H_2 is joined to a by but a single arc, and one is joined to b by but a single arc. These arcs form a cut pair in G, again a contradiction.

THEOREM 9. *Any two 2-homeomorphic basic graphs G and G' are isomorphic.*

By Theorem 4, G and G' are 2-isomorphic. If they are of nullity 1 or 2, the theorem is true; we assume they are of nullity > 2, in which case they are triply connected. By III, Theorem 4, it is seen that the only operation of type (3) possible is the trivial one of turning around a single arc, which does not alter the graph. Thus G and G' are isomorphic.

6. THEOREM 10. *Let G be an elementary graph, and let the non-separable graph G' be formed from G by an operation of type* (4). *Then G' is elementary.*

For a cut pair of arcs of G' would evidently be a cut pair of arcs of G.

THEOREM 11. *Any elementary non-basic graph G which is not a single*

† See III, p. 158. A graph is triply connected if it contains at least four vertices, and is not disconnected by the omission of any one or two vertices.

arc can be formed from a basic graph G' of the same nullity by operations of type (4).

We shall show how an elementary graph G_1 can be formed from G by the inverse of an operation of type (4). Similarly an elementary graph G_2 can be formed from G_1, etc. Obviously we arrive at a basic graph G' after a finite number of steps; the inverse of these operations carries G' into G.

As G is not basic, there is a vertex a on at least four arcs $aa_1, aa_2, \cdots, aa_m$ $(m \geqq 4)$. Take a new vertex b, replace the arcs aa_1 and aa_2 by the arcs ba_1 and ba_2, and add the arc ab, giving a graph G_1; G is formed from G_1 by an operation of type (4).

G_1 is easily seen to be non-separable. If it is not elementary, it has a cut pair of arcs ab, cd (one of these must obviously be ab, as G has no cut pair of arcs); the four vertices are distinct (Theorem 5). Dropping out these arcs gives two connected graphs H_1 and H_2 containing say b and d, a and c, respectively. b is not a cut vertex of H_1, as otherwise a would be a cut vertex of G. Hence there is a chain C_1 joining a_1 and a_2 in $H_1 - b$. Similarly there is a chain C_2 joining a_3 and a_4 in $H_2 - a$.

Form G'_1 from G by adding the new vertex b', replacing the arcs aa_1 and aa_3 by the arcs $b'a_1$ and $b'a_3$, and adding the arc ab'. G'_1 is elementary. For suppose it had a cut pair of arcs ab', ef; then these four vertices are distinct, and every chain joining a to b' in $G'_1 - ab'$ must contain ef. But one of the chains C_1, C_2, say C_1, does not contain ef; thus $aa_2 + C_1 + b'a_1$ is a chain joining a and b' in $G'_1 - ab' - ef$, a contradiction, proving the theorem.

7. THEOREM 12. *If an arc ab is removed from a basic graph G, the resulting graph G^* is non-separable.*

G^* is surely connected. If it is separable, it has a cut vertex x. Let xy_1, xy_2, (xy_3) be the two, or three, arcs of G^* on x; then one of the vertices y_1, y_2, (y_3), say y_1, is joined to none of the others by a chain in $G^* - x$. Hence a chain from y_1 to x must contain the arc xy_1, that is, xy_1 is a cut arc of G^*. Thus ab, xy_1 are a cut pair of arcs of G, contradicting Theorem 6.

THEOREM 13. *Any basic graph G of nullity > 2 can be formed from a basic graph G_1 of nullity one less by replacing two arcs of G_1 by two pairs of arcs in series, and joining the two new vertices by a new arc.*

Given the basic graph G, we shall find such a graph G_1. Remove an arc ab from G, and replace the two pairs of arcs in series that are now present by single arcs α, β. (It is easily seen that the two pairs of arcs consist of four distinct arcs.)

If the resulting non-separable † graph G_1 is not basic, it has a cut pair of arcs cd, ef; the four vertices are distinct, by Theorem 5. These two arcs were present in G, i. e. neither is α or β. For otherwise, suppose for instance $cd = \alpha$ was formed from the two arcs $ca + ad$, while ef is not β. As $G_1 - cd - ef$ is in exactly two connected pieces, the end vertices of β are connected to either c or d in this graph, say to c. Then replacing β by the two original arcs touching b and adding the vertex a and the arcs $ab + ac$ leaves the graph unconnected, and forms the graph $G - ad - ef$; thus ad, ef are a cut pair of arcs in G, a contradiction. If $cd = \alpha$ and $ef = \beta$, then either ac, be, or ac, bf are a cut pair of arcs in G. This proves the statement. Moreover, the vertices a, b, c, d, e, f are all distinct, as a and b are not in G_1 and are thus distinct from c, d, e, f.

$G - ab - cd - ef$ is in two connected pieces H_1 and H_2; say H_1 contains a, c and e, and H_2, b, d and f. Neither graph has a cut arc. For suppose H_1 say had a cut arc xy. One of the vertices a, c, e, say a, is connected to neither of the others in $H_1 - xy$, as otherwise they would all be connected, and xy would be a cut arc of G. Then a is not connected to c in $G - xy - ab$, and xy, ab are a cut pair of arcs in G, a contradiction.

The next step is to show that if $a'b'$, $c'd'$, $e'f'$ are any cut triple of arcs in G, then either H_1 or H_2 contains none of these arcs. Otherwise, one of these graphs, say H_1, would contain exactly one of the arcs, say $a'b'$. Dropping out the three arcs disconnects a' and b' in G. But as H_1 has no cut arc, a' and b' are connected in $H_1 - a'b'$, and thus in $G - a'b' - c'd' - e'f'$, a contradiction.

We can now prove the theorem easily. Having assumed that $G - ab$ did not give a basic graph, we saw it contained a cut pair of arcs cd, ef, and contained graphs H_1 and H_2 as above. Let $a'b'$ be an arc of H_1. If $G - a'b'$ does not give a basic graph when the pairs of arcs in series are replaced by single arcs, it has a cut pair of arcs $c'd'$, $e'f'$, neither arc of which lies in H_2. Hence, of the two corresponding graphs H'_1, H'_2, one of them, say H'_1, is contained in $H_1 - a'b'$. Let $a''b''$ be an arc of H'_1; if $G - a''b''$ does not give a basic graph, we find a cut pair of arcs and graphs H_1'', H_2'', with H_1'' contained in $H'_1 - a''b''$, etc. As each graph $H_1^{(i)}$ contains arcs, this process must come to an end, and we find finally a graph $G - a^{(i)}b^{(i)}$ which gives a basic graph.

8. *The construction of graphs.* We can now give a standard method for

† See Theorems 1 and 12.

the construction of graphs, forming first the basic graphs, then the elementary graphs, then any graphs.

(1) The basic graphs of nullities 1 and 2 are known. We form successively the basic graphs of nullities $3, 4, \cdots$, as in Theorem 13.

(2) From the basic graphs of a certain nullity, we form all elementary graphs of the same nullity as in Theorem 11. Any non-separable graph we can form thus is elementary (Theorem 10). A given elementary graph may, however, be derived from different basic graphs. If we wish, we can forget all graphs formed which have 2-circuits (arcs in parallel).

(3) Taking all elementary graphs of a given nullity, we form all non-separable graphs of the same nullity by operations first of type (1a) and then of type (3) (Theorem 3). If we left out elementary graphs with 2-circuits and wish to include graphs with 2-circuits now, we must add arcs in parallel with various arcs of non-separable graphs of lesser nullity. We form finally any graph by taking non-separable graphs and letting vertices coalesce in such a manner that the graphs are the components of the final graph.

9. *The classification of graphs by nullity.* To list all graphs, even all non-separable graphs with no two arcs in series, of nullities say one to five, would be a tremendous task. It is thus natural to list only graphs of some class from which all non-separable graphs can be derived without too much difficulty. The elementary graphs form such a class. Moreover, from any group of 2-homeomorphic and thus 2-isomorphic elementary graphs, we need list but one.

As the larger part of the elementary graphs have 2-circuits,† a great saving could be effected by listing none of these graphs. But then, to form all non-separable graphs from these, we would have an added operation to perform, which would increase the difficulty greatly.

We note that an elementary graph with two arcs in parallel may not remain elementary if we drop out one of these arcs; hence we cannot form all elementary graphs from elementary graphs without arcs in parallel by merely replacing single arcs by arcs in parallel.

PRINCETON UNIVERSITY.

† See the figures in R. M. Foster's paper.

2-ISOMORPHIC GRAPHS.*

By HASSLER WHITNEY.†

1. In the preceding paper we said that two graphs G and G' are 2-isomorphic if one can be transformed into the other by operations of the following two types: (2) The arrangement of the components in the graph is altered. (3) If $G = H_1 \dotplus H_2$, where H_1 and H_2 have just the vertices a and b in common and these vertices are connected in both H_1 and H_2, then H_1 is turned around at these vertices.

If G and G' are 2-isomorphic, then any circuit in one graph corresponds to a circuit in the other; for an operation of either type transforms any circuit into a circuit. It was shown in III, Theorem 2,‡ that if there is a $1-1$ correspondence between the arcs of two triply connected graphs so that circuits correspond to circuits,§ then the two graphs are isomorphic (we formerly used the term "congruent"). The question arises, what can be said about any two graphs in which circuits correspond to circuits? The answer is given in the following theorem. The phrase "strictly isomorphic (2-isomorphic)" means: "isomorphic (2-isomorphic), preserving the correspondence between the arcs of the graphs."

THEOREM. *If there is a $1-1$ correspondence between the arcs of the two graphs G and G' so that circuits correspond to circuits, then the graphs are strictly 2-isomorphic.*

In this theorem we can replace the word "circuits" by the words "subgraphs of nullity 0" or "subgraphs of nullity 1" or "cuts sets of arcs" (see the paper on topological invariants). For the first statements see the proof of III, Theorem 3; for the last, see a paper "Planar graphs."

As an example, the graphs G' and G'' of I, p. 353, are strictly 2-isomorphic.

---

* Presented to the American Mathematical Society, August 30, 1932.

† National Research Fellow.

‡ See references in the paper on topological invariants.

§ That is, to any set of arcs forming a circuit in one graph corresponds a set of arcs forming a circuit in the other.

III, Theorem 2 is slightly strengthened in the following corollary.

COROLLARY. *If G and G' satisfy the conditions of the above theorem and one of them is triply connected, then both are, and the two graphs are strictly isomorphic.*

APPLICATION TO ELECTRICAL THEORY. Let us say two graphs are electrically equivalent if there is a $1-1$ correspondence between their arcs so that if corresponding arcs of the two graphs are replaced by the same arbitrary electrical elements, then the same current will flow through corresponding elements. As R. M. Foster has stated (see reference in the last paper), two 2-isomorphic graphs are electrically equivalent. Is the converse true? If not, then it follows immediately from the above theorem that in some two electrically equivalent graphs, we can find a circuit in one corresponding to a subgraph of nullity 0 in the other. If we replace the elements of the circuit by a cell and conductors, a current will flow. But no current can flow through a network of nullity 0. Hence the two graphs are not electrically equivalent. Thus *two graphs are electrically equivalent if and only if they are 2-isomorphic, or, if and only if there is a $1-1$ correspondence between their arcs so that circuits correspond to circuits.*

2. The remainder of the paper is devoted to proving the theorem. We prove first a lemma. If X is a subgraph of G, we shall always denote by X' the corresponding subgraph of G'. We let a subgraph contain only those vertices which are on arcs of the subgraph.

LEMMA 1. *If G and G' satisfy the conditions of the theorem and H is a non-separable subgraph of G, then H' is non-separable.*

It is easily seen that if circuits correspond to circuits in two graphs, then subgraphs of rank i, nullity j, correspond to similar subgraphs. We need merely build up the two subgraphs arc by arc, and note that when corresponding arcs are added, the nullity of one graph increases if and only if the nullity of the other does.

If the lemma is false, then we can put $H' = I'_1 \dot{+} I'_2$, where I'_1 and I'_2 each contain an arc, and $R(H')$* $= R(I'_1) + R(I'_2)$ (I, Theorem 13). Hence, by the above remark, $R(H) = R(I_1) + R(I_2)$, contradicting I, Theorem 14.

Suppose now we have proved the theorem for the case that both graphs are non-separable; then it follows for the general case. For if one of the

* That is, the rank of H'.

graphs, say G, were separable, let $G_1, \cdots, G_m$ be its components. From the lemma it follows that the corresponding subgraphs $G'_1, \cdots, G'_m$ of G' are the components of G'. The conditions of the theorem are satisfied for each pair of graphs G_i, G'_i $(i = 1, \cdots, m)$, hence G_i and G'_i are strictly 2-isomorphic. Having altered each G_i by operations of type (3) so that it becomes strictly isomorphic with G'_i, we bring G into strict isomorphism with G' by operations of type (2).

3. We now assume that both G and G' are non-separable. By I, Theorem 19,* we can build up G in the following manner. Take first an arc H_0. Next add an arc or chain, which with H_0 forms a circuit H_1 (if G was not H_0). Next add an arc or suspended chain, forming with H_1 the non-separable graph H_2 (if G was not H_1), etc. The subgraph H_0 of G, being a single arc, is strictly isomorphic with H'_0 (if $G = H_0$ is a 1-circuit, so is G'). If H_1 is not strictly isomorphic with H'_1, we alter G by a number of operations of type (3) so that it becomes so. If now H_2 is not strictly isomorphic with H'_2, we alter G again, etc. Thus to prove the theorem, we need merely show that *if H is a non-separable subgraph of G strictly isomorphic with H' and A is a chain in G with just its two end vertices in H, then we can alter G by operations of type* (3) *so that $K = H \dot{+} A$ becomes strictly isomorphic with K'.*

4. The graph H, together with the arc or arcs of A, form a circuit of graphs M_1. The corresponding (non-separable) subgraphs of G' form a circuit of graphs M'_1. For, K being non-separable, so is K', by Lemma 1; hence the graphs in M'_1 are not the components of H', and some subset of them form a circuit of graphs (I, Theorem 17). But no proper subset of them do, for the resulting graph would be non-separable (I, Theorem 16), while the corresponding graph in G is not; therefore the whole set forms a circuit of graphs.

H is by hypothesis strictly isomorphic with H', while K is not strictly isomorphic with K'. Group the arcs of A with the graph H into connected

* This theorem, and with it, I, Theorem 18, can be proved most simply as follows. G contains a circuit H_1 (I, Theorems 8 and 4). If $G \neq H_1$, let a be a vertex not in H_1, and let A be a chain from a to a vertex b of H_1. By I, Theorem 6, there is a chain B from a to a vertex c of H_1 not passing through b. From A and B we pick out a chain C with just its two end vertices b and c in H_1. Put $H_2 = H_1 \dot{+} C$. If $G \neq H_2$, we find similarly a chain forming with H_2 a non-separable graph H_3, etc. We arrive finally at a graph $H_s = G$.

subgraphs $G_1, G_2, \cdots, G_n$, such that each G_i is strictly isomorphic with G'_i (hence each G'_i is connected), while no two of these graphs together have this property. The graphs $G_1, \cdots, G_n$ satisfy the conditions for being a circuit of graphs, except that some of them may be separable (they are at least connected). We shall say they form a *generalized circuit of graphs* M_2. Of course $G'_1, \cdots, G'_n$ also form a generalized circuit of graphs M'_2.

5. The proof of the theorem rests on the following two facts.

(A) *Suppose two graphs G_j and G_k have no common vertex, while G'_j and G'_k have a common vertex. Then we can alter G by operations of type* (3) *so that G_j and G_k will get a common vertex. The graphs $G_1, \cdots, G_n$ form at each step a generalized circuit of graphs.*

(B) *Suppose G_j and G_k, also G'_j and G'_k, have a common vertex.* (1) *If there are at least three graphs in M_2, let a' and b' be the vertices of G'_k joining it to G'_j and some other graph of M'_2 respectively, and let a and b be the corresponding vertices of G_k (determined by the isomorphism between G_k and G'_k). Then if a is not the vertex of G_k joining it to G_j, we can alter G by an operation of type* (3) *so that it will become so.* (2) *If M_2 contains G_j and G_k alone and $G_j \dot{+} G_k$ is not strictly isomorphic with $G'_j \dot{+} G'_k$, then we can alter G by an operation of type* (3) *so that this will be true.*

With (A) and (B) proved, we prove the theorem as follows. If the supposition of (A) holds, we bring G_k into contact with G_j. If now the supposition of (B) holds, we turn G_k around so as to touch G_j correctly. Turning G_j around also if necessary, we bring $G_j \dot{+} G_k$ into strict isomorphism with $G'_j \dot{+} G'_k$. We employ (B) until any two graphs which touch, touch correctly. Now employ (A) again if necessary, etc. K is finally brought into strict isomorphism with K', as required.

6. We prove a lemma.

LEMMA 2. *If a and b are two vertices of G, and A_1, A_2, A_3 are three chains in G joining a and b and having no other common vertices, then A'_1, A'_2, A'_3 are chains, joining two vertices of G' and having no other common vertices.*

As $A_1 \dot{+} A_2$, $A_1 \dot{+} A_3$ and $A_2 \dot{+} A_3$ are circuits, so are $A'_1 \dot{+} A'_2$, $A'_1 \dot{+} A'_3$, $A'_2 \dot{+} A'_3$. Hence each A'_i is either a chain or a set of chains. But in the latter case, there are circuits in $A'_1 \dot{+} A'_2 \dot{+} A'_3$ besides those named, which cannot be, as there are no such circuits in $A_1 \dot{+} A_2 \dot{+} A_3$ (see the proof of III, Theorem 2).

7. *Proof of* (A). Suppose the graphs $G_1, G_2, \cdots, G_n$ are named so that they lie in that order in M_2. Let $a_{k+1/2}$ be the vertex joining G_k and G_{k+1}, $(k = 1, \cdots, n)$ (putting $n + 1 = 1$). Suppose G_1 and G_i have no common vertex, while G'_1 and G'_i have a common vertex; we shall bring G_1 and G_i into contact.

The first step is to divide the graphs $G_2, \cdots, G_{i-1}$, and the graphs $G_{i+1}, \cdots, G_n$, into groups as follows. Suppose there is a chain C in G with just its two end vertices x and y in K. Suppose x lies in the graph G_p, and is not the vertex $a_{p+1/2}$, and y lies in the graph G_q, and is not the vertex $a_{q-1/2}$, and $1 < p < q < i$. Then we put the graphs $G_p, G_{p+1}, \cdots, G_q$ into the same group. If G_p and G_q, also G_q and G_r, fall into the same group, then we put all these graphs into the same group. Similarly for the graphs $G_{i+1}, \cdots, G_n$. Each group of graphs forms a subgraph of K. Let $I_2, \cdots, I_{j-1}$ be those formed from $G_2, \cdots, G_{i-1}$, and $I_{j+1}, \cdots, I_m$, those formed from $G_{i+1}, \cdots, G_n$. Put $I_1 = G_1$ and $I_j = G_i$. Then $I_1, I_2, \cdots, I_m$ *form a generalized circuit of graphs* M; say they lie in that order in M.

7′. Each I_k, formed from the graphs $G_{h_k}, G_{h_k+1}, \cdots, G_{h_{k+1}-1}$, say, is a *chain of graphs,* in that each graph G_p of the set is connected and has exactly one vertex in common with G_{p-1} and G_{p+1} (if they are in the set), and no other two have a common vertex. We shall show now that *the corresponding graphs of* M'_2 *form a chain of graphs.* This is trivial if there is just one graph in the set, which happens in particular if $k = 1$ or j. Suppose $1 \neq k \neq j$, and $h_{k+1} - h_k > 1$. The graph G_{h_k} is joined to some other graph G_{q_1} $(h_k < q_1 < h_{k+1})$ in I_k by a chain C_1 with only its end vertices x_1 and y_1 in K, where $x_1 \neq a_{h_k+1/2}$, $y_1 \neq a_{q_1-1/2}$. We shall form now three chains A_1, A_2, A_3, which have just two common vertices $c_1 \neq a_{h_k+1/2}$ and $d_1 \neq a_{q_1-1/2}$, lying in G_{h_k} and G_{q_1} respectively, and so that A_1 contains C_1 and possibly arcs of G_{h_k} or G_{q_1}, A_2 lies in $G_{h_k} \dot{+} \cdots \dot{+} G_{q_1}$, and $A_2 \dot{+} A_3$ is a circuit running around M.

If neither x_1 nor y_1 lies in H, C_1 and any circuit P running around K give such chains; then $c_1 = x_1$, $d_1 = y_1$. Suppose x_1 say, is in H; we must take care then to make $c_1 \neq a_{h_k+1/2}$. Join x_1 to $a_{h_k+1/2}$ by a chain D_1 in H. As $a_{h_k+1/2}$ is not a cut vertex of H, this graph being non-separable, there is a chain D_2 from $a_{h_k-1/2}$ to x_1 in G_{h_k} (if $a_{h_k-1/2}$ is not x_1) which does not pass through $a_{h_k+1/2}$ (see I, Theorem 6). Let c_1 be the first vertex of this chain on D_1; if D_1 contains $a_{h_k-1/2}$, let this vertex be c_1. Then A_1 is C_1 plus that

7

much of D_1 (if there is any) between x_1 and c_1, and A_2 and A_3 are two chains of a circuit P, which consists of the chain we have constructed in G_{h_k} joining $a_{h_k-1/2}$ and $a_{h_{k+1}/2}$, and a chain joining these vertices which runs around the other graphs of M_2. $d_1 = y_1$ as before.

Now by Lemma 2, A'_2 is a chain. It lies wholly in $G'_{h_k} \dotplus \cdots \dotplus G'_{q_1}$, and contains arcs of each of these graphs, as these statements are true when primes are dropped. Following from one end of A'_2 to the other, we pass through all these graphs, and hence they form a chain of graphs (remembering that $G'_1, \cdots, G'_n$ form a circuit of graphs). (We do not know in what order the graphs lie in the chain.)

If G_{q_1} is not $G_{h_{k+1}-1}$, there is a chain C_2 with just its two end vertices x_2 and y_2 in K, where $x_2 \neq a_{p_2+1/2}$ lies in a graph G_{p_2}, $y_2 \neq a_{q_2-1/2}$ lies in G_{q_2}, and $h_k \leqq p_2 \leqq q_1 < q_2 < h_{k+1}$. Hence the graphs $G'_{p_2}, \cdots, G'_{q_2}$ form a chain of graphs. It follows that the graphs $G'_{h_k}, \cdots, G'_{q_2}$ form a chain of graphs. Continuing in this manner, we see finally that $G'_{h_k}, \cdots, G'_{h_{k+1}-1}$ form a chain of graphs, as stated. From this fact we see at once that *$I'_1, \cdots, I'_m$ form a generalized circuit of graphs M'.*

8. Let $b_{k+1/2}$ be the vertex joining I_k and I_{k+1}, $k = 1, \cdots, m (m + 1 = 1)$. Put also $\alpha_k = I_k$ and $\alpha_{k+1/2} = b_{k+1/2}$, $k = 1, \cdots, m$. Suppose C is a chain in G with only its end vertices in K, joining b_k to b_l, or an inner vertex * of I_k to b_l, or an inner vertex of I_k to an inner vertex of I_l. Then we call the chain a (b_k, b_l), or an (I_k, b_l), or an (I_k, I_l) chain respectively. In any case, we can call it an (α_k, α_l) chain.

We now study what types of chains are possible in G. By symmetry, all the properties given below hold if the graphs I_1 and I_j, or the sets of graphs $\{I_p\}$ and $\{I_q\}$, $1 < p < j$, $j < q \leqq m$, are interchanged.

(a) *There is no (α_k, α_l) chain, such that $1 < k < l < j$, and for some integer p, $k < p + 1/2 < l$* (i. e. some vertex $b_{p+1/2}$ lies between α_k and α_l). This follows immediately from the definition of the graphs I_k.

(b) *There is no (I_1, I_j) chain.* For suppose there were. Then we can construct chains A_1, A_2 and A_3, with just the vertices c and d in common lying within I_1 and I_j respectively, and such that A_1 contains (I_1, I_j) and possibly arcs of I_1 or I_j, and $A_2 \dotplus A_3$ is a circuit going around M. (We use the proof in § 7', with but slight changes). By Lemma 2, A'_2 and A'_3 are

* That is, a vertex of I_k which is neither $b_{k-1/2}$ nor $b_{k+1/2}$. We shall say such a vertex lies *within* I_k.

chains in K' having exactly two common vertices c', d'. I'_1 and I'_j each contain arcs of both A'_2 and A'_3, as this is true if primes are dropped; hence I'_1 and I'_j each contain one of the vertices c', d'. By hypothesis, $I'_1 = G'_1$ and $I'_j = G'_i$ have a common vertex; consequently one of the chains A'_2, A'_3 contains arcs of I'_1 and I'_j alone. But this cannot be, as A_2 and A_3 each contain arcs of graphs I_p, $p \neq 1, j$, and (b) is established.

(c) *There is no* (I_k, I_l) *chain,* $1 < k < j$, $j < l \leqq m$. For in this case, constructing chains A_1, A_2, A_3 as above, we find that the end vertices of A'_2 and A'_3 lie within I'_k and I'_l, and one chain passes through I'_1 while the other passes through I'_j; thus I'_1 and I'_j cannot have a common vertex, a contradiction.

(d) *There are no two chains* (I_1, α_k), (I_1, α_l), *with* $2 \leqq k < j$, $j < l \leqq m$. For suppose there were. Then following the proof in § 7′, we see that the graphs $I'_1, \cdots, I'_{k'}$ (k' is the greatest integer $\leqq k$) form a chain of graphs; similarly, $I'_{l'}, \cdots, I'_m, I'_1$ (l' is the smallest integer $\geqq l$) form a chain of graphs. Hence I'_1 touches both a graph I'_p, $1 < p \leqq k'$, and a graph I'_q, $l' \leqq q \leqq m$, and thus does not touch I'_j, a contradiction.

(e) *There are no two chains* $(\alpha_{k_1}, \alpha_{l_1})$, $(\alpha_{k_2}, \alpha_{l_2})$, *where* $1 \leqq k_1 \leqq j - 1$, $2 \leqq k_2 \leqq j$, $j < l_1 \leqq m + 1/2$, $j < l_2 \leqq m + 1/2$, *and* $k_1 < k_2$, *and for some integer* q, $l_1 \leqq q \leqq l_2$. For suppose there were. Consider first Case I: either $k_1 > 1$ or $k_2 < j$, say the former. Using $(\alpha_{k_1}, \alpha_{l_1})$, we form chains A_1, A_2 and A_3, and using $(\alpha_{k_2}, \alpha_{l_2})$, we form chains B_1, B_2 and B_3, as before, where A_2 and B_2 do not pass through I_1. As A'_2 and B'_2 are chains, the graphs $I'_{k'_1}, \cdots, I'_{l'_1}$, and also the graphs $I'_{k'_2}, \cdots, I'_{l'_2}$, form chains of graphs S'_1 and S'_2 respectively, neither of which contains I'_1, where k'_s is the smallest integer $\geqq k_s$, and l'_s is the largest integer $\leqq l_s$, $s = 1, 2$. (see § 7′).

Suppose first $k'_2 > k'_1$, $l'_2 > l'_1$. Starting at $I'_{k'_1}$, which is in S'_1 but not in S'_2, pass along the graphs of S'_1 towards I'_j, which is in both S'_1 and S'_2. As $I'_{l'_2}$ is in S'_2 but not in S'_1, we have not yet passed through all the graphs of S'_2; hence we can continue from I'_j into more graphs of S'_2. This shows that I'_j touches two graphs of the sets S'_1, S'_2, and thus does not touch I'_1, a contradiction.

Suppose next $k'_2 = k'_1$, $l'_2 > l'_1$. Then $k_2 = k'_2$, an integer, and the chain $(\alpha_{k_2}, \alpha_{l_2})$ has an inner vertex of I_{k_2} as end vertex. Hence B_2 has an end vertex within I_{k_2} (we can arrange that it has, as in § 7′); it follows that B'_2 has an end vertex within I'_{k_2}, as in (b). Therefore I'_{k_2} is at one end of the chain of graphs S'_2. I'_{k_2} (which is not I'_j) lies in S'_1, so we can follow S'_1 towards I'_j. We have not yet passed into $I'_{l'_2}$, which lies in S'_2 but not in S'_1, so we can follow S'_2 further out of I'_j. Again I'_j has no vertex in common with I'_1, a contradiction.

The case $k'_2 > k'_1$, $l'_2 = l'_1$ is similar (this time $l_1 = l'_1 = q$, an integer). Suppose $k'_2 = k'_1$, $l'_2 = l'_1$. Then k_2 and l_1 are both integers, and $(\alpha_{k_2}, \alpha_{l_2})$ and $(\alpha_{k_1}, \alpha_{l_1})$ have vertices within I_{k_2} and I_{l_1} respectively. Hence I'_{k_2} is at one end of S'_2, and I'_{l_1} is at one end of S'_1. S'_1 and S'_2 contain the same graphs, and are thus the same chain; it has the distinct graphs I'_{k_2} and I'_{l_1} as ends, and thus has I'_j in its interior, so I'_j does not touch I'_1, a contradiction.

We have left Case II to consider, where $k_1 = 1$, $k_2 = j$. In this case $I'_j, \cdots, I'_{l'_2}$ and $I'_{l'_1}, \cdots, I'_m$, I'_1 form chains of graphs with I'_j and I'_1 as end graphs, and it is seen that I'_1 and I'_j have no common vertex. (e) is now proved.*

9. We return now to the proof of (A). Let us say that a chain (α_k, α_l) alternates with the vertices b_p, b_q, if the numbers k, l, p, q, are all distinct, and α_k, b_p, α_l, b_q lie in that, or the reverse, cyclic order in M. The next step is to show that there are two vertices b_{p_1}, b_{q_1}, $1 < p_1 < j$, $j < q_1 \leqq m + 1/2$, such that

(α) no chain (α_k, α_l) alternates with these vertices,

(β) I_j is in contact with one of these, and

(γ) either I_j is in contact with the other also, or it can be brought into contact with it by an operation of type (3) on G.

If there is no (I_j, α_s) chain, $s \leqq j - 1$ or $s \geqq j + 1$, then $b_{j-1/2}$ and $b_{j+1/2}$ form such a pair of vertices. Suppose there is such a chain. By (b) and (d), either $1 < s \leqq j - 1$ or $s \geqq j + 1$ for all such chains; say the latter is true (the other case is similar). Let q_1 be the smallest number such that $q_1 - 1/2$ is an integer, and for any chain (I_j, α_s) or $(b_{j-1/2}, \alpha_s)$, $s \leqq q_1$. Put $p_1 = j - 1/2$; then b_{p_1} and b_{q_1} are the required vertices. To prove (α), we note first that no (I_j, α_k) chain alternates with these vertices. Take now any number l, $j < l < q_1$. There is no (α_l, α_k) chain, $q_1 < k \leqq m + 1/2$, by (a). Suppose there were a chain (α_l, α_k), with $1 \leqq k < p_1$. By the definition of q_1, there is either a chain (I_j, α_t) or $(b_{j-1/2}, \alpha_t)$, with $q_1 - 1/2 \leqq t \leqq q_1$;

* We can state properties (*b*), (*c*), (*d*) and (*e*) in the single sentence: *It is not true that there is a chain* (a_{k_1}, a_{l_1}), *and there is a chain* (a_{k_2}, a_{l_2}), *where* $1 \leqq k_s \leqq j$, $j \leqq l_s \leqq m + 1$ $(= 1)$, $s = 1, 2$, *and for some integers* p *and* q, $k_1 \leqq p \leqq k_2$, $l_1 \leqq q \leqq l_2$.

in either case, (e) is contradicted (put $k = k_1$, $l = l_1$, $j - 1/2$ or $j = k_2$, $t = l_2$).

(β) holds; we show that (γ) holds. Suppose I_j is in contact with b_{p_1}, but not with b_{q_1}. Let J_1 be that subgraph of G containing $I_{p_1+1/2}$ ($= I_j$), $\cdots, I_{q_1-1/2}$, all vertices which can be joined to these graphs by chains not containing b_{p_1} or b_{q_1}, and all other arcs of G which touch these vertices. If J_2 is the complementary subgraph of G, then J_2 contains the graphs I_s, $s < p_1$ and $s > q_1$, as no chain alternates with b_{p_1} and b_{q_1}, and J_1 and J_2 have only these two vertices in common. Turning J_1 around at the vertices b_{p_1} and b_{q_1} brings I_j into contact with b_{q_1}, as required.—To continue the analysis, using the same notation, we should now rename all graphs and vertices α_k for $p_1 < k < q_1$, replacing subscripts k by $p_1 + q_1 - k$, so that the renamed graphs lie in cyclic order in K. I_j is then renamed $I_{p_1+q_1-j}$; it is this graph we must bring into contact with I_1.

9′. If $p_1 = 1 + 1/2$ or $q_1 = m + 1/2$, I_j is brought into contact with I_1. Suppose not. Then we shall find another pair of vertices b_{p_2} and b_{q_2} with the properties (α), (β) and (γ)—perhaps after we have performed the operation of type (3) described above—, that are nearer to I_1 in that $p_2 \leqq p_1$, $q_2 \geqq q_1$, and either $p_2 < p_1$ or $q_2 > q_1$. If one of these is $b_{1+1/2}$ or $b_{m+1/2}$, I_j can be brought into contact with I_1; if not, we find another such pair of vertices b_{p_3}, b_{q_3}, still nearer to I_1. After a finite number of steps we bring I_j into contact with I_1.

If no chain alternates with the vertices b_{p_1-1} and b_{q_1}, we can put $p_2 = p_1 - 1$, $q_2 = q_1$. I_j can be brought into contact with b_{q_2} if it is not already in contact; (α), (β) and (γ) now hold. Suppose there is a chain alternating with b_{p_1-1} and b_{q_1}; such a chain must be of the form (α_s, α_t), with $p_1 - 1/2 \leqq s \leqq p_1$, and $t > q_1$ or $t = 1$. We shall show then that no chain alternates with b_{p_1} and b_{q_1+1}, from which follows that we can put $p_2 = p_1$, $q_2 = q_1 + 1$. A chain alternating with these vertices must be of the form (α_l, α_k), with $q_1 \leqq l \leqq q_1 + 1/2$, $1 \leqq l < p_1$. But the chains (α_s, α_t) and (α_l, α_k) cannot both exist. For if $s = p_1$ and $t > q_1$, (e) is contradicted; if $s = p_1$ and $t = 1$, then $k = 1$ contradicts (d), and $k > 1$ contradicts (e). If $s = p_1 - 1/2$, then $t > q_1 + 1/2$ or $t = 1$, by (c); (d) is contradicted if $t = k = 1$, otherwise (e) is contradicted. (A) is now proved.

10. *Proof of* (B). We show first that b and a are the vertices joining G_k to G_j and another graph of M_2 respectively. If P' is a circuit running around M'_2, then that part of P' in G'_k is a chain A'_0 with end vertices a' and b'. P is a circuit; A_0 is those arcs of P lying in G_k, and is a chain with

end vertices a and b, as G_k is strictly isomorphic with G'_k. The statement now follows.

There is no chain in G joining a vertex of G_k to a vertex of the other graphs G_s which does not contain a or b. For suppose there were. We construct three chains A_1, A_2, A_3, as we have so often done, joining an inner vertex c of G_k to a vertex of K not in G_k. $A_2 \dotplus A_3$ is a circuit passing around M_2; say A_2 contains b, and A_3, a. First suppose A_2 contains arcs of G_j and G_k alone; then A'_3 contains arcs of G'_j and G'_k alone. As G_k is strictly isomorphic with G'_k and A_3 does not contain a, the arcs of A'_3 in G'_k do not contain a'. Thus A'_3 is not a chain, a contradiction. Suppose now A_2 contains arcs of other graphs besides; then A_3 contains no arcs of G_j. A_3 does not contain b; hence the arcs of A'_3 in G'_k do not contain b'. As A'_3 contains no arcs of G'_j, it is not a chain, a contradiction again.

Consequently, if we define J_1 containing G_k and J_2 containing the other graphs of M_2, as we did in proving (A), we can turn J_1, and with it, G_k, around at the vertices a and b, proving (B), (1). If the supposition of (B), (2) holds, the above operation can be performed, and $G_j \dotplus G_k$ is made strictly isomorphic with $G'_j \dotplus G'_k$. This completes the proof of the theorem.

PRINCETON UNIVERSITY,
HARVARD UNIVERSITY.

Planar Graphs [1]).

By

Hassler Whitney [2]) (Cambridge, U. S. A.).

Introduction. Kuratowski [3]) has shown that a topologica graph is planar, i. e. it can be mapped in a 1—1 continuous manner on the surface of a sphere, if and only if it contains neither of two certain graphs within it. The author has shown [4]) (I, Theorem 29) that a graph is planar if and only if it has a dual as defined in I. It is the main purpose of the present paper to give a proof of the purely combinatorial theorem (Theorem 12 that a graph has a dual if and only if it contains neither of Kuratowski's graphs as a subgraph [5]). This together with the above mentioned theorem of the author gives a proof (for graphs) of Kuratowski's theorem which involves little of a point set nature.

[1]) Presented to the American Mathematical Society, Dec. 28, 1931.

[2]) National Research Fellow.

[3]) Fund. Math. vol 15 (1930), pp. 271—283. He considers a more general point set than a graph.

[4]) We shall refer to the following papers by the author.

I. *Non-separable and planar graphs*. Trans. Amer. Math. Soc. Vol. 34 (1932), pp. 339—362.

II. *Congruent graphs and the connectivity of graphs*, Amer. Journ of Math. Vol. 54 (1932), pp. 150—168.

III. 2-*isomorphic graphs*, Amer. Journ. of Math. Vol. 55 (1933), pp. 245—254. Note that we now use the term *isomorphic* instead of *congruent*.

[5]) Kuratowski's theorem was proved independently by Orrin Frink and P. A. Smith. Prof. Frink has been kind enough to show me their (unpublished) proof; our proof (of Theorem 12) has many points in common with theirs.

In § 1 we show that, to determine whether two graphs are duals or not, it is not necessary to regard all their subgraphs, but merely a part of them. This section will not be used in the sequel. In § 2 cut sets of arcs are discussed. In §§ 3, 4 and 5 some properties of planar graphs are described which correspond to common point set theorems in the plane. The rest of the paper is devoted to the proof of Theorem 12. In this proof we need only Theorems 4, 9 and 11.

We recall the following definitions and theorems from the paper I. A *graph* G consists of a set of *vertices* $a, b, \ldots, f$, and *arcs* $ab, ac, \ldots, df$; the end vertices a and b of each arc ab must be in the graph. A *chain* consists of distinct vertices $a, b, c, \ldots, d, e$, and arcs $ab, bc, \ldots, de$. A *suspended chain* is a chain such that the first and last vertices and only those, are on at least three arcs of the graph. A *circuit* is a set of distinct vertices $a, b, c, \ldots, d, e$, and arcs $ab, bc, \ldots, de, ea$; a *k-circuit* is a circuit of k arcs. A graph is *separable* if it is the union of two graphs, each containing at least one arc, and having at most a single common vertex. A *component* of a graph is a maximal non-separable subgraph.

If a graph G has E arcs and V vertices and is in P connected pieces, then its *rank* R and *nullity* N are defined by the equations $R = V - P$, $N = E - R = E - V + P$. A *forest* is a graph containing no circuit, i. e. a graph of nullity 0. A *subgraph* H of G contains some arcs of G and those vertices which are on these arcs. The *complement* of H in G is the subgraph of G containing those arcs not in H; the complement of G in G is the null graph. Let R, R', r, r', etc. be the ranks of G, G', H, H', etc. Suppose there is a $1-1$ correspondence between the arcs of G and G' so that if H is any subgraph of G and H' is the complement of the corresponding subgraph of G', then $r' = R' - n$; then G' is a *dual* of G.

I, **Theorem** 2. *If an arc ab is added to a graph, the rank (nullity) is increased if and only if a and b were not (were) formerly connected.*

I, **Theorem** 9. *Let G be a graph of nullity* 1 *containing no isolated vertices, such that removing any arc reduces the nullity to* 0. *Then G is a circuit.*

I, **Theorem** 11. *Every non-separable subgraph of G is contained wholly in one of the components of G.*

I, **Theorem** 18. *If G is a non-separable graph of nullity* $N > 1$, *we can remove an arc or suspended chain, leaving a non-separable graph of nullity* $N - 1$.

A simple proof of this theorem is given in III, footnote on p. 247.

I, **Theorems** 20 and 21. *If* G' *is a dual of* G, *then* G *is a dual of* G', *and* $R' = N$, $N' = R$.

I, **Theorem** 23. *Let* $G_1, \ldots, G_m$ *and* $G'_1, \ldots, G'_m$ *be the components of* G *and* G' *respectively, and let* G'_i *be a dual of* G_i $(i = 1, \ldots, m)$. *Then* G' *is a dual of* G.

I, Theorem 25. *Let G and G' be dual graphs, and let $H_1, \ldots, H_m$ be the components of G. Let $H'_1, \ldots, H'_m$ be the corresponding subgraphs of G'. Then $H'_1, \ldots, H'_m$ are the components of G', and H'_i is a dual of H_i $(i = 1, \ldots, m)$.*

I, Theorem 28. *Let ab and $a'b'$ be corresponding arcs of the dual graphs G and G'. Form G_1 from G by dropping out ab, and form G'_1 from G' by dropping out $a'b'$ and letting a' and b' coalesce. Then G_1 and G'_1 are duals.*

I, Theorem 30. *If the non-separable graph G has a (non-separable) dual G', then we can map G and G' together on a sphere so that* (1) *corresponding arcs, and only such arcs, cross each other, and* (2) *inside each region of one graph there is just one vertex of the other graph.*

1. Theorem 1. *Suppose there is a 1—1 correspondence between the arcs of the two graphs G and G' so that a set of arcs in G' form a circuit if and only if the corresponding arcs of G form a circuit. Then any two corresponding subgraphs H and H' of G and G' are of the same rank and nullity.*

The conclusion holds if the word „circuit“ is replaced by „forest“ or „subgraph of nullity > 0“ or „subgraph of nullity 1“, or „cut set of arcs“ (see § 2) *or „subgraph containing no cut set of arcs“ or „subgraph containing at least one cut set of arcs“ or „subgraph containing exactly one cut set of arcs“.*

Let H and H' be any corresponding subgraphs. We shall build them up arc by arc: whenever we add an arc of H, we add also the corresponding arc of H'. To begin with, the two graphs contain no arcs, and are of rank 0, nullity 0. When we add a given arc ab to H, the nullity increases if and only if a and b were already connected, i. e. if and only if ab together with some other arcs already added form a circuit P; similarly for H'. But by hypothesis, if the circuit P is present, then the corresponding arcs of H' form a circuit P', and conversely, if P' is present, so is P. Hence the nullity of H' increases each time if and only if the nullity of H increases. It follows that H and H' are of the same nullity, and therefore also of the same rank.

The next statement follows from the easily proved fact that if any forest in one graph corresponds to a forest in the other, then also circuits correspond to circuits (use I, Theorem 9); similarly for the next two statements.

To prove the last statements, we form H and H' by dropping out the remaining arcs of G and G' one by one, dropping out corresponding arcs at the same time. The proof runs just as before, cut sets of arcs etc. taking the place of circuits etc.

Theorem 1'. *Under the same conditions as Theorem* 1, *G and G' are* 2-*isomorphic.*

This follows from the theorem of III.

Theorem 2. *Suppose there is a* 1 — 1 *correspondence between the arcs of the two graphs G and G' so that a set of arcs in G' form a cut set of arcs if and only if the corresponding arcs of G form a circuit. Then G and G' are duals.*

The conclusion holds if the words „cut set of arcs" and „circuit" are replaced by „subgraph containing no cut set of arcs" and „forest", or „subgraph containing a cut set of arcs" and „subgraph of nullity > 0*", or „subgraph containing exactly one cut set of arcs" and „subgraph of nullity* 1*".*

Suppose cut sets of arcs in G' correspond to circuits in G. Let H, of nullity n, be any subgraph of G, and let H', of rank r', be the complement of the corresponding subgraph of G'. We must show that $r' = R' - n$.

We form H and H' together: We begin with no arcs of G and all the arcs of G'. Each time we add an arc of H, we drop out the corresponding arc of G'; H and H' are then formed at the same time. The nullity of the subgraph of G increases if and only if the last arc added together with arcs already present form a circuit. The rank of the subgraph of G' decreases if and only if the last arc dropped out together with arcs already dropped out form a cut set of arcs. As circuits in G correspond to cut sets of arcs in G', the nullity of the first subgraph increases if and only if the rank of the other decreases, and the first statement follows.

The proof of the other statements follows the lines of the proof of Theorem 1.

Theorem 3. *If G' is a dual of G, then G'' is a dual of G if and only if G' and G'' are* 2-*isomorphic.*

This follows from Theorem 2 and the Theorem of III.

2. *Cut sets of arcs.* Suppose that dropping out a certain set of arcs from a graph G increases the number of connected pieces in the graph, while dropping out no proper subset of these arcs does; we then say these arcs form a *cut set of arcs.* If a single arc forms a cut set of arcs, we call it a *cut arc.* Note that no cut set of arcs contains a 1-circuit.

Theorem 4. *If G and G' are duals, then any circuit in one graph corresponds to a cut set of arcs in the other, and conversely.*

This follows from the definition of dual graphs and I, Theorem 9.

Theorem 5. *If a cut set of arcs is dropped out of a graph G, then the resulting graph G' contains two connected pieces H_1 and H_2 such that each arc of the cut set joined H_1 and H_2.*

Each arc of the cut set joined two distinct connected pieces of G', as putting it back reduces the number of connected pieces in the graph. Say an arc ab of the cut set joins the two pieces H_1 and H_2. Having put back ab, putting back any other arc cd of the cut set leaves the number of connected pieces the same; hence c and d are in the same connected piece in $G' + ab$, while they are not connected in G'. This can only be if cd joins H_1 to H_2 also.

The next two theorems are immediate consequences of this theorem.

Theorem 6. *Any two arcs of a cut set are contained in a circuit in the graph.*

Theorem 7. *Any circuit has an even number of arcs in common with any cut set of arcs.*

Theorem 6, and I, Theorem 11 give.

Theorem 8. *All the arcs of a cut set in a graph lie in a single component of the graph.*

3. *The sides of a circuit.* Suppose a topological planar graph G^*, containing a simple closed curve P^*, is mapped on the surface of a sphere. This surface is divided into two regions by P^*, the two sides of P^*, and all of G^* not in P^* lies in one of these regions.

Let G be a non-separable planar graph containing a circuit P, and let G' be a dual of G; G' is non-separable, by I, Theorem 26. Then, with reference to G', we can define the two sides of P in G as follows. If P' is the subgraph of G' corresponding to P, then P' is a cut set of arcs (Theorem 4). G' minus the arcs of P' is in two connected pieces, say H_1' and H_2'. Let the corresponding subgraphs of G be H_1 and H_2; the arcs of these two graphs, together with the vertices of these graphs which are not in P, form the *two sides of P in the graph G, with reference to the dual G'.*

To show that this definition is admissible, we must show that there is no vertex not on P which lies in both H_1 and H_2; i. e. *if an arc of* H_1 *and an arc of* H_2 *have a common vertex* c, *then* c *lies in* P.

Let C be the arcs on c; the corresponding arcs C' of G' form a circuit (Theorem 4), containing an arc of H_1' and an arc of H_2'. Hence C' contains arcs of the cut set P', and C contains arcs of P; therefore c lies in P. We can now state.

Theorem 9. *If the vertices* a *and* b *are on opposite sides of a circuit* P *in a (non-separable) graph* G, *with reference to a dual* G', *then* a *and* b *are distinct, and every chain from* a *to* b *in* G *contains a vertex of* P.

The converse of this theorem is not true: Two vertices may be on the same side of a circuit, while there is no chain joining them which does not contain vertices of the circuit.

The different duals a non-separable graph which is not triply connected may have correspond to the different ways of mapping this graph on a sphere (see I, Theorem 30, and II, Theorem 11).

4. *Boundaries.* We consider in this section a pair of dual graphs G, G', neither of which contains a 1-circuit. (Hence also neither contains a cut arc). Consider a vertex a' of one of the graphs, say G', and let A' be the arcs on a'. The corresponding arcs A of G we shall say form the *boundary of* a' *in* G. If G and G' are non-separable, and the corresponding topological graphs are mapped on a sphere as in I, Theorem 30, then A is a simple closed curve, and forms the boundary of the region of G which contains a'. If G and G' are separable, the theorem referred to does not hold in general, and this interpretation falls down.

Theorem 10. *The boundary of a vertex* a' *as above described consists of a set of circuits, each of which lies in a different component of* G[1]).

Let $a'b'$ be one of the arcs on a'. If we drop out all the arcs A', a' and b' are disconnected. If we put back as many as we can without connecting these vertices, the remaining arcs A_1' of A' form

[1]) Recall that neither graph contains a 1-circuit or cut arc.

a cut set of arcs. (If the graphs are non-separable, $A_1' = A'$). The corresponding arcs A_1 of G form a circuit; thus we see that each arc of A is contained in a circuit in A.

Consider now an arc $a'c'$ of A' not in A_1', if there is one; it lies in a different component of G' from $a'\,b'$. For otherwise, there would be a chain joining b' and c' and not passing through a', by I, Theorem 6, which together with $a'b'$ and $a'c'$ forms a circuit P'. But this circuit contains only the arc $a'\,b'$ of the cut set A_1', contrary to Theorem 7. Hence the arc of A corresponding to $a'\,c'$ lies in a different component of G from that containing the circuit A_1, by I, Theorem 25, and is contained in a circuit A_2 of A lying in this component. Continuing in this manner, we see that A consists of circuits $A_1, \ldots, A_m$, as required.

5. *Construction of dual graphs.* We prove here some theorems similar to I, Theorem 28, and II, Theorem 8. The present theorems correspond to theorems in the plane on how cross cuts may be drawn in a region, dividing it into two regions, etc.

Theorem 11. *Let G and G' be duals. Let a be a vertex of G, and let A be the arcs on a. Let A', the corresponding arcs of G', consist of two chains B' and C', each joining the vertices b' and c'. Let B and C be the corresponding arcs of G. Then:*

(1) *If G_1' is formed from G' by letting b' and c' coalesce into the vertex a', and G_1 is formed from G by replacing the vertex a by the two vertices b and c, and letting the arcs of B and C end on b and c respectively, then G_1 and G_1' are duals, preserving the correspondence between their arcs.*

(2) *If G_2' is formed from G' by adding the arc $b'\,c'$, and G_2 is formed from G_1 by adding a corresponding arc $b\,c$, then G_2 and G_2' are duals, preserving the correspondence between their arcs.*

(1) is a slight generalization of II, Theorem 8; no change in the proof of that theorem is necessary.

To prove (2), let H_2 be a subgraph of G_2, and H_2', the complement of the corresponding subgraph of G_2'. Suppose first H_2 contains bc; then H_2' does not contain $b'c'$. Form H from H_2 by dropping out the arc bc and letting the vertices b and c coalesce; G is formed similarly from G_2. Thus $r = r_2 - 1$, $R = R_2 - 1$. As H_2' is also the complement of the subgraph of G' corresponding to the subgraph H of G, and G and G' are duals, $r' = R' - n_2'$; therefore $r_2' = R_2' - n_2'$, as required. If H_2 does not contain $b\,c$ while H_2' contains $b'\,c'$, we employ the graphs G_1 and G_1' instead of G and G'.

The last theorem can be generalized in the following one; we make use of the notation of combinatorial analysis situs.

Theorem 11'. *Theorem* 11 *holds if we assume of the arcs A' merely that they contain no* 1*-circuit, provided we choose the sets of arcs B' and C' and the vertices b' and c' so that B' and C' together make up A', and they are each bounded by $a' + b'$ (mod* 2*).*

We note first that A' is a cycle (we always understand mod 2), being a sum of circuits (see Theorem 10); hence such a division of its arcs into two groups B' and C' is always possible for some two vertices b' and c'.

The theorem follows as before as soon as we have proved that b' and c' are connected in H' if and only if b and c are not connected in H_1 (see the proof of II, Theorem 8). The only difficulty is in showing that if b and c are connected in H_1, then b' and c' are not connected in H'. We shall now prove this fact.

As B' is bounded by $b' + c'$, there is a chain E' joining b' and c' in B'; similarly there is a chain F' joining b' and c' in C'. $E' + F'$ has no boundary, is thus a cycle, and consists of one or more circuits $P_1', \ldots, P_p'$ [1]). Similarly $B' - E'$ and $C' - F'$ consist of circuits $Q_1', \ldots, Q_q'$ and $R_1', \ldots, R_r'$ respectively. As the circuits of A' are in different components of G' (Theorem 10), A' can be expressed as the union of circuits in but a single manner; these circuits are therefore those above named.

Let D_1 be a chain joining b and c in H_1; the corresponding arcs in G form a circuit D, containing an arc of B and an arc of C. The corresponding subgraph D' of G' is a cut set of arcs, containing an arc of B' and an arc of C', and lying in a single component of G' (Theorem 8). Hence the cut set contains arcs of some circuit P_i', and thus contains an arc of E' and an arc of F'. The required fact now follows exactly as in II, Theorem 8, Case 2.

[1]) For these facts, see for instance Whitney, *A characterization of the closed 2-cell*, Trans. Amer. Math. Soc., vol. 35 (1933), Lemmas G and H.

6. We turn now to the main theorem of the paper.

Theorem 12. *A necessary and sufficient condition that a graph have a dual is that it contain neither of the two following types of graphs as a subgraph:*

K_1. *This graph is formed by taking five vertices, and joining each two by an arc or suspended chain.*

K_2. *This graph is formed by taking two sets of three vertices each, and joining each vertex in one set to each vertex in the other set by an arc or suspended chain.*

The necessity of the condition was proved in I, Theorem 32 [1]).

We must prove the sufficiency of the condition. We shall assume the theorem holds for all graphs containing fewer than E arcs, and shall prove it for any graph containing E arcs. As a graph of a single arc has a dual, the theorem will be proved in general.

Let G be a graph of E arcs therefore containing neither a graph K_1 nor a graph K_2 as subgraph. Suppose first G is separable. Then none of its components contain K_1 or K_2, hence each component has a dual, and hence G has a dual, by I, Theorem 23.

Suppose next G is non-separable, and suppose it contains a vertex b on but two arcs ab and bc. Dropping out the vertex b and replacing ab and bc by the single arc ac gives a graph G_1 containing neither K_1 nor K_2; it has therefore a dual G_1'. If $a'\,b'$ is the arc of G_1' corresponding to the arc ac of G_1, then adding another arc $a'\,b'$ to G_1' gives a graph G' which is a dual of G, as is easily seen (compare the proof of Theorem 11).

7. Suppose finally G is non-separable, and each vertex is on at least three arcs. By I, Theorem 18, we can drop out an arc ab, leaving a non-separable graph G_0. As G_0 contains neither K_1 nor K_2 and has fewer than E arcs, it has a non-separable dual G_0' (I, Theorem 26). Let A and B be the arcs of G_0 on a and b respectively. The corresponding subgraphs A' and B' of G_0' are circuits (Theorem 4). Suppose A' and B' have a common vertex c'; then if C' is the

[1]) We proved it by proving that neither of two certain graphs G_3 and G_4 had a dual. Note that we proved there a more general result: *There is no graph G_3' (or G_4') which, for any two numbers i and j, has the same number of subgraphs of rank i, nullity j, that a dual of G_3 (or G_4) should have.*

arcs of G_0 on c', C is a circuit in G, containing arcs of both A and B, and containing therefore a and b. A dual of G is now constructed as in Theorem 11.

The rest of the proof is concerned with the case that A' and B' have no common vertex. If we drop out just enough arcs P' from G' to disconnect A' and B', then P' is a cut set of arcs; hence the corresponding arcs P of G form a circuit.

8. We show in this section that *there are two vertices c and d in P, dividing P into the two chains C and D, such that G_0 can be expressed as the union of two graphs G_1^* and G_2^*, the first containing a and C, and the second containing b and D; moreover, G_1^* and G_2^* have only the vertices c and d in common.*

We note first that as the arcs P' separate A' and B', the vertices a and b are on opposite sides of P relative to G', and hence every chain from a to b in G_0 passes through P (Theorem 9). Say P consists of the arcs $a_1\,a_2$, $a_2\,a_3, \ldots, a_{n-1}\,a_n$, $a_n\,a_1$. If there is a chain in G_0 from a to a vertex a_i which does not contain any other vertex of P, we shall say that a_i is *accessible* from a. Let $a_{i_1}, a_{i_2}, \ldots, a_{i_p}$ be the vertices of P accessible from a, named in cyclic order, and let $a_{j_1}, a_{j_2}, \ldots, a_{j_q}$ be those accessible from b. There are at least two vertices in each set, as G_0 is non-separable.

To prove the statement, we need merely show that *for some two vertices c and d and corresponding chains C and D, all the vertices $a_{i_1}, \ldots, a_{i_p}$ lie in C, and all the vertices $a_{j_1}, \ldots, a_{j_q}$ lie in D.* For then we can let G_1^* contain C, and all those arcs of G_0, some end vertex of which can be joined to a by a chain containing neither c nor d, and let G_2^* be the complementary subgraph of G_1^* in G_0. G_1^* and G_2^* evidently have the required properties.

To find the vertices c and d, suppose first some vertex a_{j_t} does not coincide with any vertex a_{i_s}, but lies in P between two vertices $a_{i_h} = c$ and $a_{i_{h+1}} = d$. Then these last two vertices are the required vertices. For if not, then there is some vertex a_{j_u} lying in P on the other side of c and d from a_{j_t}. From a, draw chains to c and d having only their end vertices in P; from these we pick out a chain E from c to d. Using b, we find a similar chain F joining a_{j_t} and a_{j_u}. Of course E and F have no vertex in common. Using the arc $a\,b$, we find a chain in G having no vertices in P, and joining a vertex e of E and a vertex f of F. In G there are therefore two sets of vertices e, a_{j_t}, a_{j_u}, and f, c, d, there are chains joining each vertex of the first set to each vertex of the second, and no two of these chains have a common vertex, except perhaps for their end vertices. Thus G contains a graph K_2, contrary to hypothesis.

Suppose next every vertex a_{j_t} coincides with some vertex a_{i_s}. If there are only two vertices in the set $a_{j_1}, \ldots, a_{j_q}$, we can call these c and d (use the reasoning above). Otherwise let c, d and e be three of these vertices. Draw chains from a to these three vertices, each having only an end vertex on P. From these chains we can pick out a vertex f and three chains fc, fd, fe, which have only the vertex f in common, and have only their other end vertices in P. Using b, we find similar chains gc, gd, ge. Using the arc ab, we find in G a chain joining a vertex x of the first set of chains to a vertex y of the second set; this chain does not touch P. If x is f and u is g, the graph thus constructed (including P) is a graph K_1 in G, contrary to hypothesis. Suppose this is not the case; say x lies on the chain fc between f and c. From the graph we have constructed, drop out the chain de of P, and the chain gc if y is not an inner vertex of this chain, otherwise that much of gc between y and c. The resulting graph is a graph K_2 in G (the two sets of vertices are x, e, d, and f, c, g or y), again a contradiction; thus the supposition that there were three vertices in the set $a_{j_1}, \ldots, a_{j_q}$, was impossible. The statement of § 8 is now proved.

9. Form the graph G_1 from the graph G_1^* by renaming the vertices c and d c_1 and d_1 (and renaming the arcs on these vertices accordingly), and adding a new vertex e_1 and arcs $e_1 a$, $e_1 c_1$, $e_1 d_1$; form G_2 from G_2^* by renaming c and d c_2 and d_2, and adding a vertex e_2 and arcs $e_2 b$, $e_2 c_2$, $e_2 d_2$. We shall show that G_1 and G_2 have duals, and shall then reconstrust G from them, at the same time forming a dual of G.

G_1 can be constructed from G as follows. From chains in G_2^* joining b to the vertices c and d, together with the arc ab, we find in G a vertex e_1 and three chains $e_1 a$, $e_1 c$, $e_1 d$, which have only the vertex e_1 in common and only their other ends in G_1^* We now drop out all other arcs of G_2^* (note that we drop out at least one arc, namely an arc of P), replace each of the above chains by a single arc, and rename c and d c_1 and d_1. Thus we see that G_1 contains fewer arcs than G, and contains neither a graph K_1 nor a graph K_2, as G contains neither. Hence G_1 has a dual G_1'; similarly G_2 has a dual G_2'.

Let $c_1'd_1'$, $d_1'a_1'$, $a_1'c_1'$, and $c_2'd_2'$, $d_2'b_2'$, $b_2'c_2'$, be the arcs of G_1' and G_2' corresponding to the arcs $e_1 a$, $e_1 c_1$, $e_1 d_1$, and $e_2 b$, $e_2 c_2$, $e_2 d_2$, of G_1 and G_2 respectively (these arcs must form circuits). Form G_3 by letting the vertices c_1 and c_2 of G_1 and G_2 coalesce into the vertex c; form G_3' by letting the vertices a_1' and b_2' of G_1' and G_2' coalesce into the vertex a'. By I, Theorem 23, G_3 and G_3' are duals (preserving the correspondence between their arcs — we shall understand these words without mention in the future).

Evidently G_1 and G_2 are non-separable; hence the sets of arcs P_1 and P_2 of G_1 and G_2 corresponding to the sets of arcs of G_1' and G_2' on a_1' and b_2' respectively form circuits. P_1 is formed of two chains E_1 and F_1 where $E_1 = c_1 e_1 + e_1 d_1$, and $P_2 = E_2 + F_2$, where $E_2 = c_2 e_2 + e_2 d_2$. Thus in G_3, the arcs corresponding to the arcs of G_3' on a' consist of two chains, $E_1 + E_2$ and $F_1 + F_2$; each of these joins d_1 to d_2. By Theorem 11, if we let d_1 and d_2 coalesce into the vertex d, forming the graph G_4, and replace a' by the two vertices a_3' and a_4', letting the arcs corresponding to E_1 and E_2 end on a_3', and those corresponding to F_1 and F_2, on a_4', forming the graph G_4', G_4 and G_4' are duals.

By the same theorem, if we let e_1 and e_2 coalesce in G_4 into the vertex e, forming the graph G_5, and replace the vertex a_3' in G_4' by the vertices a_5' and a_6', letting $a_3'd_1'$ and $a_3'd_2'$ end on a_5', and $a_3'c_1'$ and $a_3'c_2'$, on a_6', forming the graph G_5', G_5 and G_5' are duals. By I, Theorem 28, if we drop out both arcs ec and both arcs ed and the vertex e, and replace the arcs ae and eb by the single arc ab, the resulting graph has a dual. But this graph is just the graph G, and the theorem is proved.

ON THE ABSTRACT PROPERTIES OF LINEAR DEPENDENCE.[1]

By HASSLER WHITNEY.

1. Introduction. Let $C_1, C_2, \cdots, C_n$ be the columns of a matrix $\boldsymbol{M}$. Any subset of these columns is either linearly independent or linearly dependent; the subsets thus fall into two classes. These classes are not arbitrary; for instance, the two following theorems must hold:

(a) Any subset of an independent set is independent.

(b) If $\boldsymbol{N}_p$ and $\boldsymbol{N}_{p+1}$ are independent sets of p and $p+1$ columns respectively, then $\boldsymbol{N}_p$ together with some column of $\boldsymbol{N}_{p+1}$ forms an independent set of $p+1$ columns.

There are other theorems not deducible from these; for in § 16 we give an example of a system satisfying these two theorems but not representing any matrix. Further theorems seem, however, to be quite difficult to find. Let us call a system obeying (a) and (b) a "matroid." The present paper is devoted to a study of the elementary properties of matroids. The fundamental question of completely characterizing systems which represent matrices is left unsolved. In place of the columns of a matrix we may equally well consider points or vectors in a Euclidean space, or polynomials, etc.

This paper has a close connection with a paper by the author on linear graphs;[2] we say a subgraph of a graph is independent if it contains no circuit. Although graphs are, abstractly, a very small subclass of the class of matroids, (see the appendix), many of the simpler theorems on graphs, especially on non-separable and dual graphs, apply also to matroids. For this reason, we carry over various terms in the theory of graphs to the present theory. Remarkably enough, for matroids representing matrices, dual matroids have a simple geometrical interpretation quite different from that in the case of graphs (see § 13).

The contents of the paper are as follows: In Part I, definitions of matroids in terms of the concepts rank, independence, bases, and circuits are considered, and their equivalence shown. Some common theorems are deduced (for instance Theorem 8). Non-separable and dual matroids are studied in

[1] Presented to the American Mathematical Society, September, 1934.

[2] "Non-separable and planar graphs," *Transactions of the American Mathematical Society*, vol. 34 (1932), pp. 339-362. We refer to this paper as G.

Part II; this section might replace much of the author's paper G. The subject of Part III is the relation between matroids and matrices. In the appendix, we completely solve the problem of characterizing matrices of integers modulo 2, of interest in topology.

I. Matroids.

2. Definitions in terms of rank. Let a set M of elements $e_1, e_2, \cdots, e_n$ be given. Corresponding to each subset N of these elements let there be a number $r(N)$, the *rank* of N. If the three following postulates are satisfied, we shall call this system a *matroid*.

(R_1) *The rank of the null subset is zero.*

(R_2) *For any subset N and any element e not in N,*

$$r(N+e) = r(N) + k, \qquad (k = 0 \text{ or } 1).$$

(R_3) *For any subset N and elements e_1, e_2 not in N, if $r(N+e_1) = r(N+e_2) = r(N)$, then $r(N+e_1+e_2) = r(N)$.*

Evidently *any subset of a matroid is a matroid.* In what follows, M is a fixed matroid. We make the following definitions:

$$\rho(N) = \text{number of elements in } N.$$

$$n(N) = \rho(N) - r(N) = \textit{nullity} \text{ of } N.$$

N is *independent*, or, the elements of N are independent, if $n(N) = 0$; otherwise, N, and its set of elements, are *dependent*.

Lemma 1. *For any N, $r(N) \geqq 0$ and $n(N) \geqq 0$. If $N \subset M$, then $r(N) \leqq r(M)$, $n(N) \leqq n(M)$.*

Lemma 2. *Any subset of an independent set is independent.*

e is *dependent on* N if $r(N+e) = r(N)$; otherwise e is *independent of* N.

A *base* is a maximal independent submatroid of M, i. e. a matroid B in M such that $n(B) = 0$, while $B \subset N$, $B \neq N$ implies $n(N) > 0$. See also Theorem 7. A *base complement* $A = M - B$ is the complement in M of a base B. A *circuit* is a minimal dependent matroid, i. e. a matroid P such that $n(P) > 0$, while $N \subset P$, $N \neq P$ implies $n(N) = 0$.[3]

Theorem 1. *N is independent if and only if it is contained in a base, or, if and only if it contains no circuit.*

[3] Compare G, Theorem 9.

THEOREM 2. *A circuit is a minimal submatroid contained in no base, i. e. containing at least one element from each base complement. A base is a maximal submatroid containing no circuit. A base complement is a minimal submatroid containing at least one element from each circuit.*

The above facts follow at once from the definitions. Note the reciprocal relationship between circuits and base complements. Note also that the definitions of independence and of being a circuit depend only on the given subset, while the property of being a base depends on the relationship of the subset to M.

3. Properties of rank. Our object here is to prove Theorem 3. The following definition will be useful:

$$\Delta(M, N) = r(M + N) - r(M). \tag{3.1}$$

LEMMA 3. $\Delta(M + e_2, e_1) \leqq \Delta(M, e_1)$.

Suppose first $r(M + e_1) = r(M) + 1$; then $r(M + e_1 + e_2) = r(M) + k$, $k = 1$ or 2. If $k = 2$, then $r(M + e_2) = r(M) + 1$, on account of (R_2), and the inequality holds; if $k = 1$, $r(M + e_2) = r(M) + l$, $l = 0$ or 1, and it holds again. If $r(M + e_2) = r(M) + 1$, the same reasoning applies. If finally $r(M + e_1) = r(M + e_2) = r(M)$, the inequality follows from (R_3).

LEMMA 4. $\Delta(M + N, e) \leqq \Delta(M, e)$.

If $N = e_1 + \cdots + e_p$, the last lemma gives

$$\Delta(M + N, e) \leqq \Delta(M + e_1 + \cdots + e_{p-1}, e) \leqq \cdots \leqq \Delta(M, e).$$

THEOREM 3. $\Delta(M + N_2, N_1) \leqq \Delta(M, N_1)$, *or,*

$$r(M + N_1 + N_2) \leqq r(M + N_1) + r(M + N_2) - r(M). \tag{3.2}$$

This is true if N_1 contains but a single element. For the general case, we apply the last lemma and induction, setting $N_1 = N' + e$:

$$\begin{aligned}\Delta(M + N_2, N_1) &= \Delta(M + N_2 + e, N') + \Delta(M + N_2, e)\\ &\leqq \Delta(M + e, N') + \Delta(M, e) = \Delta(M, N_1).\end{aligned}$$

(3.2) is evidently equivalent to:

$$r(M_1 + M_2) \leqq r(M_1) + r(M_2) - r(M_1 M_2). \tag{3.3}$$

4. Deduction of (I_1), (I_2) from (R_1), (R_2), (R_3). The first postulate

4

on independent sets below obviously holds if (R_1) and (R_2) hold. To prove (I_2), take N, N' as given there; then

$$r(N) = p, \qquad r(N') = p + 1.$$

We must show that for some i, $\Delta(N, e'_i) = 1$. (Then e'_i does not lie in N.) If this is not so, then on using Lemma 4 we find

$$\begin{aligned} 1 = r(N') - r(N) &\leqq \Delta(N, N') \\ &= \Delta(N, e'_1) + \Delta(N + e'_1, e'_2) + \cdots + \Delta(N + e'_1 + \cdots + e'_p, e'_{p+1}) \\ &\leqq \Delta(N, e'_1) + \Delta(N, e'_2) + \cdots + \Delta(N, e'_{p+1}) = 0, \end{aligned}$$

a contradiction.

5. Deduction of (C_1), (C_2) **from** (R_1), (R_2), (R_3). We shall need here a theorem showing how the nullity (or rank) of a matroid may be determined when we know what circuits it contains.

LEMMA 5. *Each element of a circuit is dependent on the rest of the circuit.*

If e is an element of the circuit P, then $n(P) = 1$, $n(P - e) = 0$; hence $r(P) = \rho(P) - 1 = \rho(P - e) = r(P - e)$.

LEMMA 6. *If e is dependent on P_1 but on no proper subset of P_1, then $P = P_1 + e$ is a circuit.*

As $\Delta(P_1, e) = 0$, $r(P) = r(P_1) \leqq \rho(P_1) < \rho(P)$, $n(P) > 0$, and P contains a circuit P'. If P' does not contain e, take e' in P'; then

$$\Delta(P_1 - e', e') \leqq \Delta(P' - e', e') = 0,$$

hence $r(P_1 - e') = r(P_1)$, and

$$\begin{aligned} \Delta(P_1 - e', e) &= r(P_1 - e' + e) - r(P_1 - e') \\ &\leqq r(P_1 + e) - r(P_1) = \Delta(P_1, e) = 0, \end{aligned}$$

and e is dependent on the proper subset $P_1 - e'$ of P_1, a contradiction. Therefore P' contains e. As P' is a circuit, e is dependent on the rest of P'; hence $P' = P$.

THEOREM 4. *If e is not in N, there is a circuit in $N + e$ which contains e if and only if e is dependent on N.*

Suppose $P_1 + e = P$ is a circuit, $P_1 \subset N$. Then

$$\Delta(N, e) \leqq \Delta(P_1, e) = 0,$$

and e is dependent on N. Suppose, conversely, $\Delta(N, e) = 0$. Let P_1 be a smallest subset of N on which e is dependent; then by the last lemma, $P = P_1 + e$ is a circuit. (It may be that $P = e$.)

THEOREM 5. *If N is formed element by element, then $n(N)$ is just the number of times that adding an element increases the number of circuits present.*

Say $N = e_1 + \cdots + e_p$. Then if O is the null set,

$$r(N) = \Delta(O, e_1) + \Delta(e_1, e_2) + \cdots + \Delta(e_1 + \cdots + e_{p-1}, e_p).$$

Each $\Delta(e_1 + \cdots + e_{i-1}, e_i) = 0$ or 1, and $= 0$ if and only if e_i is dependent on $e_1 + \cdots + e_{i-1}$, i.e. if and only if there is a circuit in $e_1 + \cdots + e_i$ containing e_i. The number of terms is $p = \rho(N)$, and the theorem follows.

We turn now to the proof of (C_1) and (C_2). The first is obvious. To prove the second, take P_1, P_2, e_1, e_2 as given. As

$$\Delta(P_1 - e_2, e_2) = \Delta(P_2 - e_1, e_1) = 0,$$

we have

$$\Delta(P_1 + P_2 - e_2, e_2) = \Delta(P_1 + P_2 - e_1 - e_2, e_1) = 0.$$

These equations give

$$r(P_1 + P_2 - e_1 - e_2) = r(P_1 + P_2 - e_2) = r(P_1 + P_2).$$

Using (R_2) gives

$$r(P_1 + P_2 - e_1) = r(P_1 + P_2 - e_1 - e_2);$$

hence the required circuit P_3 exists, by Theorem 4.

6. Postulates for independent sets. Let M be a set of elements. Let any subset N of M be either "independent" or "dependent." Let the two following postulates be satisfied:

(I_1) *Any subset of an independent set is independent.*

(I_2) *If $N = e_1 + \cdots + e_p$ and $N' = e'_1 + \cdots + e'_{p+1}$ are independent, then for some i such that e'_i is not in N, $N + e'_i$ is independent.*

The resulting system is equivalent to a matroid, as we now show. Given any subset N of M, we let $r(N)$ be the number of elements in a largest independent subset of N. Obviously Postulates (R_1) and (R_2) are satisfied; we must prove (R_3). Say

$$r(N+e_1)=r(N+e_2)=r(N)=r.$$

Then $r(N+e_1+e_2)=r$ or $r+1$. If it equals $r+1$, there is an independent set $N'=e'_1+\cdots+e'_{r+1}$ in $N+e_1+e_2$. Let $N''=e_1''+\cdots+e_r''$ be an independent set in N. By (I_2) there is an i such that $N''+e'_i$ is an independent set of $r+1$ elements. But $N''+e'_i$ lies in $N+e_1$ or in $N+e_2$, and hence $r(N+e_1)$ or $r(N+e_2)\geqq r+1$, a contradiction. Therefore $r(N+e_1+e_2)=r$, as required.

We have shown how to deduce either set of postulates (R) or (I) from the other. Moreover the definitions of the rank and the independence or dependence of any subset of M agree under the two systems, and hence they are equivalent.

7. Postulates for bases. Let M be a set of elements, and let each subset either be or not be a "base." We assume

(B_1) *No proper subset of a base is a base.*

(B_2) *If B and B' are bases and e is an element of B, then for some element e' in B', $B-e+e'$ is a base.*

We shall prove the equivalence of this system with the preceding one. We write here $e_1e_2\cdots$ instead of $e_1+e_2+\cdots$ for short.

THEOREM 6. *All bases contain the same number of elements.*

For suppose

$$B=e_1\cdots e_pe_{p+1}\cdots e_qe_{q+1}\cdots e_r,$$
$$B'=e_1\cdots e_pe'_{p+1}\cdots e'_q$$

are bases, with exactly $e_1,\cdots,e_p$ in common, and $r>q$. We might have $p=0$. $q>p$, on account of (B_1). By (B_2), we can replace e_{p+1} in B by an element e' of B', giving a base B_1. $e'=e'_{i_1}$ is one of the elements $e'_{p+1},\cdots,e'_q$, for otherwise B_1 would be a proper subset of B. Hence

$$B_1=e_1\cdots e_pe'_{i_1}e_{p+2}\cdots e_qe_{q+1}\cdots e_r.$$

If $q>p+1$, we replace e_{p+2} in B_1 by an element e'_{i_2} of B', giving a base B_2. Continuing in this manner, we obtain finally the base

$$B_{q-p} = e_1 \cdots e_p e'_{p+1} \cdots e'_q e_{q+1} \cdots e_r.$$

But this contains B' as a proper subset, contradicting (B_1).

We shall say a subset of M is independent if it is contained in a base. (I_1) obviously holds; we shall prove (I_2). Let N, N' be independent sets in the bases B, B'. Say

$$B = e_1 \cdots e_p e_{p+1} \cdots e_q e_{q+1} \cdots e_r e_{r+1} \cdots e_s,$$
$$B' = e_1 \cdots e_p e'_{p+1} \cdots e'_q e'_{q+1} \cdots e'_r e_{r+1} \cdots e_s,$$
$$N = e_1 \cdots e_p e_{p+1} \cdots e_q, \qquad N' = e_1 \cdots e_p e'_{p+1} \cdots e'_q e'_{q+1}.$$

Then N and N' have just $e_1, \cdots, e_p$ in common, and B and B' have just these elements and $e_{r+1}, \cdots, e_s$ in common. By (B_2), there is an element e'_{i_1} of B' such that

$$B_1 = B - e_{q+1} + e'_{i_1}$$

is a base. (This element cannot be any of $e_1, \cdots, e_p, e_{r+1}, \cdots, e_s$, by (B_1).) If i_1 is one of the numbers $p+1, p+2, \cdots, q+1$, then $N + e'_{i_1}$ is in a base B_1, as required. Suppose not; then there is a base

$$B_2 = B_1 - e_{q+2} + e'_{i_2},$$

with $i_2 \neq i_1$. If $p + 1 \leqq i_2 \leqq q + 1$, $N + e'_{i_2}$ is in a base B_2. If not, we find a base B_3, etc. We can drop out each of the $r - q$ elements $e_{q+1}, \cdots, e_r$ in turn; as there are only $r - q - 1$ elements e'_i with $i > q + 1$, we find at some point a base containing $e_1, \cdots, e_q, e'_j$ with $p + 1 \leqq j \leqq q + 1$. Then e'_j is in N', and $N + e'_j$ is in a base and is thus independent, as required.

The definitions of base and independent sets in the two systems (I) and (B) are easily seen to agree. Suppose (I_1) and (I_2) hold. (B_1) obviously holds; using (I_2), we prove that all bases contain the same number of elements; (B_2) now follows at once from (I_2). Hence the two systems are equivalent.

THEOREM 7. *B is a base in M if and only if*

$$r(B) = r(M), \qquad n(B) = 0.$$

Evidently B is a base under the given conditions. To prove the converse, we note first that there exists a base with $r(M)$ elements, as $r(M)$ is the maximum number of independent elements in M (see § 6). By Theorem 6, all bases have this many elements, and the equations follow.

THEOREM 8. *If B is a base and N is independent, then for some N' in B, $N + N'$ is a base.*

This follows from repeated application of Postulate (I_2) and the last theorem.

8. Postulates for circuits. Let M be a set of elements, and let each subset either be or not be a "circuit." We assume:

(C_1) *No proper subset of a circuit is a circuit.*

(C_2) *If P_1 and P_2 are circuits, e_1 is in both P_1 and P_2, and e_2 is in P_1 but not in P_2, then there is a circuit P_3 in $P_1 + P_2$ containing e_2 but not e_1.*

(C_2) may be phrased as follows: If the circuits P_1 and P_2 have the common element e, then $P_1 + P_2 - e$ is the union of a set of circuits.

We shall define the rank of any subset of M, and shall then show that the postulates for rank are satisfied. Let $e_1, \cdots, e_p$ be any ordered set of elements of M. Set $\Gamma_i = 0$ if there is a circuit in $e_1 + \cdots + e_i$ containing e_i, and set $\Gamma_i = 1$ otherwise (compare Theorem 5). Let the "rank" of $(e_1, \cdots, e_p)$ be

$$r(e_1, \cdots, e_p) = \sum_{i=1}^{p} \Gamma_i.$$

LEMMA 7. $r(e_1, \cdots, e_{q-2}, e_{q-1}, e_q) = r(e_1, \cdots, e_{q-2}, e_q, e_{q-1})$.

To prove this, let N be the ordered set $e_1, \cdots, e_{q-2}$, and set

$$r(N) = r, \quad r(N, e_{q-1}) = r_1, \quad r(N, e_q) = r_2,$$
$$r(N, e_{q-1}, e_q) = r_{12}, \quad r(N, e_q, e_{q-1}) = r_{21}.$$

CASE 1. There is no circuit in $N + e_{q-1}$ containing e_{q-1}, and none in $N + e_q$ containing e_q. Then

$$r_1 = r_2 = r + 1.$$

If there is a circuit in $N + e_{q-1} + e_q$ containing e_{q-1} and e_q, then

$$r_{12} = r_1 = r_2 = r_{21};$$

otherwise,

$$r_{12} = r_1 + 1 = r_2 + 1 = r_{21}.$$

CASE 2. There is a circuit P_2 in $N + e_{q-1}$ containing e_{q-1}, and a circuit P_1 in $N + e_{q-1} + e_q$ containing e_{q-1} and e_q. Then, by (C_2), there is a circuit P_3 in $N + e_q$ containing e_q. Hence

$$r_{12} = r_1 = r = r_2 = r_{21}.$$

CASE 3. There is a circuit P_2 as above, but no circuit P_1 as above. If there is a circuit P_3 as above, the last set of equations hold. Otherwise,

$$r_{12} = r_1 + 1 = r + 1 = r_2 = r_{21}.$$

CASE 4. There is a circuit in $N + e_q$ containing e_q. This case overlaps the two preceding ones; the proof above applies here also.

LEMMA 8. *The rank of any subset N is independent of the ordering of the elements of N.*

We saw above that interchanging the last two elements of any subset does not alter the rank; hence, evidently, interchanging any two adjacent elements leaves the rank unchanged. Any ordering of M may be obtained from any other by a number of interchanges of adjacent elements; the rank remains unchanged at each step, proving the lemma.

Postulates (R_1) and (R_2) are obviously satisfied. To prove (R_3), suppose $r(N + e_1) = r(N + e_2) = r(N)$. Then there is a circuit in $N + e_1$ containing e_1 and one in $N + e_2$ containing e_2; hence $r(N + e_1 + e_2) = r(N)$.

The definitions of rank and of circuits under the two systems (R), (C) agree, and hence the systems are equivalent.

9. Fundamental sets of circuits. The circuits $P_1, \cdots, P_q$ of a matroid M form a *fundamental set of circuits* if $q = n(M)$ and the elements $e_1, \cdots, e_n$ of M can be ordered so that P_i contains e_{n-q+i} but no e_{n-q+j} $(j > i)$. The set is *strict* if P_i contains e_{n-q+i} but no e_{n-q+j} $(0 < j < i$ or $j > i)$. These sets may be called sets *with respect to* $e_{n-q+1}, \cdots, e_n$.

THEOREM 9. *If $B = e_1 + \cdots + e_{n-q}$ is a base in $M = e_1 + \cdots + e_n$, then there is a strict fundamental set of circuits with respect to $e_{n-q+1}, \cdots, e_n$; these circuits are uniquely determined.*

As $r(B) = r(M)$, $\Delta(B, e_i) = 0$ $(i = n - q + 1, \cdots, n)$. Hence, by Theorem 4, there is a circuit P_i containing e_i and elements (possibly) of B. $P_{n-q+1}, \cdots, P_n$ is the required set. Suppose, for a given i, there were also a circuit $P'_i \neq P_i$. Then Postulate (C_2) applied to P_i and P'_i would give us a circuit P in B, which is impossible.

This theorem corresponds to the theorem that if a square submatrix N of a matrix M is non-singular, then N can be turned into the unit matrix by a linear transformation on the rows of M.

THEOREM 10. *If $P_1, \cdots, P_q$ form a fundamental set of circuits with* *respect to* $e_{n-q+1}, \cdots, e_n$, *then there is a unique strict set* $P'_1, \cdots, P'_q$ *with respect to* $e_{n-q+1}, \cdots, e_n$.

Set $B = M - (e_{n-q+1} + \cdots + e_n)$. The existence of $P_1, \cdots, P_q$ shows that $r(M) = r(M - e_n) = \cdots = r(B)$. Hence $\rho(B) = n - q = r(M) = r(B)$, and B is a base, by Theorem 7. Theorem 9 now applies.

Note that a matroid is not uniquely determined by a fundamental set of circuits (but see the appendix). This is shown by the following two matroids, in each of which the first two circuits form a strict fundamental set:

M, with circuits 1234, 1256, 3456;
M', with circuits 1234, 1256, 13456, 23456.

II. Separability, Dual Matroids.

10. Separable matroids. If $M = M_1 + M_2$, then $r(M) \leqq r(M_1) + r(M_2)$, on account of (3.3). If it is possible to divide the elements of M into two groups, M_1 and M_2, each containing at least one element, such that

$$r(M) = r(M_1) + r(M_2), \tag{10.1}$$

or, which is equivalent (as M_1 and M_2 have no common elements),

$$n(M) = n(M_1) + n(M_2), \tag{10.2}$$

we shall say M is *separable*; otherwise, M is *non-separable*.[4] Any single element forms a non-separable matroid. Any maximal non-separable part of M is a *component* of M.[5]

Theorem 11. *If*

$$M = M_1 + M_2, \quad r(M) = r(M_1) + r(M_2),$$
$$M'_1 \subset M_1, \quad M'_2 \subset M_2, \quad M' = M'_1 + M'_2,$$

then

$$r(M') = r(M'_1) + r(M'_2).$$

Set $M_1'' = M_1 - M_1'$, $M_2'' = M_2 - M_2'$. The relations (see Theorem 3)

$$\begin{aligned} r(M) &= \Delta(M_1 + M_2', M_2'') + \Delta(M', M_1'') + r(M') \\ &\leqq \Delta(M_2', M_2'') + \Delta(M_1', M_1'') + r(M') \\ &= r(M_2) - r(M_2') + r(M_1) - r(M_1') + r(M') \end{aligned}$$

[4] Compare G, Theorem 15.
[5] See G, § 4.

together with the fact that $r(M) = r(M_1) + r(M_2)$ show that $r(M') \geqq r(M'_1) + r(M'_2)$ and hence $r(M') = r(M'_1) + r(M'_2)$.

THEOREM 12.[6] *If* $M = M_1 + M_2$, $r(M) = r(M_1) + r(M_2)$, M' *is non-separable, and* $M' \subset M$, *then either* $M' \subset M_1$ *or* $M' \subset M_2$.

For suppose $M' = M'_1 + M'_2$, $M'_1 \subset M_1$, $M'_2 \subset M_2$, and M'_1 and M'_2 each contain an element. By the last theorem, $r(M') = r(M'_1) + r(M'_2)$, which cannot be.

THEOREM 13. *If* M_1 *and* M_2 *are non-separable matroids with a common element* e, *then* $M = M_1 + M_2$ *is non-separable.*

For suppose $M = M'_1 + M'_2$, $r(M) = r(M'_1) + r(M'_2)$. By the last theorem, $M_1 \subset M'_1$ or $M_1 \subset M'_2$, and $M_2 \subset M'_1$ or $M_2 \subset M'_1$; this shows that either M'_1 or M'_2 is void.

THEOREM 14. *No two distinct components of* M *have common elements.*

This is a consequence of the last theorem. From this follows:

THEOREM 15.[7] *Any matroid may be expressed as a sum of components in a unique manner.*

THEOREM 16.[8] *A non-separable matroid* M *of nullity* 1 *is a circuit, and conversely.*

If M_1 is a proper non-null subset of the non-separable matroid M, and $M_2 = M - M_1$, then $r(M) < r(M_1) + r(M_2)$. Hence

$$1 = n(M) > n(M_1) + n(M_2),$$

and $n(M_1) = 0$, proving that M is a circuit.

Conversely, if $M = M_1 + M_2$ is a circuit, and M_1 and M_2 each contain elements, then

$$\begin{aligned} r(M_1) + r(M_2) &= \rho(M_1) + \rho(M_2) - n(M_1) - n(M_2) \\ &= \rho(M) > r(M), \end{aligned}$$

showing that M is non-separable.

---

[6] Compare G, Lemma, p. 344.

[7] Compare G, Theorem 12.

[8] Compare G, Theorem 10.

LEMMA 9. *Let $M = M_1 + M_2$ be non-separable, and let M_1 and M_2 each contain elements but have no common elements. Then there is a circuit P in M containing elements of both M_1 and M_2.*

Suppose there were no such circuit. Say $M_2 = e_1 + \cdots + e_s$. Using Theorem 4, we see that

$$\Delta(M_1 + e_1 + \cdots + e_{i-1}, e_i) = \Delta(e_1 + \cdots + e_{i-1}, e_i) \quad (i = 1, \cdots, s),$$

and hence $r(M) = r(M_1) + r(M_2)$, a contradiction.

THEOREM 17.[9] *Any non-separable matroid M of nullity $n > 0$ can be built up in the following manner: Take a circuit M_1; add a set of elements which forms a circuit with one or more elements of M_1, forming a non-separable matroid M_2 of nullity 2 (if $n(M) > 1$); repeat this process till we have $M_n = M$.*

As $n > 0$, M contains a circuit M_1. If $n > 1$, we use the preceding lemma $n - 1$ times. The matroid at each step is non-separable, by Theorems 16 and 13.

THEOREM 18.[10] *Let $M = M_1 + \cdots + M_p$, and let $M_1, \cdots, M_p$ be non-separable. Then the following statements are equivalent:*

(1) *$M_1, \cdots, M_p$ are the components of M.*

(2) *No two of the matroids $M_1, \cdots, M_p$ have common elements, and there is no circuit in M containing elements of more than one of them.*

(3) *$r(M) = r(M_1) + \cdots + r(M_p)$.*

We cannot replace rank by nullity in (3); see G, p. 347.

(2) follows from (1) on application of Theorems 13 and 16.

To prove (1) from (2), take any M_i. If it is not a component of M, there is a larger non-separable submatroid M'_i of M containing it. By Lemma 9, there is a circuit P in M'_i containing elements of M_i and elements not in M_i; P must contain elements of some other M_j, a contradiction.

Next we prove (3) from (1). If $p > 1$, M is separable; say $M = M'_1 + M'_2$, $r(M) = r(M'_1) + r(M'_2)$. By Theorem 12, each M_i is in either M'_1 or M'_2; hence M'_1 and M'_2 are each a sum of components of M. If one of these

[9] See G, Theorem 19; also Whitney, "2-isomorphic graphs," *American Journal of Mathematics*, vol. 55 (1933), p. 247, footnote.

[10] Compare G, Theorem 17.

contains more than one component, we separate it similarly, etc. (3) now follows easily.

Finally we prove (1) from (3). Let M' be a component of M, and suppose it has an element in M_i. As

$$r(M) = r(M_i) + \sum_{j \neq i} r(M_j),$$

M' is contained in M_i, by Theorem 12; as M_i is non-separable, $M' = M_i$.

THEOREM 19.[11] *The elements e_1 and e_2 are in the same component of M if and only if they are contained in a circuit P.*

If e_1 and e_2 are both in P, they are part of a non-separable matroid, which lies in a single component of M. Suppose now e_1 and e_2 are in the same component M_0 of M, and suppose there is no circuit containing them both. Let M_1 be e_1 plus all elements which are contained in a circuit containing e_1. By Lemma 9, there is a subset M^* of $M_0 - M_1$ which forms with part of M_1 a circuit P_3. P_3 does not contain e_1. If e'_4 is an element of P_3 in M_1, there is a circuit P_1 in M_1 containing e_1 and e'_4. Let e_3 be an element of M^*. Then in $M_1 + M^*$ there are circuits P_1 and P_3 which contain e_1 and e_3 respectively, and have a common element.

Let M' be a smallest subset of M_0 which contains circuits P'_1 and P'_3 such that one contains e_1, the other contains e_3, and they have common elements. Then P'_1 and P'_3 are distinct, and $M' = P'_1 + P'_3$. Let e_4 be a common element. By Postulate (C_2), there is a circuit P_1 in $M' - e_4$ containing e_1, and a circuit P_3 in $M' - e_4$ containing e_3. By the definition of M', P_1 and P_3 have no common elements. By Postulate (C_1), P_1 is not contained in P'_1; hence it contains an element e_5 of $M' - P'_1$. P_3 does not contain e_5. As P_3 is not contained in P'_3, it contains an element e_6 of P'_1. But now P'_1 contains e_1, P_3 contains e_3, $P'_1 + P_3$ have a common element e_6, and $P'_1 + P_3$ does not contain e_5 and is thus a proper subset of M', a contradiction. This proves the theorem.

11. Dual matroids. Suppose there is a $1 - 1$ correspondence between the elements of the matroids M and M', such that if N is any submatroid of M and N' is the complement of the corresponding matroid of M', then

$$r(N') = r(M') - n(N). \tag{11.1}$$

[11] Compare D. König, *Acta Litterarum ac Scientiarum Szeged*, vol. 6, pp. 155-179, 4. (p. 159). The present theorem shows that a "glied" is the same as a component.

We say then that M' is a *dual* of M.[12]

THEOREM 20. *If M' is a dual of M, then*

$$r(M') = n(M), \qquad n(M') = r(M).$$

Set $N = M$; then $n(N) = n(M)$. In this case N' is the null matroid, and $r(N') = 0$. (11.1) now gives $r(M') = n(M)$. Also

$$n(M') = \rho(M') - r(M') = \rho(M) - n(M) = r(M).$$

THEOREM 21. *If M' is a dual of M, then M is a dual of M'.*

Take any N and corresponding N' as before. The equations

$$r(N') = r(M') - n(N), \qquad r(M') = n(M),$$
$$\rho(N) + \rho(N') = \rho(M)$$

give

$$\begin{aligned} r(N) &= \rho(N) - n(N) = \rho(N) - [r(M') - r(N')] \\ &= \rho(N) - n(M) + [\rho(N') - n(N')] \\ &= \rho(M) - n(M) - n(N') = r(M) - n(N'), \end{aligned}$$

as required.

THEOREM 22. *Every matroid has a dual.*

This is in marked contrast to the case of graphs, for only a planar graph has a dual graph (see G, Theorem 29).

Let M' be a set of elements in 1 — correspondence with elements of M. If N' is any subset of M', let N be the complement of the corresponding subset of M, and set $r(N') = n(M) - n(N)$. (R_1), (R_2), (R_3) are easily seen to hold in M', as they hold in M; hence M' is a matroid. Obviously $r(M') = n(M)$, and M' is a dual of M.

THEOREM 23. *M and M' are duals if and only if there is a 1 — 1 correspondence between their elements such that bases in one correspond to base complements in the other.*

Suppose first M and M' are duals. Let B be a base in either matroid, say in M, and let B' be the complement of the corresponding submatroid of the other matroid, M'. Then

[12] Compare G, § 8. Theorems 20, 21, 24, 25 correspond to Theorems 20, 21, 23, 25 in G. Note that *two duals of the same matroid are isomorphic*, that is, there is a 1 — 1 correspondence between their elements such that corresponding subsets have the same rank. Such a statement cannot be made about graphs. Compare H. Whitney, "2-isomorphic graphs," *American Journal of Mathematics*, vol. 55 (1933), pp. 245-254.

$$r(B') = r(M') - n(B) = r(M'),$$
$$n(B') = r(M) - r(B) = 0,$$

and B' is a base in M', by Theorem 7.

Suppose, conversely, that bases in one correspond to base complements in the other. Let N be a submatroid of M and let N' be the complement of the corresponding submatroid of M'. There is a base B' in M' with $r(N')$ elements in N', by Theorem 8. The complement in M of the submatroid corresponding to B' in M' is a base B in M with $\rho(N') - r(N') = n(N')$ elements in $M - N$, and hence with $r(M) - n(N')$ elements in N. This shows that

$$r(N) = r(M) - n(N') + k, \qquad k \geqq 0.$$

In a similar fashion we see that

$$r(N') = r(M') - n(N) + k', \qquad k' \geqq 0.$$

As B contains $r(M)$ elements and B' contains $r(M')$ elements, $r(M) + r(M') = \rho(M)$. Hence, adding the above equations,

$$k + k' = r(N) + r(N') + n(N) + n(N') - r(M) - r(M')$$
$$= \rho(N) + \rho(N') - \rho(M) = 0.$$

Hence $k = 0$, and the first equation above shows that M and M' are duals.

There are various other ways of stating conditions on certain submatroids of M and M' which will ensure these matroids being duals.[13]

THEOREM 24. *Let $M_1, \cdots, M_p$ and $M'_1, \cdots, M'_p$ be the components of M and M' respectively, and let M'_i be a dual of M_i $(i = 1, \cdots, p)$. Then M' is a dual of M.*

Let N be any submatroid of M, and let the parts of N in $M_1, \cdots, M_p$ be $N_1, \cdots, N_p$. Let N'_i be the complement in M'_i of the submatroid corresponding to N_i; then $N' = N'_i + \cdots + N'_p$ is the complement in M' of the submatroid corresponding to N. By Theorems 18 and 11 we have

$$r(N') = r(N'_1) + \cdots + r(N'_p), \qquad n(N) = n(N_1) + \cdots + n(N_p).$$

Also

$$r(M') = r(M'_1) + \cdots + r(M'_p), \qquad r(N'_i) = r(M'_i) - n(N_i);$$

adding the last set of equations gives $r(N') = r(M') - n(N)$, as required.

[13] See for instance a paper by the author "Planar graphs," *Fundamenta Mathematicae*, vol. 21 (1933), pp. 73-84, Theorem 2. Cut sets may of course be defined in terms of rank.

THEOREM 25. *Let M and M' be duals, and let $M_1, \cdots, M_p$ be the components of M. Let $M'_1, \cdots, M'_p$ be the corresponding submatroids of M'. Then $M'_1, \cdots, M'_p$ are the components of M', and M'_i is a dual of M_i $(i = 1, \cdots, p)$.*

The complement in M of the submatroid corresponding to M'_i in M' is $\sum_{j \neq i} M_j$. Hence, as M and M' are duals and the M_j $(j \neq i)$ are the components of $\sum_{j \neq i} M_j$ (see Theorem 18),

$$r(M'_i) = r(M') - n(\sum_{j \neq i} M_j) = r(M') - \sum_{j \neq i} n(M_j).$$

Adding gives

$$\sum_i r(M'_i) = pr(M') - (p-1) \sum_j n(M_j) = pr(M') - (p-1)n(M)$$
$$= pr(M') - (p-1)r(M') = r(M').$$

Therefore, by Theorem 12, each component of M' is contained in some M'_i. In the same way we see that each component of M is contained in a matroid corresponding to a component of M'; hence the components of one matroid correspond exactly to the components of the other.

Let N_i be any submatroid of M_i, and let N' and N'_i be the complements in M' and M'_i of the submatroid corresponding to N_i. The equations

$$r(M') = \sum_j r(M'_j), \qquad r(N') = r(N'_i) + \sum_{j \neq i} r(M'_j),$$
$$r(N') = r(M') - n(N_i),$$

give

$$r(N'_i) = r(M'_i) - n(N_i),$$

which shows that M'_i is a dual of M_i.

THEOREM 26. *A dual of a non-separable matroid is non-separable.*

This is a consequence of the last theorem.

III. MATRICES AND MATROIDS.

12. Matrices, matroids, and hyperplanes. Consider the matrix

$$\boldsymbol{M} = \left\| \begin{matrix} a_{11} & \cdots & a_{1n} \\ \cdot & \cdot & \cdot \\ a_{m1} & \cdots & a_{mn} \end{matrix} \right\|;$$

let its columns be $C_1, \cdots, C_n$. Any subset $\boldsymbol{N}$ of these columns forms a matrix, and this matrix has a rank, $r(\boldsymbol{N})$. If we consider the columns as abstract elements, we have a matroid M. The proof of this is simple if we consider the rank of a matrix as the number of linearly independent columns in it. (R_1) and (R_2) are then obvious. To prove (R_3), suppose $r(\boldsymbol{N} + C_1) = r(\boldsymbol{N} + C_2) = r(\boldsymbol{N})$; then C_1 and C_2 can each be expressed as a linear combination of the other columns of $\boldsymbol{N}$, and hence $r(\boldsymbol{N} + C_1 + C_2) = r(\boldsymbol{N})$. The terms independent and base carry over to matrices and agree with the ordinary definitions; a base in $\boldsymbol{M}$ is a minimal set of columns in terms of which all remaining columns of $\boldsymbol{M}$ may be expressed.

We may interpret $\boldsymbol{M}$ geometrically in two different ways; the second is the more interesting for our purposes:

(a) Let E_m be Euclidean space of m dimensions. Corresponding to each column C_i of $\boldsymbol{M}$ there is a point X_i in E_m with coördinates $a_{1i}, \cdots, a_{mi}$. The subset $C_{i_1}, \cdots, C_{i_p}$ of $\boldsymbol{M}$ is linearly independent if and only if the points $O = (0, \cdots, 0)$, $X_{i_1}, \cdots, X_{i_p}$ are linearly independent in E_m, i. e. if and only if these $p+1$ points determine a hyperplane in E_m of dimension p. A base in $\boldsymbol{M}$ corresponds to a minimal set of points $X_{i_1}, \cdots, X_{i_p}$ in E_m such that each X_j of $\boldsymbol{M}$ lies in the hyperplane determined by $O, X_{i_1}, \cdots, X_{i_p}$. Then p is the rank of $\boldsymbol{M}$.

(b) Let E_n be Euclidean space of n dimensions. Let $R_1, \cdots, R_m$ be the rows of $\boldsymbol{M}$. If $Y_1, \cdots, Y_m$ are the corresponding points of E_n: $Y_i = (a_{i1}, \cdots, a_{in})$, then the points $O, Y_1, \cdots, Y_m$ determine a hyperplane $H = H(\boldsymbol{M})$, which we shall call the *hyperplane associated with* $\boldsymbol{M}$. The dimension $d(H)$ of H is $r(\boldsymbol{M})$. Let $\boldsymbol{N} = C_{i_1} + \cdots + C_{i_p}$ be a subset of $\boldsymbol{M}$, and let E' be the p-dimensional coördinate subspace of E_n containing the x_{i_1} and ... and the x_{i_p} axes. The j-th row of $\boldsymbol{N}$ corresponds to the point Y'_j in E' with coördinates $(a_{ji_1}, \cdots, a_{ji_p})$; this is just the projection of Y_j onto E'. If H' is the hyperplane in E' determined by the points $O, Y'_1, \cdots, Y'_m$, then H' is exactly the projection of H onto E', and

$$d(H') = r(\boldsymbol{N}). \tag{12.1}$$

Let $\boldsymbol{N} = (C_{i_1}, \cdots, C_{i_p})$ be any subset of $\boldsymbol{M}$, and let E', H' correspond to $\boldsymbol{N}$. Then $\boldsymbol{N}$ *is independent if and only if*

$$d(H') = p,$$

and is a base if and only if

$$d(H') = d(H) = p.$$

THEOREM 27. *There is a unique matroid M associated with any hyperplane H through the origin in E_n.*

Let M contain the elements $e_1, \cdots, e_n$, one corresponding to each coördinate of E_n. Given any subset $e_{i_1}, \cdots, e_{i_p}$, we let its rank be the dimension of the projection of H onto the corresponding coördinate hyperplane E' of E_n. It was seen above that if $\boldsymbol{M}$ is any matrix determining H, then M is the matroid associated with $\boldsymbol{M}$.

13. Orthogonal hyperplanes and dual matroids. We prove the following theorem:

THEOREM 28. *Let H be a hyperplane through the origin in E_n, of dimension r, and let H' be the orthogonal hyperplane through the origin, of dimension $n - r$. Let M and M' be the associated matroids. Then M and M' are duals.*

We shall show that bases in one matroid correspond to base complements in the other; Theorem 23 then applies. Let

$$\boldsymbol{M} = \left\| \begin{matrix} a_{11} \cdots a_{1n} \\ \cdot \quad \cdot \quad \cdot \quad \cdot \\ a_{r1} \cdots a_{rn} \end{matrix} \right\|, \qquad \boldsymbol{M}' = \left\| \begin{matrix} b_{11} \quad \cdots b_{1n} \\ \cdot \quad \cdot \quad \cdot \quad \cdot \quad \cdot \\ b_{n-r,1} \cdots b_{n-r,n} \end{matrix} \right\|$$

be matrices determining H and H' respectively. Say the first r columns of $\boldsymbol{M}$ form a base in $\boldsymbol{M}$, i. e. the corresponding determinant A is $\neq 0$. As H and H' are orthogonal, we have for each i and j

$$a_{i1}b_{j1} + a_{i2}b_{j2} + \cdots + a_{in}b_{jn} = 0.$$

Keeping j fixed, we have a set of r linear equations in the b_{jk}. Transpose the last $n - r$ terms in each equation to the other side, and solve for b_{jk}. We find

$$b_{jk} = \frac{-1}{A} \sum_{l=r+1}^{n} b_{jl} \begin{vmatrix} a_{11} \cdots a_{1l} \cdots a_{1r} \\ \cdot \quad \cdot \quad \cdot \quad \cdot \quad \cdot \quad \cdot \quad \cdot \\ a_{r1} \cdots a_{rl} \cdots a_{rr} \end{vmatrix} = \sum_{l=r+1}^{n} c_{kl} b_{jl} \qquad (k = 1, \cdots, r).$$

This is true for each $j = 1, \cdots, n - r$, and the c_{kl} are independent of j. Thus the k-th column of $\boldsymbol{M}'$ is expressed in terms of the last $n - r$ columns. As this is true for $k = 1, \cdots, r$, the last $n - r$ columns form a base in $\boldsymbol{M}'$, as required.

14. The circuit matrix of a given matrix. Consider the matrix $\boldsymbol{M}$ of § 12. Suppose the columns $C_{i_1}, \cdots, C_{i_p}$ form a circuit, i. e. the corresponding elements of the corresponding matroid form a circuit. Then these columns are linearly dependent, and there are numbers $b_1, \cdots, b_n$ such that

$$(14.1) \qquad \begin{array}{ll} a_{i1}b_1 + \cdots + a_{in}b_n = 0 & (i = 1, \cdots, m), \\ b_j = 0 \quad (j \neq i_1, \cdots, i_p), & b_j \neq 0 \quad (j = i_1, \cdots, i_p). \end{array}$$

The b_j are all $\neq 0$ $(j = i_1, \cdots, i_p)$, for otherwise a proper subset of the columns would be dependent, contrary to the definition of a circuit. (They are uniquely determined except for a constant factor; see Lemma 11.) Suppose the circuits of $\boldsymbol{M}$ are $P_1, \cdots, P_s$. Then there are corresponding sets of numbers $b_{i1}, \cdots, b_{in}$ $(i = 1, \cdots, s)$, forming a matrix

$$\boldsymbol{M}' = \left\| \begin{matrix} b_{11} \cdots b_{1n} \\ \cdot \quad \cdot \quad \cdot \quad \cdot \\ b_{s1} \cdots b_{sn} \end{matrix} \right\|,$$

the *circuit matrix* of the matrix $\boldsymbol{M}$.

THEOREM 29. *Let $P_1, \cdots, P_q$ be a fundamental set of circuits in $\boldsymbol{M}$* (see § 9). *Then the corresponding rows of the circuit matrix $\boldsymbol{M}'$ form a base for the rows of $\boldsymbol{M}'$. Hence $r(\boldsymbol{M}') = q = n(\boldsymbol{M})$.*

Suppose the columns of $\boldsymbol{M}$ are ordered so that P_i contains C_{n-q+i} but no column C_{n-q+j} $(j > i)$. Then if the corresponding row of $\boldsymbol{M}'$ is $R'_i = (b_{i1}, \cdots, b_{in})$, we have $b_{i,n-q+i} \neq 0$ and $b_{i,n-q+j} = 0$ $(j > i)$. Hence the rows $R'_1, \cdots, R'_q$ of $\boldsymbol{M}'$ are linearly independent, and $r(\boldsymbol{M}') \geqq q$. Hence $r(\boldsymbol{M}') = n(\boldsymbol{M}) = q$, and each row of $\boldsymbol{M}'$ may be expressed in terms of $R'_1, \cdots, R'_q$.

THEOREM 30. *If $\boldsymbol{M}'$ is the circuit matrix of $\boldsymbol{M}$ and H', H are the corresponding hyperplanes, then H' is the hyperplane of maximum dimension orthogonal to H.*

This is a consequence of (14.1) and the last theorem.

THEOREM 31. *The matroids corresponding to a matrix and its circuit matrix are duals.*

This follows from the last theorem and Theorem 28.

15. On the structure of a circuit matrix. Let M be any matroid, and M', its dual. If there exists a matrix $\boldsymbol{M}$ corresponding to M, it is perhaps most easily constructed by considering it as the circuit matrix of a matrix $\boldsymbol{M}'$

5

corresponding to M'. Let H and H' be the hyperplanes corresponding to $\mathbf{M}$ and $\mathbf{M}'$. We shall say the set of numbers $(a_1, \cdots, a_n)$ is in $Z_{i_1 \ldots i_p}$ if

$$a_j \neq 0 \quad (j = i_1, \cdots, i_p), \qquad a_j = 0 \quad (j \neq i_1, \cdots, i_p).$$

If $(a_1, \cdots, a_n)$ is in H and in $Z_{i_1 \ldots i_p}$, then the columns $C_{i_1}, \cdots, C_{i_p}$ of $\mathbf{M}'$ are dependent, evidently.

LEMMA 10. *Let $(b_1, \cdots, b_n)$ be a point of H. If it is in $Z_{i_1 \ldots i_p}$, then the matroid $N' = e_{i_1} + \cdots + e_{i_p}$ is the union of a set of circuits in M'.*

Here e_i in M' corresponds to C_i in $\mathbf{M}$. We need merely show that for each i_s there is a circuit P in N' containing e_{i_s}. Let $k_1 = i_s, k_2, \cdots, k_q$ be a minimal set of numbers from $(i_1, \cdots, i_p)$ containing i_s such that there is a point $(c_1, \cdots, c_n)$ of H in $Z_{k_1 \ldots k_q}$; then $e_{k_1} + \cdots + e_{k_q}$ is the required circuit. For if it were not a circuit, there would be a proper subset $(l_1, \cdots, l_r)$ of $(k_1, \cdots, k_q)$ and a point $(d_1, \cdots, d_n)$ of H in $Z_{l_1 \ldots l_r}$. No $l_i = k_1$, on account of the minimal property of $(k_1, \cdots, k_q)$. Say $l_1 = k_t$, and set

$$a_i = d_{k_t} c_i - c_{k_t} d_i \qquad (i = 1, \cdots, n).$$

Then $(a_1, \cdots, a_n)$ is in H and in $Z_{m_1 \ldots m_u}$ with $(m_1, \cdots, m_u)$ a proper subset of $(k_1, \cdots, k_q)$ containing k_1, again a contradiction.

LEMMA 11. *If $P = e_{i_1} + \cdots + e_{i_p}$ is a circuit of M' and $(b_1, \cdots, b_n)$ and $(b'_1, \cdots, b'_n)$ are in H and in $Z_{i_1 \ldots i_p}$, then these two sets are proportional.*

For otherwise, $(c_1, \cdots, c_n)$ with $c_i = b'_{i_1} b_i - b_{i_1} b'_i$ would be a point of H in some $Z_{k_1 \ldots k_q}$ with $(k_1, \cdots, k_q)$ a proper subset of $(i_1, \cdots, i_p)$, and P would not be a circuit.

It is instructive to show directly that Postulate (C_2) holds for matrices: P_1 and P_2 are represented by rows $(b_1, \cdots, b_n)$ and $(b'_1, \cdots, b'_n)$ of $\mathbf{M}$, lying in $Z_{12 i_1 \ldots i_p}$ and $Z_{1 k_1 \ldots k_q}$ respectively, where $k_1, \cdots, k_q \neq 2$. Set $c_i = b'_1 b_i - b_1 b'_i$; then $(c_1, \cdots, c_n)$ is in H and in $Z_{2 l_1 \ldots l_r}$, with $(l_1, \cdots, l_r)$ a subset of $(i_1, \cdots, i_p, k_1, \cdots, k_q)$; the existence of P_3 now follows from Lemma 10.

THEOREM 32. *Let $\mathbf{M}$ be the circuit matrix of $\mathbf{M}'$. Let $P_1, \cdots, P_q$ form a strict fundamental set of circuits in M' with respect to $e_{n-q+1}, \cdots, e_n$, and let the first q rows in $\mathbf{M}$ correspond to $P_1, \cdots, P_q$. Let $(i_1, \cdots, i_s)$ be any set of numbers from $(1, \cdots, q)$, let $(j_1, \cdots, j_s)$ be any set from $(1, \cdots, n - q)$, and let $(i'_1, \cdots, i'_{q-s})$ be the set complementary to $(i_1, \cdots, i_s)$ in $(1, \cdots, q)$.*

Then the determinant D in $\boldsymbol{M}$ with rows $i_1, \cdots, i_s$ and columns $j_1, \cdots, j_s$ equals zero if and only if the determinant D' with rows $1, \cdots, q$ and columns $j_1, \cdots, j_s$, $n-q+i'_1, \cdots, n-q+i'_{q-s}$ equals zero, or, if and only if there exists a circuit P in $\boldsymbol{M'}$ containing none of the columns $e_{j_1}, \cdots, e_{j_s}$, $e_{n-q+i'_1}, \cdots, e_{n-q+i'_{q-s}}$.

In the matrix of the last $q = r(\boldsymbol{M})$ columns of $\boldsymbol{M}$, the terms along the main diagonal and only those are $\neq 0$. If we expand D' by Laplace's expansion in terms of the columns $n-q+i'_1, \cdots, n-q+i'_{q-s}$, we see at once that $D' = 0$ if and only if $D = 0$.

Suppose $D = 0$. Then there is a set of numbers $(\alpha_1, \cdots, \alpha_q)$, not all zero, with $\alpha_i = 0$ $(i \neq i_1, \cdots, i_s)$, such that

$$b_k = \alpha_1 b_{1k} + \cdots + \alpha_q b_{qk} = 0 \qquad (k = j_1, \cdots, j_s),$$

$(b_{i1}, \cdots, b_{in})$ being the i-th row of $\boldsymbol{M}$., $b_k = 0$ also for $k = n-q+i'_1, \cdots, n-q+i'_{q-s}$, as each term is zero for such k. The point $(b_1, \cdots, b_n)$ is in H. Any circuit given by Lemma 10 is the required circuit P.

Suppose the circuit P exists. Then it is represented by a row $(b_1, \cdots, b_n)$ in $\boldsymbol{M}$. As the first q rows of $\boldsymbol{M}$ are of rank $q = r(\boldsymbol{M})$, $(b_1, \cdots, b_n)$ can be expressed in terms of them; say $b_k = \Sigma \alpha_i b_{ik}$. As $b_k = 0$ $(k = n-q+i'_1, \cdots, n-q+i'_{q-s})$, certainly $\alpha_k = 0$ $(k = i'_1, \cdots, i'_{q-s})$. $D = 0$ now follows from the fact that $b_k = 0$ $(k = j_1, \cdots, j_s)$.

16. A matroid with no corresponding matrix.[14] The matroid M' has seven elements, which we name $1, \cdots, 7$. The bases consist of all sets of three elements except

(16.1) 124, 135, 167, 236, 257, 347, 456.

Defining rank in terms of bases, we have: Each set of k elements is of rank k if $k \leqq 2$ and of rank 3 if $k \geqq 4$; a set of three elements is of rank 2 if the set is in (16.1) and is of rank 3 otherwise. It is easy to see that the postulates for rank are satisfied. (R_3) in the case that N contains two elements is satisfied vacuously. For suppose $r(N+e_1) = r(N+e_2) = r(N) = 2$. Then $N+e_1$ and $N+e_2$ are both in (16.1); but any two of these sets have but a single element in common.

[14] After the author had noted that M' satisfies (C^*) and corresponds to no linear graph, and had discovered a matroid with nine elements corresponding to no matrix, Saunders MacLane found that M' corresponds to no matrix, and is a well known example of a finite projective geometry (see O. Veblen and J. W. Young, *Projective Geometry*, pp. 3-5).

If there exists a matrix $\boldsymbol{M}'$, corresponding to M', then let $\boldsymbol{M}$ be its circuit matrix. 123 is a base in M', and hence

$$(16.2) \qquad 124, \quad 135, \quad 236, \quad 1237$$

form a fundamental set of circuits in M'. Let R_1, R_2, R_3, R_4 be the corresponding rows of $\boldsymbol{M}$. By multiplying in succession row 1, column 2, rows 2, 3, 4, and columns 4, 5, 6, 7 by suitable constants $\neq 0$, we bring $\boldsymbol{M}$ into the following form:

$$(16.3) \qquad \boldsymbol{M} = \left\| \begin{array}{ccc|cccc} 1 & 1 & 0 & 1 & 0 & 0 & 0 \\ 1 & 0 & a & 0 & 1 & 0 & 0 \\ 0 & 1 & b & 0 & 0 & 1 & 0 \\ 1 & c & d & 0 & 0 & 0 & 1 \\ \cdot & \cdot & \cdot & \cdot & \cdot & \cdot & \cdot \end{array} \right\|;$$

a, b, c and d are $\neq 0$. We now apply Theorem 32 with

$$(i_1, \cdots, i_s; j_1, \cdots, j_s) = (1, 4; 1, 2), \quad (2, 4; 1, 3), \quad (3, 4; 2, 3),$$

i. e. using the circuits 347, 257, 167. This gives

$$\begin{vmatrix} 1 & 1 \\ 1 & c \end{vmatrix} = \begin{vmatrix} 1 & a \\ 1 & d \end{vmatrix} = \begin{vmatrix} 1 & b \\ c & d \end{vmatrix} = 0,$$

and hence $c = 1$, $a = d = b$. Using the circuit 456, with sets $(1, 2, 3; 1, 2, 3)$ gives $2a = 0$, $a = 0$, a contradiction.

In regard to this example, see the end of the paper.

APPENDIX.

MATRICES OF INTEGERS MOD 2.

We wish to characterize those matroids M corresponding to matrices $\boldsymbol{M}$ of integers mod 2,[15] i. e. matrices whose elements are all 0 or 1, where rank etc. is defined mod 2. We shall consider linear combinations, *chains:*

$$(A.1) \qquad \alpha_1 e_1 + \cdots + \alpha_n e_n \qquad (\alpha\text{'s integers mod } 2)$$

in the elements of M. The α's may be taken as 0 or 1; (A. 1) may then be interpreted as the submatroid N whose elements have the coefficient 1. Conversely, any $N \subset M$ may be written as a chain. Submatroids are added

[15] See O. Veblen, "Analysis situs," 2nd ed., *American Mathematical Society Colloquium Publications*, Ch. I and Appendix 2.

(mod 2) by adding the corresponding chains (mod 2). For instance, $(e_1 + e_2) + (e_2 + e_3) \equiv e_1 + e_3 \pmod 2$.

Any sum (mod 2) of circuits in M we shall call a *cycle* in M. N is the *true sum* of $N_1, \cdots, N_s$ if these latter have no common elements and $N = N_1 + \cdots + N_s$. We consider matroids which satisfy the following postulate:

(C^*) *Each cycle is a true sum of circuits.*

Postulate (C_2) is a consequence of (C^*). For the cycle $P_1 + P_2$ is a submatroid containing e_2 but not e_1; The existence of P_3 now follows from (C^*).

A simple example of a matroid not satisfying (C^*) is given by the matroid M' at the end of § 9.

THEOREM 33. *A circuit is a minimal non-null cycle, and conversely.*

This is proved with the aid of Postulates (C_1) and (C^*).

THEOREM 34. *Let $P_1, \cdots, P_q$ be a strict fundamental set of circuits in M with respect to $e_{n-q+1}, \cdots, e_n$. Then there are exactly 2^q cycles in M, formed by taking all sums* (mod 2) *of $P_1, \cdots, P_q$.*

First, each sum $P_{i_1} + \cdots + P_{i_s}$ (mod 2) is a cycle, containing $e_{n-q+i_1}, \cdots, e_{n-q+i_s}$ and elements (perhaps) from $B = e_1, \cdots, e_{n-q}$; obviously distinct sums give distinct cycles. Now let Q be any cycle in M; say Q contains $e_{n-q+k_1}, \cdots, e_{n-q+k_r}$ and elements (perhaps) from B. Set $Q' = P_{k_1} + \cdots + P_{k_r}$; then $Q + Q'$ is a cycle containing elements from B alone. But B is a base (see the proof of Theorem 10), and hence contains no circuits. Consequently $Q + Q'$ is the null cycle, and $Q = Q'$.

THEOREM 35. *As soon as the circuits of a strict fundamental set are known, all the circuits may be determined.*

This is a consequence of the last two theorems. It is to be contrasted with the final remark of § 9.

Remark. The word "strict" may be omitted in the last two theorems.

THEOREM 36. *Let $e_1, \cdots, e_n$ be a set of elements, and let $P_1, \cdots, P_q$ be any subsets such that P_i contains e_{n-q+i} and possibly elements from $e_1, \cdots, e_{n-q}$ alone. Then there is a unique matroid M satisfying* (C^*), *with $P_1, \cdots, P_q$ as a strict fundamental set of circuits.*

We form the 2^q cycles of Theorem 34. Those cycles which contain no other non-null cycle as a proper subset we call circuits; in particular, $P_1, \cdots, P_q$ are circuits. To prove (C*), let Q be a non-null cycle. If it is not a circuit, it contains a circuit P as a proper subset. Q and P are sums (mod 2) from $P_1, \cdots, P_q$, hence the same is true of $Q - P$, and $Q - P$ is one of the 2^q cycles. If it is not a circuit, we again extract a circuit, etc.

This theorem furnishes a simple method of constructing all matroids satisfying (C*).

We turn now to the study of matrices of integers (mod 2)

$$\boldsymbol{M} = \begin{Vmatrix} a_{11} & \cdots & a_{1n} \\ \cdot & \cdot & \cdot \\ a_{m1} & \cdots & a_{mn} \end{Vmatrix} \qquad \text{(each } a_{ij} = 0 \text{ or } 1\text{)}.$$

Any linear combination (mod 2) of the columns

$$\text{(A. 2)} \qquad \alpha_1 C_1 + \cdots + \alpha_n C_n \qquad (\alpha\text{'s integers mod } 2)$$

is a set of numbers $(\Sigma\alpha_i a_{1i}, \cdots, \Sigma\alpha_i a_{mi})$, which we call a *chain* (mod 2) in $\boldsymbol{M}$. As before, we may take each coefficient as 0 or 1, and we may consider any chain merely as a submatrix of $\boldsymbol{M}$. The chain is a *cycle* if each of the corresponding numbers is $\equiv 0$ (mod 2). The columns $C_{i_1}, \cdots, C_{i_p}$ are *independent* (mod 2) if there exists no set of integers $\alpha_1, \cdots, \alpha_n$ not all $\equiv 0$ (mod 2), with $\alpha_i = 0$ $(i \neq i_1, \cdots, i_p)$, such that $\Sigma\alpha_i C_i$ is a cycle, i. e. if no non-null subset of $C_{i_1}, \cdots, C_{i_p}$ is a cycle. Using this definition, the terms base, circuit, rank, nullity etc. (mod 2) can be defined as in Part I.

Let M be a set of elements $e_1, \cdots, e_n$ corresponding to $C_1, \cdots, C_n$ in $\boldsymbol{M}$, and let $e_{i_1} + \cdots + e_{i_p}$ be a circuit in M if and only if $C_{i_1}, \cdots, C_{i_p}$ is a circuit in $\boldsymbol{M}$. We shall show that *M is a matroid satisfying* (C*) *and the definitions of cycle in M and $\boldsymbol{M}$ agree.*

We show first that each circuit is a cycle in $\boldsymbol{M}$. If $C_{i_1}, \cdots, C_{i_p}$ is a circuit, then these columns are dependent; hence $\Sigma\alpha_i C_i$ is a cycle, with $\alpha_i = 0$ $(i \neq i_1, \cdots, i_p)$. Moreover $\alpha_i = 1$ $(i = i_1, \cdots, i_p)$, for otherwise a proper subset of $C_{i_1}, \cdots, C_{i_p}$ would be dependent. Hence $C_{i_1} + \cdots + C_{i_p}$ is a cycle. Next, any sum (mod 2) of circuits is a cycle, evidently. Next we prove (C*). Suppose $Q = C_{i_1} + \cdots + C_{i_p}$ is a cycle. Let $(k_1, \cdots, k_q)$ be a minimal subset of $(i_1, \cdots, i_p)$ such that $P = C_{k_1} + \cdots + C_{k_q}$ is a cycle; then P is a circuit. $Q - P$ is a cycle; from it we extract a circuit, just as above, etc. It follows from (C*) that the definitions of cycle in M and $\boldsymbol{M}$ agree. Theorems 33, 34 and 35 now apply to $\boldsymbol{M}$ also.

We are now ready to prove the final theorem:

THEOREM 37. *Let M be any matroid satisfying* (C*). *Suppose* $\rho(M) = n$, *and* $e_1 + \cdots + e_{n-q}$ *is a base. Then if* $\mathbf{M}_1$ *is any matrix of integers* (mod 2) *with* $n - q$ *columns which are independent* (mod 2), *columns* $C_{n-q+1}, \cdots, C_n$ *can be adjoined in a unique manner to* $\mathbf{M}_1$, *forming a matrix* $\mathbf{M}$ *of which the corresponding matroid is* M.

Let $P_1, \cdots, P_q$ be a strict fundamental set of circuits in M with respect to $e_{n-q+1}, \cdots, e_n$ (Theorem 9). Say $P_1 = e_{i_1} + \cdots + e_{i_p} + e_{n-q+1}$. Set $C_{n-q+1} \equiv C_{i_1} + \cdots + C_{i_p}$ (mod 2); this determines C_{n-q+1} as a column of 0's and 1's so that $P'_1 = C_{i_1} + \cdots + C_{i_p} + C_{n-q+1}$ is a circuit. (P'_1 is a cycle; (C*) shows that it is a single circuit, as $C_1 + \cdots + C_{n-q}$ contains no circuit.) C_{n-q+1} evidently must be chosen in this manner. We choose the remaining columns of $\mathbf{M}$ similarly. Let M' be the matroid corresponding to $\mathbf{M}$. Then $P'_1, \cdots, P'_q$ is a strict set of circuits in M'. These same sets form a strict set in M; hence, by Theorem 35, the circuits in M' correspond to those in M. Consequently $M' = M$, completing the proof.

We end by noting that the matroid M' of § 16 satisfies Postulate (C*) but corresponds to no linear graph. For letting 123 be a base and (16. 2) a fundamental set of circuits and determining the matroid as in Theorem 36, we come out with exactly M'. A corresponding matrix of integers mod 2 is constructed from (16. 3) with $a = b = c = d = 1$; we interchange rows and columns in the left-hand portion, leave out the last row and column of the right-hand portion, and interchange these two parts. (The relation $2a = 0$ is of course true mod 2.)

On the other hand, it is easily seen that if the element 7 is left out, there is a corresponding graph, which must be of the following sort: It has four vertices a, b, c, d, and the arcs corresponding to the elements $1, \cdots, 6$ are

$$ab, \quad ac, \quad ad, \quad bc, \quad bd, \quad cd.$$

There is no way of adding the required seventh arc.

The problem of characterizing linear graphs from this point of view is the same as that of characterizing matroids which correspond to matrices (mod 2) with exactly two ones in each column.

HARVARD UNIVERSITY.

A numerical equivalent of the four color map Problem.[1])

By **Hassler Whitney** in Cambridge (U. S. A.).

Mit 2 Textfiguren.

1. *Introduction.* A famous unsolved problem is the four color map problem, stated below as Problem I. The essential difficulty of the problem is due perhaps mainly to the fact that it relates two quite different types of question: What kind of a configuration is a map, and what kind of configurations can be colored in four colors; the first problem is geometric in character, and the second, combinatorial. Our purpose here is to state a problem equivalent to the four color problem which is purely combinatorial in character; this is Problem IV below.

To prove that Problems I and IV are equivalent, we prove in succession that Problems I and II, Problems II and III, and Problems III and IV below are equivalent. The first step is the only step of any real difficulty; it is caried out in another paper.[2]) The other steps, which we give here, are quite simple; for this reason, in studying the map problem, it is perhaps as useful to consider Problem II as Problem IV.

The relation between the four problems is illustrated in § 5. An example of a map is given in Fig. 1, and the example is then translated into forms corresponding to the other three problems.

In § 6 we consider further the nature of Problem II, and give two other equivalent formulations. Some questions of probability are considered in § 7; it is shown that, in one sense, probably all large maps can be colored in four colors.

---

[1]) Presented to the American Mathematical Society, Dec. 26, 1933.

[2]) A theorem on graphs, Annals of Math., vol. 32 (1931), pp. 378—390. We refer to this paper as G. In connection with the map problem, see a coming number of the Mémorial des Sciences Mathématiques, by G. D. Birkhoff.

2. *The problems.* They are as follows:

I. A map is colored by assigning to each region a color in such a manner that no two regions with a common boundary are of the same color.

Problem I. Given any map on the surface of a sphere, is it always possible to color it in four colors?

II. Draw a regular polygon; divide the inside into triangles in any manner by drawing non-intersecting diagonals; divide the outside into circular triangles in any manner by drawing non-intersecting arcs (which we also call diagonals). The result we shall call a polygonal configuration. It is colored by assigning to each vertex of the polygon a color in such a manner that two vertices which are joined by a line, either a side of the polygon or a diagonal, are of the same color.

Problem II. Can all polygonal configurations be colored in four colors?

III. Let $(p_1, q_1), (p_2, q_2), \ldots, (p_s, q_s)$ be pairs of integers. We shall say the form an n-admissible set of pairs if:

$$0 \leqq p_i, \quad p_i + 2 \leqq q_i, \quad q_i \leqq n \quad (i = 1, \ldots, s), \tag{1}$$

$$p_i < p_j < q_i < q_j \text{ is true for no } i \text{ and } j. \tag{2}$$

We say the set of pairs is complete if it is impossible to add a new pair of integers in such a manner that the above relations still hold. Any two complete n-admissible sets have the same number of pairs.

Problem III. Let $(p_1, q_1), \ldots, (p_s, q_s)$ and $(p_1', q_1'), \ldots, (p_s', q_s')$ be any two complete n-admissible sets of pairs. Is it always possible to find numbers $b_0, \ldots, b_n$, each equal to 1, 2, 3 or 4, so that

$$b_i \neq b_{i+1}, \qquad (i = 0, \ldots, n-1), \tag{3}$$

$$b_{p_i} \neq b_{q_i}, \quad b_{p_i'} \neq b_{q_i'} \qquad (i = 1, \ldots, s)? \tag{4}$$

IV. Consider the sum $a_1 + a_2 + \ldots + a_n$. If we put in as many pairs of parentheses as possible, we will have defined a definite manner of evaluating the sum. Such a sum with parentheses we call an arranged sum. For instance, $a+b+c$ may be arranged as $(a+b)+c$

or $a+(b+c)$; $a+b+c+d$ may be arranged as $((a+b)+c)+d$, $(a+(b+c))+d$, $(a+b)+(c+d)$, $a+((b+c)+d)$, or $a+(b+(c+d))$. The numbers $a_1, \ldots, a_n$ are the **terms** of the sum. Any number formed in evaluating the sum is a **partial sum**. Thus the partial sums of $(3+7)+(1+4)$ are 3, 7, 10, 1, 4, 5, 15.

Problem IV. **If an n-fold sum (any n) is expressed in any two ways as an arranged sum, is it always possible to choose the terms of the sum as integers so that no partial sum of either arranged sum is divisible by 4?**

In particular, no term may be divisible by 4; we may choose each term as 1, 2 or 3.

3. *Equivalence of Problems II and III.* Suppose the answer to Problem II is yes. Consider any two complete n-admissible sets as in Problem III. Construct a regular polygon with $n+1$ vertices $B_0, \ldots, B_n$. Corresponding to each pair (p_i, q_i) other than the pair $(0, n)$ draw the straight line segment $B_{p_i} B_{q_i}$ in the polygon. As the pairs are admissible, no two of these lines cross. We can also, corresponding to each pair (p'_i, q'_i) other than $(0, n)$, draw a curve joining $B_{p'_i}$ and $B_{q'_i}$ outside the polygon in such a manner that no two of these curves cross. As the sets of pairs are complete, the inside and outside of the polygon are divided into triangles. We have thus constructed a polygonal configuration as in Problem II. Color this in colors numbered 1, 2, 3, 4. For each $i=0, \ldots, n$, let b_i be the number of the color in which B_i is colored; then the inequalities (3), (4) all hold. Thus the answer to Problem III is yes.

Conversely, suppose Problem III is answered in the affirmative. Given any polygonal configuration, let its vertices be $B_0, \ldots, B_n$. Corresponding to each diagonal $B_{p_i} B_{q_i}$ inside the polygon define the pair (p_i, q_i), and to each diagonal $B_{p'_i} B_{q'_i}$ outside, the pair $(p'_i\, q'_i)$. Add the pair $(0, n)$ to each set. Each set of pairs is n-admissible, as no two diagonals cross, and is complete. Hence we can define numbers $b_0, \ldots, b_n$, each equal to 1, 2, 3 or 4, so that (3), (4) hold. If we color the vertices $B_0, \ldots, B_n$ correspondingly, the polygonal configuration is colored in four colors.

4. *Equivalence of Problems III and IV.* Suppose the answer to Problem III is yes. Let two arrangements of an n-fold sum $a_1+\ldots+a_n$ be given. Any pair of parentheses in the first arrangement encloses a certain number of the a's, for instance $(a_k+\ldots+a_l)$. Corresponding to

16*

this pair of parentheses define the pair of integers $(k-1, l)$; similarly for each other pair in the first arrangement. Condition (1) is obviously satisfied. Suppose that, for some i and j, $p_i<p_j<q_i<q_j$; then there would be in the sum pairs of parantheses

$$(a_{p_{i+1}}+\ldots+(a_{p_j+1}+\ldots+a_{q_i})+\ldots+a_{q_j}).$$

But these define perhaps pairs (p_i, q_j), (p_j, q_i), certainly not (p_i, q_i), (p_j, q_j). Hence (2) holds also, and the pairs are n-admissible. When we add the pair $(0, n)$, the set is evidently complete. Similarly we define a set of pairs corresponding to the second arrangement.

Define $b_0, \ldots, b_n$ so that (3) and (4) hold, and set

$$a_i=b_i-b_{i-1} \text{ or } b_i-b_{i+1}+4, \tag{5}$$

according to which number is positive and $\leqq 4$. (3) shows that each $a_i=1$, 2 or 3. Each partial sum in either arrangement

$$a_s+a_{s+1}+\ldots+a_t=b_t-b_{s-1}+4k \qquad (t>s) \tag{6}$$

(where k is some integer) corresponds to a pair of parentheses, and hence to a pair $(s-1, t)$; hence by (4), $b_{s-1}\neq b_t$, b_t-b_{s-1} is not divisible by four, and $a_s+\ldots+a_t$ is not divisible by four.

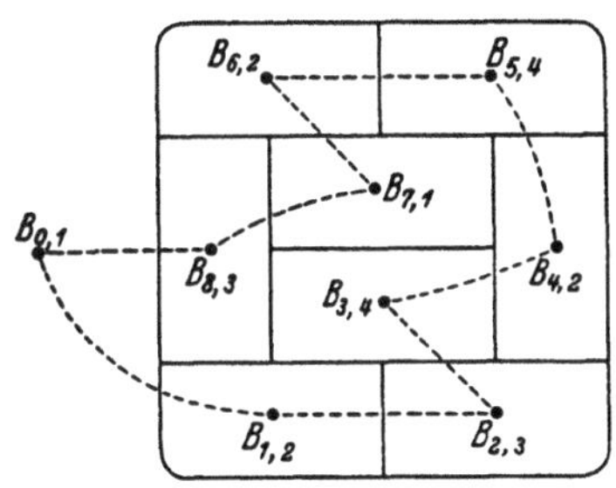

Fig. 1.

Conversely, suppose the answer to Problem IV is yes. Take two n-admissible sets of pairs as in Problem III. Consider a sum $a_1+a_2+\ldots+a_n$. Corresponding to each pair (p_i, q_i) put in a pair of parentheses $a_1+\ldots+(a_{p_i+1}, +\ldots+a_{q_i})+\ldots+a_n$. Because of (2), $a_{p_i+1}+\ldots+a_{q_i}$ is a partial sum of the resulting arranged sum. Similarly form a second arrangement from the second set of pairs. Define $a_1, \ldots, a_n$ so that no partial sum is divisible by four in either arranged sum. Set

$$b_0=1, \quad b_i=1+a_1+a_2+\ldots+a_i-4k_i \qquad (i>0), \tag{7}$$

where k_i is an integer chosen so that $b_i=1$, 2, 3 or 4. (6) now holds; it is at once seen that (3) and (4) hold.

5. *Example.* Fig. 1 (compare Fig. 3 of G) shows a map of nine regions, colored in the colors 1, 2, 3, 4. Fig. 2 shows a corresponding polygonal configuration, distorted so that all the sides of the polygon

but one lie on a line. The other arcs (diagonals inside the polygon, above, and outside, below) show what other pairs of regions in the map have a common boundary.

From Fig. 2 we at once read off two admissible sets of pairs:

$$(0,8),\ (1,8),\ (1,3),\ (3,8),\ (3,7)\ (4,7),\ (5,7);$$
$$(0,8),\ (0,6),\ (0,5),\ (0,4),\ (0,2),\ (2,4),\ (6,8).$$

Il we write the sum $a_1 + \ldots + a_8$ and put in parentheses in the two corresponding ways, we have

$$a_1+((a_2+a_3)+((a_4+(a_5+(a_6+a_7)))+a_8)),$$
$$((((a_1+a_2)+(a_3+a_4))+a_5)+a_6)+(a_7+a_8).$$

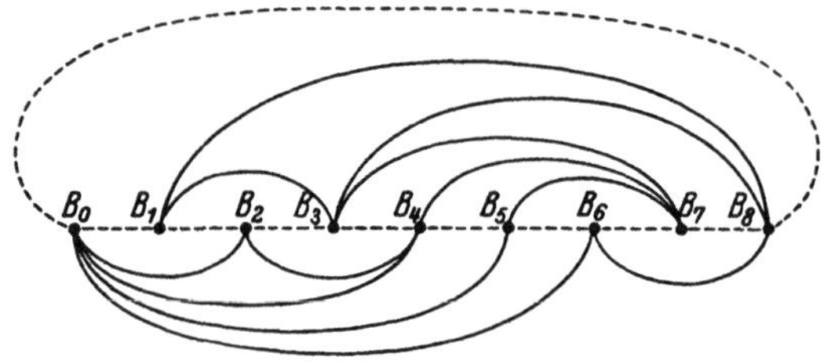

Fig. 2.

Sets of numbers satisfying the last two problems are:

b_0		b_1		b_2		b_3		b_4		b_5		b_6		b_7		b_8
1		2		3		4		2		4		2		1		3
	1		1		1		2		2		2		3		2	
	a_1		a_2		a_3		a_4		a_5		a_6		a_7		a_8.	

The first set of numbers is taken from Fig. 1; the second set is then determined by eq. (5).

6. *Other forms of Problems II.* Let P_n be a regular polygon of n sides. We define the following numbers:

τ_n = the number of ways of filling in the interior (or exterior) of P_n with diagonals, that is, the number of triangulations of P_n. Then [3])

$$\tau_n = \frac{(2n-4)!}{(n-2)!\,(n-1)!} = 2^{n-3}\,\frac{3\cdot5\cdot7\ldots(2n-5)}{3\cdot4\cdot5\ldots(n-1)}.$$

---

[3]) See G., p. 390; J. H. M. Wedderburn, Annals of Math., vol. 24 (1923), p. 121).

ν_n = the number of ways of coloring P_n in four colors. This is[4])

$$\nu_n = 3^n + (-1)^n \cdot 3.$$

δ_n = the number of ways of coloring P_n with interior triangulated in four colors. This is easily seen to be

$$\delta_n = 4 \cdot 3 \cdot 2^{n-2} = 3 \cdot 2^n.$$

For instance, in Fig. 2, leaving out the lower set of diagonals, B_0 may be colored in any one of four colors; then B_1, in any one of three; then B_8, B_3, B_2, B_7, B_4, B_5, B_6, in succession, in any one of two colors. Note that *all colorings of any triangulation are easily written down.*

Let $T_1, \ldots, T_{\tau_n}$ be the triangulutions of P_n, let $C_1, \ldots, C_{\nu_n}$ be the ways of coloring P_n in four colors, and let $D_1^i, \ldots, D_{\delta_n}^i$ be the ways of coloring T_i in four colors ($i = 1, \ldots, \tau_n$). If S_i is the set of colorings $D_1^i, \ldots, D_{\delta_n}^i$, then problem II may be stated in the following form:

Problem V. Given any n, have any two sets S_i, S_j $(i, j = 1, \ldots, \tau_n)$ *a common coloring?*

Let us set $\eta_{ij}^n = 1$ if C_j is a coloring of T_i, i. e. if C_j is a member of S_i, and set $\eta_{ij}^n = 0$ otherwise. We have then the matrix

$$\eta^n = \left\| \begin{matrix} \eta_{11}^n, & \cdots & \eta_{1\nu_n}^n \\ \cdots & \cdots & \cdots \\ \eta_{\tau_n 1}^n & \cdots & \eta_{\tau_n \nu_n}^n \end{matrix} \right\|.$$

Another form of Problem II is then:

Problem VI. *Are no two rows of η^n orthogonal?*

7. *Some probability considerations.* Suppose ξ^n is any matrix of τ_n rows and ν_n columns, with δ_n ones in each row and all other elements zero. What is the probability of no pair of rows ξ^n being orthogonal? First we shall find the probability of two given rows being orthogonal. The ones having a fixed position in the first row, there are $\binom{\nu_n}{\delta_n}$ possible arrangements of δ_n ones in the second row, and $\binom{\nu_n - \delta_n}{\delta_n}$ arrangements which make this row orthogonal to the first, i. e. such that no column has a one in both rows. The probability is therefore the ratio

[4]) See for instance Whitney, the coloring of graphs. Annals of Math., vol. 33 (1932), p. 691.

$$p_n = \binom{\nu_n - \delta_n}{\delta_n} : \binom{\nu_n}{\delta_n} = \frac{[(\nu_n - \delta_n)!]^2}{\nu_n!\,(\nu_n - 2\delta_n)!}$$

$$= \frac{(\nu_n - \delta_n)(\nu_n - \delta_n - 1)\dots(\nu_n - 2\delta_n + 1)}{\nu_n(\nu_n - 1)\dots(\nu_n - \delta_n + 1)}$$

$$= \left(1 - \frac{\delta_n}{\nu_n}\right)\left(1 - \frac{\delta_n}{\nu_n - 1}\right)\dots\left(1 - \frac{\delta_n}{\nu_n - \delta_n + 1}\right) < \left(1 - \frac{\delta_n}{\nu_n}\right)^{\delta_n}.$$

For large n we may replace ν_n and δ_n by 3^n and 2^n respectively; p_n is then less than or equal to

$$\left[1 - \left(\frac{2}{3}\right)^n\right]^{(\frac{3}{2})^n (\frac{4}{3})^n} = e^{-(\frac{4}{3})^n} = \alpha_n$$

approximately; this approaches zeio rapidly as $n \to \infty$. There are τ_n rows, and less than $\tau_n^2 < 16^n$ pairs of rows. The probability of two given rows not being orthogonal is $1 - \alpha_n$; the probability that no two rows are orthogonal is surely greater than $1 - 16^n \alpha_n$, and the probability that ξ^n is not of the desired type is consequently less that $16^n \alpha_n$, which also approaches zero when n is once large enough. Moreover, the sum

$$\sum_n 16^n e^{-(\frac{4}{3})^n}$$

converges, and hence, for N sufficiently large, the probability that there exists a map of more than N regions which cannot be colored in four colors is (from this point of view) very small. The remainder after ten terms in the sum with 16^n replaced by τ_n^2 is less than .002.

16^n could of course be replaced by a much smaller number for n not too large. Thus $\tau_{10} = 1430 = (2.07)^{10}$, $\tau_{20} = 477638700 = (2.716)^{20}$. $\tau_{100} = 1.4\,(10)^{56} = (3.64)^{100}$, $\tau_{1000} = 2\,(10)^{596} = (3.947)^{1000}$.

More accurate information could be obtained with the help of the number:

ρ_n = the maximum number of triangulations which will fit any given coloring of P_n. This is the maximum number of ones in any column of η^n. It seems quite certain that $\rho_n < (3.2)^{n-3}$ $(n \geqq 4)$; very probably $\rho_n < 3^{n-3}$ $(n \geqq 4)$. The average number of triangulations is $(2\frac{2}{3})^n$ approximately for n sufficiently large.

(Eingegangen: 12. IX. 1936.)

On reducibility in the four color problem

Hassler Whitney

November 5, 1971

Quite recently, Y. Shimamoto showed (Report, AMD 618, Brookhaven National Laboratory) that the four color map theorem would be proved if a certain configuration H of H. Heesch were proved "D-reducible." A computer program has been run (considering nearly 200,000 different colorings of the boundary C of H, each one a complicated problem); the computer said "yes." I am told that the computer was asked the wrong question.

We show here that K-reducibility of H implies K-reducibility of E, a pentagon surrounding a vertex. (Possibly the two definitions of reducibility differ.) Hence there seems no purpose in studying the more complicated graph for this purpose. The proof is obtained through a study of Shimamoto's paper.

In the diagram below, C is the heavily marked circuit, H is C with interior, and S is C with exterior. Note that S contains nine pentagons, X_i and X_i^1.

The essence of Shimamoto's paper is the following theorem.

Theorem. In any coloring of $H \cup S$, one of the pentagons of S is colored in only three colors.

We divide the proof into cases.

Case I, $col(a) \neq col(a_1)$ and $col(a) \neq col(a_1')$. Then all the vertices of X_1 are colored differently from a, and thus use only three colors.

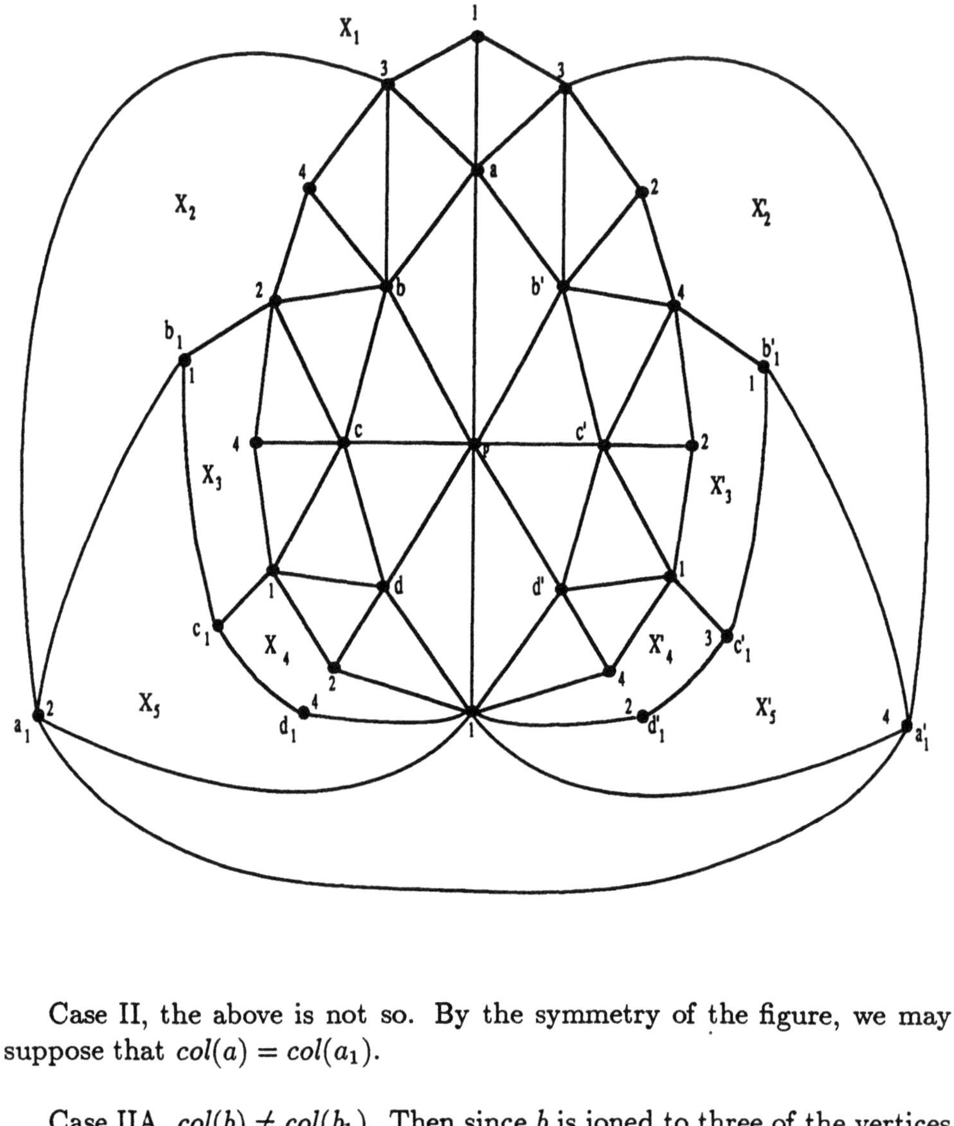

Case II, the above is not so. By the symmetry of the figure, we may suppose that $col(a) = col(a_1)$.

Case IIA, $col(b) \neq col(b_1)$. Then since b is joned to three of the vertices of X_2 and to a, ony three colors are used in X_2.

Case IIB, $col(b) = col(b_1)$. We again have subcases.

Case IIB$_1$, $col(c) \neq col(c_1)$. Then since, $col(b_1) = col(b)$, X_3 uses three colors.

Case IIB$_2$, $col(c) = col(c_1)$. Again there are two cases. If $col(d) \neq col(d_1)$, then X_4 uses three colors. Otherwise, using the vertex p shows that X_5 uses three colors, completing the proof.

We now consider what we shall call K-reducibility, defined through Kempe blocks. It is presumably the same as Heesch's D-reducibility, but I cannot vouch for this.

By a colored graph we shall always mean an assignment of one of four colors to each of the vertices of the graph, such that two vertices that are joined by an edge are of different colors. We usually call the colors 1, 2, 3, 4.

Definition. Let G be a colored graph, and let α and β be two of the colors. A Kempe α, β -block in G is a maximal connected subgraph G' of G which is colored using only the colors α and β. We say that G' is from the subgraph G_1 of G if G' contains a vertex of G_1. A Kempe α, β -chain from a to b is a path from a to b along edges of a Kempe α, β -block. We generally omit "Kempe." Note that a Kempe block may contain only one vertex.

Definition. Let Γ be a coloring of G. Then a Kempe succession of interchanges is a succession of colorings $\Gamma_1 = \Gamma, \Gamma_2, \cdots, \Gamma_n$, such that for each $i > 1, \Gamma_1$ is formed from Γ_{i-1} by interchanging two colors in a Kempe block of the coloring Γ_{i-1}. If the block contains only one vertex x, we may change its color to the other color. The succession is from the subgraph G_1 if each block used is from G_1.

Definition. Let G be as graph in the plane, and let F be one of its closed faces, with boundary C. Then if Γ is a coloring of C, we say that G is K-reducible about C for Γ if, for every graph X in F such that $J = G \cup X$ is as graph, and for every extension Γ' (if there is one) of Γ through X, there is a Kempe succession from C starting with Γ' and ending with a coloring

Γ'_n which is extendable through G. We say G is K-reducible about C if this holds for every coloring of C. If C is not mentioned, it is to be taken as the outer boundary of G.

Theorem. If H is K-reducible, then so is E.

Let X be any graph in the plane surrounding E, and let Γ be any coloring of X. If only three colors are used in the pentagon P, we may insert the vetex x and color it in the remaining color. We now suppose that all four colors are used in P.

Color S as shown in the figure. We may insert X into the position of each X_i and X'_i so that, after a permutation of Γ, the coloring agrees with that in S. This gives a colored graph S^* surrounding C; let Γ_1 be the coloring. Since H is K-reducible, there is a Kempe succession $\Gamma_2, \cdots, \Gamma_n$ from C, such that Γ_n is extendable to a coloring Γ^* of $S \cup H$.

Consider each coloring $\Gamma_k(X_i)$ or $\Gamma_k(X'_i)$, restricting to X_i or X'_i. The change from Γ_{k-1} to Γ_k uses a block B from C. Set $B_i = B \cap X_i$. Now B_i is a union of blocks from the boundary C_i of X_i. (In the figure, X_i is now the graph, not just the boundary.) Hence $\Gamma_k(X_i)$ is obtained from $\Gamma_{k-1}(X_1)$ by a Kempe succession from C_1. By the previous theorem, using Γ^*, some C_j or C'_j is now colored in three colors; say C_i is. We have found a Kempe succession in X_i from C_i, from the given coloring to a coloring extendable to the remaining vertex, completing the proof.

We end by describing briefly Shimamoto's construction, showing how S and its coloring are obtained.

We start a graph G with a 5-fold vertex x, surrounded by P (we naturally suppose that G is a minimal non-colorable graph). Color $G' = G - x$; then P uses four colors. We may put back x and four of the edges, and color x. Make another copy of this colored graph, with the quadrilateral on the outside, and fit it into the first quadrilateral, the new x, now called x', separate from x; add the edge xx'.

The dashed lines show where there is more of the graph.

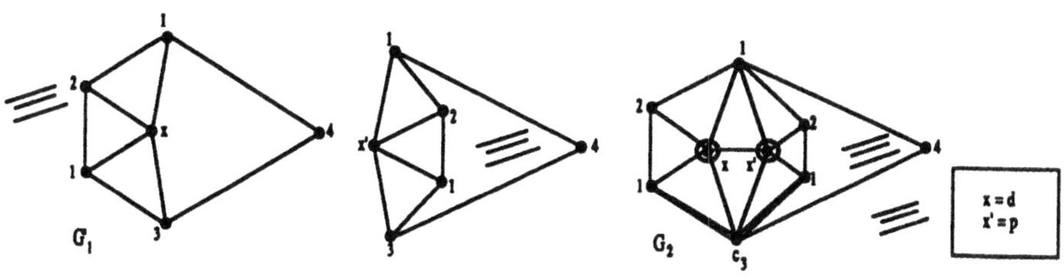

We next "split" G_2 along the pair of doubled edges, as shown. Form G_1'' from G by dropping out an edge, that one from x separating the vertices of G colored the same, and insert, after permuting colors, as shown, forming G_3.

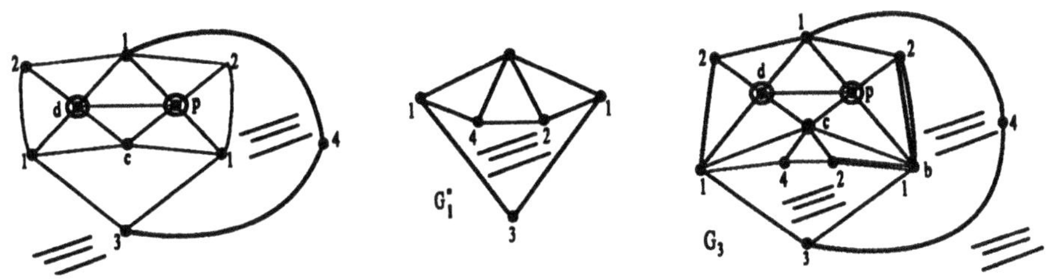

Another split along the doubled edges and insertion of G_1'', again with its coloring permuted, gives G_4. We show certain 5-fold vertices with a circle, and 6-fold vertics with a heavy dot.

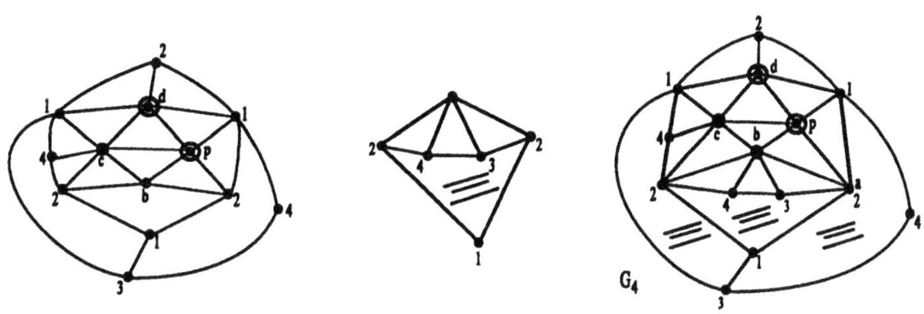

We next show G_4 flipped over and turned around. We now split as shown, along one edge and between two other edges; the top vertex, colored 3,

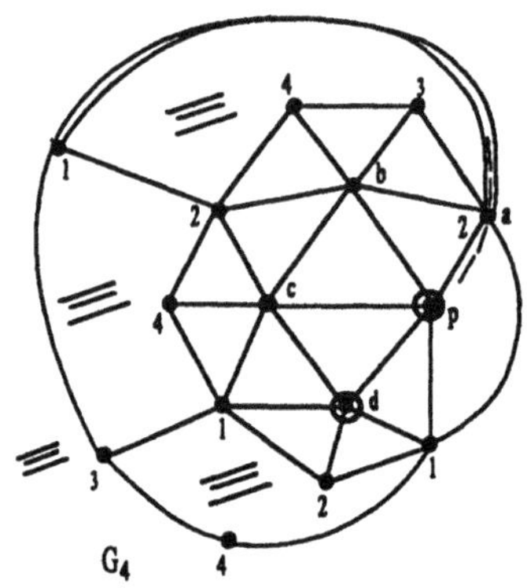

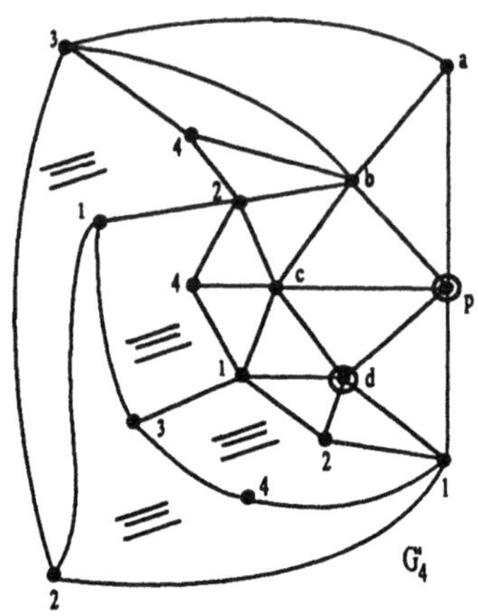

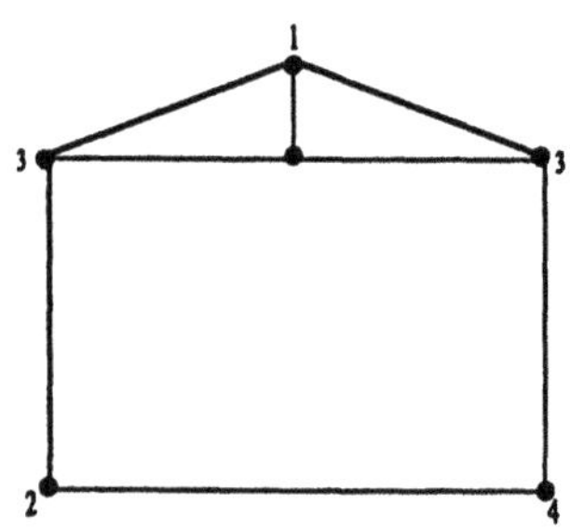

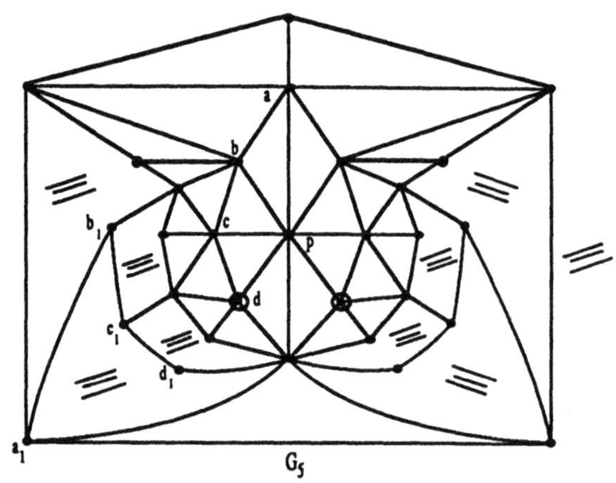

is puled way to the left. We have G'_4.

Permute the colors again in G''_1 as shown just above. Then join G'_4 in the left side of G''_1, at three vertices; the colors agree at two of them. Flip G'_4 over a vertical axis, and permute colors 2 and 4; we may fit it into G''_1 at three vertices, the colors agreeing at two of them, and so that it is also joined to the inserted G'_4 at the lower vertex. We now have the Heesch-Shimamoto graph, colored except inside C.

Note that the proof of the first theorem corresponds to different points in the construction, which were made to result in non-colorability at each step.

KEMPE CHAINS AND THE FOUR COLOUR PROBLEM

Hassler Whitney and W. T. Tutte

1. *Introduction.*

In October 1971 the combinatorial world was swept by the rumour that the notorious Four Colour Problem had at last been solved, - that with the help of a computer it had been demonstrated that any map in the plane can be coloured with at most four-colours so that no two countries with a common boundary line are given the same colour.

The first ostensible proof of the conjecture to be published was that of A. B. Kempe [4]. This appeared in 1879 and was accepted for a decade. A flaw in the argument was pointed out in 1890 by P. J. Heawood [2], whose papers on the problem span the next sixty years. After Heawood's first paper mathematicians began to suspect that the Four Colour Problem was of surpassing difficulty; perhaps it was to be ranked with Fermat's Last Theorem and the Riemann Hypothesis. G. D. Birkhoff once told one of the authors that every great mathematician had at some time attempted the Four Colour Conjecture, and had for a while believed himself successful. "Proofs" of the Conjecture are still written every now and then, and occasionally one gets published. Anyone now having a proof that he wishes to be taken seriously would be well advised to write it out clearly and in full logical detail, so that any mathematician willing to spend enough time on it will be able to check it.

This work was partly supported by a grant from the National Research Council of Canada.

UTILITAS MATHEMATICA, Vol. 2 (1972), pp. 241-281.

For the history and present status of the Four Colour Problem reference may be made to the book by O. Ore [5] and the recent article by T. L. Saaty [7].

The rumour mentioned above arose from the work of Y. Shimamoto, who claimed a proof based on the work of H. Heesch [3]. Heesch has for years been studying the "reducibility" of maps. He shows that some configurations of countries have a property that he calls "D-reducibility", and he has a method whereby a given configuration can be tested for this property. It requires much computer time. Shimamoto, on the assumption that the Four Colour Conjecture was false, showed that there must be a non-colourable map M containing a configuration H that had already passed the computer test for D-reducibility. He then arrived at a contradiction by showing that the D-reducibility of H implied the 4-colourability of M . This argument seemed to prove the Conjecture. The burden of proof was not now on a few pages of close reasoning, but on a computer!

This method of proof was greeted by the present authors (independently) first with some misgivings and then with real scepticism. It seemed to both of us that if the proof was valid it implied the existence of a much simpler proof (to be obtained by confining one's attention to one small part of M), and that this simpler proof would be so simple that its existence was incredible. The present paper is essentially the result of our attempts to give a proper mathematical form to our objection.

We found no essential flaw in Shimamoto's reasoning. (It was later set out "clearly and in full logical detail" in an article circulated by W. R. G. Haken). We therefore decided that the computer result must be wrong. (We learned later that a reprogramming of the computer had indeed given the result that H was not D-reducible). However it turned out that the basic result of Shimamoto was simply that a certain plane graph G_{12} (see Figure 12) cannot be vertex-coloured in four colours so that the boundary of each pentagonal face uses all four colours. (The reader can check this at once; see the proof near the end of Section 9).

Is there a moral to be drawn? Perhaps it is that if you are really interested in the results of a study you should give it an analysis in depth and try to understand fully its implications. In fact the strength of present-day mathematics owes much to this principle. It seems that in the present case deductions were made from the accepted D-reducibility of H without an adequate understanding of what D-reducibility is. With this understanding the deductions become incredible and are seen to constitute a proof by reductio ad absurdum that H is not in fact D-reducible.

In this paper we give a general description of this type of approach to the Four Colour Problem. We define Kempe chains, and we point out some things that can be done with Kempe chains and some things that cannot. The exposition is intended to be generally understandable, not requiring any special mathematical preparation.

2. *Preliminaries.*

First we note a dual formulation of the problem, easier to visualize and use. Consider any map M on the plane or the sphere. In each country we mark the capital; this will be a vertex of a graph G . If two countries have a common border we join their capitals by a railroad across it. This railroad is an edge of G . (The terms "vertex" and "edge" are taken from the theory of polyhedra). Colouring the map M is equivalent to colouring the vertices of G so that no two vertices joined by an edge are of the same colour. Any plane graph comes from a map in this way, provided that no edge is a "loop", i.e., joins a vertex to itself. From now on all the plane graphs that we consider are to be assumed loopless. The Four Colour Conjecture can now be formulated as follows: any plane graph can be 4-coloured. This is the formulation used in Ore's book.

The edges of a connected plane graph cut the plane into regions that we call the *faces* of G ; one of these is the *outside* face. If we project the plane stereographically onto the sphere the outside face is made to surround the North Pole. By choosing another diameter as the axis of the sphere we can arrange that any desired face contains the North Pole, so on projecting back into the plane we can arrange that any desired face becomes the outside face.

A *face-boundary* of G is the boundary of a face of G . A *circuit* is a graph defined by the set of edges and vertices of a simple closed curve. We note that a face-boundary is not necessarily a circuit; it may include an edge or vertex whose removal disconnects G . (An edge of this kind is called an *isthmus* of G). If Q is a face-boundary of G we say for short that G is *face-bounded* by Q . If Q is a circuit the remainder of G lies on one side of Q , the inside or the outside, unless G consists solely of Q . A circuit Q made up of edges of G is *vertex-separating* in G if G has a vertex inside Q and a vertex outside Q . Similarly it is *edge-separating* if G has an edge inside Q and an edge outside Q .

A *triangulation* of the plane is a plane graph G whose faces are all triangles, that is whose face-boundaries consist of three edges each. The *valency* v(A) of a vertex A is the number of edges having A as an end. Most commonly these edges go to distinct vertices, but it is not necessary to impose this as part of the definition. A k-*wheel* is a graph W formed from a circuit of k edges by adjoining a new vertex A and then joining A to each vertex of the circuit by a single new edge. A is the *hub* of the wheel, the circuit is the *rim* and the new edges are the *spokes*. In a triangulation in which at most one edge joins any two vertices (and which has at least 4 vertices) each vertex is the hub of a wheel.

A graph is k-*chromatic* if it can be coloured in k colours but not in fewer. A *full* k-colouring of a graph is one that uses exactly k colours. A 3-colouring is usually accepted as a special kind of 4-colouring, but it is not a full 4-colouring.

3. *The Euler formula.*

In the rest of this paper the symbol G is used to denote a non-null connected plane graph in which there is no loop. Let such a graph G have N_v vertices, N_e edges and N_f faces. Then the Euler formula is

$$N_v - N_e + N_f = 2 . \tag{1}$$

The formula can be proved by induction on $N = N_v + N_e + N_f$. The smallest possible value of N is 2. This occurs only when G has a single vertex, no edge and a single face (which is an "agon"). In this simple case the Euler formula holds. In the general case suppose first that G has a circuit. We drop out one edge of the circuit. This decreases N_e and N_f by one each, and leaves N_v unchanged. By the inductive hypothesis (1) holds for the resulting graph, and therefore it holds for G. If G has no circuit we can find a vertex joined to just one another. Dropping out this vertex and its edge reduces N_v and N_e each by one and leaves N_f unchanged. (It is 1). Again the formula follows for G.

We show next that for triangulations of the plane we have

$$N_f = 2N_v - 4 , \quad N_e = 3N_v - 6 . \tag{2}$$

In each face draw a new face-boundary just inside the original one. We have drawn $3N_f$ new "edges" and two of these are beside each former edge. hence $3N_f = 2N_e$. Combining this with (1) we obtain (2).

We can carry the argument further. Each vertex X is on $v(X)$ edges and each edge is on two vertices. Hence $\Sigma v(X) = 2N_e$. Let us write $v'(X) = v(X) - 6$. Then by (2) we have

$$(3) \qquad \Sigma v'(X) = \Sigma v(X) - 6N_v = -12 .$$

Hence each triangulation has vertices of valency less than 6. If each valency is at least 5 then there are at least 12 vertices of valency 5, and more according as vertices of valency greater than 6 are present.

4. Kempe chains.

Consider a fixed set $\{\alpha, \beta, \gamma, \delta\}$ of four colours. There are three *colour partitions* of this set into two pairs of colours, namely $(\alpha\beta, \gamma\delta)$, $(\alpha\gamma, \beta\delta)$ and $(\alpha\delta, \beta\gamma)$.

Let the graph be 4-coloured, and let Γ denote its 4-colouring. For each pair of colours, say $\{\alpha, \beta\}$ let $G_{\alpha\beta}$ denote the subgraph of G consisting of the vertices coloured α or β and the edges joining them. Now the "components" of a graph are its maximal connected subgraphs. We refer to the components of $G_{\alpha\beta}$ as the *Kempe chains belonging to* the unordered pair $\{\alpha, \beta\}$ in Γ . We call them also the *$\alpha\beta$-chains* of Γ . It is possible for a Kempe chain to consist of a single vertex.

A *Kempe interchange* in Γ , with respect to the colour-partition $(\alpha\beta, \gamma\delta)$ is an interchange of the colours α and β in one of the $\alpha\beta$-chains, or of the colours γ and δ in one of the $\gamma\delta$-chains. If there are m Kempe chains belonging to $\{\alpha, \beta\}$ or $\{\gamma, \delta\}$ then we obtain 2^m colourings of G by performing or not performing the interchange in each of the m chains. Each interchange leaves the $\alpha\beta$-chains and the $\gamma\delta$-chains unaltered.

We note however that an interchange in an αβ-chain of Γ must alter the system of αγ-chains for example. Some of them are destroyed or new ones appear, or both. We proceed to state an obvious but important theorem about Kempe chains.

THEOREM 4.1. *Let a plane graph* U *be face-bounded by a circuit* Q *. Let the vertices* A , B , C , D *lie in that order in* Q *. Let* U *be 4-coloured. Then if there is an* αβ-*chain of the 4-colouring containing* A *and* C *but not* B *or* D *there can be no* αβ-*chain or* γδ-*chain of the 4-colouring containing both* B *and* D *.* (See Figure 1).

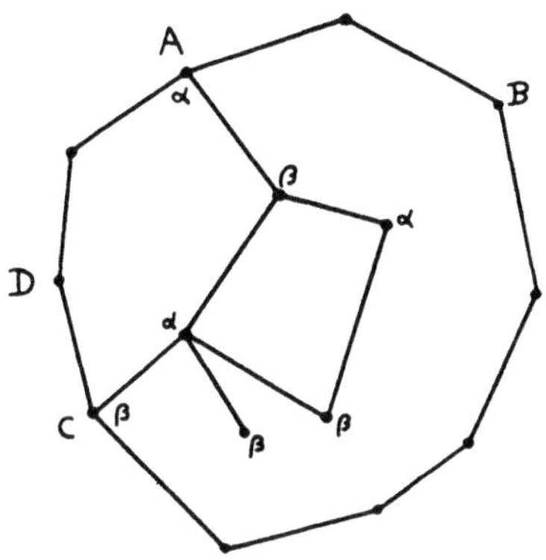

Figure 1

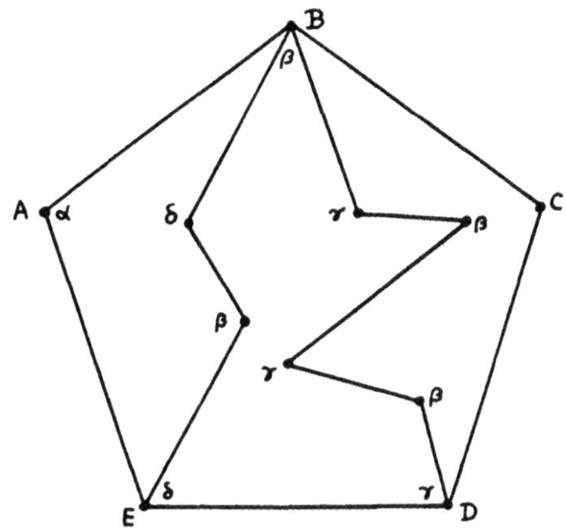

Figure 2

Theorem 4.1 has the following elementary consequence.

THEOREM 4.2. *Let a plane graph* U *be face-bounded by a quadrilateral* Q = ABCD *, and let* Γ *be a 4-colouring of* U *using all four colours in* Q *. Then we can find a Kempe interchange in* Γ *that transforms* Γ *into a 4-colouring of* U *using exactly three colours in* Q *.*

Proof. Let A , B , C and D have colours α, β, γ and δ respectively. If the αγ-chain containing A does not contain C then a Kempe interchange in it has the required effect. In the remaining case the $\beta\delta$-chain containing B does not contain D, by 4.1, and we can use the Kempe interchange in it.

5. *Elementary reductions.*

If the Four Colour Conjecture is false there must be a least integer N such that a 5-chromatic plane graph of N vertices exists. Such a graph is called *minimal*. T. L. Saaty calls N the *Birkhoff Number*, and completes his definition by saying that $N = \infty$ if the Conjecture is true. If the Four Colour Conjecture is assumed for all finite plane graphs it follows at once for infinite ones. This is an easy consequence of the Cantor "diagonal process". In this Section we assume that N is finite, and we study the properties of minimal graphs.

Obviously $N \geq 4$. In 1938 C.E. Winn showed that $N \geq 36$ [9]. In a paper published in 1970 O. Ore and J. Stemple claim to have shown that $N \geq 40$ [6]. Their numerical calculations are too lengthy for publication in a Journal, but are available in the library of the Mathematics Department at Yale University.

Besides the minimal graph, investigators of colouring problems often use the concept of a *critical graph*. For our purposes we can define a critical graph as a 5-chromatic plane graph whose proper subgraphs are all 4-colourable. A minimal graph, we shall show, is critical, but we are not entitled to assert that any critical graph must be minimal.

Most studies of the 4-colour problem have been studies of what minimal graphs must be like, with the ultimate object of showing that no such graphs exist. We give some sample properties.

THEOREM 5.1. *In a minimal graph* G *each face-boundary is a circuit.*

Proof. Suppose the boundary P of some face F is not a circuit. Let us go around it close to the boundary; we pass near some vertex X at least twice. We can go from X through F and back to X in such a way as to traverse a simple closed curve having vertices of G both inside and outside. Thus G is the union of two plane graphs G_1 and G_2 having only the vertex X in common. Each of these can be 4-coloured since each has fewer than N vertices. Having 4-coloured G_1 and G_2 we can permute the colours in G_1 so as to make the two 4-colourings agree at X . We can then combine the two 4-colourings into a 4-colouring of G . This contradiction establishes the theorem.

THEOREM 5.2. *A minimal graph* G *is a triangulation.*

Proof. Assume the contrary. Then G has a face, which may be taken as an inside face with a boundary circuit Q (see 5.1) containing four vertices A , B , C and D , in that order. Now we cannot have edges AC and BD outside Q . We may therefore assume without loss of generality that no edge joins A and C . We may therefore pull A and C together inside Q and let them become a single vertex, so obtaining a plane graph G' . This has fewer than N vertices and is thus 4-colourable. But a 4-colouring of G' obviously determines one of G , and we have a contradiction.

We can argue from 5.2 that it is sufficient to prove the Four Colour Conjecture for triangulations. If it is true for them it must be true for all plane graphs. Accordingly most papers on the subject are concerned with triangulations only (or their duals, if face-colourings are being considered).

THEOREM 5.3. *A minimal graph has no vertex of valency less than* 5.

Proof. Suppose a minimal graph G to have a vertex V of valency < 5 . Let G' be the plane graph obtained from it by deleting V and its incident edges. Since G' has fewer than N vertices it can be 4-coloured. If V is joined to at most 3 other vertices in G we can obviously extend any 4-colouring of G' to a 4-colouring of G . In the remaining case G' is bounded by a quadrilateral ABCD , by 5.2. By 4.2 we can find a 4-colouring of G' that uses only three colours in ABCD , and this can be extend as a 4-colouring of G , - a contradiction.

THEOREM 5.4. *A minimal graph* G *has no vertex-separating circuit of fewer than* 5 *vertices.*

Proof. Suppose G has such a vertex-separating circuit Q . Let Q bound the subgraphs G_1 and G_2 , whose union is G and whose intersection is Q . We can find 4-colourings Γ_1 and Γ_2 of G_1 and G_2 respectively, since each of these graphs has fewer than N vertices. If Q has at most three vertices we can permute the colours in Γ_1 so as to make Γ_1 and Γ_2 agree in Q . We can then combine Γ_1 and Γ_2 to make a 4-colouring of G .

From now on we may suppose Q to be a quadrilateral ABCD . If all four colours are used in Q in both Γ_1 and Γ_2 we can permute and combine, much as before, to obtain a 4-colouring of G . We may therefore assume, without loss of generality, that there is no 4-colouring of G_1 for which all four colours appear in Q .

Let us modify G_2 by joining A and C across the face Q . The resulting graph has fewer than N vertices and is thus 4-colourable. We deduce that G_2 has a 4-colouring with at least 3 colours in Q . Hence we can choose Γ_2 to have exactly 3 colours in Q , by 4.2. Without loss of generality we can suppose B and D to have the same colour α in Γ_2 , while A and C have distinct colours.

Considering the effect of joining A and C across Q in G_1 we find that we can choose Γ_1 so that A and C have distinct colours. Then B and D must have the same colour, by the restriction we have been able to impose on G_1 . We can now permute colours in Γ_1 so as to make it agree with Γ_2 in Q , and then combine Γ_1 and Γ_2 into a 4-colouring of G .

In every case we have found a contradiction.

By a similar but more complex argument G. D. Birkhoff showed that a minimal graph has no circuit of 5 edges separating at least two vertices from at least two others [1].

There is a similar theory of 5-colourings in which one proves, in the manner of 4.2 and 5.3, that a 5-minimal graph has no vertex of valency less than 6. It follows that 5-minimal graphs do not exist, by § 3 ; the "Five Colour Theorem" is true.

THEOREM 5.5. *A minimal graph* G *is critical.*

Proof. Let A be any edge of G . Form G' from G by deleting A . Suppose first that some edge B of G' still joins the ends of A . Then A and B make up a circuit in G , and this circuit is vertex-separating, by 5.2. But this is contrary to 5.4.

In the remaining case the edge A of G can be contracted to a single vertex so as to transform G into a new connected loopless plane graph G" with one vertex fewer. But G" has a 4-colouring, by the minimality of G , and this evidently determines a 4-colouring of G' . Hence each subgraph of G' is 4-colourable. Since A was chosen arbitrarily it follows that every proper subgraph of G is 4-colourable, that is G is critical.

6. *K-reducibility.*

Consider an edge-separating circuit Q (as defined in Section 2) in a plane graph G . It decomposes G into two plane graphs U and V , each having Q as a face-boundary. Each of U and V has one new face bounded by Q , and its other faces are faces of G . The graphs U and V have G as their union and Q as their intersection. Of course each of U and V has an edge not in Q .

Many studies of the Four Colour Problem are concerned with such figures, which we call Q-*decompositions* of G . Typically it is shown that if V has a specified structure then G cannot be minimal, whatever the structure of U . We express this property of the given V , with its boundary Q , by saying "(V, Q) is *reducible*". As far as we know at present it is not inconsistent with V being 5-chromatic, but of course this possibility has not been realized in practice.

If a plane graph V is bounded by a circuit Q of fewer than 4 vertices, then (V, Q) is reducible. For no corresponding graph G is minimal, by 5.4.

Let U be a plane graph bounded by a circuit Q . Let Γ be a 4-colouring of Q . It may happen that there is a 4-colouring Γ' of U that reduces to Γ on Q . If so we say that Γ is U-*extensible*, and that Γ' is a U-*extension* of Γ .

Now let Γ be a 4-colouring of Q , and let S be a set of 4-colourings of Q . It may happen that Γ is U-extensible and that every U-extension of Γ can be transformed, by a succession of Kempe interchanges in U , into a U-extension of some member of S . If so we say that Γ is U-*immersible* in S .

Let Γ be a 4-colouring of a circuit Q , and let S be a set of 4-colourings of Q . It may happen that Γ is U-immersible in S for every plane graph U bounded by Q and such that Γ is U-extensible. If so we say simply that Γ is *immersible* in S .

Let V be a plane graph face-bounded by a circuit Q and having at least one edge not in Q . Let S be the set of all V-extensible 4-colourings of Q . Evidently each member of S is immersible in S . We say that V is K-*reducible* with respect to Q , or, briefly, that (V, Q) is K-reducible, if every 4-colouring of Q is immersible in S.

In applications of the last definition we usually think Q as bounding an inner face of V . But of course the distinction between "inner" and "outer" is only a matter of convenience.

THEOREM 6.1. *If* (V, Q) *is K-reducible, then* (V, Q) *is reducible.*

Proof. Let V be represented as part of a Q-decomposition of a plane graph G . Then Q is edge-separating in G . Assume that G is minimal. Then, by 5.5, U has a 4-colouring Γ' , a U-extension of a 4-colouring Γ of Q . Since (V, Q) is K-reducible Γ is transformable into some 4-colouring Γ_1 of Q that is V-extensible, by a succession of Kempe interchanges in U starting with Γ' . Now Γ_1 has both a U-extension and a V-extension. Combining these we obtain a 4-colouring of G , which is a contradiction. Thus G cannot be minimal, for any U , and so (V, Q) is reducible.

If the 5-wheel could be proved reducible with respect to its rim, then the Four Colour Conjecture would be verified. For since no vertex of a minimal graph G can have valency less than 5, by 5.3, and since no vertex-separating circuit of G has fewer than 5 edges, by 5.4, it follows from §3 that G has a vertex of valency 5 and that this is the hub of a 5-wheel contained in G .

Let us consider an attack on the K-reducibility of a 5-wheel, with rim Q = ABCDE . In Figure 2 we show part of a coloured plane graph U inside Q . Actually we try to build up a graph U so that the 4-colouring of Q shown in the figure is U-extensible, but is not U-immersible in the set S of all 4-colourings of Q using only 3 distinct colours. If we can carry out the proposed construction then the 5-wheel is not K-reducible. If we can prove the construction impossible K-reducibility is established and the Four Colour Conjecture is verified. Actually we shall achieve neither of these results. But let us assume that we have found a plane graph U of the required kind, and that we have a U-extension Γ of the 4-colouring of Q shown in the diagram. We note that all full 4-colourings of Q are equivalent to within rotations and reflections of the pentagon and permutations of the four colours.

First, if the $\beta\delta$-chain containing B does not contain E then an interchange of β and δ in this chain removes β from Q . Hence we must construct U and Γ so that this chain also contains E . Similarly we must arrange that the $\beta\gamma$-chain containing B contains also D . (See Figure 2). Because of the $\beta\delta$-chain, the $\alpha\gamma$-chain containing A does not contain D ; interchange α and γ in it. Because of the $\beta\gamma$-chain the $\alpha\delta$-chain containing C does not contain E ; interchange α and δ in it. At first sight it seems that we thus remove α from Q ; the construction has been proved impossible and the Four Colour Conjecture follows.

This in fact was Kempe's proof. What is wrong? Simply that we may not be able to make both the interchanges called for This point is clarified by Figure 3. In this figure the $\alpha\gamma$-chain

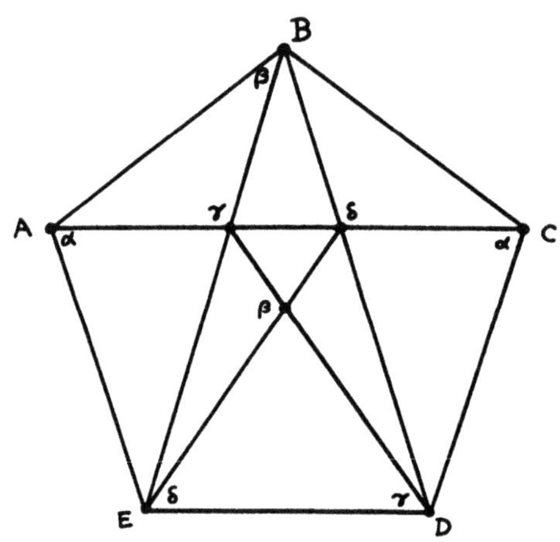

Figure 3

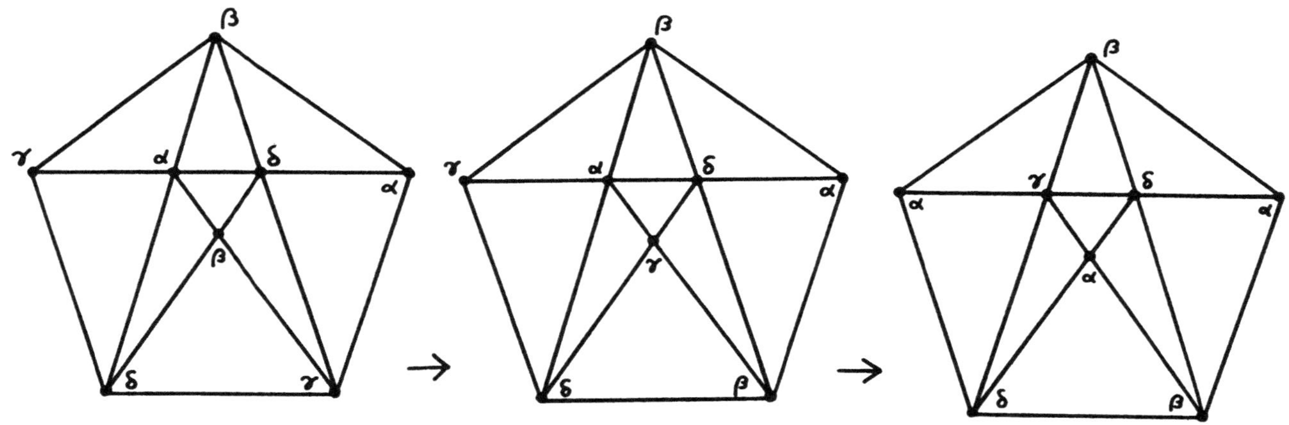

Figure 4

from A contains a vertex of the $\beta\gamma$-chain from B to D . The interchange of α and γ breaks the latter chain and sets up a new $\alpha\delta$-chain from C to E . This makes it impossible to remove α from Q by an interchange of α and δ .

In the graph shown we can remove a colour from Q by making first the suggested interchange of α and γ , and then operating on a $\beta\gamma$-chain and a $\gamma\alpha$-chain as shown in Figure 4. We ought now to complicate U so as to make this sequence of operations impossible. If the reader tries to do this we feel safe in saying that other ways of eliminating a colour from Q will appear, calling for further complication. If there is an end to this process it is not yet in sight.

7. *D-reducibility.*

The difficulty in studying K-reducibility comes from the fact that if we make a Kempe interchange, Kempe chains from other colour partitions are altered. This leads to theoretical complications that mathematicians have so far been unable to resolve.

One of the present authors sought to make a modest first step toward resolving them through a theory of "parity", whereby the 4-colourings of a triangulation can be classified as "even" or "odd". He was able to show that the parity of a 4-colouring is invariant under Kempe interchanges [8]. But this result is not strong enough to be of much help in the theory of reducibility.

We now discuss a special kind of reducibility, called *D-reducibility*, introduced by H. Heesch. Its theory avoids any consideration of the effect of a Kempe interchange on the Kempe chains of other colour partitions. Heesch gives an algorithm whereby a given configuration can be tested for D-reducibility. As a hand method it is long and tedious, but it can be programmed for a computer. Heesch advocates the construction of a catalogue of D-reducible, and otherwise reducible, configurations, hoping that ultimately it can be shown that every triangulation contains a member of the list. If so it will be proved that no triangulation is minimal, and so that the Four Colour Conjecture is true.

We define a *near-triangulation* as a plane graph N in which at most one face is non-triangular. When we speak of a near-triangulation as being bounded or face-bounded by a circuit Q it is to be understood that if there is a non-triangular face it is the one bounded by Q.

Let Γ be a 4-colouring and S a set of 4-colourings of a circuit Q. It may happen that there is a colour partition Π with the following property: if U is any near-triangulation bounded by Q and if Γ is U-extensible, then there is a U-extension Γ' of Γ that can be transformed into a U-extension of some member of S by a succession of Kempe interchanges with respect to Π. If so we say that Γ is *simply immersible* in S.

We denote the set of all 4-colourings of Q that are simply immersible in S by $f(S)$. Evidently $S \subseteq f(S)$. Then $f^2(S) = f(f(S))$ is the set of all 4-colourings of Q simply immersible in $f(S)$, and so on. A 4-colouring Γ is said to be *crudely immersible* in S if it belongs to $f^k(S)$ for some non-negative integer k. ($f^0(S) = S$). Its crude immersion in S is effected by a succession of simple immersions in sets $f^j(S)$ with decreasing j, and it is not necessary that these simple immersions shall all be associated with the same colour partition Π. We say that the set S is *dominant* if every 4-colouring of Q is crudely immersible in S.

Can one classify the dominant sets for any given circuit Q? That would seem to be an interesting question. Heesch does not solve it, but his algorithm will determine whether a given set of 4-colourings is dominant.

Let V be a near-triangulation bounded by a circuit Q and having at least one edge not in Q. We say that V is

D-reducible with respect to Q , or, briefly, that (V, Q) is D-reducible, if the set of all V-extensible 4-colourings of Q is dominant.

THEOREM 7.1. *If* (V, Q) *is D-reducible, then* (V, Q) *is reducible.*

Proof. Adjoin to V any other near-triangulation U bounded by Q so as to form a triangulation G with Q as an edge-separating circuit. Assume that G is minimal. Then U has a 4-colouring Γ' , a U-extension of a 4-colouring Γ of Q , by 5.5. Let S be the set of all V-extensible 4-colourings of Q . Then $\Gamma \in f^k(S)$ for some non-negative integer k , since (V, Q) is D-reducible. Choose Γ' , Γ and k so that k has the least possible value.

If $k > 0$ then, by the definition of $f^k(S)$, some U-extension of Γ can be transformed by a succession of Kempe interchanges, all with respect to the same colour partition, into a U-extension of some member Γ_1 of $f^{k-1}(S)$. This being contrary to the choice of k we deduce that in fact $k = 0$. This means that Γ has both a U-extension and a V-extension. Combining these extensions we obtain a 4-colouring of G .

From this contradiction we deduce that G is in fact not minimal. Thus (V, Q) is reducible.

As an example let V be a 4-wheel and Q its rim. Then S , the set of V-extensible 4-colourings of Q , consists of the non-full 4-colourings of Q . By 4.2 each of the remaining 4-colourings of Q is in f(S) . Thus S is dominant and (V, Q) is D-reducible. On the other hand we have the following theorem.

THEOREM 7.2. *Let* Q = ABCDE *be a pentagon, and let* S *be the set of all 4-colourings of* Q *using only 3 distinct colours. Then* S *is not dominant.*

Proof. Let Γ be any full 4-colouring of Q . Adjusting the notation we can represent it by Figure 5.

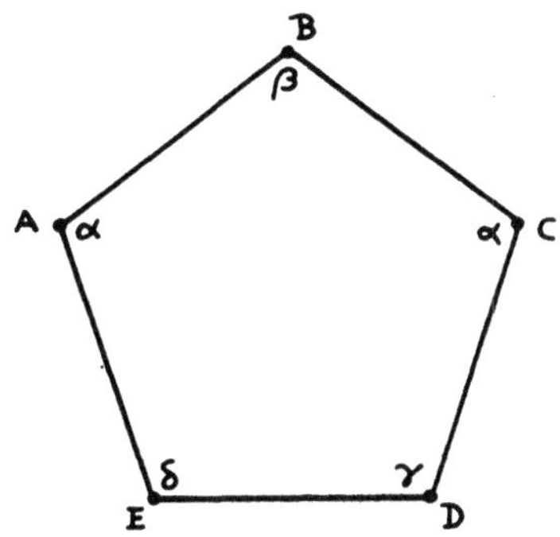

Figure 5

Consider the three colour partitions in turn. In the case of (αβ, γδ) whatever U we take, bounded by Q, the vertices A, B and C are in one αβ-chain and the vertices D and E are in one δγ-chain. Interchanges in this partition yield only full 4-colourings of Q. In the case of (αγ, βδ) let us form U by adding an edge BD and an edge BE. Any Kempe interchange leaves both colours β and δ on BE, and both colours α and γ on CD. Again, only full 4-colourings of Q can be derived. By the symmetry of the figure, the same proof applies to the case (αδ, βγ). We conclude that Γ is not in f(S). In fact f(S) must be identical with S, and so S is not dominant.

COROLLARY. *The 5-wheel is not D-reducible.*

THEOREM 7.3. *Let* U *be a near-triangulation bounded by a pentagon* Q, *and let* U *be 4-colourable. Suppose that no 4-colouring of* Q *using only three distinct colours is U-extensible. Then every full 4-colouring of* Q *is U-extensible.*

We prove this by "crude chaining", that is we do not need to consider the effect of our Kempe interchanges on the Kempe chains of other colour partitions.

Proof. We may assume that the full 4-colouring of Q shown in Figure 5 is U-extensible. There is a $\beta\delta$-chain in U from B to E , since otherwise an interchange would eliminate β from Q . An interchange on one $\alpha\gamma$-chain interchanges the colours of C and D , leaving A coloured α . We have thus moved the pair of vertices distinguished by a common colour two steps counter-clockwise round the pentagon. Repetition of this process gives five distinct full 4-colourings of Q , all U-extensible, from which all the full 4-colourings of Q can be derived by permutations of the four colours. The theorem follows.

8. *Chromodendra.*

The definitions of K-reducibility and D-reducibility suffer from one disadvantage. They require the consideration of an infinity of possible near-triangulations U bounded by Q . In the case of K-reducibility we may hope to overcome this disadvantage by proving general theorems about the interrelations of Kempe chains of colourings of U , but so far no suitable general theorems have been established. In the case of D-reducibility we shall show, following Heesch, that the possible near-triangulations U can be classified under a finite number of cases, and that each case can be dealt with by a finite argument. The question of whether a given V , bounded by Q , is D-reducible can thus be reduced to a finite problem.

Let Q be a circuit of n edges, and let Γ be a 4-colouring of Q . Of course Γ has its Kempe chains, these being subgraphs of Q . If α and β are any two of the four colours we define an $\alpha\beta$-*cluster* of Γ as a non-null set of Kempe chains of Γ belonging to the colour-pair $\{\alpha, \beta\}$. A *clustering* of Γ , with respect to the colour-partition $(\alpha\beta, \gamma\delta)$, is a family of $\alpha\beta$-clusters and $\gamma\delta$-clusters of Γ such that each Kempe chain of Γ belonging to $\{\alpha, \beta\}$ or $\{\gamma, \delta\}$ is contained in exactly one of them. A vertex of G is said to be *included* in a cluster C if it is a vertex of one of the Kempe chains making up C.

A clustering Z of Γ, with respect to $(\alpha\beta, \gamma\delta)$, is said to be *admissible* if it has the following properties.

(i) Let C_1 and C_2 be any two distinct members of Z. Let a_1 and b_1 be any two vertices included in C_1, and let a_2 and b_2 be any two vertices included in C_2. Then a_1 and b_1 do not separate a_2 from b_2 in Q.

(ii) Let C_1 be any cluster in Z and let L be any arc in Q such that no vertex of L is included in C_1 but each end of L is a vertex adjacent to a vertex that is included in C_1. Then there is a cluster C_2 in Z such that each end of L is included in C_2.

Two clusters C_1 and C_2 in a clustering Z are said to be *adjacent* in Z if there is an edge A of Q for which one end is included in C_1 and one in C_2. We then say that A is a *joining edge* of the two clusters. Evidently adjacent clusters are associated with complementary colour-pairs.

THEOREM 8.1. *Let* C_1 *and* C_2 *be adjacent clusters in an admissible clustering* Z. *Then* C_1 *and* C_2 *have exactly two joining edges* A *and* B. *Moreover the deletion of* A *and* B *decomposes* Q *into two disjoint connected graphs* L_1 *and* L_2 *such that* L_1 *contains all the vertices of* Q *included in* C_1, *and* L_2 *all those included in* C_2.

Proof. C_1 and C_2 have one joining edge A. Let its ends included in C_1 and C_2 be a_1 and a_2 respectively. Follow along Q from a_2, away from a_1, until we reach a last vertex b_2 not in C_1; the next edge B has its other end b_1 in C_1. This gives L_1 and L_2. (L_1 is an arc, unless $a_1 = b_1$). By (ii), b_2 is in C_2. By construction L_2 has no vertices in C_1, and by (i) L_1 has no vertices in C_2. Thus (8.1) follows.

It is convenient to represent an admissible clustering Z of Γ , with respect to (αβ, γδ) by a graph χ called its *chromodendron*. The vertices of χ represent the member-clusters of Z , and two vertices are joined by an edge if and only if the two member-clusters corresponding are adjacent. Two vertices are not to be joined by more than one edge.

If two vertices are adjacent in Q they are included in the same cluster, or in adjacent clusters, of Z . From this observation we deduce that χ is connected. From 8.1 we deduce that each edge of χ is an isthmus. Thus,

THEOREM 8.2. *Every chromodendron is a tree.*

The chromodendra of the admissible clusterings of Γ , with respect to (αβ, γδ) will be called the chromodendra of Γ , with respect to (αβ, γδ) . If n is reasonably small there is no difficulty in making a list of all such structures for a given Γ and a given colour-partition.

Let Γ and $Γ_1$ be 4-colourings of the circuit Q and let χ be a chromodendron of Γ with respect to (αβ, γδ) . We say that Γ is *simply* χ-*transformable* into $Γ_1$ if it is transformed into $Γ_1$ by interchanging the two colours in each of the Kempe chains of Γ belonging to a single cluster, this cluster being represented by a vertex of χ . All the Kempe chains of this cluster are affected, but no other αβ-chains or γδ-chains. We note that after the operation χ remains as a chromodendron of $Γ_1$, with respect to (αβ, γδ) , and we can consider the application of a second simple χ-transformation to $Γ_1$. We say that Γ is χ-*transformable* into a 4-colouring Γ' of Q if Γ can be changed into Γ' by a succession of simple χ-transformations, all of course referring to the same chromodendron χ . We also express this by saying that χ *admits* a transformation of Γ into Γ' .

We proceed to relate the theory of chromodendra to that of D-reducibility.

Let U be a near-triangulation bounded by a circuit Q . Let Γ be a 4-colouring of Q and let Γ' be a U-extension of Γ . Let K be a Kempe chain of Γ' belonging to the colour-pair $\{\alpha, \beta\}$. Suppose it to have at least one vertex in common with Q . Then the intersection of K with Q is a union of one or more $\alpha\beta$-chains of Γ . The $\alpha\beta$-chains of Γ contained in $K \cap Q$ thus constitute an $\alpha\beta$-cluster of Γ . We call this the *residue* of K in Γ . Kempe chains of Γ' having no vertex in common with Q are considered to have no residues in Γ . Figure 6 shows Q bounding U , with the colourings Γ and Γ' . The full lines represent the edges of the $\alpha\beta$-chains and $\gamma\delta$-chains of Γ' . There are two $\alpha\beta$-chains. One of them has a residue consisting of three

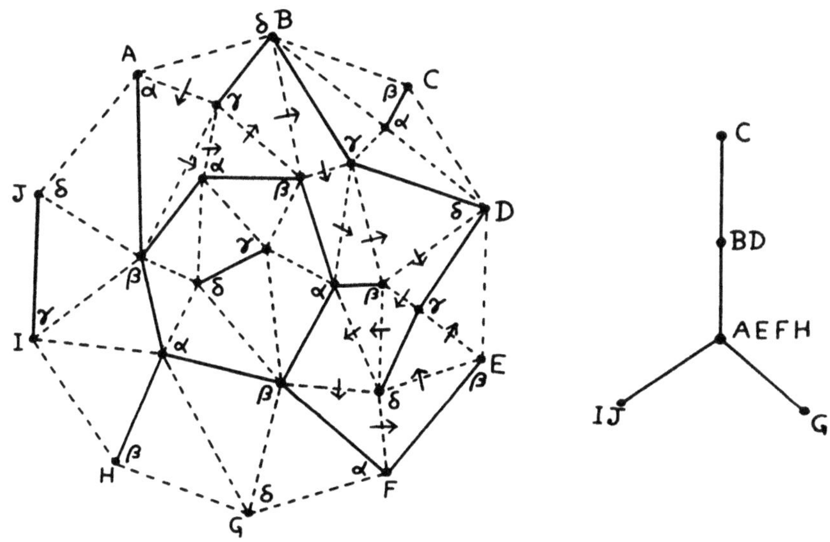

Figure 6

$\alpha\beta$-chains of Γ . These have the vertex-sets $\{A\}$, $\{H\}$ and $\{E, F\}$. The second $\alpha\beta$-chain of Γ' has a residue consisting of a single Kempe chain of Γ , this having the vertex-set $\{C\}$. One of the $\gamma\delta$-chains of Γ' is separated from Q by an $\alpha\beta$-chain, and so has no residue in Γ . There are three others. Two of them have residues consisting of a single Kempe chain of Γ each. The vertex-sets are $\{G\}$ and $\{I, J\}$. The residue of the third consists of two Kempe chains of Γ . The vertex sets are $\{B\}$ and $\{D\}$. The arrows marked on some of the edges are intended to clarify part of the following proof.

THEOREM 8.3. *Let* U *be a near-triangulation bounded by* Q *, let* Γ *be a 4-colouring of* Q *and let* Γ' *be a U-extension of* Γ . *Let* Z *be the family of residues in* Γ *of the* $\alpha\beta$*-chains and* $\gamma\delta$*-chains of* Γ' . *Then* Z *is an admissible clustering of* Γ *with respect to* $(\alpha\beta, \gamma\delta)$.

Proof. That Z is a clustering of Γ follows from the fact that each vertex of Q belongs to exactly one $\alpha\beta$-chain or $\gamma\delta$-chain of Γ' .

By 4.1, Z satisfies Condition (i) for an admissible clustering.

To prove Condition (ii) for Z let C_1 be any cluster in Z and let L be an arc in Q , with ends B and D say, such that no vertex of L is included in C_1 but B and D are adjacent in Q to vertices A and E respectively (not necessarily distinct) that are included in C_1 . Let C_2 be the member of Z that includes B . We have to show that C_2 also includes D .

Let C_1 and C_2 be residues of Kempe chains K_1 and K_2 of Γ' respectively. We may suppose K_1 to belong to $\{\alpha, \beta\}$ and K_2 to $\{\gamma, \delta\}$.

Let us define a *crossing edge* of U as an edge with one end in K_1 and one in K_2 . Let a *crossing triangle* of U be a face, other than the one bounded by Q , that is incident with a vertex of K_1 and a vertex of K_2 . We assume in this definition that only one face of U can be bounded by Q . If two are then Q is a triangle identical with U , and the required result is trivial. Evidently the following rules hold.

(iii) *if* T *is a crossing triangle then each vertex of* T *is in* K_1 *or* K_2 .

(iv) *Each crossing triangle is incident with exactly two crossing edges.*

Consider the crossing edge AB of Q . It is incident with a single crossing triangle T_1 , and this has a unique second crossing edge t_1 . If t_1 is not an edge of Q it is incident with a second crossing triangle T_2 , and this has a second crossing edge t_2 . If t_2 is not an edge of Q it is incident with a second crossing triangle T_3 , and so on. Such a sequence is indicated in Figure 6. T_1 is the triangle incident with AB . An arrow leads from this triangle into T_2 , another arrow from T_2 into T_3 , and so on.

Suppose some triangle is repeated in the sequence $T_1, T_2, \ldots$. Let us say $T_i = T_j$, where $i < j$ and j has the least value consistent with this condition. Now t_{j-1} is either t_{i-1} or t_i . If $t_{j-1} = t_{i-1}$, then $T_{j-1} = T_{i-1}$, contrary to the choice of j . Hence $t_{j-1} = t_i$, $T_{j-1} = T_{i+1}$ and, by the choice of j , $j - 1 = i + 1$. But now $t_{i+1} = t_{j-1} = t_i$, which is impossible.

We conclude that in fact no triangle is repeated in the sequence. The sequence therefore terminates with a triangle T_m such that t_m is an edge of Q .

The crossing edge t_m has one end X_1 in K_1 and one end X_2 in K_2 . Hence X_1 and X_2 are included in the clusters C_1 and C_2 respectively. Following along L from B we cannot reach X_1 before reaching E . Applying (i) to A , E , B and

X_2 we find that X_2 cannot lie beyond E . Hence $X_2 = D$, C_2 includes D , and the proof is complete.

We refer to Z as the admissible clustering of Γ , with respect to $(\alpha\beta, \gamma\delta)$, *induced by* Γ' . Similarly the corresponding chromodendron χ is the *chromodendron of* Γ , with respect to $(\alpha\beta, \gamma\delta)$, *induced by* Γ' . An example of such a chromodendron is shown in Figure 6.

We need the following converse to 8.3.

THEOREM 8.4. *Let* Γ *be a 4-colouring of a circuit* Q , *and let* χ *be a chromodendron of* Γ *with respect to* $(\alpha\beta, \gamma\delta)$. *Then we can construct a near-triangulation* U *bounded by* Q , *and a* U-*extension* Γ' *of* Γ *such that* Γ' *induces* χ *by way of the residues of its* $\alpha\beta$-*chains and* $\gamma\delta$-*chains.*

Proof. Let Z be the admissible clustering of Γ corresponding to χ .

Let n be the number of Kempe chains of Γ belonging to $\{\alpha, \beta\}$ and $\{\gamma, \delta\}$ respectively. For the construction of U in the cases n = 1 and n = 2 see Figures 7 and 8 respectively.

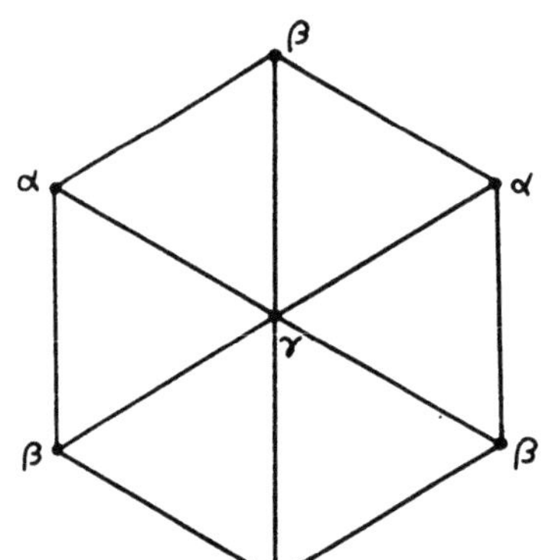

Figure 7

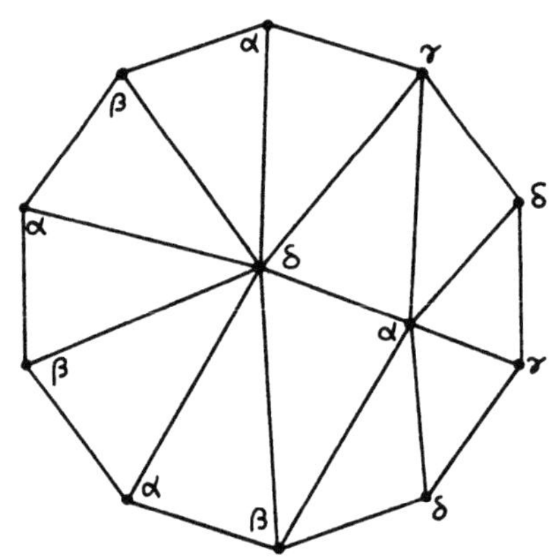

Figure 8

Let us assume as an inductive hypothesis that the theorem is true whenever n is less than some integer $q \geq 3$, and let us consider the case $n = q$. We can suppose q to be even, since Kempe chains of Γ belonging to $\{\alpha, \beta\}$ and $\{\gamma, \delta\}$ occur alternately in Q.

Evidently we can find vertices A and B of Q, each coloured α or β, and vertices C and D of Q, each coloured γ or δ, such that A and B separate C and D in Q. It may happen that A and B are included in the same cluster of Z. But suppose not. Then there is an arc M in Q that contains B but no vertex included in the same cluster as A, and has the maximum number of vertices consistent with this condition. We can take C and D to be the ends of M since these ends will not be coloured α or β. Then C and D are each adjacent to a vertex included in the same cluster as A, since otherwise M could be extended. It follows from Condition (ii) that C and D are included in a common cluster of Z.

Adjusting the notation we can assert that Q has two vertices A and B, each coloured α or β, belonging to different $\alpha\beta$-chains J and K respectively of Q, but included in the same cluster W of Z. We join A and B by an arc L inside Q as the first step in the construction of U. If A and B have different colours L consists of a single edge. Otherwise L consists of two edges, and their common end is given a colour α or β different from the colour of A and B. (See Figure 9).

Now Q is the union of two arcs L_1 and L_2 with common ends A and B but otherwise disjoint. We write Q_1 and Q_2 for the circuits $L \cup L_1$ and $L \cup L_2$ respectively. The colours already assigned determine 4-colourings Γ_1 and Γ_2 of Q_1 and Q_2 respectively.

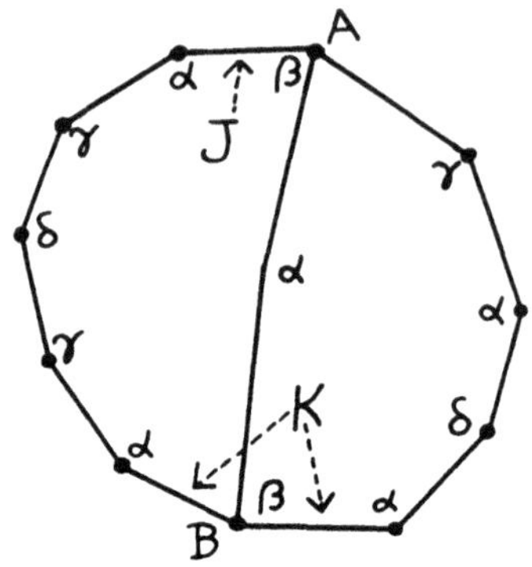

Figure 9

Let j be 1 or 2 . The Kempe chains of Γ_j belonging to {α, β} or {γ, δ} are the correspondingly coloured Kempe chains of Γ contained in L_j , together with an αβ-chain H_j which is the union of L with $J \cap L_j$ and $K \cap L_j$. Let us write n_j for the number of αβ-chains and γδ-chains of Γ_j . Then evidently $n_j < q$. By Condition (i) each cluster of Z includes vertices of only one of the arcs L_1 and L_2 , except for the cluster W . Hence Γ_j has a clustering Z_j consisting of those clusters of Z that include only vertices of L_j , together with a cluster W_j defined as follows. The members of W_j are those members of W , other than J and K , that are contained in L_j , together with H_j . It can now be shown that Z_j is an admissible clustering of Γ_j . The proof is quite straightforward, and to save space we leave it to the reader.

Since $n_j < q$ it follows by the inductive hypothesis that we can triangulate the inside of Q_j to form a near-triangulation U_j bounded by Q_j , arranging that Γ_j has a U_j-extension Γ_j' the residues of whose αβ-chains and γδ-chains are the clusters of Z_j .

Combining U_1 and U_2 we obtain a near-triangulation U bounded by Q, and we can combine Γ_1' and Γ_2' to obtain a U-extension Γ' of Γ. The $\alpha\beta$-chains and $\gamma\delta$-chains of Γ' are those of Γ_1' and Γ_2', except that those containing L are replaced by their union, an $\alpha\beta$-chain of Γ' whose residue is the cluster W. Thus Z is induced by Γ'.

The theorem is now established for the case $n = q$. It follows in general by induction.

The next theorem reduces to finiteness the problem of the dominance of a given set of 4-colourings of a circuit.

THEOREM 8.5. *Let* Γ *be a 4-colouring of a circuit* Q, *and let* S *be a set of 4-colourings of* Q. *Then* Γ *is simply immersible in* S *if and only if the following condition holds: There exists a colour-partition* Π *such that, for each chromodendron* χ *of* Γ *with respect to* Π, Γ *is* χ*-transformable into a member of* S.

Proof. Suppose Γ is simply immersible in S. Then there is a colour-partition Π with the following property: if U is any near-triangulation bounded by Q and if Γ is U-extensible then there is a U-extension Γ' of Γ that can be transformed into a U-extension of a member of S by a succession of Kempe interchanges with respect to Π. By 8.4 we can choose U and Γ' so that Γ' induces χ, where χ is an arbitrary chromodendron of Γ with respect to Π. In Q the succession of Kempe interchanges reduces to a succession of simple χ-transformations, taking Γ into a member of S. Thus the stated condition holds.

Conversely suppose the condition to hold. Let U be any near-triangulation bounded by Q such that Γ has a U-extension Γ'. Let Γ' induce the clustering Z of Γ, corresponding to a chromodendron χ. There exists a succession of simple χ-transformations changing Γ into a member of S. Each affects a single cluster of Z, and each can be effected by a

Kempe interchange applied to the Kempe chain of Γ' having that cluster as its residue. In this way we construct a succession of Kempe interchanges in U , with respect to Π , transforming Γ' into a U-extension of a member of S . Thus Γ is simply immersible in S .

Let us consider how to test for dominance a set S of 4-colourings of a circuit Q . Consider any 4-colouring Γ of Q not in S . Let $J_1(\Gamma)$, $J_2(\Gamma)$ and $J_3(\Gamma)$ be the sets of all chromodendra of Γ with respect to $(\alpha\beta, \gamma\delta)$, $(\alpha\gamma, \beta\delta)$ and $(\alpha\delta, \beta\gamma)$ respectively. We begin by determining these three sets. Then for each $\chi \in J_1(\Gamma)$ we determine all the 4-colourings Γ'' of Q such that χ admits a transformation of Γ into Γ'' . It may happen that for each such χ one of the 4-colourings Γ'' is in S . If so we have $\Gamma \in f(S)$, by 8.5. If not we repeat the procedure with $J_2(\Gamma)$, and then if necessary with $J_3(\Gamma)$. If for some i each chromodendron of $J_i(\Gamma)$ admits a transformation of Γ into a member of S then $\Gamma \in f(S)$, but otherwise Γ is not in f(S) . Repeating the whole procedure for each 4-colouring Γ not in S we complete the determination of the set f(S) . Then we can find $f^2(S)$ in the same way, and so on. Eventually the process must terminate with a set $f^k(S)$ identical with $f^{k+1}(S)$. (A set T such that $f(T) = T$ will be called *closed*. Perhaps the term "D-closed" would be better; but we shall not make use of other kinds of reducibility in this connection.) If $f^k(S)$ includes all the 4-colourings of Q then S is dominant; otherwise it is not.

This is the procedure applied by Heesch to the set S of V-extensible 4-colourings of Q to determine if a pair (V, Q) is D-reducible, that is if S is dominant. Let us call it *Heesch's algorithm*.

It can be shown that if (V, Q) is D-reducible then (V, Q) is K-reducible. The proof uses the fact that in the second paragraph of the proof of 8.5 Γ' can be any U-extension of Γ .

We leave it to the reader to bridge the gap due to the fact that we have defined K-reducibility in terms of general planar graphs, and D-reducibility only in terms of near-triangulations.

Perhaps we should take note here of some other kinds of reducibility defined by Heesch [3]. If V is a near-triangulation bounded by Q then (V, Q) is said to be A-*reducible* if the following condition holds: there is a near-triangulation W bounded by Q , having fewer vertices than V , and such that every W-extensible 4-colouring of Q is in the set S of V-extensible 4-colourings of Q . To prove that (V, Q) is then reducible consider a supposedly minimal graph G separated into V and another near triangulation U by the circuit Q . The union of U and W , we deduce, is 4-colourable. Hence there is a 4-colouring of Q that is U-extensible and W-extensible, and therefore V-extensible. This implies that G has a 4-colouring and we have a contradiction.

B-reducibility is similarly defined except that it requires the W-extensible 4-colourings of Q to be simply immersible in S , not necessarily all in S . The definition of C-reducibility replaces "simply immersible" by "crudely immersible" . In each case the proof of reducibility is a straightforward generalization of the one we have sketched in the case of A-reducibility. Heesch gives examples from the literature of these three kinds of reducibility.

In the theory of D-reducibility we try to make as much progress as possible without considering the effect of a Kempe interchange on the Kempe chains of other colour-partitions. Perhaps the Four Colour Problem can be settled in this way, but perhaps the drastic simplification rejects essential information. The authors would feel much more confident in the power of the theory if it could be used to prove that the Birkhoff number exceeds some reasonable number, - say 40, or even 20.

9. *A construction for D-irreducible pairs.*

We now give a method for constructing D-irreducible pairs (V, Q) . Of course an attempt to prove the Four Colour Conjecture uses, rather, D-reducible pairs; thus the present construction puts difficulties in the way of the method. It shows directions in which the method need not be pursued.

The authors feel tempted to construct an abstract theory of dominant sets. They have defined a *closed* set of 4-colourings of a circuit Q towards the end of Section 8. The definition amounts to saying that a set T is closed if no member of its complement is simply immersible in T . A set R can be called *open* if its complement is closed, that is if no member of R is simply immersible in the complement of R . Here is an example of a theorem from the abstract theory of dominant sets.

THEOREM 9.1. *Let* J *be the set of all 4-colourings of a circuit* Q . *Then each non-null open subset of* J *meets every dominant subset of* J .

Proof. Let S be a dominant subset of J , and T a non-null open subset of J . Assume that their intersection is null. There is a least integer k such that $f^k(S)$ meets T . Moreover $k > 0$. Choose $\Gamma \in T \cap f^k(S)$. Then Γ is simply immersible in $f^{k-1}(S)$ and therefore it is simply immersible in the complement of T . This is contrary to the hypothesis that T is open.

Let us define a *pentatriangulation* as a plane graph in which each face is either a triangle or a pentagon. A *near-pentatriangulation* is a plane graph in which each face with at most one exception is a triangle or a pentagon. When we say that a near-pentatriangulation is bounded or face-bounded by a circuit Q we shall imply that Q is the face-boundary of the exceptional face, if there is one.

A P-*colouring* of a pentatriangulation G is a 4-colouring of G such that all four colours appear on the boundary of each pentagonal face. A P-colouring of a near-pentatriangulation G bounded by a circuit Q is defined in the same way, with the understanding that the four-colour condition may be relaxed for Q if Q is a pentagon and not the whole of G .

THEOREM 9.2. *Let* U *be a near-pentatriangulation bounded by a circuit* Q . *Let* T *be the set of all 4-colourings of* Q *having* U-*extensions that are* P-*colourings of* U . *Then* T *is open.*

Proof. Let Γ be a 4-colouring of Q having a U-extension Γ' that is a P-colouring of U .

We convert U into a near-triangulation bounded by Q as follows. Consider any pentagonal face P_j of U satisfying the four-colour condition. Let its face-boundary be Q_j . In Γ' one colour, say α , is repeated in Q_j . (See Figure 5). Call the vertices coloured α "special", and call the vertex between them the "apex". In each face P_j we join the apex to the two non-adjacent vertices of Q_j . We observe that Γ' is preserved as a 4-colouring of the resulting near-triangulation U' . We note also that no succession of Kempe interchanges in Γ' and U' , provided they are all with respect to the same colour-partition, can transform Γ' into a 4-colouring of U' in which only three colours appear on one of the circuits Q_j . This means that Γ is not simply immersible in the complement of T . We deduce that T is open.

THEOREM 9.3. *Let a circuit* Q *separate a pentatriangulation* G *into a near-pentatriangulation* U *and a near-triangulation* V , *both bounded by* Q . *Suppose* U *to have a* P-*colouring, but* G *to have no* P-*colouring. Then* (V, Q) *is* D-*irreducible.*

Proof. Assume the contrary. Let S be the set of V-extensible 4-colourings of Q , and T the set of 4-colourings of Q having U-extensions that are P-colourings of U . Then S is dominant. T is non-null by hypothesis, and open by 9.2. Hence S and T have a common member Γ , by 9.1. Combining a V-extension and a suitable U-extension of Γ we can obtain a P-colouring of G , contrary to hypothesis.

We go on to describe a construction due to Shimamoto.

Suppose there is a critical graph containing a 5-wheel, any minimal graph for example. Dropping out the hub and spokes of this wheel gives a plane graph G* that we call a "chromatic obstacle". It is bounded by the pentagonal rim of the wheel. It is 4-colourable, but all four colours must appear on the rim.

In the construction as described by Shimamoto at the time of the rumours new critical graphs are built up from smaller ones. It is found that a critical graph obtained in this way must have the following structure. There is a pentatriangulation X with no P-colouring, and the critical graph is obtained from X by filling each face with a chromatic obstacle. It is possible therefore to describe the construction as one in which new pentatriangulations without P-colourings are obtained from old ones. We prefer to describe it in this way and so to avoid any use of such hypothetical figures as critical graphs and chromatic obstacles.

The principal step in the construction is based on the following fact. Suppose Q bounds U and V , the whole forms G , and the conditions of 9.3 hold. Let A , X , B be three consecutive vertices of Q . Let us cut along the arc AXB and open it out into a quadrilateral AXBX' , with X still on the boundary of V . Finally let us fill the new quadrilateral with three triangles and a pentagon, as shown in Figure 10.

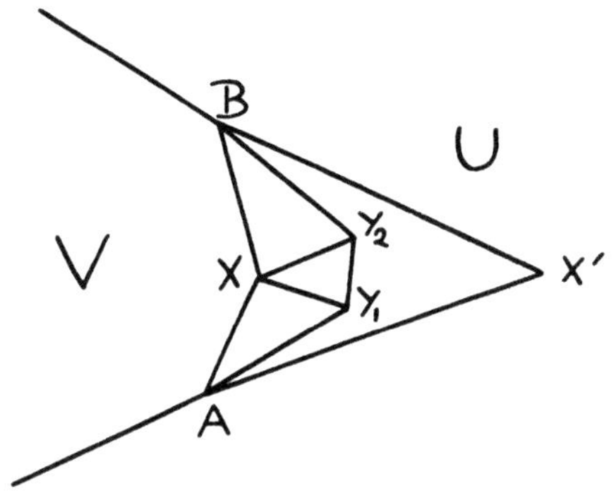

Figure 10

Let us now replace the arc AXB in Q by the arc AY_1Y_2B , to form a circuit Q' . There is a near-triangulation V' defined by the faces of G in V together with the three new triangles. This is separated by Q; from a near-pentatriangulation U' defined by the faces of G in U together with the new pentagon AY_1Y_2BX' . The new pentatriangulation that is the union of U' and V' we denote by G' . It is appropriate to call this operation Shimamoto's First Construction. The construction is reminiscent of one used by Hajós for critical graphs [5, Section 11.4].

The conditions of 9.3 hold with Q', U' and V' replacing Q , U and V respectively. We know U has a P-colouring, and we can extend this as a P-colouring of U' by assigning appropriate colours to the new vertices Y_1 and Y_2 , both divalent in U' . If G' has a P-colouring it is clear that X and X' must have the same colour in it, and so we can derive a P-colouring of G , which is impossible. We have derived a new D-irreducible pair (V', Q') . We can now drop the primes and repeat the construction.

As our starting point we can take G to be the 5-wheel and Q its rim. U is the circuit Q , and V is the 5-wheel itself. U has two pentagonal faces, both bounded by Q . It is thus a near-pentatriangulation bounded by Q . In a P-colouring of U each face must have all four colours in its boundary. Evidently the conditions of 9.3 hold. Thus 9.3 shows that the 5-wheel is D-irreducible, as we have seen already in 7.2, Corollary.

Figure 11 shows the results of some successive applications of Shimamoto's Construction to the 5-wheel. In each diagram the circuit Q is shown by arrows. An asterisk indicates the vertex X that is to be split.

For the final step we start with the last diagram of Figure 11 and omit the broken edges. We take two copies of the resulting graph and identify their arcs AKL . We then introduce new vertices R , A_1' and A_2' , and join from them as shown in Figure 12.

This is an example of Shimamoto's Second Construction.

Let us show that the graph G_{12} of Figure 12 satisfies the conditions of 9.3 with Q as the circuit Q_{12} indicated by the arrows, V as the near-triangulation, which we shall also call H , inside Q_{12} , and U as the near-pentatriangulation U_{12} outside Q_{12} .

We prove that G_{12} has no P-colouring as follows: If it did, then using Σ shows that A has the same colour as A_1' or A_2' ; say $col(A) = col(A_1')$. Using Σ_B now shows that $col(B) = col(B')$. Using Σ_C and Σ_D in turn shows that $col(C) = col(C')$ and $col(D) = col(D')$. Noting that K is joined to all of A , B , C , D and L now shows that Σ' uses only three colours, a contradiction.

It remains to construct a P-colouring for U_{12} , shown completely in Figure 12. Ignoring R and the edge $A_1'A_1'$ we find that this graph consists of two isomorphic parts with only the vertex L in common. From a P-colouring of the graph U of the

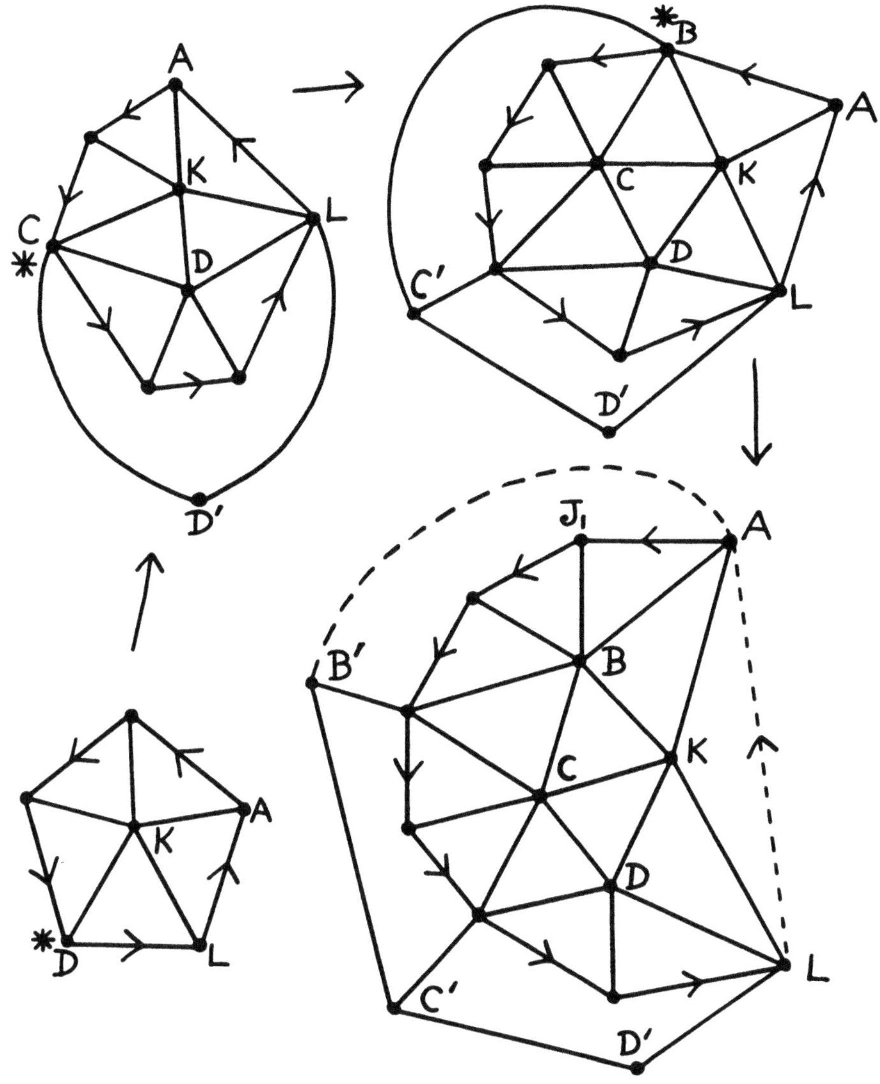

Figure 11

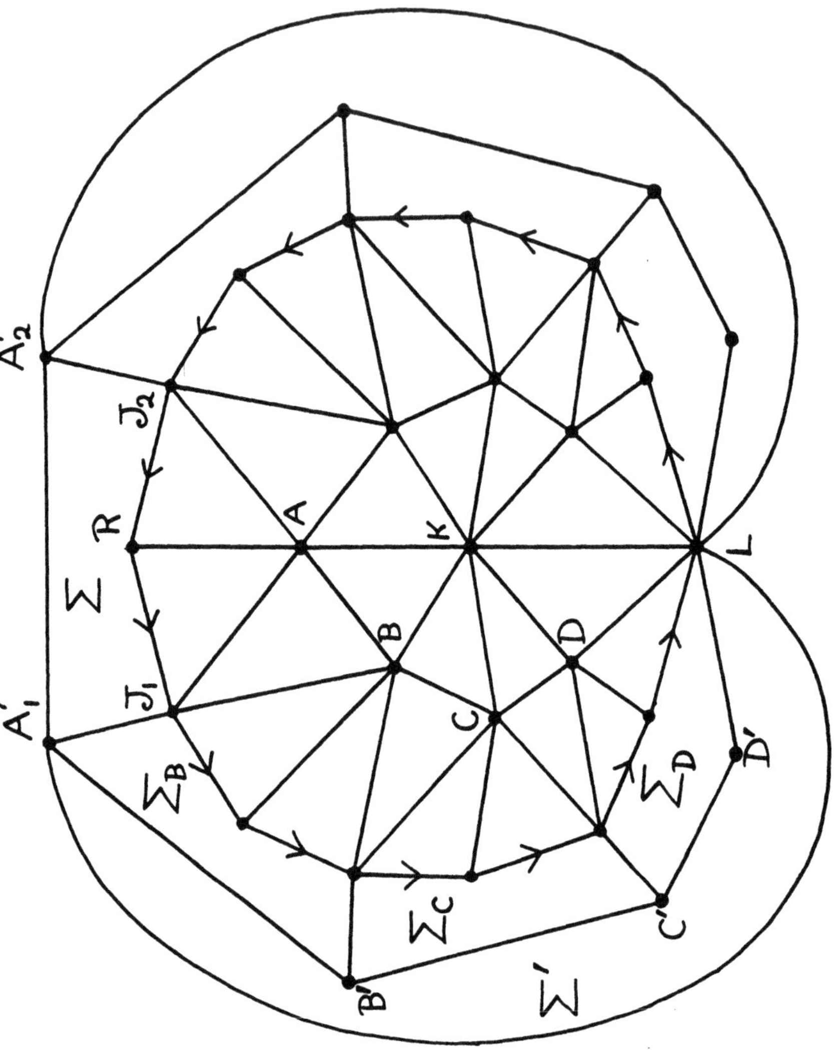

Figure 12

last diagram of Figure 11 we can, for each part, derive a 4-colouring that obeys the four colour rule for pentagonal faces of G_{12} bounded by circuits in the part considered. After a permutation in one of the two parts we can combine the two 4-colourings to obtain a 4-colouring of U_{12}, less R, such that A_1' and A_2' have different colours and at least three colours occur among the four vertices A_1', A_2', J_1 and J_2. We can now colour R so as to complete a P-colouring of U_{12}.

We can now apply 9.3 to obtain

THEOREM 9.4. *The near-triangulation* H *is* D-*irreducible.*

10. *Concluding remarks.*

The configuration H of 9.4 is the one mentioned in Section 1. At first it was thought to have been proved D-reducible by a computer programmed to apply Heesch's Algorithm. In the original form of Shimamoto's Construction the pentagonal faces of G_{12} were supposed to be filled with chromatic obstacles. Then U_{12} was a near-triangulation and it could be proved to have a 4-colouring. Arguments resembling those of Section 9 led to a contradiction corresponding to our 9.4. It seemed that this could only be resolved by supposing that U_{12} was impossible, i.e., that no chromatic obstacles existed and the Four Colour Conjecture was true. Now it is clear that Shimamoto had discovered not a proof of the Conjecture but a construction for D-irreducible configurations.

To the present authors the supposed D-reducibility of H meant that any 4-colouring of U_{12} could be converted into one extendable to all G_{12} by crude chaining applied to U_{12}. Somehow in the course of this chaining the four colours on the boundary of one of the supposed chromatic obstacles would reduce to three. By isolating the effect of the chaining on this one chromatic obstacle we should obtain a proof by crude chaining of the reducibility of the 5-wheel. Yet it seemed clear that our 7.3, a well-known result, was the best that could be expected along this line. It was to rigorize this objection that we introduced P-colourings and worked out their theory as given in Section 9.

In this report on the present state of the Four Colour Problem there is little for which we claim originality, apart from the recognition of the true meaning of Shimamoto's Construction. We have tried to clarify the theory for ourselves, and we dare to hope that we may thereby have clarified it for others.

It now seems to us that the next step in the theory of D-reducibility should be an attempt to classify the minimal dominant sets for the smaller circuits. To test (V, Q) for D-reducibility we would then determine the set S of V-extensible 4-colourings of Q and check it against a list of dominant sets of Q to see if it contained one of them. Should this method prove feasible it might eliminate much repetitive work.

The authors wish to propose the following Conjecture for consideration.

CONJECTURE. *Let* S *and* T *be non-null open sets of 4-colourings of a circuit* Q . *Then they have a common element.*

This conjecture implies the Four Colour Conjecture. For suppose a minimal graph G is separated by a circuit Q into near-triangulations U and V . By 9.2, specialized to near-triangulations, we can take S and T to be the sets of U-extensible and V-extensible 4-colourings of Q respectively. Their common elements, being both U-extensible and V-extensible, give rise to 4-colourings of G , contrary to the assumption of minimality. Yet the conjecture is concerned only with the properties of circuits.

REFERENCES

[1] G. D. Birkhoff, *The Reducibility of Maps*, Amer. J. Math., *35* (1913), 115-128.

[2] P. J. Heawood, *Map-colour Theorem*, Quart. J. Math. Oxford, Ser. *24* (1890), 322-338.

[3] H. Heesch, *Untersuchungen zum Vierfarbenproblem*, Hochschulskripten 810/810a/810b, Mannheim 1969.

[4] A. B. Kempe, *On the Geographical Problem of the Four Colours*, Amer. J. Math., *2* (1879), 193-200.

[5] O. Ore, *The Four-color Problem*, Academic Press, New York, 1967.

[6] O. Ore and J. Stemple, *Numerical Calculations on the Four-color Problem*, J. Combinatorial Theory, *8* (1970), 65-78.

[7] T. L. Saaty, *Thirteen Colorful Variations on Guthrie's Four-color Conjecture*, Amer. Math. Monthly, *79* (1972), 2-43.

[8] W. T. Tutte, *On the Four Colour Conjecture*, Proc. London Math. Soc., *50* (1948), 137-149.

[9] C. E. Winn, *On the Minimum Number of Polygons in an Irreducible Map*, Amer. J. Math., *62* (1940), 406-416.

Institute of Advanced Studies, Princeton.
University of Waterloo

Received September 1, 1972.

Chapter 2

Differentiable Functions and Singularities

ANALYTIC EXTENSIONS OF DIFFERENTIABLE FUNCTIONS DEFINED IN CLOSED SETS*

BY

HASSLER WHITNEY†

I. Differentiable functions in closed sets

1. Introduction. Let A be a closed set, bounded or unbounded, in euclidean n-space E, and let $f(x)$ be a function defined and continuous in A. It is well known that this function can be extended so as to be continuous throughout E.‡ If A satisfies certain conditions, the solution of the Dirichlet problem is a function harmonic in $E-A$ and taking on the given boundary values in A. Two questions which arise are the following: Is there always a function differentiable, or perhaps analytic, in $E-A$, and taking on the given values in A? If the given function $f(x)$ is in some sense differentiable in A, can the extension $F(x)$ be made differentiable to the same order throughout E?

These questions are answered in the affirmative in Theorem I. We use a definition of the derivatives of a function in a general set which arises naturally from a consideration of Taylor's formula. In Part II, a differentiable extension of $f(x)$ is found, whether $f(x)$ is differentiable to finite or infinite order. Part III is devoted to some general approximation theorems. It is well known that a continuous function in a bounded closed set can be approximated uniformly (together with any finite number of derivatives) by polynomials; we show that functions defined in *open* sets may be approximated (together with derivatives) by *analytic* functions, the approximation being closer and closer as we approach the boundary of the set. This theorem, together with the results of Part II, furnish an immediate proof of Theorem I. In Part IV we give some extensions of Theorem I; in particular, we show that

* Presented to the Society, December 29, 1932; received by the editors March 29, 1933, and, after revision, May 2, 1933.

† National Research Fellow.

‡ See references in a paper by P. Urysohn, Mathematische Annalen, vol. 94 (1925), p. 293, footnote 51.

A continuous extension the author has not seen in the literature may be given as follows; we assume for simplicity that A is bounded. Let $h(r)$ $(r\geqq 0)$ be a continuous and monotone increasing function such that $h(0)=0$, and if x and y are any two points of A whose distance apart is r_{xy}, then $|f(x)-f(y)|\leqq h(r_{xy})$. For any points x of E and y of A, set $H(x, y)=f(y)-h(r_{xy})$; then if x is in A, $H(x,y)\leqq f(x)$. The continuous extension of $f(x)$ is $F(x)$, which at each point x of E equals the maximum of $H(x, y)$ as y varies over A.

the extension of $f(x)$ may be made analytic at the isolated points of A. Theorem III includes all preceding results but Lemma 7.

2. **Notations.** We shall write all equations involving n variables as if there were but a single variable present. For instance, we write

$$f_0(x) \quad \text{for} \quad f_{0\ldots 0}(x_1, \cdots, x_n),$$

$$D_k f(x') \text{ for } \frac{\partial^{k_1+\cdots+k_n}}{\partial x_1^{k_1} \cdots \partial x_n^{k_n}} f(x_1', \cdots, x_n'),$$

$$\binom{k}{l} \text{ for } \binom{k_1}{l_1} \cdots \binom{k_n}{l_n},$$

etc. For any n-fold subscript k, we put

$$\sigma_k = k_1 + \cdots + k_n.$$

Note that $\sigma_{k+l} = \sigma_k + \sigma_l$. r_{xy} will always denote the distance between x and y (unless x and y are complex). As an example, (3.1) below is short for

$$f_{k_1 \cdots k_n}(x_1', \cdots, x_n') = \sum_{\substack{l_1+\cdots+l_n \\ \leqq m-(k_1+\cdots+k_n)}} \frac{f_{k_1+l_1,\cdots,k_n+l_n}(x_1, \cdots, x_n)}{l_1! \cdots l_n!} (x_1' - x_1)^{l_1} \cdots (x_n' - x_n)^{l_n} + R_{k_1 \cdots k_n}(x_1', \cdots, x_n'; x_1, \cdots, x_n).$$

3. **Differentiable functions in subsets of E.** Let $f(x)$ be defined in the set A, and let m be an integer $\geqq 0$. We say $f(x) = f_0(x)$ *is of class C^m in A in terms of the functions $f_k(x)$ $(\sigma_k \leqq m)$* if the functions $f_k(x)$ are defined in A for all n-fold subscripts k with $\sigma_k \leqq m$, and

$$f_k(x') = \sum_{\sigma_l \leqq m-\sigma_k} \frac{f_{k+l}(x)}{l!} (x' - x)^l + R_k(x'; x) \tag{3.1}$$

for each $f_k(x)$ $(\sigma_k \leqq m)$, where $R_k(x'; x)$ has the following property. Given any point x^0 of A and any $\epsilon > 0$, there is a $\delta > 0$ such that if x and x' are any two points of A with $r_{xx^0} < \delta$ and $r_{x'x^0} < \delta$, then

$$| R_k(x'; x) | \leqq r_{xx'}^{m-\sigma_k} \epsilon. \tag{3.2}$$

One might define the derivatives of a function at the points of a set B, when the function is defined in a larger set A. We shall not do this here.

If $m = 0$, (3.1) and (3.2) state merely that $f(x)$ is continuous. Note that the conditions are satisfied automatically at all isolated points of A, no matter how the $f_k(x)$ are defined there.

It is easily seen that the $f_k(x)$ are continuous in a neighborhood of each point of A, and are thus bounded there. From this we prove that if $f(x)$ is of class C^m in A in terms of the $f_k(x)$ $(\sigma_k \leqq m)$, then it is of class $C^{m'}$ in A $(m' < m)$ in terms of the $f_k(x)$ $(\sigma_k \leqq m')$.

Any function we shall say is of class C^{-1} in A. $f(x)$ is of class C^∞ in A in terms of the $f_k(x)$ (defined for all k) if it is of class C^m in A in terms of the $f_k(x)$ $(\sigma_k \leqq m)$ for each m.

Suppose $f(x)$ is defined throughout the region R, and is of class C^m in terms of the $f_k(x)$ $(\sigma_k \leqq m)$. Then putting $x = (x_1, \cdots, x_n)$, $x' = (x_1, \cdots, x_h + \Delta x_h, \cdots, x_n)$, (3.1) gives

$$(3.3) \qquad f_{k_1 \cdots k_n}(x') = f_{k_1 \cdots k_n}(x) + f_{k_1, \cdots, k_h+1, \cdots, k_n}(x)\Delta x_h + R^{(h)}_{k_1 \cdots k_n}(x'; x)$$

(provided $\sigma_k < m$), where $R^{(h)}_{k_1 \cdots k_n}(x'; x)/\Delta h \to 0$ as $\Delta h \to 0$, which shows that

$$(3.4) \qquad \frac{\partial}{\partial x_h} f_{k_1 \cdots k_n}(x) = f_{k_1, \cdots, k_h+1, \cdots, k_n}(x) \qquad (\sigma_k < m)$$

in R; thus in this case, $f(x)$ is of class C^m in the ordinary sense, and the $f_k(x)$ are the partial derivatives of $f(x)$. The converse is true, by Taylor's Theorem.

4. The main theorem of the present paper is the following:

THEOREM I. *Let A be a closed subset of E, and let $f(x) = f_0(x)$ be of class C^m (m finite or infinite) in A in terms of the $f_k(x)$ $(\sigma_k \leqq m)$. Then there is a function $F(x)$ of class C^m in E in the ordinary sense, such that*

(1) *$F(x) = f(x)$ in A,*

(2) *$D_k F(x) = f_k(x)$ in A $(\sigma_k \leqq m)$,*

(3) *$F(x)$ is analytic in $E - A$.*†

Of course (2) includes (1).

No such theorem holds if we leave out the uniformity condition on $R_k(x'; x)$, i.e. if we assume merely that for any x and $\epsilon > 0$ there is a $\delta > 0$ such that if $r_{xx'} < \delta$, then $|R_k(x'; x)| < r_{xx'}^{m-\sigma_k}\epsilon$. The following example shows this. Let A be the set of points (using one variable) $x = 0$, $1/2^s$ and $1/2^s + 1/2^{2s}$ $(s = 1, 2, \cdots)$. Set $f(x) = 0$ at $x = 0$ and $1/2^s$, and $f(x) = 1/2^{2s}$ at the remaining points. Set $f_1(x) \equiv 0$ in A. The above condition is satisfied, but there is no extension of $f(x)$ which has a continuous first derivative.

5. The following lemma will be needed; its proof is elementary.

† It is seen from the proof in §16 that $F(x)$ is analytic in a complex region with the following property. If x is a point of $E - A$ distant 3ρ from A, then the region contains all points within a distance ρ of x.

LEMMA 1. *Let $w(z)$ be a continuous function of one variable defined throughout an interval containing z_0, let A^* be a closed set in this interval, and let w_0' be a fixed number. Suppose that for every $\epsilon>0$ there is a $\delta>0$ such that*

(1) *if z is in A^* and $|z-z_0|<\delta$, then*

$$|w(z) - w(z_0)/(z - z_0) - w_0'| < \epsilon;$$

(2) *if z is not in A^* and $|z-z_0|<\delta$, then the derivative $w'(z)$ exists and*

$$|w'(z) - w_0'| < \epsilon.$$

Then $w(z)$ has a derivative at z_0, and $w'(z_0)=w_0'$.

II. DIFFERENTIABLE EXTENSIONS

6. **The functions $\psi_k(x'; x)$.** We shall make use of functions defined as follows for x in A and x' in E (m finite):

$$\psi_k(x'; x) = \sum_{\sigma_l \leqq m-\sigma_k} \frac{f_{k+l}(x)}{l!}(x' - x)^l \qquad (\sigma_k \leqq m); \tag{6.1}$$

$\psi_k(x'; x)$ is the value at x' of the polynomial of degree at most $m-\sigma_k$ which approximates the function $f_k(x)$ to the $(m-\sigma_k)$th order at x. Keeping x fixed, it is a polynomial in x', given by Taylor's formula in terms of its value and derivatives at x. In terms of these functions, (3.1) becomes

$$f_k(x') = \psi_k(x'; x) + R_k(x'; x) \qquad (\sigma_k \leqq m). \tag{6.2}$$

The lth derivative of the function of x' $\psi_k(x'; x)$ at x' is $\psi_{k+l}(x'; x)$; if we express $\psi_k(x''; x)$ by Taylor's formula in terms of its value and derivatives at x', we obtain

$$\begin{aligned}\psi_k(x''; x) &= \sum_l \frac{\psi_{k+l}(x'; x)}{l!}(x'' - x')^l \\ &= \sum_l \frac{(x'' - x')^l}{l!} \sum_j \frac{f_{k+l+j}(x)}{j!}(x' - x)^j.\end{aligned}$$

The definition of $\psi_k(x''; x')$ in conjunction with this identity gives, for any points x and x' in A and x'' in E,

$$\begin{aligned}\psi_k(x''; x') &= \sum_l \frac{f_{k+l}(x')}{l!}(x'' - x')^l \\ &= \sum_l \frac{(x'' - x')^l}{l!}\left[\sum_j \frac{f_{k+l+j}(x)}{j!}(x' - x)^j + R_{k+l}(x'; x)\right] \\ &= \psi_k(x''; x) + \sum_l \frac{R_{k+l}(x'; x)}{l!}(x'' - x')^l.\end{aligned} \tag{6.3}$$

7. **The function $\Theta(x)$.** Let R be the region given by the inequalities $|x_h|<1$ $(h=1, \cdots, n)$, let R' be R minus the origin, and let R^* be the boundary of R. Define the functions θ, θ', Θ as follows:

$$\theta(x) = 2(1 - x_1^2) \cdots (1 - x_n^2) - 1 \text{ in } R', \tag{7.1}$$

$$\theta'(x) = \frac{\theta(x)}{1 - \theta^2(x)} \qquad \text{in } R', \tag{7.2}$$

$$\Theta(x) = \begin{cases} e^{\theta'(x)} & \text{in } R', \\ 0 & \text{in } E - R. \end{cases} \tag{7.3}$$

It is seen that $-1<\theta(x)<+1$, $\theta(x)\to+1$ as $x\to0$, and $\theta(x)\to-1$ as $x\to R^*$; hence $\theta'(x)\to+\infty$ as $x\to0$ and $\theta'(x)\to-\infty$ as $x\to R^*$. Consequently $\Theta(x)\to+\infty$ to infinite order as $x\to0$ and $\Theta(x)\to0$ to infinite order as $x\to R^*$; also $\Theta(x)$ is of class C^∞ for $x\neq0$. If $\Theta'(x)=1/\Theta(x)$ in R' and $\Theta'(x)=0$ for $x=0$, then $\Theta'(x)$ is of class C^∞ in R.

8. **The subdivision of $E-A$.** Divide E into n-cubes of side 1, and let K_0 be the set of all these cubes whose distances from A are at least $6n^{1/2}$ (if there are any). In general, having constructed the cubes of K_{s-1}, divide each cube which is now present but is not in $K_0+\cdots+K_{s-1}$ into 2^n cubes of side $1/2^s$, and let K_s be the set of all these cubes whose distances from A are at least $6n^{1/2}/2^s$ (if there are any).

The distance from any cube C of K_s to A is $<18n^{1/2}/2^s$ $(s\geqq1)$; for it lies in a cube C' of the previous subdivision which does not belong to K_{s-1}, and whose distance from A is therefore $<6n^{1/2}/2^{s-1}$.

Any cube C of K_s is separated from any cube C' of K_{s+2} by at least four cubes of K_{s+1}. For the distance from C to A is $\geqq12n^{1/2}/2^{s+1}$, the distance from any point of C' to A is $<9n^{1/2}/2^{s+1}$, and the diameter of any cube of K_{s+1} is $n^{1/2}/2^{s+1}$.

9. **The functions $\phi_\nu(x)$.** We introduce the following definitions:

$y^1, y^2, \cdots$ is the set of all vertices of cubes of $K_0+K_1+\cdots$, arranged in a sequence.

r_ν is the distance from y_ν to A $(\nu=1, 2, \cdots)$.

x^ν is a fixed point of A whose distance from y^ν is r_ν.

b_ν is the length of side of the largest cube of $K_0+K_1+\cdots$ with y^ν as a vertex.

I_ν is the set of points x for which $|x_h-y_h^\nu|\leqq b_\nu$ $(h=1, \cdots, n)$; B_ν is its boundary.

$$\pi_\nu(x) = \Theta\left(\frac{x_1 - y_1^\nu}{b_\nu}, \cdots, \frac{x_n - y_n^\nu}{b_\nu}\right) \quad \text{in} \quad E - y^\nu;$$

$$\pi_\nu'(x) = \Theta'\left(\frac{x_1 - y_1^\nu}{b_\nu}, \cdots, \frac{x_n - y_n^\nu}{b_\nu}\right) \quad \text{in} \quad I_\nu - B_\nu.$$

$$\phi_\nu(x) = \begin{cases} \dfrac{\pi_\nu(x)}{\sum_\lambda \pi_\lambda(x)} & \text{in} \quad E - A,\ x \neq y^1, y^2, \cdots, \\ 1, & x = y^\nu, \\ 0, & x = y^\mu\ (\mu \neq \nu). \end{cases}$$

Suppose y^* is a given point of $E-A$, distant δ_* from A (or from a given point x^0 of A), and suppose y^* lies in the cube C of K_s. Then if I_ν, with center y^ν, has points in common with C, and y^ν is distant d_ν from A (or from x^0),

$$\delta_*/2 \leqq d_\nu < 2\delta_*. \tag{9.1}$$

To prove this, say C' is a largest cube with y^ν as a vertex, and C' is in K_t; then $t \geqq s-1$. The diameter of C' is $n^{1/2}/2^t$; hence y^ν is distant at most $n^{1/2}/2^t \leqq 2n^{1/2}/2^s$ from any point of I_ν. As the diameter of C is $n^{1/2}/2^s$, y^ν is distant at most $3n^{1/2}/2^s$ from y^*. But $\delta_* \geqq 6n^{1/2}/2^s$, and the inequalities follow.

Each function $\pi_\nu(x)$ is >0 in $I_\nu - B_\nu - y^\nu$ and only there; it approaches ∞ and 0 to infinite order as x approaches y^ν and B_ν respectively. Each point x of $E-A$ is interior to some cube I_ν, hence $\pi_\nu(x) > 0$ for some ν, and $\sum \pi_\lambda(x) > 0$ in $E-A$, justifying the definition of $\phi_\nu(x)$. Note that $\phi_\nu(x)$ is $\neq 0$ in $I_\nu - B_\nu$ and only there; also

$$\sum_\nu \phi_\nu(x) = 1 \quad \text{in} \quad E - A. \tag{9.2}$$

We shall show that $\phi_\nu(x)$ is of class C^∞ in $E-A$. This is obvious at points $x \neq y^\nu$. Consider a small neighborhood U_λ of y^λ, $\lambda \neq \nu$. $\pi_\lambda'(x)$ is of class C^∞ in U_λ; hence the same is true of $\phi_\nu = \pi_\lambda' \pi_\nu / (1 + \pi_\lambda' \sum_{\mu \neq \lambda} \pi_\mu)$ in U_λ. Similarly $\phi_\nu = 1/(1 + \pi_\nu' \sum_{\mu \neq \nu} \pi_\mu)$ is of class C^∞ in a small neighborhood U_ν of y^ν; the statement follows.

10. **The derivatives of the $\phi_\nu(x)$.** Consider two (closed) cubes C and C' of $K_0 + K_1 + \cdots$, and let J and J' be those sets I_ν with points in C and C' respectively. We shall say *C and C' are of the same type* if the sets in J' can be brought into coincidence with the sets in J by a translation and stretching of the axes, that is, if the structure of the subdivision about C' is the same as that about C. There are but a finite number, say d, of possible types of cubes, and for some number c, there are at most c sets I_ν with points in any given cube C.

Take a fixed cube C of K_0 and a fixed k. As each $\phi_\nu(x)$ is of class C^∞, $D_k\phi_\nu(x)$ is bounded in C; there are only a finite number of these functions $\neq 0$ in C, and hence they are uniformly bounded:

$$|D_k\phi_\nu(x)| < N_k(C) \quad \text{in} \quad C \qquad (\nu = 1, 2, \cdots).$$

Consider now any cube C' of any K_s, and let C be a (perhaps hypothetical) cube of K_0 of the same type as C'. If $I_{\lambda_1'}, \cdots, I_{\lambda_t'}$ are the sets I_λ with points in C', let $I_{\lambda_1}, \cdots, I_{\lambda_t}$ be the corresponding sets with points in C; the latter set of sets is carried into the former by a translation of the axes and a stretching by a factor $1/2^s$. Each function ϕ_{λ_q} corresponding to I_{λ_q} goes thereby into the function

$$\phi_{\lambda_q'}(x) = \phi_{\lambda_q}[y^{\lambda_q} + 2^s(x - y^{\lambda_q'})]$$

corresponding to $I_{\lambda_q'}$. Therefore, differentiating σ_k times with respect to x,

$$D_k\phi_{\lambda_q'}(x) = 2^{s\sigma_k}D_k\phi_{\lambda_q}[y^{\lambda_q} + 2^s(x - y^{\lambda_q'})]$$

for x in C', and hence

$$|D_k\phi_\nu(x)| < 2^{s\sigma_k}N_k(C) \quad \text{in} \quad C' \qquad (\nu = 1, 2, \cdots),$$

as $\phi_\nu(x) = 0$ in C' for $\nu \neq \lambda_1', \cdots, \lambda_t'$. Now the constants $N_k(C)$ take on at most d distinct values for a fixed k; if we let N_k be the largest of these, we can state: *Given any n-fold set of numbers k, there is a number N_k such that if C is any cube of K_s, then*

$$(10.1) \qquad |D_k\phi_\nu(x)| < 2^{s\sigma_k}N_k \quad \textit{in} \quad C \qquad (\nu = 1, 2, \cdots).$$

11. **A differentiable extension of $f(x)$, m finite.** We are now in a position to prove, for m finite,

LEMMA 2. *Under the conditions of Theorem* I, *there is a function $g(x)$ of class C^∞ in $E-A$, having the properties* (1) *and* (2) *of Theorem* I.

For each ν ($\nu = 1, 2, \cdots$) there are functions $\phi_\nu(x)$ and $\psi(x; x^\nu) = \psi_0(x; x^\nu)$; we put

$$(11.1) \qquad g(x) = \begin{cases} \sum_\nu \phi_\nu(x)\psi(x; x^\nu) & \text{in } E - A, \\ f(x) & \text{in } A. \end{cases}$$

As the $\phi_\nu(x)$ and $\psi(x; x^\nu)$ are of class C^∞ in $E-A$, the same is true of $g(x)$. The function $g(x) = f(x)$ is of class C^m at all inner points of A, by §3. It remains to show that $D_kg(x)$ exists, equals $f_k(x)$, and is continuous, at all boundary points of A, for $\sigma_k \leqq m$.

Take a fixed boundary point x^0 of A, and any ϵ, $0<\epsilon<1$. Take

$$\eta < \epsilon/\{2c[(m+2)!]^n(108n^{1/2})^m N\} \text{ and } \eta < \epsilon/6,$$

where N is the largest of the numbers N_k for $\sigma_k \leqq m$. Take $M>|f_k(x)|$ ($\sigma_k\leqq m$, x in A and $r_{xx^0}\leqq 1$), and take

$$\delta < \epsilon/\{6(m+1)^n M\} \text{ and } \delta < 1$$

so small that (3.2) holds at the point x^0 with ϵ replaced by η. Take now any point y^* of $E-A$ within a distance $\delta/4$ of x^0; we shall show that

$$(11.2) \qquad |D_k g(y^*) - f_k(x^0)| < \epsilon \qquad (\sigma_k \leqq m).$$

Say the distance from y^* to A is $\delta_*/4$ (then $\delta_*<\delta$), and let x^* be a point of A distant $\delta_*/4$ from y^*. Consider the sum in (6.1) with x' and x replaced by x^* and x^0 respectively; as each l_h is $\leqq m$, it contains at most $(m+1)^n$ terms. If we take the term with $l_1=\cdots=l_n=0$ to the other side, there is in each remaining term a factor $(x_h^*-x_h^0)^{l_h}$ with $l_h>0$. As each $|x_h^*-x_h^0|$ is $<\delta<1$, we find

$$|\psi_k(x^*;x^0) - f_k(x^0)| < (m+1)^n M\delta < \epsilon/6.$$

But also $|R_k(x^*;x^0)|<\eta<\epsilon/6$; hence, using (6.2),

$$|f_k(x^*) - f_k(x^0)| < \epsilon/3.$$

Similarly we see that $|\psi_k(y^*;x^*)-f_k(x^*)|<\epsilon/6$; therefore

$$(11.3) \qquad |\psi_k(y^*;x^*) - f_k(x^0)| < \epsilon/2 \qquad (\sigma_k \leqq m).$$

Say y^* lies in the cube C of K_s, and let $I_{\lambda_1}, \cdots, I_{\lambda_t}$ be those sets I_λ with points in C. Each corresponding point y^{λ_q} is distant $<\delta/2$ from x^0, by (9.1), and hence each corresponding point x^{λ_q} is distant $<\delta$ from x^0. As the same is true of x^*, (3.2) gives

$$(11.4) \qquad |R_k(x^\nu;x^*)| \leqq r_{x^\nu x^*}^{m-\sigma_k}\eta \qquad (\nu=\lambda_1,\cdots,\lambda_t).$$

Set

$$\zeta_{\nu;k}(x) = \psi_k(x;x^\nu) - \psi_k(x;x^*) \qquad (\nu=\lambda_1,\cdots,\lambda_t);$$

then as $r_{x^\nu x^*}<\delta_*$ and $|x_h-x_h^\nu|<\delta_*$ for x in C, $|(x-x^\nu)^l|<\delta_*^{\sigma_l}$, and (6.3) and (11.4) give

$$(11.5) \qquad |\zeta_{\nu;k}(x)| < (m+1)^n\delta_*^{m-\sigma_k}\eta \text{ in } C \qquad (\nu=\lambda_1,\cdots,\lambda_t).$$

Using (9.2), we see that

$$(11.6) \qquad g(x) = \psi(x;x^*) + \sum_{s=1}^{t}\phi_{\lambda_s}(x)\zeta_{\lambda_s;0}(x) \text{ in } C.$$

As $D_k\psi(x; x^\nu)=\psi_k(x; x^\nu)$ and therefore $D_k\zeta_{\nu;0}(x)=\zeta_{\nu;k}(x)$,

$$D_k g(x) = \psi_k(x; x^*) + \sum_{s=1}^{t} \sum_l \binom{k}{l} D_l\phi_{\lambda_s}(x)\zeta_{\lambda_s;k-l}(x) \text{ in } C.$$

(10.1) and (11.5) give, as $t \leqq c$ (see §10) and

$$\binom{k_h}{l_h} \leqq m!,$$

$$(11.7) \qquad \left| D_k g(x) - \psi_k(x; x^*) \right| < \sum_l c[(m+1)!]^n 2^{s\sigma_l} N\delta_*^{\,m-\sigma_k+\sigma_l}\eta \text{ in } C.$$

Now the distance from C to A is $>\delta_*/6$; also, as C is in K_s, this distance is $<18n^{1/2}/2^s$. Hence $18n^{1/2}/2^s>\delta_*/6$, or, $2^s<108n^{1/2}/\delta_*$. This gives, as $\sigma_k \leqq m$ and $\delta_*<1$,

$$(11.8) \quad \left| D_k g(x) - \psi_k(x; x^*) \right| < c[(m+2)!]^n(108n^{1/2})^m N\delta_*^{\,m-\sigma_k}\eta < \epsilon/2$$

in C, and in particular, at y^*. This inequality together with (11.3) gives (11.2), as required.

The proof can now be completed with the aid of Lemma 1. (11.2) with $k=0$ shows that $g(x)$ is continuous throughout E. Take any number $k=(k_1, \cdots, k_n)$ with $\sigma_k<m$, and put $k'=(k_1, \cdots, k_h+1, \cdots, k_n)$. Assuming that $D_k g(x)$ is continuous in E, we shall show that $D_{k'}g(x)$ exists and is continuous in E. Take any boundary point $x^0=(x_1^0, \cdots, x_n^0)$ and put $z_0=x_h^0$, $w(z)=w(x_h)=D_k g(x_1^0, \cdots, x_h, \cdots, x_n^0)$, $w_0'=f_{k'}(x^0)$. Let A^* be the set of points of A for which $x_p=x_p^0$ $(p\neq h)$. (3.3) with $x=x^0$ and $\Delta x_h=x_h-x_h^0$, and (11.2) with k replaced by k', show that the conditions of the lemma are fulfilled; hence $\partial w(z_0)/\partial x_h=D_{k'}g(x^0)$ exists and equals $f_{k'}(x^0)$. (11.2) shows that $D_{k'}g(x^0)$ is continuous at x^0. Therefore $g(x)$ is of class C^m in E.

12. A differentiable extension of $f(x)$, m infinite. We now prove Lemma 2 for the case $m=\infty$. For any given m, let $\psi_{m;k}(x'; x)$ $(\sigma_k \leqq m)$ be the function given by the right hand side of (6.1). Choose the axes so that the origin falls on a point of A. Let S_p be the set of all points of E whose distances from the origin are $\leqq 2^p$, $p=1, 2, \cdots$. Let M_p be the maximum of $|f_k(x)|$ for $\sigma_k \leqq p$ and x in $A\cdot S_p$, and let $N^{(p)}$ be the maximum of N_k for $\sigma_k \leqq p$. Choose for each positive integer p a number δ_p such that

$$\delta_p < 1/\left\{2^{2p+1}c[(p+2)!]^n(36n^{1/2})^p N^{(p)} M_{p+1}\right\}, \ \delta_p < \delta_{p-1}/2.$$

The extension $g^{(\infty)}(x)$ of $f(x)$ is determined as follows. Given any number ν, determine the number γ_ν so that $\delta_{\gamma_\nu+1} \leqq r_\nu < \delta_{\gamma_\nu}$ (see §9); set $\gamma_\nu=0$ if $r_\nu>\delta_1$.

Put

$$(12.1)\qquad g^{(\infty)}(x) = \begin{cases} \sum_\nu \phi_\nu(x)\psi_{\gamma_\nu;0}(x;\, x^\nu) & \text{in } E - A, \\ f(x) & \text{in } A. \end{cases}$$

Given any fixed k, we shall find an inequality similar to (11.2) for $D_k g^{(\infty)}(x)$. Let $g^{(m)}(x)$ be the extension of $f(x)$ of class C^m given by Lemma 2 ($m=1, 2, \cdots$). Given any boundary point x^0 of A and any $\epsilon>0$, choose $p \geqq \sigma_k+2$ so that x^0 lies in S_p and so that $1/2^p<\epsilon$. Take $\delta<\delta_p$ so that (11.2) with g replaced by $g^{(\sigma_k)}$ will hold for our given k and any y^* of $E-A$ within δ of x^0; we show next that for any such y^*,

$$(12.2)\qquad \left| D_k g^{(\infty)}(y^*) - D_k g^{(\sigma_k)}(y^*) \right| < \epsilon.$$

Choose q so that $\delta_{q+1} \leqq \delta_* < \delta_q$, where δ_* is the distance from y^* to A; then $q \geqq p$. Define C, K_s, $I_{\lambda_1}, \cdots, I_{\lambda_t}$ as in §11. Note that for $\nu =$ any λ_h, $\delta_{\gamma_\nu+1} \leqq r_\nu < 2\delta < 2\delta_p < \delta_{p-1}$, hence $\gamma_\nu+1>p-1$, and thus $\gamma_\nu>p-2\geqq\sigma_k$. Set

$$(12.3)\qquad \xi_\nu(x) = \psi_{\gamma_\nu;0}(x;\, x^\nu) - \psi_{\sigma_k;0}(x;\, x^\nu) \qquad (\nu = \lambda_1, \cdots, \lambda_t);$$

using (12.1) and (11.1), we see that

$$(12.4)\qquad g^{(\infty)}(x) = g^{(\sigma_k)}(x) + \sum_{u=1}^{t} \phi_{\lambda_u}(x)\xi_{\lambda_u}(x) \text{ in } C.$$

Now $D_j\xi_\nu(x) = \psi_{\gamma_\nu;j}(x;\, x^\nu) - \psi_{\sigma_k;j}(x;\, x^\nu)$. If we replace k by j in (6.1), then those and only those terms in the sum with $\sigma_l \leqq m-\sigma_j$ occur. Replacing m by γ_ν and σ_k successively and subtracting, we have

$$(12.5)\qquad D_j\xi_\nu(x) = \sum_{\sigma_l=\sigma_k-\sigma_j+1}^{\gamma_\nu-\sigma_j} \frac{f_{j+l}(x^\nu)}{l!}(x - x^\nu)^l \text{ in } C.$$

Now $r_\nu > \delta_*/2$, by (9.1), hence $r_\nu > \delta_{q+2}$, and thus $\gamma_\nu \leqq q+1$ ($\nu = \lambda_1, \cdots, \lambda_t$); there are therefore less than $(q+2)^n$ terms in the sum, and in each term, $\sigma_j+\sigma_l \leqq q+1$. It follows that $|f_{j+l}(x^\nu)|<M_{q+1}$ in each term. Also $|x_h - x_h^\nu| < 2\delta_* < 2\delta_q$ and $\sigma_l \geqq \sigma_k - \sigma_j + 1$ in each term; hence

$$\left| D_j\xi_\nu(x) \right| < (q+2)^n M_{q+1} 2^{q+1} \delta_*^{\sigma_k-\sigma_j} \delta_q$$

in C. This with (12.4) gives

$$\left| D_k g^{(\infty)}(x) - D_k g^{(\sigma_k)}(x) \right| \leqq \sum_{u=1}^{t} \sum_j \binom{k}{j} \left| D_{k-j}\phi_{\lambda_u}(x) \right| \left| D_j\xi_{\lambda_u}(x) \right|$$

$$< c \sum_j \binom{k}{j} 2^{s(\sigma_k-\sigma_j)} N^{(\sigma_k)} (q+2)^n M_{q+1} 2^{q+1} \delta_*^{\sigma_k-\sigma_j} \delta_q$$

in C. Now the distance from C to A is $>\delta_*/2$ and is $<18n^{1/2}/2^s$; hence $2^s<36n^{1/2}/\delta_*$. Also $\sigma_k<p\leqq q$; therefore

$$
\begin{aligned}
(12.6)\qquad |D_kg^{(\infty)}(x)-D_kg^{(\sigma_k)}(x)| &< c[(q+2)!]^n(36n^{1/2})^qN^{(q)}M_{q+1}2^{q+1}\delta_q \\
&< 1/2^q<\epsilon
\end{aligned}
$$

in C, and in particular, at y^*, proving (12.2). Using (11.2), we find $|D_kg^{(\infty)}(y^*)-f_k(x^0)|<2\epsilon$ for any point y^* of $E-A$ within δ of x^0. Again we can apply Lemma 1 and show that $D_kg^{(\infty)}(x)$ exists and is continuous throughout E. As this is true for every k, the proof is complete.

13. We prove next a combined extension and approximation theorem.

LEMMA 3. *Let $f(x)$ be of class C^m (m finite) in E, with $D_kf(x)=f_k(x)$ ($\sigma_k\leqq m$) there, and let $f_k(x)$ ($m<\sigma_k\leqq m'$, $m'>m$ finite or infinite) be defined in the closed set A so that $f(x)$ (considered now only in A) is of class $C^{m'}$ there. Then for an arbitrary $\epsilon>0$ there is a function $g(x)$ which is of class C^m in E, of class $C^{m'}$ in a neighborhood of A, and equals $f(x)$ outside another neighborhood of A, such that*

$$(13.1)\qquad D_kg(x)=f_k(x)\ \text{in } A \qquad (\sigma_k\leqq m'),$$

and

$$(13.2)\qquad |D_kg(x)-D_kf(x)|<\epsilon\ \text{in } E \qquad (\sigma_k\leqq m).$$

Let $f'(x)$ be the extension of class $C^{m'}$ of the values of $f(x)$ in A given by the last lemma, and put $\zeta(x)=f'(x)-f(x)$; then $\zeta(x)$ is of class C^m in E, and

$$D_k\zeta(x)=0\ \text{in } A \qquad (\sigma_k\leqq m).$$

Set $\eta=\epsilon/\{c[(m+1)!]^n(36n^{1/2})^mN\}$ ($N=\max N_k$ for $\sigma_k\leqq m$). As $\zeta(x)$ is of class C^m and $D_k\zeta(x)$ vanishes in A ($\sigma_k\leqq m$), we can find an open set R containing A so that if y is any point of $R-A$, at a distance δ from A, then

$$|D_k\zeta(y)|<\eta\delta^{m-\sigma_k} \qquad (\sigma_k\leqq m).$$

Let $\nu_1,\nu_2,\cdots$ be those numbers such that I_{ν_p} lies wholly in R ($p=1,2,\cdots$). We set

$$(13.3)\qquad g(x)=f(x)+\zeta(x)\sum_{p=1}^{\infty}\phi_{\nu_p}(x)\ \text{in } E-A,$$

and $g(x)=f(x)$ in A. As $\sum\phi_{\nu_p}(x)=1$ in an open set surrounding A, $g(x)=f'(x)$ there. As $\sum\phi_{\nu_p}(x)=0$ in $E-R$, $g(x)=f(x)$ there. The statements about the class of $g(x)$ are true. To show that (13.2) holds, let y be a point of $R-A$, distant δ from A; then, defining C, K_s, $I_{\lambda_1},\cdots,I_{\lambda_t}$ as in the previous lemma, we have

$$| D_k g(y) - D_k f(y) | \leq \sum_p \sum_l \binom{k}{l} | D_l \phi_{\nu_p}(y) | \, | D_{k-l} \zeta(y) |$$
$$< c \sum_l (m!)^n 2^{\nu \sigma_l} N \delta^{m-\sigma_k+\sigma_l} \eta < \epsilon \qquad (\sigma_k \leq m).$$

14. We close this section with a theorem concerning the isolated points of A. Define α_p as follows:

$$(14.1) \qquad \alpha_p = \begin{cases} m \text{ if } m \text{ is finite} \\ p \text{ if } m \text{ is infinite} \end{cases} \qquad (p = 1, 2, \cdots).$$

LEMMA 4. *Consider the closed set $A = A' + a_1 + a_2 + \cdots$, where $a_1, a_2, \cdots$ are isolated points (then A' is closed), and let m be finite or infinite. Let $f_k(x)$ be defined in A' for $\sigma_k \leq m$ and at each a_ν for all k, so that $f(x)$ is of class C^m in A in terms of the $f_k(x)$ $(\sigma_k \leq m)$. Then there is a function $g'(x)$ of class C^∞ in $E - A'$ and of class C^m in E, such that*

$$(14.2) \qquad D_k g'(x) = f_k(x) \text{ in } A' \text{ for } \sigma_k \leq m \text{ and at each } a_\nu \text{ for all } k.$$

Let $g(x)$ be the extension of $f(x)$ of class C^m given by Lemma 2. Let $U_1, U_2, \cdots$ be neighborhoods of $a_1, a_2, \cdots$, chosen so that each is at a positive distance from each other and from A'. If m is finite, we alter $g(x)$ in U_1, next in U_2, etc., by means of the last lemma†, so that the new function $g'(x)$ will take on the required derivatives at $a_1 + a_2 + \cdots$, and so that

$$(14.3) \qquad | D_k g'(x) - D_k g(x) | < 1/p \text{ in } U_p \quad (\sigma_k \leq \alpha_p, \; p = 1, 2, \cdots).$$

(14.2) is an immediate consequence of this inequality and Lemma 1.

III. Approximation theorems

15. We prove first the following extension of the Weierstrass approximation theorem.‡

LEMMA 5. *Let $g(x)$ be of class C^m in E (m finite), and let S be a bounded closed set in E.§ Then for each $\epsilon > 0$ there exists a function $G(x)$ analytic in E and such that*

$$(15.1) \qquad | D_k G(x) - D_k g(x) | < \epsilon \text{ in } S \qquad (\sigma_k \leq m).$$

Let R_b be the set of points distant at most b from the origin ($b \geq 0$). Consider the n-tuple integral

† We use the last lemma with A replaced by a_ν and m' by ∞.

‡ Compare de la Vallée Poussin, *Cours d'Analyse*, vol. II, 2d edition, 1912, pp. 126–137.

§ It is sufficient that $g(x)$ be defined over S, for we can then extend its definition over E, by Lemma 2.

$$(15.2)\qquad \Phi(b) = T\int_{R_b} e^{-r_{0y}^2}\,dy = T\int\cdots\int e^{-(y_1^2+\cdots+y_n^2)}dy_1\cdots dy_n,$$

where T is chosen so that $\Phi(\infty)=1$; then $0\leqq\Phi(b)\leqq 1$ for all b. If we replace y by κy and b by κb, we see that

$$(15.3)\qquad \Phi(\kappa b) = T\kappa^n\int_{R_b} e^{-\kappa^2 r_{0y}^2}\,dy.$$

Let $v(x)$ be a function $\equiv 1$ in S, $\equiv 0$ outside some neighborhood of S, and of class C^∞ in E, such that $D_k v(x)=0$ in S for all k. (Such a function may be found for instance by the aid of Lemma 2.) Put $g'(x)=v(x)g(x)$, and

$$(15.4)\qquad G(x) = T\kappa^n\int_E g'(y)e^{-\kappa^2 r_{xy}^2}\,dy,$$

where κ will be chosen later; $G(x)$ is analytic in E. As r_{xy} is a function of $y-x$ alone, differentiating under the integral sign gives

$$D_kG(x) = T\kappa^n\int_E g'(y)D_k^{(x)}e^{-\kappa^2 r_{xy}^2}dy = (-1)^{\sigma_k}T\kappa^n\int_E g'(y)D_k^{(y)}e^{-\kappa^2 r_{xy}^2}\,dy,$$

where $D_k^{(x)}$ and $D_k^{(y)}$ denote differentiation with respect to x and y respectively. Integrating by parts σ_k times gives

$$(15.5)\qquad D_kG(x) = T\kappa^n\int_E D_kg'(y)e^{-\kappa^2 r_{xy}^2}\,dy.$$

As $\Phi(\infty)=1$, we see that

$$(15.6)\qquad D_kG(x) - D_kg'(x) = T\kappa^n\int_E [D_kg'(y) - D_kg'(x)]e^{-\kappa^2 r_{xy}^2}\,dy.$$

Take M so large that

$$(15.7)\qquad |D_kg'(x)| < M \text{ in } E \qquad (\sigma_k \leqq m).$$

The functions $D_kg'(x)$ are uniformly continuous in E; hence there is a $\delta>0$ such that

$$(15.8)\qquad |D_kg'(y) - D_kg'(x)| < \epsilon/2 \qquad (r_{xy}<\delta,\ \sigma_k\leqq m).$$

Take κ so large that

$$(15.9)\qquad 1 - \Phi(\kappa\delta) < \epsilon/(4M).$$

For a given x, let U consist of all points within δ of x; then if J_1 and J_2 are formed by replacing the domain of integration on the right hand side of

(15.6) by U and $E-U$ respectively, we have, using (15.3),

$$|J_1| < T\kappa^n \int_U \frac{\epsilon}{2} e^{-\kappa^2 r_{xy}^2}\, dy = \frac{\epsilon}{2}\Phi(\kappa\delta) < \frac{\epsilon}{2},$$

$$|J_2| < T\kappa^n \int_{E-U} 2Me^{-\kappa^2 r_{xy}^2}\, dy = 2M[1 - \Phi(\kappa\delta)] < \frac{\epsilon}{2},$$

and hence $|D_kG(x) - D_kg'(x)| < \epsilon$ in $E(\sigma_k \leqq m)$, which gives (15.1).

$G(x)$ may of course be replaced by a polynomial if desired.

16. The above lemma can be generalized as follows.

LEMMA 6. *Let R be an open set and let $R_1, R_2, \cdots$ be bounded open sets (some of which may be void) whose sum is R, such that each $\overline{R}_p = R_p$ plus boundary is in R_{p+1}. Then if $g(x)$ is defined and of class C^m (m finite or infinite) in R, and $\epsilon_1 \geqq \epsilon_2 \geqq \cdots$ are given positive numbers, there is an analytic function $G(x)$ defined in R such that*

$$(16.1)\qquad |D_kG(x) - D_kg(x)| < \epsilon_p \text{ in } R - R_p \qquad (\sigma_k \leqq \alpha_p,\ p = 1, 2, \cdots).$$

α_p is defined in (14.1). Note that, if $R_1, \cdots, R_q$ are void, then

$$(16.2)\qquad |D_kG(x) - D_kg(x)| < \epsilon_q \text{ in } R \qquad (\sigma_k \leqq \alpha_q).$$

Consider the closed set $\overline{R}_{p-1} + (\overline{R}_{p+1} - R_p) + (E - R_{p+2}) = Q_p' + Q_p + Q_p''$; if in Lemma 2 we replace A by this set and $f(x)$ by a function $\equiv 1$ in Q_p and $\equiv 0$ in $Q_p' + Q_p''$, we find a function $u_p(x)$ for each p, of class C^∞ in E, such that

$$(16.3)\qquad u_p(x) = \begin{cases} 1 \text{ in } Q_p, \\ 0 \text{ in } Q_p' + Q_p''; \end{cases} \qquad D_ku_p(x) = 0 \text{ in } Q_p' + Q_p + Q_p''\ (\sigma_k > 0).$$

(If R_{p+1} is void, we put $u_p(x) \equiv 0$; if R_{p+1} is not void but R_{p-1} is void, we have $u_p(x) = 0$ in Q_p'' and $=1$ in $\overline{R}_{p+1}$.) Let $Z_p \geqq 1$ be such a number that

$$(16.4)\qquad |D_ku_p(x)| < Z_p \text{ in } E \qquad (\sigma_k \leqq \alpha_p,\ p = 1, 2, \cdots).$$

We define successively analytic functions $G_1(x), G_2(x), \cdots$, by the following formula:

$$(16.5)\quad G_p(x) = T\kappa_p^n \int_E u_p(y)[g(y) - \{G_1(y) + \cdots + G_{p-1}(y)\}]e^{-\kappa_p^2 r_{xy}^2}\, dy.$$

(For $p = 1$, the factor in brackets is simply $g(y)$.) κ_p is chosen so that, if we set

$$(16.6)\qquad H_p(x) = u_p(x)[g(x) - \{G_1(x) + \cdots + G_{p-1}(x)\}],$$

then

$$(16.7)\quad |D_kG_p(x) - D_kH_p(x)| < \beta_p' = \epsilon_{p+1}/\{2^{p+2}[(\alpha_{p+1}+1)!]^n Z_{p+1}\} \quad \text{in } \bar{R}_{p+1} \quad (\sigma_k \leqq \alpha_{p+1})$$

(see Lemma 5); we shall restrict κ_p further later. Remembering the definition of $u_p(x)$, we see that (16.7) with (16.6) gives

$$(16.8)\quad |D_kg(x) - D_k\{G_1(x) + \cdots + G_p(x)\}| < \beta_p' < \epsilon_p/2 \text{ in } Q_p(\sigma_k \leqq \alpha_{p+1}).$$

Differentiating $H_p(x)$ and using (16.4) and (16.8) with p replaced by $p-1$, we see that (compare the derivation of (11.7))

$$|D_kH_p(x)| < [(\alpha_p + 1)!]^n Z_p\beta_{p-1}' = \epsilon_p/2^{p+1} \text{ in } Q_{p-1} \qquad (\sigma_k \leqq \alpha_p).$$

As $u_p(x)$ and its derivatives are 0 in R_{p-1}, this holds in R_{p-1} also; hence, using (16.7), we have

$$(16.9)\qquad |D_kG_p(x)| < \epsilon_p/2^p \text{ in } \bar{R}_p \qquad (\sigma_k \leqq \alpha_p).$$

We set now

$$(16.10)\qquad G(x) = G_1(x) + G_2(x) + \cdots;$$

this is the desired approximation to $g(x)$. To prove this, we see first from (16.9) that $D_k[G_1(x)+\cdots+G_p(x)]$ converges uniformly in any bounded closed subset of R $(\sigma_k \leqq m)$; hence $G(x)$ is defined in R, and

$$(16.11)\qquad D_kG(x) = D_kG_1(x) + D_kG_2(x) + \cdots \text{ in } R \qquad (\sigma_k \leqq m).$$

Next (16.9) shows that

$$(16.12)\quad \begin{aligned} |D_kG_{p+1}(x) + D_kG_{p+2}(x) + \cdots| &< \epsilon_{p+1}/2^{p+1} + \epsilon_{p+2}/2^{p+2} + \cdots \\ &\leqq \epsilon_p(1/2^2 + 1/2^3 + \cdots) = \epsilon_p/2 \text{ in } \bar{R}_{p+1} \qquad (\sigma_k \leqq \alpha_{p+1}); \end{aligned}$$

this with (16.8) gives $|D_kG(x) - D_kg(x)| < \epsilon_p$ in $Q_p(\sigma_k \leqq \alpha_p)$, proving (16.1).

It remains to be shown that $G(x)$ is analytic in R. To this end we extend the definition of each $G_p(x)$ to complex values of $x = (x_1' + ix_1'', \cdots, x_n' + ix_n'')$, using (16.5) still. Consider the analytic function of x

$$r_{xy}^2 = \sum (y_h - x_h)^2 = \sum [(y_h' - x_h') + i(y_h'' - x_h'')]^2;$$

as $y_h'' = 0$ in (16.5), the domain of integration being real,

$$\Re(r_{xy}^2) = \sum [(y_h' - x_h')^2 - x_h''^2].$$

Take any point x^0 of R and let U be the complex region of radius ρ about x^0, where ρ is so small that the real points in the complex region of radius 3ρ about x^0 lie in some R_q; we take q so that $3\rho^2 > 1/2^q$. Now if $p > q$, x is in U, and y is in $R - R_{p-1}$, then $\sum x_h''^2 < \rho^2$ and $\sum (y_h' - x_h')^2 \geqq 4\rho^2$, and hence

$$\Re(r_{xy}^2) > 3\rho^2.$$

Also $H_p(y)$ vanishes in R_q and in $E-R_{p+2}$ for $p>q$; therefore if M_p' is the maximum of $|H_p(y)|$ (note that $H_p(y)$ is determined before we determine κ_p) and V_p is the volume of R_p $(p=1, 2, \cdots)$,

$$(16.13)\qquad \begin{aligned}|G_p(x'+ix'')| &< T\kappa_p^n \int_{R_{p+2}-R_{p-1}} M_p' e^{-3\kappa_p^2\rho^2} dy \\ &< T\kappa_p^n\, e^{-\kappa_p^2/2^p} M_p' V_{p+2}\end{aligned}$$

for x in U and $p>q$. Hence if we choose κ_p successively for $p=1, 2, \cdots$, so that this quantity is $<1/2^p$ (and so that (16.7) holds), then the series in (16.10), when defined for complex values of x, converges uniformly in a complex neighborhood of any point of R. Therefore the function $G(x)$ is analytic in R, completing the proof.

17. The numbers κ_p as chosen above depend not only on the functions $u_p(x)$ but also on the function $g(x)$. Under certain restrictions, we can take them independent of $g(x)$, as follows.

LEMMA 7. *Let the open sets $R_1, R_2, \cdots$, the numbers $\epsilon_1, \epsilon_2, \cdots$, and the functions $u_1(x), u_2(x), \cdots$ be given as in Lemma 6; let $\Delta_1(r), \Delta_2(r), \cdots$ be a sequence of positive continuous functions defined for $r>0$, such that $\Delta_p(r)\to 0$ as $r\to 0$ and $\Delta_{p+1}(r)\geqq\Delta_p(r)$; let a be a point of R, and M a positive number. Then there is a sequence of numbers $\kappa_1, \kappa_2, \cdots$, with the following property. If $g(x)$ is any function of class C^m defined in R such that $|g(a)|\leqq M$ and*

$$(17.1)\qquad |D_k g(x') - D_k g(x)| < \Delta_p(r_{xx'}) \text{ in } \overline{R}_p \qquad (\sigma_k \leqq \alpha_p,\ p = 1, 2, \cdots),$$

and if $G(x)$ is defined in terms of $g(x)$ as in the previous lemma, using the above numbers κ_p, then $G(x)$ is analytic in R and (16.1) *holds.*

As the u's and their derivatives are uniformly continuous in E, there are functions $\Gamma_p(x)$ of the same sort as the Δ's above such that

$$(17.2)\qquad |D_k u_p(x') - D_k u_p(x)| < \Gamma_p(r_{xx'}) \text{ in } E \ (\sigma_k \leqq \alpha_{p+1},\ p = 1, 2, \cdots).$$

The conditions on $g(x)$ imply that for some M_1'', $|D_k g(x)|<M_1''$ in $\overline{R}_3$ $(\sigma_k\leqq\alpha_2)$.† Say

$$|D_k u_p(x)| < Z_p' \text{ in } E \qquad (\sigma_k \leqq \alpha_{p+1},\ p = 1, 2, \cdots).$$

Then as $u_1(x)=0$ in $R-R_3$, we have

$$|D_j u_1(x')D_l g(x') - D_j u_1(x) D_l g(x)| \leqq \Gamma_1(r_{xx'})M_1'' + Z_1'\Delta_3(r_{xx'})$$

† If d_3 is the diameter of $\overline{R}_3$, then $|g(x)|<M+\Delta_1(d_3)$ in $\overline{R}_3$. Now take any $k'=(k_1, \cdots, k_h-1, \cdots, k_n)$ and $k=(k_1, \cdots, k_n)(0<\sigma_k\leqq\alpha_2)$. Let $x'x''$ be a line segment parallel to the x_h-axis and lying wholly in $\overline{R}_3$; set $r=|x_h''-x_h'|$. As $|D_{k'}g(x'')-D_{k'}g(x')|<\Delta_1(r)$, the law of the mean gives, for some point x^* of $x'x''$, $|D_k(x^*)|<\Delta_1(r)/r$. Hence $|D_k(x)|<\Delta_1(r)/r+\Delta_2(d_3)$ in $\overline{R}_3$ $(0<\sigma_k\leqq\alpha_2)$.

for $\sigma_j \leqq \alpha_2$ and $\sigma_l \leqq \alpha_2$ and any x and x' in E. Hence if we put

$$\Delta_1^*(r) = [(\alpha_2 + 1)!]^n [\Gamma_1(r) M_1'' + Z_1' \Delta_3(r)],$$

we shall have, on differentiating $H_1(x) = u_1(x) g(x)$,

$$(17.3) \qquad | D_k H_1(x') - D_k H_1(x) | \leqq \Delta_1^*(r_{xx'}) \text{ in } E \qquad (\sigma_k \leqq \alpha_2).$$

Also $| D_k H_1(x) | < [(\alpha_2+1)!]^n Z_1' M_1''$ in $\overline{R}_3$ and $=0$ in $E - R_3$ $(\sigma_k \leqq \alpha_2)$; thus inequalities corresponding to (15.7) and (15.8) hold for $H_1(x)$. Hence if we take $\delta_1 > 0$ so that

$$\Delta_1^*(r) < \beta_1'/2 \qquad (r < \delta_1),$$

and take κ_1 so that

$$1 - \Phi(\kappa_1 \delta_1) < \beta_1' / \{4[(\alpha_2 + 1)!]^n Z_1' M_1''\},$$

then if we form $G_1(x)$ for any admissible $g(x)$ by means of (16.5), (16.7) will hold with $p = 1$; we take κ_1 large enough so that the right hand side of (16.13) with $p = 1$ will be $< 1/2$.

If we differentiate (16.5) with $p = 1$ σ_k times $(\sigma_k \leqq m)$, we derive an equation similar to (15.5); forming this for $x = x$ and $x = x'$ and subtracting, we find (changing y to $y + x' - x$ in one equation)

$$(17.4) \quad D_k G_1(x') - D_k G_1(x) = T \kappa_1^n \int_E [D_k H_1(y + x' - x) - D_k H_1(y)] e^{-\kappa_1^2 r_{xy}^2} dy.$$

This with (17.3), (15.3), and the definition of $\Phi(\infty)$ gives

$$(17.5) \qquad | D_k G_1(x') - D_k G_1(x) | \leqq \Delta_1^*(r_{xx'}) \text{ in } E \qquad (\sigma_k \leqq \alpha_2).$$

Assume now we have defined functions $\Delta_p^*(r)$ and have chosen numbers κ_p so that

$$(17.6) \qquad | D_k G_p(x') - D_k G_p(x) | \leqq \Delta_p^*(r_{xx'}) \text{ in } E \qquad (\sigma_k \leqq \alpha_{p+1}),$$

so that (16.7) holds, and so that the quantity in (16.13) is $< 1/2^p$, for $p < q$. Then for any admissible $g(x)$, $g(x) - \{G_1(x) + \cdots + G_{q-1}(x)\}$ satisfies the same kind of conditions as $g(x)$; hence, just as before, we find a function $\Delta_q^*(r)$ so that an inequality similar to (17.3) holds for $D_k H_q(x)$ in E $(\sigma_k \leqq \alpha_{q+1})$. Also $H_q(x)$ is bounded properly; hence we can choose κ_q so that (16.7) holds for any admissible $g(x)$ with p replaced by q, and so that (16.13) with $p = q$ is $< 1/2^q$. From this we show, as before, that (17.6) holds with p replaced by q. We can thus continue finding functions $\Delta_p^*(r)$ and numbers κ_p indefinitely. We put finally $G(x) = G_1(x) + G_2(x) + \cdots$, and show, just as in Lemma 6, that $G(x)$ has the required properties. This ends the proof.

IV. Analytic extensions

18. **Proof of Theorem I.** Let $g(x)$ be the extension of $f(x)$ of class C^m given by Lemma 2. Set $R=E-A$ and define $R_1, R_2, \cdots, \alpha_1, \alpha_2, \cdots$, and numbers $\epsilon_1, \epsilon_2, \cdots$, approaching zero as in §16. Define $G(x)$ in $E-A$ as in Lemma 6, and set $F(x)=G(x)$ in $E-A$, $F(x)=f(x)$ in A. That $F(x)$ is of class C^m in E and property (2) holds follows from (16.1) and Lemma 1, just as in §11; the other facts are obvious.

19. **The functions $\omega_{\nu k}(x)$.** In the next sections we shall discuss the analyticity of the extension of $f(x)$ at the isolated points of A. Let R be an open set, let $a_1, a_2, \cdots$ be points of R having no limit point in R, let $m_1, m_2, \cdots$ be corresponding integers $\geqq 0$, and let m be an integer $\geqq -1$ or ∞. We assume that if $a_{\nu_1}, a_{\nu_2}, \cdots$ is any sequence of points a_ν approaching the boundary of R, then

$$\liminf_{s\to\infty} m_{\nu_s} \geqq m. \tag{19.1}$$

Choose about each a_ν a neighborhood U_ν lying, with its boundary, in R, so that no two have common points. Define the numbers $\rho(\nu; k)$ so that when $(\nu; k)$ runs through the values $(1; k)$, $\sigma_k \leqq m_1$; $(2; k)$, $\sigma_k \leqq m_2$; etc.; then $\rho(\nu; k)$ runs through the values $1, 2, 3, \cdots$. Let $\rho'(\nu; k)$ equal one plus the largest of the numbers $m_1, \cdots, m_\nu, \rho(\nu; k)$.

Take any positive integer s, and consider all neighborhoods U_ν such that $\rho'(\nu k) \geqq s$ for some k $(\sigma_k \leqq m_\nu)$; let R_s be the set of all points of R whose distances from these neighborhoods and from the boundary of R are $>1/s$, and whose distances from the origin are $<s$. Then R_s is a bounded open set, $\overline{R}_s$ lies in R_{s+1} $(s=1, 2, \cdots)$, $R_1+R_2+\cdots=R$, and U_ν lies in $R-R_{\rho'(\nu k)}$ $(\sigma_k \leqq m_\nu)$. By Lemma 2, there are functions $\omega_{\nu k}(x)$ of class C^∞ in E, defined for $\sigma_k \leqq m_\nu$, $\nu=1, 2, \cdots$, such that

$$D_l\omega_{\nu k}(a_\nu) = \begin{cases} 1, & l = k \\ 0, & l \neq k \end{cases} \quad (\sigma_l \leqq m_\nu); \ \omega_{\nu k}(x) = 0 \text{ in } E - U_\nu. \tag{19.2}$$

Choose for each ν a positive number $\beta_\nu < 1/\nu$ so that $\beta_\nu \geqq \beta_{\nu+1}$, and

$$\beta_\nu \left| D_l\omega_{\nu k}(x) \right| < 1/\nu[(m_\nu + 1)^n] \text{ in } E \quad (\sigma_k \leqq m_\nu, \sigma_l \leqq m_\nu, \nu = 1, 2, \cdots). \tag{19.3}$$

Now let $f_{\nu k}$ be any set of numbers, defined for $\sigma_k \leqq m_\nu$, $\nu=1, 2, \cdots$, satisfying the condition

$$\left| f_{\nu k} \right| \leqq \beta_\nu \qquad (\sigma_k \leqq m_\nu, \nu = 1, 2, \cdots). \tag{19.4}$$

Set

$$(19.5)\qquad c_{\rho(\nu k)} = f_{\nu k}, \quad \omega'_{\rho(\nu k)}(x) = \omega_{\nu k}(x) \quad (\sigma_k \leqq m_\nu,\ \nu = 1, 2, \cdots).$$

Take any $s=\rho(\nu k)$. As $\omega_s'(x)=\omega_{\nu k}(x)=0$ in $R-U_\nu$, and $R_s=R_{\rho(\nu k)}$ is in $R_{\rho'(\nu k)}$ which has no points in common with U_ν,

$$(19.6)\qquad \omega_s'(x) = 0 \text{ in } R_s \qquad (s = 1, 2, \cdots).$$

20. **The transformation** L. Define functions $u_1(x)$, $u_2(x)$, $\cdots$ as in Lemma 6. Consider any function

$$(20.1)\qquad g(x) = \lambda_1\omega_1'(x) + \lambda_2\omega_2'(x) + \cdots \qquad (|\lambda_s| \leqq 1, s = 1, 2, \cdots);$$

such functions and λ's we shall call *admissible*. Set

$$(20.2)\qquad \epsilon_s = \beta_s/2^{s+1} \qquad (s = 1, 2, \cdots).$$

There are, obviously, functions $\Delta_1(r)$, $\Delta_2(r)$, $\cdots$, so that (17.1) holds for any such $g(x)$; hence, by Lemma 7, we can define numbers κ_1, κ_2, $\cdots$, so that if $G(x)$ is defined in terms of $g(x)$ as in Lemma 6, then $G(x)$ is analytic in R and (16.1) holds. *In using Lemma* 6, *we replace* α_p *by* p.

We note here a certain property of $G(x)$: *If* $g(x)$ *is admissible and*

$$(20.3)\qquad \textit{if } g(x) = 0 \textit{ in } R_s, \textit{ then } |D_kG(x)| < \epsilon_{s-1}/2^{s-2} \textit{ in } \overline{R}_{s-1} \qquad (\sigma_k \leqq s - 1).$$

As $u_p(x)=0$ in $R-R_s$ $(p\leqq s-2)$, $u_p(x)g(x)=0$ in E $(p\leqq s-2)$. Using (16.5), we see in succession that $G_1(x)\equiv 0$, $\cdots$, $G_{s-2}(x)\equiv 0$. This with (16.9) and (16.11) gives

$$|D_kG(x)| \leqq \sum_{p=s-1}^{\infty} |D_kG_p(x)| < \sum_{p=s-1}^{\infty} \epsilon_p/2^p \leqq \epsilon_{s-1}/2^{s-2}$$

in $\overline{R}_{s-1}$ $(\sigma_k\leqq s-1)$, as required.

Given any admissible $g(x)$, let $Lg(x)$ be the corresponding function $G(x)$. It follows easily from the definition of $G(x)$ that L is linear:

$$(20.4)\qquad L[\lambda_1g_1(x) + \lambda_2g_2(x)] = \lambda_1Lg_1(x) + \lambda_2Lg_2(x).$$

We show now that for admissible numbers λ,

$$(20.5)\qquad L\sum_{s=1}^{\infty}\lambda_s\omega_s'(x) = \sum_{s=1}^{\infty}\lambda_sL\omega_s'(x).$$

To prove this, take any point x^0 of R, in the set R_q, and any $\epsilon>0$. Take $q'\geqq q$ so that $1/2^{q'-s}<\epsilon$. (19.6) and (20.3), for $s=q'+1$, $q'+2$, $\cdots$, give, as $\lambda_s\omega_s'(x)$ is admissible,

$$\Big|\sum_{s=q'+1}^{\infty}\lambda_sL\omega_s'(x)\Big| = \Big|\sum_{s=q'+1}^{\infty}L\lambda_s\omega_s'(x)\Big| < \sum_{s=q'+1}^{\infty}1/2^{s-2} = 1/2^{q'-2} < \epsilon/2$$

in $R_{q'}$, and in particular, at x^0. As

$$\sum_{s=q'+1}^{\infty} \lambda_s \omega_s'(x) = 0 \text{ in } R_{q'+1}$$

and is admissible,

$$\left| L \sum_{s=q'+1}^{\infty} \lambda_s \omega_s'(x^0) \right| < 1/2^{q'-1} < \epsilon/2.$$

Moreover

$$L \sum_{s=1}^{\infty} \lambda_s \omega_s'(x) = \sum_{s=1}^{q'} \lambda_s L\omega_s'(x) + L \sum_{s=q'+1}^{\infty} \lambda_s \omega_s'(x);$$

hence

$$\left| L \sum_{s=1}^{\infty} \lambda_s \omega_s'(x^0) - \sum_{s=1}^{\infty} \lambda_s L\omega_s'(x^0) \right|$$

$$\leqq \left| L \sum_{s=q'+1}^{\infty} \lambda_s \omega_s'(x^0) \right| + \left| \sum_{s=q'+1}^{\infty} \lambda_s L\omega_s'(x^0) \right| < \epsilon,$$

which proves (20.5).

We prove two inequalities. Take any $(\nu; k)$, $(\mu; l)$ $(\sigma_k \leqq m_\nu, \sigma_l \leqq m_\mu)$; then

$$(20.6) \qquad \left| D_k L\omega_{\mu l}(a_\nu) - D_k \omega_{\mu l}(a_\nu) \right| < \epsilon_{\rho(\nu k)},$$

$$(20.7) \qquad \left| D_k L\omega_{\mu l}(a_\nu) - D_k \omega_{\mu l}(a_\nu) \right| < \epsilon_{\rho(\mu l)}.$$

The first follows from (16.1) when we note that a_ν is in $R - R_{\rho'(\nu k)}$, and $\epsilon_{\rho'(\nu k)} < \epsilon_{\rho(\nu k)}$, and $\sigma_k \leqq m_\nu < \rho'(\nu k)$ (recall that α_p was replaced by p in using Lemma 6). We now prove the second. As $\omega_{\mu l}(x) = 0$ in $R_{\rho'(\mu l)}$ and $\rho'(\mu l) \geqq \rho(\mu l) + 1$, (20.3) gives

(a) $\left| D_{k'} L\omega_{\mu l}(x) - D_{k'} \omega_{\mu l}(x) \right| < \epsilon_{\rho(\mu l)}$ in $\overline{R}_{\rho'(\mu l) - 1}$ $\qquad (\sigma_{k'} \leqq \rho'(\mu l) - 1)$.

Also (16.1) gives

(b) $\left| D_{k'} L\omega_{\mu l}(x) - D_{k'} \omega_{\mu l}(x) \right| < \epsilon_{\rho(\mu l)}$ in $R - R_p$ $\qquad (\sigma_{k'} \leqq p, \ p \geqq \rho'(\mu l) - 1)$.

Say a_ν is in $R_{p+1} - R_p$. As a_ν is not in $R_{\rho'(\nu k)}$, $\rho'(\nu k) \leqq p$, and $\sigma_k \leqq m_\nu \leqq \rho'(\nu k) - 1 \leqq p - 1$. If $p \geqq \rho'(\mu l) - 1$, (20.7) follows directly from (b). If $p < \rho'(\mu l) - 1$, then a_ν is in $R_{\rho'(\mu l) - 1}$, and $\sigma_k \leqq p - 1 < \rho'(\mu l) - 1$, and (a) applies.

21. **An infinite system of linear equations.** We prove here

LEMMA 8. *Suppose* η_s *and* c_s $(s = 1, 2, \cdots)$, *and* γ_{st} $(s, t = 1, 2, \cdots)$, *are given, so that* $1 \geqq \eta_s \geqq \eta_{s+1} > 0$ $(s = 1, 2, \cdots)$, $|c_s| \leqq 1$, *and*

$$(21.1) \qquad |\gamma_{st}| < \eta_s / 2^{t+1} \qquad (s, t = 1, 2, \cdots).$$

Then there are numbers λ_s $(s=1, 2, \cdots)$ *such that*

$$(21.2)\qquad \sum_{t=1}^{\infty}(\gamma_{st}+\delta_{st})\lambda_t = \sum_{t=1}^{\infty}\gamma_{st}\lambda_t + \lambda_s = c_s \qquad (s = 1, 2, \cdots),$$

and

$$(21.3)\qquad |\lambda_s - c_s| \leqq \eta_s \qquad (s = 1, 2, \cdots).$$

Using the method of successive approximations, put

$$(21.4)\qquad \lambda_{1s} = c_s, \quad \lambda_{ps} = -\sum_{t=1}^{\infty}\gamma_{st}\lambda_{p-1,t} \qquad (p = 2, 3, \cdots).$$

It is readily proved by induction that

$$(21.5)\qquad |\lambda_{ps}| < \eta_s/2^{p-1} \qquad (p = 2, 3, \cdots).$$

Hence the series $\lambda_{1s}+\lambda_{2s}+\cdots$ converges to a limit λ_s $(s=1, 2, \cdots)$, and

$$\sum_{t=1}^{\infty}(\gamma_{st}+\delta_{st})\lambda_t = \sum_{p=1}^{\infty}\sum_{t=1}^{\infty}(\gamma_{st}+\delta_{st})\lambda_{pt} = \sum_{p=1}^{\infty}(\lambda_{ps}-\lambda_{p+1,s}) = c_s,$$

$$|\lambda_s - c_s| = \Big|\sum_{p=2}^{\infty}\lambda_{ps}\Big| \leqq \sum_{p=2}^{\infty}\eta_s/2^{p-1} = \eta_s.$$

22. We are now ready to prove

LEMMA 9. *Let* R, m, a_ν, $m_\nu(\nu=1, 2, \cdots)$ *be defined as in* §19. *Then there are numbers* $\beta_\nu>0$ $(\nu=1, 2, \cdots)$ *with the following property. Given any set of numbers* $f_{\nu k}$ *defined for* $\sigma_k\leqq m_\nu$, $\nu=1, 2, \cdots$, *such that* (19.4) *holds, there exists a function* $G(x)$ *analytic in* R, *such that*

$$(22.1)\qquad D_kG(a_\nu) = f_{\nu k} \qquad (\sigma_k \leqq m_\nu, \quad \nu = 1, 2, \cdots),$$

and such that if we set $G(x)\equiv 0$ *in* $E-R$, *then* $G(x)$ *is of class* C^m *in* E, *and*

$$(22.2)\qquad D_kG(x) = 0 \text{ in } E - R \qquad (\sigma_k \leqq m).$$

We define the $\omega_{\nu k}(x)$ and the β_ν as in §19. Now take any $f_{\nu k}$ satisfying (19.4), and define the $c_{\rho(\nu k)}$ by (19.5). Define the ϵ_s and the transformation L as in §20. Set

$$(22.3)\qquad \eta_{\rho(\nu k)} = \beta_\nu \qquad (\sigma_k \leqq m_\nu, \nu = 1, 2, \cdots),$$

and

$$(22.4)\qquad \gamma_{st} = \gamma_{\rho(\nu k)\rho(\mu l)} = D_kL\omega_{\mu l}(a_\nu) - \delta_{st} = D_kL\omega_{\mu l}(a_\nu) - D_k\omega_{\mu l}(a_\nu).$$

Let $u=\rho(\theta j)$ be the larger of the two numbers $s=\rho(\nu k)$, $t=\rho(\mu l)$. Then using (20.6) or (20.7) according as $u=s$ or $u=t$, we find (as $\beta_{\rho(\theta j)}\leqq\beta_{\rho(\nu k)}\leqq\beta_\nu$)

$$|\gamma_{st}| < \epsilon_{\rho(\theta_j)} \leqq \beta_\nu/2^{u+1} \leqq \eta_s/2^{t+1}.$$

Also $|c_s| = |f_{\nu k}| \leqq \beta_\nu < 1$. Therefore the equations (21.2) have a solution $\lambda_1, \lambda_2, \cdots$, and

$$(22.5) \qquad \sum_{t=1}^{\infty}(\gamma_{st} + \delta_{st})\lambda_t = \sum_{\mu,l}\lambda_{\rho(\mu l)}D_kL\omega_{\mu l}(a_\nu) = c_s = c_{\rho(\nu k)} = f_{\nu k}.$$

By (22.8) below, the λ's are admissible (§20), and we can define the analytic function $G(x)$ in R by the equation

$$(22.6) \qquad G(x) = L\sum_{t=1}^{\infty}\lambda_t\omega_t'(x).$$

(20.5) and (22.5) give

$$(22.7) \qquad D_kG(a_\nu) = D_k\sum_{t=1}^{\infty}\lambda_tL\omega_t'(a_\nu) = \sum_{\mu,l}\lambda_{\rho(\mu l)}D_kL\omega_{\mu l}(a_\nu) = f_{\nu k}.$$

(19.6) and (20.3) show that the last sum above is uniformly convergent in any $\bar{R}_p$; hence the termwise differentiation is permissible.

Set $G(x) \equiv 0$ in $E-R$; we must show that $G(x)$ is of class C^m in E. (This is trivial if $m=-1$.) First note that, by (19.4) and (21.3),

$$(22.8) \qquad |\lambda_{\rho(\nu k)}| \leqq |c_{\rho(\nu k)}| + \eta_{\rho(\nu k)} = |f_{\nu k}| + \beta_\nu \leqq 2\beta_\nu;$$

this with (19.3) gives (replacing ν, k and l by μ, l and k)

$$(22.9) \qquad |\lambda_{\rho(\mu l)}D_k\omega_{\mu l}(x)| < 2/[\mu(m_\mu+1)^n] \text{ in } E \qquad (\sigma_k, \sigma_l \leqq m_\mu, \mu = 1, 2, \cdots).$$

Now take any boundary point x^0 of R, any integer $m' \leqq m$, and any $\epsilon > 0$. Take $q \geqq m'$ so that $\epsilon_q < \epsilon/2$. Take $\delta > 0$ so that R_q has no points within δ of x^0, and so that if ν is any number such that U_ν has points within δ of x^0, then $m_\nu \geqq m'$ and $2/\nu < \epsilon/2$ (see (19.1)). Consider any point y of R within δ of x^0, and take any k, $\sigma_k \leqq m'$. Either $D_k\omega_{\mu l}(y) = 0$ for all μ, l, or else for some μ, y lies in U_μ, in which case there are at most $(m_\mu+1)^n$ such numbers $\neq 0$, and $2/\mu < \epsilon/2$, and $m_\mu \geqq m'$. Thus if we replace x by y in (22.9) and sum over μ and l, we find

$$\left|D_k\sum_{t=1}^{\infty}\lambda_t\omega_t'(y)\right| = \left|\sum_{\mu,l}\lambda_{\rho(\mu l)}D_k\omega_{\mu l}(y)\right| < \epsilon/2 \qquad (\sigma_k \leqq m').$$

As y is in $R-R_q$, replacing α_q by $q \geqq m'$ in (16.1) gives

$$\left|D_kL\sum_{t=1}^{\infty}\lambda_t\omega_t'(y) - D_k\sum_{t=1}^{\infty}\lambda_t\omega_t'(y)\right| < \epsilon_q < \epsilon/2 \qquad (\sigma_k \leqq m').$$

This with the last inequality gives

$$|D_kG(y)| < \epsilon \text{ in } R \text{ if } r_{x^0y} < \delta \qquad (\sigma_k \leqq m');$$

the proof is completed with the aid of Lemma 1.

23. **Functions analytic at the isolated points of A.** Lemmas 4, 6 and 9 lead directly to the following theorem.

THEOREM II. *Let A be a closed set in E, and let $a_1, a_2, \cdots$ be isolated points of A. Set $A' = A - (a_1 + a_2 + \cdots)$. Let m be an integer $\geqq -1$ or ∞, and let the integers $m_\nu \geqq 0$, $\nu = 1, 2, \cdots$, satisfy (19.1). Let $f_k(x)$ be defined for x in $A'(\sigma_k \leqq m)$, and for $x = a_\nu$ $(\sigma_k \leqq m_\nu)$, so that $f(x)$ is of class C^m in A. Then there is a function $F(x)$ of class C^m in E such that*

(1) *$F(x) = f(x)$ in A,*
(2) *$D_kF(x) = f_k(x)$ in A' for $\sigma_k \leqq m$ and at each a_ν for $\sigma_k \leqq m_\nu$,*
(3) *$F(x)$ is analytic in $E - A'$.*

We asked that $f(x)$ be of class C^m in A, while $f_k(a_\nu)$ may not be defined for certain values of ν and k $(\sigma_k \leqq m)$. We require merely that after setting $f_k(a_\nu) = 0$ $(\sigma_k > m_\nu)$, $f(x)$ shall be of class C^m in A.

A special case of interest is $m = -1$. The m_ν and the $f_k(a_\nu)$ are then unrestricted. A' may be void, in which case $f(x)$ is analytic throughout E. A' may of course contain isolated points.

To prove the theorem, set $R = E - A'$ and determine the open sets R_ν and the numbers β_ν $(\nu = 1, 2, \cdots)$, as in §19. Let $g'(x)$ be the extension of $f(x)$ of class C^m in E and of class C^∞ in $E - A'$ given by Lemma 4 (setting $f_k(a_\nu) = 0$ for $\sigma_k > m_\nu$). Let $G'(x)$ be the analytic function in R given by Lemma 6 (with α_p replaced by p) such that

$$(23.1) \qquad |D_kG'(x) - D_kg'(x)| < \beta_p \text{ in } R - R_p \qquad (\sigma_k \leqq p),$$

and set $G'(x) = f(x)$ in A'. $G'(x)$ is of class C^m in E, and

$$(23.2) \qquad D_kG'(x) = f_k(x) \text{ in } A' \qquad (\sigma_k \leqq m)$$

(see §18). Set

$$(23.3) \qquad f_{\nu k} = D_kg'(a_\nu) - D_kG'(a_\nu) \qquad (\sigma_k \leqq m_\nu, \nu = 1, 2, \cdots).$$

As a_ν lies in $R - R_{\rho'(\nu k)}$ and $\rho'(\nu k) > m_\nu$, (23.1) gives $|f_{\nu k}| < \beta_{\rho'(\nu k)} < \beta_\nu (\sigma_k \leqq m_\nu)$. Thus the conditions of Lemma 9 are satisfied, and there is a function $G(x)$ analytic in R, $\equiv 0$ in A', of class C^m in E, and such that (22.1) and (22.2) hold. Set

$$(23.4) \qquad F(x) = G'(x) + G(x);$$

then $F(x)$ is our required function. It is of class C^m in E as the same is true of $G'(x)$ and $G(x)$; it is analytic in $R=E-A'$ as the same is true of $G'(x)$ and $G(x)$; it equals $f(x)$ in A' as $G'(x)=f(x)$ and $G(x)=0$ there. (22.1), (23.3) and (14.2) show that $D_kF(a_\nu)=D_kg'(a_\nu)=f_k(a_\nu)$ $(\sigma_k\leqq m_\nu)$; (22.2) and (23.2) show that $D_kF(x)=f_k(x)$ in A', completing the proof.

24. **An extension-approximation theorem.** We prove here

THEOREM III.† *Let A be closed, and let $A_{-1}, A_0, A_1, \cdots$ be closed subsets of A such that each A_s lies in A_{s+1}. Let $a_{s1}, a_{s2}, \cdots$ be points of A_s-A_{s-1} which are isolated points of A, and set $A'=A-\sum a_{st}$. Let B_{-1} be void, and let $B_0, B_1, \cdots$ be sets whose sum B lies in $E-A$, such that each B_s lies in B_{s+1}, such that each set B_s-B_{s-1} has limit points in $B-B_{s-1}+A_s$ only, and such that each set $A+B-B_s$ is closed. Let $f_k(x)$ be defined in each set $\Gamma_s=A+B-(A_{s-1}+B_{s-1})$ for $\sigma_k\leqq s$ $(s=0, 1, \cdots)$ so that $f(x)=f_0(x)$ is of class C^s in Γ_s in terms of the $f_k(x)$ for each s. Let $\epsilon(x)$ be a continuous function, positive in $E-A'$ and zero in A'. Then there is a function $F(x)$ defined in $E-A_{-1}$ such that*

(1) *$F(x)$ is of class C^s in $E-A_{s-1}$ $(s=0, 1, \cdots)$,*
(2) *$D_kF(x)=f_k(x)$ in $A-A_{s-1}$ $(\sigma_k\leqq s, s=0, 1, \cdots)$,*
(3) *$|D_kF(x)-f_k(x)|<\epsilon(x)$ in $B-B_{s-1}$ $(\sigma_k\leqq s, s=0, 1, \cdots)$,*
(4) *$F(x)$ is analytic in $E-A'$.*

Any number of sets A_s, B_s may be void; any of the points a_{st} may not exist. Note that if $A_\infty=A-(A_{-1}+A_0+\cdots)$, then $F(x)$ is of class C^∞ at all points of A_∞. Theorem I for m finite is obtained by letting B and $A_{-1}, \cdots, A_{m-1}$ be void, and setting $A=A_m$; and for m infinite, by letting B and every A_s be void. Theorem II is obtained similarly; we arrange the a_{st} in a sequence $a_1, a_2, \cdots$, and set $m_\nu=s$ if a_ν is in A_s-A_{s-1}. Lemma 6 is obtained by setting $A=A_{-1}=E-R$, $B_s=R_{s+1}$ $(s=0, 1, \cdots)$, and taking $\epsilon(x)$ so that $\epsilon(x)\leqq\epsilon_s$ in $R-R_s$.

We turn now to the proof. Take a subdivision of the open set $E-(A+B)$ as in §8, let $y^{0\nu}$ $(\nu=1, 2, \cdots)$ be the vertices of the cubes, and let $x^{0\nu}$ be a point of Γ_0 whose distance from $y^{0\nu}$ is not more than twice the distance from $y^{0\nu}$ to Γ_0. Define the functions $\phi_{0\nu}(x)$ in $E-(A+B)$ as in §9, and define $g_0'(x)$ by (11.1), using the functions $\phi_{0\nu}(x)$ and $\psi_{0;0}(x; x^{0\nu})=f(x^{0\nu})$, and replacing $E-A$ and A by $E-(A+B)$ and Γ_0 respectively. Then $g_0'(x)$ is defined throughout $E-A_{-1}$, and is easily seen to be a continuous extension of $f(x)$. Let $g_0(x)$ be a function $=g_0'(x)$ in $A-A_{-1}+\Gamma_1$ and analytic in the open set $E-(A+\Gamma_1)$ so that

† A special case of this theorem has been proved by A. Besikowitsch, *Über analytische Funktionen mit vorgeschriebenen Werten ihrer Ableitungen*, Mathematische Zeitschrift, vol. 21 (1924), pp. 111–118.

$$|g_0(x) - g_0'(x)| < \theta_1(x)/4 \text{ in } E - (A + \Gamma_1), \tag{24.1}$$

where $\theta_p(x) = \min [\epsilon(x), \text{distance from } x \text{ to } A+\Gamma_p]$ $(p=1, 2, \cdots)$. Then $g_0(x)$ is continuous in $E-A_{-1}$.

We shall now define in succession functions $g_1(x), g_2(x), \cdots$, with the following properties:

(a) $g_p(x)$ is defined in $E-A_{-1}$, is of class C^s in $E-A_{s-1}$ $(s=0, \cdots, p)$, and is analytic in $E-(A+\Gamma_{p+1})$.

(b) $D_k g_p(x) = f_k(x)$ in $A - A_{s-1} + \Gamma_{p+1}$ $(\sigma_k \leqq s,\ s=0, \cdots, p)$.

(c) $|D_k g_p(x) - D_k g_{p-1}(x)| < \epsilon(x)/2^{p+2}$ in $B_{p-1} - B_{s-1}$ $(\sigma_k \leqq s,\ s=0, \cdots, p-1)$.

(d) $|D_k g_p(x) - f_k(x)| < \epsilon(x)/2^{p+2}$ in $B_p - B_{p-1}$ $(\sigma_k \leqq p)$.

Assuming $g_0(x), \cdots, g_{p-1}(x)$ are defined, we shall define $g_p(x)$. Consider any point of Γ_p; it is at a positive distance from the closed set A_{p-1}, and hence we can enclose it in an open set lying at a positive distance from A_{p-1}. We thus enclose Γ_p in an open set Γ_p' containing no points of A_{p-1}, and having no limit points in A_{p-1} other than limit points of Γ_p. Take a subdivision of the open set $E-(A+\Gamma_p)$, let $y^{p\nu}$ $(\nu=1, 2, \cdots)$ be the vertices of the cubes, and let $x^{p\nu}$ be a point of Γ_p whose distance from $y^{p\nu}$ is not more than twice the distance from $y^{p\nu}$ to Γ_p $(\nu=1, 2, \cdots)$. Define the functions $\phi_{p\nu}(x)$ in $E-(A+\Gamma_p)$ as in §9 and define $\psi_{p;k}(x'; x)$ by (6.1) $(\sigma_k \leqq p)$, replacing m by p. Remembering that $f(x)$ is of class C^p in Γ_p, set

$$g_p'(x) = \sum_\nu \phi_{p\nu}(x)\psi_{p;0}(x; x^{p\nu}) \text{ in } \Gamma_p' - \Gamma_p, \tag{24.2}$$

and set $g_p'(x) = g_{p-1}(x)$ in Γ_p. From the proof in §11 it is seen that $g_p'(x)$ is an extension of class C^p of the values of $f(x)$ in Γ_p.

Set $\zeta_p(x) = g_p'(x) - g_{p-1}(x)$ in Γ_p'; then $\zeta_p(x)$ is of class C^{p-1} in Γ_p', and

$$D_k\zeta_p(x) = 0 \text{ in } \Gamma_p \qquad (\sigma_k \leqq p-1). \tag{24.3}$$

Set $\eta_p = 1/\{2^{p+4}c[(p+1)!]^n(36n^{1/2})^p N^{(p)}\}$ $(p=0, 1, \cdots)$, where $N^{(p)} = \max N_k$ for $\sigma_k \leqq p$. Let K_{p-1} be the set of points of B_{p-1} for which

$$|D_k\zeta_p(x)| \geqq \eta_{p-1}\epsilon(x)\delta_x^{p-1-\sigma_k} \text{ for some } k \qquad (\sigma_k \leqq p-1),$$

where δ_x is the distance from x to Γ_p, or 1 if that is smaller. Each point of $A - A_{p-1}$ is at a positive distance from K_{p-1}, as B_{p-1} has no limit points in $A - A_{p-1}$, and each point of $B - B_{p-1}$ is at a positive distance from K_{p-1} on account of (24.3), as $\epsilon(x) > 0$ in B; hence each point of Γ_p is at a positive distance from K_{p-1}, and we can enclose Γ_p in an open set Γ_p'' which lies in Γ_p' and contains no points of K_{p-1}. Now

(24.4) $$|D_k\zeta_p(x)| < \eta_{p-1}\epsilon(x)\delta_x^{p-1-\sigma_k} \text{ in } \Gamma_p'' \cdot B_{p-1} \qquad (\sigma_k \leqq p-1).$$

We can also take Γ_p'' so that if $\rho_p(x)$ is the distance from x to A_{p-1}, then

(24.5) $$|D_k\zeta_p(x)| < \eta_{p-1}\rho_p(x)\delta_x^{p-1-\sigma_k} \text{ in } \Gamma_p'' \qquad (\sigma_k \leqq p-1).$$

Let I_{ν_t} be those sets I_ν of the subdivision of $E-(A+\Gamma_p)$ lying wholly in Γ_p'' $(t=1, 2, \cdots)$, and set

(24.6) $$g_p''(x) = g_{p-1}(x) + \zeta_p(x)\sum_{t=1}^{\infty}\phi_{\nu_t}(x) \text{ in } E-(A+\Gamma_p),$$

and $g_p''(x)=g_{p-1}(x)=f(x)$ in $A-A_{-1}+\Gamma_p$. Then $g_p''(x)$ is of class C^p in $E-A_{p-1}$, and with the help of (24.4) we find

(24.7) $$|D_kg_p''(x) - D_kg_{p-1}(x)| < \epsilon(x)/2^{p+3} \text{ in } B_{p-1} \qquad (\sigma_k \leqq p-1)$$

(see §13). Also (24.5) gives

(24.8) $$|D_kg_p''(x) - D_kg_{p-1}(x)| < \rho_p(x)/2^{p+3} \text{ in } E-A \qquad (\sigma_k \leqq p-1),$$

and hence $g_p''(x)$ is of class C^s in $E-A_{s-1}$ $(s=0, \cdots, p)$, as the same is true of $g_{p-1}(x)$ $(s=0, \cdots, p-1)$.

Finally let $g_p(x)$ be an analytic function such that

(24.9) $$|D_kg_p(x) - D_kg_p''(x)| < \theta_{p+1}(x)/2^{p+3} \text{ in } E-(A+\Gamma_{p+1}) \quad (\sigma_k \leqq p);$$

set $g_p(x)=g_{p-1}(x)=f(x)$ in $A-A_{-1}+\Gamma_{p+1}$. Then $g_p(x)$ has all the required properties. (c) is a direct consequence of the above inequality and (24.7); (d) follows from (24.9) and the fact that $D_kg_p''(x)=f_k(x)$ in $\Gamma_p(\sigma_k \leqq p)$; (a) and (b) follow with the aid of Lemma 1.

Set

(24.10) $$g(x) = \lim_{p\to\infty} g_p(x) \text{ in } E-A_{-1}.$$

By (24.8) and (24.9), $g(x)$ exists and is of class C^∞ in $E-A$. Let x be any point of any A_s-A_{s-1}; by (a), $g_p(x)$ is of class C^s in a neighborhood of x for $p \geqq s$, and by (24.8) and (24.9), the same is true of $g(x)$. The same argument, using (b), shows that

(24.11) $$D_kg(x) = f_k(x) \text{ in } A-A_{s-1} \quad (\sigma_k \leqq s,\ s=0, 1, \cdots).$$

Finally (c), (d), (24.1) and the definition of $g_0'(x)$ show that

(24.12) $$|D_kg(x) - f_k(x)| < \epsilon(x)/2 \text{ in } B-B_{s-1} \quad (\sigma_k \leqq s,\ s=0, 1, \cdots).$$

We have now found an extension $g(x)$ with all the properties but (4). It is replaced by an analytic extension $F(x)$ just as in §23; we must be careful merely to make

(24.13) $\quad |D_kF(x) - D_kg(x)| < \epsilon(x)/2$ in $B - B_{s-1}$ $\quad (\sigma_k \leqq s, s = 0, 1, \cdots)$.

Let $a_1, a_2, \cdots$ be the a_{st} arranged in a sequence, and set $m_\nu = s$ if a_ν is in $A_s - A_{s-1}$. Set $R = E - A'$. Let R_p consist of those points of the R_p of §19 whose distances from the closed set $A + B - B_{p-1}$ are $> 1/p$ $(p = 1, 2, \cdots)$ Every point of $E - A'$ lies in some R_p, as $B - (B_0 + B_1 + \cdots)$ is void. Take the β_ν (§19) small enough so that if $|\lambda_{\rho(\nu k)}| \leqq 2\beta_\nu$ (see (22.8)) and $g^*(x) = \sum \lambda_t \omega_t'(x)$, then

(24.14) $\quad |D_kg^*(x)| < \epsilon(x)/8$ in $R_{p+1} - R_p$ $\quad (\sigma_k \leqq p)$.

Let ϵ_s' be one eighth the lower bound of $\epsilon(x)$ for x in R_{s+1} $(s = 1, 2, \cdots)$, or 1 if that is smaller. Replace the ϵ_s of (20.2) by min $(\epsilon_s, \epsilon_s')$. Then for any such $g^*(x)$, (16.1) gives

(24.15) $\quad |D_kLg^*(x)| < \epsilon_p + |D_kg^*(x)| < \epsilon(x)/4$ in $R_{p+1} - R_p$ $\quad (\sigma_k \leqq p)$.

Replace the $g'(x)$ of §23 by the present $g(x)$, and determine $G'(x)$ so that

(24.16) $\quad |D_kG'(x) - D_kg(x)| < \min\, [\beta_p, \epsilon(x)/4]$ in $R - R_p$ $\quad (\sigma_k \leqq p)$,

and, in particular, in $B - B_{p-1}$. Now if $G(x)$ and $F(x)$ are determined as in §23, then (24.13) and hence property (3) hold; the other properties are easily verified, and the proof of the theorem is complete.

Harvard University,
Cambridge, Mass.

DERIVATIVES, DIFFERENCE QUOTIENTS, AND TAYLOR'S FORMULA*

BY HASSLER WHITNEY†

1. *Introduction*. Let $f(x)$ be defined in the closed interval I. If $f(x)$ has a continuous mth derivative, it can be expanded in a Taylor's formula with $m+1$ terms plus remainder; the mth difference quotient of $f(x)$ approaches $d^m f(x)/dx^m$ uniformly. If $f(x)$ is a polynomial of degree at most $m-1$, then the mth difference quotient is identically zero. The object of the present note is to prove converses of these theorems. The results hold also in an open interval, as they hold in every closed subinterval.

2. *Difference Quotients*. Given a function $f(x)$ defined in I, the pth difference quotient is defined by the equations $\Delta_h^0 f(x) = f(x)$, and

$$(1)\quad \begin{aligned} \Delta_h^p f(x) &= \frac{1}{h^p}\sum_{i=0}^{p}(-1)^{p-i}\binom{p}{i} f(x+ih) \\ &= \frac{1}{h}[\Delta_h^{p-1} f(x+h) - \Delta_h^{p-1} f(x)] \end{aligned}$$

for $p>0$. We say $\Delta_h^p f(x) \to f_p(x)$ uniformly in I if for every $\epsilon>0$ there is a $\delta>0$ such that $|\Delta_h^p f(x) - f_p(x)| < \epsilon$ for every x in I and every h, $|h|<\delta$.‡

(a) Suppose $f_0(x), f_1(x), \cdots, f_m(x)$ are defined in I, and

$$(2)\ f_0(x+h) = f_0(x) + \frac{1}{1!}f_1(x)h + \cdots + \frac{1}{m!}f_m(x)h^m + R(x,h),$$

where $R(x,h)/h^m \to 0$ uniformly in I as $h\to 0$. If we form the pth difference quotient, we find§

* Presented to the Society, June 23, 1933.

† National Research Fellow.

‡ We always suppose that the values of x under consideration (here, $x+ih$ for $i=0,\cdots,p$) lie in I.

§ If we solve the linear equations (setting $0^0=1$)

$$0^j u_0 + 1^j u_1 + \cdots + p^j u_p = p!\delta_{jp}, \qquad (j=0,\cdots,p),$$

we find $u_i = (-1)^{p-i}\binom{p}{i}$; hence $\sum_{i=0}^{p}(-1)^{p-i}\binom{p}{i} i^j = 0$, $(j<p)$, and $=p!$, $(j=p)$.

$$(3)\quad \Delta_h^p f_0(x) = \frac{1}{h^p}\sum_{i=0}^{p}(-1)^{p-i}\binom{p}{i}\left[\sum_{j=0}^{m}\frac{f_j(x)}{j!}(ih)^j + R(x,ih)\right]$$
$$= \sum_{j=0}^{m}\frac{f_j(x)}{j!}h^{j-p}\sum_{i=0}^{p}(-1)^{p-i}\binom{p}{i}i^j + \frac{1}{h^p}R_p(x,h)$$
$$= f_p(x) + \sum_{j=p+1}^{m}\frac{f_j(x)}{j!}h^{j-p}\sum_{i=0}^{p}(-1)^{p-i}\binom{p}{i}i^j$$
$$+ \frac{1}{h^p}R_p(x,h),$$

where

$$(4)\qquad R_p(x,h) = \sum_{i=0}^{p}(-1)^{p-i}\binom{p}{i}R(x,ih).$$

From (2) we see that $f_0(x)$ is continuous; hence $\Delta_h^p f_0(x)$ is continuous (h fixed, $p=1,\cdots,m$). Setting $p=m$ in (3), we see that $\Delta_h^m f_0(x)\to f_m(x)$ uniformly; hence $f_m(x)$ is continuous. Having proved that $\Delta_h^q f_0(x)\to f_q(x)$ uniformly and $f_q(x)$ is continuous, ($q=m, m-1,\cdots,p+1$), (3) shows that $\Delta_h^p f_0(x)\to f_p(x)$ uniformly, and hence $f_p(x)$ is continuous, ($p=1,\cdots,m$).

(b) Suppose $f(x)=f_0(x)$ and $d^pf(x)/dx^p$ exists and equals the continuous function $f_p(x)$, ($p=1,\cdots,m$). Then (2) holds, by Taylor's theorem, and hence $\Delta_h^p f(x)\to d^pf(x)/dx^p$ uniformly, ($p=1,\cdots,m$).

(c) Suppose $f(x)=a_0+a_1x+\cdots+a_mx^m$ is a polynomial of degree at most m. Then a Taylor's formula (2) holds with $f_m(x)\equiv m!a_m$ and $R(x,h)\equiv 0$, and using (3) with $p=m$, we have $\Delta_h^m f(x)\equiv m!a_m$. If $f(x)$ is of degree at most $m-1$, $\Delta_h^m f(x)\equiv 0$.

3. *Rolle's Theorem for Difference Quotients.* We shall prove first the following lemma.

LEMMA 1. *If $k_0<k_1<\cdots<k_m$ are integers, h is >0, and*

$$(5)\qquad (-1)^{m-i}f(x+k_ih)\geqq 0,\quad (i=0,\cdots,m),$$

then there is an integer $k_0^{(m)}$ ($k_0\leqq k_0^{(m)}<k_m$) such that

$$(6)\qquad \Delta_h^m f(x+k_0^{(m)}h)\geqq 0.$$

As $f(x+k_mh)\geqq 0$ and $f(x+k_{m-1}h)\leqq 0$, there is a $k_{m-1}^{(1)}$,

($k_{m-1} \leqq k_{m-1}^{(1)} < k_m$), such that $\Delta_h^1 f(x+k_{m-1}^{(1)}h) \geqq 0$. In this manner we find numbers $k_0^{(1)}, \cdots, k_{m-1}^{(1)}$, ($k_{i-1} \leqq k_{i-1}^{(1)} < k_i$), such that

$$(-1)^{m-1-i}\Delta_h^1 f(x + k_i^{(1)}h) \geqq 0, \quad (i = 0, \cdots, m-1).$$

By the same method, using $\Delta_h^1 f(x)$ in place of $f(x)$, we find numbers $k_0^{(2)}, \cdots, k_{m-2}^{(2)}, (k_{i-1}^{(1)} \leqq k_{i-1}^{(2)} < k_i^{(1)})$, such that

$$(-1)^{m-2-i}\Delta_h^2 f(x + k_i^{(2)}h) \geqq 0, \quad (i = 0, \cdots, m-2).$$

Continuing in this manner we find, finally, the required number $k_0^{(m)}$.

4. THEOREM 1. *Let $f(x)$ be measurable in the closed interval I. A necessary and sufficient condition that $f(x)$ be a polynomial of degree at most $m-1$ is that $\Delta_h^m f(x) \to 0$ uniformly.**

The theorem is not true without the measurability condition. For the discontinuous functions defined by G. Hamel† are seen to have the property $\Delta_h^r f(x) \equiv 0$.

The necessity of the condition follows from §2 (c); we must prove the sufficiency. We shall prove first the following lemma.

LEMMA 2. *Let $f(x)$ satisfy the conditions of the theorem. Let $r_1, r_2, \cdots, r_m$ be rational numbers. Take a fixed number a and a fixed t. Let $P(x)$ be the polynomial of degree at most $m-1$ such that $f(a+r_it) = P(a+r_it)$, $(i=1, \cdots, m)$. Then if r is any rational number, $f(a+rt) = P(a+rt)$.*

Suppose that the lemma is not true. Then for some rational r_0, $f(a+r_0t) \neq P(a+r_0t)$. Let

$$Q(x) = c_0 + c_1x + \cdots + c_mx^m, \quad (c_m \neq 0),$$

be the polynomial of degree m such that $Q(a+r_it) = f(a+r_it)$, $(i=0, \cdots, m)$. Set $\phi(x) = f(x) - Q(x)$; then $\phi(a+r_it) = 0$, $(i=0, \cdots, m)$, and by §2 (c),

$$\Delta_h^m \phi(x) = \Delta_h^m f(x) - m!c_m \to -m!c_m$$

uniformly as $h \to 0$. Hence we can take a $\delta > 0$ so that $\Delta_h^m \phi(x) < 0$ or > 0 for all x in I and $h < \delta$ according as $c_m > 0$ or $c_m < 0$. As

* The case that $f(x)$ is continuous and $\Delta_h^m f(x) \equiv 0$ has been considered by Anghelutza, Mathematica (Cluj), vol. 6 (1932), pp. 1–7.

† Mathematische Annalen, vol. 60(1905), p. 461, equation (2).

the r_i are rational, there is a number r such that $rt<\delta$ and the numbers $a+krt$ (k an integer) include the numbers $a+r_it$. By Lemma 1, there are integers k' and k'' such that $\Delta_{rt}^m\phi(a+k'rt)\geqq 0$ and $\Delta_{rt}^m\phi(a+k''rt)\leqq 0$, a contradiction, proving the lemma.

5. *Proof of Theorem* 1. If we define $P(x)$ as in the lemma, a and t being rational, then $f(x)=P(x)$ at all rational points of I. If $f(x)$ is continuous, it follows that $f(x)\equiv P(x)$ in I. To complete the proof of the theorem, we must show that if $f(x)\not\equiv P(x)$, then $f(x)$ is not measurable.

Suppose there is a number a in I such that $f(a)\neq P(a)$. If $Q(x)$ is the polynomial of degree at most $m-1$ such that $f(x)=Q(x)$ at points $a, a+r_2, \cdots, a+r_m$, ($r_2, \cdots, r_m$ rational), then $f(x)=Q(x)$ at all points $a+r$, (r rational), by the lemma. Set $\sigma=|Q(a)-P(a)|$, and take $\delta>0$ so that

$$(7)\qquad |P(x)-P(a)|<\frac{\sigma}{4},\qquad |Q(x)-Q(a)|<\frac{\sigma}{4},\qquad (|x-a|\leqq\delta).$$

Take $\eta>0$ so that if $R(x)$ is any polynomial of degree at most $m-1$ such that $|R(i)|\leqq\eta$, $(i=1,\cdots,m)$, then $|R(0)|<\sigma/4$.* By a change of variable, we see then that for any x, h, and y_0, *if* $|R(x+ih)-y_0|\leqq\eta$, $(i=1,\cdots,m)$, *then*

$$|R(x)-y_0|<\sigma/4.$$

If $f(x)$ is measurable, there is a number y_0 such that the set of points E in $(a-\delta, a+\delta)$ (and in I) for which $|f(x)-y_0|\leqq\eta$ is of positive measure. Let I' be a closed subinterval of $(a-\delta, a+\delta)$ such that if $E_m=E\cdot I'$, then†

$$(8)\qquad \text{meas}\,(E_m)>\frac{m^2}{m^2+1}\,\text{meas}\,(I').$$

Either $|y_0-P(a)|\geqq\sigma/2$ or $|y_0-Q(a)|\geqq\sigma/2$, say the latter. Let b be a number in I' such that $f(b)=Q(b)$, (take $b-a$ rational); then, using (7), $|f(b)-y_0|>\sigma/4$.

Let S be the set of numbers s such that $b+ms$ is in E_m, and E_i the set of numbers $b+is$, where s is in S, $(i=1,\cdots,m-1)$.

* That this can be done follows easily from the fact that $\Delta_1^m R(0)=0$; see §2 (c).

† See Hobson, *Functions of a Real Variable*, vol. 1, 3d ed., 1923, p. 194.

Set $E_i' = E_i \cdot E_m$, and let S_i be the set of numbers s in S such that $b+is$ is in E_i', $(i=1, \cdots, m)$. There is no number s_0 in every S_i. For if there were, we would have simultaneously

$$(9)\quad |f(b + is_0) - y_0| \leqq \eta,\ (i = 1, \cdots, m),\ |f(b) - y_0| > \sigma/4.$$

Hence by our choice of η, if $R(x)$ is the polynomial of degree at most $m-1$ such that $R(b+is_0) = f(b+is_0)$, $(i=1, \cdots, m)$, then $R(b) \neq f(b)$. But this contradicts Lemma 2.

Consequently, every number s of S is in some set $S-S_i$, and hence for some j, meas $(S-S_j) \geqq$ meas $(S)/m$. Therefore

$$\text{meas } (E_j - E_j') \geqq \frac{\text{meas } (E_j)}{m} \geqq \frac{\text{meas } (E_m)}{m^2}.$$

As E_m and $E_j - E_j'$ have no common points, this with (8) gives

$$(10)\qquad \text{meas } [E_m + (E_j - E_j')] \geqq \text{meas } (E_m)\left[1 + \frac{1}{m^2}\right] > \text{meas } (I').$$

But this contradicts the fact that $E_m + E_j$ lies in I'. Hence $f(x)$ is not measurable, and the theorem is proved.

THEOREM 2. *Let $f(x)$ be measurable in the closed interval $I = (\alpha, \beta)$. A necessary and sufficient condition that $d^m f(x)/dx^m$ exist and equal the continuous function $f_m(x)$ is that $\Delta_h^m(f(x) \to f_m(x)$ uniformly in I.*

* A direct proof of this theorem may be given as follows, as suggested by Birkhoff. The function $f(kh)$, $(h=1/p$, k and p integral), may be expressed in terms of $\Delta_h^m f(x)$ by the formula

$$(12)\quad f(kh) = \sum_{i=0}^{m-1} a_i (kh)^i + \frac{h^m}{(m-1)!} \sum_{l=0}^{k-1} (k-l-1)(k-l-2) \cdots (k-l-m+1)\Delta_h^m f(lh),$$

where the a_i are determined so that $f(sph) = f(s)$, $(s=0, \cdots, m-1)$. We solve a set of linear equations with a Vandermonde determinant, and find

$$(13)\quad a_i = \sum_{s=0}^{m-1} \eta_{is}\Bigg[f(s) - \frac{h^m}{(m-1)!} \sum_{l=0}^{sp-1} (sp-l-1) \cdots (sp-l-m+1)\Delta_h^m f(lh)\Bigg],$$

The necessity of the condition was proved in §2 (b). To prove the sufficiency, we show first that $f_m(x)$ is continuous. Take δ so small that $|\Delta_h^m f(x) - f_m(x)| < \epsilon/2$ for $|h| < \delta$. Take any two points x_1 and x_2 of I such that $x_2 - x_1 = mh$, $|h| < \delta$. Then $|\Delta_h^m f(x_1) - f_m(x_1)| < \epsilon/2$, $|\Delta_{-h}^m f(x_2) - f_m(x_2)| < \epsilon/2$. But $\Delta_{-h}^m f(x_2) = \Delta_h^m f(x_1)$; hence $|f_m(x_2) - f_m(x_1)| < \epsilon$.

Define the m-fold integral

$$(11) \quad g(x) = \int_a^x \cdots \int_a^{x_3} \int_a^{x_2} f_m(x_1) dx_1 dx_2 \cdots dx_m;$$

then $d^m g(x)/dx^m = f_m(x)$ in I, and by §2 (b), $\Delta_h^m g(x) \to f_m(x)$ uniformly in I. Set $\phi(x) = f(x) - g(x)$. Then $\Delta_h^m \phi(x) \to 0$ uniformly; hence $\phi(x)$ is a polynomial of degree at most $m-1$ (Theorem 1), and therefore $f(x)$ has a continuous mth derivative in I, and

$$\frac{d^m}{dx^m} f(x) = \frac{d^m}{dx^m} \phi(x) + \frac{d^m}{dx^m} g(x) = f_m(x).$$

THEOREM 3. *A necessary and sufficient condition that the function $f(x) = f_0(x)$ defined in I have a continuous mth derivative is that there exist functions $f_1(x), \cdots, f_m(x)$ such that* (2) *holds. In this case, $d^p f(x)/dx^p = f_p(x)$, $(p = 1, \cdots, m)$.*

The necessity of the condition is a consequence of Taylor's formula. To prove the sufficiency, §2 (a) shows that $\Delta_h^p f(x) \to f_p(x)$ uniformly as $h \to 0$; consequently†, by Theorem 2, $d^p f(x)/dx^p$ exists and equals $f_p(x)$, $(p = 1, \cdots, m)$.

HARVARD UNIVERSITY

---

where the η_{is} are numerical coefficients. If in the above equations we replace p by $p_j = 2^j p$, h by $h_j = h/2^j$, and k by $k_j = 2^j k$, and set $r = k_j h_j = kh$, then as $\Delta_h^m f(x) \to f_m(x)$ uniformly, passing to the limit $j = \infty$ gives

$$(14) \quad \begin{aligned} (r) = \sum_{i=0}^{m-1} r^i \sum_{s=0}^{m-1} \eta_{is} \left[f(s) - \frac{1}{(m-1)!} \int_0^s (s-t)^{m-1} f_m(t) dt \right] \\ + \frac{1}{(m-1)!} \int_0^r (r-t)^{m-1} f_m(t) dt, \end{aligned}$$

for any rational $r > 0$. If $f(x)$ is continuous, this is true for all $r > 0$ and, the theorem follows immediately on differentiating. (The restriction $r > 0$ may of course be replaced by the restriction $r > -c$ for any c.)

† A direct proof of this theorem is given by the author in a paper entitled *Differentiable functions defined in closed sets I*, Theorem IV.

DIFFERENTIABLE FUNCTIONS DEFINED IN CLOSED SETS. I†

BY

HASSLER WHITNEY‡

1. **Introduction.** In a recent paper§ the author has shown that if a function $f(x)$ defined in a closed set A in n-space E satisfies certain conditions involving Taylor's formula (in finite form), i.e. if it is "of class C^m in A," then its definition can be extended over E so that it will have continuous partial derivatives through the mth order. In this paper we restrict ourselves to the one-dimensional case. (For the above theorem in this case, see §4.) Let $x_0, \cdots, x_m$ be distinct points of A. If $P(x)=c_0+\cdots+c_mx^m$ is the polynomial of degree at most m such that $P(x_i)=f(x_i)(i=0, \cdots, m)$, the mth difference quotient of $f(x)$ at these points is $\Delta_{0\ldots m}f=\Delta^m f(x)=m!c_m$. The main object of this paper is to prove (see §§2 and 3 for definitions)

THEOREM I. *A necessary and sufficient condition that $f(x)$ be of class C^m in A is that $\Delta^m f(x)$ converge in A.*

This theorem furnishes a direct definition of the differentiability of a function; the former definition (see §3) involved the existence of other functions $f_1(x), \cdots, f_m(x)$.

The necessity of the condition is easily proved. The definition of $f(x)$ being extended over the x-axis E, consider any $m+1$ points $x_0, \cdots, x_m$ $(x_0<x_1<\cdots<x_m)$. Define $P(x)$ as above. As $f(x_i)-P(x_i)=0(i=0, \cdots, m)$ there is a point $x'(x_0<x'<x_m)$ such that $(d^m/dx^m)[f(x')-P(x')]=0$. But $d^mP(x)/dx^m\equiv m!c_m=\Delta_{0\ldots m}f$; hence $\Delta_{0\ldots m}f=d^mf(x')/dx^m$. Therefore if $x_0, \cdots, x_m$ are in A and are sufficiently near a point x^* of A, $\Delta_{0\ldots m}f$ $=d^mf(x')/dx^m=d^mf(x^*)/dx^m$ approximately, and $\Delta^mf(x)$ converges in A (in fact, in E). This may be proved also from (2.6) for $s=m$.

We note that, for $f(x)=f_0(x)$ to be of class C^m in a general closed set A, it is not sufficient that there exist functions $f_s(x)$ $(s=1, \cdots, m)$ in A such that $df_s(x)/dx=f_{s+1}(x)$ there. As an example, set $f_0(0)=0$ and $f_0(x)=1/2^{2i}$ $(1/2^i\leqq x\leqq 3/2^{i+1}, i=1, 2, \cdots)$, and set $f_1(x)\equiv 0$ and $f_2(x)\equiv 0$ in the same point set A.

The majority of the paper is devoted to the proof of Theorem I. In the

† Presented to the Society, October 28, 1933; received by the editors July 27, 1933.

‡ National Research Fellow.

§ *Analytic extensions of differentiable functions defined in closed sets*, these Transactions, vol. 36 (1934), pp. 63–89; this paper will be referred to as A.E.

last section we study Taylor's formula in finite form, when it holds in closed sets, and when its validity implies differentiability of the given function.

2. **Difference quotients.**† If $x_0, \cdots, x_m$ are distinct numbers, set‡

$$(2.1)\quad u_{ij} = x_j - x_i, \quad r_{ij} = |u_{ij}|, \quad \overset{i}{\alpha}_{01\cdots m} = \frac{1}{u_{0i}\cdots u_{i-1,i}u_{i+1,i}\cdots u_{mi}}.$$

Given a function $f(x)$, we define the mth difference quotient by the formula

$$(2.2)\quad \Delta^m f(x) = \Delta(x_0, x_1, \cdots, x_m; f) = \Delta_{01\cdots m}f = m! \sum_{i=0}^{m} \overset{i}{\alpha}_{01\cdots m} f(x_i).$$

In particular, $\Delta_0 f = f(x_0)$, $\Delta_{01}f = [f(x_1) - f(x_0)]/(x_1 - x_0)$. $\Delta_{0\cdots m}$ is symmetric in the points $x_0, \cdots, x_m$.

If $i \geqq 2$,

$$\frac{1}{u_{01}}(\overset{i}{\alpha}_{12\cdots s} - \overset{i}{\alpha}_{02\cdots s}) = \frac{1}{u_{01}}\left(\frac{1}{u_{1i}\cdots} - \frac{1}{u_{0i}\cdots}\right) = \frac{1}{u_{01}} \frac{u_{0i} - u_{1i}}{u_{0i}u_{1i}\cdots} = \overset{i}{\alpha}_{012\cdots s};$$

hence

$$(2.3)\quad \begin{aligned} \frac{s}{u_{01}}(\Delta_{12\cdots s} - \Delta_{02\cdots s}) &= \frac{s!}{u_{01}}\Big[-\overset{0}{\alpha}_{02\cdots s}f(x_0) + \overset{1}{\alpha}_{12\cdots s}f(x_1) \\ &\qquad + \sum_{i\geqq 2}(\overset{i}{\alpha}_{12\cdots s} - \overset{i}{\alpha}_{02\cdots s})f(x_i)\Big] \\ &= s! \sum_{i=0}^{s} \overset{i}{\alpha}_{012\cdots s} f(x_i) = \Delta_{012\cdots s}. \end{aligned}$$

Suppose $*$ is a set of subscripts containing neither 0, 1, nor 2; then for some m,

$$\Delta_{012*} = \frac{m}{u_{01}}(\Delta_{12*} - \Delta_{02*}) = \frac{m}{u_{02}}(\Delta_{12*} - \Delta_{01*}).$$

Solving for Δ_{01*}, we find

$$(2.4)\quad \Delta_{01*} = \frac{u_{02}}{u_{01}}\Delta_{02*} + \frac{u_{21}}{u_{01}}\Delta_{21*},$$

which may be written as follows: $u_{01}\Delta_{01*} + u_{12}\Delta_{12*} + u_{20}\Delta_{20*} = 0$.

Let $x_0, \cdots, x_s$ be distinct numbers. If we solve the equations $\sum_{i=0}^{s}$

† Compare Nörlund, *Differenzenrechnung*, Berlin, 1924, pp. 8–9. It is seen that $\Delta_{01\cdots m} = m! \cdot [x_0x_1\cdots x_m]$.

‡ In the equations below, the numbers $0, 1, \cdots$, when appearing as subscripts, are to be considered as variables. Thus, as a particular case of (2.1), $\overset{0}{\alpha}_{023} = 1/(u_{20}u_{30})$; in the second equation of §6, $\Sigma_j 1/u_{j'i} = 1/u_{0'i} + \cdots$. Without this notation, the equations would often get quite cumbersome.

$(x_i-x)^j z_i=\delta_{js}(j=0,\cdots,s)$, x being any fixed number, we find $z_i=\alpha_{0\cdots s}^i$. Hence

$$\sum_{i=0}^{s}\alpha_{0\cdots s}^{i}(x_i-x)^j=0 \qquad (j=0,\cdots,s-1),$$
$$(2.5)\qquad \sum_{i=0}^{s}\alpha_{0\cdots s}^{i}(x_i-x)^s=1.$$

Suppose $f(x)=f_0(x),\cdots,f_m(x)$, $R(x',x)=R_0(x',x)$ satisfy (3.1) below for $s=0$. Then (2.2) and (2.5) give

$$(2.6)\qquad \begin{aligned}\Delta_{0\cdots s}f&=s!\sum_{i=0}^{s}\alpha_{0\cdots s}^{i}\left[\sum_{j=0}^{m}\frac{f_j(x)}{j!}(x_i-x)^j+R(x_i,x)\right]\\&=f_s(x)+s!\sum_{j=s+1}^{m}\frac{f_j(x)}{j!}\sum_{i=0}^{s}\alpha_{0\cdots s}^{i}(x_i-x)^j+s!\sum_{i=0}^{s}\alpha_{0\cdots s}^{i}R(x_i,x).\end{aligned}$$

If $f(x)=c_0+\cdots+c_mx^m$ is a polynomial of degree at most m, then (3.1) is satisfied with $f_m(x)\equiv m!c_m$ and $R_s(x',x)\equiv 0$. Setting $s=m$ in (2.6) gives

$$(2.7)\qquad \Delta_{0\cdots m}f\equiv m!c_m.$$

We say $\Delta^m f(x)$ *converges in the set* A if for each point x of A and every $\epsilon>0$ there is a $\delta>0$ such that if $x_0,\cdots,x_m$, $x_{0'},\cdots,x_{m'}$ are any two sets of distinct points of A, all within δ of x, then

$$|\Delta_{0\cdots m}-\Delta_{0'\cdots m'}|<\epsilon.$$

$\Delta^m f(x)$ of course converges at all isolated points of A. We say $\Delta^m f(x)\to f_m(x)$ in A if $|\Delta_{0\cdots m}-f_m(x)|<\epsilon$ whenever $x_0,\cdots,x_m$ are in A and within δ of x. Evidently *if* $\Delta^m f(x)\to f_m(x)$ *in* A, *then* $f_m(x)$ *is continuous in the set of limit points of* A.

Differentiable functions

3. Definition of differentiable functions. Let $f(x)=f_0(x)$ be defined in the closed set A. We say $f(x)$ *is of class* C^m *in* A (see A. E.) if there exist functions $f_1(x),\cdots,f_m(x)$, $R(x',x)=R_0(x',x),\cdots,R_m(x',x)$ in A such that

$$(3.1)\qquad f_s(x')=\sum_{i=s}^{m}\frac{f_i(x)}{(i-s)!}(x'-x)^{i-s}+R_s(x',x)\qquad (s=0,\cdots,m),$$

and for each s, each point x of A, and every $\epsilon>0$ there is a $\delta>0$ such that

$$(3.2)\qquad \left|\frac{R_s(x'',x')}{(x''-x')^{m-s}}\right|<\epsilon \qquad (x',x''\text{ in }A;\ |x'-x|,\ |x''-x|<\delta).$$

If $f_i(x),\cdots,f_m(x)$, $R_i(x',x)$ satisfy (3.1) and (3.2) for $s=i$, we say $f_i(x)$

can be expanded in a Taylor's formula to the $(m-i)$th order locally uniformly in terms of $f_i(x), \cdots, f_m(x)$. If $f(x)$ is defined throughout an open interval and has a continuous mth derivative there, then it is of class C^m, by Taylor's theorem.

4. **Extension of differentiable functions.** *If $f_0(x)$ is of class C^m in terms of $f_0(x), \cdots, f_m(x)$ in A, then the definitions of these functions can be extended throughout E so they will be continuous and so that $df_s(x)/dx = f_{s+1}(x)$ there* $(s=0, \cdots, m-1)$ (see A. E., Lemma 2). As the proof can be given more simply in the one-dimensional case, we give it here. We can assume A is unbounded on both sides; otherwise, take a point a beyond A on either side, and set $f_s(x) \equiv 0$ $(s=0, \cdots, m)$ beyond a.

For each interval (a, b) of $E-A$, let $P(x)$ be the polynomial of degree at most $2m+1$ such that

$$\frac{d^s}{dx^s}P(a) = f_s(a), \qquad \frac{d^s}{dx^s}P(b) = f_s(b) \qquad (s = 0, \cdots, m); \tag{4.1}$$

we set

$$f_s(x) = \frac{d^s}{dx^s}P(x) \text{ in } (a, b). \tag{4.2}$$

$df_s(x)/dx = f_{s+1}(x)$ $(s=0, \cdots, m-1)$ in $E-A$; we must show that this holds also at any point x_0 of A.

Suppose each $f_{s+1}(x)$ is continuous in E. Then given x_0 in A and $\epsilon>0$, take $\delta>0$ so small that

$$|f_{s+1}(x') - f_{s+1}(x_0)| < \frac{\epsilon}{2} \qquad (|x' - x_0| < \delta).$$

By (3.1) and (3.2), we can also take δ so small that if a is in A, $|a-x_0| < \delta$, and

$$f_s(a) = f_s(x_0) + f_{s+1}(x_0)(a - x_0) + R'(a, x_0),$$

then $|R'(a, x_0)/(a-x_0)| < \epsilon/2$. Now take any point x within δ of x_0. If x is in A, set $a=x$; otherwise, let a be the end point nearest x_0 of the interval of $E-A$ containing x. Now for some x', $a \leqq x' \leqq x$,

$$f_s(x) = f_s(a) + f_{s+1}(x')(x - a).$$

Adding this to the last equation and dividing by $x - x_0$, we find

$$\frac{f_s(x) - f_s(x_0)}{x - x_0} = f_{s+1}(x_0) + [f_{s+1}(x') - f_{s+1}(x_0)]\frac{x - a}{x - x_0} + \frac{R'(a, x_0)}{x - x_0}.$$

As $|x'-x_0|<\delta$, $|x-a|\leqq|x-x_0|$ and $|x-x_0|\geqq|a-x_0|$,

$$\left|\frac{f_s(x)-f_s(x_0)}{x-x_0}-f_{s+1}(x_0)\right|<\epsilon \qquad (|x-x_0|<\delta),$$

as required. (We have given here the details of A. E., Lemma 1.)

We must prove still that each $f_s(x)$ is continuous at each point x_0 of A; it is of course true in $E-A$. As $f_s(x)$ is continuous in A, it is sufficient to prove that for every $\epsilon>0$ there is a $\delta>0$ such that if (a, b) is any interval of $E-A$ lying within δ of x_0, then

$$|f_s(x)-f_s(a)|<\epsilon \qquad (a\leqq x\leqq b).$$

Take $\epsilon'<\epsilon/[2(m+1)^2K]$, where K is a number to be determined later. Let M be the maximum of $|f_i(x)|$ in A ($|x-x_0|\leqq 1$, $i=0,\cdots,m$). Take $\delta<\epsilon/(2mM)$ and <1 so small that (3.2) holds with ϵ replaced by ϵ' for any x, x' within δ of x_0. Now take any interval (a, b) of $E-A$ lying within δ of x_0. In (a, b), $f(x)$ equals

$$P(x)=\sum_{i=0}^{m}\frac{f_i(a)}{i!}(x-a)^i+\sum_{i=m+1}^{2m+1}\frac{\gamma_i}{i!}(x-a)^i,$$

where the γ_i are determined by the relations

$$\frac{d^s}{dx^s}P(b)=\sum_{i=s}^{m}\frac{f_i(a)}{(i-s)!}(b-a)^{i-s}+\sum_{i=m+1}^{2m+1}\frac{\gamma_i}{(i-s)!}(b-a)^{i-s}=f_s(b);$$

hence

$$\sum_{i=m+1}^{2m+1}\frac{\gamma_i}{(i-s)!}(b-a)^{i-s}=f_s(b)-\sum_{i=s}^{m}\frac{f_i(a)}{(i-s)!}(b-a)^{i-s}=R_s(b, a).$$

Solving for the γ_i, we find

$$\gamma_i=\sum_{j=0}^{m}K_{ij}\frac{R_j(b, a)}{(b-a)^{i-j}},$$

where the K_{ij} depend on m alone. Set $K=\max|K_{ij}|$; then

$$|\gamma_i|\leqq\sum_{j=0}^{m}\frac{K}{|b-a|^{i-m}}\left|\frac{R_j(b, a)}{(b-a)^{m-j}}\right|<\frac{(m+1)K}{|b-a|^{i-m}}\epsilon'.$$

Now if x is any point in (a, b), then $|x-a|\leqq|b-a|$, and

$$
\begin{aligned}
|f_s(x) - f_s(a)| &= \left| \frac{d^s}{dx^s} P(x) - f_s(a) \right| \\
&= \left| \sum_{i=s+1}^{m} \frac{f_i(a)}{(i-s)!}(x-a)^{i-s} + \sum_{i=m+1}^{2m+1} \frac{\gamma_i}{(i-s)!}(x-a)^{i-s} \right| \\
&< mM|x-a| + (m+1)K\epsilon' \sum_{i=m+1}^{2m+1} \frac{|x-a|^{i-s}}{|b-a|^{i-m}} \\
&< mM\delta + (m+1)^2 K |b-a|^{m-s}\epsilon' < \epsilon,
\end{aligned}
$$

as required.

Theorem I, A perfect

5. A succession of lemmas culminates in Lemma 7, which is the sufficiency part of Theorem I for perfect sets.

Lemma 1. *Let A be a closed set, and let $\Delta^s f(x)$ converge on A. Then we can define $f_s(x)$ on the set of limit points A^* of A so that the following is true. Given x in A^* and $\epsilon > 0$, we can choose a $\delta > 0$ so that if $x_0, \cdots, x_s$ is any set of distinct points of A lying within δ of x, then $|\Delta_{0\ldots s} - f_s(x)| < \epsilon$.*

The proof is simple.

Lemma 2. *If $\Delta^s f_0(x)$ converges in the perfect set A, then for each point x of A and every $\epsilon > 0$ there is a $\delta > 0$ such that*

$$
|\Delta_{0\cdots t-1,t\cdots s-1} - \Delta_{0'\cdots(t-1)',t\cdots s-1}| < \epsilon \tag{5.1}
$$

$(0 \leqq t \leqq s)$ whenever all the points concerned lie within δ of x.

This is trivial if $t=0$. We assume it holds for numbers $0, \cdots, t-1$, and shall prove it for t. Given a point x_s distinct from all former points, the equations

$$
\Delta_{0\cdots t-1,t\cdots s-1,s} = \frac{s}{u_{0s}} (\Delta_{1\cdots t-1,t\cdots s} - \Delta_{0\cdots t-1,t\cdots s-1}),
$$

$$
\Delta_{0'\cdots(t-1)',t\cdots s-1,s} = \frac{s}{u_{0's}} (\Delta_{1'\cdots(t-1)',t\cdots s} - \Delta_{0'\cdots(t-1)',t\cdots s-1})
$$

give

$$
\begin{aligned}
\Delta_{0\cdots t-1,t\cdots s-1} - \Delta_{0'\cdots(t-1)',t\cdots s-1} &= (\Delta_{1\cdots t-1,t\cdots s} - \Delta_{1'\cdots(t-1)',t\cdots s}) \\
&\quad - \frac{1}{s}(u_{0s}\Delta_{0\cdots t-1,t\cdots s} - u_{0's}\Delta_{0'\cdots(t-1)',t\cdots s}).
\end{aligned} \tag{5.2}
$$

As $\Delta_{0\ldots s}$ converges, we can take $M>0$ and $\delta' < \epsilon/(4M)$ so that $|\Delta_{0\ldots s}| < M$ whenever $x_0, \cdots, x_s$ are within δ' of x. By induction, we can take $\delta < \delta'$ so small that the first term on the right in (5.2) is in absolute value $<\epsilon/2$ whenever all points concerned are within δ of x. Now given the points $x_0, \cdots, x_{s-1}$,

$x_{0'}, \cdots, x_{(t-1)'}$ within δ of x, let x_s be another such point; then (5.2) gives (5.1).

The lemma with $t=s$ shows that $\Delta^{s-1}f_0(x)$ converges in A.

LEMMA 3. *If $\Delta^m f_0(x)$ converges in the perfect set A, then there are continuous functions $f_1(x), \cdots, f_m(x)$ in A such that $\Delta^s f_0(x)\to f_s(x)$ in A $(s=1,\cdots,m)$.*

We prove this successively for $s=m, m-1, \cdots, 1$ with the help of Lemmas 1 and 2.

6. We proceed to the following lemma.

LEMMA 4. *If $\Delta^p g(x)\to g_p(x)$ and $\Delta^1 g(x)\to g_1(x)$ in the perfect set A, then $\Delta^{p-1}g_1(x)\to g_p(x)$ in A.*

Set $q=p-1$. If we apply the relation $\Delta^1\phi(x_i)\to d\phi(x_i)/dx_i$ to $\alpha^i_{0\cdots i\cdots q}$ as a differentiable function of x_i, we find

$$\frac{\alpha^{i'}_{0\cdots i'\cdots q}-\alpha^i_{0\cdots q}}{u_{ii'}} = -\alpha^i_{0\cdots q}\sum_{j\neq i}\frac{1}{u_{ji}}+\epsilon(x_{i'}),$$

where $\epsilon(x_{i'})\to 0$ as $x_{i'}\to x_i$. Hence

$$\begin{aligned}\alpha^i_{0\cdots q}\sum_j\frac{1}{u_{j'i}} &= -\frac{\alpha^{i'}_{0\cdots i'\cdots q}}{u_{ii'}}+\alpha^i_{0\cdots q}\sum_{j\neq i}\left(\frac{1}{u_{j'i}}-\frac{1}{u_{ji}}\right)+\epsilon(x_{i'})\\ &= -\frac{\alpha^{i'}_{0\cdots i'\cdots q}}{u_{ii'}}+\zeta_i(x_{0'},\cdots,x_{q'}),\end{aligned}$$

where $\zeta_i(x_{0'},\cdots,x_{q'})\to 0$ as $x_{j'}\to x_j (j=0,\cdots,q)$. Consider the $2q$ points $x_0, x_{0'}, \cdots, x_q, x_{q'}$. We have

$$\Delta_{0\cdots jj'\cdots q}g = p!\left[\sum_{i=0}^{q}\frac{\alpha^i_{0\cdots q}}{u_{j'i}}g(x_i)+\frac{\alpha^{j'}_{0\cdots j'\cdots q}}{u_{jj'}}g(x_{j'})\right],$$

$$\begin{aligned}\frac{1}{p}\sum_{j=0}^{q}\Delta_{0\cdots jj'\cdots q}g &= (p-1)!\left[\sum_{i=0}^{q}\alpha^i_{0\cdots q}g(x_i)\sum_{j=0}^{q}\frac{1}{u_{j'i}}+\sum_{i=0}^{q}\frac{\alpha^{i'}_{0\cdots i'\cdots q}}{u_{ii'}}g(x_{i'})\right]\\ &= q!\sum_{i=0}^{q}\left[\alpha^{i'}_{0\cdots i'\cdots q}\frac{g(x_{i'})-g(x_i)}{u_{ii'}}+g(x_i)\zeta_i(x_{0'},\cdots,x_{q'})\right].\end{aligned}$$

As $\Delta^1 g(x)\to g_1(x)$, this gives, letting $x_{j'}\to x_j$ $(j=0,\cdots,q)$,

$$\frac{1}{p}\lim\sum_{j=0}^{q}\Delta_{0\cdots jj'\cdots q}g = q!\sum_{i=0}^{q}\alpha^i_{0\cdots q}g_1(x_i)=\Delta_{0\cdots q}g_1. \tag{6.1}$$

Now given a point x of A and an $\epsilon>0$, take $\delta>0$ so that if $x_0, \cdots, x_j, x_{j'}, \cdots, x_q$ are within δ of x, then

$$|\Delta_{0\cdots jj'\cdots q}g - g_p(x)| < \epsilon.$$

Then if $x_0, \cdots, x_q$ are within δ of x, we find by adding points $x_{0'}, \cdots, x_{q'}$ within δ of x and letting $x_{i'}\to x_i$ $(i=0, \cdots, q)$ that

$$|\Delta_{0\cdots q}g_1 - g_p(x)| < \epsilon,$$

as required.

LEMMA 5. *If $\Delta^m f_0(x)$ converges in the perfect set A, then there are continuous functions $f_1(x), \cdots, f_m(x)$ such that $\Delta_p f_q(x)\to f_{p+q}(x)$ in A.*

This follows from Lemmas 3 and 4.

7. We now present the two final lemmas needed for the proof of Theorem I when A is perfect.

LEMMA 6. *Let $g(x)=g_0(x), \cdots, g_s(x)$ be defined in the perfect set A, and suppose $\Delta^s g(x)\to g_s(x)$. If $g(x)$ can be expanded in a Taylor's formula to the $(s-1)$th order in terms of $g_0(x), \cdots, g_{s-1}(x)$, then it can be expanded in a Taylor's formula to the sth order locally uniformly in terms of $g_0(x), \cdots, g_s(x)$.*

Given a point x of A and an $\epsilon>0$, take $\delta>0$ so that

$$|\Delta_{0\cdots s} - g_s(x_0)| < \frac{s!}{2^s}\frac{\epsilon}{3} \tag{7.1}$$

whenever $x_0, \cdots, x_s$ are within δ of x (recall that $g_s(x)$ is continuous, by §2). Take any two points x_0 and x_s of A within δ of x; we must show that $|R^{(s)}(x_s, x_0)|/r_{0s}^s<\epsilon$.

Take δ' so small that if $|x'-x_0|<\delta'$, then

$$\left|\frac{R^{(s-1)}(x', x_0)}{(x'-x_0)^{s-1}}\right| < \frac{r_{0s}}{2^{2s}s}\frac{\epsilon}{3},$$

where $R^{(s-1)}(x', x_0)=g(x')-\sum_{j=0}^{s-1}g_j(x_0)(x'-x_0)^j/j!$. Take $M>|g_s(x_0)|$. Take a point x_{s-1} in A within δ' of x_0 and so close to x_0 that

$$\frac{r_{0,s-1}}{r_{0s}} < \frac{s!}{2^s M}\frac{\epsilon}{3} \quad \text{and} \quad < \frac{1}{2},$$

and (if $s>2$) take in succession points $x_{s-2}, \cdots, x_1$ in A so that

$$r_{0,t-1} < \tfrac{1}{2}r_{0t} \qquad (t = 2. \cdots, s-1); \tag{7.2}$$

let these points lie within δ of x. Then if $i<s$,

$$\left|\alpha_{0\ldots s}^{i}R^{(s-1)}(x_i, x_0)\right| < \frac{1}{r_{si}}\,\frac{r_{0i}^{s-1}}{r_{0i}\cdots r_{i-1,i}r_{i+1,i}\cdots r_{s-1,i}}\,\frac{r_{0s}}{2^{2s}s}\,\frac{\epsilon}{3} < \frac{1}{2^s s}\,\frac{\epsilon}{3}\,.$$

Now

$$\frac{1}{s!}\Delta_{0\ldots s} = \alpha^s g(x_s) + \sum_{i=0}^{s-1}\alpha^i\left[\sum_{j=0}^{s-1}\frac{g_j(x_0)}{j!}u_{0i}^j + R^{(s-1)}(x_i, x_0)\right]$$

$$= \alpha^s g(x_s) - \alpha^s\sum_{j=0}^{s-1}\frac{g_j(x_0)}{j!}u_{0s}^j + \sum_{i=0}^{s-1}\alpha^i R^{(s-1)}(x_i, x_0),$$

on account of (2.5). Therefore

$$\begin{aligned}R^{(s)}(x_s, x_0) &= g(x_s) - \sum_{j=0}^{s}\frac{g_j(x_0)}{j!}u_{0s}^j \\ &= \frac{\Delta_{0\ldots s}}{s!\alpha^s} - \frac{g_s(x_0)u_{0s}^s}{s!} - \sum_{i=0}^{s-1}\frac{\alpha^i}{\alpha^s}R^{(s-1)}(x_i, x_0),\end{aligned} \tag{7.3}$$

and as $r_{is}/r_{0s} \leqq (r_{0s}+r_{0i})/r_{0s} = 1 + r_{0i}/r_{0s}$,

$$\frac{\left|R^{(s)}(x_s, x_0)\right|}{r_{0s}^s} \leqq \frac{r_{0s}\cdots r_{s-1,s}}{s!r_{0s}^s}\left|\Delta_{0\ldots s} - g_s(x_0)\right| + \frac{\left|g_s(x_0)\right|}{s!}\left|\frac{r_{0s}\cdots r_{s-1,s}}{r_{0s}} - 1\right|$$

$$+ \sum_{i=0}^{s-1}\frac{r_{0s}\cdots r_{s-1,s}}{r_{0s}}\left|\alpha^i R^{(s-1)}(x_i, x_0)\right| \,.$$

$$< \frac{2^s}{s!}\,\frac{s!}{2^s}\,\frac{\epsilon}{3} + \frac{M}{s!}\left[\left(1 + \frac{r_{00}}{r_{0s}}\right)\cdots\left(1 + \frac{r_{0,s-1}}{r_{0s}}\right) - 1\right] + s\cdot 2^s\frac{1}{2^s s}\,\frac{\epsilon}{3} < \epsilon,$$

as required.

LEMMA 7. *If $\Delta^m f(x)$ converges in the perfect set A, then $f_0(x) = f(x)$, $f_1(x), \cdots, f_m(x)$ can be defined in A so that $f(x)$ is of class C^m in A in terms of the $f_s(x)$ $(s = 0, \cdots, m)$.*

We define $f_1(x), \cdots, f_m(x)$ by means of Lemma 5. Taylor's formula for each $f_s(x)$ holds to the 0th order, as $f_s(x)$ is continuous (see §2). We prove in succession that it holds to the kth order for $k = 1, \cdots, m-s$. This completes the proof of the lemma, and therefore of Theorem I for the case that A is perfect.

P-SETS AND Q-SETS

8. We shall prove a lemma which will be needed in the next part. Let $A' = a_1, a_2, \cdots$ be a set of isolated points, at least $m+1$ in number. With each point a_i we shall associate m other points $a_{i_1}, \cdots, a_{i_m}$; these m points

together with a_i we say form the *Q-set* $Q(a_i)$. Take a Q-set Q_i, and let $a_{j_1}, \cdots, a_{j_s}$ be all those points such that $Q(a_{j_t})=Q_i$; these points form the *P-set* P_i corresponding to Q_i. Each point of P_i is in Q_i. Each point a_i lies in just one P-set $P(a_i)$, as a_i is associated with just one Q-set $Q(a_i)$; however, a_i may lie in several Q-sets. Let $\delta(Q_i)$ be the greatest distance between pairs of points of Q_i.

LEMMA 8. *The P-sets and Q-sets may be so chosen that for any two points* a_i *and* a_j,

$$\text{(8.1)} \qquad \textit{if} \quad \frac{\delta(Q(a_i)) + \delta(Q(a_j))}{|a_j - a_i|} > 2m, \quad \textit{then } P(a_i) = P(a_j).$$

We first associate sets of points with certain of the limit points of the points $a_1, a_2, \cdots$ as follows. Let c_i be a point such that there is a sequence of points of A' approaching it from one side, say the left, while there is a nearest point of A' to c_i on the other side of c_i. Let μ equal $m+1$, or the number of points a_j between c_i and the next limit point c_k to the right of c_i if that number is smaller, and let $a_{j_1}, \cdots, a_{j_\mu}$ be the points nearest c_i on the right (counting from left to right). Let τ be the smallest of the numbers $|a_j - a_{j_s}|$ $(s, t=1, \cdots, \mu)$ which are $> |a_{j_1} - c_i|$, if there are such. Let $a_1(c_i)$ be a point of A' to the left of c_i such that

$$\text{(8.2)} \qquad |c_i - a_1(c_i)| < |a_{j_1} - c_i|, \quad \text{and} \quad |a_{j_1} - a_1(c_i)| < \tau$$

if τ is defined. Let $a_2(c_i), \cdots, a_m(c_i)$ be points of A' lying between c_i and $a_1(c_i)$.

We now define the Q-sets. Given a point a_i, we associate another point with it as follows. Suppose, Case I, there is a point a_j whose distance from a_i is less than or equal to the distance from any other a_k to a_i; then we associate a_j with a_i, or that one of the pair a_j, a_k which lies to the left of a_i, if their distances from a_i are the same. Suppose, Case II, there is no such point. Then there is a limit point c_i nearer a_i than any point a_j. If there are two such points, we consider that one c_i on the left. The point we associate with a_i is then $a_1(c_i)$.

Suppose now we have associated a number of points with a_i, forming the set of points S. We associate the next point in a fashion much the same as above. If Case II has not occurred in associating the other points of S with a_i, we again have two cases to consider. Case I, there is a nearest point a_j to the set S; we then associate this point with S (or the point a_k, as above). Case II, there is none; then take the point c_i as above, and associate $a_1(c_i)$ with S. At any time we employ Case II, we immediately associate also the points $a_2(c_i)$, $a_3(c_i)$, $\cdots$ with S, till we have the required $m+1$ points $Q(a_i)$.

Note that the point we associate with S does not depend on which point a_i of S we started with. Also if Case I has occurred each time in forming the subset S of $Q(a_i)$, then there is no point a_k not in S which lies between two points of S.

9. To prove that (8.1) holds take any two points a_i and a_j; set $r_{ij}=|a_j-a_i|$.

(1) Suppose there are at most a finite number of points of A' between a_i and a_j. If $\delta(Q(a_i))+\delta(Q(a_j))>2mr_{ij}$, then either $\delta(Q(a_i))>mr_{ij}$ or $\delta(Q(a_j))>mr_{ij}$, say the former. Then there is a first time when, on adding a point a_k to a set S in forming $Q(a_i)$, the distance from a_k to S is $>r_{ij}$.

(a) In forming S from a_i, Case I has occurred each time. For if Case II had occurred, say in adding the point $a_1(c_i)$ to the subset S_1 of S, then a_k would be some $a_s(c_i)$; but the distance from a_k to S is then at most the distance from $a_s(c_i)$ to $a_1(c_i)$ which is less than the distance from $a_1(c_i)$ to S_1 which is by hypothesis $\leqq r_{ij}$.

(b) There is no point a_s whose distance from S is $\leqq r_{ij}$. For suppose there were; then Case II must occur in adding $a_k=a_1(c_i)$ to S, and c_i is nearer S than any point a_r. (If Case I occurred, a_s or a nearer point, not a_k, would be added to S.) Say c_i lies to the left of S. Let a_p and a_q be the left and right-hand end points of S respectively. As there is a point a_s distant $\leqq r_{ij}$ from S, $|a_p-c_i|<r_{ij}$. Suppose a_j is not in S. As there are no limit points between a_i and a_j, a_j lies to the right of S, and hence there is a first point a_r to the right of S. Then as a_i is in S, $|a_r-a_q|\leqq r_{ij}$. But as a_q and a_r are among the first $m+1$ (or μ) points to the right of c_i, and $|a_p-c_i|<|a_r-a_q|$, (8.2) gives $|a_p-a_k|<|a_r-a_q|\leqq r_{ij}$, a contradiction; therefore a_j is in S. As a_j is in S and $|a_p-c_i|<r_{ij}$, (8.2) gives $|a_p-a_k|<r_{ij}$, again a contradiction.

(c) S contains a_j. For otherwise (b) would be contradicted.

(d) In forming $Q(a_j)$, the points of S are chosen first. For suppose not. Then after perhaps adding some points of S to a_j, forming the set S', we choose a point a_l not in S. By (b), the distance from a_l to S is $>r_{ij}$. As there is a point in S whose distance from S' is at most r_{ij}, a_l must have been chosen under Case II; then the distance from some point c_j to S' is $<r_{ij}$. But then as c_j is a limit point of points a_s, there is a point a_s whose distance from S is $<r_{ij}$, a contradiction.

Now in forming both $Q(a_i)$ and $Q(a_j)$, the points of S are chosen first. As the remaining points chosen depend only on S, $Q(a_i)$ and $Q(a_j)$ must coincide; hence a_i and a_j lie in the same P-set.

(2) Suppose there is a limit point of isolated points b between a_i and a_j. In forming $Q(a_i)$, the set S at any step is at a distance $\leqq|b-a_i|$ from b; hence in adding the next point a_k to S, its distance from S is $\leqq|b-a_i|$ if Case I

occurs, and is $<2|b-a_i|$ if Case II occurs, by (8.2). Therefore $\delta(Q(a_i)) < 2m|b-a_i|$. Similarly $\delta(Q(a_j)) < 2m|b-a_j|$. Adding,

$$\delta(Q(a_i)) + \delta(Q(a_j)) < 2m(|b - a_i| + |b - a_j|) = 2mr_{ij},$$

completing the proof.

Remark. Given a point a_i, if there exist m points $a_{i_1}, \cdots, a_{i_m}$ such that the m intervals between $a_i, a_{i_1}, \cdots, a_{i_m}$ are all $\leqq \rho$, or if there exists a point a not in $Q(a_i)$ within ρ of a_i, then $\delta(Q(a_i)) < 2m\rho$. This follows from the proof in (2).

THEOREM I, A CLOSED

Each isolated point of A is enclosed in an interval; this gives a perfect set B. The definition of $f(x)$ is extended over B. With the help of Lemma 8 it is shown that $\Delta^m f(x)$ now converges over B. By Lemma 7, $f(x)$ is of class C^m in B; hence the same is true in A.

10. **The sets A' and B.** Let A_1 be the set of isolated points of the closed set A, let A_2 be the set of limit points of isolated points, and let A_3 be the remaining points of A. Let A' consist of A_1, together with certain other points as follows. A_1+A_2 being closed, let I be any open interval of $E-(A_1+A_2)$ containing points of A_3. If an end point a_i of I is in A_1, then there is, in I, a nearest point $a_1(a_i)$ of A_3 to a_i. We associate this point with a_i, and also points $a_2(a_i), \cdots, a_m(a_i)$ of A_3 in I, chosen so that

$$(10.1) \qquad |a_s(a_i) - a_1(a_i)| < |a_1(a_i) - a_i| \qquad (s = 2, \cdots, m).$$

A' is a set of isolated points; we may name them $a_1, a_2, \cdots$. A' is contained in A_1+A_3.

For each point a_i of A_1, let $d(a_i)$ be its distance from the rest of A, and let B_i be a closed interval of length $d(a_i)/2$, with a_i as center. Let the perfect set B be A plus all of these intervals. Arrange the points of A' into P-sets and Q-sets so as to obey Lemma 8. For each P-set P_i, let the corresponding P'-set P_i' contain the points of P_i, together with the points of any intervals B_j there may be which enclose points of P_i.

Given any set S of $m+1$ points in B, we shall define its *complexity* $\sigma(S)$ as follows. If all the points of S are in A, set $\sigma(S)=0$. If S contains $p>0$ points in $B-A$, and all these points lie in a single P'-set P_i', let q be the number of remaining points of S which do not lie in the corresponding Q-set Q_i, and set $\sigma(S)=pq$. The complexity of S is in this case certainly $\leqq m^2$. If S contains p points in $B-A$, and these points do not all lie in the same P'-set, set $\sigma(S)=m^2+p-1$. The complexity of any set S is $\leqq m^2+m$.

11. The following lemma together with Lemma 7 gives Theorem I.

LEMMA 9. *Let $f(x)$ be defined in the closed set A so that $\Delta^m f(x)$ converges in A. Then A can be enclosed in a perfect set B, the definition of $f(x)$ can be extended over B, and $f_m(x)$ can be defined in B, so that $\Delta^m f(x) \to f_m(x)$ in B.*

Define the sets A', B etc. as above. We may assume there are at least $m+1$ points in A'. Define $f_m(x)$ at each point of A_2+A_3 as in Lemma 1. Take a fixed interval B_i with center a_i; we define $f(x)$ and $f_m(x)$ over B_i as follows. Let $Q_i = Q(a_i)$ be the corresponding Q-set. Let

$$R_i(x) = \gamma_0 + \cdots + \gamma_m x^m \tag{11.1}$$

be the polynomial of degree at most m such that $R_i(x) = f(x)$ at each point of Q_i; then $\Delta(Q_i) = m!\gamma_m$, by (2.7). Set

$$f(x) = R_i(x), \quad f_m(x) = m!\gamma_m \text{ in } B_i. \tag{11.2}$$

The same polynomial $R_i(x)$ is used in defining $f(x)$ and $f_m(x)$ over each interval of the P'-set P_i' corresponding to Q_i; hence if S is any set of $m+1$ points such that all of its points in $B-A$ lie in P_i', and all remaining points lie in Q_i, then $f(x) = R_i(x)$ at each point of S, and therefore, by (2.7), $\Delta(S) = m!\gamma_m = \Delta(Q_i)$.

Each point x of A_3 is at a positive distance from $B-A$; by the definition of $f_m(x)$, $\Delta^m f(x) \to f_m(x)$ at such points. Each point x of $B-(A_2+A_3)$ is in an interval B_i; hence near x, $f(x)$ is a polynomial, and $\Delta^m f(x) \to f_m(x)$ there also. It remains to show that for each point x of A_2 and every $\epsilon > 0$ there is a $\delta > 0$ such that if S is any set of $m+1$ points of B within δ of x, then

$$|\Delta(S) - f_m(x)| < \epsilon. \tag{11.3}$$

By Lemma 1, we can take $\delta' > 0$ so that

$$|\Delta(S_0) - f_m(x)| < \frac{\epsilon}{(8m+8)^{m^2+m}} \tag{11.4}$$

for any set S_0 of $m+1$ points of A lying within δ' of x. Set $\delta = \delta'/(4m+2)$. We shall prove the following:

(A) *If S is any set in B, of complexity $\sigma(S) = \sigma$, composed of sets of points S_1 in $B-A$ and S_2 in A, and if S_1 lies within δ of x and S_2 lies within δ' of x, then*

$$|\Delta(S) - f_m(x)| < \epsilon_\sigma = \frac{\epsilon}{(8m+8)^{m^2+m-\sigma}}. \tag{11.5}$$

As $\sigma \leqq m^2+m$, (11.3) follows.

12. We note first that *if b_i is in some interval of the P'-set P_i', and b_i lies within δ of x, then Q_i lies within δ' of x.* Say a_i is the center of B_i; then a_i lies

within 2δ of x, a limit point of points of A'. Hence $\delta(Q(a_i)) < 4m\delta$, by the remark at the end of §9, and $Q_i = Q(a_i)$ lies within $(4m+2)\delta = \delta'$ of x.

We shall prove (A) first for $\sigma = 0$, then for $\sigma > 0$, using induction. Suppose $\sigma = 0$. If S is in A, the fact follows from (11.4). If S contains points of $B-A$, then all these points lie in a single P'-set P_i', and the rest of S lies in the corresponding Q-set Q_i; hence $\Delta(S) = \Delta(Q_i)$. Q_i lies within δ' of x; hence (11.4) holds with S_0 replaced by Q_i or by S, and therefore (11.5) holds.

Now suppose (11.5) is proved for all sets S' with $\sigma(S') < \sigma$; we shall prove it for any set S with $\sigma(S) = \sigma$. Suppose first $\sigma > m^2$; then the points of S in $B-A$ lie in at least two P'-sets. Let P_i' and P_j' be two of these sets, let b_i and b_j be points of S (in $B-A$) in P_i' and P_j' respectively, and let a_i and a_j be the centers of the corresponding intervals. Let a_k be a point of $Q(a_i)$ not lying in S. If $S' = S - b_i - b_j$, then, by (2.4),

$$(12.1)\quad \Delta(S) = \Delta(b_i, b_j, S') = \frac{a_k - b_i}{b_j - b_i}\Delta(b_i, a_k, S') + \frac{b_j - a_k}{b_j - b_i}\Delta(a_k, b_j, S').$$

The sets $S' + b_i + a_k$ and $S' + b_j + a_k$ each contain fewer points of $B-A$ than S; hence their complexities are each $< \sigma$. Also $Q(a_i)$ and therefore a_k lie within δ' of x. Therefore, by induction,

$$(12.2)\quad |\Delta(b_i, a_k, S') - f_m(x)| < \epsilon_{\sigma-1}, \quad |\Delta(a_k, b_j, S') - f_m(x)| < \epsilon_{\sigma-1}.$$

As a_i and a_j lie in distinct P-sets, $\delta(Q(a_i)) + \delta(Q(a_j)) \leqq 2mr_{ij}$, by (8.1). As $|b_i - a_i| \leqq r_{ij}/4$ and $|b_j - a_j| \leqq r_{ij}/4$, $|b_j - b_i| \geqq r_{ij}/2$. As a_k and a_j lie in $Q(a_i)$ and $Q(a_j)$ respectively, $|a_j - a_k| \leqq \delta(Q(a_i)) + \delta(Q(a_j)) + r_{ij} \leqq (2m+1)r_{ij}$; hence $|b_j - a_k| < (2m+2)r_{ij}$. Also $|a_k - b_i| \leqq \delta(Q(a_i)) + |a_i - b_i| < (2m+2)r_{ij}$; hence

$$(12.3)\quad \left|\frac{a_k - b_i}{b_j - b_i}\right| < 4m + 4, \quad \left|\frac{b_j - a_k}{b_j - b_i}\right| < 4m + 4.$$

This with (12.2) and (12.1) gives

$$\begin{aligned}|\Delta(S) - f_m(x)| &< \left|\frac{a_k - b_i}{b_j - b_i}\right| |\Delta(b_i, a_k, S') - f_m(x)| \\ &\quad + \left|\frac{b_j - a_k}{b_j - b_i}\right| |\Delta(a_k, b_j, S') - f_m(x)| \\ &< (8m + 8)\epsilon_{\sigma-1} = \epsilon_\sigma,\end{aligned}$$

as required.

Suppose now $0 < \sigma \leqq m^2$; then the points of S in $B-A$ lie in a single P'-set P_i', and there are points of S not in $P_i' + Q_i$. Let b_i be a point of S in $B-A$, let a be a point of S not in $P_i' + Q_i$, and let a_k be a point of Q_i which is not in S. If $S' = S - b_i - a$, the sets $S' + b_i + a_k$ and $S' + a + a_k$ each have a smaller complexity than S. a_k lies within δ' of x, and hence, by induction, (12.2) holds with b_j replaced by a. Let a_i be the center of the interval B_i containing b_i.

Suppose, (1), $a=a_j$ is in A'. Then $|a_j-b_i|>r_{ij}/2$. As a_k is in $Q_i=Q(a_i)$ while a_j is not, $|a_k-b_i|<|a_k-a_i|+r_{ij}<(2m+1)r_{ij}$, by the remark, and $|a_j-a_k|<(2m+1)r_{ij}$. Hence (12.3) holds with b_j replaced by $a_j=a$, and (11.5) follows just as before. Suppose, (2), a is in $A-(A'+A_2)$. From a, move toward a_i to the first point a' in A_1+A_2. If a' is in A_1, move back to the first point $a_1(a')$ in A_3. Then $|a_1(a')-a'|\leqq|a-a_i|$ and $|a_s(a')-a_1(a')|<|a_1(a')-a'|\leqq|a-a_i|$ $(s=2,\cdots,m)$, by (10.1). Hence $\delta(Q_i)<2m|a-a_i|$, by the remark, and $|a_k-b_i|<(2m+1)|a-a_i|$, and $|a-a_k|<(2m+1)|a-a_i|$. As $|a-b_i|>|a-a_i|/2$, (12.3) and (11.5) follow, as before. If a' is in A_2, there are m points of A' nearer a_i than a, and again $\delta(Q_i)<2m|a-a_i|$ and (11.5) follows. Suppose finally, (3), a is in A_2. Again we must have $\delta(Q_i)<2m|a-a_i|$ and (11.5) follows. This completes the proof of (A), therefore of Lemma 9, and therefore of Theorem I.

TAYLOR'S FORMULA

13. **Conditions under which Taylor's formula is valid.** Taylor's formula for $f(x)$ may hold to the mth order in certain closed sets even if $f(x)$ is not of class C^m (see §14). We find here a difference quotient condition equivalent to the validity of Taylor's formula, at least for perfect sets.

LEMMA 10. *If $f(x)=f_0(x)$ can be expanded in a Taylor's formula to the mth order locally uniformly in terms of $f_0(x),\cdots,f_m(x)$ in the closed set A, then these functions are continuous in A.*

It is apparent from (3.1) and (3.2) with $s=0$ that $f_0(x)$ is continuous. Take any s, $0<s\leqq m$. We shall assume $f_j(x)$ is continuous for $s<j\leqq m$, if there are such values of j, and shall prove that $f_s(x)$ is continuous.

Let $x_0,\cdots,x_s$ be distinct points of A. If we subtract (2.6) with x replaced by x_0 from the same equation with x replaced by x_1, we find

$$(13.1)\quad \begin{aligned} f_s(x_1)-f_s(x_0) &= s!\sum_{j=s+1}^{m}\left[\frac{f_j(x_0)}{j!}\sum_{i=0}^{s}\alpha^i u_{0i}^j-\frac{f_j(x_1)}{j!}\sum_{i=0}^{s}\alpha^i u_{1i}^j\right]\\ &\quad+s!\sum_{i=0}^{s}\alpha^i[R(x_i,x_0)-R(x_i,x_1)]. \end{aligned}$$

Given any limit point x_0 of A and any $\epsilon>0$, take $\delta<\epsilon/[2^{s+3}(s+1)mM]$ (if $s<m$) and $<1/2$ so small that (3.2) holds with x and ϵ replaced by x_0 and $\epsilon/[2^{s+2}(s+1)!]$ respectively, where $M=\max|f_j(x')|$ $(|x'-x_0|\leqq 1,\ s<j\leqq m)$. If $s>1$, take a point x_s of A within δ of x_0, and take points $x_{s-1},\cdots,x_2$ of A so that $r_{0i}<r_{0,i+1}/3$ $(i=2,\cdots,s-1)$. Now take any point x_1 within δ

of x_0, so that $r_{01}<r_{02}/3$ if $s>1$. From (13.1) we see that $|f_s(x_1)-f_s(x_0)|<\epsilon$, as required (see the proof of Lemma 6).

Let $x_0, \cdots, x_s$ be an ordered set of points. We say they form an (x_0, ρ)-set $(\rho>1)$, if

$$r_{0,i-1} < \frac{r_{0i}}{\rho} \qquad (i = 1, \cdots, s). \tag{13.2}$$

THEOREM II. *Let $f(x)=f_0(x), \cdots, f_m(x)$ be defined in the closed set A. A necessary condition that a Taylor's expansion for $f(x)$ should hold to the mth order locally uniformly in terms of $f_0(x), \cdots, f_m(x)$ is that for each (or some) $\rho>1$, each s $(0\leqq s\leqq m)$, each point x of A, and each $\epsilon>0$, there exist a $\delta>0$, such that if $x_0, \cdots, x_s$ is any (x_0, ρ)-set of points lying within δ of x, then*

$$|\Delta_{0\cdots s}f - f_s(x)| < \epsilon.$$

By the last lemma, the $f_i(x)$ are continuous. Take M so that $|f_i(x')|<M$ for $|x'-x|<1$. Take $\delta<\epsilon(\rho-1)^s/[2(s+1)mM\rho^s]$ and <1 so that $|f_s(x')-f_s(x)|<\epsilon/2$ $(|x'-x|<\delta)$, and so that (3.2) holds with ϵ replaced by $\epsilon(\rho-1)^s/[2(s+1)!\rho^s]$. Now take any (x_0, ρ)-set of points $x_0, \cdots, x_s$ lying within δ of x. Then

$$\frac{r_{0i}}{r_{ki}} < \frac{\rho}{\rho - 1}$$

for $k\neq i$. For if $k<i$, then $r_{0k}\leqq r_{0,i-1}<r_{0i}/\rho$, hence $r_{ki}\geqq r_{0i}-r_{0k}>r_{0i}(1-1/\rho)$, and $r_{0i}/r_{ki}<1/(1-1/\rho)=\rho/(\rho-1)$; if $k>i$, then $r_{0k}\geqq r_{0,i+1}>\rho r_{0i}$, hence $r_{ki}\geqq r_{0k}-r_{0i}>r_{0i}(\rho-1)$, and $r_{0i}/r_{ki}<1/(\rho-1)<\rho/(\rho-1)$. Replacing x by x_0 in (2.6) gives immediately $|\Delta_{0\cdots s}f-f_s(x_0)|<\epsilon/2$; hence $|\Delta_{0\cdots s}f-f_s(x)|<\epsilon$.

THEOREM III. *If A is perfect, then the condition in Theorem* II *is also sufficient.*

We shall prove successively for $s=0, \cdots, m$ that $f(x)$ can be expanded in a Taylor's formula to the sth order locally uniformly in terms of $f_0(x), \cdots, f_s(x)$. Evidently $f_0(x)$ is continuous; hence this is true for $s=0$. The proof for a general s follows the proof of Lemma 6; we need merely be careful to choose $x_{s-1}, \cdots, x_1$ so that $r_{0,t-1}<r_{0t}/\rho$ $(t=2, \cdots, s)$.

14. **Taylor's formula and differentiability.** We shall say the set A has the property Z_ρ at the point $x(\rho>1)$ if there is an $\eta>0$ such that corresponding to any two points x_0 and x_1 of A within η of x, points $x_2, \cdots, x_s$ of A can be found such that

$$\frac{1}{\rho} < \left| \frac{x_i - x_j}{x_1 - x_0} \right| < \rho \qquad (i, j = 0, \cdots, s;\ i \neq j); \tag{14.1}$$

then $r_{ij}/r_{kl}<\rho^2$ for $i\neq j$, $k\neq l$. This condition is satisfied for instance by Cantor's set. s is any number $\leqq m$, m fixed.

THEOREM IV.* *Let A be a closed set having the property Z_ρ for some $\rho=\rho(x)$ at each point x, and let $f(x)=f_0(x),\cdots,f_m(x)$ be defined in A. A necessary and sufficient condition that $f(x)$ be of class C^m in terms of $f_0(x),\cdots,f_m(x)$ is that Taylor's formula for $f(x)$ should hold to the mth order locally uniformly in terms of $f_0(x),\cdots,f_m(x)$.*

In short, in this case, Taylor's formula for $f_0(x)$ implies Taylor's formula for each $f_s(x)$.

The necessity of the condition being trivial, we turn to the sufficiency. By Lemma 10, $f_m(x)$ is continuous. It remains to prove that for any s, $0<s<m$, $f_s(x)$ may be expanded in a Taylor's formula to the $(m-s)$th order locally uniformly in terms of $f_s(x),\cdots,f_m(x)$. We shall prove this for s, assuming it for numbers $s+1,\cdots,m$.

Let $x_0,\cdots,x_s$ be distinct points of A. Set

$$(14.2)\quad H_j=\sum_{i=0}^{s}\overset{i}{\alpha}\overset{j}{u}_{0i},$$

$$(14.3)\quad H_j'=\sum_{i=0}^{s}\overset{i}{\alpha}\overset{j}{u}_{1i}=\sum_{i=0}^{s}\overset{i}{\alpha}(u_{0i}-u_{01})^{j}=\sum_{i=0}^{s}\overset{i}{\alpha}\sum_{l}(-1)^{j-l}\binom{j}{l}\overset{l}{u}_{0i}\overset{j-l}{u}_{01}$$
$$=\sum\nolimits_l(-1)^{j-l}\binom{j}{l}H_l\overset{j-l}{u}_{01},$$

where $\sum_l$ means summation over all values of l. We can write (if $s<m$)

$$\sum_{j=s+1}^{m}\frac{f_j(x_1)}{j!}H_j'=\sum_{j=s+1}^{m}\frac{1}{j!}\sum_{k=j}^{m}\frac{f_k(x_0)}{(k-j)!}\overset{k-j}{u}_{01}\sum\nolimits_l(-1)^{j-l}\binom{j}{l}H_l\overset{j-l}{u}_{01}+R$$
$$=\sum_{k=s+1}^{m}\frac{f_k(x_0)}{k!}\sum\nolimits_l\overset{k-l}{u}_{01}H_l\sum_{j=s+1}^{k}(-1)^{j-l}\binom{k}{j}\binom{j}{l}+R,$$

where

$$R=\sum_{j=s+1}^{m}\frac{1}{j!}H_j'R_j(x_1,x_0).$$

Now if $k\geqq l>s$, then on replacing j by $k-j$ we find

* For the special case that A is a closed interval, see a paper by the author, *Derivatives, difference quotients and Taylor's formula*, Bulletin of the American Mathematical Society, vol. 40 (1934), pp. 89–94; Theorem III.

$$\sum_{j=s+1}^{k}(-1)^{j-l}\binom{k}{j}\binom{j}{l}=\sum_{j}(-1)^{k-j-l}\binom{k}{j}\binom{k-j}{l}$$
$$=(-1)^{k-l}\binom{0}{k-l}=\delta_{kl},^{*}$$

and if $k>l=s$,

$$\sum_{j=s+1}^{k}(-1)^{j-l}\binom{k}{j}\binom{j}{l}=\delta_{ks}-(-1)^{s-s}\binom{k}{s}\binom{s}{s}=-\binom{k}{s}.$$

Therefore, as $H_l=0$ $(l<s)$ and $H_s=1$,

$$\sum_{j=s+1}^{m}\frac{f_j(x_1)}{j!}H_j'=\sum_{k=s+1}^{m}\frac{f_k(x_0)}{k!}\left[-\binom{k}{s}u_{01}^{k-s}+H_k\right]+R. \tag{14.4}$$

Putting this in (13.1) gives

$$\begin{aligned} f_s(x_1)=&\sum_{k=s}^{m}\frac{f_k(x_0)}{(k-s)!}u_{01}^{k-s}-\sum_{j=s+1}^{m}\frac{s!}{j!}\sum_{i=0}^{s}\alpha^i u_{1i}^{j}R_j(x_1,x_0)\\ &+s!\sum_{i=0}^{s}\alpha^i[R(x_i,x_0)-R(x_i,x_1)]. \end{aligned} \tag{14.5}$$

Given a point x of A and an $\epsilon>0$, take ρ and η corresponding to x, and take $\delta'<\eta$ so that (3.2) holds with δ and ϵ replaced by δ' and $\epsilon/[3m(s+1)!\rho^{2m}]$ and with s taking on the values $0, s+1, \cdots, m$. Set $\delta=\delta'/(2\rho)$. Now if x_0 and x_1 are points of A within δ of x, we can add points $x_2, \cdots, x_s$ of A so that (14.1) holds and these points will lie within δ' of x. Then

$$\left|\frac{\alpha^i u_{1i}^{j}}{u_{01}^{m-s}}R_j(x_1,x_0)\right|=\frac{r_{1i}^{j}}{r_{0i}\cdots r_{i-1,i}r_{i+1,i}\cdots r_{si}r_{01}^{j-s}}\frac{|R_j(x_1,x_0)|}{r_{01}^{m-j}}<\frac{\epsilon}{3m(s+1)!},$$

and similarly for the other remainder terms. Therefore $|R_s(x_1,x_0)|/r_{01}^{m-s}<\epsilon$, as required.

COROLLARY. *If $m\leqq 2$, Theorem* IV *holds for all closed sets.*

The only value of s we may need in the above proof is $s=1$; the condition Z_ρ is satisfied trivially if $s=1$.

Example. Theorem IV does not hold for all closed sets, as we now show, using $m=3$. Set $a_i=1/2^i$, $b_i=1/2^{2i}$, $c_i=1/2^{3i}$; $b_i'=a_i+b_i$, $c_i'=a_i+c_i$, $d_i=a_i+b_i-c_i (i=1, 2, \cdots)$. Let A be the set of points $0, a_i, c_i', d_i, b_i'$. Set $f_0(0)=f_1(0)=f_2(0)=f_3(0)=0$,

* See Netto, *Lehrbuch der Combinatorik*, Leipzig, 1927, §158, (27).

$$f_0(a_i) = 0, \quad f_0(c_i') = 0, \quad f_0(d_i) = 0, \quad f_0(b_i') = b_i^2 c_i,$$
$$f_1(a_i) = 0, \quad f_1(c_i') = 0, \quad f_1(d_i) = b_i^2 - b_i c_i, \quad f_1(b_i') = b_i^2 + b_i c_i,$$
$$f_2(a_i) = 0, \quad f_2(c_i') = 0, \quad f_2(d_i) = 2b_i, \quad f_2(b_i') = 2b_i,$$
$$f_3(a_i) = 0, \quad f_3(c_i') = 0, \quad f_3(d_i) = 0, \quad f_3(b_i') = 0.$$

As $\Delta(0, a_i, c_i', d_i) = 0$, while $\Delta(a_i, c_i', d_i, b_i') = 3!\, b_i^2 c_i / [b_i(b_i - c_i)c_i] \to 6$ as $i \to \infty$, $\Delta^3 f_0(x)$ does not converge at $x = 0$, and hence $f_0(x)$ is not of class C^3, by Theorem I. However, Taylor's formula holds for $f_0(x)$ to the third order locally uniformly. For a calculation shows that $R(x, y) = 0$ whenever x and y are chosen from the points a_i, c_i', d_i, b_i', except that $R(b_i', a_i) = R(b_i', c_i') = b_i^2 c_i$, $R(c_i', d_i) = R(c_i', b_i') = b_i c_i (b_i - 2c_i)$; hence if x and y are chosen in any manner from the points a_i, c_i', d_i, b_i', $R(y, x)/(y - x)^3 \to 0$ as $i \to \infty$. Suppose now x_i and y_j are chosen from a_i, c_i', d_i, b_i', and from a_j, c_j', d_j, b_j' respectively, $j \neq i$ (or $x_i = 0$ or $y_j = 0$). If k is the larger of the numbers i, j, then

$$|R(y_j, x_i)| < 2b_k^2 c_k + (b_k^2 + b_k c_k)(a_k + b_k) + b_k(a_k + b_k)^2,$$

and as $|y_j - x_i|^3 \geq a_k^3/8$, $R(y_j, x_i)/(y_j - x_i)^3 \to 0$ as $i, j \to \infty$ $(j \neq i)$. Hence for some $\delta > 0$, if x and y are any two points of A within δ of 0, $|R(y, x)/(y - x)^3| < \epsilon$. This is true also at each isolated point of A; hence Taylor's formula is valid.

Note that we may increase A to a perfect set by adding the intervals between a_i and c_i' and between d_i and b_i', and giving the obvious definitions of $f_0(x), \cdots, f_3(x)$ there. In this example, Taylor's formula holds to the required order for neither $f_1(x)$ nor $f_2(x)$.

HARVARD UNIVERSITY,
CAMBRIDGE, MASS.

ANNALS OF MATHEMATICS
Vol. 35, No. 3, July, 1934

DERIVATIVES, DIFFERENCE QUOTIENTS AND TAYLOR'S FORMULA II[1]

BY HASSLER WHITNEY

(Received September 12, 1934)

1. **Introduction.** In Paper I[2] we gave conditions under which a function of one variable $f(x)$ is a polynomial of degree at most $m - 1$, or has a continuous m^{th} derivative, in terms of the uniform convergence of the mth difference quotient. Here we shall give similar results in two dimensions; the extension to n dimensions is immediate. The functions involved are always supposed defined in a certain (open) region R of space.

2. **Partial difference quotients.** Let $f(x, y)$ be defined in R. If i and j are two non-negative integers and h is $\neq 0$, the partial difference quotient $\Delta_h^{i\,j} f(x, y)$ is defined by the equation

$$\Delta_h^{i\,j} f(x,y) = \frac{1}{h^{i+j}} \sum_{k=0}^{i} \sum_{l=0}^{j} (-1)^{i+j-k-l} \binom{i}{k}\binom{j}{l} f(x + kh, y + lh). \tag{1}$$

Thus for instance

$$\Delta_h^{2\,0} f(x,y) = \frac{1}{h^2}[f(x + 2h, y) - 2f(x + h, y) + f(x, y)],$$

$$\Delta_h^{1\,1} f(x,y) = \frac{1}{h^2}[f(x + h, y + h) - f(x, y + h) - f(x + h, y) + f(x, y)].$$

It is easily seen that

$$\Delta_h^{i+1,\,j} f(x,y) = \frac{1}{h}[\Delta_h^{i\,j} f(x + h, y) - \Delta_h^{i\,j} f(x, y)], \tag{2}$$

with a similar equation for $\Delta_h^{i,\,j+1} f(x, y)$.

(a) Suppose the functions $f_{ij}(x, y)$ are defined in R for $i + j \leqq m$. Suppose

$$f_{00}(x + \xi, y + \eta) = \sum_{i+j \leqq m} \frac{f_{ij}(x, y)}{i!\, j!} \xi^i \eta^j + R(x, y; \xi, \eta), \tag{3}$$

where for each point (x^*, y^*) of R and each $\epsilon > 0$ there is a $\delta > 0$ such that if (x, y) and $(x + \xi, y + \eta)$ are any two points within δ of (x^*, y^*), then

$$| R(x, y; \xi, \eta) | \leqq (\xi^2 + \eta^2)^{m/2} \epsilon. \tag{4}$$

---

[1] Presented to the American Mathematical Society, December 27, 1933.

[2] Derivatives, difference quotients and Taylor's formula I, this Bulletin, vol. 40 (1934), pp. 89–94.

We find as in Paper I for $i + j \leqq m$

$$\Delta_h^{i\,j} f_{00}(x, y) = \frac{1}{h^{i+j}} \sum_{k,\,l} (-1)^{i+j-k-l} \binom{i}{k}\binom{j}{l} \left[\sum_{p+q \leqq m} \frac{f_{pq}(x, y)}{p!\, q!} (kh)^p (lh)^q + R(x, y; kh, lh) \right]$$

(5)
$$= f_{ij}(x, y) + \sum{}' \frac{f_{i+p,\, j+q}(x, y)}{(i + p)!\, (j + q)!} h^{p+q} \sum_{k,\,l} (-1)^{i+j-k-l} \binom{i}{k}\binom{j}{l} k^{i+p} l^{j+q} + \frac{1}{h^{i+j}} \sum_{k,\,l} (-1)^{i+j-k-l} \binom{i}{k}\binom{j}{l} R(x, y; kh, lh),$$ [3]

where in Σ' those and only those terms occur in which

$$p \geqq 0, \qquad q \geqq 0, \qquad 0 < p + q \leqq m - i - j.$$

(3) and (4) show that $f_{00}(x, y)$ is continuous. As before, we see in succession that $\Delta_h^{i\,j} f_{00}(x, y) \to f_{ij}(x, y)$ uniformly in the neighborhood of any point of R and $f_{ij}(x, y)$ is continuous ($i + j = m, m - 1, \cdots, 1$).

(b) Suppose $f(x, y) = f_{00}(x, y)$ and $\partial^{i+j} f(x, y)/\partial x^i\, \partial y^j$ exists and equals the continuous function $f_{ij}(x, y)$ ($i + j \leqq m$). Then (3) and (4) hold, by Taylor's theorem, and hence $\Delta_h^{i\,j} f(x, y) \to f_{ij}(x, y)$ uniformly ($i + j \leqq m$).

(c) If $f(x, y) = \Sigma_{i+j \leqq m}\, a_{ij}\, x^i y^j$ is a polynomial of degree at most m, then a Taylor's formula (3) holds with $f_{ij}(x, y) \equiv i!\, j!\, a_{ij} (i + j = m)$ and $R(x, y; \xi, \eta) \equiv 0$. Hence $\Delta_h^{i\,j} f(x, y) \equiv i!\, j!\, a_{ij} (i + j = m)$. If $f(x, y)$ is of degree at most $m - 1$, $\Delta_h^{i\,j} f(x, y) \equiv 0$ $(i + j = m)$.

3. **Functions with given m^th partial derivatives.** Suppose the continuous functions $f_{ij}(x, y)$ are defined for $i + j = m$ throughout an open square S with center at (a, b). Under what conditions does there exist a function $f_{00}(x, y)$ in R such that

(6)
$$\frac{\partial^{i+j}}{\partial x^i\, \partial y^j} f_{00}(x, y) = f_{ij}(x, y) \qquad (i + j = m)?$$

The answer is given by the following lemma.

---

[3] We have here used the fact that

$$\sum_{k,\,l} (-1)^{i+j-k-l} \binom{i}{k}\binom{j}{l} k^p l^q = \sum_{k=0}^{i} (-1)^{i-k} \binom{i}{k} k^p \cdot \sum_{l=0}^{j} (-1)^{j-l} \binom{j}{l} l^q,$$

which equals zero if $p < i$ or $q < j$, and equals $i!\, j!$ if $p = i$ and $q = j$; we then changed p and q to $i + p$ and $j + q$ respectively.

LEMMA 1. *Let the* $f_{ij}(x, y)$ *be given as above. If for any* k, l *such that* $k + l = m - 1$,

$$\int_x^{x'} f_{k+1,\,l}(s, y)\, ds + \int_y^{y'} f_{k,\,l+1}(x', s)\, ds + \int_{x'}^{x} f_{k+1,\,l}(s, y')\, ds + \int_{y'}^{y} f_{k,\,l+1}(x, s)\, ds = 0, \tag{7}$$

provided the path of integration lies in S, *then there exists a function* $f_{00}(x, y)$ *in* S *satisfying* (6).[3a]

The necessity of the condition merely reflects the fact that $\partial^{k+l} f_{00}(x, y)/\partial x^k\, \partial y^l$ is a single valued function. We prove the sufficiency. For any k and l, $k + l < m$, we define the function f_{kl} along the line $y = b$ by the equation

$$f_{kl}(x, b) = \int_a^x \cdots \int_a^{s_3} \int_a^{s_2} f_{m-l,\,l}(s_1, b)\, ds_1\, ds_2 \cdots ds_{m-k-l}. \tag{8}$$

We now extend the definitions of these functions throughout S:.

$$f_{kl}(x, y) = \sum_{p=0}^{m-k-l-1} \frac{1}{p!} f_{k,\,l+p}(x, b)(y - b)^p + \int_b^y \cdots \int_b^{s_3} \int_b^{s_2} f_{k,\,m-k}(x, s_1)\, ds_1\, ds_2 \cdots ds_{m-k-l}. \tag{9}$$

If we show that

$$\frac{\partial}{\partial x} f_{kl}(x, y) = f_{k+1,\,l}(x, y), \qquad \frac{\partial}{\partial y} f_{kl}(x, y) = f_{k,\,l+1}(x, y), \tag{10}$$

(6) will follow. (7) gives

$$\int_b^{s_2} [f_{k,\,m-k}(x', s_1) - f_{k,\,m-k}(x, s_1)]\, ds_1 = \int_x^{x'} [f_{k+1,\,m-k-1}(s, s_2) - f_{k+1,\,m-k-1}(s, b)]\, ds;$$

hence

$$\int_b^y \cdots \int_b^{s_3} \int_b^{s_2} [f_{k,\,m-k}(x', s_1) - f_{k,\,m-k}(x, s_1)]\, ds_1\, ds_2 \cdots ds_{m-k-l} = \int_x^{x'} \int_b^y \cdots \int_b^{s_3} [f_{k+1,\,m-k-1}(s, s_2) - f_{k+1,\,m-k-1}(s, b)]\, ds_2 \cdots ds_{m-k-l}\, ds.$$

[3a] In n dimensions we must assume such a formula holds where the contour is any rectangle with sides parallel to any two coordinate axes.

On dividing both sides by $x' - x$ and letting x' approach x, we find the derivative with respect to x of the integral in (9). Differentiating (8) gives

$$\frac{\partial}{\partial x} f_{kl}(x, b) = f_{k+1, l}(x, b).$$

Hence, differentiating (9) with respect to x,

$$\begin{aligned}\frac{\partial}{\partial x} f_{kl}(x, y) &= \sum_{p=0}^{m-k-l-1} \frac{1}{p!} f_{k+1, l+p}(x, b)(y-b)^p \\ &\quad + \int_b^y \cdots \int_b^{s_3} [f_{k+1, m-k-1}(x, s_2) - f_{k+1, m-k-1}(x, b)]\, ds_2 \cdots ds_{m-k-l} \\ &= \sum_{p=0}^{m-k-l-1} \frac{1}{p!} f_{k+1, l+p}(x, b)(y-b)^p \\ &\qquad - \frac{1}{(m-k-l-1)!} f_{k+1, m-k-1}(x, b)(y-b)^{m-k-l-1} \\ &\qquad + \int_b^y \cdots \int_b^{s_2} f_{k+1, m-(k+1)}(x, s_1)\, ds_1 \cdots ds_{m-(k+1)-l} \\ &= f_{k+1, l}(x, y).\end{aligned}$$

The other equation follows immediately on differentiating (9) with respect to y.

4. **Difference quotients and contour integrals.** We prove the following lemma.

LEMMA 2. *If $\Delta_h^{ij} f(x, y)$ approaches the continuous function $f_{ij}(x, y)$ uniformly in the neighborhood of any point of R for $i + j = m$, then (7) is satisfied in any square S lying wholly in R.*

We first derive an identity in difference quotients. Let x and y be fixed numbers, let p and q be fixed integers, take $h \neq 0$, and set $x^* = x + ph$, $y^* = y + qh$. Replace x by $x + kh$ in (2), and sum. This gives

$$h \sum_{k=0}^{p-1} \Delta_h^{i+1, j} f(x + kh, y) = \Delta_h^{ij} f(x^*, y) - \Delta_h^{ij} f(x, y).$$

Similarly we find

$$h \sum_{k=0}^{p-1} \Delta_h^{i+1, j} f(x + kh, y^*) = \Delta_h^{ij} f(x^*, y^*) - \Delta_h^{ij} f(x, y^*),$$

$$h \sum_{k=0}^{q-1} \Delta_h^{i, j+1} f(x, y + kh) = \Delta_h^{ij} f(x, y^*) - \Delta_h^{ij} f(x, y),$$

$$h \sum_{k=0}^{q-1} \Delta_h^{i, j+1} f(x^*, y + kh) = \Delta_h^{ij} f(x^*, y^*) - \Delta_h^{ij} f(x^*, y).$$

These equations give

$$(11)\quad \begin{aligned} &h\sum_{k=0}^{p-1}\Delta_h^{i+1,j}f(x+kh,y)+h\sum_{k=0}^{q-1}\Delta_h^{i,j+1}f(x^*,y+kh)\\ &\quad -h\sum_{k=0}^{p-1}\Delta_h^{i+1,j}f(x+kh,y^*)-h\sum_{k=0}^{q-1}\Delta_h^{i,j+1}f(x,y+kh)=0. \end{aligned}$$

The functions $f_{ij}(x, y)$ are uniformly continuous in S. Hence, given x, x', y, y', and an $\epsilon > 0$, we can take a $\delta > 0$ so that if $|h| < \delta$, $|y' - y^*| < \delta$, and $x + ph \leqq x' < x + (p + 1)h$, then

$$\left| \int_x^{x'} f_{i+1,j}(s, y')\, ds - h\sum_{k=0}^{p-1} f_{i+1,j}(x+kh, y^*) \right| < \epsilon.$$

As $\Delta_h^{i+1,j} f(x, y) \to f_{i+1,j}(x, y)$ uniformly in $S(i + 1 + j = m)$, we can take $\delta' < \delta$ so small that if $|h| < \delta'$, then

$$\left| h\sum_{k=0}^{p-1} f_{i+1,j}(x+kh, y^*) - h\sum_{k=0}^{p-1}\Delta_h^{i+1,j} f(x+kh, y^*) \right| < hp\epsilon \leqq |x' - x|\epsilon.$$

We derive similar inequalities corresponding to the other three terms of (11). As ϵ is arbitrarily small, (11) together with these inequalities gives (7), as required.

5. **Difference quotients and polynomials.** Theorem I of Paper I generalizes to two (or n) dimensions as follows:

THEOREM I. *Let $f(x, y)$ be a function defined in the square S and measurable in each variable separately. A necessary and sufficient condition that $f(x, y)$ be a polynomial of degree at most $m - 1$ is that $\Delta_h^{m0} f(x, y) \to 0$ and $\Delta_h^{0m} f(x, y) \to 0$ uniformly in S.*

The necessity of the condition follows from 2(c); we must prove the sufficiency. Take a fixed y_0, and consider $f(x, y_0)$ as a function of x. $\Delta_h^{m0} f(x, y_0) = \Delta_h^m f(x, y_0) \to 0$ uniformly, and hence $f(x, y_0)$ is a polynomial of degree at most $m - 1$ in x, by Paper I, Theorem I. Similarly $f(x_0, y)$ is a polynomial of degree at most $m - 1$ in y; it follows that $f(x, y)$ is a polynomial of degree at most $m - 1$ in x and y.[4] This is true if S is replaced by a region R.

[4] Choose m numbers $y_0, \cdots, y_{m-1}$; then

$$f(x, y_k) = a_{k0} + a_{k1}x + \cdots + a_{k,m-1}x^{m-1} \qquad (k = 0, \cdots, m - 1).$$

But also

$$f(x, y) = b_0(x) + b_1(x)y + \cdots + b_{m-1}(x)y^{m-1}.$$

Setting $y = y_k(k = 0, \cdots, m - 1)$ in this equation and using the first equations, we find that the b's are polynomials in x.

6. **Existence of derivatives.** We are now ready to prove the fundamental theorem:

THEOREM II. *Let $f(x, y)$ be a function defined in R and measurable in each variable separately. A necessary and sufficient condition that $\partial^{i+j} f(x, y)/\partial x^i \partial y^j$ exist and equal the continuous function $f_{ij}(x, y)(i + j = m)$ is that*

$$\Delta_h^{ij} f(x, y) \to f_{ij}(x, y)$$

uniformly in the neighborhood of any point of $R(i + j = m)$.

If $f(x, y)$ is continuous, then $f_{ij}(x, y)$ is continuous, by 2(a).

The necessity of the condition was proved in 2(b); we must prove the sufficiency. Consider any square S in R. By Lemmas 1 and 2 there is a function $g(x, y)$ in S such that $\partial^{i+j} g(x, y)/\partial x^i \partial y^j = f_{ij}(x, y)$ $(i + j = m)$; hence, by 2(b), $\Delta_h^{ij} g(x, y) \to f_{ij}(x, y)$ uniformly in S. Set $\phi(x, y) = f(x, y) - g(x, y)$; then $\Delta_h^{ij} \phi(x, y) \to 0$ uniformly in S, and $\phi(x, y)$ is a polynomial of degree at most $m - 1$, by Theorem I. Therefore $\partial^{i+j} f(x, y)/\partial x^i \partial y^j = \partial^{i+j} g(x, y)/\partial x^i \partial y^j = f_{ij}(x, y)$ in S, and hence in R.

THEOREM III. *A necessary and sufficient condition that $f(x, y) = f_{00}(x, y)$ have continuous m^{th} partial derivatives in R is that there exist functions $f_{ij}(x, y)$ in R $(i + j \leqq m)$ such that* (3) *and* (4) *hold. In this case, $\partial^{i+j} f(x, y)/\partial x^i \partial y^j = f_{ij}(x, y)$ $(i + j \leqq m)$.*

The necessity of the condition is a consequence of Taylor's formula. To prove the sufficiency, 2(a) shows that $\Delta_h^{ij} f(x, y) \to f_{ij}(x, y)$ uniformly $(i + j \leqq m)$; hence, by Theorem II, $\partial^{i+j} f(x, y)/\partial x^i \partial y^j$ exists and equals $f_{ij}(x, y)$ $(i + j \leqq m)$.

HARVARD UNIVERSITY.

Annals of Mathematics
Vol. 35, No. 3, July, 1934

FUNCTIONS DIFFERENTIABLE ON THE BOUNDARIES OF REGIONS[1]

By Hassler Whitney

(Received September 12, 1934)

1. Introduction. Let the function $f(x_1, \cdots, x_n)$ be defined in the bounded region[2] R of n-space E, and suppose f has continuous m^{th} partial derivatives in R, i.e. f "is of class C^m" in R. If B is the boundary of R, how shall we decide whether f is of class C^m in $R + B$? If the derivatives of f take on boundary values on B, it would be natural to define the derivatives on B as the limit of their values in R. But it is easy to construct a region R and a function f such that the kth partial derivatives of $f (0 < k \leqq m)$ are continuous in $R + B$, whereas at a certain boundary point P of B, f is not continuous;[3] it seems unreasonable in this case to say that f is of class C^m in $R + B$.

If it is possible to extend the definition of f throughout a region containing $R + B$ so that it has continuous m^{th} partial derivatives there, we may then surely say that f is of class C^m in $R + B$; this is the definition we shall use. We show in this note that, for certain regions, for a function to be of class C^m in the closed region, it is sufficient that the mth partial derivatives be continuous on the boundary.

We shall use a one-dimensional notation, as in a paper by the author AE.[4] Thus $f_k(x) = f_{k_1 \cdots k_n}(x_1, \cdots, x_n)$, $(x' - x)^l = (x_1' - x_1)^{l_1} \cdots (x_n' - x_n)^{l_n}$, $l! = l_1! \cdots l_n!$, etc. We set $\sigma_k = k_1 + \cdots + k_n$. $r_{xx'}$ is the distance between the two points x and x'. The fundamental definition is: $f(x) = f_0(x)$ is of class C^m in A if functions $f_k(x)$ and $R_k(x'; x)$ $(\sigma_k \leqq m)$ exist in A such that

$$f_k(x') = \sum_{\sigma_l \leqq m - \sigma_k} \frac{f_{k+l}(x)}{l!} (x' - x)^l + R_k(x'; x) \tag{1}$$

for each $k (\sigma_k \leqq m)$, and R_k has the following property. Given the point x^0 of A and an $\epsilon > 0$, there is a $\delta > 0$ such that if x and x' are any two points of A within δ of x^0, then

$$| R_k(x'; x) | \leqq r_{xx'}^{m - \sigma_k} \epsilon. \tag{2}$$

[1] Presented to the American Mathematical Society, December 27, 1933.

[2] The restriction that R be bounded is made merely for simplicity; the theorem holds equally well without it.

[3] We give an example with $m = 1$. In polar coordinates, let S be the spiral $r = \theta^{-\frac{1}{2}}$ $(\theta \geqq 0)$. We let R be a narrow region about S, and set $f(x, y) = F(r, \theta) = \log \theta$ for those values of r which are approximately $\theta^{-\frac{1}{2}}$. Differentiating, we find $\partial f/\partial x = -\theta^{-\frac{1}{2}} \sin \theta$ and $\partial f/\partial y = \theta^{-\frac{1}{2}} \cos \theta$ approximately in R. Hence $\partial f/\partial x$ and $\partial f/\partial y \to 0$ at the origin; but $f \to \infty$ there.

[4] Analytic extensions of differentiable functions defined in closed sets, Transactions of the American Mathematical Society, Vol. 36 (1934), pp. 63–89.

If $f(x)$ has continuous m^{th} partial derivatives in R, it is of class C^m in R, as we see by setting $f_{k_1 \ldots k_n} = \partial^{k_1 + \cdots + k_n} f / \partial x_1^{k_1} \cdots \partial x_n^{k_n}$ and using Taylor's formula. The converse also is true.

LEMMA 1. *If $f(x)$ is of class C^m in the closed set A, then its definition can be extended throughout E so it will be of class C^m there.*

This is proved in AE, Lemma 2.

2. The property P. We shall say a point set A has the property P if there is a number ω such that any two points x and y of A are joined by a curve C in A of length $L \leqq \omega\, r_{xy}$.

LEMMA 2. *If R has the property P, so has $R + B$.*

Let x and y be points of $R + B$, and suppose first that y is in B while x is in R. Let $y_1, y_2, \cdots$ be a sequence of points of R such that $r_{y_s y} < 2^{-s-2}\, r_{xy}$. As $r_{xy_1} < 3r_{xy}/2$, there is a curve C_0 joining x and y_1 of length $< 3\omega\, r_{xy}/2$. In general, $r_{y_s y_{s+1}} < 2^{-s-1}\, r_{xy}$, and hence there is a curve C_s joining y_s and y_{s+1} of length at most $2^{-s-1}\,\omega\, r_{xy}$. From the curves $C_0, C_1, C_2, \cdots$ we pick out a curve C joining x to y; its length is

$$L < 3\omega r_{xy}/2 + 2^{-2}\,\omega r_{xy} + 2^{-3}\,\omega r_{xy} + \cdots = 2\omega r_{xy}.$$

If both x and y are in B, we take a point x' with $r_{x'x} < r_{xy}/2$, find curves joining it to x and y, of lengths $L_1 < \omega r_{xy}$ and $L_2 < 3\omega r_{xy}$ respectively, as above; from these we pick out a curve C joining x and y, of length $L < 4\omega r_{xy}$. Thus the number $\omega' = 4\omega$ will do for $R + B$. Evidently any number $\omega' > \omega$ will do.

3. A remainder formula. We prove here that *if (1) holds on the rectifiable curve C of length L with end points x^* and x', then $R_k(x'; x^*)(\sigma_k \leqq m)$ is given by the Stieltjes integral*[4a]

$$(3) \qquad R_k(x'; x^*) = -\sum_{\sigma_l = m - \sigma_k} \frac{1}{l!} \int_0^L [f_{k+l}(x(s)) - f_{k+l}(x^*)]\, d(x' - x(s))^l,$$

where s denotes the length of that part of C between x^* and the variable point $x(s)$. If we carry out the differentiation indicated, we may write this as a sum of contour integrals

$$(4) \qquad R_k(x'; x^*) = \sum_{\sigma_l = m - \sigma_k} \sum_{\sigma_j = 1} \int_{x^*}^{x'} [f_{k+l}(x) - f_{k+l}(x^*)] \frac{(x' - x)^{l-j}}{(l-j)!} (dx)^j.$$

In the case $n = 2$, $k = 0$, this equation, written out in full, is

$$(5) \qquad \begin{aligned} R_{00}(x', y'; x^*, y^*) = \sum_{i+j=m} \int_{(x^*, y^*)}^{(x', y')} & [f_{ij}(x, y) - f_{ij}(x^*, y^*)] \\ & \cdot \left[\frac{(x' - x)^{i-1} (y' - y)^j}{(i-1)!\, j!} dx + \frac{(x' - x)^i (y' - y)^{j-1}}{i!\,(j-1)!} dy \right]. \end{aligned}$$

[4a] An example shows that no such formula holds for all curves.

To prove (3), let $x^0 = x^*, x^1, \cdots, x^p = x'$ be the end points of a subdivision of C. If in equation (6.3) of AE we subtract $f_k(x'')$ from both sides and change x, x', x'' to x^{i-1}, x^i, x' respectively, we find

$$R_k(x'; x^{i-1}) - R_k(x'; x^i) = \sum_{\sigma_l \leqq m - \sigma_k} \frac{R_{k+l}(x^i; x^{i-1})}{l!} (x' - x^i)^l. \tag{6}$$

As $R_k(x'_h; x^i_h) = 0$, summing over i gives

$$R_k(x'; x^*) = \sum_{\sigma_l \leqq m - \sigma_k} \frac{1}{l!} \sum_{i=1}^{p} R_{k+l}(x^i; x^{i-1})(x' - x^i)^l. \tag{7}$$

We show first that as the norm of the subdivision tends to zero, the terms on the right with $\sigma_l < m - \sigma_k$ tend to zero. By (2), given an $\eta > 0$, we can take the norm so small that

$$| R_j(x^i; x^{i-1}) | < r_{x^{i-1} x^i}\, \eta \qquad (\sigma_j < m).$$

As $| x'_h - x^i_h | \leqq L$, we find for $\sigma_l < m - \sigma_k$

$$\left| \sum_{i=1}^{p} R_{k+l}(x^i; x^{i-1})(x' - x^i)^l \right| < \eta L^{\sigma_l} \sum_{i=1}^{p} r_{x^{i-1} x^i} \leqq \eta L^{\sigma_l + 1}.$$

as η is arbitrary, the statement is proved.

Now take any l with $\sigma_l = m - \sigma_k$. We have

$$\sum_{i=1}^{p} R_{k+l}(x^i; x^{i-1})(x' - x^i)^l = \sum_{i=1}^{p} [f_{k+l}(x^i) - f_{k+l}(x^{i-1})]\, (x' - x^i)^l$$

$$= - \sum_{i=1}^{p-1} [f_{k+l}(x^i) - f_{k+l}(x^*)]\, [(x' - x^{i+1})^l - (x' - x^i)^l].$$

Putting this in (7) and passing to the limit as the norm of the subdivision tends to zero, we obtain (3).

LEMMA 3. *Let $f(x)$ be of class C^m on the curve C of length L with end points x^* and x'. If*

$$| f_k(x) - f_k(x^*) | < \epsilon \qquad (\sigma_k = m, x \text{ on } C), \tag{8}$$

then

$$| R_k(x'; x^*) | < n(m + 1)^n L^{m - \sigma_k} \epsilon^{5} \qquad (\sigma_k \leqq m). \tag{9}$$

[5] The numerical factor $n(m + 1)^n$ may of course be replaced by the factor

$$\sum_{\sigma_l = m - \sigma_k} \sum_{\sigma_j = 1} \frac{1}{(l - j)!} = \frac{n^{m - \sigma_k}}{(m - \sigma_k - 1)!}.$$

This follows at once from (4) when we note that the first sum contains at most $(m+1)^n$ terms, the second sum contains n terms, and

$$\left| \int_{x^*}^{x'} \frac{(x'-x)^{l-j}}{(l-j)!} (dx)^j \right| < \int_0^L L^{\sigma_l - 1}\, ds = L^{\sigma_l}.$$

4. Functions of class C^m in $R + B$. One more lemma will lead us to the main theorem of the paper.

LEMMA 4. *If R has the property P and $f_l(x)$ is uniformly continuous in $R(\sigma_l = m)$, then $f_k(x)$ is also $(\sigma_k < m)$.*

Assuming this is true for values of k such that $\sigma_k > s$ $(0 \leqq s < m)$, we shall prove it for any k with $\sigma_k = s$. By hypothesis, $f_{k+l}(x)$ is uniformly continuous for $\sigma_l > 0$; from (1), we see that it is sufficient to show that for every $\epsilon > 0$ there is a $\delta > 0$ such that if x' and x^* are any two points of R with $r_{x'x^*} < \delta$, then $| R_k(x'; x^*) | < \epsilon$. Take $\delta' < 1$ so that $| f_k(x) - f_k(x^*) | < \epsilon/[n(m+1)^n]$ if $r_{xx^*} < \delta'$ and $\sigma_k = m$. Set $\delta = \delta'/\omega$, and take any two points x', x^* of R with $r_{x'x^*} < \delta$. There is a curve C joining them of length $L \leqq \omega r_{x'x^*} < \delta'$; if x is on C, then $r_{xx^*} < \delta'$. As $L < 1$, Lemma 3 gives $| R_k(x'; x^*) | < \epsilon$, as required.

THEOREM. *Let the region R have the property P, and let $f(x_1, \cdots, x_n)$ be of class C^m in R. If $\partial^{k_1+\cdots+k_n} f / \partial x_1^{k_1} \cdots \partial x_n^{k_n}$ $(k_1 + \cdots + k_n = m)$ can be defined on the boundary B of R so that it is continuous in $R + B$, then the definition of f can be extended throughout space so that it is of class C^m there.*

By the last lemma, $f_k(x)$ can be defined in B so that it is continuous in $R + B$ $(\sigma_k \leqq m)$. We must show that $f_0(x) = f(x)$ is of class C^m in $R + B$; the theorem then follows from Lemma 1. Given an $\epsilon > 0$, take $\delta' < 1$ so that $| f_k(x) - f_k(x^*) | < \epsilon/[n(m+1)^n \omega'^m]$ $(\sigma_k = m)$ if x and x^* are in $R + B$ and $r_{xx^*} < \delta'$; set $\delta = \delta'/\omega'$. Now let y' and y^* be any two points of $R + B$ with $r_{y'y^*} < \delta$. By Lemma 2, there is a curve C' joining them of length $L' \leqq \omega' r_{y'y^*} < \delta'$; all of C' except possibly its end points lies in R. If x' and x^* are interior points of C', and C, of length L, is that part of C' joining them, then, as $L < L'$, Lemma 3 gives

$$\frac{| R_k(x'; x^*) |}{r_{y'y^*}^{m-\sigma_k}} < n(m+1)^n \frac{L^{m-\sigma_k}}{r_{y'y^*}^{m-\sigma_k}} \frac{\epsilon}{n(m+1)^n \omega'^m} < \epsilon.$$

As $f_k(x)$ is continuous in $R + B$, (1) shows that $R_k(x'; x^*)$ is also. Hence, letting $x' \to y'$ and $x^* \to y^*$ in the above inequality, we obtain (2) for $R_k(y'; y^*)$, as required.

HARVARD UNIVERSITY.

Reprinted from DUKE MATHEMATICAL JOURNAL
Vol. 1, No. 4, December, 1935

A FUNCTION NOT CONSTANT ON A CONNECTED SET OF CRITICAL POINTS

BY HASSLER WHITNEY

1. **Introduction.** Let $f(x_1, \cdots, x_n)$ be a function of class C^m (i.e., with continuous partial derivatives through the mth order) in a region R. Any point at which all its first partial derivatives vanish is called a *critical point* of f. Suppose every point of a connected set A of points in R is a critical point. It is natural to suspect then that f is a constant on A. But this need not be so. We construct below an example with $n = 2$, $m = 1$, A = an arc. The example may be extended to the case $n = n$, $m = n - 1$, A = an arc. The arc and the function on the arc are easily defined. The extension of the function through the rest of the plane or space is given by a theorem of the author.[1]

The question settled in this paper was raised implicitly in a paper of W. M. Whyburn.[2] It is brought up by his definition of critical sets as the maximal connected subsets of the set of critical points on which the function takes a single critical value. Theorem 2 of Whyburn's paper shows that an example of the type given in the present paper can be constructed only by using critical sets which have points that cannot be joined in these sets by rectifiable arcs. It would be interesting to discover how far from rectifiable a closed set must be to be a set of critical points of some function but not a critical set of the function. It may be remarked that any closed set may be a critical set.[3]

For fixed n and m large enough, $m \geqq [(n - 3)^2/16 + n]$, where $[n]$ is the integral part of n, f must be constant on any connected critical set, as shown by M. Morse and A. Sard in an unpublished paper.

The example shows that it is in general impossible to express the values of a function $f(x_1, \cdots, x_n)$ along a curve which is not rectifiable by means of an integral of a function of partial derivatives of f of order $\leqq n - 1$ along the curve.[4]

2. **The arc.** Let Q be a square of side 1 in the plane. Let Q_0, Q_1, Q_2, Q_3 be squares of side $1/3$ lying interior to Q in cyclical order, each a distance $1/12$

Received May 16, 1935; presented to the American Mathematical Society, October 26, 1935. The example given here was discovered in 1932 while the author was a National Research Fellow.

[1] H. Whitney, *Analytic extensions of differentiable functions defined in closed sets*, Transactions of the American Mathematical Society, vol. 36 (1934), pp. 63–89, Lemma 2. We refer to this paper as AE.

[2] W. M. Whyburn, Bull. Amer. Math. Soc., vol. 35 (1929), pp. 701–708.

[3] See A. Ostrowski, Bull. des Sciences Math., Feb. (1934), pp. 64–72.

[4] For such an expression (using partial derivatives of any desired order) along a rectifiable curve, see H. Whitney, *Functions differentiable on the boundaries of regions*, Annals of Mathematics, vol. 35 (1934), pp. 482–485, (1) and (3).

from the boundary of Q. Let q and q' be the centers of the sides of Q along Q_0, Q_1, and along Q_3, Q_0. Let q_i and q_i' be centers of adjacent sides of Q_i ($i = 0, 1, 2, 3$) so that q_{i-1}' and q_i face each other ($i = 1, 2, 3$), and q_0 is near q, q_3' is near q'. Let A_0 be a line joining q and q_0, let A_i join q_{i-1}' and q_i ($i = 1, 2, 3$), and let A_4 join q_3' and q'.

Suppose we have constructed squares $Q_{i_1 \cdots i_t}$, points $q_{i_1 \cdots i_t}$, $q_{i_1 \cdots i_t}'$, and lines $A_{j_1 \cdots j_t}$ (each $i_k = 0, 1, 2, 3$; each $j_k = 0, 1, 2, 3, 4$) for $t < s$. By taking a square $Q_{i_1 \cdots i_{s-2}}$, shrinking it to a third its size, and turning it around and upside down if necessary, we may place it in $Q_{i_1 \cdots i_{s-1}}$ so that $q_{i_1 \cdots i_{s-2}}$ and $q_{i_1 \cdots i_{s-2}}'$ go into $q_{i_1 \cdots i_{s-1}}$

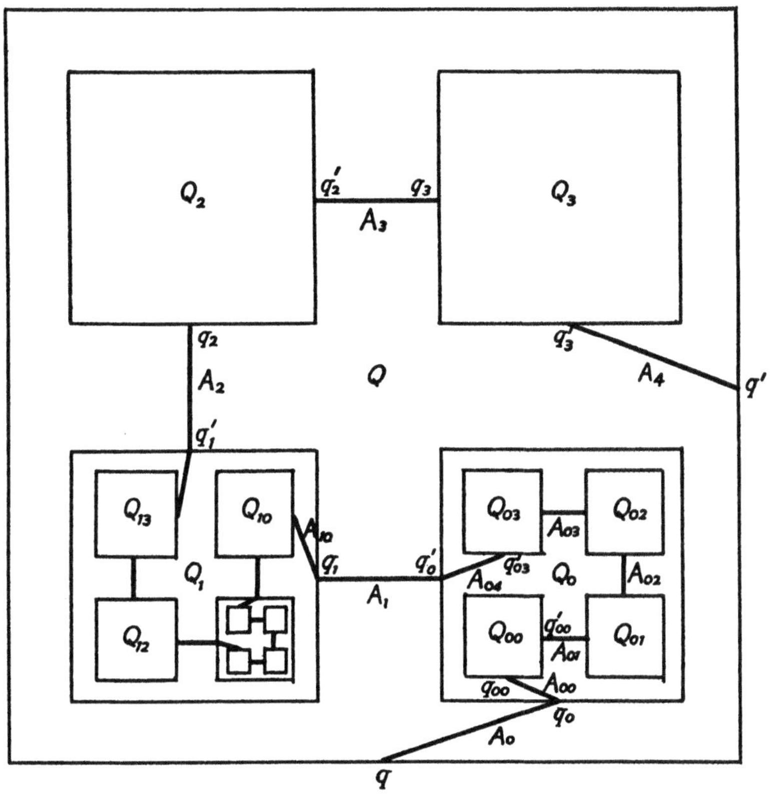

and $q_{i_1 \cdots i_{s-1}}'$, and thus construct squares $Q_{i_1 \cdots i_s}$, etc. We continue this process indefinitely. Let $Q_{i_1 i_2 \cdots}$ be the point common to Q, Q_{i_1}, $Q_{i_1 i_2}$, $\cdots$ for each $(i_1, i_2, \cdots)$.

The line segments $A_{i_1 \cdots i_s}$ together with the points $Q_{i_1 i_2 \cdots}$ form an arc A. It may be represented as the topological image of the segment (0, 1) by letting $A_{i_1 \cdots i_s}$ correspond to the segment

$$\left(\frac{2i_1 + 1}{9} + \cdots + \frac{2i_{s-1} + 1}{9^{s-1}} + \frac{2i_s}{9^s}, \frac{2i_1 + 1}{9} + \cdots + \frac{2i_{s-1} + 1}{9^{s-1}} + \frac{2i_s + 1}{9^s}\right),$$

and letting $Q_{i_1 i_2 \cdots}$ correspond to the number

$$\frac{2i_1 + 1}{9} + \frac{2i_2 + 1}{9^2} + \cdots .$$

3. **The function** $f(x, y)$. We first define $f(x, y)$ along the arc A as follows:

$$\text{on } A_{i_1 \cdots i_s}, \qquad f = \frac{i_1}{4} + \cdots + \frac{i_s}{4^s};$$

$$\text{at } Q_{i_1 i_2 \cdots}, \qquad f = \frac{i_1}{4} + \frac{i_2}{4^2} + \cdots .$$

f increases from 0 to 1 as we run along A from q to q'. Set $f_{00}(x,y) = f(x,y)$, $f_{10}(x,y) = f_{01}(x,y) = 0$ on A. We shall show that f_{00} *is of class* C^1 *on* A *in terms of* (f_{00}, f_{10}, f_{01}) (see AE). It will follow from Lemma 2 of AE that the definition of $f(x,y)$ may be extended over the plane (in particular, over Q) so that f is of class C^1; also $\partial f/\partial x = f_{10} = 0$, $\partial f/\partial y = f_{01} = 0$ on A, and hence each point of A is a critical point of f.

As f_{10} and f_{01} are continuous in A, we need merely prove that for each $\epsilon > 0$ there is a $\delta > 0$ such that if (x,y) and (x',y') are points of A whose distance apart is $r < \delta$, then

$$|f(x',y') - f(x,y)| < r\epsilon, \tag{1}$$

(see AE (3.1) and (3.2)).

The proof rests on the following two facts.

(a) If (x,y) and (x',y') are points of A in $Q_{i_1 \cdots i_s}$, then

$$|f(x',y') - f(x,y)| \leqq 1/4^s . \tag{2}$$

(b) If (x,y) and (x',y') are points of A separated by some point $Q_{i_1 i_2 \cdots}$, and if $Q_{i_1 \cdots i_s}$ is the smallest square containing them both, then

$$r > \frac{1}{12} \frac{1}{3^{s+1}} . \tag{3}$$

Assume (a) and (b) are true. Given $\epsilon > 0$, choose s_0 and δ so that

$$36 \left(\frac{3}{4}\right)^{s_0} < \epsilon, \qquad \delta < \frac{1}{12} \frac{1}{3^{s_0+1}} . \tag{4}$$

Now let (x,y) and (x',y') be any two points of A distant $r < \delta$. If no point $Q_{i_1 i_2 \cdots}$ separates them, then $f(x',y') = f(x,y)$, and (1) holds. Otherwise, let $Q_{i_1 \cdots i_s}$ be the smallest square containing them both. (b) gives

$$\frac{1}{12} \frac{1}{3^{s+1}} < r < \delta < \frac{1}{12} \frac{1}{3^{s_0+1}}, \quad \therefore s > s_0 .$$

Hence

$$\frac{|f(x',y') - f(x,y)|}{r} \leqq \frac{12 \cdot 3^{s+1}}{4^s} = 36 \left(\frac{3}{4}\right)^s < \epsilon .$$

It remains to prove (a) and (b). (a) is obvious from the definition of f. To prove (b), we consider three cases: neither point is in any square $Q_{i_1 \cdots i_s i_{s+1}}$; each point is in such a square; one point is in such a square, and the other is not. In the first two cases we see that $r > 1/(2 \cdot 3^{s+1})$; in the third case, (3) holds.

4. **Generalization to higher dimensions.** We shall indicate how the corresponding example is constructed for $n = 3$, $m = 2$; the generalization to higher dimensions is obvious.

Let Q be a cube of side 1. Let $Q_0, \cdots, Q_7$ be cubes of side 2/5 arranged in Q so that Q_i is adjacent to Q_{i-1}. Let q and q' be the centers of faces of Q which are adjacent and adjacent respectively to Q_0 and Q_7. Define q_i and q'_i $(i = 0, \cdots, 7)$ as before, and similarly for $A_0, \cdots, A_8$. The process is continued indefinitely, as before. A is again an arc. We set

$$\text{on } A_{i_1 \cdots i_s}, \quad f = \frac{i_1}{8} + \cdots + \frac{i_s}{8^s},$$

$$\text{at } Q_{i_1 i_2 \cdots}, \quad f = \frac{i_1}{8} + \frac{i_2}{8^2} + \cdots .$$

Set $f_{000} = f$, $f_{\alpha\beta\gamma} = 0$ $(\alpha + \beta + \gamma = 1 \text{ or } 2)$ on A. To prove that $f = f_{000}$ is of class C^2 in terms of these functions we need merely show that

$$\Delta = | f(x', y', z') - f(x, y, z) | < r^2 \epsilon \text{ on } A \qquad (r < \delta) . \tag{5}$$

Note that f varies by at most $1/8^s$ in any $Q_{i_1 \cdots i_s}$. Now if the smallest cube containing (x,y,z) and (x',y',z') is $Q_{i_1 \cdots i_s}$, r is of the order $(2/5)^s$, while $\Delta \leqq 1/8^s$; hence Δ/r^2 is of the order $(25/32)^s$, which is $< \epsilon$ for r small enough and hence s large enough.

Note that *all partial derivatives of* f (of order $\leqq n - 1 = 2$) *vanish on* A.

HARVARD UNIVERSITY.

DIFFERENTIABLE FUNCTIONS DEFINED IN ARBITRARY SUBSETS OF EUCLIDEAN SPACE*

BY

HASSLER WHITNEY

1. Introduction. In a former paper† we studied the differentiability of a function defined in closed subsets of Euclidean n-space E. We consider here the differentiability "about" an arbitrary point of a function defined in an arbitrary subset of E. We show in Theorem 1 that any function defined in a subset A of E which is differentiable about a subset B of E may be extended over E so that it remains differentiable about B. This theorem is a generalization of AE Lemma 2. We show further that any function of class C^m about a set B is of class C^{m-1} about an open set B' containing B. In the second part of the paper we consider some elementary properties of differentiable functions, such as: the sum or product of two such functions is such a function.‡ We end with the theorem that differentiability is a local property.§

2. Definitions and elementary properties. We use a one-dimensional notation as in AE. Thus $f_k(x) = f_{k_1 \cdots k_n}(x_1, \cdots, x_n)$, $x^l = x_1^{l_1} \cdots x_n^{l_n}$, $l! = l_1! \cdots l_n!$, $D_k f(x) = \partial^{k_1 + \cdots + k_n} f(x)/\partial x_1^{k_1} \cdots \partial x_n^{k_n}$, etc.; we set $\sigma_k = k_1 + \cdots + k_n$, r_{xy} = distance from x to y. We always set $f(x) = f_0(x)$. Suppose the functions $f_k(x)$ for $\sigma_k \leqq m$ are defined in the subset A of Euclidean n-space E. Define $R_k(x'; x)$ for x, x' in A by

* Presented to the Society, January 2, 1936; received by the editors October 26, 1935.

† *Analytic extensions of differentiable functions defined in closed sets*, these Transactions, vol. 36 (1934), pp. 63–89. We refer to this paper as AE. See also *Functions differentiable on the boundaries of regions*, Annals of Mathematics, vol. 35 (1934), pp. 482–485, and *Differentiable functions defined in closed sets*, I, these Transactions, vol. 36 (1934), pp. 369–387, which we refer to as F and D respectively.

P. Franklin in Theorem 1 of a paper *Derivatives of higher order as single limits*, Bulletin of the American Mathematical Society, vol. 41 (1935), pp. 573–582, has given a necessary and sufficient condition for the existence of a continuous mth derivative. We remark that this theorem is exactly the special case of Theorem I of D obtained by letting $f(x)$ be defined in an interval. It is also a special case of Theorem 2 of the author's *Derivatives, difference quotients, and Taylor's formula*, Bulletin of the American Mathematical Society, vol. 40 (1934), pp. 89–94 (see also Errata, p. 894). For his assumption is easily seen to imply the needed uniformity condition; it also implies at once that $f(x)$ is continuous, so that no considerations of measurability are necessary. His Theorem 2 should be compared with Theorems II and III of D.

‡ If the set is closed, these theorems may be proved by first extending the functions throughout E.

§ For the case of one variable this follows from D, Theorem I.

$$(1)\qquad f_k(x') = \sum_{\sigma_l \leqq m-\sigma_k} \frac{f_{k+l}(x)}{l!}(x'-x)^l + R_k(x';x).$$

Let x^0 be an arbitrary point of E. If for each k $(\sigma_k \leqq m)$ and every $\epsilon > 0$ there is a $\delta > 0$ such that

$$(2)\qquad |R_k(x';x)| \leqq r_{xx'}^{m-\sigma_k}\epsilon \text{ if } x,\ x' \text{ in } A,\ r_{xx^0} < \delta,\ r_{x'x^0} < \delta,$$

we shall say that $f(x)$ is *of class* C^m *in* A *about* x^0 *in terms of the* $f_k(x)$, or, $f(x)$ is $(C^m, A, x^0, f_k(x))$. If this is true for each x^0 in B, we say $f(x)$ is $(C^m, A, B, f_k(x))$, and replace "about x^0" by "about B." We say $f(x)$ (defined in A) is *of class* C^m *in* A *about* B, or, $f(x)$ is (C^m, A, B), if there exist functions $f_k(x)$ $(\sigma_k \leqq m)$ defined in A such that $f(x)$ is $(C^m, A, B, f_k(x))$. If $B = A$ in the last two definitions, we leave out the words "about B"; this is in agreement with the previous definitions. We say $f(x)$ is $(C^\infty, A, B, f_k(x))$ if $f(x)$ is $(C^m, A, B, f_k(x))$ for each m. Any function defined in A is (C^{-1}, A, E).

Remark. We might define in an obvious manner such relations as (C^m, A, x^0), (C^∞, A, B). To study them would require a study of the different possible definitions of the $f_k(x)$ if $f(x)$ is (C^m, A, B). The $f_k(x)$ are not in general determined by $f(x)$. Thus if $A = B$ is the x_1-axis, only the $f_k(x)$ with $k_2 = \cdots = k_n = 0$ are determined by $f(x)$. It is not obvious for what point sets A the $f_k(x)$ are all determined by $f(x)$.

If $f(x)$ is $(C^m, A, B, f_k(x))$ $(m \geqq 0)$, then the $f_k(x)$ are continuous at each point of B;* that is, the $f_k(x)$ may be defined in $B - B \cdot A$ so that this will be true. To show this, take x^0 in B, set $\epsilon = 1$, and choose δ so that (2) holds for any k $(\sigma_k \leqq m)$. Take x in A within δ of x^0 (if there is such a point); then (1) and (2) show that $f_k(x')$ is bounded for x' in A within δ of x^0 $(\sigma_k \leqq m)$. Now let $\{x^i\}$ be any sequence of points of A, $x^i \to x^0$; (1) and (2) show that $\{f_k(x^i)\}$ is a regular sequence.

If A is open and $f(x)$ is $(C^m, A, A, f_k(x))$, then $D_k f(x)$ exists and equals $f_k(x)$ in A $(\sigma_k \leqq m)$. (See AE.) If x^0 is an isolated point of A or is at a positive distance from A, then $f(x)$ is $(C^m, A, x^0, f_k(x))$ for any $f_k(x)$. If $f(x)$ is $(C^m, A, B, f_k(x))$ [or (C^m, A, B)], and A' is in A, B' is in B, then $f(x)$ is $(C^m, A', B', f_k(x))$ [or (C^m, A', B')]. Also $f(x)$ is (C^0, A, B) if and only if it is continuous at each point of B. If $f(x)$ is $(C^m, A, B, f_k(x))$, then it is $(C^{m'}, A, B, f_k(x))$ for all $m' < m$; a stronger theorem is proved in Theorem 2. If $f(x)$ is $(C^m, A, B, f_k(x))$, then $f_k(x)$ is $(C^{m-\sigma_k}, A, B, f_l(x))$.

3. **Extension theorems.** We prove here a theorem which gives the maximum range of differentiability of a function, and a theorem about the still larger range of differentiability of a function to an order one less.

* Or better, "continuous in A about B."

THEOREM 1. *If $f(x)$ is $(C^m, A, B, f_k(x))$* (m finite or infinite), then the $f_k(x)$ may be extended throughout E so that $f(x)$ is $(C^m, E, B, f_k(x))$.*†

We note, conversely, that if $f(x)$ is not $(C^m, A, x^0, f_k(x))$, then no extension of $f(x)$ will be so. We remark also that $f(x)$ may be made analytic in $E-\bar{A}$ ($\bar{A}=A$ plus limit points).

To prove the theorem, we first extend the $f_k(x)$ through $\bar{A}-A$ as follows: Take any x^0 in $\bar{A}-A$. Let $f_k(x^0)$ be the upper limit of $f_k(x^i)$ for sequences $\{x^i\}$, $x^i \to x^0$, x^i in A, if this is finite; otherwise, set $f_k(x^0)=0$. Next we extend the $f_k(x)$ throughout $E-\bar{A}$ by the method of AE Lemma 2. We shall assume in the proof that m is finite. If $m=\infty$, we prove $C^{m'}$ for every integer m'. The only alteration needed in the proof is that AE §12 should be used; but this makes no essential change.

As $E-\bar{A}$ is open, $f(x)$ is $(C^m, E, E-\bar{A}, f_k(x))$; we must show that $f(x)$ is $(C^m, E, B\cdot\bar{A}, f_k(x))$. Take a fixed point x^0 in $B\cdot\bar{A}$. Let us say (k, ϵ, A_1, A_2) holds if there is a $\delta>0$ such that (2) holds whenever x is in A_1, x' is in A_2, and $r_{xx^0}<\delta$, $r_{x'x^0}<\delta$. We must prove (k, ϵ, E, E) for each k ($\sigma_k \leqq m$) and each $\epsilon>0$.

First we prove $(k, \epsilon, \bar{A}, \bar{A})$. Set $\epsilon'=\epsilon/[2(m+1)^n]$, and let δ be the smallest of the δ's given by (l, ϵ', A, A) for $\sigma_l \leqq m$. Let U be the spherical neighborhood of x^0 of radius δ; then $f_l(x)$ is bounded in $U\cdot A$ ($\sigma_l \leqq m$). Given x, x' in $U\cdot\bar{A}$, choose sequences $\{x^i\}$, $\{x'^i\}$ of points of $U\cdot A$, with $x^i\to x$, $x'^i \to x'$. Suppose first $\sigma_k=m$. Then we may take these sequences so that $f_k(x^i)\to f_k(x)$, $f_k(x'^i)\to f_k(x')$, and the desired inequality for $R_k(x'; x)$ follows from that for $R_k(x'^i; x^i)$. Suppose now that $\sigma_k<m$. Relations (1) and (2) with k, x', x replaced by l, x^i, x^j show that for any such $\{x^i\}$, $\{f_l(x^i)\}$ is a regular sequence ($\sigma_l<m$); hence $f_l(x^i)\to f_l(x)$, and similarly $f_l(x'^i)\to f_l(x')$ ($\sigma_l<m$). Relation (1) now shows that for i large enough, $\Delta = R_k(x'; x) - R_k(x'^i; x^i)$ differs as little as we please from

$$-\sum_{\sigma_l=m-\sigma_k} \frac{f_{k+l}(x) - f_{k+l}(x^i)}{l!}(x'-x)^l.$$

As $|f_j(x)-f_j(x^i)| \leqq \epsilon'$ ($\sigma_j=m$) and $|(x'-x)^l| \leqq r_{xx'}^{\sigma_l}$, $|\Delta| \leqq (m+1)^n\epsilon' r_{xx'}^{m-\sigma_k}$ for i large enough; the inequality again follows.

Next we prove $(k, \epsilon, \bar{A}, E-\bar{A})$. Set $\epsilon'=\epsilon/[2\cdot 4^m(m+1)^n]$, and define η in terms of ϵ' and then δ as in AE §11, using $(k, \eta, \bar{A}, \bar{A})$. Take x in $\bar{A}$ and x' in $E-\bar{A}$, each within $\delta/4$ of x^0. By AE (6.3) and the equation following (11.6),

* Or merely locally (C^m, A, B); see Theorem 6.

† If $A=B$ is closed, then B may be replaced by E; the present proof then gives a proof of AE Lemma 2 which makes no use of AE Lemma 1.

$$R_k(x'; x) = D_kf(x') - \psi_k(x'; x)$$
$$= \sum_l \frac{R_{k+l}(x^*; x)}{l!}(x' - x^*)^l + \sum_{s=1}^{t}\sum_l \binom{k}{l} D_l\phi_{\lambda_s}(x')\zeta_{\lambda_s;k-l}(x'),$$

where x^* is a point of $\bar{A}$ distant $\delta_*/4$ from x', $\delta_*/4$ being the distance from x' to $\bar{A}$. As $r_{x^*x} \leqq 2r_{xx'}$, $r_{x'x^*} \leqq 2r_{xx'}$, and $\delta_* \leqq 4r_{xx'}$, we find with the help of AE (11.8)

$$\left| R_k(x'; x) \right| \leqq (m+1)^n (2r_{xx'})^{m-\sigma_k}\eta + (4r_{xx'})^{m-\sigma_k}\epsilon'/2 < r_{xx'}^{m-\sigma_k}\epsilon.$$

Next we prove $(k, \epsilon, E-\bar{A}, \bar{A})$. As is easily seen from AE (6.3) or by F (6) with x^{i-1}, x^i replaced by x, x',

$$R_k(x'; x) = \sum_l \frac{R_{k+l}(x; x')}{l!}(x' - x)^l.$$

Set $\epsilon' = \epsilon/(m+1)^n$, and take the smallest δ given by $(k+l, \epsilon', \bar{A}, E-\bar{A})$ for $\sigma_l \leqq m - \sigma_k$. The required inequality now follows at once.

Finally we must show $(k, \epsilon, E-\bar{A}, E-\bar{A})$. Set $\epsilon' = \epsilon/[2n(m+1)^n]$, and take δ smaller than the $\delta/4$ given by AE §11 with ϵ replaced by ϵ' and smaller than the δ's given by $(k+l, \epsilon', \bar{A}, E-\bar{A})$ and $(k+l, \epsilon', E-\bar{A}, \bar{A})$ for $\sigma_l \leqq m - \sigma_k$. Now take x and x' in $E-\bar{A}$ within δ of x^0; we must consider two cases. Case I: The line segment $S = xx'$ lies wholly in $E-\bar{A}$. By AE (11.2), $|f_l(y) - f_l(x')| < 2\epsilon'$ for y on S $(\sigma_l \leqq m)$; the desired inequality now follows from F, Lemma 3. Case II: There is a point x^* of $\bar{A}$ on S. From AE (6.3), or F (6) with x^{i-1}, x^i replaced by x, x^*, we find

$$R_k(x'; x) = R_k(x'; x^*) + \sum_l \frac{R_{k+l}(x^*; x)}{l!}(x' - x^*)^l,$$

and the inequality again follows.

THEOREM 2. *If $f(x)$ is $(C^m, A, B, f_k(x))$ (m finite), then there is an open set B' containing B such that $f(x)$ is $(C^{m-1}, A, B', f_k(x))$.*

For each x in B, let $\delta(x)$ be the largest of the numbers δ for which (2) holds for all k $(\sigma_k \leqq m)$ with ϵ replaced by 1. Let $U(x)$ be the set of all points x' within $\delta(x)$ of x; then B' is the sum of all $U(x)$. The set B' is open. To prove $(C^{m-1}, A, B', f_k(x))$, take any x^0 in B' and any $\epsilon > 0$. For some x^* in B, $r_{x^*x^0} < \delta(x^*)$. There is an M such that $|f_k(y)| < M$ for y in $A \cdot U(x^*)$ $(\sigma_k \leqq m)$.† Let δ be the smaller of $\delta(x^*) - r_{x^*x^0}$ and $\epsilon/[2(m+1)^nM+2]$. Now take any x and x' in A within δ of x^0. We are interested in the remainders

† For the proof, see the paragraph following the remark.

$$R_k'(x';x) = \sum_{\sigma_l = m - \sigma_k} \frac{f_{k+l}(x)}{l!}(x'-x)^l + R_k(x';x)$$

with $\sigma_k < m$. As $r_{xx'} < 2\delta$,

$$\left| R_k'(x';x) \right| \leqq (m+1)^n M r_{xx'}^{m-\sigma_k} + r_{xx'}^{m-\sigma_k} < r_{xx'}^{m-1-\sigma_k}\epsilon.$$

COROLLARY. *If $f(x)$ is of class C^m in any given point set about B, then it may be extended through an open set B' containing B so that it is of class C^{m-1} in B' and of class C^m in B' about B.*

4. **Composite functions, etc.** We prove here three theorems.

THEOREM 3. *If f and g are of class C^m in A about B, then so are $f+g$ and $f-g$, with*

$$(f \pm g)_k = f_k \pm g_k. \tag{3}$$

This is obvious.

THEOREM 4. *If f and g are of class C^m in A about B, then so is fg, and f/g if $g \neq 0$. The derivatives are given by the ordinary formulas. Thus*

$$(fg)_k = \sum_l \binom{k}{l} f_l g_{k-l}. \tag{4}$$

We might prove this theorem directly, but it follows from Theorem 5: fg and f/g are functions (of two variables) of class C^∞ of the functions f and g. (The condition B in A is obtained by using Theorem 1.)

THEOREM 5. *Let A and B be subsets of n-space E_n, and let A' and B' be subsets of ν-space E_ν. Let $f^i(x)$ be $(C^m, A, B, f_k^i(x))$ $(i=1, \cdots, \nu)$, and let $g(y)$ be $(C^m, A', B', g_k(y))$ (m finite or infinite). Suppose B is in A, x in A implies*

$$y = (y_1, \cdots, y_\nu) = (f^1(x), \cdots, f^\nu(x)) = f(x)$$

in A', and x in B implies $f(x)$ in B'. Then the function

$$h(x) = g(f^1(x), \cdots, f^\nu(x)) = g(f(x))$$

is $(C^m, A, B, h_k(x))$; the $h_k(x)$ are given by the ordinary formulas (9) *for derivatives.*

As a consequence of this theorem, the definition of being of class C^m is independent of the coordinate system chosen. If the condition x in A [or B] does not imply $f(x)$ in A' [or B'], we may apply the theorem to any subset A_1 [or B_1] of A [or B] for which it does. We shall suppose m is finite; if $m = \infty$, we merely apply the reasoning below for each positive integer.

Suppose first $u^1(x), \cdots, u^\nu(x)$ are functions of class C^m in an open set Γ of E_n, suppose $v(y)$ is of class C^m in an open set Γ' of E_ν, and suppose x in Γ implies $u(x)$ in Γ'. Letting R'^i, S' denote remainders for u^i, v, Taylor's formula gives

$$(5) \qquad u_k^i(x') = D_k u^i(x') = \sum_{\sigma_l \leqq m-\sigma_k} \frac{u_{k+l}^i(x)}{l!}(x'-x)^l + R_k'^i(x';x),$$

$$(6) \qquad v_k(y') = D_k v(y') = \sum_{\sigma'_l \leqq m-\sigma'_k} \frac{v_{k+l}(y)}{l!}(y'-y)^l + S_k'(y';y),$$

certain inequalities on the $R_k'^i$ and S_k' being satisfied. We have set $\sigma_k' = k_1 + \cdots + k_\nu$. Set $w(x) = v(u(x))$; then (5) and (6) with $k=0$ give

$$(7) \qquad w(x') = \sum_t \frac{v_t(u(x))}{t!}\left\{\sum_{\sigma_j \geqq 1} \frac{u_j(x)}{j!}(x'-x)^j + R'(x';x)\right\}^t + S'(u(x'); u(x)),$$

where $S' = S_0'$. Also, by Taylor's formula,

$$(8) \qquad w_k(x') = \sum_l \frac{w_{k+l}(x)}{l!}(x'-x)^l + T_k'(x';x).$$

Subtract (8) with $k=0$ from (7); then as R'^i, S', and T' all approach 0 to the mth order as $x' \to x$,† we may equate coefficients of $(x'-x)^k$ for $\sigma_k \leqq m$.‡ Thus we find polynomials

$$P_k(u_p^i, v_q) \qquad (\sigma_p \leqq \sigma_k,\ \sigma_q' \leqq \sigma_k;\ \sigma_k \leqq m)$$

such that, for any x in Γ,

$$(9) \qquad w_k(x) = P_k(u_p^i(x), v_q(u(x))).$$

Using (8) gives for $w_k(x')$

$$(10) \qquad w_k(x') = \sum_l \frac{P_{k+l}(u_p^i(x), v_q(u(x)))}{l!}(x'-x)^l + T_k'(x';x).$$

We may also evaluate it by replacing x by x' in (9) and using (5) and (6). (In (6) we replace y' by $u(x')$ and use (5) again.) Each variable in the resulting polynomial P_k consists of a polynomial in quantities R', S', and other quantities; if we multiply out and collect all terms with an R' or an S' as a factor, we obtain

† This is clear for S' if $m=0$; if $m>0$, then

$$S'/r_{xx'}^m = [S'/|u(x')-u(x)|^m] \cdot [|u(x')-u(x)|/r_{xx'}]^m,$$

where $|y'-y| = r_{yy'}$, and the last factor is bounded in $U \cdot A$.

‡ This is easily proved in succession for $\sigma_k = 0, 1, \cdots$ on letting $x' \to x$.

$$w_k(x') = P_k\Bigg[\sum_s \frac{u^i_{p+s}(x)}{s!}(x'-x)^s,\tag{11}$$
$$\sum_t \frac{v_{q+t}(u(x))}{t!}\Bigg\{\sum_{\sigma_j\geqq 1}\frac{u_j(x)}{j!}(x'-x)^j\Bigg\}^t\Bigg] + Q_k,$$

where Q_k is a polynomial containing an R' or an S' as a factor in each term. It must be understood that $\sum u^i_{p+s}(x)(x'-x)^s/s!$ appears as the variable in the position of u_p^i, etc., in $P_k(u_p^i, v_q)$.

We now prove: *If* u_k^i $(\sigma_k \leqq m; i=1, \cdots, \nu)$, v_k $(\sigma_k' \leqq m)$ *are any numbers, then*

$$P_k^*(x; u_p^i, v_q) = P_k\Bigg[\sum_s \frac{u^i_{p+s}}{s!}x^s, \sum_t \frac{v_{q+t}}{t!}\Bigg\{\sum_{\sigma_j\geqq 1}\frac{u_j}{j!}x^j\Bigg\}^t\Bigg]\tag{12}$$
$$- \sum_l \frac{P_{k+l}(u_p^i, v_q)}{l!}x^l,$$

considered as a polynomial in x, *contains no terms of degree* $\leqq m-\sigma_k$. To prove this, define the polynomials

$$u^i(x) = \sum_{\sigma_l\leqq m}\frac{u_l^i}{l!}x^l, \qquad v(y) = \sum_{\sigma_l'\leqq m}\frac{v_l}{l!}(y-u_0)^l;\tag{13}$$

then $u_k^i(0) = D_k u^i(0) = u_k^i$, $v_k(u_0) = D_k v(u_0) = v_k$. Set $w(x) = v(u(x))$. Replacing x', x by x, 0 in (10) and (11) and putting in (12) gives, as $Q_k = 0$ in this case,

$$P_k^*(x; u_p^i, v_q) = T_k'(x; 0).\tag{14}$$

As $T_k' \to 0$ to the $(m-\sigma_k)$th order as $x\to 0$, P_k^* cannot contain any terms of degree $\leqq m-\sigma_k$.

We return now to the functions $f^i(x)$, $g(y)$, $h(x)$. Set $h_k(x) = P_k(f_p^i(x), g_q(f(x)))$. The formulas (10) and (11) hold equally well for the f^i, g, h. Hence using (10), (11), and (12), we find for the remainder for $h_k(x)$

$$T_k(x'; x) = P_k^*(x'-x; f_p^i(x), g_q(f(x))) + Q_k.\tag{15}$$

To show that $h(x)$ is $(C^m, A, B, h_k(x))$, take any x^0 in B, and set $y^0 = f(x^0)$. As $f(x)$ is continuous in A about B, for each neighborhood V of y^0 there is a neighborhood $U(V)$ of x^0 such that x in $U(V)\cdot A$ implies $f(x)$ in $V\cdot A'$. As y^0 is in B', we may take V so that the $g_k(y)$ are bounded in $V\cdot A'$. We may take U in $U(V)$ so small that the $f_k(x)$ are bounded in $U\cdot A$. Because of the property of P_k^*, we may obviously take δ small enough so that P_k^* satisfies an inequality of the nature of (2). Moreover each term in Q_k contains an $R_p(x'; x)$ or an $S_q(u(x'); u(x))$ with $\sigma_p \leqq \sigma_k$ or $\sigma_q' \leqq \sigma_k$; as each such remainder satisfies an inequality (2) (see a recent footnote) and all other quantities entering into Q_k are bounded, we may take δ small enough so that Q_k also satisfies an inequality (2). Hence the same is true of T_k, and the theorem is proved.

5. **Differentiability a local property.** Our object is to prove

THEOREM 6. *Let $f(x)$ be locally (C^m, A, B) (m finite or infinite). For each point x^0 of B there is a neighborhood U of x^0 and functions $f_k^{(x^0)}(x)$ defined in $U\cdot A$ such that $f(x)$ is $(C^m, U\cdot A, U\cdot B, f_k^{(x^0)}(x))$.*† *Then $f(x)$ is (C^m, A, B). If the $f_k^{(x^0)}(x)$ for $\sigma_k \leqq p$ are independent (at any x for which they are defined) of x^0, then these functions may be included among the $f_k(x)$ $(\sigma_k \leqq m)$.*

We may take each neighborhood U as an open n-cube, so small that the $f_k^{(x^0)}(x)$ are bounded in U. A finite or denumerable number of them, $C_1, C_2, \cdots$, cover B; we may take them so that any one touches at most a finite number of the others, and so that any boundary point of any C_i is interior to some C_j.‡ By hypothesis, to each i there correspond functions $f_k^i(x)$, $\sigma_k \leqq m$, such that $f(x)$ is $(C^m, C_i\cdot A, C_i\cdot B, f_k^i(x))$. In each C_i we define the function $\pi_i(x)$ as it was defined in I_i in AE §9; set

$$\phi_i(x) = \pi_i(x)/\sum_j \pi_j(x) \tag{16}$$

in $C_1+C_2+\cdots$. Set $g^i(x)=\phi_i(x)f(x)$ in $C_i\cdot A$. By Theorem 4, $g^i(x)$ is $(C^m, C_i\cdot A, C_i\cdot B)$, and

$$g_k^i(x) = \sum_l \binom{k}{l} D_l\phi_i(x)f_{k-l}^i(x). \tag{17}$$

As the $f_k^i(x)$ are bounded in $C_i\cdot A$ and the $D_l\phi_i(x)\to 0$ to infinite order as x approaches the boundary of C_i (see AE §9), the latter statement is true also of the $g_k^i(x)$. Hence, evidently, if we set $g_k^i(x)=0$ in $A-C_i\cdot A$, $g^i(x)$ is $(C^m, A, B, g_k^i(x))$. Set

$$f_k(x) = g_k^1(x) + g_k^2(x) + \cdots, \tag{18}$$

which in any $C_i\cdot A$ is a finite sum; this reduces to $f(x)$ for $k=0$. Theorem 3 shows at once that $f(x)$ is $(C^m, A, B, f_k(x))$. (Given x^0 in B, to apply Theorem

† Note that "$f(x)$ is locally $(C^\infty, \cdots)$" is not the same statement as "$f(x)$ is locally $(C^m, \cdots)$ for each m."

‡ Let $C^1, C^2, \cdots$ be a denumerable set of the cubes which cover B. Express each C^i as the sum of a denumerable number of cubes C_j^i with the following properties: Each C_j^i is, with its boundary, interior to C^i; the diameter of C_j^i, $\delta(C_j^i)$, is $<1/i$; $\delta(C_j^i)\to 0$ as $j\to\infty$; the cubes C_j^i approach the boundary of C^i as $j\to\infty$. Now drop out all cubes C_j^i which are interior to larger cubes C_l^k; the remaining cubes $C_1, C_2, \cdots$ still cover B. To each cube C_j^i corresponds a number $\eta>0$ such that any point set of diameter $<\eta$ having points in common with C_j^i lies interior to some C_k^i; using this fact, it is easily seen that any C_i has points in common with but a finite number of the C_j.

3, we choose δ so small that the points within δ of x^0 lie in but a finite number of the C_i.)

To prove the second statement, let $f_l'(x)$ denote the common value of $f_l^i(x)$ for $\sigma_l \leqq p$. Differentiating $\sum \phi_i = 1$ gives

$$\sum_i D_l\phi_i(x) = \begin{cases} 1 \text{ if } l = 0, \\ 0 \text{ if } \sigma_l > 0. \end{cases} \tag{19}$$

Define the $f_k(x)$ as before. Take any k with $\sigma_k \leqq p$; then (17) and (18) give

$$f_k(x) = \sum_l \binom{k}{l} f_{k-l}'(x) \sum_i D_l\phi_i(x) = f_k'(x)$$

in $C_1 + C_2 + \cdots$. It does not matter how $f_k(x)$ is defined outside this set.

The second statement in the theorem does not hold for an arbitrary set of $f_k(x)$, at least using the above method. To see this, take $n = m = 2$, $A = B$ = the interval $(-1, 1)$ of the x_1-axis, $C_1 = C_2$ = the square with corners $(\pm 1, \pm 1)$; set $f = 0$,

$$f_{10}^1 = f_{20}^1 = f_{11}^1 = f_{02}^1 = 0, \qquad f_{01}^1 = 1,$$

and $f_{ij}^2 = -f_{ij}^1$ on A. Also set

$$\phi_1(x, y) = \tfrac{1}{2} + \tfrac{3}{4}x - \tfrac{1}{4}x^3, \qquad \phi_2(x, y) = \tfrac{1}{2} - \tfrac{3}{4}x + \tfrac{1}{4}x^3.$$

(Though ϕ_1 and ϕ_2 are not the functions defined above, they have the necessary properties.) We find on A

$$g_{11}^1(x, y) = g_{11}^2(x, y) = \tfrac{3}{4} - \tfrac{3}{4}x^2, \qquad f_{11}(x, y) = \tfrac{3}{2} - \tfrac{3}{2}x^2 \neq 0.$$

Harvard University,
Cambridge, Mass.

DIFFERENTIABILITY OF THE REMAINDER TERM IN TAYLOR'S FORMULA

By Hassler Whitney

1. **Introduction.** Taylor's formula with exact remainder may be written

$$f(x) = f(0) + \frac{f'(0)}{1!}x + \cdots + \frac{f^{(n-1)}(0)}{(n-1)!}x^{n-1} + \frac{1}{n!}x^n f_n(x), \tag{1}$$

where

$$f_n(x) = \frac{n}{x^n}\int_0^x (x-t)^{n-1} f^{(n)}(t)\, dt \qquad (x \neq 0). \tag{2}$$

The main object of the paper (Theorem 1) is to show that if f is of class C^{n+p} (i.e., it has continuous derivatives through the order $n + p$), then f_n is of class C^p, but not necessarily of higher class. The condition that f be of class C^{n+p} may be expressed purely in terms of f_n (Theorem 2). A generalization is given to several dimensions (Theorem 3). The functions are always assumed defined in a neighborhood of the origin. The second theorem will be used in the following note; the results of both papers are essential in studying the singularities of certain mappings (see the next following paper).

2. **Properties of the remainder term.** We prove

THEOREM 1. *Let $1 \leq n \leq n + p$, let f be of class C^{n+p} (in a region about the origin), and let* (1) *and* (2) *hold. Set $f_n(0) = f^{(n)}(0)$. Then f_n is of class C^p, and of class C^{n+p} for $x \neq 0$. Its derivatives are, for $k = 0, \cdots, p$,*

$$f_n^{(k)}(x) = \frac{n}{x^{n+k}}\int_0^x (x-t)^{n-1} t^k f^{(n+k)}(t)\, dt \qquad (x \neq 0), \tag{3}$$

$$f_n^{(k)}(0) = \frac{n!k!}{(n+k)!} f^{(n+k)}(0). \tag{4}$$

Furthermore,

$$\lim_{x \to 0} x^k f_n^{(p+k)}(x) = 0 \qquad (k = 1, \cdots, n). \tag{5}$$

See also Corollary 2 below.

We require two formulas, assuming ϕ differentiable.

$$\int_0^x t^{k+1}\phi'(t)\, dt = x^{k+1}\phi(x) - (k+1)\int_0^x t^k \phi(t)\, dt; \tag{6}$$

Received July 28, 1942; presented to the American Mathematical Society September 8, 1942.

also, for $n \geq 2$,

$$0 = \int_0^x \frac{d}{dt} [(x - t)^{n-1} t^{k+1} \phi(t)] \, dt$$

$$= \int_0^x [(n - 1)(x - t)^{n-2} t^k \{(x - t) - x\} \phi(t) + (k + 1)(x - t)^{n-1} t^k \phi(t) + (x - t)^{n-1} t^{k+1} \phi'(t)] \, dt,$$

$$\int_0^x (x - t)^{n-1} t^{k+1} \phi'(t) \, dt = -(n + k) \int_0^x (x - t)^{n-1} t^k \phi(t) \, dt + (n - 1)x \int_0^x (x - t)^{n-2} t^k \phi(t) \, dt. \tag{7}$$

Formula (3) with $k = 0$ is merely (2). We prove (3) by induction on k. Suppose it holds for a given k. If, first, $n = 1$, then differentiating (3) gives, with the help of (6),

$$f_1^{(k+1)}(x) = -\frac{k+1}{x^{k+2}} \int_0^x t^k f^{(k+1)}(t) \, dt + \frac{1}{x^{k+1}} x^k f^{(k+1)}(x)$$

$$= \frac{1}{x^{k+2}} \int_0^x t^{k+1} f^{(k+2)}(t) \, dt,$$

as required. Next, for $n \geq 2$, we obtain also the required formula

$$f_n^{(k+1)}(x) = -\frac{n(n+k)}{x^{n+k+1}} \int_0^x (x - t)^{n-1} t^k f^{(n+k)}(t) \, dt$$

$$+ \frac{n(n-1)}{x^{n+k}} \int_0^x (x - t)^{n-2} t^k f^{(n+k)}(t) \, dt$$

$$= \frac{n}{x^{n+k+1}} \int_0^x (x - t)^{n-1} t^{k+1} f^{(n+k+1)}(t) \, dt.$$

Next we prove a well-known formula about the B function. Setting $f(x) = x^{n+k}$ gives

$$f_n(x) = n! x^k, \qquad f_n^{(k)}(x) = n! k!, \qquad f^{(n+k)}(x) = (n + k)!;$$

substituting these equations in (3) gives

$$\frac{n}{x^{n+k}} \int_0^x (x - t)^{n-1} t^k \, dt = \frac{n! k!}{(n + k)!}. \tag{8}$$

It also follows that

$$\left| \frac{n}{x^{n+k}} \int_0^x (x - t)^{n-1} t^k \eta(t) \, dt \right| \leq \frac{n! k!}{(n + k)!} \max | \eta(t) |. \tag{9}$$

Now we determine the limit of $f_n^{(k)}(x)$ as $x \to 0$. Because of (9), we may replace $f^{(n+k)}(t)$ by $f^{(n+k)}(0)$ in (3); using (8), we have

$$\lim_{x\to 0} f_n^{(k)}(x) = \frac{n!k!}{(n+k)!} f^{(n+k)}(0). \tag{10}$$

It is now an elementary argument to show in succession that f_n is continuous, of class $C^1, \cdots$, and that (4) holds.

Because of (1), f_n is of class C^{n+p} for $x \neq 0$. Differentiating (3) with $k = p$ gives, for $x \neq 0$,

$$\begin{aligned} f_n^{(p+k)}(x) &= \sum_{j=0}^{k} \binom{k}{j} \frac{d^{k-j}}{dx^{k-j}} \left(\frac{n}{x^{n+p}}\right) \frac{d^j}{dx^j} \int_0^x (x-t)^{n-1} t^p f^{(n+p)}(t)\,dt \\ &= \sum_{j=0}^{k} \frac{A_{pnkj}}{x^{n+p+k-j}} \int_0^x (x-t)^{n-j-1} t^p f^{(n+p)}(t)\,dt \qquad (k \le n-1), \end{aligned} \tag{11}$$

$$f_n^{(n+p)}(x) = \sum_{j=0}^{n-1} \frac{A_{pnj}}{x^{2n+p-j}} \int_0^x (x-t)^{n-j-1} t^p f^{(n+p)}(t)\,dt + \frac{A_{pn}}{x^n} f^{(n+p)}(x), \tag{12}$$

for some constants A. Using (8) and (9), we find in either case, for some constants B,

$$\lim_{x\to 0} x^k f_n^{(p+k)}(x) = B_{pnk} f^{(n+p)}(0) \qquad (k = 0, \cdots, n).$$

To find the B_{pnk}, take $f(x) = x^{n+p}$; then $f_n(x) = n!x^p$, and

$$x^k f_n^{(p+k)}(x) = 0 \qquad (k = 1, \cdots, n).$$

Since $f^{(n+p)}(0) = (n+p)!$, $B_{pnk} = 0$ for $k > 0$, and (5) is proved.

Example. Set $f(x) = |x|^{n+p+\frac{1}{2}}$. Then f is of class C^{n+p}, and $f_n(x) = \pm n! \,|x|^{p+\frac{1}{2}}$ is of class C^p but not of higher class.

COROLLARY 1. *Let $F(x)$ be continuous, and let $x^nF(x)$ be of class C^{n+p}. Then $x^iF(x)$ is of class C^{p+i} for $i = 0, \cdots, n$.*

Set $f(x) = x^nF(x)$. Expand f as in (1). Setting $x = 0$ shows that $f(0) = 0$. Dividing by x and letting $x \to 0$ then shows that $f'(0) = 0$, etc. Hence, $f_n(x) = n!F(x)$, and F is of class C^p, by Theorem 1. If we apply this result with $F(x)$ replaced by $x^iF(x)$ and n replaced by $n - i$, the corollary follows. From this follows

COROLLARY 2. *With the notations of Theorem 1, $x^if_n(x)$ is of class C^{p+i} for $1 \le i \le n \le n + p$.*

3. **A converse theorem.** We shall show that some of the conditions in the last theorem are sufficient to make f of class C^{n+p}.

THEOREM 2. Let $f_n(x)$ be of class C^p, and of class C^{n+p} for $x \neq 0$, and let

$$\lim_{x\to 0} x^k f_n^{(p+k)}(x) \qquad (k = 0, \cdots, n) \tag{13}$$

exist. Set $f(0) = a_0$, *and*

$$f(x) = a_0 + \frac{a_1}{1!}x + \cdots + \frac{a_{n-1}}{(n-1)!}x^{n-1} + \frac{1}{n!}x^n f_n(x) \qquad (x \neq 0). \tag{14}$$

Then f is of class C^{n+p}, and $f^{(i)}(0) = a_i$ $(i = 0, \cdots, n-1)$, $f^{(n)}(0) = f_n(0)$.

Remark. The limits in (13) are 0 for $k > 0$, by Theorem 1.

Set $F(x) = x^n f_n(x)/n!$. It is clearly sufficient to show that F is of class C^{n+p}, $F^{(i)}(0) = 0$ for $i = 0, \cdots, n-1$,m and $F^{(n)}(0) = f_n(0)$. Differentiating F gives, for $x \neq 0$,

$$\begin{aligned} F^{(i)}(x) &= \sum_{j=0}^{i} \binom{i}{j} \frac{d^{i-j}}{dx^{i-j}} \left(\frac{x^n}{n!}\right) f_n^{(j)}(x) \\ &= \sum_{j=\max(0,i-n)}^{i} \binom{i}{j} \frac{x^{n-i+j}}{(n-i+j)!} f_n^{(j)}(x). \end{aligned} \tag{15}$$

If we show that $\lim_{x\to 0} F^{(i)}(x)$ exists for $i = 0, \cdots, n+p$, it will follow that F is of class C^{n+p}. Take a fixed i. Each term in (15) with $j \leq p$ is of course continuous at $x = 0$. If $j = p + k$, $k > 0$, then

$$\lim_{x\to 0} x^{n-i+j} f_n^{(j)}(x) = \lim_{x\to 0} x^{n+p-i} \lim_{x\to 0} [x^k f_n^{(p+k)}(x)],$$

which exists, by (13). Hence, $F(x)$ is of class C^{n+p}. That $F^{(i)}(0) = 0$ for $i < n$ is obvious from (15); $F^{(n)}(0) = f_n(0)$ is (4) of Theorem 1 with $k = 0$; and the theorem is proved.

Example. Let (for $x \neq 0$)

$$f(x) = \tfrac{1}{2} x^2 f_2(x), \qquad f_2(x) = x \sin (x^{-2});$$

then f_2 is continuous, and is of class C^2 for $x \neq 0$; but $f(x)$ is not even of class C^1.

4. **Two or more dimensions.** We shall prove

THEOREM 3. *Let $f(x, y)$ be of class C^{n+p}, $n \geq 1$, $p \geq 0$. Then*

$$f(x, y) = g_0(y) + \cdots + \frac{x^{n-1}}{(n-1)!} g_{n-1}(y) + \frac{x^n}{n!} g_n(x, y), \tag{16}$$

$$g_i(y) = \frac{\partial^i}{\partial x^i} f(0, y) \qquad (i = 0, \cdots, n-1), \tag{17}$$

$$g_n(x, y) = \frac{n}{x^n} \int_0^x (x - t)^{n-1} \frac{\partial^n}{\partial t^n} f(t, y)\, dt \qquad (x \neq 0); \tag{18}$$

the g_i are of class C^{n+p-i} for $i = 0, \cdots, n$; for $k + l \leq p$,

$$\frac{\partial^{k+l}}{\partial x^k \partial y^l} g_n(x, y) = \frac{n}{x^{n+k}} \int_0^x (x - t)^{n-1} t^k \frac{\partial^{n+k+l}}{\partial t^{n+k} \partial y^l} f(t, y)\, dt \qquad (x \neq 0). \tag{19}$$

$$\frac{\partial^{k+l}}{\partial x^k \partial y^l} g_n(0, y) = \frac{n!k!}{(n+k)!} \frac{\partial^{n+k+l}}{\partial x^{n+k} \partial y^l} f(0, y). \tag{20}$$

Clearly the theorem may be generalized to more dimensions. Also, we may expand each g_i to the order $n - i$, giving a complete expansion of f, with remainder terms.

First, for any fixed y (in the given neighborhood of the origin), expanding f in terms of x gives (16), (17) and (18). The g_i for $i < n$ are obviously of class C^{n+p-i}. Differentiating (18) l times with respect to y gives (19) with $k = 0$. Now take y fixed again, and apply the proof of (3) to the function $\partial^l g_n / \partial y^l$ of x; this proves (19). Next, we prove

$$\lim_{\substack{x \to 0 \\ y \to y'}} \frac{\partial^{k+l}}{\partial x^k \partial y^l} g_n(x, y) = \frac{n!k!}{(n+k)!} \frac{\partial^{n+k+l}}{\partial x^{n+k} \partial y^l} f(0, y'),$$

exactly as (10) was proved. As before, the fact that g_n is of class C^p, together with the formula (20), follows at once.

Example. Let

$$f(x, y) = \frac{x \mid y \mid^{7/2}}{x^2 + y^2}, \qquad (x, y) \neq (0, 0), \qquad f(0, 0) = 0.$$

Set $r^2 = x^2 + y^2$, and let P_i, Q_i and R_i denote polynomials of degree i. Then, for $y > 0$, it is easily seen that

$$f = \frac{P_4 y^{\frac{1}{2}}}{r^2}, \qquad \frac{\partial f}{\partial x} = \frac{Q_5 y^{\frac{1}{2}}}{r^4}, \qquad \frac{\partial f}{\partial y} = \frac{Q_5' y^{\frac{1}{2}}}{r^4},$$

$$\frac{\partial^2 f}{\partial x^2} = \frac{R_6 y^{\frac{1}{2}}}{r^6}, \qquad \frac{\partial^2 f}{\partial x \partial y} = \frac{R_6' y^{\frac{1}{2}}}{r^6}, \qquad \frac{\partial^2 f}{\partial y^2} = \frac{R_6'' y^{\frac{1}{2}}}{r^6}.$$

Since $2 \mid xy \mid \leq r^2, f(x, 0) = 0$, and all the above $\rightarrow 0$ as $y \rightarrow 0$; proceed similarly for $y < 0$. Thus it is clear that f is of class C^2.

Expanding f as in (16) gives

$$f(x, y) = 0 + x \mid y \mid^{3/2} - x^2 \frac{x \mid y \mid^{3/2}}{x^2 + y^2};$$

thus $g_2(x, y) = -2x \mid y \mid^{3/2}/r^2$ for $r \neq 0$. Since

$$\frac{\partial}{\partial x} g_2(0, y) = -2 \frac{\mid y \mid^{3/2}}{y^2} = -2 \mid y \mid^{-\frac{1}{2}} \qquad (y \neq 0),$$

g_2 is not of class C^1.

HARVARD UNIVERSITY.

DIFFERENTIABLE EVEN FUNCTIONS

By Hassler Whitney

An even function $f(x) = f(-x)$ (defined in a neighborhood of the origin) can be expressed as a function $g(x^2)$; $g(u)$ is determined for $u \geq 0$, but not for $u < 0$. We wish to show that g may be defined for $u < 0$ also, so that it has roughly half as many derivatives as f. A similar result for odd functions is given.

Theorem 1. *An even function $f(x)$ may be written as $g(x^2)$. If f is analytic, of class C^∞ or of class C^{2s}, g may be made analytic, of class C^∞ or of class C^s, respectively.*

If $f(x) = \sum a_i x^i$ is even and analytic, then $a_i = 0$ for i odd, and we may set $g(u) = \sum a_{2i} u^i$, which is analytic.

Suppose f is even and of class C^{2s}. Then Taylor's formula gives

$$f(x) = a_0 + a_1 x^2 + \cdots + a_{s-1} x^{2s-2} + x^{2s}\phi(x). \tag{1}$$

By Theorem 1 of [3], ϕ is even and continuous, and of class C^{2s} for $x \neq 0$, and

$$\lim_{x \to 0} x^k \phi^{(k)}(x) = 0 \qquad (k = 1, \cdots, 2s). \tag{2}$$

Set $\psi(u) = \psi(-u) = \phi(u^{\frac{1}{2}})$, and

$$g(u) = a_0 + a_1 u + \cdots + a_{s-1} u^{s-1} + u^s \psi(u). \tag{3}$$

Then $g(x^2) = f(x)$. To show that g is of class C^s, it is sufficient to show, by Theorem 2 of [3], that

$$\lim_{u \to 0} u^k \psi^{(k)}(u) \qquad (k = 0, \cdots, s) \tag{4}$$

exists.

If we differentiate $\psi(x^2) = \phi(x)$ $(x > 0)$, a simple proof by induction shows that, for some constants α_{ki},

$$\phi^{(k)}(x) = \sum_{1 \leq i \leq \frac{1}{2}k} \alpha_{ki} x^{k-2i} \psi^{(k-i)}(x^2) + 2^k x^k \psi^{(k)}(x^2). \tag{5}$$

Solving these equations in succession gives, for some β_{ki},

$$2^k x^k \psi^{(k)}(x^2) = \phi^{(k)}(x) + \sum_{1 \leq i \leq k-1} \beta_{ki} x^{-i} \phi^{(k-i)}(x). \tag{6}$$

Hence,

$$x^{2k} \psi^{(k)}(x^2) = \sum_{0 \leq i \leq k-1} \beta'_{ki} x^{k-i} \phi^{(k-i)}(x),$$

and (4) for $x > 0$ follows from (2). Since $\psi(-u) = \psi(u)$, the theorem is proved for this case.

Received July 28, 1942; presented to the American Mathematical Society September 8, 1942.

Suppose finally that f is even and of class C^∞. Then, for each s, there is a function $g_s(u)$ of class C^s such that $f(x) = g_s(x^2)$. Since $g(u) = g_s(u)$ is independent of s for $u \geq 0$, $g(u)$ is of class C^∞ for $u \geq 0$. Hence [1], [2; Theorem I] it may be extended so as to be of class C^∞ for all u.

Example 1. Let $f(x) = | x |^{2s+1+\frac{1}{2}}$. Then f is of class C^{2s+1}, while $g(u) = u^{s+3/4}$ for $u > 0$, and g cannot be of class C^{s+1}.

Example 2. Let $f(x)$ be Cauchy's function $\exp(-1/x^2)$. Then g exists, of class C^∞; for $u > 0$, $g(u) = \exp(-1/u)$. Of course we cannot define g by this formula for $u < 0$; but we may set $g(u) = \exp(-1/|u|)$ for $u \neq 0$.

THEOREM 2. *An odd function $f(x)$ may be written as $xg(x^2)$. If f is analytic, of class C^∞ or of class C^{2s+1}, g may be made analytic, of class C^∞ or of class C^s, respectively.*

Again the analytic case is trivial. If f is odd and of class C^{2s+1}, then expanding gives

$$f(x) = a_0 x + \cdots + a_{s-1}x^{2s-1} + x^{2s+1}\phi(x) = xF(x);$$

applying the proof of the last theorem to the even function $F(x)$ gives the statement. (That F is of class C^{2s} follows from Corollary 1 of the preceding paper.) The case that f is of class C^∞ goes as before.

Remark. Since g is constructed in a definite fashion, the theorems hold for functions of several variables which are even in one of them. (The case that f is of class C^∞ offers no further difficulty.) The reference above to [2] is to take care of this case.

BIBLIOGRAPHY

1. H. WHITNEY, *Analytic extensions of differentiable functions defined in closed sets*, Transactions of the American Mathematical Society, vol. 36(1934), pp. 63–89.
2. H. WHITNEY, *Derivatives, difference quotients, and Taylor's formula* II, Annals of Mathematics, vol. 35(1934), pp. 476–485.
3. H. WHITNEY, *Differentiability of the remainder term in Taylor's formula*, this Journal, vol. 10(1943), pp. 153–158.

HARVARD UNIVERSITY.

THE GENERAL TYPE OF SINGULARITY OF A SET OF $2n - 1$ SMOOTH FUNCTIONS OF n VARIABLES

By Hassler Whitney

1. **Introduction.** Let a region R of n-space E^n, or more generally, of a differentiable n-manifold, be mapped differentiably into m-space E^m. If $m \geq 2n$, it is always possible [1; 818], [3], by a slight alteration of the mapping function f (letting also any finite number of derivatives change arbitrarily slightly), to obtain a mapping f^* which is everywhere regular. That is, for any p in R, and any set of independent vectors $u_1, \cdots, u_n$ in R at p, f^* carries these vectors into independent vectors. Here, vector equals the vector in "tangent space" equals the differential. As a consequence, some neighborhood U of p is mapped by f in a one-one way. The object of this paper is to determine what can be obtained by slight alterations of f in case $m = 2n - 1$. It turns out that any singularities may be made into a fixed kind. (It will be shown in other papers that any smooth n-manifold may be imbedded in $(2n)$-space, and may be immersed (self-intersections allowed) in $(2n - 1)$-space.)

There are two main theorems in the paper, roughly:

(a) We may alter f arbitrarily slightly, forming f^*, for which the singular points (points where f^* is not regular) are isolated, and such that a certain condition (C) below holds at each singular point. (The self-intersection may also be made simple; cf. [3; 655, (D)].)

(b) Let f^* satisfy the condition mentioned. Then for any singular point p, we may choose coördinate systems $x_1, \cdots, x_n$ in a neighborhood of p and $y_1, \cdots, y_{2n-1}$ in a neighborhood of $f(p)$ such that f^* is given exactly by the equations (4.2). Here, f^* must have many derivatives.

Remark. As a consequence, there is a slight deformation of E^{2n-1} which carries $f(U)$ (U a neighborhood of p) into the set of points given by (4.2).

The transformations in (b) may lower the class of f^* considerably; but if f^* is of class C^∞, or analytic, the transformations will be also. The condition mentioned in (a) is the following:

(C) There is a direction through p with the following properties: (C_1) f^* maps any vector in this direction into the null vector in E^{2n-1}, but maps any other vector at p into a non-null vector. (C_2) If $g(p')$ is the derivative of $f^*(p')$ in the direction given above, for p' near p, then there is no vector in E^{2n-1} which is the image both of a vector under f^* and a vector $\neq 0$ under g, both at p.

We may phrase the second condition as follows:

(C_2') Suppose a coördinate system is chosen in which the given vector is in the x_1-direction. Then

Received July 28, 1942; presented to the American Mathematical Society September 8, 1942.

$$(1.1)\qquad \frac{\partial f^*}{\partial x_2}, \quad \cdots, \quad \frac{\partial f^*}{\partial x_n}, \quad \frac{\partial^2 f^*}{\partial x_1^2}, \quad \frac{\partial^2 f^*}{\partial x_1 \partial x_2}, \quad \cdots, \quad \frac{\partial^2 f^*}{\partial x_1 \partial x_n},$$

taken at p, are independent vectors.

We shall show first that the two conditions (C_2) and (C_2') are equivalent, and independent of the coördinate system chosen (with the same or opposite direction of the x_1-axis). Next we show that (C) implies regularity near p, so that p is an isolated singularity. (This fact follows also from (b); a proof is given here to help in understanding condition (C).) We then study the typical singularity mentioned in (b). Next we prove (b), and finally (a). The proof of (a) uses methods found in [3]; the other proofs are straightforward analysis.

2. **Equivalence of** (C_2) **and** (C_2'). We use f in place of f^*. Having chosen a coördinate system as in (C_2'), we note that $g(p') = \partial f(p')/\partial x_1$. If e_i is the unit vector in R at p in the direction of x_i, then by definition, f carries it into the vector $\partial f(p)/\partial x_i$. Hence, for any vectors $u = (u_1, \cdots, u_n)$ and $v = (v_1, \cdots, v_n)$ at p, f and g carry these into

$$f(u) = \sum_{i=1}^{n} u_i \frac{\partial f(p)}{\partial x_i} = \sum_{i=2}^{n} u_i \frac{\partial f(p)}{\partial x_i},$$

$$g(v) = \sum_{i=1}^{n} v_i \frac{\partial^2 f(p)}{\partial x_1 \partial x_i}.$$

Suppose (C_2') holds. Then if $f(u) = g(v)$, since the vectors (1.1) are independent, the coefficients $u_2, \cdots, u_n, v_1, \cdots, v_n$ are all 0, and $v = 0$, proving (C_2). Suppose conversely that (C_2) holds. If there is a linear relation between the vectors (1.1) with coefficients, say $u_2, \cdots, u_n, -v_1, \cdots, -v_n$, not all 0, then defining u and v as above, with $u_1 = 0$, we have $f(u) = g(v)$, and $u \neq 0$ or $v \neq 0$. By (C_2), $v = 0$, and hence $f(u) = 0$. By (C_1), u is in the x_1-direction, that is, $u_2 = \cdots = u_n = 0$, which is a contradiction.

We shall show that condition (C_2'), and hence (C), is independent of the coördinate system employed. Take two systems, each with the first axis (at p) in the given direction; then $\partial x_k/\partial x_1' |_p = a\delta_{1k}$, $a \neq 0$. Let T^{n-1} be the plane (in E^{2n-1}) of all directional derivatives of f at p, i.e., all vectors $f(u)$. Let $u \sim v$ denote $u - v \in T^{n-1}$. If we suppose that (C_2') holds in the first but not in the second system, we have, for some $v' = (v_1', \cdots, v_n') \neq 0$, if $g' = \partial f/\partial x_1'$,

$$g'(v') = \sum_i v_i' \frac{\partial^2 f}{\partial x_1' \partial x_i'}\bigg|_p = \sum_i v_i' \frac{\partial}{\partial x_1'} \sum_j \frac{\partial f}{\partial x_j} \frac{\partial x_j}{\partial x_i'}\bigg|_p$$

$$\sim \sum_i v_i' \sum_j \sum_k \frac{\partial^2 f}{\partial x_k \partial x_j} \frac{\partial x_k}{\partial x_1'} \frac{\partial x_j}{\partial x_i'}\bigg|_p$$

$$= a \sum_j \left[\sum_i v_i' \frac{\partial x_j}{\partial x_i'}\right] \frac{\partial^2 f}{\partial x_1 \partial x_j}\bigg|_p \sim 0.$$

Hence, all $v_j = \sum_i v'_i \partial x_j / \partial x'_i$ are 0. But the Jacobian $| \partial x_j / \partial x'_i | \neq 0$ at p, so that the columns of $\| \partial x_j / \partial x'_i \|$ are independent, which contradicts the assumption that not all the v'_i are 0.

3. **Regularity near a singular point.** Let (C) hold for f at p; we shall show that f is regular in the rest of a neighborhood of p. Using (C'_2), let $T^{n-1}(p')$ and $T^n(p')$ be the planes determined by the $\partial f / \partial x_i$ $(i > 1)$ and the $\partial^2 f / \partial x_1 \partial x_i$ respectively, taken at p'. Then in a neighborhood U_0 of p, they are of the dimensions shown, and if $p' \in U_0$ and T'^n is any n-plane sufficiently near $T^n(p)$, T'^n and $T^{n-1}(p')$ have only one point in common. Since the vectors $\partial g(p)/\partial x_1, \cdots, \partial g(p)/\partial x_n$ determine $T^n(p)$, we can take $U_1 \subset U_0$ so that for any $p'' \in U_1$, there is a plane T'^n as near $T^n(p)$ as required above which contains $g(p'') - g(p)$; it follows that if $p'' \neq p$, $g(p'') - g(p)$ is in no $T^{n-1}(p')$ $(p' \in U_0)$. Now, since $g(p) = \partial f(p)/\partial x_1 = 0$, we have in particular: for any $p' \in U_1$, $p' \neq p$, $g(p')$ is not in $T^{n-1}(p')$; that is, $\partial f(p')/\partial x_1$ is independent of $\partial f(p')/\partial x_2, \cdots, \partial f(p')/\partial x_n$. Thus f is regular in U_1 except at p, as required.

4. **A typical singularity.** As is well known, a mapping of a projective plane into 3-space may be obtained by replacing a piece of the surface of a sphere by a "cross cap". A cross cap (or rather the top of one) may be described as follows: Let $T(y)$ be the plane in E^3 perpendicular to the y-axis at a given y. Take a parabola in $T(y_0)$ for $y_0 < 0$; say $z = \pm y_0 x^{\frac{1}{2}}$. As we let y increase, pull in the two sides of the parabola until they coincide (and thus form a half ray) at $y = 0$ and become a parabola reversed in direction for $y > 0$. The locus of these parabolas forms the top of the cross cap. A sphere about the origin cuts the cross cap in a curve in the form of a bent figure 8.

Using coördinates y_1, y_2, y_3 in E^3, the cross cap may be represented parametrically by the equations

$$y_1 = x_1^2, \qquad y_2 = x_2, \qquad y_3 = x_1 x_2. \tag{4.1}$$

Generalizing this, let us map E^n into E^{2n-1} by:

$$\begin{aligned} y_1 &= x_1^2, & & \\ y_i &= x_i & & (i = 2, \cdots, n), \\ y_{n+i-1} &= x_1 x_i & & (i = 2, \cdots, n). \end{aligned} \tag{4.2}$$

The matrix $\| \partial y_i / \partial x_j \|$, transposed, is

$$\left\| \begin{matrix} 2x_1 & 0 & 0 & \cdots & 0 & x_2 & x_3 & \cdots & x_n \\ 0 & 1 & 0 & \cdots & 0 & x_1 & 0 & \cdots & 0 \\ 0 & 0 & 1 & \cdots & 0 & 0 & x_1 & \cdots & 0 \\ \cdots & \cdots & \cdots & \cdots & \cdots & \cdots & \cdots & \cdots & \cdots \\ 0 & 0 & 0 & \cdots & 1 & 0 & 0 & \cdots & x_1 \end{matrix} \right\|$$

If we call the mapping $y = f(x)$, f is clearly regular except at the origin; only $\partial f/\partial x_1$ is 0 there. Also, the matrix $\| \partial^2 y_i/\partial x_1 \partial x_j \|$, transposed, has a 2 in the upper left corner, a diagonal of $n - 1$ ones in the lower right, and zero elsewhere. Combining the two matrices, with the first row $\partial y_1/\partial x_j$ omitted, we obtain a diagonal matrix, except for an interchange of rows, at the origin. Hence, the vectors (1.1) are independent, and (C) holds.

To determine the self-intersections, suppose $f(x'_1, \cdots, x'_n) = f(x_1, \cdots, x_n)$, where the points are distinct. Then $x'_i = x_i$ for $i > 1$, since $y'_i = y_i$; hence, $x'_1 \neq x_1$, and, since $y'_1 = y_1$, $x'_1 = -x_1 \neq 0$. Since $y'_{n+i-1} = y_{n+i-1}$ and $x_1 \neq 0$, $x_i = 0$ $(i > 1)$. Thus,

$$f(\alpha, 0, \cdots, 0) = f(-\alpha, 0, \cdots, 0),$$

and there are no other self-intersections.

Examining the matrix $\| \partial y_i/\partial x_j \|$, we see that at any such self-intersection, the two tangent planes have only the y_1-axis in common.

5. **Proof of** (b). We shall prove:

THEOREM 1. *Let f be a mapping of class C^s, $s = 4r + 8$, $r \geq 1$, of the region R of E^n into E^{2n-1}, and let* (C) *hold at the origin (assumed in R). Then there are curvilinear coördinate systems about* 0 *and* $f(0)$, *of class C^r, in terms of which f has the form* (4.2). *If f is analytic, or of class C^∞, so are the coördinate systems.*

Probably s need not be taken so large in terms of r. We shall consider the case where f is of class C^s; in the other cases, the transformations employed are clearly analytic or of class C^∞, respectively.

As a first step, choose coördinates $x_1, \cdots, x_n$ in E^n as in (C'_2); since the vectors (1.1) are independent, we may choose oblique axes $y_1, \cdots, y_{2n-1}$ in E^{2n-1} so that these vectors are unit vectors (except for $\partial^2 f/\partial x_1^2$) on the axes of

$$y_2, \quad \cdots, \quad y_n, \quad y_1, \quad y_{n+1}, \quad \cdots, \quad y_{2n-1},$$

respectively. Now take each y_k, expand it in terms of x_1 to the third order, expand the coefficient of x_1^i in terms of x_2 to the order $3 - i$, etc. This gives

$$y_k = A^k + \sum_i B_i^k x_i + \sum_{i \leq j} C_{ij}^k x_i x_j + \sum_{h \leq i \leq j} D_{hij}^k (x_1, \cdots, x_n) x_h x_i x_j .$$

(Many of the D_{hij}^k are actually functions of fewer variables.) The A^k, B_i^k and C_{ij}^k are the y_k and first and second derivatives (except for a factor) at the origin. By the choice of the y-coördinate system, therefore, many of these may be determined, and we find

$$\begin{aligned} y_1 &= x_1^2 + R_1(x_1, \cdots, x_n), \\ y_i &= x_i + R_i(x_1, \cdots, x_n) \qquad (i = 2, \cdots, n), \\ y_{n+i-1} &= x_1 x_i + R_{n+i-1}(x_1, \cdots, x_n) \qquad (i = 2, \cdots, n), \end{aligned} \tag{5.1}$$

where

$$R_k = \sum_{2 \le i \le j} C_{ij}^k x_i x_j + \sum_{h \le i \le j} D_{hij}^k(x_1, \cdots, x_n) x_h x_i x_j, \tag{5.2}$$

the R_k being of class C^s.

Set $x_1' = x_1$, and

$$x_i' = x_i + R_i(x_1, \cdots, x_n) \qquad (i = 2, \cdots, n). \tag{5.3}$$

Then $\partial x_i' / \partial x_j |_0 = \delta_{ij}$, so that this is a transformation of coördinates of class C^s near the origin. Furthermore, $\partial x_i / \partial x_j' |_0 = \delta_{ij}$, so that

$$x_i = x_i' + \sum_{j \le k} a_{jk}^i(x_1', \cdots, x_n') x_j' x_k'.$$

Substituting in (5.1) and dropping primes gives

$$\begin{aligned} y_1 &= x_1^2 + R_1(x_1, \cdots, x_n), \\ y_i &= x_i \qquad (i = 2, \cdots, n), \\ y_{n+i-1} &= x_1 x_i + R_{n+i-1}(x_1, \cdots, x_n) \qquad (i = 2, \cdots, n), \end{aligned} \tag{5.4}$$

where the new R_k have the same form (5.2).

Next we simplify the form of y_1. Since

$$\left.\frac{\partial y_1}{\partial x_1}\right|_0 = 0, \qquad \left.\frac{\partial^2 y_1}{\partial x_1^2}\right|_0 = 2,$$

we may solve $\partial y_1 / \partial x_1 = 0$ in a neighborhood of the origin, obtaining a function $x_1 = \phi(x_2, \cdots, x_n)$, of class C^{s-1}. Then, by definition,

$$2\phi(x_2, \cdots, x_n) + \frac{\partial}{\partial \phi} R_1(\phi(x_2, \cdots, x_n), x_2, \cdots, x_n) = 0. \tag{5.5}$$

Set

$$x_1' = x_1 - \phi(x_2, \cdots, x_n), \qquad x_i' = x_i \qquad (i > 1). \tag{5.6}$$

For each $(x_2, \cdots, x_n)$ near $(0, \cdots, 0)$, expand y_1 in terms of x_1'. Since $\partial y_1 / \partial x_1' = 0$ for $x_1' = 0$, this gives

$$y_1 = \psi_0(x_2, \cdots, x_n) + x_1'^2 \psi_2(x_1', x_2, \cdots, x_n); \tag{5.7}$$

ψ_0 and ψ_2 are of class C^{s-1} and C^{s-3}, respectively, by [2; Theorem 3]. Differentiating (5.5) gives

$$\left[2 + \left.\frac{\partial^2 R_1}{\partial x_1^2}\right|_0\right] \left.\frac{\partial \phi}{\partial x_i}\right|_0 + \left.\frac{\partial^2 R_1}{\partial x_1 \partial x_i}\right|_0 = 0 \qquad (i > 1).$$

Since $\partial^2 R_1 / \partial x_1 \partial x_j |_0 = 0$ for all j, this gives $\partial \phi / \partial x_i |_0 = 0$; hence $\partial x_i' / \partial x_j |_0 = \delta_{ij}$. It follows that the Jacobian of the transformation is 1 at the origin; also $\phi(x_2', \cdots) = \sum \phi_{ij}(x_2', \cdots) x_i' x_j'$, so that the y_i are given by the same kinds of expressions in terms of $x_1', x_2, \cdots$ as in terms of $x_1, x_2, \cdots$.

Next, set

$$\text{(5.8)}\quad \begin{aligned} x_1'' &= x_1'[\psi_2(x_1', \cdots, x_n')]^{\frac{1}{2}}, \qquad x_i'' = x_i \qquad (i > 1),\\ y_1' &= y_1 - \psi_0(y_2, \cdots, y_n), \qquad y_i' = y_i \qquad (i > 1). \end{aligned}$$

Now

$$2 = \frac{\partial^2 y_1}{\partial x_1^2}\bigg|_0 = \frac{\partial^2 y_1}{\partial x_1'^2}\bigg|_0 = 2\psi_2\bigg|_0;$$

hence $\psi_2 |_0 = 1$, and the transformations to the x_i'' and y_i' are allowable and are of class C^{s-3} in a neighborhood of the origin. Dropping primes again, we have the same equations as (5.4), with R_1 missing; the R_k are of class C^{s-3}, and have again the form (5.2). To prove this for y_{n+i-1}, we note that if $x_1' = \theta(x_1'', \cdots, x_n)$, then $\partial\theta/\partial x_1'' |_0 = 1$, and hence

$$\frac{\partial^2}{\partial x_1''\partial x_j''}[\theta x_i + R_{n+i-1}'(\theta, \cdots, x_n)]\bigg|_0 = \delta_{ij} \qquad (i = 1, j \geq 1).$$

Next, set $y_i' = y_i$ $(i = 1, \cdots, n)$, and

$$\text{(5.9)}\quad y_k' = y_k - R_k(0, y_2, \cdots, y_n) \qquad (k = n+1, \cdots, 2n-1).$$

Setting $k = n + i - 1$ and using (5.2), we find

$$\begin{aligned} y_k' = x_1x_i &+ \sum_{i \leq l} D_{1il}^k(x_1, \cdots, x_n)x_1x_jx_l \\ &+ \sum_{2 \leq h \leq i \leq l} [D_{hil}^k(x_1, x_2, \cdots) - D_{hil}^k(0, x_2, \cdots)]x_hx_jx_l ; \end{aligned}$$

if we expand the first bracketed term D in terms of x_1, and drop primes, we have

$$\text{(5.10)}\quad \begin{aligned} y_1 &= x_1^2, \qquad y_i = x_i \qquad (i = 2, \cdots, n),\\ y_{n+i-1} &= x_1[x_i + R_{n+i-1}(x_1, \cdots, x_n)] \qquad (i = 2, \cdots, n), \end{aligned}$$

where (for $k = n + i - 1$)

$$\text{(5.11)}\quad R_k(x_1, \cdots, x_n) = \sum_{i \leq l} E_{jl}^k(x_1, \cdots, x_n)x_jx_l .$$

The new R_k are of class C^{s-4}; the E_{jl}^k are of class C^{s-6}.

Our next job is to move the curve of self-intersection over onto the y_1-axis; this will result in the term in x_1^2 in (5.11) dropping out.

Examples. Take $n = 2$. If $y_3 = x_1(x_2 + x_1^2)$, then the curve $x_2 = -x_1^2$ maps into the curve of self-intersection, which is $(x_1^2, -x_1^2, 0)$ in E^3. If $y_3 = x_1(x_2 + x_1^2 + x_1x_2)$, the curve is $x_2 = -x_1^2$ again, mapping onto $(x_1^2, -x_1^2, -x_1^4)$.

As in §4, if the distinct points $(x_1', x_2', \cdots)$ and $(x_1, x_2, \cdots)$ go into the same point, then $x_i' = x_i$ for $i > 1$, and $x_1' = -x_1 \neq 0$. Using $y_k' = y_k$ for $n + i - 1 = k > n$ gives

$$\text{(5.12)}\quad \begin{aligned} x_1[x_i + R_k(x_1, x_2, \cdots)] &= -x_1[x_i + R_k(-x_1, x_2, \cdots)],\\ x_i &= -\tfrac{1}{2}[R_k(x_1, x_2, \cdots) + R_k(-x_1, x_2, \cdots)]. \end{aligned}$$

If we write these in the form $x_i + \frac{1}{2}[\ \ ] = 0$, then since $\partial R_k/\partial x_j |_0 = 0$, the functional matrix is $|| \delta_{ij} ||$ at the origin; hence, we may solve these for $x_2, \cdots, x_n$ in terms of x_1. The resulting functions are obviously even and of class $C^{s-4} = C^{4r+4}$; hence, by the theorem of the preceding paper, the solutions may be written as $\phi_i(x_1^2)$, where the ϕ_i are of class C^{2r+2}. Then

$$\phi_i(x_1^2) = -\tfrac{1}{2}[R_k(x_1, \phi_2(x_1^2), \cdots) + R_k(-x_1, \phi_2(x_1^2), \cdots)]. \tag{5.13}$$

Also, for $k > n$, we may define

$$\phi_k(x_1^2) = \tfrac{1}{2}x_1[R_k(x_1, \phi_2(x_1^2), \cdots) - R_k(-x_1, \phi_2(x_1^2), \cdots)]. \tag{5.14}$$

(Then $y_k = \phi_k(x_1^2)$ for all k on the curve of intersection.)

Make the transformation

$$\begin{aligned} x_1' &= x_1, \quad & x_i' &= x_i - \phi_i(x_1^2) \quad & (i &= 2, \cdots, n), \\ y_1' &= y_1, \quad & y_i' &= y_i - \phi_i(y_1) \quad & (i &= 2, \cdots, 2n-1). \end{aligned} \tag{5.15}$$

These have non-vanishing Jacobians near the origins, and are of class C^{2r+2}. Now $y_1' = x_1'^2$, $y_i' = x_i'$ $(i = 2, \cdots, n)$, and using (5.13) and (5.14), we find for $n + i - 1 = k > n$,

$$\begin{aligned} y_k' &= x_1'[x_i' + \phi_i(x_1'^2) + R_k(x_1', x_2, \cdots)] - \phi_k(x_1'^2) \\ &= x_1'[x_i' - \tfrac{1}{2}\{R_k(x_1', \phi_2(x_1'^2), \cdots) + R_k(-x_1', \phi_2(x_1'^2), \cdots)\} \\ &\qquad + R_k(x_1', x_2' + \phi_2(x_1'^2), \cdots) \\ &\qquad\qquad - \tfrac{1}{2}\{R_k(x_1', \phi_2(x_1'^2), \cdots) - R_k(-x_1', \phi_2(x_1'^2), \cdots)\}], \end{aligned}$$

or

$$y_k' = x_1'[x_i' + R_k'(x_1', x_2', \cdots)], \tag{5.16}$$

where the R_k', of class C^{2r+2}, are given by

$$R_k'(x_1', x_2', \cdots) = R_k(x_1', x_2' + \phi_2(x_1'^2), \cdots) - R_k(x_1', \phi_2(x_1'^2), \cdots).$$

Since $R_k'(x_1', 0, \cdots) = 0$, expanding, for each x_1', in terms of $x_2', \cdots, x_n'$ to the first order and dropping primes from the x_i' gives

$$R_k'(x_1, x_2, \cdots) = \sum_{1<j} x_j R_{kj}(x_1, x_2, \cdots). \tag{5.17}$$

Using the definition of R_k' and (5.11), we find

$$R_{kj}(0, 0, \cdots) = \left.\frac{\partial R_k'}{\partial x_j}\right|_0 = \left.\frac{\partial R_k}{\partial x_j}\right|_0 = 0;$$

therefore,

$$R_{kj}(x_1, x_2, \cdots) = \sum_l x_l R_{kjl}(x_1, x_2, \cdots). \tag{5.18}$$

Putting in (5.16) (with primes dropped) gives, if $k = n + i - 1$,

$$(5.19)\qquad \begin{aligned} y_1 &= x_1^2, \qquad y_i = x_i &&(i = 2, \cdots, n),\\ y_k &= x_1[x_i + \sum_{j=2}^{n} x_j R_{kj}(x_1, \cdots, x_n)] &&(i = 2, \cdots, n). \end{aligned}$$

The R_{kj} are of class C^{2r+1}.

Next, by the theorems of the preceding paper, we may write

$$(5.20)\qquad \begin{aligned} P_{kj}(x_1^2, x_2, \cdots) &= \tfrac{1}{2}[R_{kj}(x_1, x_2, \cdots) + R_{kj}(-x_1, x_2, \cdots)],\\ x_1 Q_{kj}(x_1^2, x_2, \cdots) &= \tfrac{1}{2}[R_{kj}(x_1, x_2, \cdots) - R_{kj}(-x_1, x_2, \cdots)], \end{aligned}$$

the P_{kj} and Q_{kj} being of class C^r. This gives

$$y_k = x_1[x_i + \sum_{j=2}^{n} x_j P_{kj}(x_1^2, x_2, \cdots)] + x_1^2 \sum_{j=2}^{n} x_j Q_{kj}(x_1^2, x_2, \cdots).$$

The next to the last transformation is: $y_i' = y_i$ $(i \leq n)$,

$$(5.21)\qquad y_k' = y_k - y_1 \sum_{j=2}^{n} y_j Q_{kj}(y_1, y_2, \cdots) \qquad (k > n).$$

The Jacobian is 1 at the origin. Now, dropping primes again, we have

$$(5.22)\qquad \begin{aligned} y_1 &= x_1^2, \qquad y_i = x_i &&(i = 2, \cdots, n),\\ y_{n+i-1} &= x_1[x_i + \sum_{j=2}^{n} x_j P_{n+i-1,j}(x_1^2, x_2, \cdots)] &&(i = 2, \cdots, n). \end{aligned}$$

We are now ready for the final transformation. Define functions

$$F_{kl}(u_1, \cdots, u_n) \qquad (k, l = n + 1, \cdots, 2n - 1),$$

for each fixed k, as the solutions of the linear equations

$$(5.23)\qquad \sum_{h=n+1}^{2n-1} [\delta_{n+i-1,h} + P_{hi}(u_1, \cdots)]F_{kh}(u_1, \cdots) = P_{ki}(u_1, \cdots),$$

for $i = 2, \cdots, n$. By (5.20) and (5.18), the determinant $| \delta_{n+i-1,h} + P_{hi} |$ is near 1 in a neighborhood of the origin; hence, the F_{kh} are defined and of class C^r there, and $F_{kh}(0, \cdots, 0) = 0$. Set

$$(5.24)\qquad \begin{aligned} y_i' &= y_i &&(i = 1, \cdots, n),\\ y_k' &= y_k - \sum_{h=n+1}^{2n-1} F_{kh}(y_1, \cdots, y_n) y_h &&(k = n + 1, \cdots, 2n - 1). \end{aligned}$$

Since, for $k \geq n + 1$,

$$\frac{\partial y_k'}{\partial y_h} = \delta_{kh} - F_{kh} \qquad (h > n),$$

and the $F_{kh}\,|_0 = 0$, the Jacobian $\neq 0$ at the origin, and so the transformation is allowable, and of class C^r.

We find the y'_k in terms of the x_i : Writing P_{kj} for $P_{kj}(x_1^2, x_2, \cdots)$, etc., and taking $k = n + i - 1$, $i > 1$, we have

$$\begin{aligned}\frac{1}{x_1} y'_k &= x_i + \sum_{j=2}^{n} P_{kj}x_j - \sum_{h=n+1}^{2n-1} F_{kh}[x_{h-n+1} + \sum_{j=2}^{n} P_{hj}x_j] \\ &= x_i + \sum_{j=2}^{n} x_j[P_{kj} - F_{k,n+j-1} - \sum_{h=n+1}^{2n-1} F_{kh}P_{hj}] \\ &= x_i .\end{aligned}$$

This proves the theorem.

6. **Proof of** (a). The other main theorem is:

THEOREM 2. *Let f be a mapping of class C^2 of a region R of E^n into E^{2n-1}. Then arbitrarily close to f (together with first and second derivatives) there is a mapping f^* satisfying* (C) *at each singular point. f^* may be made analytic.*

Remark. It will be clear from the proof that the theorem holds equally well with R and E^{2n-1} replaced by smooth manifolds M^n and M^{2n-1}.

Let $R_1, R_2, \cdots$ be rectangular regions in R, and choose R'_i with $\overline{R}'_i \subset R_i$, so that the R'_i cover R. Suppose we have shown how to deform f slightly so that (C) holds at any singular point in a given $\overline{R}'_\alpha$. Then carrying out deformations successively, we make it hold in $\overline{R}'_1, \overline{R}'_2, \cdots$, each time making the deformation so slight that the property is not disturbed in any preceding $\overline{R}'_1$, and so that the limit will be a mapping with the property in every $\overline{R}'_i$ and hence in R. (Since for any f_0 and $\eta(p)$, the property of f satisfying (C) at any singular point is an $(f_0, 2, \eta)$-property, as in [3; §7], the statement follows at once from [3; Lemma 12].) We consider only functions f of class C^r, $r \geq 2$; the final function may be made analytic, by [3; Lemma 9].

If f is not of class C^3, approximate to it by a function f^{**} of class C^3; thus we may suppose f is of class C^3 in the first place.

Let $\lambda(p)$ be a function of class C^∞ in R, $\equiv 1$ in $\overline{R}'_\alpha$, $\equiv 0$ in $R - R_\alpha$. (Use [3; Lemma 11], or construct it directly.) We shall define certain kinds of mappings G_{ij}, H_{ij}, with which f may be combined to give the required function satisfying (C) at all singular points in $\overline{R}'_\alpha$.

First, by the proof of [3; Lemma 18], we may suppose also that $\partial f/\partial x_2, \cdots, \partial f/\partial x_n$ are independent in $\overline{R}'_\alpha$.

Set $m = 2n - 1$. Let $v_1, \cdots, v_m$ be the unit vectors in E^m. Let $\overline{E} = E^{2mn}$ be a Euclidean space, with unit vectors (in a definite order)

$$v_{ij}, \qquad w_{ij} \qquad\qquad (i = 1, \cdots, n; j = 1, \cdots, m).$$

Corresponding to any mapping $g(p)$ of class C^3 of $\overline{R}_\alpha$ into E^m such that $\partial g(p)/\partial x_2, \cdots, \partial g(p)/\partial x_n$ are independent for each $p = (x_1, \cdots, x_n)$, define $\overline{g}^2(p), \cdots,$

$\bar{g}^n(p)$ as the numbers such that

$$g^{\odot}(p) = \frac{\partial g(p)}{\partial x_1} - \sum_{1<i} \bar{g}^i(p) \frac{\partial g(p)}{\partial x_i}$$

is orthogonal to $\partial g(p)/\partial x_2, \cdots, \partial g(p)/\partial x_n$. If $g_1(p), \cdots, g_m(p)$ are the coördinates of $g(p)$, define the corresponding mapping $g^*(p)$ of $\bar{R}_\alpha$ into $\bar{E}$ by

$$g^* = \sum_j g_j^{\odot} v_{1j} + \sum_{i,j;1<i} \frac{\partial g_j}{\partial x_i} v_{ij}$$

$$+ \sum_j \left[\frac{\partial^2 g_j}{\partial x_1^2} - 2\sum_{1<i} \bar{g}^i \frac{\partial^2 g_j}{\partial x_1 \partial x_i} + \sum_{1<i,h} \bar{g}^h \bar{g}^i \frac{\partial^2 g_j}{\partial x_h \partial x_i}\right] w_{1j}$$

$$+ \sum_{i,j;1<i} \left[\frac{\partial^2 g_j}{\partial x_1 \partial x_i} - \sum_{1<h} \bar{g}^h \frac{\partial^2 g_j}{\partial x_h \partial x_i}\right] w_{ij} .$$

Then the $\bar{g}^i$ and $g^{\odot}$ are of class C^2 and g^* is of class C^1.

Given g as above, if we take a fixed $p_0 = (\bar{x}_1, \cdots, \bar{x}_n)$ and make the change of variables

$$x_1 = x_1' + \bar{x}_1, \qquad x_i = x_i' - \bar{g}^i(p_0)x_1' + \bar{x}_i \qquad (i > 1),$$

we find

$$g^*(p_0) = \sum_{i,j} \frac{\partial g_j(p_0)}{\partial x_i'} v_{ij} + \sum_{i,j} \frac{\partial^2 g_j(p_0)}{\partial x_1' \partial x_i'} w_{ij} .$$

Remark. One could carry out the proof, using all $\partial g/\partial x_i$ and $\partial^2 g/\partial x_h \partial x_i$ independently; then $\bar{E}$ will have $m[n + \frac{1}{2}n(n+1)]$ dimensions.

For any sets of numbers β_{ij}, γ_{ij} $(i = 1, \cdots, n; j = 1, \cdots, m)$, set

$$f_{\beta\gamma}(p) = f(p) + \lambda(p)[\sum_{i,j} \beta_{ij} x_i v_j + \sum_{i,j} \gamma_{ij} x_1 x_i v_j].$$

If the β_{ij} and γ_{ij} are small enough, $\partial f_{\beta\gamma}/\partial x_2, \cdots, \partial f_{\beta\gamma}/\partial x_n$ are independent in $\bar{R}_\alpha$, so that the $\bar{f}^i_{\beta\gamma}(p)$ etc. are defined. To find $f^*_{\beta\gamma}(p_0)$, make the change of variables described above, differentiate $f_{\beta\gamma}$ with respect to these variables, and put in the second relation above for $f^*_{\beta\gamma}$; we find (at $p_0 \in \bar{R}'_\alpha$)

$$f^*_{\beta\gamma} = \sum_j \left[\frac{\partial f_j}{\partial x_1'} + \beta_{1j} - \sum_{1<i} \bar{f}^i_{\beta\gamma}\beta_{ij} + 2\gamma_{1j}x_1 + \sum_{1<i} \gamma_{ij}x_i - \sum_{1<i} \bar{f}^i_{\beta\gamma}\gamma_{ij}x_1\right] v_{1j}$$

$$+ \sum_{i,j;1<i} \left[\frac{\partial f_j}{\partial x_i'} + \beta_{ij} + \gamma_{ij}x_1\right] v_{ij}$$

$$+ \sum_j \left[\frac{\partial^2 f_j}{\partial x_1'^2} + 2\gamma_{1j} - 2\sum_{1<i} \bar{f}^i_{\beta\gamma}\gamma_{ij}\right] w_{1j} + \sum_{i,j;1<i} \left[\frac{\partial^2 f_j}{\partial x_1' \partial x_i'} + \gamma_{ij}\right] w_{ij} .$$

It follows that (all at p_0)

$$\frac{\partial f^*_{\beta\gamma}}{\partial \beta_{1j}} = v_{1j}, \qquad \frac{\partial f^*_{\beta\gamma}}{\partial \beta_{ij}} = -\bar{f}^i_{\beta\gamma} v_{1j} + v_{ij} \qquad (i > 1),$$

$$\frac{\partial f^*_{\beta\gamma}}{\partial \gamma_{1j}} = 2x_1 v_{1j} + 2w_{1j},$$

$$\frac{\partial f^*_{\beta\gamma}}{\partial \gamma_{ij}} = x_i v_{1j} - \bar{f}^i_{\beta\gamma} x_1 v_{1j} + x_1 v_{ij} - 2\bar{f}^i_{\beta\gamma} w_{1j} + w_{ij} \qquad (i > 1).$$

Clearly the $f^*_{\beta\gamma}$ form a $(2mn)$-parameter family of mappings of $\bar{R}'_\alpha$ into $\bar{E}$ (see [3; §18]).

We wish to find, for arbitrarily small β_{ij} and γ_{ij}, a mapping $f_{\beta\gamma}$ with the property (C) at each singular point in $\bar{R}'_\alpha$. We shall construct sets $S_k \subset \bar{E}$, $k \geq 1$, and show that if a mapping g of $\bar{R}'_\alpha$ into E^m does not satisfy (C) at all singular points, then $g^*(p) \in \sum S_k$ for some $p \in \bar{R}'_\alpha$. Take $(t_{ij}, u_{ij}) \in \bar{E}$, and let t_i and u_i be the vectors in E^m with components t_{ij} and u_{ij}, respectively. Then (t, u) is in S_k if the following hold:

(a) $t_2, \cdots, t_n$ are independent;

(b) if T^{n-1} is the plane in E^m determined by them, then $t_1 \in T^{n-1}$;

(c) the vectors $t_1, \cdots, t_n, u_1, \cdots, u_n$ determine exactly a plane T^{m-k}.

Now take any mapping g, such that, at some singular point p, (C) does not hold, but $\partial g(p)/\partial x_1, \cdots, \partial g(p)/\partial x_n$ are independent. Let t_{ij} and u_{ij} be the coördinates of the corresponding $g^*(p)$:

$$g^*(p) = \sum_{i,j} t_{ij} v_{ij} + \sum_{i,j} u_{ij} w_{ij},$$

i.e., $g^*(p) = (t, u)$. Since p is a singular point of g, and the $\partial g(p)/\partial x_i$ are independent for $i = 2, \cdots, n$, $\partial g(p)/\partial x_1$ lies in the plane T^{n-1} determined by them. Hence, if we change coördinates to the x'_i as above,

$$\frac{\partial g(p)}{\partial x'_1} = g^{\odot}(p) = 0.$$

Since (C) does not hold, the vectors $\partial g(p)/\partial x'_i$ $(i > 1)$ and $\partial^2 g(p)/\partial x'_1 \partial x'_i$ are dependent, and therefore determine a plane T^{m-k} with $k > 0$. But the expressions for $g^*(p)$ show that these vectors are exactly the t_i $(i > 1)$ and the u_i; hence $g^*(p) \in S_k$.

Next we show that S_k is an analytic manifold, and determine its dimension. Let (t^*, u^*) be any point of S_k. We shall show that the nearby points (t, u) of S_k are obtained by varying the t_{ij} and u_{ij} under certain restrictions; if we introduce d independent parameters describing them in an analytic fashion, we will have an analytic regular mapping of a d-manifold, and the statement will follow. First, $t_2, \cdots, t_n$ may be varied (of course slightly enough to keep them independent) in any fashion; there are thus $(n - 1)m$ parameters so far. Next, keeping these fixed, vary t_1, keeping it in the plane T^{n-1}; $n - 1$ parameters are added. Next, we consider the positions of the plane T^{m-k}. Given T^{m-1}, we determined them by naming a plane $T^{n-k} \subset T^{m-k}$ orthogonal to T^{m-1}, since $T^{m-1} \subset T^{m-k}$ and $(n - 1) + (n - k) = m - k$. Now T^{n-k} is any plane in the maximal plane T^n orthogonal to T^{n-1}; it may be varied by varying each of a set of $n - k$ vectors ξ_i determining it, so that $\Delta\xi_i$ is in the plane $T^k \subset T^n$ orthogonal to T^{n-k}; thus it is determined by $k(n - k)$ parameters; see also [3; §24]. Finally, vary $u_1, \cdots, u_n$ in any manner such that, with T^{m-1}, they still determine T^{m-k} Since they can vary (slightly) in any manner, so long as they remain in T^{m-k}, $n(m - k)$ more parameters are introduced. Therefore,

$$\dim (S_k) = d = (n - 1)m + (n - 1) + k(n - k) + n(m - k)$$

$$= 4n^2 - 3n - k^2.$$

Consequently [3; §17], $\sum S_k$ is a denumerable sum of compact sets, each of zero $(4n^2 - 3n)$-extent. Therefore, using [3; Lemma 16] (see also [3; Lemma 12]), since $\bar{R}'_\alpha$ is of finite n-extent, and the $f_{\beta\gamma}$ form a $(4n^2 - 2n)$-parameter family, there is an arbitrarily small (β, γ), for which $f_{\beta\gamma}(p)$ is not in any S_k for any $p \in \bar{R}'_\alpha$; then $f_{\beta\gamma}$ is the required mapping. This completes the proof of the theorem. (The proof should be compared with that in [3; 677–679].)

Bibliography

1. H. Whitney, *Analytic coördinate systems and arcs in a manifold*, Annals of Mathematics, vol. 38(1937), pp. 809–818.
2. H. Whitney, *Differentiability of the remainder term in Taylor's formula*, this Journal, vol. 10(1943), pp. 153–158.
3. H. Whitney, *Differentiable manifolds*, Annals of Mathematics, vol. 37(1936), pp. 645–680.

Harvard University.

ON THE EXTENSION OF DIFFERENTIABLE FUNCTIONS

HASSLER WHITNEY

The author has shown previously how to extend the definition of a function of class C^m defined in a closed set A so it will be of class C^m throughout space (see [1]).[1] Here we shall prove a uniformity property: If the function and its derivatives are sufficiently small in A, then they may be made small throughout space. Besides being bounded, we assume that A has the following property:

(P) There is a number ω such that any two points x and y of A are joined by an arc in A of length less than or equal to ωr_{xy} (r_{xy} being the distance between x and y).

This property was made use of in [2]; its necessity in the theorem is shown by two examples below.

A second theorem removes the boundedness condition in the first theorem, and weakens the hypothesis (*P*); its proof makes use of the proof of the first theorem. We remark that in each theorem, as in [1], the extended function is a linear functional of its values in A.

The proof of Theorem I is obtained by examining the proof in [1]; hence we assume that the reader has this paper before him, and we shall follow its notations closely.

THEOREM I. *Let A be a bounded closed set in n-space E with the property* (P), *and let m be a positive integer. Then there is a number α with the following property. Let $f(x)$ be any function of class C^m in A, with derivatives $f_k(x)$ $(\sigma_k = k_1 + \cdots + k_n \leqq m)$. Suppose*

$$|f_k(x)| < \eta \qquad (x \in A, \sigma_k \leqq m).$$

Then $f(x)$ may be extended throughout E so that

$$|f_k(x)| < \alpha\eta \qquad (x \in E, \sigma_k \leqq m).$$

Let d be the diameter of A, or 1 if this is larger, and let R be a spherical region of radius $2d$ with its center at a point of A. Set $f(x) = 0$ in $E - R$. Then the extension of f in $R - A$ given in [1] will be shown to have the property, using

$$\alpha = 2n(m!)^n(m+1)^{3n}(433n^{1/2}d\omega)^m cN,$$

where N and c are as given in [1, §§11, 12]. Note that $433 = 4 \cdot 108 + 1$.

Presented to the Society, September 13, 1943; received by the editors November 27, 1943.

[1] Numbers in brackets refer to the references cited at the end of this paper.

Set $B=A\cup(\bar{R}-R)$. We show first that for any points x', x'' of B,

$$|R_k(x';x'')|<\beta r_{x'x''}^{m-\sigma_k}\eta,\qquad \beta=2n(m+1)^n\omega^m.$$

Suppose first that x' and x'' are in A. Let C be a curve in A joining them, of length less than or equal to $\omega r_{x'x''}$. The inequality is then a consequence of [2, Lemma 3]. Suppose next that one of the points is in A, and the other is in $\bar{R}-R$ (the case that both are in $\bar{R}-R$ is trivial). By [1, (3.1)], since $r_{x'x''}\geqq d\geqq 1$,

$$|R_k(x';x'')|\leqq\eta+\sum_{\sigma_l\leqq m-\sigma_k}\eta r_{x'x''}^{\sigma_l}\leqq r_{x'x''}^{m-\sigma_k}\eta\left[1+\sum_{\sigma_l\leqq m-\sigma_k}1\right]$$

$$\leqq(m+1)^n r_{x'x''}^{m-\sigma_k}\eta.$$

Now take any x in $R-B$. Let $\delta^*/4$ be the distance from x to B, and let x^* be a point of B distant $\delta^*/4$ from x. Say x is in the cube C of the set of cubes K_s; let $I_{\lambda_1},\cdots,I_{\lambda_t}$ be those I_λ with points in C (see [1, §11]). Now y^ν is the center of I_ν, and x^ν is a nearest point of B to y^ν. As noted in [1, (9.1)], $r_{y^\nu x^*}$ and $r_{y^\nu x^\nu}$ each lie between $\delta^*/8$ and $\delta^*/2$. Since $r_{xy^\nu}<\delta^*/2$, we have

$$r_{x^\nu x^*}<\delta^*,\qquad r_{xx^\nu}<\delta^*.$$

The definition of ζ in [1, §11] together with [1, (6.3)] gives

$$\zeta_{\nu;k}(x)=\psi_k(x;x^\nu)-\psi_k(x;x^*)=\sum_{\sigma_l\leqq m-\sigma_k}\frac{R_{k+l}(x^\nu;x^*)}{l!}(x-x^\nu)^l.$$

Hence

$$|\zeta_{\nu;k}(x)|<(m+1)^n\beta r_{x^\nu x^*}^{m-\sigma_k-\sigma_l}r_{xx^\nu}^{\sigma_l}\eta<(m+1)^n\beta\delta^{*m-\sigma_k}\eta.$$

Following [1, §11] still, we find

$$|D_kg(x)-\psi_k(x;x^*)|<c\sum_{\sigma_l\leqq m-\sigma_k}(m!)^n2^{s\sigma_l}N(m+1)^n\beta\delta^{*m-\sigma_k+\sigma_l}\eta.$$

As in [1], $2^s<108n^{1/2}/\delta^*$; hence

$$|D_kg(x)-\psi_k(x;x^*)|<c(m!)^nN(m+1)^{2n}(108n^{1/2})^m\beta\delta^{*m-\sigma_k}\eta.$$

Moreover, since $r_{xx^*}<3d$, [1, (6.1)] gives

$$|\psi_k(x;x^*)|<3^m(m+1)^nd^m\eta.$$

Since $\delta^*\leqq 4d$ and $f(x)=g(x)$ in $R-B$, the theorem follows.

We turn now to the second theorem. We shall say A *satisfies* (P) *locally* if for each $x\in A$ there is a neighborhood U of x and a number

ω such that any two points y and z of $A \cap U$ are joined by an arc in A of length not greater than ωr_{zy}.

THEOREM II. *Let A be a closed subset of an open set R in E, satisfying (P) locally, and let m be a positive integer. Then for any continuous function $\epsilon(x)$ defined and greater than 0 in R there is a continuous function $\delta(x)$ defined and greater than 0 in A with the following property. Let $f(x)$ be any function of class C^m in A, such that*

$$|f_k(x)| < \delta(x) \qquad (x \in A, \sigma_k \leqq m).$$

Then $f(x)$ may be extended throughout R so that

$$|f_k(x)| < \epsilon(x) \qquad (x \in R, \sigma_k \leqq m).$$

REMARKS. The preceding theorem is easily seen to be a consequence of this one. The present theorem holds if E is replaced by a differentiable manifold M, in which a fixed set of coordinate systems (each one intersecting but a finite number of others) is used to measure the size of derivatives. To show this, we imbed M in a Euclidean space E' (see [3, Theorem 1]), giving $A \subset R \subset R' \subset E'$ (R' open in E'; we let R' contain no points of the limit set of M), extend f throughout R' (see the proof of [3, Lemma 4]), and consider its values in R.

To prove the theorem, we begin by choosing spherical regions $U_1, U_2, \cdots$, each $\overline{U}_i \subset R$, with the following properties:

(a) Each U_i is in a neighborhood U as described above.

(b) Each $\overline{U}_i$ intersects but a finite number of other $\overline{U}_j$.

(c) If U_i is of radius ρ_i, and U_i' is the concentric region of radius $\rho_i/2$, then $R = \sum U_i'$.

Let $\psi^i(x)$ be a function of class C^m in E such that

$$\psi^i(x) > 0 \qquad (x \in U_i'),$$
$$\psi^i(x) = 0 \qquad (x \in E - U_i').$$

Set

$$\phi^i(x) = \psi^i(x)/\sum \psi^j(x) \qquad (x \in R);$$

then $\phi^i(x)$ is of class C^m in R, and

$$\phi^i(x) = 0 \qquad (x \in R - U_i'),$$
$$\sum \phi^i(x) = 1 \qquad (x \in R).$$

The extension of $f(x)$ is defined as follows. Set

$$f^i(x) = \phi^i(x) f(x) \qquad (x \in A),$$
$$f^i(x) = 0 \qquad (x \in R - U_i).$$

Then f^i is of class C^m in $A \cup (R - U_i)$. Extend it as in [1] (using a fixed subdivision of U_i-A; we could set $f^i(x)=0$ in $E-R$) to be of class C^m in R (or E). (Note that if $A\cap U_i'=0$, then $f^i(x)=0$, $x\in R$.) Set

$$f(x)=\sum f^i(x) \qquad (x\in R).$$

Then f is an extension of class C^m of its values in A. We must show that it satisfies the condition of smallness.

Choose $a_i\geqq 1$ so that

$$\left|\overset{i}{\phi}_k(x)\right|\leqq a_i \qquad (x\in R,\ \sigma_k\leqq m),$$

then if $|f_k(x)|<\eta$ $(x\in A\cap U_i')$,

$$\left|\overset{i}{f}_k(x)\right|=\left|\sum_{\sigma_l\leqq\sigma_k}\overset{i}{\phi}_l(x)f_{k-l}(x)\right|\leqq(m+1)^n a_i\eta \qquad (x\in A).$$

By the choice of U_i, there is an ω_i such that any x' and x'' in $A\cap U_i$ are joined by an arc in A of length not greater than $\omega_i r_{x'x''}$. Set $\sigma_i=\max(1,2/\rho_i)$. If R_k^i is the remainder for f_k^i, we shall show that for any x' and x'' in $A\cup(R-U_i)$,

$$\left|\overset{i}{R}_k(x';x'')\right|<2n(m+1)^{2n}\omega_i^m a_i\sigma_i^m r_{x'x''}^{m-\sigma_k}\eta.$$

If x' and x'' are both in U_i, we apply [2, Lemma 3]. If $x'\in R-U_i$ and $x''\in U_i'$, or vice versa, then $r_{x'x''}\geqq\rho_i/2$, and the proof in the preceding theorem applies; we consider separately the cases $\rho_i/2\geqq 1$, $\rho_i/2<1$, using $r_{x'x''}\geqq 1$ and $\sigma_i r_{x'x''}\geqq 1$ respectively. If $x'\in R-U_i$ and $x''\in R-U_i'$, or vice versa, $R_k^i=0$, since $\phi_l^i(x')=\phi_l^i(x'')=0$. The proof of the preceding theorem now shows that for some α_i, if

$$\left|f_k(x')\right|<\eta \qquad (x'\in A\cap U_i',\ \sigma_k\leqq m),$$

then

$$\left|\overset{i}{f}_k(x)\right|<\alpha_i\eta \qquad (x\in R,\ \sigma_k\leqq m).$$

(We may set $\alpha_i=1$ if $A\cap U_i'=0$.)

Given $\epsilon(x)$, we determine $\delta(x)$ as follows. For each $x\in R$ there is a set of numbers $\lambda_1,\cdots,\lambda_s$, $s=s(x)$, such that

$$x\in\text{each } U_{\lambda_j}', \qquad x\in\text{no other } U_i'.$$

Because of (b), s is finite. Set $\alpha(x)=\alpha_{\lambda_1}+\cdots+\alpha_{\lambda_s}$.

We can clearly choose a continuous function $\beta(x)$ in R such that

$$\alpha(x)<\beta(x) \qquad (x\in R).$$

We may now choose a continuous function $\delta(x')>0$ in A such that for any $x' \in A$, if $U'_{\mu_1}, \cdots, U'_{\mu_t}$ are the U'_j containing x', then

$$\delta(x') \leqq \epsilon(x)/\beta(x) \quad (x \in U'_{\mu_1} \cup \cdots \cup U'_{\mu_t}).$$

Now take any f of class C^m in A, with $|f_k(x)| < \delta(x)$ $(x \in A, \sigma_k \leqq m)$; the extension of f through R has been defined. Take any $x \in R$; define $\lambda_1, \cdots, \lambda_s$ as above. Then

$$|f_k(x')| < \delta(x') \leqq \epsilon(x)/\beta(x) \qquad (x' \in A \cap U'_{\lambda_j}, \sigma_k \leqq m),$$

and hence

$$|f_k^{\lambda_j}(x)| < \alpha_{\lambda_j}\epsilon(x)/\beta(x) \qquad (\sigma_k \leqq m).$$

Since $f_k(x) = f_k^{\lambda_1}(x) + \cdots + f_k^{\lambda_s}(x)$,

$$|f_k(x)| < \alpha(x)\epsilon(x)/\beta(x) < \epsilon(x)$$

for $\sigma_k \leqq m$, which completes the proof.

EXAMPLES. (1) Let A consist of a point, together with a sequence of points approaching it. Letting $f(x) = 1$ at a finite number of points of the sequence, and $f(x) = 0$ in the rest of A shows (with $m = 1$) that the theorem fails here.

(2) Let A be the closed region of the plane defined by (a) $x^2 + y^2 \leqq 1$, and (b) either $x \leqq 0$ or $|y| \geqq x^{3/2}$. Let $f(x, y) = 0$ if $x \leqq 0$, and set

$$f(x, y) = \begin{cases} \gamma x^2/(1 + \gamma^2 x^2) & \text{if } x \geqq 0,\ y > 0, \\ -\gamma x^2/(1 + \gamma^2 x^2) & \text{if } x \geqq 0,\ y < 0. \end{cases}$$

We see easily that f is of class C^1 in A. (It would not be if, in (b), we used $|y| \geqq x^2$.) The maximum $\partial f/\partial x$ occurs at $x = 1/(3^{1/2}\gamma)$, and has the value $9/(8 \cdot 3^{1/2})$. Set

$$p = (1/3^{1/2}\gamma,\ 1/3^{3/4}\gamma^{3/2}), \qquad q = (1/3^{1/2}\gamma,\ -1/3^{3/4}\gamma^{3/2}).$$

Then

$$\frac{f(p) - f(q)}{r_{pq}} = \frac{2\gamma/3\gamma^2}{1 + \gamma^2/3\gamma^2} \div \frac{2}{3^{3/4}\gamma^{3/2}} = \frac{3^{3/4}}{4}\gamma^{1/2}.$$

Hence, in any extension of f through the plane, we must have $|\partial f/\partial y| \geqq 3^{3/4}\gamma^{1/2}/4$ at some point (between p and q); yet $|f|$, $|\partial f/\partial x|$ and $|\partial f/\partial y|$ are uniformly bounded for all $\gamma > 1$. Taking γ arbitrarily large shows that the conclusion of the theorem does not hold.

REFERENCES

1. H. Whitney, *Analytic extensions of differentiable functions defined in closed sets*, Trans. Amer. Math. Soc. vol. 36 (1934) pp. 63–89.

2. ———, *Functions differentiable on the boundaries of regions*, Ann. of Math. vol. 35 (1934) pp. 482–485.
3. ———, *Differentiable manifolds*, Ann. of Math. vol. 37 (1936) pp. 645–680.
4. ———, *Differentiable functions defined in arbitrary subsets of Euclidean space*, Trans. Amer. Math. Soc. vol. 40 (1936) pp. 309–317. Further references are given here.
5. H. O. Hirschfeld, *Continuation of differentiable functions through the plane*, Quart. J. Math. Oxford Ser. vol. 7 (1936) pp. 1–15.
6. M. R. Hestenes, *Extension of the range of a differentiable function*, Duke Math. J. vol. 8 (1941) pp. 183–192.

HARVARD UNIVERSITY

ON IDEALS OF DIFFERENTIABLE FUNCTIONS.*

By HASSLER WHITNEY.

1. Introduction. We shall consider functions of class C^r defined in an open subset G of Euclidean n-space E^n, that is, functions with continuous partial derivatives through the order r in G. (We take r finite.) A set I of such functions forms an *ideal* if, whenever f_1, f_2 are in I and ϕ_1, ϕ_2, are of class C^r in G, $\phi_1 f_1 + \phi_2 f_2$ is in I. We shall call I an *r-ideal.*

Let I be an r-ideal in G. Given any point $x = (x_1, \cdots, x_n)$ in G, consider all sets of numbers

$$a_{\alpha_1 \ldots \alpha_n} \qquad \text{(each } \alpha_i \geqq 0,\ \alpha_1 + \cdots + \alpha_n \leqq r)$$

such that for each such set there is a function f in I with

$$\frac{\partial^{\alpha_1 + \cdots + \alpha_n} f(x)}{\partial x_1^{\alpha_1} \cdots \partial x_n^{\alpha_n}} = a_{\alpha_1 \ldots \alpha_n}.$$

These sets form the *local ideal* $\mathfrak{J}_r(I, x)$ *of* I *at* x.

It was conjectured by Laurent Schwartz (personal communication) that an ideal is determined by its set of local ideals, provided that the ideal is closed (we use the topology described below). The main object of this paper is to prove this conjecture (see Theorem I). There is a rather obvious generalization of the theorem to the case where E^n is replaced by a manifold of class C^r.

The case $r = 0$ is much simpler than the case $r > 0$; we treat it separately in **3**.

In the last section we consider briefly the following problem: What sets of local ideals can be the set of local ideals of an r-ideal? A satisfactory answer to this question seems quite difficult to find.

The topology we shall use is determined as follows. Let f be a function of class C^r in G. Then to any compact subset A of G (i. e. bounded closed subset A of E^n lying in G), and any $\epsilon > 0$, the set of all functions g of class C^r in G such that

$$\left| \frac{\partial^{\alpha_1 + \cdots + \alpha_n}(g - f)}{\partial x_1^{\alpha_1} \cdots \partial x_n^{\alpha_n}} \right| < \epsilon \text{ in } A \qquad (\alpha_1 + \cdots + \alpha_n \leqq r)$$

---

* Received November 25, 1947.

form a "special neighborhood" of f. A set of functions forms an open set if it can be expressed as the union of a set of special neighborhood of functions.

2. The principal theorems. The main theorem is:

THEOREM 1. *Any two closed r-ideals in G with the same sets of local ideals are identical.*

This is an immediate consequence of the following approximation theorem. For the statement of this theorem, we need some more definitions.

We say that a set of functions forms a *locally finite set of functions in G* if the following are true:

(a) Each function is identically zero outside some compact subset of G.

(b) Each point of G is in a neighborhood such that at most a finite number of the functions fail to vanish identically in this neighborhood.

Because of the Heine-Borel theorem, (b) holds with neighborhoods of points replaced by arbitrary compact subsets of G.

If $f_1, f_2, \cdots$ form a locally finite set, and the f_i are of class C^r in G, clearly $f_1 + f_2 + \cdots$ exists and is of class C^r in G.

If I is an r-ideal in G, we define its *locally finite completion* I^* as follows. A function f is in I^* if there is an infinite series of functions of I which converges to f, these functions forming a locally finite set.

It is easily seen that I^* is an ideal. It is not hard to show that $I \subset I^*$; we do not need this fact.

THEOREM 2. *Let I be an r-ideal in G, and let F be a function of class C^r in G such that at each x in G, the values of F and its derivatives form a set of numbers in $\mathfrak{J}_r(I, x)$. Then for each positive continuous function $\epsilon(x)$ in G there is a function f^* in I^* such that*

$$\left| \frac{\partial^{\alpha_1+\cdots+\alpha_n}[F(x) - f^*(x)]}{\partial x_1^{\alpha_1} \cdots \partial x_n^{\alpha_n}} \right| < \epsilon(x), \quad x \in G, \quad \alpha_1 + \cdots + \alpha_n \leqq r.$$

We prove Theorem 1 with the help of Theorem 2 as follows. Let I and I' be two closed r-ideals in G as stated. To show that they are identical, we show that each is contained in the other; for instance, that $I' \subset I$. Take any function F in I'. To prove $F \varepsilon I$ it is sufficient to find a function f of I in an arbitrary special neighborhood U of F, since I is closed. Let U be defined by the compact subset A of G and the number $\epsilon > 0$. Set $\epsilon(x) \equiv \epsilon$, and choose f^* by Theorem 2; say $f^* = g_1 + g_2 + \cdots$. As noted above, at most a finite number of these are $\not\equiv 0$ in A; hence

$$f^* = g_1 + \cdots + g_m \text{ in } A \text{ for some } m.$$

Set $f = g_1 + \cdots + g_m$ in G. Then $f \varepsilon I \cap U$, as required.

Most of the paper is devoted to the proof of Theorem 2. The methods used are similar to those employed in extending differentiable functions.[1]

REMARK. The hypothesis of closure in Theorem 1 is necessary, as shown by the first two examples of **4.**

3. The case $r = 0$. This case is described in full by the following theorem.

THEOREM 3. *A closed ideal I of continuous functions in G is determined by the set of points A where all the functions vanish. This set is closed in G. Any closed subset of G can occur.*

The set of points where one function of I vanishes is closed, since the function is continuous. The set A is the intersection of such sets, and hence is closed.

Let A be any closed subset of G. Let $f(x)$ be the distance from any point x of G to A; this is continuous in G, and vanishes in A only. Let I be the ideal generated by f, i. e. the set of all functions ϕf, ϕ continuous. Clearly A is the set of points where all functions of I vanish.

To prove the first part of the theorem, we prove Theorem 2 for this case. Let, then, I and A be as in Theorem 3, and let F be a continuous function such that $F(x) = 0$ in A. Let $\epsilon(x)$ be any positive continuous function in G.

To each x in $G - A$ there corersponds a function f_x in I such that $f_x(x) \neq 0$. By multiplying by a suitable continuous function, we may replace f_x by a function f'_x in I such that for some spherical neighborhood U_x of x (we take $\bar{U}_x \subset G - A$),

$$f'_x(x') > 0 \text{ in } U_x, \qquad f'_x(x') = 0 \text{ in } G - U_x.$$

It is easy to choose $x^1, x^2, \cdots$ so that the spheres $U_{x^1}, U_{x^2}, \cdots$ cover $G - A$, and each intersects but a finite number of the others. Set

$$g'_i(x) = f'_{x^i}(x) / \sum_j f'_{x^j}(x) \text{ in } U_{x^i}, g'_i(x) = 0 \text{ in } G - U_{x^i}.$$

Since the numerator is $\equiv 0$ in $G - U_{x^i}$ and the denominator is > 0 in $G - A$, g'_i is continuous. In fact, we may write $g'_i = \phi_i f'_{x^i}$, where

[1] See H. Whitney, "Analytic extensions of differentiable functions defined in closed sets," *Transactions of the American Mathematical Society*, vol. 36 (1934), pp. 63-89. We refer to this paper as AE.

12

$\phi_i = 1/\Sigma f'_{x^j}$ in U_{x^i} and $\phi_i = 0$ outside a larger neighborhood of x^i, showing that g'_i is in I. Note that

$$0 \leqq g'_i(x) \leqq 1, \qquad \sum_i g'_i(x) = 1 \text{ in } G - A.$$

Let U be a neighborhood of A such that

$$|F(x)| < \epsilon(x) \text{ in } U.$$

Let $\lambda_1, \lambda_2, \cdots$ be those numbers such that for any $k = \lambda_j$, U_{x^k} has points in $G - U$. Set

$$g_i(x) = F(x) g'_{\lambda_i}(x) \qquad (i = 1, 2, \cdots).$$

Then these form a locally finite set of functions in G, each lying in I. If $f^* = \Sigma g_i$, clearly $f^* = F(x)$ in $G - U$, while for $x \varepsilon U$,

$$|F(x) - f^*(x)| = |F(x)|\,|1 - \Sigma g'_{\lambda_i}(x)| \leqq |F(x)| < \epsilon(x),$$

completing the proof.

4. Examples. We give first two examples of ideals which are not closed.

Example 1. Let I be the set of all functions of one variable, of class C^r, each of which vanishes except in some bounded set. Let $\phi(x) \equiv 1$; ϕ is not in I. Given any bounded set A, let f be a function $\equiv 1$ in A and $\equiv 0$ outside some neighborhood of A; f is in I. Hence, if $\bar{I}$ is the closed ideal, ϕ is in $\bar{I}$. Any function F of class C^r equals $F\phi$, and hence is in $\bar{I}$.

Example 2. Let I consist of all functions of class C^r which vanish in some neighborhood of the origin. Then $\bar{I}$ contains all functions of class C^r which vanish together with their derivatives of order $\leqq r$, at the origin.

The next three examples serve to illustrate what local ideals may be like.

Example 3. $n = 2$, $r = 2$. Let I consist of all functions of class C^r which vanish on the x-axis; we consider the local ideal $\mathfrak{J}_r(I, O)$ of I at the origin O. For f in I, the only necessary relations between derivatives at the origin are

$$f(O) = 0, \qquad \partial f(O)/\partial x = 0, \qquad \partial^2 f(O)/\partial x^2 = 0.$$

Consider the polynomials (which are in I)

$$P_1(x, y) = y, \qquad P_2(x, y) = xy, \qquad P_3(x, y) = y^2.$$

Any function f in I has the same value and derivatives of order $\leqq 2$ at O

as those of some linear combinations of these polynomials; these polynomials (or rather, their sets of values and derivatives) form a base for $\mathfrak{J}(I, O)$.

Note that the local ideal is generated by $P_1 = y$; for $P_2 = xP_1$, $P_3 = yP_1$, and thus P_1 being in I implies that P_2 and P_3 are also.

A new ideal would be obtained by adding the condition $\partial f/\partial y = 0$ at the origin.

EXAMPLE 4. $n = 2$, $r = 2$. Let I consist of all functions of class C^2 which vanish on both axes. Then for any $x^0 \neq O$ on either axis, $\mathfrak{J}(I, x^0)$ is like the local ideal of Example 3. At the origin, $\mathfrak{J}(I, O)$ consists of the set of values 0 alone, except that $\partial^2 f/\partial x \partial y$ may be $\neq 0$; hence the polynomial xy forms a base for $\mathfrak{J}_r(I, O)$.

If we wish the functions to vanish on the x-axis and the line $y = x$, transforming axes (compare Example 5 below) shows that the local ideal is then generated by $xy - y^2$.

EXAMPLE 5. $n = 2$, $r = 2$. Let I now be the set of functions of class C^2 which vanish on the curve $y = x^2$. If we use the variables x', y' in Example 3, we obtain Example 5 by the transformation $x' = x$, $y' = y - x^2$. Calculating derivatives of $f \in I$ gives, referring to Example 3,

$$\partial f(O)/\partial x' = \partial f(O)/\partial x = 0, \qquad \partial^2 f(O)/\partial x'^2 = 2\,\partial f(O)/\partial y + \partial^2 f(O)/\partial x^2 = 0.$$

The relations defining $\mathfrak{J}_r(I, O)$ are obtained by setting these expressions and $f(O)$ equal to 0. The polynomials

$$P_1 = y - x^2, \qquad P_2 = xy, \qquad P_3 = y^2$$

form a base for $\mathfrak{J}_r(I, O)$, and the local ideal is generated by P_1. As far as $\mathfrak{J}_r(I, O)$ is concerned, P_2 and P_3 are equivalent to xP_1 and yP_1 respectively (see **6**).

The next two examples illustrate the dependence of a local ideal on neighboring values of the functions.

EXAMPLE 6. $n = 2$, $r = 1$. Let A be a point set defined as follows. It contains the origin, and the closed intervals $1/2^{i+1} \leqq x \leqq 1/2^i$ $(y = 0)$ for $i = 1, 3, 5, \cdots$. The omitted intervals $(i = 2, 4, \cdots)$ are replaced by the circular arcs with the same end points, each arc lying in the half plane $y \geqq 0$ and being of radius 1. Let I consist of all functions of class C^1 which vanish on A.

Any function in I has vanishing partial derivatives at the ends of each interval; hence the same is true at the origin. This is in spite of the fact that the direction between any two points of A sufficiently near to O is as nearly in the x-direction as we please.

EXAMPLE 7. $n=2$, $r=1$. Let us round off the corners in the point set A of Example 6. Then I contains a function f such that $\partial f(O)/\partial y=1$. If we change the original A sufficiently slightly, $\partial f/\partial y$ cannot satisfy a Lipschitz condition.

The last two examples show how the functions in ideals may lack differentiability properties.

EXAMPLE 8. $n=2$, $r=1$. Let I be generated by

$$f(x,y)=[1+(x^2+y^2)^{1/2}]y;$$

then I consists of all functions ϕf, where ϕ is of class C^1. Since $f=0$ only on the x-axis, and $\partial f/\partial y\neq 0$ there, the local ideals are the same as for the ideal I' generated by $f'(x,y)=y$. Since each of f, f' is obtained from the other by multiplying by a function which is not of class C^1, $I\neq I'$.

EXAMPLE 9. $n=3$, $r=1$. Let I be generated by the two functions

$$f(x,y,z)=y+[1+(x^2+y^2+z^2)^{1/2}]z,$$
$$g(x,y,z)=y^2+z^2.$$

Then the set of points where all functions of I vanish is the x-axis. Differentiating shows that

$$\partial f/\partial z=(1+|x|)(\partial f/\partial y) \text{ on the } x\text{-axis.}$$

Since this is trivially true with f replaced by g, we see at once that this relation holds for all functions of I. It follows that there is no function of class C^2 in I for which $\partial f/\partial y\neq 0$ at the origin. The same is of course true in the closed ideal. (A similar remark applies to the last example.)

5. Notations. We shall commonly use single letters to denote n-fold quantities; for $\alpha=(\alpha_1,\cdots,\alpha_n)$, write

$$[\alpha]=\alpha_1+\cdots+\alpha_n.$$

We use D_α for $\partial^{[\alpha]}/\partial x_1^{\alpha_1}\cdots\partial x_n^{\alpha_n}$ etc. Thus the right hand side of the second displayed relation in **6**, written out in full, is

$$\sum_{\alpha_1+\cdots+\alpha_n\leqq r}\frac{a_{\alpha_1\ldots\alpha_n}}{\alpha_1!\cdots\alpha_n!}(x_1-x_1^0)^{\alpha_1}\cdots(x_n-x_n^0)^{\alpha_n}.$$

We shall use ρ_{xy} or $\rho(x,y)$ for the distance between x and y; similarly for $\rho_{xA}=\rho(x,A)$ where A is a point set. We set $\rho(x,A)=\infty$ if A is void.

By $U_\delta(x)$, or $U_\delta(A)$, we denote the set of points within δ of the point x or the point set A.

Let f be a function of class C^r. Its *r-value* $V_r(f, x)$ at x is the set of numbers consisting of its value and partial derivatives of order $\leqq r$ at x. Two functions f, g are r-equivalent at x if they have the same r-value there; we write

$$f \approx_r g \text{ at } x, \quad \text{or,} \quad V_r(f, x) = V_r(g, x).$$

We shall call the *grade* of a function f at x^0 the smallest number i such that some partial derivative of f of order i is $\neq 0$ at x^0, provided there is such an i. The grade of a polynomial at the origin is then the lowest degree of its terms if the polynomial is not $\equiv 0$. The grade of an r-value is that of any corresponding function.

6. Local ideals as linear spaces. Given a point x^0, any set a of numbers $\{a_\alpha\}$, $[\alpha] \leqq r$, considered as associated with x^0, is the r-value of a uniquely defined polynomial P of degree $\leqq r$. P must satisfy the condition

$$V_r(P, x^0) = a, \text{ i. e. } D_\alpha P(x^0) = a_\alpha, \ [\alpha] \leqq r.$$

P is given by the formula

$$P(x) = \sum_{[\alpha] \leqq r} (a_\alpha / \alpha!)\,(x - x^0)^\alpha$$

The addition and multiplication of r-values corresponds to the same for polynomials, provided we drop out terms of degree $> r$ when they arise, i. e. reduce mod polynomials of grade $> r$. We may thus represent the r-value by the corresponding polynomial.

As an illustration of the multiplication, at the end of Example 5, **4**, we noted that

$$x(y - x^2) = xy - x^3 \approx_2 xy, \qquad y(y - x^2) \approx_2 y^2, \text{ at } O.$$

We note in passing that x and $x + xy$ generate the same local ideal at the origin: $\mathfrak{J}_r(x, O) = \mathfrak{J}_r(x + xy, O)$. For if x is in an ideal, so is yx, and hence also $x + xy$. Conversely, if $x + xy$ is in an ideal, so are

$$x + xy, \qquad y(x + xy) = xy + xy^2, \cdots, xy^{r-1} + xy^r;$$

hence so is the alternating sum, which is $x \pm xy^r \approx_r x$.

For each point x^0, consider the classes of functions of C^r, two functions being in the same class if they are r-equivalent at x^0. These add and multiply like r-values; hence they may be used to represent r-values. If we are in a differentiable manifold, they may be used to define r-values and local ideals.

In the proofs of the theorems of this paper, it is addition and multiplication by constants in local ideals which play the essential role. With these operations, a local ideal forms a finite dimensional linear space. A *base* in a local ideal is then a set of r-values in the local ideal such that any r-value in the local ideal can be expressed uniquely as a linear combination of these (with numerical coefficients). A set of r-values forms a base if and only if they are independent and their number equals the dimension of the ideal (as a linear space).

The set of all r-values forms the *unit* local ideal $\mathfrak{J}_r{}^1(x^0)$ at x^0; the set containing the zero r-value (i. e. $a_\alpha = 0$ for each α) only forms the *zero* local ideal $\mathfrak{J}_r{}^0(x^0)$ at x^0. Any local ideal at x^0 contains the second and is contained in the first. If an ideal contains an r-value a with $a_{0\ldots 0} \neq 0$, it is the unit ideal, clearly.

7. On the structure of local ideals. The results of this section will not be needed in the proof of the theorems of the paper.

Consider a point x^0 and an integer s, $0 \leqq s \leqq r$. The set of all r-values $\{a_\alpha\}$ such that $a_\alpha = 0$ if $[\alpha] \neq s$ forms a subspace of $\mathfrak{J}_r{}^1(x^0)$; its dimension is easily seen to be $\binom{n-1+s}{n-1}$. These subspaces are independent and generate $\mathfrak{J}_r{}^1(x^0)$; the dimension of the whole space is $\binom{n+r}{n}$.

Now take any local r-ideal H at x^0. Let H^s be the subset of H consisting of all r-values in H of grade $\geqq s$. Let $\bar{H}^s$ denote the local s-ideal at x^0, formed from H^s by considering a_α with $[\alpha] \leqq s$ only (i. e. by reducing mod r-values of grade $> s$). Then $\bar{H}^s$ is a linear space; let us call its dimension the s-dimension $\dim_s(H)$ of H.

We can relate bases in the $\bar{H}^s$ to a base in H as follows.

LEMMA 7a. *Let H be a local r-ideal at x^0. For each s, $0 \leqq s \leqq r$, let $P_1{}^s, \cdots, P_{d_s}{}^s$ be r-values in H^s such that, when reduced mod r-values of grade $> s$, they form a base in $\bar{H}^s$. Then the $P_i{}^s$ form a base in H, and $d_s = \dim_s(H)$.*

The proof is straightforward, and may be omitted.

8. The sets A_λ, B_λ. We begin by associating with I certain point sets, as follows. Define:

$$A_\lambda = \text{all } x \text{ such that } \dim \mathfrak{J}_r(I, x) = \lambda,$$

$$B_\lambda = \sum_{\mu \geqq \lambda} A_\mu.$$

LEMMA 8a. *The B_λ are open in G.*

Take $x \varepsilon B_\lambda$; say $x \varepsilon A_\mu$, $\mu \geqq \lambda$. Now $\mathfrak{I}_r(I,x)$ is a linear space of dimension μ. Choose μ r-values in it which form a base. For each of these r-values, choose a function in I with this r-value at x^0. Let $f_1, \cdots, f_\mu$ be these functions. Then $V_r(f_1,x^0), \cdots, V_r(f_\mu, x^0)$ are independent. Since the f_i are of class C^r, $V_r(f_i,x^0)$ is a continuous function of x^0; it follows that $V_r(f_1,x), \cdots, V_r(f_\mu,x)$ are independent, x near x^0. This shows that for some neighborhood U of x^0, $\dim \mathfrak{I}_r(I,x) \geqq \mu$ for $x \varepsilon U$, i. e., $U \subset B_\mu \subset B_\lambda$, proving that B_λ is open.

9. Local functions. Let I and F be as in the theorem. For each $x \varepsilon G$ there is a function f in I such that $V_r(f,x) = V_r(F,x)$; it follows from Taylor's formula with remainder that for each $\epsilon > 0$ there is a $\delta > 0$ such that

$$| D_\alpha[f(y) - F(y)] | \leqq \rho_{xy}{}^{r-[\alpha]}\epsilon, \qquad \rho_{xy} < \delta, \ [\alpha] \leqq r.$$

However, δ may depend on x; no uniformity condition holds in general. But if we restrict ourselves to the individual sets A_λ, we can prove a uniformity condition, as shown by the lemmas below.

LEMMA 9a. *For any $x \varepsilon A_\lambda$ and any $\epsilon > 0$ there is a $\delta > 0$ with the following property. For each y in A_λ within δ of x there is a function in I such that if $[\alpha] \leqq r$,*

$$(9.1) \qquad | D_\alpha[f(z) - F(z)] | \leqq \rho_{yz}{}^{r-[\alpha]}\epsilon, \qquad \rho_{xz} < \delta,$$

$$(9.2) \qquad D_\alpha f(y) = D_\alpha F(y).$$

First choose functions $f_1, \cdots f_\lambda$ in I such that $V_r(f_1,x), \cdots, V_r(f_\lambda,x)$ form a base in $\mathfrak{I}_r(I,x)$. Let U be a neighborhood of x such that the r-values of $f_1, \cdots f_\lambda$ are independent at any $y \varepsilon U$. Now take any $y \varepsilon U \cap A_\lambda$. Then the r-values of $f_1, \cdots f_\lambda$ form a base there; hence there are uniquely defined numbers $u_1(y), \cdots, u_\lambda(y)$ such that

$$V_r[u_1(y)f_1 + \cdots + u_\lambda(y)f_\lambda, y] = V_r(F,y), \qquad y \varepsilon U \cap A_\lambda.$$

If we set

$$(9.3) \qquad g_y(z) = u_1(y)f_1(z) + \cdots + u_\lambda(y)f_\lambda(z),$$

this means that

$$(9.4) \qquad D_\alpha g_y(y) = D_\alpha F(y), \qquad y \varepsilon U \cap A_\lambda, \ [\alpha] \leqq r.$$

Choose $\delta_1 > 0$ and N so that

$$| u_i(y) | \leqq N, \qquad y \varepsilon A_\lambda, \ \rho_{xy} < \delta_1.$$

Set $\epsilon_1 = \epsilon/(1+\lambda N)$. By Taylor's formula with remainder, there is a $\delta_2 \leqq \delta_1$ such that, for each k and α,

$$| D_\alpha f_k(z) - \sum_\beta (D_{\alpha+\beta} f_k(y)/\beta!)(z-y)^\beta | \leqq \rho_{yz}{}^{r-[\alpha]}\epsilon_1, \qquad \rho_{xy}, \rho_{xz} < \delta_2.$$

Since $D_\gamma g_y(z) = \sum_k u_k(y) D_\gamma f_k(z)$, this gives

$$\begin{aligned} &| D_\alpha g_y(z) - \sum_\beta (D_{\alpha+\beta} g_y(y)/\beta!)(z-y)^\beta | \\ &\quad = | \sum_k u_k(y) [D_\alpha f_k(z) - \sum_\beta (D_{\alpha+\beta} f_k(y)/\beta!)(z-y)^\beta] | \\ &\quad \leqq \lambda N \rho_{yz}{}^{r-[\alpha]}\epsilon_1, \qquad \rho_{xy}, \rho_{xz} < \delta_2. \end{aligned}$$

Again by Taylor's formula, there is a $\delta \leqq \delta_2$ such that

$$|D_\alpha F(z) - \sum_\beta (D_{\alpha+\beta} F(y)/\beta!)(z-y)^\beta | \leqq \rho_{yz}{}^{r-[\alpha]}\epsilon_1, \qquad \rho_{xy}, \rho_{xz} < \delta.$$

Using (9.4) with $\alpha+\beta$ in place of α and comparing the last two inequalities gives

$$| D_\alpha g_y(z) - D_\alpha F(z) | \leqq (1+\lambda N)\rho_{yz}{}^{r-[\alpha]}\epsilon_1, \qquad \rho_{xy}, \rho_{xz} < \delta.$$

Now take any y in A_λ within δ of x. Set $f(z) = g_y(z)$. Then f is in I, and (9.1) and (9.2) follow from the last relation and (9.4).

LEMMA 9b. *For each positive continuous function $\epsilon(x)$ in A_λ there is a positive continuous function $\eta(x)$ in A_λ with the following property. For each $x \in A_\lambda$ there is a function f in I such that*

$$(9.5) \qquad | D_\alpha[f(y) - F(y)] | \leqq \rho_{xy}{}^{r-[\alpha]}\epsilon(x), \qquad \rho_{xy} < \eta(x), \ [\alpha] \leqq r,$$

$$(9.6) \qquad D_\alpha f(x) = D_\alpha F(x), \qquad [\alpha] \leqq r,$$

Note that y need not lie in A_λ.

Define the positive continuous function $\epsilon_1(x)$ in A_λ by

$$\epsilon_1(x) = \underset{y \in U_\theta(x) \cap A_\lambda}{\text{g.l.b.}} \epsilon(y), \qquad \theta = \rho(x, E - B_\lambda)/2.$$

For any x in A_λ, let $\eta(x)$ be the upper bound of numbers η with the following property. There is a point $x' \in A_\lambda$ such that, using x' and $\epsilon = \epsilon_1(x')$ in Lemma 9a, there is a δ satisfying the condition of the lemma, and also the conditions

$$(9.7) \qquad U_{2\eta}(x) \subset U_\delta(x'), \qquad \delta < \rho(x', E - B_\lambda)/2.$$

Certainly $\eta(x) > 0$. For if we take $x' = x$, we may apply Lemma 9a to find a number $\eta \leqq \delta/2$. To show that $\eta(x)$ is continuous, it is sufficient to show that for any x_1 and x_2 in A_λ,

$$\eta(x_2) \geqq \eta(x_1) - \rho(x_1, x_2)/2.$$

Take any $\theta > 0$, take $\eta > \eta(x_1) - \theta$, and find an x' and δ corresponding to x_1 and η; then clearly x' and δ will also do for x_2, with η replaced by $\eta - \rho_{x_1x_2}/2$, which is arbitrarily near $\eta(x_1) - \rho_{x_1x_2}/2$.

Now take any $x \varepsilon A_\lambda$. Set $\eta = \eta(x)/2$. Using the definition of $\eta(x)$, find x' and δ correspondingly. By (9.7), x is within δ of x'; using $y = x$, let f be the corresponding function in I. Then (9.2) becomes (9.6), while (9.1) gives

$$| D_a[f(z) - F(z)] | \leqq \rho_{xz}{}^{r-[a]}\epsilon_1(x'), \qquad \rho_{x'z} < \delta.$$

Now take any y within $\eta(x)$ of x. Then $y \varepsilon U_{2\eta}(x) \subset U_\delta(x')$; hence the above inequality holds with $z = y$. Moreover,

$$\rho(x, x') < \delta < \rho(x', E - B_\lambda)/2;$$

therefore $\epsilon_1(x') \leqq \epsilon(x)$, and (9.5) is proved.

10. The functions $\phi_\nu(x)$. In each of the next two sections we shall need a process of interpolation between functions of class C^r. For a given closed subset P of E, the functions will be defined in the open set $E - P$ (or in part of it). We follow essentially AE, §§ 7-10.

We shall use a "standard subdivision" of $E - P$, defined as follows. Divide E into n-cubes, each of side 1; let K_0 be the set of all these cubes whose distances from P are at least $6n^{1/2}$ (if there are any). In general, K_s is formed by subdividing all the cubes into cubes of side $1/2^s$, and choosing all those (if any) which do not lie in any of $K_0, \cdots, K_{s-1}$, and whose distances from P are at least $6n^{1/2}/2^s$.

It is easily seen that the distance from any cube Q of K_s to P is $< 18n^{1/2}/2^s$ if $s \geqq 1$, and that any cube of K_s is separated from any cube of K_{s+2} by at least four cubes of K_{s+1}.

Let $y^1, y^2, \cdots$ denote the vertices of all the cubes of $K_0 + K_1 + \cdots$, arranged in a sequence. For each ν, let x_ν be a point of P such that $\rho(x^\nu, y^\nu) = \rho(y^\nu, P)$. Let J_ν be the cube with center y^ν which is just large enough to contain all cubes Q of $K_0 + K_1 + \cdots$ with y^ν as a vertex.

Let Q_0 be a fixed n-cube. Let $\phi'(x)$ be a function of class C^r in E such that

$$\phi'(x) > 0 \text{ within } Q_0, \qquad \phi'(x) = 0 \text{ outside } Q_0.$$

By translating and contracting, we obtain functions $\phi'_\nu(x)$, positive within J_ν and zero outside J_ν. Setting

$$\phi_\nu(x) = \phi'_\nu(x)/\Sigma_\mu\phi_\mu(x)$$

gives functions of class C^r in E, such that ϕ_ν is positive within J_ν and is zero outside J_ν, and such that

$$\sum_\nu \phi_\nu(x) = 1, \qquad x \varepsilon E - P. \tag{10.1}$$

If J_ν is of side σ, then any derivative of order t of ϕ'_ν is $1/\sigma^t$ times the same derivative of ϕ' at a corresponding point. The definition of ϕ_ν in terms of the ϕ'_μ depends on the relative arrangements of the cubes J_μ about J_ν; there are but a finite number of arrangements. From these facts, we see easily (compare AE, § 10) that for some number $N \geqq 1$,

$$| D_\alpha\phi_\nu(x) | < 2^{s[\alpha]}N, \qquad x \varepsilon Q,\ Q \text{ in } K_s.$$

We shall replace this by another inequality. Take any $x \varepsilon Q$, Q in K_s, $s \geqq 1$. Since $6n^{1/2}/2^s \leqq \rho(Q,P) < 18n^{1/2}/2^s$, and diam $(Q) = n^{1/2}/2^s$, we have

$$\rho(x,P) \leqq \rho(Q,P) + n^{1/2}/2^s \leqq 7\rho(Q,P)/6 < 21n^{1/2}/2^s.$$

Therefore

$$| D_\alpha\phi_\nu(x) | < (21n^{1/2})^r N\rho_{xP}{}^{-[\alpha]} \tag{10.2}$$

if $x \varepsilon Q$, Q in K_s, $s \geqq 1$, $[\alpha] \leqq r$.

We note that N depends on n and r only, and thus is independent of the shape of P.

We shall need some further properties of the cubes J_ν. Let x be any point of $E - P$, let x lie in the cube Q of K_s, and let x lie within J_ν; then [2]

$$\rho(x,P)/2 < \rho(y^\nu,P) < 2\rho(x,P), \tag{10.3}$$

$$\operatorname{diam}(J_\nu) \leqq 4n^{1/2}/2^s \leqq 2\rho(x,P)/3. \tag{10.4}$$

To prove these, let Q' be a largest cube of $K_0 + K_1 + \cdots$ with y^ν as vertex; say Q' is in K_t. Then $t \geqq s - 1$, and

$$\operatorname{diam}(Q') = n^{1/2}/2^t \leqq 2n^{1/2}/2^s;$$

from which (using the definition of J_ν) the first part of (10.4) follows. It follows from this that $\rho(x,y^\nu) \leqq 2n^{1/2}/2^s$. But also

$$\rho(x,P) \geqq \rho(Q,P) \geqq 6n^{1/2}/2^s;$$

hence $\rho(x,y^\nu) \leqq \rho(x,P)/3$, and (10.3) and the rest of (10 4) follow.

---

[2] Correction to AE, (9.1): If y^* is distant δ_* from a given point x^0 of A, we prove $d_v < 2\delta_*$ only. Note that $d_\nu = r_\nu$.

We note finally that there is a number c such that at most c sets J_ν contain points of any cube Q of $K_0 + K + \cdots$.

11. The functions in the A_λ. In **9** we showed that we could find functions approximating to F locally, with a uniformity condition holding throughout A_λ. In the present section we find a single function approximating to F throughout A_λ, in fact throughout an open subset R_λ of B_λ which contains A_λ. This function will be obtained by interpolating between the functions defined locally.

When a positive continuous function $\zeta(x)$ is given in A_λ, we shall let R_λ denote the set of all x such that for some $y \in A_\lambda$, $\rho(x, y) < \zeta(y)$; R'_λ will denote the same, using $\zeta(y)/2$. It will be clear, with the ζ to be used, that $R_\lambda \subset B_\lambda$.

LEMMA 11a. *Let A_λ be non-void, and let $\epsilon(x)$ be a positive continuous function in B_λ. Then there is a positive continuous function $\eta(x)$ in A_λ such that for each positive continuous function $\zeta(x) \leqq \eta(x)$ in A_λ there is a function f of class C^r in B_λ, given by* (11.10), *the $f_{x\nu}$ being in I, such that if $[\alpha] \leqq r$, then*

$$|D_\alpha[f(x) - F(x)]| \leqq \epsilon(x), \qquad x \in R_\lambda, \tag{11.1}$$

$$|D_\alpha[f(x) - F(x)]| \leqq \rho_{xA_\lambda}{}^{r-[\alpha]}\epsilon(x), \qquad x \in R_\lambda - R'_\lambda. \tag{11.2}$$

In the application in **12**, we shall use $\zeta = \eta$.

Using the numbers N and c introduced in **10**. and recalling that $\rho(x, A) = \infty$ if A is void, set

$$\epsilon_1(x) = \epsilon(x)/\{c[(r+1)!]^n(126n^{1/2})^r N\}, \tag{11.3}$$

$$\epsilon_2(x) = \text{g.l.b.}\, \epsilon_1(y), \qquad \rho(x, y) \leqq \min[\rho(x, E - B_\lambda)/2, 1]. \tag{11.4}$$

Using $\epsilon_2(x)$ in A_λ, Lemma 9b gives a positive continuous function in A_λ, which we shall call $\eta_1(x)$; we take

$$\eta_1(x) \leqq \min[\rho(x, E - B_\lambda)/2, 1]. \tag{11.5}$$

Define $\eta_2(x)$ and $\eta(x)$ in A_λ by

$$\eta_2(x) = \text{g.l.b.}\, \eta_1(y), \qquad \rho(x, y) \leqq \min[\rho(x, E - B_\lambda)/2, 4], \tag{11.6}$$

$$\eta(x) = \min[\eta_2(x)/6, \rho(x, E - B_\lambda)/16, 1/2]. \tag{11.7}$$

Take any $\zeta(x) \leqq \eta(x)$, and define R_λ, R'_λ correspondingly.

Choose a finite or denumerable set A'_λ of points in A_λ, with the following properties:

(a) A'_λ has no limit points in B_λ.

(b) for each $y \varepsilon A_\lambda$ there is a $z \varepsilon A'_\lambda$ such that

$$\rho(y,z) < \zeta(y)/2, \qquad \zeta(z) > 3\zeta(y)/4.$$

The proof of existence of A'_λ is quite elementary.

We prove three properties of A'_λ:

(c) For each $x \varepsilon R_\lambda$ there is a $z \varepsilon A'_\lambda$ such that

$$\rho(x,z) < 2\zeta(z).$$

(d) $$\rho(x, A'_\lambda) < \rho(x, E - B_\lambda)/7, \qquad x \varepsilon R_\lambda.$$

(e) $$\rho(x, A'_\lambda) < 2\rho(x, A_\lambda), \qquad x \varepsilon R_\lambda - R'_\lambda.$$

To prove (c), take $x \varepsilon R_\lambda$, choose $y \varepsilon A_\lambda$ so that $\rho(x,y) < \zeta(y)$, and choose z by (b). Then

$$\rho(x,z) \leqq \rho(x,y) + \rho(y,z) < 3\zeta(y)/2 < 2\zeta(z),$$

as required. To prove (d) (which is trivial if $B_\lambda = E$), given $x \varepsilon R_\lambda$, find y and z as before. Now

$$\rho(x,y) < \zeta(y) \leqq \eta(y) \leqq \rho(y, E - B_\lambda)/16,$$

hence

$$\begin{aligned}\rho(y, E - B_\lambda) &\leqq \rho(x,y) + \rho(x, E - B_\lambda)\\ &\leqq \rho(y, E - B_\lambda)/16 + \rho(x, E - B_\lambda),\end{aligned}$$

and it follows that

$$\rho(y, E - B_\lambda) \leqq 16\rho(x, E - B_\lambda)/15.$$

Using an inequality above, this gives

$$\begin{aligned}\rho(x, A'_\lambda) &\leqq \rho(x,z) < 3\zeta(y)/2 < 2\eta(y)\\ &\leqq \rho(y, E - B_\lambda)/8 \leqq 2\rho(x, E - B_\lambda)/15,\end{aligned}$$

which proves (d). To prove (e), let y be a point of $E - B_{\lambda+1}$ such that $\rho(x,y) = \rho(x, E - B_{\lambda+1})$. Since $E - B_{\lambda+1} = A_\lambda \cup (E - B_\lambda)$, and $\rho(x, A_\lambda) \leqq \rho(x, A'_\lambda) < \rho(x, E - B_\lambda)$, by (d), it follows that

$$y \varepsilon A_\lambda, \qquad \rho(x,y) = \rho(x, A_\lambda).$$

Since x is not in R'_λ, $\rho(x,y) \geqq \zeta(y)/2$. Choose z by (b). Then

$$\rho(x, A'_\lambda) \leqq \rho(x,z) \leqq \rho(x,y) + \rho(y,z)$$
$$< \rho(x,y) + \zeta(y)/2 \leqq 2\rho(x,y) = 2\rho(x, A_\lambda),$$

proving (e).

Take now the standard subdivision of the set $B_\lambda - A'_\lambda$; this set is open, because of (a). To each vertex y^ν corresponds a cube J_ν and a function ϕ_ν, as in **10**. We shall, however, use only certain ones of these; we denote by $y^1, y^2, \cdots$ those vertices of the subdivision such that

$$\rho(y^\nu, A'_\lambda) < \rho(y^\nu, E - B_\lambda),$$

and let x^ν, J_ν and ϕ_ν correspond. Then $x^\nu \varepsilon A'_\lambda$.

By the choice of $\eta_1(x)$, for each $z \varepsilon A'_\lambda$ there is a function $f_z(x)$ in I such that, for $[\alpha] \leqq r$,

$$(11.8) \qquad | D_\alpha[f_z(x) - F(x)] | \leqq \rho_{zx}{}^{r-[\alpha]}\epsilon_2(z), \qquad \rho_{zx} < \eta_1(z),$$

$$(11.9) \qquad D_\alpha f_z(z) = D_\alpha F(z).$$

We define f in B_λ by the formulas

$$(11.10) \qquad f(x) = \begin{cases} \Sigma\phi_\nu(x) f_{x^\nu}(x), & x \varepsilon B_\lambda - A'_\lambda, \\ F(x), & x \varepsilon A'_\lambda. \end{cases}$$

Clearly f is of class C^r in $B_\lambda - A'_\lambda$. From (11..1) we could deduce at once that f is of class C^{r-1} in B_λ. It requires but little further proof to show that f is of class C^r in B_λ; the necessary details may be found in AE, § 11. (Near any $z \varepsilon A'_\lambda$, use (11.1) with $\epsilon(x)$ replaced by an arbitrary $\epsilon > 0$.)

We turn now to the proof (11.1) and (11.2). Take any x in $R_\lambda - A'_\lambda$. Say x is in the cube Q of K_s, and in the interiors of the cubes $J_1, \cdots, J_\gamma$, to use a simple notation. Then $\gamma \leqq c$ (see **10**).

We begin by showing that (the y^i denoting for the moment all vertices of $K_0 + K_1 + \cdots$)

$$(11.11) \qquad \rho(y^i, A'_\lambda) < \rho(y^i, E - B_\lambda), \qquad i = 1, \cdots, \gamma.$$

Set $P = E - B_\lambda + A'_\lambda$. Because of (d),

$$(11.12) \qquad \rho(x, P) = \rho(x, A'_\lambda).$$

Using (10.4) gives (since y^i is the center of J_i)

$$(11.13) \qquad \rho(x, y^i) \leqq \rho(x, P)/3 = \rho(x, A'_\lambda)/3.$$

Hence

$$\rho(y^i, A'_\lambda) \leqq \rho(x, y^i) + \rho(x, A'_\lambda) \leqq 4\rho(x, A'_\lambda)/3.$$

Moreover, using (11.13) and (d) again,

$$\rho(y^i, E - B_\lambda) \geqq \rho(x, E - B_\lambda) - \rho(x, y^i) \geqq 6\tfrac{2}{3}\rho(x, A'_\lambda).$$

These inequalities give (11.11).

It follows from (11.11) that the y^i actually are properly named; further, that the sum in (11.10) contains all ϕ_ν corresponding to $J_1, \cdots, J_\gamma$. Consequently, (10.1) holds at x. Subtracting $F(x)$ from both sides of (11.10) gives therefore (with the present notation)

$$f(x) - F(x) = \sum_{i=1}^{\gamma} \phi_i(x)[f_{x^i}(x) - F(x)]. \tag{11.14}$$

With the same notations again, (11.11), (10.3) and (11.12) give

$$\rho(y^i, x^i) = \rho(y^i, P) < 2\rho(x, P) = 2\rho(x, A'_\lambda).$$

This, with (11.13), gives

$$\rho(x, x^i) < 3\rho(x, A'_\lambda). \qquad i = 1, \cdots, \gamma. \tag{11.15}$$

Using (c), choose $z \in A'_\lambda$ so that $\rho(x, z) < 2\zeta(z)$. Now

$$\rho(x, x^i) < 3\rho(x, A'_\lambda) \leqq 3\rho(x, z),$$

and hence, using (11.7)

$$\rho(z, x^i) < 4\rho(x, z) < 8\zeta(z) \leqq 8\eta(z)$$
$$\leqq \min[\rho(z, E - B_\lambda)/2, 4].$$

Consequently, by (11.6), $\eta_2(z) \leqq \eta_1(x^i)$, from which follows, using (11.7) again,

$$\rho(x, x^i) < 3\rho(x, z) < 6\zeta(z) \leqq \eta_2(z) \leqq \eta_1(x^i).$$

We may therefore apply (11.8), giving

$$| D_\alpha[f_{x^i}(x) - F(x)] | \leqq \rho_{x^i x}{}^{r-[\alpha]}\epsilon_2(x^i).$$

Moreover, using (11.5),

$$\rho(x, x^i) < \eta_1(x^i) \leqq \min[\rho(x^i, E - B_\lambda)/2, 1];$$

hence, by (11.4) $\epsilon_2(x^i) \leqq \epsilon_1(x)$. It follows that

$$| D_\alpha[f_{x^i}(x) - F(x)] | \leqq \rho_{x^i x}{}^{r-[\alpha]}\epsilon_1(x), \qquad i = 1, \cdots, \gamma. \tag{11.16}$$

Differentiating (11.14) gives

$$D_\alpha[f(x)-F(x)]=\sum_{i=1}^{\gamma}\sum_{\beta}\binom{\alpha}{\beta}D_\beta\phi_i(x)D_{\alpha-\beta}[f_{x^i}(x)-F(x)].$$

Applying (10.2), (11.12) and (11.16) gives

$$|D_\alpha[f(x)-F(x)]|\leqq\sum_{i=1}^{\gamma}\sum_{\beta}\binom{\alpha}{\beta}(21n^{1/2})^rN\rho_{xA'\lambda}{}^{-[\beta]}\ \rho_{x^ix}{}^{r-[\alpha-\beta]}\epsilon_1(x).$$

(Using an inequality proved below, $\rho(Q,A'_\lambda)\leqq\rho(x,A'_\lambda)\leqq 1<6n^{1/2}$; hence $s\geqq 1$, and the use of (10.2) was legitimate.) Now $\gamma\leqq c$ (see **10**). Each β_k is one of $0,\cdots,r$; hence there are at most $(r+1)^n$ terms in the second sum. Each binomial coefficient is at most $\alpha_1!\cdots\alpha_n!\leqq(r!)^n$. Consequently, using (11.15),

$$(11.17)\qquad |D_\alpha[f(x)-F(x)]|\leqq c[(r+1)!]^n(63n^{1/2})^rN\rho_{xA'\lambda}{}^{r-[\alpha]}\epsilon_1(x),$$

for $x\in R_\lambda-A'_\lambda$, $[\alpha]\leqq r$.

To prove (11.1), take any $x\in R_\lambda-A'_\lambda$. Choosing z as before and applying (c) and (11.7) gives

$$\rho(x,A'_\lambda)\leqq\rho(x,z)<2\zeta(z)\leqq 2\eta(z)\leqq 1;$$

(11.1) is thus an immediate consequence of (11.17) and (11.3), for such x. For $x\in A'_\lambda$, it follows by continuity. (Actually, $D_\alpha f(x)=D_\alpha F(x)$ there.)

To prove (11.2), take $x\in R_\lambda-R'_\lambda$. Then (11.17), property (e) and (11.3) give (11.2) at once. This completes the proof of the lemma.

12. The function f^*. We are given the r-ideal I in G, the function F of class C^r in G such that $V_r(F,x)\in\mathfrak{J}_r(I,x)$ for each $x\in G$, and the positive continuous function $\epsilon(x)$ in G. Define the sets A_λ etc. as in **8.** Let $\lambda_1>\lambda_2>\cdots>\lambda_h$ be those numbers such that each A_{λ_i} is non-void. Set

$$(12.1)\qquad \rho_{x\lambda}=\rho(x,A_\lambda),\qquad \bar\rho_{x\lambda}=\rho(x,E-B_\lambda),$$

$$(12.2)\qquad a=1/\{3c[(r+1)!]^n(21n^{1/2})^rN\},$$

and for each λ, let $\epsilon_\lambda(x)$ be the positive continuous function

$$(12.3)\qquad \epsilon_\lambda(x)=\tfrac{1}{2}\min[1,\bar\rho_{x\lambda}{}^r]a^{\lambda+1}\epsilon(x),\qquad x\in B_\lambda.$$

For each i, we apply Lemma 11a to the set B_{λ_i}, with the function $\epsilon_{\lambda_i}(x)$. This gives us a function $\eta_{\lambda_i}(x)$ in A_{λ_i}; set $\zeta_{\lambda_i}(x)=\eta_{\lambda_i}(x)$. There is then a function f_{λ_i} in B_{λ_i}, given by (11.10), such that for $[\alpha]\leqq r$,

$$(12.4)\qquad |D_\alpha[f_{\lambda_i}(x)-F(x)]|\leqq\epsilon_{\lambda_i}(x),\qquad x\in R_{\lambda_i},$$

(12.5) $\quad | D_a[f_{\lambda_i}(x) - F(x)] | \leqq \rho_{x\lambda_i}{}^{r-[a]}\epsilon_{\lambda_i}(x), \qquad x \varepsilon R_{\lambda_i} - R'_{\lambda_i}.$

The following relations are immediate consequences:

(12.6) $\quad | D_a[f_{\lambda_i}(x) - F(x)] | \leqq \tfrac{1}{2}\epsilon(x), \qquad x \varepsilon R_{\lambda_i},$

(12.7) $\quad | D_a[f_{\lambda_i}(x) - F(x)] | \leqq \tfrac{1}{2}a^{\lambda_{i+1}}\bar{\rho}_{x\lambda_i}{}^{r-[a]}\epsilon(x), \qquad x \varepsilon R_{\lambda_i},$

(12.8) $\quad | D_a[f_{\lambda_i}(x) - F(x)] | \leqq \tfrac{1}{2}a^{\lambda_{i+1}}\rho_{x\lambda_i}{}^{r-[a]}\epsilon(x), \qquad x \varepsilon R_{\lambda_i} - R'_{\lambda_i}.$

(Consider separately the cases $\bar{\rho}_{x\lambda_i} \leqq 1$, $\bar{\rho}_{x\lambda_i} > 1$.)

Take the standard subdivision of each $B_{\lambda_i} - A_{\lambda_i} = B_{\lambda_{i+1}}$. We shall use only those cubes J_ν which have points not in R_{λ_i}; call these $J_{\lambda_i 1}, J_{\lambda_i 2}, \cdots$. Let $\phi_{\lambda_i j}$ correspond to $J_{\lambda_i j}$, and set

(12.9) $$\phi^*_{\lambda_i}(x) = \sum_j \phi_{\lambda_i j}(x).$$

We show that

(12.10) $\quad \phi^*_{\lambda_i}(x) = 1, \qquad x \varepsilon B_{\lambda_i} - R_{\lambda_i},$

(12.11) $\quad \phi^*_{\lambda_i}(x) = 0, \qquad x \varepsilon R'_{\lambda_i} - A_{\lambda_i}.$

The first relation is clear; see (10.1). To prove the other, we need merely show that no $J_{\lambda_i j}$ has points in R'_{λ_i}. Suppose that this is false; let it have points $x \varepsilon R'_{\lambda_i}$ and $z \varepsilon B_{\lambda_i} - R_{\lambda_i}$. By definition of R'_{λ_i}, there is a point $y \varepsilon A_{\lambda_i}$ such that $\rho(x,y) < \zeta_{\lambda_i}(y)/2$. Also, following the proof of (d), **11**, and using (11.7),

$$\bar{\rho}_{y\lambda_i} \leqq 16\bar{\rho}_{x\lambda_i}/15,$$

$$\rho_{x\lambda_i} \leqq \rho_{xy} < \zeta_{\lambda_i}(y)/2 < \eta_{\lambda_i}(y) \leqq \bar{\rho}_{y\lambda_i}/16 < \bar{\rho}_{x\lambda_i}.$$

Hence, setting $P = E - B_{\lambda_i} + A_{\lambda_i}$, $\rho(x,P) = \rho_{x\lambda_i}$, and (10.4) gives, if $x \varepsilon J_{\lambda_i j}$,

$$\operatorname{diam}(J_{\lambda_i j}) < \rho(x,P) = \rho_{x\lambda_i} \leqq \rho(x,y) < \zeta_{\lambda_i}(y)/2.$$

It follows that $\rho(z,y) \leqq \operatorname{diam}(J_{\lambda_i j}) + \rho(x,y) < \zeta_{\lambda_i}(y)$, so that $z \varepsilon R_{\lambda_i}$, a contradiction.

We note in passing that the proof above (taking $x \varepsilon R_{\lambda_i}$), with the fact $A_{\lambda_{i+1}} \subset E - B_{\lambda_i}$, shows that

(12.12) $\quad \rho_{x\lambda_i} < \bar{\rho}_{x\lambda_i} \leqq \rho_{x\lambda_{i+1}}, \qquad x \varepsilon R_{\lambda_i}.$

Note also that

(12.13) $\quad B_{\lambda_1} = A_{\lambda_1} = R_{\lambda_1}, \qquad B_{\lambda_i} - A_{\lambda_i} = B_{\lambda_{i-1}} \qquad (i > 1).$

We now define functions $g_{\lambda_1}, \cdots, g_{\lambda_k}$, using induction, as follows:

(12.14) $$g_{\lambda_1}(x) = f_{\lambda_1}(x), \qquad x \in B_{\lambda_1},$$

and for $i > 1$,

(12.15) $$g_{\lambda_i}(x) = f_{\lambda_i}(x), \qquad x \in A_{\lambda_i},$$

(12.16) $$g_{\lambda_i}(x) = f_{\lambda_i}(x) + \phi^*{}_{\lambda_i}(x)[g_{\lambda_{i-1}}(x) - f_{\lambda_i}(x)], \qquad x \in B_{\lambda_i} - A_{\lambda_i}.$$

Because of (12.11), $g_{\lambda_i} = f_{\lambda_i}$ in R'_{λ_i}. It follows that g_{λ_i} is of class C^r in R'_{λ_i}. The same holds in $B_{\lambda_i} - A_{\lambda_i}$, because of (12.16); hence it is of class C^r in B_{λ_i}.

We now prove an inequality. Take any $x \in G$. Let

(12.17) $$x \in R_{\mu_1} \cap \cdots \cap R_{\mu_k}, \qquad \mu_1 > \cdots > \mu_k,$$

and suppose that x is in no other R_j (R_j is defined only for $j = \lambda_1, \cdots, \lambda_h$). Then for $i = 1, \cdots, k$,

(12.18) $$| D_\alpha[g_{\mu_i}(x) - f_{\mu_i}(x)] | \leqq \tfrac{1}{2} a^{\mu_i} \rho_{x\mu_i}{}^{r-[\alpha]} \epsilon(x).$$

Take first $i = 1$. Since

$$B_{\mu_1} - A_{\mu_1} = \sum_{\lambda > \mu_1} A_\lambda \subset \sum_{\lambda > \mu_1} R_\lambda,$$

x is not in $B_{\mu_1} - A_{\mu_1}$. Also $x \in R_{\mu_1} \subset B_{\mu_1}$. Hence $x \in A_{\mu_1} \subset R'_{\mu_1}$, and the left hand side of (12.18) is 0.

Now take any $i > 1$. Since the relation is trivial for $x \in R'_{\mu_i}$, we may suppose $x \in R_{\mu_i} - R'_{\mu_i}$. Using induction, and the facts $\mu_{i-1} > \mu_i$, $\rho_{x\mu_{i-1}} \leqq \rho_{x\mu_i}$ (see (12.12)), we have

$$\begin{aligned} | D_\alpha[g_{\mu_{i-1}}(x) - f_{\mu_{i-1}}(x)] | &\leqq \tfrac{1}{2} a^{\mu_{i-1}} \rho_{x\mu_{i-1}}{}^{r-[\alpha]} \epsilon(x) \\ &\leqq \tfrac{1}{2} a^{\mu_i+1} \rho_{x\mu_i}{}^{r-[\alpha]} \epsilon(x) \end{aligned}$$

Next, (12.7) and (12.12) give

$$| D_\alpha[f_{\mu_{i-1}}(x) - F(x)] | \leqq \tfrac{1}{2} a^{\mu_{i-1}+1} \bar{\rho}_{x\mu_{i-1}}{}^{r-[\alpha]} \epsilon(x) \leqq \tfrac{1}{2} a^{\mu_i+1} \rho_{x\mu_i}{}^{r-[\alpha]} \epsilon(x).$$

Also, using (12.8),

$$| D_\alpha[f_{\mu_i}(x) - F(x)] | \leqq \tfrac{1}{2} a^{\mu_i+1} \rho_{x\mu_i}{}^{r-[\alpha]} \epsilon(x).$$

These three relations give

(12.19) $$| D_\alpha[g_{\mu_{i-1}}(x) - f_{\mu_i}(x)] | \leqq \tfrac{3}{2} a^{\mu_i+1} \rho_{x\mu_i}{}^{r-[\alpha]} \epsilon(x).$$

From (12.16) and (12.10) we have

(12.20) $$g_{\lambda_j}(y) = g_{\lambda_{j-1}}(y), \qquad y \in B_{\lambda_j} - R_{\lambda_j}.$$

13

Since x is in no R_{λ_j} with $\mu_{i-1} > \lambda_j > \mu_i$, it follows, if $\mu_i = \lambda_m$, that $g_{\lambda_{m-1}}(x) = g_{\mu_{i-1}}(x)$. Putting this in (12.16), with i replaced by m, and differentiating gives

$$D_\alpha[g_{\mu_i}(x) - f_{\mu_i}(x)] = \sum_\beta \binom{\alpha}{\beta} D_\beta \phi^*_{\mu_i}(x) D_{\alpha-\beta}[g_{\mu_{i-1}}(x) - f_{\mu_i}(x)].$$

As in **11**, there are at most c terms in the sum in (12.9), for fixed x. Applying (10.2) and (12.19) and recalling that $\rho(x,P) = \rho_{x\mu_i}$ here, we find (as in **11**)

$$| D_\alpha[g_{\mu_i}(x) - f_{\mu_i}(x)] | \leqq c[(r+1)!]^n(21n^{1/2})^r N \tfrac{3}{2}\rho_{x\mu_i}{}^{r-[\alpha]} a^{\mu_i+1}\epsilon(x);$$

using (12.2) gives (12.18).

Noting that $B_{\lambda_k} = G$, our required function is

$$f^*(x) = g_{\lambda_k}(x), \qquad x \varepsilon G. \tag{12.21}$$

We must show that

$$| D_\alpha[f^*(x) - F(x)] | \leqq \epsilon(x), \qquad x \varepsilon G,\ [\alpha] \leqq r. \tag{12.22}$$

Take any $x \varepsilon G$; apply (12.17) and (12.18). Since x is in no R_{λ_i} with $\lambda_i < \mu_k$, (12.20) gives $g_{\mu_k}(x) = g_{\lambda_k}(x) = f^*(x)$. Choose $y \varepsilon A_{\mu_k}$ so that $\rho(x,y) < \zeta_{\mu_k}(y)$. Then

$$\rho_{x\mu_k} \leqq \rho(x,y) < \zeta_{\mu_k}(y) = \eta_{\mu_k}(y) < 1.$$

Also $a < 1$. Hence, using (12.18) with $i = k$, we have

$$| D_\alpha[f^*(x) - f_{\mu_k}(x)] | \leqq \tfrac{1}{2}\epsilon(x).$$

Combining this with (12.6) (with $\lambda_i = \mu_k$) gives (12.22).

13. Completion of the proof of Theorem 2. We must still show that f^* is in I^*. To this end, we shall replace the functions f_{λ_i} and g_{λ_i} by functions f'_{λ_i} and g'_{λ_i} in I^*. To begin with, for each i, let ψ_{λ_i} be a function of class C^r in G such that

$$\psi_{\lambda_i}(x) = 1, \qquad x \varepsilon G - \sum_{j>i} R'_{\lambda_j}, \tag{13.1}$$

$$\psi_{\lambda_i}(x) = 0, \qquad x \varepsilon R''_{\lambda_i}, \tag{13.2}$$

where R''_{λ_i} is a neighborhood of $\sum_{j>i} A_{\lambda_j} = G - B_{\lambda_i}$ in G. (We may construct such a function for instance by using the standard subdivision of B_{λ_i}, and adding together all ϕ_ν such that the corresponding J_ν does not lie entirely in some R'_{λ_j} with $j > i$.) We may clearly take $R''_{\lambda_k} \subset R''_{\lambda_j}$ for $k > j$. We now set

$$(13.3) \qquad f'_{\lambda_i}(x) = \psi_{\lambda_i}(x) f_{\lambda_i}(x), \qquad x \varepsilon B_{\lambda_i},$$

$$(13.4) \qquad f'_{\lambda_i}(x) = 0, \qquad x \varepsilon G - B_{\lambda_i}.$$

Since $f'_{\lambda_i} = 0$ in R''_{λ_i}, f'_{λ_i} is of class C^r in G. Because of (13.1),

$$(13.5) \qquad f'_{\lambda_i}(x) = f_{\lambda_i}(x), \qquad x \varepsilon G - \sum_{j>i} R'_{\lambda_j}.$$

We shall show that each f'_{λ_i} is in I^*. We cannot use the expression (11.10) for f_{λ_i}, since it would not satisfy condition (b) of **2** at the points of A'_{λ_i}. This, however, is easily remedied. Because of (a), **11**, each $z \varepsilon A'_{\lambda_i}$ is in a neighborhood U_z such that for each $y^\nu \varepsilon U_z$, $x^\nu = z$ (see **10**). Let $\Sigma^{(z)}$ denote the sum in (11.10) over those terms $\phi_\nu f_x{}^\nu$ such that the corresponding J_ν lies entirely in U_z. Set

$$(13.6) \qquad \bar{f}_z(x) = \Sigma^{(z)} \phi_\nu(x) f_x{}^\nu(x) = [\Sigma^{(z)} \phi_\nu(x)] f_z(x), \qquad x \varepsilon B_{\lambda_i} - z,$$

and $\bar{f}_z(z) = f_z(z) = F(z)$. Since $\Sigma^{(z)} \phi_\nu \equiv 1$ in a neighborhood of z (excepting z) and $\equiv 0$ outside U_z, we may set $\bar{f}_z(x) = 0$ in $G - B_{\lambda_i}$, giving a function $\bar{f}_z$ in I (since f_z is in I).

If we set $\phi'_\nu(x) = \phi_\nu(x) f_x{}^\nu(x)$ in B_{λ_i} and $= 0$ in $G - B_{\lambda_i}$, the functions ϕ'_ν are defined and of class C^r in G, and are in I.

Let $z^1, z^2, \cdots$ denote the points of A'_{λ_i}. We may suppose the U_{z^m} have no common points. If Σ' denotes the sum of those terms corresponding to no z^m, then

$$(13.7) \qquad f_{\lambda_i}(x) = \Sigma' \phi'_\nu(x) + \sum_m \bar{f}_{z^m}(x) \qquad x \varepsilon B_{\lambda_i}.$$

Arrange the functions $\psi_{\lambda_i}\phi'_\nu$, $\psi_{\lambda_i}\bar{f}_{z^m}$, using the ν appearing in Σ', in a sequence; call these functions $f^*_{\lambda_i 1}, f^*_{\lambda_i 2}, \cdots$. Then

$$(13.8) \qquad f'_{\lambda_i}(x) = \sum_k f^*_{\lambda_i k}(x), \qquad x \varepsilon G;$$

this follows from (13.7) and (13.3) for $x \varepsilon B_{\lambda_i}$, and from (13.2) and (13.4) for $x \varepsilon G - B_{\lambda_i}$.

The functions $f^*_{\lambda_i k}$ are in I, and (a), **2**, holds; we shall prove (b). Take any $x \varepsilon G$. If first $x \varepsilon B_{\lambda_i}$, it is clear from the definition of the functions ϕ'_ν and $\bar{f}_{z^m}$ that at most a finite number of the functions ϕ'_ν (ν in Σ'), $\bar{f}_{z^m}$, are $\neq 0$ in some neighborhood of x; hence this holds also for the $f^*_{\lambda_i k}$. If next $x \varepsilon G - B_{\lambda_i}$, then $x \varepsilon R''_{\lambda_i}$, and (13.2) shows that all the $f^*_{\lambda_i k}$ vanish in R''_{λ_i}. Thus (b) holds, and we have shown that f'_{λ_i} is in I^*.

We now define the g'_{λ_i}. Set $\phi^*_{\lambda_i}(x)=0$ in A_{λ_i}. From (12.11) we see that $\phi^*_{\lambda_i}$ is now of class C^r in B_{λ_i}. Define

$$g'_{\lambda_1}(x)=f'_{\lambda_1}(x), \qquad x \varepsilon G, \tag{13.9}$$

and for $i>1$,

$$g'_{\lambda_i}(x)=f'_{\lambda_i}(x)+\phi^*_{\lambda_i}(x)[g'_{\lambda_{i-1}}(x)-f'_{\lambda_i}(x)], \qquad x \varepsilon B_{\lambda_i}, \tag{13.10}$$

and $g'_{\lambda_i}(x)=0$, $x \varepsilon G-B_{\lambda_i}$. We see at once that $g'_{\lambda_i}(x)=0$ in R''_{λ_i}; it follows that g'_{λ_i} is of class C^r in G.

Let us show that

$$g'_{\lambda_i}(x)=g_{\lambda_i}(x), \qquad x \varepsilon G-\sum_{j>i} R'_{\lambda_j}. \tag{13.11}$$

For $i=1$, this follows from (13.9), (13.5) and (12.14). Now take $i>1$. If $x \varepsilon A_{\lambda_i}$, $\phi^*_{\lambda_i}(x)=0$, and (13.10), (13.5) and (12.15) apply. Now take $x \varepsilon B_{\lambda_i}-A_{\lambda_i}$. If first $x \varepsilon R'_{\lambda_i}$, then (13.10), (12.11), (13.5) and (12.16) give the result. If not, then $x \varepsilon B_{\lambda_i}-\sum_{j>i-1} R'_{\lambda_j}$. Using induction gives $g'_{\lambda_{i-1}}(x)=g_{\lambda_{i-1}}(x)$, and (13.11) now follows from (13.10), (13.5) and (12.16).

We now prove that each g'_{λ_i} is in I^*. For $i=1$, it follows from (13.9). For $i>1$, we use induction. Note that if we replace $\phi^*_{\lambda_i}$ by a function $\phi^{**}_{\lambda_i}$ which equals $\phi^*_{\lambda_i}$ in $G-R''_{\lambda_i}$ and which goes to zero together with derivatives as we approach $G-B_{\lambda_i}$, we may set $\phi^{**}_{\lambda_i}=0$ in $G-B_{\lambda_i}$, making it of class C^r in G, and we may use $\phi^{**}_{\lambda_i}$ in place of $\phi^*_{\lambda_i}$ in (13.10), since $g'_{\lambda_{i-1}}=f'_{\lambda_i}=0$ in R''_{λ_i}. Now (13.10), with $\phi^{**}_{\lambda_i}$, holds in G. Using this, and the facts that f'_{λ_i} and $g'_{\lambda_{i-1}}$ are in I^* and that I^* is an ideal, shows that g'_{λ_i} is in I^*.

Finally, setting $i=h$ in (13.11) and using (12.21) gives

$$f^*(x)=g_{\lambda_h}(x)=g'_{\lambda_h}(x), \qquad x \varepsilon G. \tag{13.12}$$

Since g'_{λ_h} is in I^*, f^* is in I^*, and the proof of Theorem 2 is complete.

14. Restrictions on the local ideals. We consider here the problem: What sets of local r-ideals are the local ideals of an r-ideal? A good answer to the question seems quite difficult to give, especially for $r \geqq 2$. For a better understanding of the problem, we shall give two theorems, which do hardly more than show wherein the difficulty lies.

The first theorem says that each r-value of each local ideal must be realized by a function, each of whose r-values lies in the corresponding local ideal.

THEOREM 4. *Given a set of local r-ideals $H(x)$ for each x in an open subset G of E^n, there is an r-ideal I in G such that*

$$(14.1)\qquad \mathfrak{J}_r(I, x) = H(x), \qquad x \in G,$$

if and only if the following holds. For each x^0 in G and for each r-value $a^0 \in H(x^0)$ there exists a function f of class C^r in G such that

$$(14.2)\qquad V_r(f, x^0) = a^0; \qquad V_r(f, x) \in H(x), \qquad x \in G.$$

To prove the necessity, suppose that I exists. Given $a^0 \in H(x^0)$, since $\mathfrak{J}_r(I, x^0) = H(x^0)$, there is a function f in I such that $V_r(f, x^0) = a^0$; clearly (14.2) holds.

To prove the sufficiency, suppose the condition satisfied. Let I consist of all functions f of class C^r in G such that $V_r(f, x) \in H(x)$ for all x. Since the $H(x)$ are local ideals, I is clearly an ideal. Certainly $\mathfrak{J}_r(I, x) \subset H(x)$, all x. To prove $H(x) \subset \mathfrak{J}_r(I, x)$, all x, take any x^0 and any $a^0 \in H(x^0)$. Choose f so as to satisfy (14.2). Then f is in I, and $V_r(f, x^0) = a^0$, so that $H(x^0) \subset \mathfrak{J}_r(I, x^0)$, completing the proof.

The next theorem is obtained by applying the last theorem, not to all points of G, but to all the points where the local ideals $H(x)$ are non-trivial. Recall (**6**, **7**) that the unit local r-ideal $\mathfrak{J}_r^1(x)$ consists of all r-values at x, and that for any other local ideal $H(x)$, $a_{0\ldots 0} = 0$ for each r-value $a \in H(x)$, that is, for any f with $V_r(f, x) \in H(x)$, $f(x) = 0$. Thus the set A of points x where $H(x) \neq \mathfrak{J}_r^1(x)$ is the set of points at which any function f in any corresponding ideal is required to vanish. The following theorem gives conditions on the local ideals at points of A; there are of course no conditions at other points of G.

THEOREM 5. *Given the $H(x)$, there is a corresponding I as in Theorem 3 if and only if the following holds. Let A be the set of points x in G for which $H(x) \neq \mathfrak{J}_r^1(x)$; we assume A is closed. Then for each x^0 in A and for each $a^0 \in H(x^0)$ there is a neighborhood $U \subset G$ of x^0 and there are r-values $a(x) \in H(x)$ for $x \in U \cap A$ such that $a(x^0) = a^0$, and so that the following is true. For each pair of points x, x' in $U \cap A$, and for each $\alpha = (\alpha_1, \cdots, \alpha_n)$, $[\alpha] \leqq r$, define $R_\alpha(x', x)$ by the relation*

$$(14.3)\qquad a_\alpha(x') = \sum_{[\beta] \leqq r-[\alpha]} (a_{\alpha+\beta}(x)/\beta!)(x'-x)^\beta + R_\alpha(x', x).$$

Then for each $x^* \in U \cap A$ and $\epsilon > 0$ there is a $\delta > 0$ such that

$$(14.4)\qquad |R_\alpha(x', x)| \leqq \rho_{xx'}{}^{r-[\alpha]}\epsilon \text{ if } \rho_{xx^*}, \rho_{x'x^*} < \delta, \qquad x, x' \in U \cap A, [\alpha] \leqq r.$$

EXAMPLE. Suppose that A consists of isolated points only. Then the condition is obviously always true.

First suppose that I exists. Given x^0 and $a^0 \varepsilon H(x^0)$, choose f in I so that $V_r(f, x^0) = a^0$, and set $a(x) = V_r(f, x)$, $x \varepsilon G$. Then the R_a are simply the remainders in Taylor's formula, and (14.4) is well known.

Now suppose that the condition holds. By Theorem I of AE, we may extend the values of the $a_\alpha(x)$ ($[\alpha] \leqq r$) through U so that $a_0(x) = a_{0\ldots 0}(x)$ is of class C^r in U in terms of the $a_\alpha(x)$. (Though Theorem I of AE is stated for E only, it clearly holds for open sets U. Actually, we could apply Theorem III of AE directly.) Since $H(x) = \mathfrak{J}_r{}^1(x)$, $x \varepsilon U - A$, $a(x) \varepsilon H(x)$ for such x also. Let $\phi(x)$ be a function of class C^r in G which is $\equiv 0$ outside a neighborhood U' of x^0 with $\bar{U}' \subset U$, and is $\equiv 1$ in a neighborhood U'' of x^0; set

$$f(x) = \phi(x) a_0(x) \text{ in } U, \qquad f(x) = 0 \text{ in } G - U.$$

Then f is of class C^r in G, and since $a(x) \varepsilon H(x)$ ($x \varepsilon U$) and the $H(x)$ are local ideals, $V_r(f, x) \varepsilon H(x)$ for all $x \varepsilon G$. Clearly $V_r(f, x^0) = a(x^0) = a^0$; thus the condition of the last theorem is proved, and the present theorem follows.

Remark. In the condition of the last theorem, we could drop the hypothesis that A is closed, if we assume that (14.4) holds for all x^* in U instead of all x^* in $U \cap A$; for we could then prove that A is closed (or we could use Theorem 1 of H. Whitney, "Differentiable functions defined in arbitrary subsets of Euclidean space," *Transactions of the American Mathematical Society*, vol. 40 (1936), pp. 307-317).

HARVARD UNIVERSITY.

ON TOTALLY DIFFERENTIABLE AND SMOOTH FUNCTIONS

HASSLER WHITNEY

1. Introduction. H. Rademacher has proved that a function of n variables satisfying a Lipschitz condition is totally differentiable a.e. (almost everywhere) (see, for instance, Saks, [6, pp. 310-311]). It was discovered by H. Federer (though not stated as a theorem; see [2, p. 442]) that if f is totally differentiable a.e. in the bounded set P, then there is a closed set $Q \subset P$ with the measure $|P - Q|$ as small as desired, such that f is smooth (continuously differentiable) in Q; that is, the values of f in Q may be extended through space so that the resulting function g is smooth there.

Theorem 1 of the present paper strengthens the latter theorem by showing that f is approximately totally differentiable a.e. in P if and only if Q exists with the above property. The rest of the paper gives further theorems in the direction of Federer's Theorem, as follows.

Suppose the domain of definition of f were a bounded open set P. Then in applying the part (a) $\longrightarrow$ (c) of Theorem 1, we might alter f in a set $P - Q$ which included a neighborhood of the boundary of P. In applications, it might be important to keep the values of f in most of a subset close to the boundary of P, or in most of some other subset. That such can be done follows from Theorem 2.

If f satisfies a Lipschitz condition, Theorem 3 shows that g may be made to satisfy a Lipschitz condition also, with a constant which equals a number ρ_n (depending on the number n of variables only) times the constant for f; in the case of one variable, we may take $\rho_1 = 1$.

If we weaken the assumption on f, assuming only that it is measurable, then Lusin's Theorem shows that we can alter f on a set of arbitrarily small measure, giving a continuous function g. In the other direction, suppose we assume that f (defined in an open set) has continuous mth partial derivatives, and that these derivatives are totally differentiable a.e. Then Theorem 4 shows that we may alter f on a set of arbitrarily small measure, giving a function g which has continuous partial derivatives of order $m + 1$. For the case of one variable, this is essentially a theorem of Marcinkiewicz, [5, Theorem 3].

Examples show that the hypotheses in the theorems cannot be materially

Received February 20, 1950.

Pacific J. Math. 1 (1951), 143-159.

weakened without altering the conclusions. For instance, define a function ϕ of one variable as follows. Let $\phi_0(t)$ be the distance from t to the nearest integer. Using any sufficiently large integer a, set

$$\phi_i(t) = 2^i \phi_0(a^i t)/a^i, \qquad \phi(t) = \sum_{i=0}^{\infty} \phi_i(t) .$$

Then ϕ satisfies a Lipschitz condition of order $1 - \alpha$, for any $\alpha > 0$; but Property (c) of Theorem 1 is not true for it. If $\Phi(t) = \int_0^t \phi(s)\,ds$, then Φ is smooth, and its first derivative satisfies a Lipschitz condition of order $1 - \alpha$; but the conclusion of Theorem 4 (with $m = 1$) fails.

2. The theorem for bounded sets. Let $x = (x_1, \cdots, x_n)$ denote points of n-space E^n. With the unit vectors $e_1, \cdots, e_n$ of a coordinate system, any vector v can be written in the form $\Sigma v_i e_i$. The length of v is $|v| = (\Sigma v_i^2)^{1/2}$; $|y - x|$ is the distance from x to y. Given n numbers $f_1(x), \cdots, f_n(x)$, set

$$F(x) \cdot v = \sum f_k(x) v_k ; \tag{2.1}$$

this is linear in v. If $f(x)$, $f(y)$, and the $f_k(x)$ are defined, set

$$e(x,y) = \frac{f(y) - f(x) - F(x) \cdot (y - x)}{|y - x|} \tag{2.2}$$

for $y \neq x$, and $e(x,x) = 0$. Let $S_z\,[\alpha(z)]$ denote the set of elements z with the property α. Given f, and so on, as above, set

$$H(x,\epsilon) = S_y[e(x,y) < \epsilon] . \tag{2.3}$$

The measurable function f defined in the set P is a.t.d. (approximately totally differentiable) at $x \in P$ in terms of the $f_k(x)$ (see [6, p. 300]) if for each $\epsilon > 0$ the set $H(x, \epsilon)$ has x as a point of density. (Any standard definition of density points may be used for the purposes of this paper.) If this holds, then x is a point of density of P, and the $f_k(x)$ are uniquely determined; if x is a point of density in the direction of each axis, then the $f_k(x)$ are the approximate partial derivatives of f at x. The f_k are measurable (see [6, p. 299]).

THEOREM 1. *Let f be measurable in the bounded set P. Then the four following conditions are equivalent:*

(a) *The function f is a. t. d. a. e. in P.*

(b) *The function f is approximately derivable with respect to each variable a. e. in P.*

(c) *For each $\epsilon > 0$ there is a closed set $Q \subset P$ such that $|P - Q| < \epsilon$ and f is smooth in Q.*

(d) *There is a sequence of disjoint closed sets $Q_1, Q_2, \cdots$ in P such that $|P - Q_1 \cup Q_2 \cup \cdots| = 0$ and f is smooth in each Q_i.*

REMARK. If f is assumed totally differentiable a.e. in P, the proof that (c) holds is simplified; see [2, p. 442].

Proof of Theorem 1. For the equivalence of (a) and (b), see [6, pp. 300-303]. Note that (b) is an obvious consequence of (d). We shall prove the equivalence of (a), (c) and (d).

Suppose (c) holds. We choose the disjoint closed sets $Q_1, Q_2, \cdots$ in succession so that f is smooth in each and $|P_i| \leq |P|/2^i$, where

$$P_i = P - Q_1 \cup \cdots \cup Q_i ,$$

as follows. Having found $Q_1, \cdots, Q_{i-1}$, choose a closed set Q_i' so that f is smooth in Q_i' and $|P - Q_i'| \leq |P|/2^{i+1}$. Let $U_\delta(A)$ denote the δ-neighborhood of the set A. For small enough δ, we may use

$$Q_i = Q_i' - U_\delta(Q_1 \cup \cdots \cup Q_{i-1}) .$$

Thus (d) holds.

Suppose (d) holds. Let Q_i^* be the set of points of density of Q_i, and set $Q^* = Q_1^* \cup Q_2^* \cup \cdots$. Then $|P - Q^*| = 0$. Take any $x \in Q^*$; say $x \in Q_i^*$. Since f is smooth in Q_i and x is a point of density of Q_i, f (considered now in P) is a.t.d. at x. Thus f is a.t.d. at all points of Q^*, and (a) holds.

Now given (a), we must prove (c). There is a number $a > 0$ with the following property. For any points x, y, and number r with $|y - x| \leq r$, we have

$$|U_r(x) \cap U_r(y)| \geq 2a\, |U_r(x)| .$$

For $x \in P$, set $V_i = |U_{1/i}(x)|$, and

$$\psi_i(x, \eta) = |U_{1/i}(x) - H(x, \eta)| , \tag{2.4}$$

$$\phi_i(x) = \text{g.l.b. } S_\eta[\psi_i(x, \eta) < aV_i] . \tag{2.5}$$

Since $e(x,y)$ is measurable in the pair of variables x, y, it follows that $\psi_i(x, \eta)$ is measurable for fixed η. Also, as a function of η, $\psi_i(x, \eta)$ is monotone and continuous on the left; hence

$$\phi_i(x) < \zeta \quad \text{if and only if} \quad \psi_i(x, \zeta) < aV_i \,. \tag{2.6}$$

Therefore ϕ_i is measurable.

Let Q_1 be the set of points where f is a.t.d.; then $f_1, \cdots, f_n$ are defined in Q_1. Given $x \in Q_1$ and $\epsilon' > 0$, we may choose $\delta > 0$ so that

$$\psi_i(x,\epsilon') < aV_i \quad \text{if} \quad 1/i < \delta \,;$$

using (2.6) shows that

$$\lim_{i\to\infty} \phi_i(x) = 0\,, \qquad x \in Q_1 \,. \tag{2.7}$$

By Lusin's and Egeroff's theorems, there is a closed set $Q \subset Q_1$ such that $|Q_1 - Q| < \epsilon$, the f_k are continuous in Q, and $\phi_i(x) \longrightarrow 0$ uniformly in Q. We now prove that for each $\epsilon' > 0$ there is a $\delta > 0$ such that

$$e(x,y) < \epsilon' \quad \text{if} \quad x,y \in Q, \quad |y-x| < \delta \,. \tag{2.8}$$

Setting $\epsilon_1 = \epsilon'/6$, we may choose δ so that

$$|F(y)\cdot v - F(x)\cdot v| \le \epsilon_1 |v| \quad \text{if} \quad x,y \in Q, \quad |y-x| < 2\delta\,, \tag{2.9}$$

$$\phi_i(x) < \epsilon_1 \quad \text{if} \quad x \in Q, \quad 1/(i+1) < \delta \,. \tag{2.10}$$

Now take any $x, y \in Q$ with $|y - x| < \delta$. Let j be the largest integer such that $1/j \ge |y-x|$, and set

$$R = U_{1/j}(x) \cap U_{1/j}(y)\,; \qquad \text{then} \quad |R| \ge 2aV_j \,.$$

Since $1/(j+1) < |y-x| < \delta$, (2.10) and (2.6) give

$$\psi_j(x, \epsilon_1)\,, \quad \psi_j(y, \epsilon_1) < aV_j \,.$$

Hence there is a point z in R in neither corresponding set; that is,

$$|z-x|, \quad |z-y| < 1/j\,; \qquad e(x,z)\,, \quad e(y,z) < \epsilon_1 \,.$$

Since $F(x)\cdot v$ is linear in v and $|z-x|, |z-y| < 2|y-x|$, we have

$$\begin{aligned} e(x,y)\,|y-x| &= |f(y) - f(x) - F(x)\cdot(y-x)| \\ &\le |f(z) - f(x) - F(x)\cdot(z-x)| \\ &\quad + |f(z) - f(y) - F(y)\cdot(z-y)| + |[F(y) - F(x)]\cdot(z-y)| \\ &\le \epsilon_1[|z-x| + |z-y| + |z-y|] < \epsilon'\,|y-x| \end{aligned}$$

if $y \neq x$, proving (2.8).

This fact, together with the continuity of the $f_k(x)$ in Q, shows that f is "of class C^1 in terms of the f_k in Q", as defined in [7] (the definition is given after (6.3), below). Hence, by [7, Lemma 2], we may extend f to be smooth in E^n, completing the proof. (The extension is described in Section 4, below; by use of the results of that section, it is not hard to show that f has the required properties.)

3. The theorem for unbounded sets. We remove the restriction of boundedness in Theorem 1, and give more information about the set in which f may be left unaltered.

THEOREM 2. *Let $A_1, A_2, \cdots$ be open sets in E^n such that each has points in common with, at most, a finite number of the others, and let $\epsilon_1, \epsilon_2, \cdots$ be positive numbers. Let K be the set of points x such that there is a sequence of distinct sets $A_{\mu_1}, A_{\mu_2}, \cdots$ and a sequence of points $x_1, x_2, \cdots$ with $x_i \in A_{\mu_i}$ and $x_i \longrightarrow x$. Let P be a measurable subset of $E^n - K$, and let f be a. t. d. a. e. in P in terms of the f_k. Then there is a set $Q \subset P$ such that Q is closed in $E^n - K$ and $|(P - Q) \cap A_i| < \epsilon_i$, and there is a smooth function g in $E^n - K$ such that $g(x) = f(x)$ and $\partial g(x)/\partial x_k = f_k(x)$ in Q.*

REMARKS. Clearly K is closed and $K \cap A_i = 0$ for all i. If $Q^* \subset P$, Q^* is closed in $E^n - K$, and for some positive continuous functions $\delta_1(x), \delta_2(x), \cdots$ in Q^*,

$$e(x,y) \leq 1/2^i \qquad \text{if } x \in Q^*, \qquad |y-x| < \delta_i(x),$$

the proof shows that we may make $Q \supset Q^*$. For instance, we may make Q contain any given set of points of P in which f is totally differentiable and which has no accumulation points in $E^n - K$. On the other hand, we must expect to drop out a neighborhood of the set of points where f is not totally differentiable. Further, we cannot in general keep in Q any given closed set where f is approximately totally differentiable, as is shown by the following example (in one variable):

$$f(t) = t^2 \sin(i/t)(t \neq 0), \; f(0) = 0 .$$

Proof of Theorem 2. For each pair (k, l) of positive integers, let $U_{k,l}$ be the set of points x satisfying the conditions (with a fixed x_0 in E^n)

$$k-1 < |x - x_0| < k+1, \qquad 1/(l-1) > \operatorname{dist}(x, K) > 1/(l+1) \;;$$

for $k = 1$ or $l = 1$, we drop out the first inequalities. If K is void, the index l is not needed, and the situation is simpler. The $U_{k,l}$ are bounded open sets covering $E^n - K$, and each one touches at most eight others. Arrange them in a sequence $U_1', U_2', \cdots$.

For each i, let $\lambda_{i,1}, \lambda_{i,2}, \cdots$ be the (finite or infinite) set of numbers such that $U'_{\lambda_{i,k}} \cap A_i \neq 0$. Since the $\bar{U}_j'$ are compact and in $E^n - K$, each touches at most a finite number of the A_i; hence for given j, there is at most a finite number of values of i such that $\lambda_{i,k} = j$ for some k. Let ϵ_j' be the smallest of the numbers $\epsilon_i/2^k$, using these values of i and corresponding k.

Considering f and the f_k in $P \cap U_j'$ alone, apply the proof of Theorem 1 to find a closed set $Q_j \subset P \cap U_j'$ such that $|P \cap U_j' - Q_j| < \epsilon_j'$, and such that f is of class C^1 in terms of the f_k in Q_j. Set

$$V_j = U_j' - Q_j, \qquad V = \bigcup_j V_j, \qquad Q = E^n - K - V.$$

Then V is open, Q is closed in $E^n - K$, and $Q \cap U_j' \subset Q_j$. Now

$$(P - Q) \cap A_i = P \cap V \cap A_i = \bigcup_j (P \cap V_j \cap A_i),$$
$$|P \cap V_j| = |P \cap U'_j - Q_j| < \epsilon'_j.$$

Since $V_j \subset U_j'$, $P \cap V_j \cap A_i$ is void unless $j = \lambda_{i,k}$ for some k. Hence

$$|(P - Q) \cap A_i| \leq \sum_j |P \cap V_j \cap A_i| < \sum_k \epsilon'_{\lambda_{i,k}} \leq \sum_k \epsilon_i/2^k = \epsilon_i.$$

Since each U_j' is open and $Q \cap U_j' \subset Q_j$, f is clearly of class C^1 in terms of the f_k in Q. Hence, as before, we may extend the values of f in Q through $E^n - K$, as required. (We are applying [7, Lemma 2] in an open set; the change required in the proof is very simple. Or we could use [7, Theorem III].)

4. The theorem for Lipschitz functions. The following theorem has two parts, corresponding to the two theorems above.

THEOREM 3. *For each positive integer n there is a number ρ_n* (we may take $\rho_1 = 1$) *with the following properties.*

(a) *Let f be defined and satisfy a Lipschitz condition in the bounded closed set $P \subset E^n$:*

$$|f(y) - f(x)| \leq N|y - x|, \qquad x, y \in P. \tag{4.1}$$

Then for each $\epsilon > 0$ there is a closed set $Q \subset P$ such that $|P - Q| < \epsilon$, and there is a smooth function g in E^n satisfying a Lipschitz condition (see (4.15)) *with the constant $\rho_n N$, such that $g = f$ in Q.*

(b) *Let the* A_i, ϵ_i, *and* K *be as in Theorem* 2. *Let* P *be closed in* $E^n - K$ *(it may have accumulation points in* K*). Let* f *be defined in* P *and satisfy* (4.1). *Then there is a set* $Q \subset P$ *which is closed in* P *(and hence in* $E^n - K$*) such that* $|(P - Q) \cap A_i| < \epsilon_i$; *and there is a function* g *satisfying* (4.15) *in* E^n *which is smooth in* $E^n - Q^*$, *where* $Q^* = \bar{Q} - Q$, *such that* $g = f$ *in* Q.

(c) *We may take* Q [*in either* (a) *or* (b)] *so that* f *is totally differentiable in* Q *in terms of functions* $f_1, \cdots, f_n$; *we may then take* g *so that* $\partial g/\partial x_k = f_k$ *in* $Q (k = 1, \cdots, n)$.

(d) *Given a positive continuous function* $\eta(x)$ in $E^n - K$ [*in* E^n, *for case* (a)], *we may make*

$$(4.2) \qquad |g(x) - f(x)| < \eta(x)\,, \qquad x \in P\,.$$

REMARKS. It is no restriction to take P closed (or closed in $E^n - K$). For if P is not closed, it is easily seen that we may extend f (uniquely) over $\bar{P}$ so that it is continuous there; then (4.1) now holds in $\bar{P}$. (We can in fact extend f to satisfy (4.1) in E^n; see [3] or [4].) Note that, in (b), $Q^* \subset \bar{P} - P$; if K is void, then Q^* is void, and g is smooth in E^n. As an immediate consequence of (4.15), we have

$$(4.3) \qquad |\Sigma v_k \partial g(x)/\partial x_k| \leq \rho_n N \quad \text{if} \quad x \in E^n - Q^*\,, \ |v| = 1\,.$$

The hypothesis of total differentiability a.e. in P, together with

$$|\Sigma v_k f_k(x)| \leq N|v|$$

where the f_k are defined, is not enough to give the theorem (unless, for instance, $P = E^n$), as simple examples show. (Compare the examples in H. Whitney [8].) If we wish to prove (4.3) rather than (4.15), the proof may be slightly simplified; of course (4.15) follows from (4.3) if $Q^* = 0$ (hence if $K = 0$). See also the remarks following Theorem 2.

Proof of Theorem 3. To prove the theorem, we first note that (a) is contained in (b); use $A_1 = E^n$, $\epsilon_1 = \epsilon$. Next, (d) will follow at once from (4.1) and (4.15) if we make sure that each point of P is sufficiently close to some point of Q; this will clearly be the case if, in applying the proof of Theorem 2, we take the ϵ'_j small enough. Also, just as in Theorem 2, (c) will hold. It remains to show that we can obtain the properties in (b), using the proof of Theorem 2. We do this here, except for showing that we can make $\rho_1 = 1$.

We must examine the proof of [7, Lemma 2]. First, since f is totally differentiable a.e. in P [6, p. 311], we may choose Q as in the proof of Theorem 2;

recall that the f_k are continuous in Q. We shall use a cubical subdivision of $E^n - \bar{Q}$, essentially as in [7]. For each integer s (in [7], only $s \geq 0$ was used), let K_s' be the set of all cubes of edge length $1/2^s$, the coordinates of whose corners are integral multiples of $1/2^s$. Let K_s'' consist of the cubes of K_s' whose distances from $\bar{Q}$ are at least $6n^{1/2}/2^s$. Let K_s consist of the cubes of K_s'' which are not in cubes of K_{s-1}''. Take any cube $C \in K_s$; suppose $C \subset C'$, $C' \in K_{s-1}'$. Then $\operatorname{dist}(C', \bar{Q}) < 6n^{1/2}/2^{s-1}$. Therefore, clearly

$$6n^{1/2}/2^s \leq \operatorname{dist}(C, \bar{Q}) < 13n^{1/2}/2^s, \qquad C \in K_s. \tag{4.4}$$

Take $C \in K_s$, $C' \in K_{s+2}$. Then each point of C' is within

$$n^{1/2}/2^{s+2} + 13n^{1/2}/2^{s+2}$$

from $\bar{Q}$; hence

$$\operatorname{dist}(C, C') \geq (5/2)n^{1/2}/2^s, \qquad C \in K_s, \quad C' \in K_{s+2}. \tag{4.5}$$

Let $y^1, y^2, \cdots$ be the set of all corners of all cubes of all K_s. Choose $x^\nu \in \bar{Q}$ with $|x^\nu - y^\nu| = \operatorname{dist}(y^\nu, \bar{Q})$. Let b_ν be the largest length of edge of any cube of any K_s with y^ν as a corner, and let I_ν be the cube defined by

$$|x_i - y_i^\nu| \leq b_\nu \ (i = 1, \cdots, n).$$

Let ϕ_0' be a smooth function which is positive within a fixed unit cube and is zero outside; by a translation and similarity transformation, define ϕ_ν', positive within I_ν and zero outside. Set $\phi_\nu = \phi_\nu'/\Sigma\phi_\lambda'$; then ϕ_ν is positive within I_ν and zero outside, and $\Sigma\phi_\nu = 1$ in $E^n - \bar{Q}$. Since there is at most some fixed number of shapes of cubes (of some K_s, and K_{s+1} perhaps) forming any I_ν, there is clearly a number $M_n \geq 1$ with the following property (compare [7, Section 10]): taking $|v| = 1$,

$$|\Sigma v_i \,\partial\phi_\nu/\partial x_i| < 2^s M_n \quad \text{if} \quad \phi_\nu(y) \neq 0 \quad \text{for some} \quad y \in C \in K_s. \tag{4.6}$$

Extend f to be continuous in $\bar{P}$ (if $\bar{P} \neq P$); (4.1) still holds. For any $x^* \in \bar{Q}$ and any $x \in E^n$, set

$$\psi(x; x^*) = f(x^*) + \sum f_i(x^*)(x_i - x_i^*); \tag{4.7}$$

this is the value at x of the linear function approximating to f at x^*. Then set

$$g(x) = \sum \phi_\nu(x) \ \psi(x; x^\nu), \qquad x \in E^n - \bar{Q}. \tag{4.8}$$

It is not hard to show that if $g = f$ in $\bar{Q}$, then g is smooth in $E^n - Q^*$ and $\partial g/\partial x_i = f_i$ in Q; see the proof of [7, Lemma 2]. We must still prove (4.15).

Take first any x and x' in $E^n - \bar{Q}$; say for definiteness that

$$x \in C \in K_s, \quad x' \in C' \in K_{s'}, \quad s \geq s' . \tag{4.9}$$

Let x^* be a point of $\bar{Q}$ nearest to x. Since $\Sigma\phi_\nu(x) = 1$, we may write

$$g(x) = f(x^*) + \sum_{\nu} \phi_\nu(x)[f(x^\nu) - f(x^*)] + \sum_{\nu, i} \phi_\nu(x) f_i(x^\nu)(x_i - x_i^\nu) .$$

Hence

$$g(x') - g(x) = \sum_{\nu} [\phi_\nu(x') - \phi_\nu(x)][f(x^\nu) - f(x^*)] \tag{4.10}$$

$$+ \sum_{\nu, i} [\phi_\nu(x') - \phi_\nu(x)] f_i(x^\nu)(x_i' - x_i^\nu) + \sum_{\nu, i} \phi_\nu(x) f_i(x^\nu)(x_i' - x_i) .$$

We shall find a bound for each non-zero term. First we show that

$$|\phi_\nu(x') - \phi_\nu(x)| \; |f(x^\nu) - f(x^*)| \leq 64 N M_n n^{1/2} \; |x' - x| . \tag{4.11}$$

Consider first any ν such that $\phi_\nu(x) \neq 0$. Then by (4.6),

$$|\phi_\nu(x') - \phi_\nu(x)| < 2^s M_n |x' - x| . \tag{4.12}$$

Also, since

$$|x^* - x| \leq \text{diam}(C) + \text{dist}(C, \bar{Q}) \leq 14 n^{1/2}/2^s ,$$

$$|y^\nu - x| \leq 2\,\text{diam}(C) \leq 2 n^{1/2}/2^s ,$$

$$|y^\nu - x^\nu| \leq |x^* - y^\nu| \leq 16 n^{1/2}/2^s ,$$

we have

$$|x^\nu - x^*| \leq 32 n^{1/2}/2^s ,$$

and hence

$$|f(x^\nu) - f(x^*)| \leq 32 N n^{1/2}/2^s .$$

These relations give (4.11). Next consider any ν such that $\phi_\nu(x') \neq 0$. Then (using inequalities like those above) we obtain

$$|x^\nu - x^*| \le |x^\nu - x'| + |x' - x| + |x - x^*|$$

$$\le 18n^{1/2}/2^{s'} + |x' - x| + 14n^{1/2}/2^s \le 32n^{1/2}/2^{s'} + |x' - x| .$$

In the present case, (4.12) holds with s'. Suppose first that

$$|x' - x| \le 32n^{1/2}/2^{s'} .$$

Then

$$|f(x^\nu) - f(x^*)| \le N \cdot 64n^{1/2}/2^{s'} ,$$

and (4.11) follows. If $|x' - x| > 32n^{1/2}/2^{s'}$, then $|x^\nu - x^*| < 2\,|x' - x|$, and since $|\phi_\nu(x') - \phi_\nu(x)| \le 2$ and $4N < 64NM_n n^{1/2}$, (4.11) follows again.

Next we show that

$$(4.13) \qquad |\phi_\nu(x') - \phi_\nu(x)|\; |f_i(x^\nu)|\; |x_i' - x_i^\nu| \le 40NM_n n^{1/2}|x' - x| .$$

We may suppose that $\phi_\nu(x') \neq 0$, in which case $|x' - x^\nu| \le 18n^{1/2}/2^{s'}$, or $\phi_\nu(x) \neq 0$, in which case $|x' - x^\nu| \le 18n^{1/2}/2^s + |x' - x|$; in either case,

$$|x_i' - x_i^\nu| \le |x' - x^\nu| \le 18n^{1/2}/2^{s'} + |x' - x| .$$

First suppose that $|x' - x| \le 2n^{1/2}/2^{s'}$. Then, by (4.5), $s \le s' + 1$. Hence, using (4.6) with s or s' we get

$$|\phi_\nu(x') - \phi_\nu(x)| \le 2^{s'+1}M_n|x' - x| ;$$

since $|f_i(x^\nu)| \le N$, (4.13) follows. Next suppose that $|x' - x| > 2n^{1/2}/2^{s'}$. Then $|\phi_\nu(x') - \phi_\nu(x)| \le 2$, and $|x'_i - x_i^\nu| < 10|x' - x|$, giving (4.13) again.

Finally, we have

$$(4.14) \qquad |\phi_\nu(x)\, f_i(x^\nu)(x_i' - x_i)| \le N|x' - x| .$$

There is clearly a number c_n such that for any x, there are at most c_n values of ν such that $\phi_\nu(x) \neq 0$. In the three groups of terms in (4.10), there are at most $2c_n$, $2c_n n$, and $c_n n$ non-zero terms respectively. Hence, by (4.11), (4.13) and (4.14), we have

$$|g(x') - g(x)| \leq c_n N M_n n^{1/2}(128 + 80n + n)|x' - x|,$$

which gives

$$(4.15) \qquad |g(x') - g(x)| \leq \rho_n N |x' - x|, \qquad \rho_n = 209\, c_n M_n n^{3/2}.$$

If x and x' are in $\bar{Q}$, (4.15) follows from (4.1), since $g = f$ (or the extended f) in $\bar{Q}$. Suppose finally that $x \in \bar{Q}$, $x' \in E^n - \bar{Q}$ (or *vice versa*). Let x'' be the last point of the segment xx' in $\bar{Q}$. Then (4.1) holds for x and x'', and (4.15) holds for x''' and x', with x''' in $x''x'$ and arbitrarily close to x''; hence (4.15) holds in all cases, and the proof is complete.

5. Lipschitz functions of one variable. We must prove Theorem 3, with $n = 1$, $\rho_1 = 1$. The proof is elementary in nature; we do not need [7]. Find a closed subset Q_1 of $E^1 - K$ (or of E^1, in case (a)) as in the proof of Theorem 2 (or Theorem 1, if we are only using (a)). Now (4.1) holds in Q_1, f_1 is continuous in Q_1, and f is smooth in terms of f_1 in Q_1 (see Section 3, above), that is, for each $x \in Q_1$ and each $\epsilon' > 0$ there is a $\delta > 0$ such that

$$(5.1) \qquad |f(x'') - f(x') - (x'' - x') f_1(x')| \leq \epsilon' |x'' - x'|$$
$$\text{if } |x'' - x|, \quad |x' - x| < \delta, \quad x', x'' \in Q_1.$$

Let primes on functions denote differentiation. We shall find a set Q which is closed in $E^1 - K$, with $|Q_1 - Q| < \epsilon^*$ in case (a) or $|(Q_1 - Q) \cap A_i| < \epsilon_i^*$ in case (b), and a function g which satisfies (4.1) in E^1 and is smooth in $E^1 - K$, and such that $g = f$ and $g' = f_1$ in Q; for ϵ^* or the ϵ_i^* small enough, Q and g have the required properties.

Let $I_1, I_2, \cdots$ be the closed intervals whose interiors fill out $E^1 - Q_1 \cup K$. Extend f through E^1 so that (4.1) holds there; see [3] or [4]. Set $g_0 = f$ in $Q_1 \cup K$, and let g_0 be linear in the I_k, so that g_0 is continuous in the closed intervals. Then g_0 is continuous in E^1, and satisfies (4.1) there.

We shall need the following lemma.

LEMMA 1. *Let ϕ be defined and satisfy* (4.1) *in the closed interval* $[a^*, b]$, *and let ϕ be linear in the subinterval* $[a, b]$. *Then there is an arbitrarily small* *interval* $[a', a'']$ *about* a *such that* $\phi'(a')$ *exists, and there is a function* ψ *in* $[a^*, b]$ *which equals* ϕ *in* $[a^*, a'] \cup [a'', b]$ *and is smooth in* $[a', b]$, *and is such that* $\psi'(a') = \phi'(a')$, *and for* x *in* $[a', a'']$, $\psi'(x)$ *lies between* $\phi'(a')$ *and* $\phi'(a'')$.

We use the notation $\Delta\phi(x, y) = [\phi(y) - \phi(x)]/(y - x)$. If ϕ is linear in some interval $[x_0, b]$ with $x_0 < a$, we may set $\psi = \phi$. If not, we may choose $c < a$ arbitrarily close to a so that $\Delta\phi(c, a) \neq \Delta\phi(a, b)$. Suppose for definiteness that $\Delta\phi(c, a) < \Delta\phi(a, b)$. Take $a'' > a$ arbitrarily close to a. Because of (4.1), ϕ is absolutely continuous, and $\phi(a'') - \phi(c) = \int_c^{a''} \phi'(x)\,dx$. Hence there is a point a' in $[c, a]$ such that $\phi'(a')$ exists and

$$\phi'(a') < \Delta\phi(a', a'') < \Delta\phi(a'', b) = \phi'(a'') ;$$

that is, the tangents at a' and a'' intersect at a point x' between a' and a''. Using these tangents except near x', and smoothing near x', gives the required ψ.

We return to the theorem. Let $x_1, x_2, \cdots$ be the set of end points of the intervals I_k. Let I_1'' be an interval about x_1, of length $< \epsilon_1''$ for some ϵ_1'' (see below), with one end point interior to an interval I_k with x_1 as end point. Apply the lemma (or the lemma with x replaced by $-x$) to find an interior interval $I_1' = [a_1', a_1'']$ about x_1, and using $\phi = g_0$ in I_1'', define $\psi = g$ in I_1'. We may require that neither a_1' nor a_1'' is any x_i. In general, having found disjoint intervals $I_1', \cdots, I_{j-1}'$, let x_h be the first point of the sequence which is in none of these, and let I_j'' be an interval about x_h, of length $< \epsilon_j''$, disjoint from the preceding I_i'. Apply the lemma as before to find I_j', and define g in I_j'. Set $g = g_0$ elsewhere in E^1. Let Q be the set of points of Q_1 interior to no I_j'. For small enough ϵ_j'', the inequalities with ϵ^* or ϵ_i^* hold. We shall show that g is smooth in $E^1 - K$ and $g' = f_1$ in Q; the other properties of g are clear.

Clearly g' is continuous in a neighborhood of any point interior to an I_j' or an I_k, that is, in $E^1 - K \cup Q$. Now take any $x \in Q$; we shall show that $g'(x) = f_1(x)$ and g' is continuous at x, considering only points $x' \geq x$ for which $g'(x')$ is defined. The same fact holds for $x' \leq x$, and this will complete the proof. By definition of g, this is true if x is the left hand end point of some I_j'; suppose this is not the case.

Given $\epsilon' > 0$, choose δ so that (5.1) holds, and so that $|f_1(x') - f_1(x)| < \epsilon'$ for $x' \in Q_1$, $|x' - x| < \delta$. Choose $y > x$ in Q_1 within δ of x. Now any difference quotient of f, with points in $[x, y] \cap Q_1$, is within $2\epsilon'$ of $f_1(x)$; hence clearly any difference quotient of g_0 in $[x, y]$ is within $2\epsilon'$ of $f_1(x)$. Hence, for any x' in $[x, y]$ such that $g_0'(x')$ exists, $|g_0'(x') - f_1(x)| < 2\epsilon'$. Because of the last property in the lemma, $|g'(x') - f_1(x)| < 2\epsilon'$ if $g'(x')$ exists. Since $g(x') - g(x) = \int_x^{x'} g'(t)\,dt$, this shows that $g'(x)$ exists (as a right hand derivative) and equals $f_1(x)$, and proves the required continuity.

6. Functions with totally differentiable *m*th partial derivatives. We shall prove a theorem corresponding to (a) $\longrightarrow$ (c) in Theorem 1; the extension to the case corresponding to Theorem 2 is clear.

THEOREM 4. *Let f and its partial derivatives of order $\leq m$ be defined in a bounded open set $P \subset E^n$, and let each mth partial derivative be totally differentiable a.e. in P. Then for each $\epsilon > 0$ there is a closed set $Q \subset P$ such that $|P - Q| < \epsilon$, and there is a function g with continuous $(m+1)$th partial derivatives in E^n such that all partial derivatives of f of order $\leq m+1$ exist in Q and equal those of g there. In particular, $g = f$ in Q.*

Because of Theorem 1, we may suppose $m \geq 1$. We use the notation of [7]; thus

$$\frac{f_k(x)}{k!}(x' - x)^k = \frac{f_{k_1\cdots k_n}(x_1,\cdots,x_n)}{k_1!\cdots k_n!}(x_1' - x_1)^{k_1}\cdots(x_n' - x_n)^{k_n}$$

[do not confuse with the earlier $f_k(x)$], $\sigma_k = k_1 + \cdots + k_n$, and so on. Also

$$f_{k_1\cdots k_n} = \partial^{\sigma_k} f/\partial x_1^{k_1}\cdots\partial x_n^{k_n}$$

where defined.

Take any k with $\sigma_k = m - 1$, and any integers i and j. Since $\partial f_k/\partial x_i$ and $\partial f_k/\partial x_j$ are defined in P and are totally differentiable a.e. in P, it follows that their partial derivatives $\partial^2 f_k/\partial x_i \partial x_j$ and $\partial^2 f_k/\partial x_j \partial x_i$ exist a.e. in P; by a theorem of Currier [1], these are equal a.e. in P. Where this is so for all i, j, it is clear that we may define f_k with $\sigma_k = m + 1$ uniquely. Let P' be the subset of P in which the f_k exist for $\sigma_k \leq m + 1$, and each f_k ($\sigma_k = m$) is totally differentiable in terms of the f_l ($l_i \geq k_i$, $\sigma_l = m + 1$); then $|P - P'| = 0$. As seen in Section 2, the f_k are measurable.

As in [7], let $\psi_k(x'; x)$, for $\sigma_k \leq m$, be the value at x' of the polynomial of degree at most $m - \sigma_k$ which has the same value and partial derivatives of order $\leq m - \sigma_k$ at x as f_k. Then

$$\psi_k(x'; x) = \sum_{\sigma_l \leq m - \sigma_k} \frac{f_{k+l}(x)}{l!}(x' - x)^l .$$

Let $R_k(x'; x)$ be the corresponding remainder in Taylor's expansion:

$$R_k(x'; x) = f_k(x') - \psi_k(x'; x), \qquad \sigma_k \leq m. \tag{6.2}$$

Define ψ_k' and R_k' similarly for $x \in P'$, with m replaced by $m + 1$. We shall say a remainder R_k' is *of order* m' *at* x^0 if the following is true. For each $\epsilon' > 0$ there is a $\delta > 0$ such that

$$|R_k'(x; x^0)| \leq \epsilon' |x - x^0|^{m'} \quad \text{if} \quad |x - x^0| < \delta. \tag{6.3}$$

Recall from [7] that in a closed set, f is of class C^{m+1} in terms of the $f_k(\sigma_k \leq m + 1)$ if and only if each R_k' is of order $m + 1 - \sigma_k$ uniformly in a neighborhood of each point.

With the help of Lemma 2 below, we prove Theorem 4 as follows. By Lusin's Theorem, there is a closed set $Q' \subset P'$ with $|P' - Q'| < \epsilon/2$ such that each $f_k(x)$ with $\sigma_k = m + 1$ is continuous in Q'; that is, $R'_k(\sigma_k = m + 1)$ is of order 0 in Q'. For each integer i and each $x^0 \in Q'$, let $\delta_i(x^0)$ be the upper bound of numbers $\delta \leq 1$ such that (6.3) holds with $m' = m - \sigma_k + 1$, $\epsilon' = 1/2^i$, for all k with $\sigma_k \leq m + 1$. Then by the lemma, $\delta_i(x) > 0$ in Q'. As in Section 2, we see that the $\delta_i(x)$ are measurable. Find sets Q_i as in Section 2, and set $Q = Q_1 \cap Q_2 \cap \cdots$. Then clearly f is of class C^{m+1} in Q in terms of the f_k, and hence [7, Lemma 2] f may be extended from Q over E^n so that

$$\partial^{\sigma_k} f / \partial x_1^{k_1} \cdots \partial x_n^{k_n} = f_k$$

in Q. This extension is the required g. There remains to prove

LEMMA 2. *Let P be open, let*

$$\partial^{\sigma_k} f / \partial x_1^{k_1} \cdots \partial x_n^{k_n} = f_k$$

in P for $\sigma_k \leq m$, let $f_k(x^0)$ be defined for $\sigma_k = m + 1$, and let the $f_k(\sigma_k = m)$ be totally differentiable in terms of the $f_l(l_i \geq k_i, \sigma_l = m + 1)$ at x^0. Define $R_k'(x; x^0)$ as above. Then R_k' is of order $m - \sigma_k + 1$ at x^0 if $\sigma_k \leq m - 1$.

Note that the hypothesis shows that R_k' for $\sigma_k = m$ is of order 1 at x^0.

Suppose we have proved Lemma 2 for the case that $f_k(x^0) = 0$ for all k, $\sigma_k \leq m + 1$. Then it holds for the general case. For set

$$\bar{f}_k(x) = f_k(x) - \psi_k'(x; x^0) = R_k'(x; x^0) \qquad (\sigma_k \leq m + 1);$$

then $\bar{f}_k(x^0) = 0$. Also, since $\bar{\psi}_k'(x; x^0) = 0$ (using the $\bar{f}_l$), $\bar{R}_k'(x; x^0) = R_k'(x; x^0)$.

Since the $\bar{R}'_k$ are of order $m - \sigma_k + 1$ at x^0 for $\sigma_k = m$, the lemma shows that this is true also for $\sigma_k < m$. Thus $R'_k = \bar{R}'_k$ is of order $m - \sigma_k + 1$ at x^0, as required.

We shall need Taylor's Theorem with exact remainder:

LEMMA 3. *Let ϕ be a function of one variable such that $\phi^{(h)} = d^h\phi/dx^h$ exists for $h \leq m'$ in an interval and is bounded. Then*

$$\phi(t_1) = \sum_{h=0}^{m'} \frac{\phi^{(h)}(t_0)}{h!}(t_1 - t_0)^h$$

$$+ \frac{1}{(m'-1)!}\int_{t_0}^{t_1} (t_1 - s)^{m'-1}[\phi^{(m')}(s) - \phi^{(m')}(t_0)]ds .$$

Since $\phi^{(m')}$ is bounded, $\phi^{(m'-1)}$ satisfies a Lipschitz condition; hence for any smooth α, $\beta = \alpha\phi^{(m'-1)}$ is absolutely continuous, and

$$\int_a^b (d\beta/dt)dt = \beta(b) - \beta(a) .$$

Therefore the usual proof applies.

We return to Lemma 2, assuming $f_k(x^0) = 0$ ($\sigma_k \leq m + 1$). Set

$$x^i = (x_1, \cdots, x_i, x^0_{i+1}, \cdots, x^0_n) ;$$

then $x^n = x$. Take any $i > 0$, and any k with $\sigma_k \leq m - 1$. Set $m' = m - \sigma_k$, $k(i) = (k_1, \cdots, k_i + m', \cdots, k_n)$, and

$$x^i(s) = (x_1, \cdots, x_{i-1}, s, x^0_{i+1}, \cdots, x^0_n) .$$

Then $x^i(x^0_i) = x^{i-1}$, $x^i(x_i) = x^i$. For some $\delta_1 > 0$, the $f_k(x)$ ($\sigma_k \leq m$) are bounded for $|x - x^0| < \delta_1$. Lemma 3 gives

$$f_k(x^i) = \psi_k(x^i; x^{i-1})$$

$$+ \frac{1}{(m'-1)!}\int_{x^0_i}^{x_i} (x_i - s)^{m'-1}\{f_{k(i)}[x^i(s)] - f_{k(i)}(x^{i-1})\}\, ds .$$

Since $f_l(x^i(s)) = R'_l(x^i(s); x^0)$, and so on, the definition of R_k gives

$$R_k(x^i; x^{i-1}) = \frac{1}{(m'-1)!}\int_{x^0_i}^{x_i} (x_i - s)^{m'-1}$$

$$\times \{R'_{k(i)}[x^i(s); x^0] - R'_{k(i)}(x^{i-1}; x^0)\}\, ds .$$

For a certain ϵ^* chosen below, choose $\delta \leq \delta_1$ so that

$$|R_l^{\prime}(x'; x^0)| \leq \epsilon^* |x' - x^0| \qquad \text{if } \sigma_l = m, \quad |x' - x^0| < \delta .$$

Then if $|x - x^0| < \delta$, using $|x^i(s) - x^0| \leq |x - x^0|$ for $x_i^0 \leq s \leq x_i$, and so on, gives

$$|R_k(x^i; x^{i-1})| \leq \frac{2\epsilon^* |x - x^0|}{(m' - 1)!} \int_{x_i^0}^{x_i} (x_i - s)^{m'-1} ds \leq \frac{2\epsilon^* |x - x^0|^{m'+1}}{m'!} .$$

Now in [7, (6.3)], subtract $f_k(x'')$ from both sides, and change x, x', x'' to x^{i-1}, x^i, x respectively; this gives

$$R_k(x; x^{i-1}) - R_k(x; x^i) = \sum_{\sigma_l \leq m - \sigma_k} \frac{R_{k+l}(x^i; x^{i-1})}{l!} (x - x^i)^l .$$

Hence,

$$|R_k(x; x^{i-1}) - R_k(x; x^i)| \leq 2\epsilon^* |x - x^0|^{m - \sigma_k + 1} \sum_{\sigma_l \leq m - \sigma_k} \frac{1}{(m - \sigma_k - \sigma_l)!\, l!} .$$

Let A_k denote the sum, and let A be the largest A_k. Since $R_k^{\prime}(x; x^0) = R_k(x; x^0)$ for the case at hand, adding the inequalities for $i = 1, \cdots, n$ gives

$$|R_k^{\prime}(x; x^0)| \leq 2nA\, \epsilon^* |x - x^0|^{m - \sigma_k + 1} .$$

Given $\epsilon' > 0$, set $\epsilon^* = \epsilon'/(2nA)$, and choose δ accordingly; this inequality then completes the proof.

REFERENCES

1. A. E. Currier, *Proof of the fundamental theorems on second order cross partial derivatives*, Trans. Amer. Math. Soc. 35 (1933), 245-253.

2. H. Federer, *Surface Area II*, Trans. Amer. Math. Soc. 55 (1944), 438-456.

3. M. D. Kirszbraun, *Über die zusammenziehenden und Lipschitzschen Transformationen*, Fund. Math. 22 (1934), 77-108.

4. E. J. MacShane, *Extension of range of functions*, Bull. Amer. Math. Soc. 40 (1934), 837-842.

5. J. Marcinkiewicz, *Sur les séries de Fourier*, Fund. Math. 27 (1936), 38-69.

6. S. Saks, *Theory of the integral,* Second Edition, Monografje Matematyczne, Warszawa-Lwow, 1937.

7. H. Whitney, *Analytic extensions of differentiable functions defined in closed sets,* Trans. Amer. Math. Soc. 36 (1934), 63-89.

8.————, *On the extension of differentiable functions*, Bull. Amer. Math. Soc. 50 (1944), 76-81.

HARVARD UNIVERSITY

ANNALS OF MATHEMATICS
Vol. 62, No. 3, November, 1955

ON SINGULARITIES OF MAPPINGS OF EUCLIDEAN SPACES. I. MAPPINGS OF THE PLANE INTO THE PLANE

BY HASSLER WHITNEY

(Received May 27, 1955)

1. Introduction

Let f_0 be a mapping of an open set R in n-space E^n into m-space E^m. Let us consider, along with f_0, all mappings f which are sufficiently good approximations to f_0. By the Weierstrass Approximation Theorem, there are such mappings f which are analytic; in fact, (see [5], Lemma 6) we may make f approximate to f_0 throughout R arbitrarily well, and if f_0 is r-smooth (i.e., has continuous partial derivatives of orders $\leqq r$), r finite, we may make corresponding derivatives of f approximate to those of f_0.

Supposing f is smooth (i.e., 1-smooth), the Jacobian martix $\mathbf{J}$ of f is defined (using fixed coordinate systems); we say the point $p \in R$ is a *regular point* or a *singular point* of f according as $\mathbf{J}$ is of maximum rank (i.e., of rank min (n, m)) or of lesser rank. In general we cannot expect f to be free of singular points. A fundamental problem is to determine what sort of singularities any good approximation f to f_0 must have; what sort of sets they occupy, what f is like near such points, what topological properties hold with reference to them, etc.

Some special cases of this problem have been studied, as follows:

(a) For $m = 1$, we have a real valued function in R. It was shown by M. Morse, in Theorem 16 of [3], that f may be chosen so that the singular points (called critical points here) are isolated, the "Hessian" being non-zero at each; see §12 below. Moreover, each critical point may be assigned a "type number"; topological relations among these were given by Morse in [2].

(b) If $m \geqq 2n$, we may find an f with no singular points; see (a) and (b) of Theorem 2 in [5].

(c) If $m = 2n - 1$, we may obtain an f with isolated singular points; see [9]. For each such point $p \in R$, coordinate systems $(x_1, \cdots, x_n)$ in E^n and $(u_1, \cdots, u_m)$ in E^m may be chosen in which f, near p, has the form

$$(1.1) \qquad u_1 = x_1^2, \qquad u_i = x_i, \qquad u_{n+i-1} = x_1 x_i \qquad (i = 2, \cdots, n).$$

The singularities are studied from a topological point of view in [10].

(d) Some beginnings have been made for other pairs of values (n, m), by N. Wolfsohn, [11], but no complete description of the singularities exists in these cases.

Thus the smallest pair of values for which the problem is open is the pair (2, 2), i.e., for mappings of the plane into the plane; it is this case we treat here. In this case, there can be "folds" lying along curves, and isolated "cusp" points on the folds. Some topological properties (using folds and branches) are given by Tucker in [4].

The nature of the singularities may be seen by considering deformations of mappings of E^1 into E^1. The function $u = x^2$ defines a mapping of E^1 into E^1, with a "fold" at $x = 0$. If we also set $v = y$, we have a mapping of E^2 into E^2, with a "fold" along $x = 0$ in the (x, y)-plane; the deformation (as y changes) leaves all points fixed. The pair of equations

$$(1.2) \qquad u = x^2, \qquad v = y,$$

are the normal form for a fold; see Theorem 15A.

Now consider a mapping of E^1 into E^1, with two folds; for instance, if $u = x - x^3$, there are folds at $x = -1$ and at $x = 1$. As y changes, let us "pull out" the image so that the folds come together and disappear; this may be accomplished by setting

$$(1.3) \qquad u = xy - x^3, \qquad v = y,$$

and letting y move from 1 to -1. These equations are the normal form for a cusp point; see Theorem 16A.

The early study of cusp points was carried out in collaboration with Deborah T. Haimo and Norman Z. Wolfsohn.

2. Summary of results

The paper is divided into four Parts. In part A we give some general properties of mappings, and consider in particular derivatives $\nabla_V f(p)$ with respect to a vector V. For a "good" mapping f, the set of singular points (where the numerical Jacobian J is 0) will consist of smooth curves; we call these "general folds" of f. In terms of derivatives $\nabla_V f$, $\nabla_V \nabla_V f$ of f in the direction of tangent vectors V to the curve $J = 0$, we define whether or not a point of a general fold C is a "fold point" or a "cusp point" of f. If all singular points are fold or cusp points, we call f "excellent". In this case, the cusp points are isolated on the general folds. The image of each general fold is a smooth curve, except for cusps at the images of the cusp points. This is the reason for the term "cusp".

Part B is devoted to the proof that arbitrarily near any mapping f_0 there is an excellent mapping f. The general method was set forth in [6], and was used there and in [9] and [11]. We give a somewhat simplified account, which should be sufficient for all applications of the general nature we consider here. To illustrate the method, the application to critical point theory is given.

The definition of fold and cusp points is given independently of any coordinate system. In Part C we show that about any fold or cusp point coordinate systems may be introduced in E and in E' so that the transformation is given exactly by (1.2) or (1.3).

Returning to critical point theory, suppose that p is an isolated critical point of the real function ϕ_0, and ϕ is a good approximation to ϕ_0, together with first derivatives. Then ϕ may have several critical points near p; but there are definite relations between the numbers of critical points of each type, as shown by Morse. If we do not require the first derivatives to approximate, the restrictions on the numbers of critical points of different types disappear.

The situation for mappings of the plane into the plane is more complex, as we see in Part D. Under general approximations, we may introduce new folds and cusps in any region. We cannot eliminate a fold from a given region with an arbitrarily good approximation (not considering derivatives); however, a cusp point may be removed, it being replaced by a pair of fold curves running out of the region. Using 1-approximations (where first derivatives also approximate), we may introduce folds and cusps near a given fold. An arc cutting across a former fold must now cut across an odd number of folds. Cusps can no longer be eliminated; near a cusp for the old mapping there is a cusp for the new mapping. Under 2-approximations, folds and cusps remain as before, except that near a cusp, an even number of new cusps may be introduced on the fold curve through the cusp, close to the cusp. Under 3-approximations, no changes in the nature of the folds and cusps is possible.

There is one general invariance theorem about numbers of cusps. If two excellent mappings agree in a neighborhood of a simple closed curve (where there are no cusp points), then they have the same number of cusp points inside, counted modulo 2. Hence if one of the mappings has exactly one cusp point inside, the other has cusp points inside. For the proof, we derive a formula for the number of cusp points inside, mod 2. It uses both the manner in which the curve crosses folds and the manner in which its image crosses itself. (In place of the latter, one could use the rotation number of the image curve.) As an application, it is shown that there is no excellent mapping of the projective plane into the plane with only regular and fold points.

Part A. Generalities

3. Some preliminaries

By *Euclidean n-space* E^n we mean a metric affine n-dimensional space, though the metric is often not needed; it may be *oriented* in the usual manner. Let *arithmetic n-space* $\mathfrak{A}^n$ be the space of all n-tuples $(x_1, \cdots, x_n)$ of real numbers. A *coordinate system* χ in E^n is a one-one smooth mapping of an open subset of $\mathfrak{A}^n$ onto an open subset of E^n, with non-vanishing Jacobian; it is *r-smooth* if the mapping χ is (see §1). With a coordinate system, the point $(x_1, \cdots, x_n)$ of $\mathfrak{A}^n$ is also used to denote the image point in E^n.

Let f be a smooth mapping of an open set $R \subset E^n$ into E^m, and let V be a vector in E^n. Then for any point $p \in R$ and small real number t, $p + tV$ is in R, and $f(p + tV) - f(p)$ is a vector in E^m. The *derivative* of f with respect to V at p is the vector in E^m

$$\nabla_V f(p) = \lim_{t \to 0+} \frac{1}{t} [f(p + tV) - f(p)]. \tag{3.1}$$

The mapping $\nabla f(p)$ of vectors in E^n into vectors in E^m is linear. In particular, if $e_1, \cdots, e_n$ are the unit vectors of a coordinate system at p, then the usual notation is

$$(3.2) \qquad f_{x_i}(p) = \frac{\partial f(p)}{\partial x_i} = \nabla_{e_i} f(p).$$

If $\bar{e}_1, \cdots, \bar{e}_n$ are the unit vectors in $\mathfrak{A}^n$, then $e_i(p) = \nabla_{\bar{e}_i}\chi(x_1, \cdots, x_n)$.

If $(u_1, \cdots, u_m)$ is a coordinate system in E^m, then the $\nabla_V u_i$ are the components of $\nabla_V f(p)$:

$$(3.3) \qquad \nabla_V f(p) = \nabla_V(u_1, \cdots, u_m) = (\nabla_V u_1, \cdots, \nabla_V u_m).$$

Suppose $V(p)$ is a smooth vector function in E^n, and f is as above. Then $(\nabla_V f)(p) = \nabla_{V(p)} f(p)$ is a function in R, whose values are vectors in E^m, and we may form the second derivative $\nabla_W \nabla_V f(p)$, for any vector W. It is easily seen that

$$(3.4) \qquad \nabla_W \nabla_V f(p) = \nabla_W \nabla_{V_0} f(p) + \nabla_{W'} f(p), \qquad V_0 = V(p),$$
$$W' = \nabla_W V(p).$$

If $V_1, \cdots, V_n$ are the components of V:

$$V = (V_1, \cdots, V_n) = V_1 e_1 + \cdots + V_n e_n,$$

then since $\nabla f(p)$ is linear,

$$(3.5) \qquad \nabla_V f(p) = V_1 f_{x_1}(p) + \cdots + V_n f_{x_n}(p).$$

If $n = m = 2$, and we use coordinate systems (x, y) and (u, v), then applying this twice gives (letter subscripts always denoting derivatives)

$$(3.6) \qquad \nabla_W \nabla_V f(p) = W_1(V_1 f_x + V_2 f_y)_x + W_2(V_1 f_x + V_2 f_y)_y.$$

Let ϕ be a 2-smooth parametrized curve in E^n: ϕ maps $\mathfrak{A}^1$ into E^n. The tangent vector is

$$(3.7) \qquad \frac{d\phi(t)}{dt} = \nabla_{\bar{e}_1} \phi(t).$$

LEMMA 3a. *Let f be a 2-smooth mapping of E^n into E^m, let $V(p)$ be a smooth vector function in E^n, and let ϕ be a 2-smooth parametrized curve in E^n such that*

$$(3.8) \qquad \frac{d\phi(t)}{dt} = V(\phi(t)) \neq 0.$$

Then

$$\left.\begin{aligned} &(3.9) \qquad \frac{d}{dt}(f\phi)(t) = \nabla_V f(p), \\ &(3.10) \qquad \frac{d^2}{dt^2}(f\phi)(t) = \nabla_V \nabla_V f(p), \end{aligned}\right\} \quad p = \phi(t).$$

Since $\nabla\phi(t)$ maps $\bar{e}_1$ into $V(p)$ and $\nabla f(p)$ maps $V(p)$ into $\nabla_V f(p)$, the composite mapping $\nabla(f\phi)(t)$ maps $\bar{e}_1$ into $\nabla_V f(p)$; this is (3.9). If we replace the mapping f of E^n into E^m by the mapping $F(p) = \nabla_V f(p)$ of E^n into vectors in E^m, and use (3.9), we obtain (3.10):

$$\frac{d^2}{dt^2}(f\phi)(t) = \frac{d}{dt}(F\phi)(t) = \nabla_V F(p) = \nabla_V \nabla_V f(p).$$

Let f and g be mappings of R into E^m, and let $\varepsilon(p)$ be a positive continuous function in R. We say g is an *ε-approximation* to f if

$$| g(p) - f(p) | < \varepsilon(p), \qquad \text{all } p \in R.$$

If f and g are r-smooth, we say g is an *(r, ε)-approximation* to f if this inequality holds, and also the similar inequalities for all partial derivatives of orders $\leqq r$, using fixed coordinate systems. (The definition depends on the coordinate systems used.) We speak of *general approximations* and *r-approximations* in the two cases.

4. The singularities

Let f be a smooth mapping of the open set $R \subset E = E^2$ into $E' = E'^2$. We say f is *regular* at p if

$$\text{(4.1)} \qquad \nabla_V f(p) \neq 0 \quad \text{whenever} \quad V \neq 0;$$

otherwise, f is *singular* at p. With coordinate systems (x, y) in R and (u, v) in E', the *Jacobian* is

$$\text{(4.2)} \qquad J = u_x v_y - u_y v_x ;$$

p is a regular or singular point according as $J(p) \neq 0$ or $J(p) = 0$.

Let f be 2-smooth. We say p is a *good* point if either $J(p) \neq 0$ or $\nabla J(p) \neq 0$ (i.e. at least one of $J_x(p)$, $J_y(p)$ is $\neq 0$). We say f is *good* if every point of R is good.

LEMMA 4a. *Let f be good in R. Then for each $p \in R$, the image space $H(p)$ of $\nabla f(p)$ is of dimension 2 or 1, according to whether p is a regular or a singular point.*

Clearly $d_p = \dim(H(p)) = 2$ if p is regular, and $d_p \leqq 1$ if p is singular. Suppose $d_p = 0$. Then $u_x = u_y = v_x = v_y = 0$ at p, and differentiating (4.2) shows that $J_x(p) = J_y(p) = 0$, a contradiction.

LEMMA 4b. *Let f be good in R. Then the singular points of f form smooth curves in R.*

Let p be singular. Then $J(p) = 0$, and $\nabla J(p) \neq 0$. Hence the solutions of $J = 0$ near p lie on a smooth curve, by the implicit function theorem.

The above curves we call the *general folds* of f.

Now let f be 3-smooth and good. Take any singular point p, and let $\phi(t)$ be a 2-smooth parametrization of the general fold C through p, with $\phi(0) = p$. Note that df/dt means $d(f\phi)/dt$. We say p is a *fold point* of f if

$$\text{(4.3)} \qquad df/dt \neq 0 \quad \text{at} \quad p,$$

and p is a *cusp point* of f if

$$\text{(4.4)} \qquad df/dt = 0, \qquad d^2f/dt^2 \neq 0 \quad \text{at} \quad p.$$

Thus p is a fold point of f (assumed good) if the image of C near p is a smooth curve with non-zero tangent vector at p, and p is a cusp point if the tangent vector is zero at p but is becoming non-zero at a positive rate as we move away from p on C. Clearly the definitions are independent of the parametrization chosen for C.

It is an immediate result of the definition that *cusp points are isolated.*

We say p is an *excellent point* of f (assumed good) if it is a regular, fold, or cusp point, and f is *excellent* if each point of R is excellent.

Using 3-smooth coordinate systems about p and about $f(p)$, define the smooth vector function

$$V(p) = (-J_y(p), J_x(p)). \tag{4.5}$$

Then by (3.5), $\nabla_{V(p)}J(p) = 0$ at each singular point; hence at all points of the general fold, *$V(p)$ is tangent to the fold.*

Given the general fold point p of f, we may therefore choose the parametrization ϕ so that (3.8) holds. Applying Lemma 3a shows that, assuming f is good, *p is a fold point if and only if*

$$\nabla_V f(p) \neq 0, \tag{4.6}$$

and *p is a cusp point if and only if*

$$\nabla_V f(p) = 0, \qquad \nabla_V \nabla_V f(p) \neq 0. \tag{4.7}$$

5. The structure of f near a cusp point

By means of Theorems 15A and 16A, we may study f near a fold or cusp point directly through the corresponding equations (1.2) and (1.3). We make a direct study of folds and cusps here. We assume f is excellent.

We shall use a certain system of curves defined by f in an open set $R' \subset R$. For each $p \in R$, consider the vectors $V' = \nabla_V f(p)$ as a function of vectors V with $|V| = 1$; we let R' contain p if the vectors V' are not all of the same length. For any $p \in R'$, there will be a pair of opposite directions at p such that for V in these directions, $|V'|$ is a minimum. (For V in the perpendicular directions, $|V'|$ will be a maximum.) Now R' is filled up by smooth curves in these directions; we call these the *curves of minimum* ∇F.

For any $p \in R'$ and vector $V \neq 0$, $\nabla_V f(p) = 0$ if and only if p is a singular point and V is tangent to the curve of minimum ∇f there; see Lemma 4a.

Now consider any general fold curve C. If a curve of minimum ∇f cuts C at a positive angle at p, then for the tangent vector $V(p)$, $\nabla_V f(p) \neq 0$, and hence p is a fold point. Now suppose C is tangent to a curve of minimum ∇f at p. Then p is not a fold point, and hence is a cusp point, since f is excellent. Set

$$V^* = \nabla_V \nabla_V f(p); \tag{5.1}$$

then $V^* \neq 0$. Since $\nabla_V f(p) = 0$, $\nabla_V f(p')$ is approximately in the direction of $\pm V^*$ for p' on C near p; by Lemma 4a and continuity, it follows that $\nabla_W f(p)$

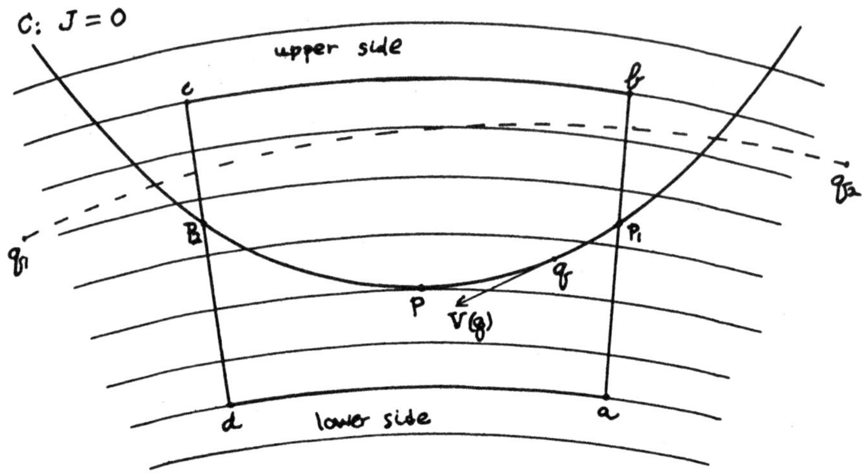

FIG. 1. General fold, cusp at p. Curves of minimum ∇f shown.

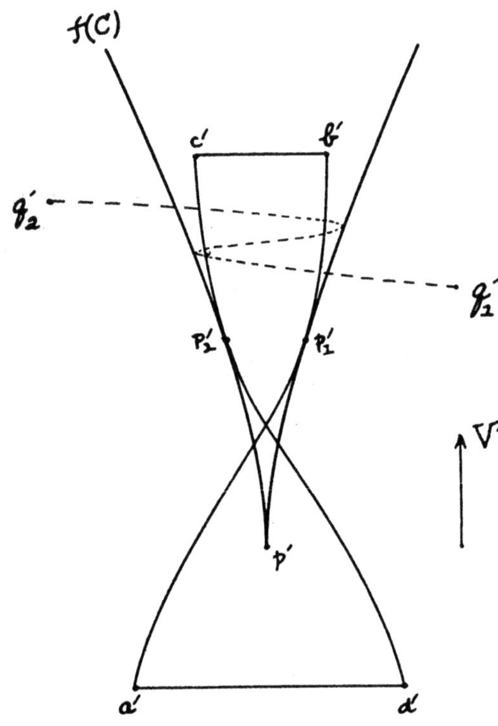

FIG. 2. Image of general fold and of box about cusp point shown.

is a multiple of V^*, for all vectors W. As we move along the general fold through p, $\nabla_V f(p')$ changes from a negative to a positive multiple of V^* (approximately); hence $V(p')$ cuts the curves of minimum ∇f in opposite senses on the two sides of p. Therefore the curves of minimum ∇f lying on one side of C cut C on both sides of p. We call this side of C the *upper side*, and the other, the *lower side*.

The image of C clearly has a cusp at $f(p)$, pointing in the direction of $-V^*$. For any vector W not tangent to C at p, $\nabla_W f(p)$ is a positive or negative multiple of V^* according as W points into the upper or lower side of C.

With coordinate systems, the Jacobian J has opposite signs on the two sides of C; but the signs depend of course on the chosen orientations of E and E'. If the cusp is given by (1.3), we find, for any $q = (x, y)$ on C,

$$\begin{aligned} &J = u_x = y - 3x^2, \qquad V(q) = (-1, -6x), \\ &\nabla_V f(q) = (-y - 3x^2, -6x), \qquad \nabla_V \nabla_V f(q) = (12x, 6). \end{aligned} \tag{5.2}$$

Hence $V^* = (0, 6)$, and the positive y-axis lies in the upper part of C for the cusp at (0, 0). The image of the general fold $J = 0$ is, by (5.2) and (1.3),

$$u = 2x^3, \; v = 3x^2, \quad \text{hence} \quad 27u^2 = 4v^3. \tag{5.3}$$

6. Some special coordinate systems

Let p be a singular point of the good mapping f. Then there is a vector in E mapped into 0 by $\nabla f(p)$; let the x-axis be in the direction of this vector. The unit vector in the y-direction is mapped into a vector $\neq 0$; let this be the unit vector in the v-direction in E'. Then we have

$$u_x = u_y = v_x = 0, \qquad v_y = 1, \text{ at } p; \tag{6.1}$$

these equations characterize coordinate systems with the above properties at p and at $f(p)$. (The coordinate systems of Theorems 15A and 16A have these properties.) Using such a coordinate system, we shall find the conditions that p be a fold or a cusp point.

Using (4.5), (4.2) and (6.1) gives

$$V = (-J_y, J_x) = (-u_{xy}, u_{xx}) \text{ at } p. \tag{6.2}$$

Hence, by (3.3), (3.5) and (6.1),

$$\nabla_V f(p) = (0, V_1 v_x + V_2 v_y) = (0, u_{xx}). \tag{6.3}$$

Therefore, by (4.6), *the condition that p be a fold point is*

$$J_x(p) = u_{xx}(p) \neq 0. \tag{6.4}$$

Now suppose $J_x = u_{xx} = 0$ at p. Differentiating (4.2) twice gives

$$J_{xx} = u_{xxx} - 2u_{xy}v_{xx} \text{ at } p.$$

Applying (3.6) and (6.2) gives

$$\nabla_V \nabla_V v = -J_y(-J_y v_x + J_x v_y)_x = -u_{xy}(u_{xxx} - 3u_{xy}v_{xx})$$

at p. Similarly $\nabla_V \nabla_V u = 0$ at p. Therefore

$$\nabla_V \nabla_V f = (0, -u_{xy}(u_{xxx} - 3u_{xy}v_{xx})) \text{ if } \nabla_V f(p) = 0, \tag{6.5}$$

and, by (4.7), *the condition that p be a cusp point is*

$$u_{xx} = 0,\ u_{xy} \neq 0, \qquad u_{xxx} - 3u_{xy}v_{xx} \neq 0, \text{ at } p. \tag{6.6}$$

With (1.3), we have (6.1), and also $u_{xx} = 0$, $u_{xy} = 1$, $u_{xxx} = -6$, $v_{xx} = 0$ at the origin; hence this is a cusp point. Since $f_x = (u_x, v_x) = (0, 0)$ at all points of the general fold, the curves of minimum ∇f are parallel to the x-axis at these points, and hence cut the general fold (other than at the origin) at a positive angle. Therefore the remaining singular points are all fold points.

We consider what happens to (1.3) under the transformation of coordinates

$$x = x', \qquad y = y' + ax'^2. \tag{6.7}$$

First take $a = 1$. Then (1.3) becomes

$$u = x'y', \qquad v = y' + x'^2, \tag{6.8}$$

in which the cubic term in the expression for u has been replaced by a quadratic term in the expression for v. Of course $u_{x'x'x'} - 3u_{x'y'}v_{x'x'} \neq 0$ at (0, 0). With $a = 3$, (1.3) becomes

$$u = x'y' + 2x'^3, \qquad v = y' + 3x'^2; \tag{6.9}$$

here $J = y'$, and the general fold is the x'-axis.

7. Removal of a branch point

In complex variables, the mapping $w = z^2$ has an isolated singularity at the origin. Using $z = x + iy$, $w = u + iv$, make a slight alteration of the mapping (near the origin), by setting

$$u = x^2 - y^2 - 4\varepsilon x, \qquad v = 2xy. \tag{7.1}$$

Now

$$J = 4x^2 - 8\varepsilon x + 4y^2, \tag{7.2}$$

and the singularities are on the circle C through the origin, with center at $(\varepsilon, 0)$.

A tangent vector to the circle is, as in (6.2),

$$V(x, y) = 8(-y, x - \varepsilon), \tag{7.3}$$

and we find at once

$$\nabla_V f = 16(-2xy + 3\varepsilon y, x^2 - y^2 - \varepsilon x) \tag{7.4}$$

on C. This vanishes at the three points

$$(0, 0), \quad \left(\frac{3}{2}\varepsilon, \frac{3^{1/2}}{2}\varepsilon\right), \quad \left(\frac{3}{2}\varepsilon, -\frac{3^{1/2}}{2}\varepsilon\right). \tag{7.5}$$

These points are cusp points, and the remaining points of C are fold points.

PART B. IMPROVING MAPPINGS

8. The method to be used

We consider mappings f of an open set $R \subset E = E^n$ into $E' = E^m$, near a given mapping f_0. There are certain properties we wish f to have, expressible in terms of partial derivatives, say of orders 1 through r. With fixed coordinate systems in E and E', each component of f has partial derivatives of each order k; hence the total number of partial derivatives considered is

$$(8.1)\quad N = m\left[\binom{n}{1} + \binom{n+1}{2} + \cdots + \binom{n+r-1}{r}\right] = m\left[\binom{n+r}{r} - 1\right].$$

Arranging these partial derivatives in a fixed order, any set of partial derivatives is an ordered set of N numbers, and hence a point of $\bar{E} = \mathfrak{A}^N$. There will be a certain "bad" set $S \subset \bar{E}$; a point q is a "bad" point, i.e., $q \in S$, if it corresponds to a set of partial derivatives which we do not wish f to have at any point of R.

Given f, take any $p \in R$. The set of partial derivatives of f at p defines a point $\bar{f}(p) \in \bar{E}$; thus f defines a mapping $\bar{f}$ of R into $\bar{E}$. Now f satisfies the required properties provided no $\bar{f}(p)$ lies in S, i.e., $\bar{f}(R) \cap S = 0$. It turns out in general that S is the union of a set of smooth manifolds $S_0, \cdots, S_\mu$.

We alter f_0 by adding to it a polynomial of degree $\leqq r$, whose coefficients form a set α of very small numbers. Let f_α be the resulting mapping; there is a corresponding mapping $\bar{f}_\alpha$ of R into $\bar{E}$. Given any compact piece T of S and any compact piece Q of R, we show that we can choose α so that $\bar{f}_\alpha(Q) \cap T = 0$, simply by looking at the dimensions of the S_i. An elementary type of reasoning now shows that we can find f with $\bar{f}(R)$ separated from S; this f is the required mapping.[1]

9. Separation of sets in $\bar{E}$

Let R and R^* be open sets in E and in $\mathfrak{A}^\nu$ respectively, and let

$$(9.1)\qquad G(p, \alpha) = g_\alpha(p) = g_p^*(\alpha)$$

be a smooth mapping of the Cartesian product $R \times R^*$ into $\bar{E}$. For each $p \in R$, g_p^* is a smooth mapping of R^* into $\bar{E}$; $\nabla g_p^*(\alpha)$ maps vectors in $\mathfrak{A}^\nu$ into vectors in $\bar{E}$. If the image vector space is of dimension ν for each p and α, we say that the g_α form a *ν-parameter family* of mappings of R into $\bar{E}$.

We say the subset Q of a Euclidean space is of *finite μ-extent* if there is a number A with the following property. For any integer h there are sets $Q_1, \cdots, Q_a$ such that

$$(9.2)\qquad Q = Q_1 \cup \cdots \cup Q_a, \qquad \operatorname{diam}(Q_j) < 1/2^h \text{ (all } j), \qquad a \leqq 2^{\mu h}A.$$

LEMMA 9a. *Let the g_α form a ν-parameter family of mappings of R into $\bar{E}$, let Q and T be compact subsets of R and of $\bar{E}$ of finite K-extent and of finite M-extent re-*

[1] For more general results, see [6].

spectively, and suppose $K + M < \nu$. Then arbitrarily near any α_0 there is an α such that

$$(9.3) \qquad g_\alpha(Q) \cap T = 0.$$

Since G is smooth, there is a spherical neighborhood U_0 of α_0 and a number L such that

$$(9.4) \qquad | g_\alpha(p') - g_\alpha(p) | \leqq L \, | p' - p |, \qquad \alpha \in \bar{U}_0, \quad p, p' \in Q.$$

Since the g_α form a ν-parameter family, no g_p^* carries any non-zero vector into the vector 0. Hence there is a $\gamma > 0$ such that

$$(9.5) \qquad | \nabla_V g_p^*(\alpha) | \geqq 2\gamma \text{ if } | V | = 1, \qquad \alpha \in \bar{U}_0, \quad p \in Q.$$

We may choose a spherical neighborhood $U_1 \subset U_0$ of α_0 so that

$$(9.6) \qquad | \nabla_V g_p^*(\alpha') - \nabla_V g_p^*(\alpha) | \leqq \gamma \text{ if } | V | = 1, \quad \alpha, \alpha' \in U_1, \quad p \in Q.$$

We now show that

$$(9.7) \qquad | g_{\alpha'}(p) - g_\alpha(p) | \geqq \gamma \, | \alpha' - \alpha | \text{ if } \alpha, \alpha' \in U_1, \qquad p \in Q.$$

For if $V = \alpha' - \alpha$, then

$$(9.8) \qquad g_{\alpha'}(p) - g_\alpha(p) = \int_0^1 \nabla_V g_p^*(\alpha + tV) \, dt,$$

and hence, using (9.6),

$$| g_{\alpha'}(p) - g_\alpha(p) - \nabla_V g_p^*(\alpha) |$$
$$= \left| \int_0^1 [\nabla_V g_p^*(\alpha + tV) - \nabla_V g_p^*(\alpha)] \, dt \right| \leqq \gamma \, | V | ;$$

now (9.7) follows with the help of (9.5).

Choose A and B so that (9.2) holds for Q, with A, and for T, with B, respectively. Choose s and h so that, if ζ is the radius of U_1,

$$L < 2^s, \qquad AB/\gamma^\nu 2^{\nu h - Ks - \nu} < \zeta^\nu.$$

Choose expressions for Q and T:

$$Q = Q_1 \cup \cdots \cup Q_a, \ \operatorname{diam}(Q_j) < 1/2^{s+h}, \qquad a \leqq 2^{K(s+h)} A,$$
$$T = T_1 \cup \cdots \cup T_b, \ \operatorname{diam}(T_k) < 1/2^h, \qquad b \leqq 2^{Mh} B.$$

Let Z_{jk} be the set of all $\alpha \in U_1$ such that

$$g_\alpha(Q_j) \cap T_k \neq 0.$$

Suppose α and α' are in Z_{jk}. Then there are points $p, p' \in Q_j$ such that $g_\alpha(p)$ and $g_{\alpha'}(p')$ are in T_k, and hence are distant apart $< 1/2^h$. Using (9.4) (with α'), we have therefore

$$| g_\alpha(p) - g_{\alpha'}(p) | < 1/2^h + L \operatorname{diam}(Q_j) < 2/2^h.$$

Using (9.7) gives $|\alpha' - \alpha| < 2/2^h\gamma$. Thus

$$\operatorname{diam}(Z_{jk}) \leqq 1/2^{h-1}\gamma.$$

Letting ρ_ν be the volume of the unit ball in ν-space, it follows that Z_{jk} lies in a ball P_{jk} of radius $1/2^{h-1}\gamma$, which has volume $\rho_\nu/(2^{h-1}\gamma)^\nu$. Since $K + M \leqq \nu - 1$, we have

$$\sum \operatorname{vol}(P_{jk}) = ab\rho_\nu/2^{h\nu-\nu}\gamma^\nu \leqq AB\rho_\nu/2^{h-Ks-\nu}\gamma^\nu < \rho_\nu\zeta^\nu,$$

which is the volume of U_1. Therefore there is an α in U_1 which is in no Z_{jk}. For this α, $g_\alpha(Q) \cap T = 0$, as required.

10. The family of mappings f_α

Returning to the considerations of §8, we take first the case $n = m = r = 2$; use coordinates (x, y) in E and (u, v) in E'. The axes in $\bar{E} = \mathfrak{A}^{10}$ correspond to the partial derivatives

$$(u_x, u_y, u_{xx}, u_{xy}, u_{yy}, v_x, v_y, v_{xx}, v_{xy}, v_{yy}).$$

Let e_u and e_v be the unit vectors in E'. Given f_0 and any set of numbers $\alpha = (\alpha_1, \cdots, \alpha_1', \cdots) \in \mathfrak{A}^{10}$, set

$$\begin{aligned} f_\alpha(x, y) = f_0(x, y) &+ (\alpha_1 x + \alpha_2 y + \alpha_{11}x^2 + \alpha_{12}xy + \alpha_{22}y^2)e_u \\ &+ (\alpha_1' x + \alpha_2' y + \alpha_{11}'x^2 + \alpha_{12}'xy + \alpha_{22}'y^2)e_v . \end{aligned} \tag{10.1}$$

Let us write $f_\alpha(x, y) = (u^\alpha, v^\alpha)$. Then the new partial derivatives are

$$u_x^\alpha = u_x^0 + \alpha_1 + 2\alpha_{11}x + \alpha_{12}y, \qquad u_y^\alpha = u_y^0 + \alpha_2 + \alpha_{12}x + 2\alpha_{22}y,$$

$$u_{xx}^\alpha = u_{xx}^0 + 2\alpha_{11}, \qquad u_{xy}^\alpha = u_{xy}^0 + \alpha_{12}, \qquad u_{yy}^\alpha = u_{yy}^0 + 2\alpha_{22},$$

and similarly for the partial derivatives of v^α. Therefore the mapping $\bar{f}_\alpha$, at any $p \in E$, is given by

$$\begin{aligned} \bar{f}_\alpha = \bar{f}_0 + (&\alpha_1 + 2\alpha_{11}x + \alpha_{12}y, \alpha_2 + \alpha_{12}x + 2\alpha_{22}y, 2\alpha_{11}, \alpha_{12}, 2\alpha_{22}, \\ &\alpha_1' + 2\alpha_{11}'x + \alpha_{12}'y, \alpha_2' + \alpha_{12}'x + 2\alpha_{22}'y, 2\alpha_{11}', \alpha_{12}', 2\alpha_{22}'). \end{aligned} \tag{10.2}$$

Writing $\bar{f}_p^*(\alpha) = \bar{f}_\alpha(p)$, we now consider the partial derivatives of $\bar{f}_p^*$, in the directions of the axes in $\mathfrak{A}^{10}$. From (10.2) we see that the first few of these are

$$\begin{aligned} \partial\bar{f}_\alpha/\partial\alpha_1 &= (1, 0, 0, 0, \cdots), \qquad \partial\bar{f}_\alpha/\partial\alpha_2 = (0, 1, 0, 0, \cdots), \\ \partial\bar{f}_\alpha/\partial\alpha_{11} &= (2x, 0, 2, 0, \cdots), \cdots . \end{aligned} \tag{10.3}$$

These vectors are the images under $\nabla\bar{f}_p^*(\alpha)$ of the unit vectors in $\mathfrak{A}^{10}$. Since the matrix of components has non-zero diagonal elements and zeros only to the right, the image vectors are independent. This shows that the $\bar{f}_\alpha$ form a 10-parameter family.

Turning to the general case, we prove

LEMMA 10a. *The f_α form an N-parameter family of mappings of R into $\bar{E}$.*

Here, $\alpha \in \mathfrak{A}^N$ is a set of numbers

$$(10.4)\qquad \alpha_{i;\lambda_1\cdots\lambda_h};\qquad 1 \leqq i \leqq m, 1 \leqq h \leqq r, \qquad 1 \leqq \lambda_1 \leqq \cdots \leqq \lambda_h \leqq n,$$

ordered as follows. Take first those with $i = 1$, then with $i = 2$, etc. For each i, take those with $h = 1$, then with $h = 2$, etc. For each h, use lexicographic ordering. (This agrees with the ordering used above.)

With unit vectors $e_1, \cdots, e_m$ in E', define f_α by

$$(10.5)\qquad f_\alpha(p) = f_0(p) + \sum_{i=1}^m [\sum_{h=1}^r \sum_{\lambda_1 \leqq \cdots \leqq \lambda_h} \alpha_{i;\lambda_1\cdots\lambda_h} x_{\lambda_1} \cdots x_{\lambda_h}] e_i .$$

The components of $f_\alpha(p)$ are

$$(10.6)\qquad u_{i;\alpha} = u_{i;0} + \sum_{h=1}^r \sum_{\lambda_1 \leqq \cdots \leqq \lambda_h} \alpha_{i;\lambda_1\cdots\lambda_h} x_{\lambda_1} \cdots x_{\lambda_h} .$$

Taking partial derivatives, we see that there are integers $K_{\lambda_1\cdots\lambda_h} > 0$ such that

$$(10.7)\qquad \frac{\partial^h u_{i;\alpha}}{\partial x_{\lambda_1} \cdots \partial x_{\lambda_h}} = \frac{\partial^h u_{i;0}}{\partial x_{\lambda_1} \cdots \partial x_{\lambda_h}} + K_{\lambda_1 \cdots \lambda_h} \alpha_{i;\lambda_1\cdots\lambda_h} + \cdots ,$$

where the remaining terms contain factors $\alpha_{i;\mu_1\cdots\mu_{h'}}$ with $h' > h$. These derivatives are the components of $\bar{f}_\alpha(p) = \bar{f}_p^*(\alpha)$. Differentiating these in the directions of the unit vectors in $\mathfrak{A}^N$, we have

$$\frac{\partial \bar{f}_\alpha(p)}{\partial \alpha_{i;\lambda_1\cdots\lambda_h}} = (\cdots, K_{\lambda_1\cdots\lambda_h}, 0, 0, \cdots);$$

the matrix of components is triangular, with the $K_{\lambda_1\cdots\lambda_h}$ as diagonal elements. Hence these vectors are independent, showing that the image vector space is of dimension N. This proves the lemma.

11. Transformation away from manifolds

A *smooth manifold* S' in $\bar{E} = \mathfrak{A}^N$ is a point set with the following property. For each point $p \in S'$ there is a neighborhood U of p such that $S' \cap U$ is exactly the set of points in U which satisfy a set of independent equations. That is, there are real smooth functions $\Phi_1, \cdots, \Phi_\delta$ in U whose gradients are independent in U, such that $\Phi_1(p) = \cdots = \Phi_\delta(p) = 0$ for $p \in U$ if and only if $p \in S'$. Equivalently using a coordinate system, U can be chosen so that δ of the coordinates are given as smooth functions of the remaining $M = N - \delta$ coordinates. We call M and δ the *dimension* and *defect* of S' respectively.

By a *manifold collection* S of dimension M and defect $\delta = N - M$ we mean a set of disjoint manifolds $S_0, \cdots, S_\mu$, such that $M = \max\{\dim(S_i)\}$, and such that each point set $S_0 \cup \cdots \cup S_i$ is closed in $\bar{E}$. The "bad" sets we shall use will be manifold collections.

THEOREM 11A. *Let f_0 be a mapping of the open set $R \subset E^n$ into E^m, let S be a manifold collection of dimension M and defect δ in $\bar{E} = \mathfrak{A}^N$* (see §8), *let $\varepsilon(p)$ be a positive continuous function in R, and suppose*

$$(11.1)\qquad n + M < N, \quad \text{or,} \quad n < \delta.$$

Then there is an ε-approximation f to f_0 in R such that for the corresponding mapping $\bar{f}$ into $\bar{E}$,

$$\bar{f}(R) \cap S = 0. \tag{11.2}$$

Moreover, if f_0 is s-smooth, we may require f to be an (s, ε)-approximation to f_0.

There is clearly a sequence of open balls $U_1, U_2, \cdots$ in E^n with the following properties. Let U_i' be the open ball concentric with U_i and of twice the radius. Then the closures $\bar{U}_i'$ lie in R, the U_i cover R, and each $\bar{U}_i$ touches but a finite number of $\bar{U}_j'$. Set $r' = \max(r, s)$. There is a polynomial $\Phi(t)$ $(1 \leqq t \leqq 2)$ such that $\Phi(1) = 1$ and $\Phi(2) = 0$, and such that all its derivatives of orders $\leqq r'$ vanish at these points. With the help of Φ we may clearly construct r'-smooth functions $\phi_1, \phi_2, \cdots$ in $E = E^n$ (we could make them ∞-smooth) such that

$$\phi_i = 1 \text{ in } \bar{U}_i, \qquad \phi_i = 0 \text{ in } E - U_i'. \tag{11.3}$$

Say $S = S_0 \cup \cdots \cup S_\mu$ as above. As noted in §1, we may suppose f_0 is r'-smooth. We shall find mappings $f_0, f^0, f^1, \cdots, f^\mu$ such that each is an arbitrarily good approximation to the last (through the order r'), and

$$\bar{f}^h(R) \cap S_h^* = 0, \qquad S_h^* = S_0 \cup \cdots \cup S_h. \tag{11.4}$$

Then $f = f^\mu$ is the required mapping.

Suppose f^h has been found; we show how to find f^{h+1}. (The same proof gives f^0.) We shall find mappings f^{hk} and numbers ε^{hk} $(k = 1, 2, \cdots)$ such that (setting $f^{h0} = f^h$)

$$f^{hk} = f^{h,k-1} \text{ in } R - U_k', \qquad f^{hk}\ (s, \varepsilon^{hk})\text{-approximates } f^{h,k-1}, \tag{11.5}$$

$$\bar{f}^{hk}(R) \cap S_h^* = 0, \qquad \bar{f}^{hk}(\bar{U}_k^*) \cap S_{h+1}^* = 0, \qquad U_k^* = U_1 \cup \cdots \cup U_k. \tag{11.6}$$

By the finiteness property of the U_i, all the f^{hk} will be the same in any $\bar{U}_i$ from some point on; hence

$$f^{h+1} = \lim_{k\to\infty} f^{hk} \tag{11.7}$$

will exist and be r'-smooth, and (11.4) for $h + 1$ will hold. With small enough numbers ε^{hk}, f^{h+1} will approximate to f^h as required.

Having $f^{h,k-1}$, we show how to find f^{hk}. Using coordinate systems in E^n and in E^m and notations as before, define f_α by (10.5) (with $f^{h,k-1}$ in place of f_0) and set

$$f^{hk}(p) = f^{h,k-1}(p) + \phi_k(p)[f_\alpha(p) - f^{h,k-1}(p)]. \tag{11.8}$$

Then (11.5) will hold if α is small enough. Hence we need merely obtain (11.6).

Since S_h^* is closed, $A = \bar{f}^{h,k-1}(\bar{U}_k')$ is compact, and the first part of (11.6) holds with $k - 1$, the sets S_h^* and A are at a positive distance apart, and hence the relation will hold with k provided we take α small enough. Since the second part of (11.6) holds with $k - 1$, we will have $\bar{f}^{hk}(\bar{U}_{k-1}^*) \cap S_{h+1}^* = 0$ for α small enough. Therefore, since $\phi_k = 1$ in $\bar{U}_k$, we need merely find α so that

$$\bar{f}_\alpha(\bar{U}_k) \cap S_{h+1} = 0. \tag{11.9}$$

There is a bounded open set Z containing $\bar{f}^{h,k-1}(\bar{U}_k)$, whose closure $\bar{Z}$ does not touch S_h^*. Set $T = S_{h+1} \cap \bar{Z}$. Using a small α, we need merely obtain (11.9) with T in place of S_{h+1}.

Clearly T is closed (since S_{h+1}^* is); we show that it is of finite M-extent. Since S_{h+1} is a smooth manifold, there are points $p_1, \cdots, p_j$ in it and portions $H_1, \cdots, H_j$ of it covering T, such that each H_i has a smooth projection π_i into the tangent plane P_i to S_{h+1} at p_i. Now the inverse π_i^{-1} expands distances by at most a certain factor L_i. Since $\pi_i(H_i)$ is in a bounded part of the M_{h+1}-dimensional plane P_i ($M_{h+1} \leqq M$), it is clearly of finite M-extent; hence so is H_i, and so is T.

Also $\bar{U}_k$ is of finite n-extent. By Lemma 10a, the f_α form an N-parameter family in R. Because of (11.1), we may apply Lemma 9a, giving $\bar{f}_\alpha(\bar{U}_k) \cap T = 0$. This proves (11.9), and the theorem follows.

12. Application to critical point theory

As an example of the method above, we consider the case $m = 1$; the mapping f of $R \subset E^n$ into E^m is then simply a real function in R. For each $p \in R$, the mapping $\nabla_V f(p)$ is a real valued linear function of vectors V, i.e., is a "covector" at p; this gives a mapping $F(p)$ of R into the linear space (of dimension n) of covectors. The point p is a *critical point* of f if $F(p) = 0$, i.e., $\nabla_V f(p) = 0$ (all V).

To insure that a critical point p be isolated, we may require that $\nabla F(p)$ carry non-zero vectors into non-zero covectors; that is, $\nabla_W F(p) \neq 0$ if $W \neq 0$. Then for any $p' \neq p$ near p, we will have $F(p') \neq 0$, so that p' is not a critical point.

Let $(x_1, \cdots, x_n)$ be a coordinate system in E^n, with unit vectors $e_1, \cdots, e_n$. Then in terms of components,

$$F(p) = \left(\frac{\partial f}{\partial x_1}, \cdots, \frac{\partial f}{\partial x_n}\right). \tag{12.1}$$

Also

$$\nabla_{e_i} F(p) = \frac{\partial F(p)}{\partial x_i} = \left(\frac{\partial^2 f}{\partial x_i \partial x_1}, \cdots, \frac{\partial^2 f}{\partial x_i \partial x_n}\right), \tag{12.2}$$

so that the condition that $\nabla F(p)$ carry non-zero vectors into non-zero vectors, or, independent vectors into independent vectors, is that the determinant (the Hessian) $| \partial^2 f/\partial x_i \partial x_j |$ be $\neq 0$.

THEOREM 12A.[2] *Arbitrarily near any f_0 there is an f such that the Hessian is $\neq 0$ at all critical points; the critical points of f are isolated.*

Define $\bar{E}$ as in §8, with $r = 2$; then

$$\bar{f}(p) = \left(\frac{\partial f}{\partial x_1}, \cdots, \frac{\partial f}{\partial x_n}, \frac{\partial^2 f}{\partial x_1^2}, \frac{\partial^2 f}{\partial x_1 \partial x_2}, \cdots, \frac{\partial^2 f}{\partial x_n^2}\right), \tag{12.3}$$

[2] Another proof may be given as follows. First approximate to f_0 by an analytic function g; set $f(x_1, \cdots, x_n) = g(x_1, \cdots, x_n) + \alpha_1 x_1 + \cdots + \alpha_n x_n$ with arbitrarily small α. See Morse, [3], Theorem 16.

and $\bar{E} = \mathfrak{A}^N$, $N = n + n(n+1)/2$. The "bad" set S in $\bar{E}$ is the set of sets $\alpha = (\alpha_1, \cdots, \alpha_{11}, \cdots)$ such that, setting $\alpha_{ji} = \alpha_{ij}$ for $j > i$,

$$(12.4) \qquad \alpha_1 = \cdots = \alpha_n = 0, \qquad D = \begin{vmatrix} \alpha_{11} & \cdots & \alpha_{1n} \\ \cdot & \cdot\ \cdot\ \cdot & \cdot \\ \alpha_{n1} & \cdots & \alpha_{nn} \end{vmatrix} = 0.$$

Let S_i be the subset of S in which the rank of the matrix of D is i; then $S = S_0 \cup \cdots \cup S_{n-1}$. We show that S_i is a smooth manifold of dimension

$$(12.5) \qquad M_i = \binom{i+1}{2} + i(n-i) = \binom{n+1}{2} - \binom{n-i+1}{2}.$$

Given $\alpha \in S_i$, say the rows $A_{\lambda_1}, \cdots, A_{\lambda_i}$ of D are independent. Then, as is well known, the principal minor with these rows and columns is $\neq 0$. (For if not, then in the submatrix containing these rows, there would be a column, say the kth, independent of the columns numbered $\lambda_1, \cdots, \lambda_i$. Since D is symmetric, the (complete) k^{th} row would then be independent of the rows $A_{\lambda_1}, \cdots, A_{\lambda_i}$, and D would be of rank $>i$, a contradiction.)

Now take any $\alpha' \in S_i$ near α; then the rows $A'_{\lambda_1}, \cdots, A'_{\lambda_i}$ are independent. Each other row is dependent on these, and has a unique expression

$$(12.6) \qquad A'_k = a_{k\lambda_1}A'_{\lambda_1} + \cdots + a_{k\lambda_i}A'_{\lambda_i}.$$

This gives

$$(12.7) \qquad \alpha'_{\lambda_j k} = \alpha'_{k\lambda_j} = a_{k\lambda_1}\alpha'_{\lambda_1\lambda_j} + \cdots + a_{k\lambda_i}\alpha'_{\lambda_i\lambda_j}.$$

These equations can be solved for the $a_{k\lambda_j}$ (k different from any λ_h), giving these as fixed analytic functions of the elements of the rows $A'_{\lambda_1}, \cdots, A'_{\lambda_i}$. The $\binom{i+1}{2}$ elements $\alpha'_{\lambda_h\lambda_j}$ with $\lambda_h \leqq \lambda_j$ may be chosen arbitrarily, and also the $\alpha'_{\lambda_h k}$ ($k \neq$ any λ_j); the remaining elements of the A'_{λ_h} are determined by symmetry, and the remaining rows A'_k by (12.6). Hence S_i is a smooth manifold, and its dimension is the number of independent elements, given by (12.5).

For any sequence of elements $\alpha^{(1)}, \alpha^{(2)}, \cdots$ of S_i, with a limit α^*, we have rank $(D^*) \leqq i$; hence $\alpha^* \in S_j$ with $j \leqq i$, showing that $S_0 \cup \cdots \cup S_i$ is closed. Therefore S is a manifold collection, of dimension

$$(12.8) \qquad M = \max(M_i) = \binom{n+1}{2} - 1.$$

Since $n + M < N$, Theorem 11A shows that we may choose f near f_0 so that $\bar{f}(R) \cap S = 0$. This is the desired function.

13. Obtaining excellent mappings

We are now ready for the basic theorem.

THEOREM 13A. *Let f_0 be a mapping of $R \subset E = E^2$ into $E' = E'^2$. Then ar-*

bitrarily near f_0 there is an excellent mapping f. If f_0 is r-smooth and $\varepsilon(p)$ is a positive continuous function in R, we may make f an (r, ε)-approximation to f_0.

As noted in §1, we may and shall suppose that f_0 is 3-smooth. Because of Theorem 11A, we need merely find the corresponding bad set S in $\bar{E}$, and show that it is a manifold collection of defect >2 in $\bar{E}$.

For a better understanding, we prove (a) and (b) before proving (c).

(a) We may obtain f in R such that ∇f is of rank $\geqq 1$ at all points of R.

With coordinate systems (x, y) in E and (u, v) in E', let $\bar{E} = \mathfrak{A}^4$ correspond to the partial derivatives (u_x, u_y, v_x, v_y); then we must avoid $\bar{f}(p) = (0, 0, 0, 0)$. Thus the bad set consists of the origin alone in $\bar{E}$. This is a manifold collection of defect $4 > 2$; hence f may be obtained.

(b) We may obtain a good f in R.

In this case, take $\bar{E} = \mathfrak{A}^{10}$ as in §8. The set $S^{(b)}$ of bad points corresponds to the equations

$$J = 0, \qquad J_x = J_y = 0. \tag{13.1}$$

Let $S_0^{(b)}$ be the subset of $\bar{E}$ corresponding to $u_x = u_y = v_x = v_y = 0$, or better, $\alpha_1 = \alpha_2 = \alpha_1' = \alpha_2' = 0$. Since the first equations imply (13.1), $S_0^{(b)} \subset S^{(b)}$. Also $S_0^{(b)}$ is of defect 4 in $\bar{E}$. Let $S_1^{(b)}$ be the rest of $S^{(b)}$; we show that it is a smooth manifold of defect 3. Differentiating (4.2) gives $J_x = u_{xx}v_y + \cdots$. Let $\bar{J}$, $\bar{J}_x$, $\bar{J}_y$ be the corresponding real valued functions in $\bar{E}$. Then

$$\bar{J} = \alpha_1\alpha_2' - \alpha_2\alpha_1', \tag{13.2}$$

$$\bar{J}_x = \alpha_{11}\alpha_2' + \alpha_1\alpha_{12}' - \alpha_{12}\alpha_1' - \alpha_2\alpha_{11}', \tag{13.3}$$

$$\bar{J}_y = \alpha_{12}\alpha_2' + \alpha_1\alpha_{22}' - \alpha_{22}\alpha_1' - \alpha_2\alpha_{12}'. \tag{13.4}$$

Differentiating these functions with respect to $\alpha_1, \alpha_2, \alpha_{11}, \cdots$ in turn shows that the gradients are

$$\begin{aligned}
\nabla\bar{J} &= (\alpha_2', \ -\alpha_1', \ 0, \ 0, \ 0, \ -\alpha_2, \ \alpha_1, \ 0, \ 0, \ 0),\\
\nabla\bar{J}_x &= (\alpha_{12}', \ -\alpha_{11}', \ \alpha_2', \ -\alpha_1', \ 0, \ -\alpha_{12}, \ \alpha_{11}, \ -\alpha_2, \ \alpha_1, \ 0),\\
\nabla\bar{J}_y &= (\alpha_{22}', \ -\alpha_{12}', \ 0, \ \alpha_2', \ -\alpha_1', \ -\alpha_{22}, \ \alpha_{12}, \ 0, \ -\alpha_2, \ \alpha_1).
\end{aligned}$$

For any point of $S_1^{(b)}$, clearly these gradients are independent. For instance, if $\alpha_1 \neq 0$, then the 7th, 9th and 10th columns are independent. Hence $S_1^{(b)}$ is of defect 3, as stated. Since $S_0^{(b)}$ and $S^{(b)}$ are closed, Theorem 11A applies.

(c) To prove the theorem, we consider $\bar{E} = \mathfrak{A}^{18}$, corresponding to all partial derivatives of orders $\leqq 3$. The bad set S corresponds to the three equations

$$J = 0, \qquad \nabla_V f = 0, \qquad \nabla_V \nabla_V f = 0, \tag{13.5}$$

the last two equations being vector equations and hence corresponding to a pair of equations apiece.

Define S_0 and S_1 like the $S_0^{(b)}$ and $S_1^{(b)}$ above; these are of defect 4 and 3 respectively. ($\nabla \bar{J}$, $\nabla \bar{J}_x$ and $\nabla \bar{J}_y$ are as before, with extra zeros.) Set $S_2 = S - (S_0 \cup S_1)$. We show that S_2 is of defect 3.

Take any point

$$\alpha = (\alpha_1, \alpha_2, \alpha_{11}, \alpha_{12}, \alpha_{22}, \alpha_{111}, \alpha_{112}, \alpha_{122}, \alpha_{222}, \alpha_1', \cdots)$$

of S_2. Since α is not in S_0 or in S_1, we have

$$(13.6) \qquad \alpha_1 \text{ or } \alpha_2 \text{ or } \alpha_1' \text{ or } \alpha_2' \text{ is } \neq 0, \qquad \bar{J}_x \neq 0 \text{ or } \bar{J}_y \neq 0.$$

Say α_2' is $\neq 0$. (The other cases are similar.) We show that the functions in $\bar{E}$ corresponding to the equations

$$(13.7) \qquad J = 0, \qquad \nabla_V v = 0, \qquad \nabla_V \nabla_V v = 0,$$

have independent gradients. (The two remaining equations, corresponding to $\nabla_V u = \nabla_V \nabla_V u = 0$, are consequences of these, since ∇f is of rank 1 when $J = 0$, and $\alpha_2' \neq 0$ corresponds to $v_y \neq 0$.)

The first equation is (13.2), and its gradient is $(\alpha_2', -\alpha_1', \cdots)$, which is $\neq 0$ since $\alpha_2' \neq 0$.

The second equation corresponds to

$$-J_y v_x + J_x v_y = 0;$$

see (3.5) and (4.5). By (13.3) and (13.4), it is

$$F_2(\alpha) = -(\alpha_{12}\alpha_2' + \cdots)\alpha_1' + (\alpha_{11}\alpha_2' + \cdots)\alpha_2' = \alpha_{11}(\alpha_2')^2 + \cdots,$$

and $\partial F_2/\partial \alpha_{11} = (\alpha_2')^2 \neq 0$. Hence ∇F_2 is independent of ∇F_1.

By (3.6) and (4.5), the third equation $F_3 = 0$ corresponds to

$$-J_y(-J_y v_x + J_x v_y)_x + J_x(-J_y v_x + J_x v_y)_y = 0.$$

We shall find where u_{xxx} and u_{xxy} appear. The above equation is, by the formulas for J_x and J_y (see (13.3) and (13.4)),

$$\begin{aligned} -J_y[-(u_{xy}v_y + \cdots)v_x + (u_{xx}v_y - u_{xy}v_x + \cdots)v_y]_x & \\ + J_x[\cdots + (u_{xx}v_y + \cdots)v_y]_y & \\ = [-J_y v_y^2]u_{xxx} + [2J_y v_x v_y + J_x v_y^2]u_{xxy} + \cdots &= 0, \end{aligned}$$

and

$$\partial F_3/\partial \alpha_{111} = -\bar{J}_y(\alpha_2')^2, \qquad \partial F_3/\partial \alpha_{112} = 2\bar{J}_y \alpha_1' \alpha_2' + \bar{J}_x(\alpha_2')^2.$$

Since $\alpha_2' \neq 0$ and one of $\bar{J}_x$, $\bar{J}_y$ is $\neq 0$, one of these is $\neq 0$; hence ∇F_3 is $\neq 0$, in fact independent of ∇F_1 and ∇F_2. Therefore S_2 is of defect 3. Also S_0, $S_0 \cup S_1$ and S are closed. Hence S is a manifold collection of defect 3, and again Theorem 11A applies. This completes the proof.

PART C. NORMAL FORMS

14. Topics related to Taylor's formula

We shall give some results from [7] and [8], and a few consequences. Part of (a) appears in Morse, [2], §2. Some work of Levi[3] touches on this also. We shall use a special notation for partial derivatives:

$$\phi_{\lambda_1\cdots\lambda_n}(x_1, \cdots, x_n) = \frac{\partial^{\sigma_\lambda}}{\partial x_1^{\lambda_1} \cdots \partial x_n^{\lambda_n}} \phi(x_1, \cdots, x_n), \quad \sigma_\lambda = \lambda_1 + \cdots + \lambda_n .$$

(a) Let $\phi(x_1, \cdots, x_n)$ be r-smooth in a neighborhood of $(0, \cdots, 0)$, and suppose $1 \leqq s \leqq r$. Then we may write

$$\begin{aligned} \phi(x_1, x_2, \cdots, x_n) &= \psi^0(x_2, \cdots, x_n) + \cdots + \frac{x_1^{s-1}}{(s-1)!} \psi^{s-1}(x_2, \cdots, x_n) \\ &\quad + \frac{x_1^s}{s!} \psi^s(x_1, x_2, \cdots, x_n), \end{aligned} \tag{14.1}$$

where

$$\psi^i(x_2, \cdots, x_n) = \phi_{i0\cdots0}(0, x_2, \cdots, x_n) \qquad (i = 0, \cdots, s-1), \tag{14.2}$$

$$\psi^s(x_1, x_2, \cdots, x_n) = \frac{s}{x_1^s} \int_0^{x_1} (x_1 - t)^{s-1} \phi_{s0\cdots0}(t, x_2, \cdots, x_n) dt \ (x_1 \neq 0), \tag{14.3}$$

$$\psi^s(0, x_2, \cdots, x_n) = \phi_{s0\cdots0}(0, x_2, \cdots, x_n). \tag{14.4}$$

Each function ψ^i is $(r - i)$-smooth $(i = 0, \cdots, s)$. Moreover, for $\sigma_\lambda \leqq r - s$,

$$\psi^s_{\lambda_1\cdots\lambda_n}(x_1, x_2, \cdots) = \frac{s}{x_1^{s+\lambda_1}} \int_0^{x_1} (x_1 - t)^{s-1} t^{\lambda_1} \phi_{s+\lambda_1,\lambda_2\cdots}(t, x_2, \cdots)\, dt \qquad (x_1 \neq 0), \tag{14.5}$$

$$\psi^s_{\lambda_1\cdots\lambda_n}(0, x_2, \cdots) = \frac{s!\lambda_1!}{(s + \lambda_1)!} \phi_{s+\lambda_1,\lambda_2\cdots}(0, x_2, \cdots), \tag{14.6}$$

$$x_1^{s-i} \psi^s_{\lambda_1\cdots\lambda_n}(x_1, \cdots, x_n) \text{ is continuous if } 0 \leqq i = r - \sigma_\lambda \leqq s. \tag{14.7}$$

Hence the last relation holds also if $0 \leqq i \leqq r - \sigma_\lambda \leqq s$.

(b) As a corollary, suppose $1 \leqq s \leqq r$, $\phi(x_1, \cdots, x_n)$ is r-smooth, and

$$\phi_{i0\cdots0}(0, x_2, \cdots, x_n) = 0 \qquad (i = 0, 1, \cdots, s-1). \tag{14.8}$$

Then there is an $(r - s)$-smooth function $\Phi(x_1, \cdots, x_n)$ such that

$$\phi(x_1, \cdots, x_n) = x_1^s \Phi(x_1, \cdots, x_n). \tag{14.9}$$

(c) For a converse, suppose $1 \leqq s \leqq r$, $\Phi(x_1, \cdots, x_n)$ is $(r - s)$-smooth every-

[3] E. E. Levi, in [1] discusses differentiability of the function ψ^s in (14.1), letting the point about which ϕ is expanded also be variable. His statement that our ψ^s is r-smooth is of course not true at $(x_1, \cdots, x_n) = (0, \cdots, 0)$.

where and is r-smooth for $x_1 \neq 0$, and (14.7) holds with Φ in place of ψ^s. If we set

$$\phi(x_1, \cdots, x_n) = x_1^s \Phi(x_1, \cdots, x_n), \tag{14.10}$$

then ϕ is r-smooth.

(d) We say an s-smooth function $\phi(x_1, \cdots, x_n)$ is of *order* $\geqq s$, or ord$(\phi) \geqq ts$, if ϕ and all its partial derivatives of orders $<s$ vanish at the origin. Clearly then $\phi_{\lambda_1 \cdots \lambda_n}(x_1, \cdots, x_n)$ is of order $\geqq s - \sigma_\lambda$ if $\sigma_\lambda < s$.

(e) Suppose $\phi(x_1, \cdots, x_n)$ is r-smooth, and $s \leqq r$. Expand ϕ as in (14.1). Now expand each $\psi^i (i < s)$ in terms of x_2, replacing s by $s - i$. Expand again, in terms of x_3, etc. This gives

$$\begin{aligned} \phi(x_1, \cdots, x_n) &= \textstyle\sum_{\sigma_\lambda < s} a_{\lambda_1 \cdots \lambda_n} \dfrac{x_1^{\lambda_1} \cdots x_n^{\lambda_n}}{\lambda_1! \cdots \lambda_n!} \\ &\quad + \textstyle\sum_{\sigma_\lambda = s} \dfrac{x_1^{\lambda_1} \cdots x_n^{\lambda_n}}{\lambda_1! \cdots \lambda_n!} F_{\lambda_1 \cdots \lambda_n}(x_1, \cdots, x_n), \end{aligned} \tag{14.11}$$

$$a_{\lambda_1 \cdots \lambda_n} = \phi_{\lambda_1 \cdots \lambda_n}(0, \cdots, 0), \; F_{\lambda_1 \cdots \lambda_n}(0, \cdots, 0) = \phi_{\lambda_1 \cdots \lambda_n}(0, \cdots, 0). \tag{14.12}$$

(Actually many of the $F_{\lambda_1 \cdots \lambda_n}$ are functions of fewer variables.) The $F_{\lambda_1 \cdots \lambda_n}$ are $(r - s)$-smooth, and each term in the last sum is of order $\geqq s$.

(f) Let $\phi(x_1, \cdots, x_n)$ be $(s + 1)$-smooth and of order $\geqq s$. Then clearly $x_i \phi$ is of order $\geqq s + 1$.

(g) Suppose ϕ is r-smooth and of order $\geqq s$, $s \leqq r$, and

$$\phi(0, x_2, \cdots, x_n) = 0, \qquad \text{all } x_2, \cdots, x_n. \tag{14.13}$$

Then there is an $(r - 1)$-smooth function Φ, of order $\geqq s - 1$, such that

$$\phi(x_1, \cdots, x_n) = x_1 \Phi(x_1, \cdots, x_n). \tag{14.14}$$

That the $(r - 1)$-smooth function Φ exists follows from (b). Now expand Φ as in (14.11), into a polynomial of degree $\leqq s - 2$ and a remainder term of order $\geqq s - 1$. Using $\phi = x_1 \Phi$ and the fact that ϕ is of order $\geqq s$ shows that the polynomial is 0; hence Φ is of order $\geqq s - 1$.

(h) Let $\Phi(x, y)$ and $\phi(x)$ be r-smooth, let Φ be of order $\geqq s$, and suppose

$$\phi(0) = 0; \qquad \Phi(x, \phi(x)) = \Phi_{01}(x, \phi(x)) = 0, \qquad \text{all } x. \tag{14.15}$$

Then there is an $(r - 2)$-smooth function $\Psi(x, y)$ of order $\geqq s - 2$ such that

$$\Phi(x, y) = [y - \phi(x)]^2 \Psi(x, y). \tag{14.16}$$

For set $\Phi'(x, y) = \Phi(x, y + \phi(x))$; then

$$\Phi'(x, 0) = \Phi'_{01}(x, 0) = 0.$$

With two applications of (*g*), we find $\Phi'(x, y) = y^2 \Phi^*(x, y)$, Φ^* being $(r - 2)$-smooth. Setting $\Psi(x, y) = \Phi^*(x, y - \phi(x))$ gives (14.16). We see easily that Φ' is of order $\geqq s$, and Φ^* and Ψ are of order $\geqq s - 2$.

(i) Suppose ϕ is r-smooth near $(0, \cdots, 0)$, $2 \leqq s \leqq r$, and

$$x_1' = \phi(x_1, \cdots, x_n) = x_1 + R(x_1, \cdots, x_n), \qquad \operatorname{ord}(R) \geqq s. \tag{14.17}$$

Then we may solve for x_1 in terms of $x_1', x_2, \cdots, x_n$ near the origin, giving

$$x_1 = x_1' + R'(x_1', x_2, \cdots, x_n), \quad \operatorname{ord}(R') \geqq s, \quad R' \ r\text{-smooth}. \tag{14.18}$$

First, the solution exists, since $\phi_{10\cdots0}(0, \cdots, 0) \neq 0$. Differentiating

$$x_1 = \phi(x_1, \cdots, x_n) + R'(\phi(x_1, \cdots, x_n), x_2, \cdots, x_n)$$

gives

$$1 = \phi_{10\cdots0}[1 + R'_{10\cdots0}], \qquad 0 = \phi_{20\cdots0}[1 + R'_{10\cdots0}] + \phi_{10\cdots0}^2 R'_{20\cdots0},$$

etc. Since $R_{i0\cdots0}(0, \cdots, 0) = 0$ $(i < s)$, we find $R'_{i0\cdots0}(0, \cdots, 0) = 0$ $(i < s)$, as required.

(j) Let $\phi(x_1, \cdots, x_n)$ be $2r$-smooth, and even in x_1:

$$\phi(-x_1, x_2, \cdots, x_n) = \phi(x_1, x_2, \cdots, x_n).$$

Then there is an r-smooth function $\psi(x_1, \cdots, x_n)$ such that

$$\phi(x_1, x_2, \cdots, x_n) = \psi(x_1^2, x_2, \cdots, x_n).$$

The proof uses (14.7) and (c); see [8].

(k) Let ϕ be $(2r + 1)$-smooth, and odd in x_1: $\phi(-x_1, x_2, \cdots) = -\phi(x_1, x_2, \cdots)$. Then there is an r-smooth function ψ such that

$$\phi(x_1, x_2, \cdots, x_n) = x_1\psi(x_1^2, x_2, \cdots, x_n).$$

15. Normal form for folds

We prove

THEOREM 15A. *Let p be a fold point of the r-smooth mapping f of E into E', with $r \geqq 3$. Then $(r - 3)$-smooth coordinate systems (x, y), (u, v) may be introduced about p and $f(p)$ respectively, in terms of which f takes the form*

$$u = x^2, \qquad v = y. \tag{15.1}$$

To start with, introduce coordinates as in §6, with origins at p, $f(p)$ respectively. Now expand u and v as functions of x and y to the second order, as in (14.11). Then, because of (14.12) and (6.1), we have

$$u = R(x, y), \qquad v = y + S(x, y), \qquad \operatorname{ord}(R) \geqq 2, \quad \operatorname{ord}(S) \geqq 2, \tag{15.2}$$

R and S being r-smooth.

Set

$$x' = x, \qquad y' = y + S(x, y); \tag{15.3}$$

since the Jacobian of the transformation is 1 at the origin, this is an allowable

r-smooth change of coordinates. By (i) of §14, we have

$$y = y' + S'(x', y'), \qquad \text{ord}(S') \geqq 2.$$

Now

$$u = R(x', y' + S'(x', y')) = R'(x', y'), \qquad \text{ord}(R') \geqq 2.$$

The formulas for u and v in the new coordinate systems are, dropping primes,

$$(15.4) \qquad u = R(x, y), \qquad v = y, \qquad \text{ord}(R) \geqq 2.$$

Since $v_x = 0$ (all x, y), the Jacobian is $J = u_x$. Since p is a fold point, and the new coordinates satisfy (6.1), (6.4) gives $J_x(0, 0) \neq 0$. By the implicit function theorem, we may solve $J = 0$ near the origin, giving an $(r - 1)$-smooth function ϕ such that

$$(15.5) \qquad J = u_x = 0 \text{ is given by } x = \phi(y) \text{ near } (0, 0).$$

Set

$$(15.6) \qquad x = x' + \phi(y'), \qquad u' = u - R(\phi(v), v),$$

and $y = y'$, $v' = v$. These are $(r - 1)$-smooth changes of coordinates. Now

$$u' = R(x' + \phi(y'), y') - R(\phi(y'), y'),$$

and hence $u'(0, y') = 0$. Also

$$u'_{x'}(0, y') = R_x(\phi(y'), y') = u_x(\phi(y'), y') = 0.$$

Also, since $u_{xx}(0, 0) = J_x(0, 0) \neq 0$,

$$u'_{x'x'}(0, 0) = R_{xx}(\phi(0), 0) = u_{xx}(0, 0) \neq 0.$$

Therefore, expanding u' as in (14.1) and using (14.2) and (14.4) gives

$$(15.7) \qquad u' = x'^2\Phi(x', y'), \qquad \Phi(0, 0) \neq 0;$$

Φ is $(r - 3)$-smooth.

We may define the $(r - 3)$-smooth function

$$(15.8) \qquad \Psi(x', y') = [\Phi(x', y')]^{\frac{1}{2}}$$

near $(0, 0)$. Set

$$(15.9) \qquad x^* = x'\Psi(x', y'), \qquad y^* = y'.$$

This is an $(r - 3)$-smooth transformation of coordinates, which gives $u' = x^{*2}$, $v' = y^*$, as required.

16. Preliminary transformations about cusps

Our object is to prove

THEOREM 16A. *Let p be a cusp point of the r-smooth mapping f of E into E', with $r \geqq 12$. Then $(r/2 - 5)$-smooth coordinate systems may be introduced about p and $f(p)$, in terms of which f takes the form* (1.3).

We may start with the coordinate systems satisfying (15.4) about the cusp point p. Expanding R as in (e) of §14 gives

$$u = R(x, y) = Ax^2 + Bxy + Cy^2 + R'(x, y), \qquad \text{ord}(R') \geqq 3.$$

Since (6.1) holds, (6.6) gives

$$A = \tfrac{1}{2}u_{xx}(0, 0) = 0, \qquad B = u_{xy}(0, 0) \neq 0.$$

Set

$$(16.1) \qquad x = x', \qquad y = y'/B, \qquad u = u', \qquad v = v'/B.$$

Then $v' = y'$, and

$$u' = u = x'y' + Cy'^2/B^2 + R'(x', y'/B),$$

or, dropping primes,

$$(16.2) \qquad u = xy + Ey^2 + R(x, y), \qquad v = y, \qquad \text{ord}(R) \geqq 3.$$

Expanding R again gives

$$u = xy + Ey^2 + Ax^3 + Bx^2y + Cxy^2 + Dy^3 + R', \qquad \text{ord}(R') \geqq 4.$$

Set

$$(16.3) \qquad u' = u - Ev^2 - Cuv - (D - EC)v^3, \qquad v' = v;$$

this gives, dropping primes again,

$$(16.4) \qquad u = xy + Ax^3 + Bx^2y + R, \qquad v = y, \qquad \text{ord}(R) \geqq 4.$$

Set

$$(16.5) \qquad x = x' - Bx'^2, \qquad y = y';$$

then, dropping primes, $v = y$, and

$$(16.6) \qquad u = xy + Ax^3 + R, \qquad \text{ord}(R) \geqq 4.$$

Since (6.1) holds still, (6.6) gives

$$A = \tfrac{1}{6}u_{xxx}(0, 0) \neq 0.$$

Set $a = |A|^{-\frac{1}{3}}$, $\sigma = -A/|A|$, and

$$(16.7) \qquad x = ax', \qquad y = \sigma y', \qquad u = \sigma a u', \qquad v = \sigma v';$$

this gives, dropping primes,

$$(16.8) \qquad u = xy - x^3 + R, \qquad v = y, \qquad \text{ord}(R) \geqq 4.$$

LEMMA 16a. *Suppose* $m \geqq 3$, *and*

$$(16.9) \qquad u = [x + S(x, y)]y - x^3 + Kx^m + T(x, y),$$

$$\text{ord}(S) \geqq m - 1, \quad \text{ord}(T) \geqq m + 1,$$

S and T being s-smooth. Then the s-smooth transformation

$$x' = x + S(x, y) \tag{16.10}$$

gives

$$u = x'y - x'^3 + Kx'^m + U(x', y), \qquad \text{ord}(U) \geqq m + 1. \tag{16.11}$$

By (i) of §14, inverting (16.10) gives $x = x' + S'(x', y)$, $\text{ord}(S') \geqq m - 1$. We now find

$$u = x'y - [x'^3 + 3x'^2S'(x', y) + \cdots] + K(x' + S')^m + T(x' + S'(x', y), y),$$

and using (f) of §14, (16.11) follows.

Returning to (16.8), expand R:

$$R(x, y) = 2Ax^4 + (Bx^3 + Cx^2y + Dxy^2 + Ey^3)y + R', \qquad \text{ord}(R') \geqq 5.$$

Applying the lemma with $m = 4$, we may transform coordinates so that we have, dropping primes,

$$u = xy - x^3 + 2Ax^4 + R, \qquad \text{ord}(R) \geqq 5. \tag{16.12}$$

Set

$$\begin{aligned} x' &= \frac{x}{1 + Ax}, \qquad y' = (1 + Ax)y - A[x^3 - 2Ax^4 - R(x, y)], \\ & \qquad\qquad u' = u, \qquad v' = v + Au. \end{aligned} \tag{16.13}$$

Then $v' = y'$. Also, using

$$x = \frac{x'}{1 - Ax'} = x' + Ax'^2 + A^2x'^3 + \cdots,$$

$$xy = (1 + Ax)x'y = x'y' + x'A(x^3 - 2Ax^4 - R(x, y)),$$

and $y = y' + S(x', y')$, $\text{ord}(S) \geqq 2$ (see (i) of §14), we find

$$\begin{aligned} u' = x'y' + x'A[(x' + Ax'^2 + \cdots)^3 - 2A(x' + \cdots)^4 - R(x' + \cdots, y' + S)] \\ - (x' + Ax'^2 + \cdots)^3 + 2A(x' + \cdots)^4 + R(x' + \cdots, y' + S), \end{aligned}$$

which, dropping primes, reduces to

$$u = xy - x^3 + R, \qquad \text{ord}(R) \geqq 5. \tag{16.14}$$

Expand R again, and get rid of the terms with y as a factor by the lemma; we obtain

$$u = xy - x^3 + Ax^5 + R, \qquad \text{ord}(R) \geqq 6. \tag{16.15}$$

Set

$$(16.16)\qquad x' = x - \frac{A}{6}xy - \frac{A}{3}x^3 + \frac{A^2}{12}x^3y, \qquad y' = y,$$
$$u' = u - \frac{A}{6}uv, \qquad v' = v.$$

We find

$$x'y' - x'^3 = (x - \frac{A}{6}xy + \frac{A}{6}x^3)y - x^3 + Ax^5 + R^*,$$

and dropping primes, $v = y$, and

$$(16.17)\qquad u = xy - x^3 + R, \qquad \operatorname{ord}(R) \geqq 6.$$

Expand R again, and use the lemma; then $u = xy - x^3 + 4Ax^6 + R$. Set

$$(16.18)\qquad x' = x - Ax^2y - Ax^4, \qquad u' = u - Au^2,$$

and $y' = y$, $v' = v$. We find, dropping primes,

$$(16.19)\qquad u = xy - x^3 + R(x, y), \qquad v = y, \qquad \operatorname{ord}(R) \geqq 7,$$

R being r-smooth.

17. Transformation of the curve $J = 0$

From (16.19), we have

$$(17.1)\qquad J = u_x = y - 3x^2 + R_x(x, y);$$

J and R_x are $(r - 1)$-smooth. We can solve $J = 0$ for y, obtaining $y = \phi(x)$, ϕ being $(r - 1)$-smooth. We have

$$\phi(x) - 3x^2 + R_x(x, \phi(x)) = 0,$$

showing that $\phi(x) - 3x^2$ is of order $\geqq 6$. Hence

$$\phi(0) = \phi_x(0) = 0, \qquad \phi_{xx}(0) = 6, \qquad d^k\phi(0)/dx^k = 0 \qquad (k = 3, 4, 5).$$

By (a) of §14, we may write

$$(17.2)\qquad \phi(x) = 3x^2H(x),$$

H being $(r - 3)$-smooth. Differentiating shows that

$$(17.3)\qquad H(0) = 1, \qquad H_x(0) = H_{xx}(0) = H_{xxx}(0) = 0.$$

Therefore we may define $K(x)$ and $L(x)$ by

$$(17.4)\qquad H(x) = [K(x)]^2, \qquad xK(x) = x + L(x).$$

We find that (17.3) holds also for K, and hence that L is of order $\geqq 5$; L is $(r-3)$-smooth.

Define the $(r-3)$-smooth change of coordinates

$$x' = x + L(x), \qquad y' = y. \tag{17.5}$$

By (i) of §14,

$$x = x' + M(x'), \qquad \operatorname{ord}(M) \geqq 5.$$

Putting into (16.19) shows that

$$u = x'y' - x'^3 + S(x', y'), \qquad \operatorname{ord}(S) \geqq 6. \tag{17.6}$$

By (17.2), (17.4) and (17.5),

$$\phi(x) = 3[xK(x)]^2 = 3x'^2. \tag{17.7}$$

Since $u_{x'} = u_x x_{x'}$, $u_{x'} = 0$ when $u_x = 0$, that is, when $y' = \phi(x) = 3x'^2$. Using (17.6) shows therefore that $S_{x'}(x', 3x'^2) = 0$. Consequently, changing notation,

$$u = xy - x^3 + R, \qquad \operatorname{ord}(R) \geqq 6, \qquad R_x(x, 3x^2) = 0; \tag{17.8}$$

the coordinate system and R are $(r-3)$-smooth.

18. Moving the image of $J = 0$

We shall choose new coordinates in E' so that the image of the curve $J = 0$ will be $27u^2 = 4v^3$, as it would be if we had $R = 0$ in (17.8).

Since $R(x, y)$ is of order $\geqq 6$, so is $R(x, 3x^2)$, and we can define the $(r-5)$-smooth function R' by

$$R(x, 3x^2) = x^2R'(x), \qquad \operatorname{ord}(R') \geqq 4. \tag{18.1}$$

Split R' into its even and odd parts, by setting

$$S'(x) = \tfrac{1}{2}[R'(x) + R'(-x)], \qquad T'(x) = \tfrac{1}{2}[R'(x) - R'(-x)]. \tag{18.2}$$

By (j) and (k) of §14, there are $(r/2-3)$-smooth functions S, T such that

$$S'(x) = S(x^2), \qquad T'(x) = xT(x^2); \tag{18.3}$$

differentiating shows that

$$\operatorname{ord}(S) \geqq 2, \qquad \operatorname{ord}(T) \geqq 1. \tag{18.4}$$

Near $t = 0$, set

$$\beta(3t) = -\frac{2tS(t)}{2 + T(t)}, \quad \gamma(3t) = -\frac{T(t)}{2 + T(t)}; \tag{18.5}$$

then

$$\operatorname{ord}(\beta) \geqq 3, \qquad \operatorname{ord}(\gamma) \geqq 1. \tag{18.6}$$

Since $R(x, 3x^2) = x^2[S(x^2) + xT(x^2)]$, we have

$$\beta(3x^2) + [2x^3 + R(x, 3x^2)]\gamma(3x^2)$$
$$= -\frac{x^2}{2 + T(x^2)}[2S(x^2) + \{2x + S(x^2) + xT(x^2)\}T(x^2)]$$
$$= -x^2[S(x^2) + xT(x^2)] = -R(x, 3x^2).$$

Set

$$u' = u[1 + \gamma(v)] + \beta(v), \qquad v' = v. \tag{18.7}$$

Then the value of u' on $J = 0$ is

$$u'(x, 3x^2) = [2x^3 + R(x, 3x^2)][1 + \gamma(3x^2)] + \beta(3x^2) = 2x^3. \tag{18.8}$$

Define R^* by $u' = xy - x^3 + R^*(x, y)$. Then

$$R^*(x, y) = R(x, y) + [xy - x^3 + R(x, y)]\gamma(y) + \beta(y),$$

which is of order $\geqq 3$. Because of (18.8), $R^*(x, 3x^2) = 0$. Also, by (17.8),

$$u'_x = [1 + \gamma(y)]u_x = 0 \text{ if } y = 3x^2.$$

Hence $R^*_x(x, 3x^2) = 0$. Since $R^*(x, 3x^2) = 0$, we have also $R^*_y(x, 3x^2) = 0$. Consequently, changing the notation, $v = y$, and

$$u = xy - x^3 + R, \qquad \operatorname{ord}(R) \geqq 3, \qquad R(x, 3x^2) = R_y(x, 3x^2) = 0; \tag{18.9}$$

R and the coordinate systems are $(r/2 - 3)$-smooth.

19. The final transformation

By (18.9) and (h) of §14, there is an $(r/2 - 5)$-smooth function $S(x, y)$ such that

$$R(x, y) = [y - 3x^2]^2 S(x, y). \tag{19.1}$$

Since R is of order 3, S is of order 1, and hence $S(0, 0) = 0$. Therefore, by the implicit function theorem, there is an $(r/2 - 5)$-smooth function $\alpha(x, y)$ near $(0, 0)$ such that

$$[1 - 3x\alpha - (y - 3x^2)\alpha^2]\alpha = S; \tag{19.2}$$

we have $\alpha(0, 0) = 0$. Set

$$x' = x + (y - 3x^2)\alpha(x, y), \qquad y' = y. \tag{19.3}$$

This is an $(r/2 - 5)$-smooth change of coordinates, and

$$[x'y - x'^3] - [xy - x^3 + R(x, y)]$$
$$= (y - 3x^2)[y - 3x^2 - 3x(y - 3x^2)\alpha - (y - 3x^2)^2\alpha^2]\alpha - R$$
$$= (y - 3x^2)^2\{[1 - 3x\alpha - (y - 3x^2)\alpha^2]\alpha - S\} = 0;$$

therefore $u = x'y' - x'^3$, $v = y'$, and the proof of Theorem 16A is complete.

PART D. TOPOLOGICAL CONSIDERATIONS

20. Insertion of folds

Let f be an excellent mapping of the open set $R \subset E$ into E'. We describe certain approximations g to f which have the singularities of f, and also further singularities.

(a) ARBITRARY APPROXIMATIONS. We show that for any smooth curve C in R which touches no general fold, we may introduce two new folds, one at C and one near C.

For each $p \in C$, let $p_t(-1 \leqq t \leqq 1)$ denote the points of a line segment S_p approximately perpendicular to C at p, with $p_0 = p$; we may choose these segments so that they fill out a neighborhood U of C which touches no general fold of f. We alter f to form g in U as follows. As t runs from -1 to 1, let $g(p_t)$ run along $f(S_p)$ from $f(p_{-1})$ to $f(p)$ (for $t = 0$), then back a little, then on through $f(p)$ to $f(p_1)$. We may clearly do this so that g is as smooth as desired, if f and C are. Now C is a fold for g, and so is a curve C', consisting of the points $p_{1/2}$ for instance. We may let $g = f$ in $R - U$. With U small enough, g is an arbitrarily good approximation to f.

The curve C might be a simple closed curve, or it might be an arc with its two ends on the boundary of R. We could let C and C' be arbitrarily small concentric circles. Then we may define a smooth one-one mapping h of a neighborhood U' of these circles into 3-space E^3 such that projecting into E' gives g; $h(U')$ is "mushroom" shaped.

(b) APPROXIMATIONS WITH FIRST DERIVATIVES. Let C_0 be a curve of fold points of f (without cusps); it may be the whole or a part of a complete general fold. We show that we may define g to be an arbitrarily good approximation to f, together with first derivatives, so that there is a new pair of folds near C_0. If C_0 is closed or has its ends on the boundary of R, there will be no new cusps for g; otherwise, the new folds will meet in a pair of cusp points for g.

We may let p_t denote points of a neighborhood of C_0, as in (a), so that the image of each S_p under f is an arc folded over on itself, the fold occurring at p. (For the method of proof, compare the early part of §15.) Let $g(p_t) = f(p_t)$ for $-1 \leqq t \leqq 0$; as t runs from 0 to 1, let $g(p_t)$ move along $f(S_p)$ towards $f(p_1)$, then back a little, and then forward again to $f(p_1)$. Clearly two new folds are obtained.

We show that we may make g approximate to f near a given point p of C_0; then it is easy to see that the approximation is possible near all of C_0.

By Theorem 15A, we may choose coordinates so that f, near p, is given by $u = x^2$, $v = y$. We may define an ∞-smooth function $\phi(t)$ so that $\phi(-t) = \phi(t)$, and

$$(20.1) \quad \phi(0) = 1, \qquad \phi(t) = 0 \ (|t| \geqq 1), \qquad 0 \leqq \phi'(t) \leqq \phi'(-\tfrac{1}{2}) = \alpha \ (t < 0).$$

Given $\varepsilon > 0$, define g by

$$(20.2) \qquad u = x^2 + \frac{10\varepsilon^2}{\alpha}\phi\left(\frac{x - 2\varepsilon}{\varepsilon}\right), \qquad v = y.$$

Then

$$u_x = 2x + \frac{10\varepsilon}{\alpha}\phi'\left(\frac{x - 2\varepsilon}{\varepsilon}\right), \qquad | u_x - 2x | \leqq 10\varepsilon \text{ (all } x),$$

so that g approximates to f as required. Also

$$u_x(x) > 0 \ (0 < x \leqq 2\varepsilon \text{ or } x \geqq 3\varepsilon), \qquad u_x(5\varepsilon/2) = -5\varepsilon < 0,$$

so that there are two numbers x_1, x_2 between 2ε and 3ε with $u_x(x_1) = u_x(x_2) = 0$; these define the points of the new folds.

21. Insertion of cusps

We consider several types of approximation, as in §20.

(a) We show that we may insert a pair of nearby arcs where the new function g will have fold points, as in (a) of §20, and run them together to give two new cusps. That the folds join in cusps will be evident from the description of cusps in §5.

Suppose that near the regular point p, f is given by $u = x$, $v = y$. Define ϕ as in §20, and define g by

$$u = x + \frac{2\varepsilon}{\alpha}\phi\left(\frac{x}{\varepsilon}\right)\phi\left(\frac{y}{\varepsilon}\right), \qquad v = y. \tag{21.1}$$

Then g is an arbitrarily good approximation to f, and $g = f$ outside a small neighborhood of p. Since $u_x(\varepsilon/2, 0) = -1$, it is clear that there are two folds cutting the x-axis. We see easily that these come together in two cusps if ϕ is of sufficiently simple shape.

(b) With coordinates about the fold point p of f, as in (b) of §20, define g by

$$u = x^2 + \frac{10\varepsilon^2}{\alpha}\phi\left(\frac{x - 2\varepsilon}{\varepsilon}\right)\phi\left(\frac{y}{\varepsilon}\right), \qquad v = y. \tag{21.2}$$

Clearly we obtain two new folds, joined at two cusps points, and g is an arbitrarily good approximation to f, together with first derivatives.

(c) Let p be a cusp point of f; choose coordinates as in Theorem 16A, so that f is given by $u = xy - x^3$, $v = y$ near p. Define g near p by setting

$$u = xy - x^3\left[1 - 2\phi\left(\frac{x}{\varepsilon}\right)\phi\left(\frac{y}{\varepsilon}\right)\right], \qquad v = y. \tag{21.3}$$

Then

$$u_x = u_y = u_{xx} = 0, \qquad u_{xy} = 1, \qquad u_{xxx} = 6, \qquad \text{at } p,$$

and hence, by the criterion (6.6), p is a cusp point. The curve C of general fold: $J = u_x = 0$ coincides with the original $C_0 : y = 3x^2$, for $| x | \geqq \varepsilon$; it contains p, and by symmetry, is in the x-direction there. Since (for small ε) $J_y > 0$ throughout, C has a single piece joining the parts of C_0 where $| x | \geqq \varepsilon$.

By (4.4)

$$(21.4)\qquad \nabla_V f = (-J_y u_x + J_x u_y,\ -J_y v_x + J_x v_y) = J_x(u_y,\ 1) \text{ on } C;$$

hence $\nabla_V f = 0$ is equivalent to $J_x = u_{xx} = 0$ on C. At points of C where $x \leqq -\varepsilon$, $u = xy - x^3$, and $u_{xx} > 0$; at points where $x \geqq \varepsilon$, $u_{xx} < 0$. On the other hand, since $u_{xx} = 0$ and $u_{xxx} = 6 > 0$ at p, u_{xx} has the same sign as x for $x \neq 0$, $|x|$ small. Therefore, as x runs from $-\varepsilon$ to ε, if we run along C, J_x changes sign at least three times. With the function ϕ of simple shape, J_x will change sign exactly three times, and $\nabla_V f$ will have a non-zero derivative as we move along C at each point where J_x changes sign; that is, g will have three cusp points. We have thus introduced two new cusps, the three cusps lying on a single general fold curve.

Differentiating (21.3) shows that, if $\bar{u} = xy - x^3$ and $\Delta = u - \bar{u}$, then each term in Δ_x and Δ_y has the factor x^2 or x^3/ε, and each factor in Δ_{xx}, Δ_{xy} and Δ_{yy} has the factor x, x^2/ε or x^3/ε^2; since $\Delta = 0$ for $|x| \geqq \varepsilon$, it follows that g is an arbitrarily good approximation of f, together with first and second derivatives.

22. Eliminating cusps

Let f be excellent in R, and let p be a cusp point, on the general fold C. Suppose there is a smooth curve A which moves from p into the lower side of C (see §5), and without touching any other general folds, reaches the boundary of R. Then there is an arbitrarily good approximation g to f which agrees with f outside a neighborhood U of A, and for which the part of the fold near p is replaced by a pair of folds going along near A to the boundary of R, without cusp points.

This may be seen simply as follows. Choose coordinates as in Theorem 16A. Now note that each line $y = a > 0$ is mapped by f so as to fold over on itself twice; the lines $y = a \leqq 0$ have no such folds. We need merely insert such folds near the negative y-axis, to join the above folds. These can be extended down along all of A.

23. Invariance under general approximations

We saw in §22 that cusps may be eliminated from regions by arbitrarily good approximations. This is not true of folds:

THEOREM 23A. *Let p be a fold point of the excellent mapping f. Then for any neighborhood U of p, each sufficiently good approximation g to f which is excellent has a fold point in U.*

Since p is a fold point, there are points p_1 and p_2 in U where the Jacobian J has opposite signs. Let U_i be a circular neighborhood of $p_i (i = 1, 2)$ which touches no fold, and let U_i' be an interior circular neighborhood. For a sufficiently good approximation g to f, if g_t is the deformation of g into f:

$$g_t(q) = g(q) + t[f(q) - g(q)] \qquad (0 \leqq t \leqq 1),$$

then the image of the boundary ∂U_i does not touch the image under f of U_i':

$$g_t(q) \neq f(q'), \qquad q \in \partial U_i, \qquad q' \in U_i', \qquad 0 \leqq t \leqq 1.$$

Hence, by a standard theorem in algebraic topology, $g(U_i)$ and $f(U_i)$ each cover $f(U_i')$ the same algebraic number of times. For f, this number is ± 1. Hence there is a point $p_i' \in U_i$ such that the Jacobian of g at p_i' is of the same sign as the Jacobian of f in U_i. Now the Jacobians of g at p_1' and at p_2' are of opposite sign; consequently the segment $p_1'p_2'$ contains a singular point of g, and since g is excellent, there is a fold point of g in U.

REMARK. It is easy to see that we may eliminate folds from a neighborhood of an arc by an arbitrarily good approximation g to f.

24. Invariance under 1-approximations

THEOREM 24A. *If Q is a bounded closed set in which f is non-singular, then any sufficiently good 1-approximation g to f is non-singular in Q.*

This is obvious, since the Jacobian involves only first derivatives.

The following theorem strengthens Theorem 23A.

THEOREM 24B. *Let the arc A have end points p_1 and p_2 where f is non-singular. Then for any sufficiently good 1-approximation g to f which is excellent, any arc A' from p_1 to p_2 which cuts only fold points of f and g cuts the same number of folds* (mod 2) *for each.*

This is clear, since the Jacobians of f and of g have the same sign at each p_i.

THEOREM 24C. *Let p be a cusp point of f. Then for any neighborhood U of p, each sufficiently good 1-approximation g to f which is excellent has a cusp point in U.*

As seen in §5, there is a curve $A = p_1p_2p_3p_4$ of minimum ∇f in U which cuts the fold C through p at the points p_2 and p_3; the open arc p_2p_3 lies in the upper part of C, and the open arcs p_1p_2 and p_3p_4 lie in the lower part. There is an arc B from p_1 to p_4 in the lower part of C, lying in U, such that A and B bound a region R' filled by curves of minimum ∇f. Now for any sufficiently good 1-approximation g to f, there will be an arc A^* of minimum ∇g, near A, which will bound, with part of B, a region R^* filled by curves of minimum ∇g; also g will be non-singular in B, and there will be fold points of g in R^*. The set Q of fold and cusp points of g in the closure $\bar{R}^*$ is a closed set; there is a "lowest curve" D of minimum ∇g in $\bar{R}^*$ which touches Q, say at p^*. Since p^* is not in B, $p^* \in R^*$. Now p^* is a singular point of g. Also, by definition of D, the general fold of g through p^* does not cross the curve D, and hence is tangent to D; therefore p^* is not a fold point of g. It follows that p^* is a cusp point of g.

25. Invariance under 2- and 3-approximations

THEOREM 25A. *For any bounded closed set Q in which the only singularities of f are fold points, any sufficiently good 2-approximation g to f which is excellent has only folds in Q, of the same character as f.*

We need not be more explicit here. Let p be a fold point of f in Q, and let A be a short segment perpendicular to the fold, with center at p. Since $J^{(f)}$ is of opposite signs at the two ends of A, the same will be true of $J^{(g)}$; hence $J^{(g)}$ will vanish somewhere on A. Since f is good, the directional derivative of $J^{(f)}$ in the direction of A is $\neq 0$; hence the same is true for g, and g has just one general fold cutting A. Thus the general folds of g are like those of f in Q, if the 2-approximation is good enough. Since the directions of curves of minimum ∇g and of general folds for g are nearly parallel to the similar curves for f, the conditions for fold points will be satisfied at all the general fold points of g in Q, for a good approximation. Hence g will have no cusp points in Q.

THEOREM 25B. *Let U be a (small) neighborhood of the cusp point p of f. Then for any sufficiently good 2-approximation g to f which is excellent, there will be a cusp point p' of g in U, on a general fold C'; there will be no other general folds of g in U, and the number of cusp points of g on C' in U will be odd.*

By the proof above, there will be a unique general fold C' of g in U. At two points p_1, p_2 of the general fold C of f, on opposite sides of p, the curves of minimum ∇f cut C in opposite senses; the same will be true, using g, for similar points p_1', p_2' of C'. Hence there will be an odd number of cusps of g between these points; there will be no cusps in $C' \cap U$ outside these points.

THEOREM 25C. *With U, p, f as in the last theorem, any sufficiently good 3-approximation g to f has a unique general fold in U, with a unique cusp point on it.*

There is a unique C' as in the last theorem, with a cusp point p'. Since $\nabla_V \nabla_V f(p) \neq 0$, the similar relation $\nabla_{V'} \nabla_{V'} g(p') \neq 0$ holds; we see easily that $\nabla_{V'} g$ is in opposite directions on opposite sides of p' on C', and hence p' is the only cusp of g in U.

26. Counting folds and cusps

Let the arc A join the points p_1 and p_2, and cut through a finite number of folds. Then this number is even or odd, according to whether the Jacobian of f has the same or opposite signs at p_1 and p_2. Thus the number (mod 2) of crossings of folds by an arc is determined by the mapping in a neighborhood of the boundary of the arc.

Let D be a region of the plane, whose boundary A touches no cusps of f. We wish to determine the number (mod 2) of cusps of f in D, in terms of the properties of f in a neighborhood U of A. We may think of U as a narrow "ribbon" about A. We will find a formula, involving the self-crossings of A under f and the manner in which A crosses folds of f. We let both E and E' be oriented.

We cannot assign signs to cusps and count their algebraic number. For as seen in §21, we may alter f inside D so a pair of new cusp points is introduced; these two points cannot be distinguished (for instance, we may deform f inside D so as to interchange them); hence we cannot assign them opposite signs.

We consider any D which is *normal relative to f*, as follows. The boundary A of D consists of a finite set $A_1, A_2, \cdots$ of smooth arcs, joined at points where f is regular. Each A_i touches no cusp of f, and cuts any fold in a direction which is parallel neither to the fold nor to the curve of minimum ∇f at that point. Moreover, f is one-one in A, except that there may be isolated pairs of points p_1, p_2 interior to the A_i such that neighborhoods of these points in A are mapped into arcs cutting each other at a positive angle.

A "self-crossing" of A is a pair p_1, p_2 of points of A as above. Let $M'_f(A)$ be the number of these.

Consider a fold crossing of A, at $p \in A_i$. Let A' be a small arc of A_i containing p, oriented in the direction from $J < 0$ to $J > 0$. We may choose a coordinate system (x, y) in E, oriented like E, so that $J = 0$, $J < 0$ or $J > 0$ according to whether $x = 0$, $x < 0$ or $x > 0$ respectively, and so that the curve of minimum ∇f through p is the x-axis. Then the part A^+ of A' in $J > 0$ is either in the first quadrant $(y > 0)$ or in the fourth $(y < 0)$; we call the crossing *positive* or *negative* correspondingly.

Suppose we choose coordinates (u, v), oriented like E', in a neighborhood of $f(p)$, so that the image of $J = 0$ is on $u = 0$ and the image of $J \neq 0$ is in $u > 0$. Then clearly the crossing is positive or negative according to whether $f(A^+)$ is on the upper side (greater v) or lower side of the image of the rest of A'.

Let $M''_f(A)$ denote the algebraic number of fold crossings. We shall prove

THEOREM 26A. *Let D be normal relative to f, and have the boundary A. Then the number of cusps of f in D is*

$$M_f(A) \equiv M'_f(A) + \tfrac{1}{2}M''_f(A) \qquad \text{(mod 2)}. \tag{26.1}$$

As an immediate consequence, we have

THEOREM 26B. *Let f and g be two excellent mappings which agree in a neighborhood of the simple closed curve A, which contains no cusp point. Then they have the same number* (mod 2) *of cusp points inside A.*

27. A formula about crossings

If B is an arc joining the points p_1, p_2 of the circle C in E, then the number of crossings of B with C (mod 2) depends on whether B goes out of or into the interior of C at each p_i. The formula is a special case of Theorem 27A, which will be used in the proof of Theorem 26A.

Let $q_1, \cdots, q_\mu$ be points of E, and let $A_{ij}(j = 1, \cdots, \nu_i; i = 1, 2, 3)$ be smooth arcs joining pairs of the points q_k, such that no two of these touch except perhaps at an end of each. Consider each A_{ij} as a 1-chain (mod 2), and set $A_i = \sum_j A_{ij}$. Assume these have the same boundary mod 2:

$$\partial A_1 = \partial A_2 = \partial A_3 = B \qquad \text{(mod 2)}. \tag{27.1}$$

B consists of a subset of the q_k, each with the coefficient 1 (mod 2); q_k is in B if and only if the number of A_{ij} (i fixed) ending on q_k is odd.

Let f be a mapping of the union A of the A_i into E, with the following properties.

(a) f is smooth in each A_{ij}.

(b) At any q_k, the images under f of those A_{ij} which end at q_k go out from $f(q_k)$ in either the same or the opposite cyclic order as these A_{ij} go out from q_k. We say f is *order preserving* or *order reversing* at q_k in these two cases respectively.

(c) If f maps the distinct points p_1, p_2 of A into the same point q, then neither is any q_k, and small subarcs of A containing these points map into arcs cutting each other at a positive angle at q. No third point of A maps into q.

Let N be the number of points q_k of B where f is order reversing. Let N_{ij} be the number of crossings of arcs of $f(A_i)$ with arcs of $f(A_j)$.

THEOREM 27A. *With the above hypotheses,*

$$N_{12} + N_{13} + N_{23} \equiv N \pmod{2}. \tag{27.2}$$

We prove this by making successive alterations of f, till the mapping becomes the identity, for which (27.2) is trivially true.

To start with, we may clearly deform f, without changing any of the above terms, so that the new mapping, say f_1 sends each q_k into itself. Using another deformation, we obtain a mapping f_2 such that at each q_k, the A_{ij} ending at q_k are mapped into arcs which go almost vertically upward from q_k, but slightly to the right of the vertical.

Suppose f_2 is order reversing at q_k. We alter f_2 as follows. A reflection across the vertical direction sends each $f_2(A_{ij})$ at q_k into an arc moving upwards slightly left of vertical. Let $f_2'(A_{ij})$ move upwards in this direction, and then let it cross over the vertical line to rejoin its former position further up. We may clearly choose f_2' so that each arc from q_k now crosses over each other arc once just above q_k, and no other crossings are added.

We find the effect of this change at each q_k on (27.2). Suppose first that q_k is in B. Since f_2' is order preserving at q_k, N is changed (mod 2). Let λ_i be the number of arcs of A_i ending at $q_k (i = 1, 2, 3)$. Since q_k is in $B = \partial A_i$, λ_i is odd. Now N_{ij} is increased by $\lambda_i \lambda_j \equiv 1 \pmod{2}$; hence so is $N_{12} + N_{13} + N_{23}$. Thus (27.2) holds for f_2 if and only if it holds for f_2'. Now suppose q_k is not in B; then N is unchanged. In this case, each λ_i is even; hence N_{ij} is changed by $\lambda_i \lambda_j \equiv 0 \pmod{2}$, so (27.2) is again unchanged. Let f_3 be f_2, altered in this manner at each q_k where f_2 is order reversing.

We may clearly deform f_3 into f_4 so that the arcs A_{ij} and $f_4(A_{ij})$ now go out from each q_k in the same direction, without altering any term of (27.2). We may also suppose that for any pair A_{ij} and A_{kl}, $f_4(A_{ij})$ and A_{kl} intersect only as in (c).

Let f_5 be the identity. We obtain this from f_4 by replacing $f_4(A_{ij})$ by A_{ij} for each A_{ij} in turn. Consider any such step; say f_4'' is obtained from f_4' by replacing $f_4'(A_{1j})$ by A_{1j}. Since $\partial A_2 + \partial A_3 = 0 \pmod{2}$, $H = f_4'(A_2) + f_4'(A_3)$ is expressible as a sum of circuits (simple closed curves). The same is true for $f_4'(A_{1j}) + A_{1j}$. Hence $f_4'(A_{1j}) + A_{1j}$ crosses H an even number of times, and therefore $f_4'(A_{1j})$ crosses H the same number (mod 2) of times as A_{1j}. Therefore $N_{12} + N_{13}$ is unchanged (mod 2), and (27.2) holds for f_4'' if and only if it held for f_4'. Since it holds for f_5, it holds for f_4, hence for f_3, for f_2, for f_1, and for f. The theorem is now proved.

28. Additivity of $M_f(A)$

Let D_1 and D_2 be normal relative to f, as in §26. We suppose the boundary A_i of D_i is made up of smooth arcs, and that any arc of A_1 and any arc of A_2 either are disjoint, or have just an end in common, or are identical. Then we can define $D_3 = D_1 + D_2 \pmod 2$; it is made up of parts which are in exactly one of D_1, D_2, and its boundary A_3 is made up of arcs from A_1 and from A_2. Assuming also that D_3 is normal relative to f, we prove

$$M_f(A_3) \equiv M_f(A_1) + M_f(A_2) \pmod 2. \tag{28.1}$$

Let A_1' denote the arcs in A_2 but not in A_1, and hence in A_3; A_2', the arcs in A_1 but not in A_2, and hence in A_3; and A_3', the arcs in both A_1 and A_2, and hence not in A_3. Now A_2' and A_3' together make up the arcs of A_1; hence $\partial A_2' + \partial A_3' = \partial A_1 = 0 \pmod 2$, and $\partial A_2' = \partial A_3' \pmod 2$. Similarly $\partial A_1' = \partial A_3' \pmod 2$. The conditions of §27 hold, using the A_i', and therefore so does (27.2).

Set

$$M' = \sum_{i=1}^3 M_f'(A_i), \qquad M'' = \sum_{i=1}^3 M_f''(A_i). \tag{28.2}$$

Since A_1 is composed of the arcs of A_2' and A_3',

$$M_f'(A_1) = N_{22} + N_{23} + N_{33},$$

with similar expressions for $M_f'(A_2)$ and $M_f'(A_3)$. Therefore

$$M' = N_{12} + N_{13} + N_{23} + 2(N_{11} + N_{22} + N_{33}). \tag{28.3}$$

Since each arc A' appearing in any A_i' is in exactly two of A_1, A_2, A_3, each fold crossing in A_i' contributes to exactly two of the terms in M''. Therefore $\frac{1}{2}M''$ is (mod 2) the total number of fold crossings of arcs of the A_i'. Take any such arc $A' = q_k q_l$. Clearly the number $M^*(A')$ of fold crossings in A' is one (mod 2) if and only if the Jacobian of f is of opposite signs at q_k and at q_l, i.e. if and only if the number of these vertices (q_k and q_l) where $J < 0$ is $\equiv 1 \pmod 2$. Hence $\frac{1}{2}M''$ is the total number (mod 2) of q_k at which $J < 0$, counting each q_k the number μ_k of times it appears as an end of an arc A'. Since μ_k is odd only for the points of B, $\frac{1}{2}M''$ is the number (mod 2) of points of B where $J < 0$. Now f reverses the cyclic order of the arcs at a vertex q_k if and only if $J < 0$ there. Consequently $\frac{1}{2}M'' \equiv N \pmod 2$.

Using (28.3) and (27.2) now gives $M' \equiv \frac{1}{2}M'' \pmod 2$; hence $\sum_{i=1}^3 M_f(A_i) \equiv 0 \pmod 2$, proving (28.1).

29. Proof of Theorem 26A

Let D be normal relative to f. We may cut D into subregions $D_1, \cdots, D_m$, so that the conditions for normality hold for the set of all boundary arcs of the D_i. We shall do this in such a manner that we can prove the theorem for each D_i; then, because of the additivity of M_i (§28), it holds also for D.

Let $p_1, \cdots, p_s$ be the cusps inside D. We choose regions $D_1, \cdots, D_s$ about these points, each in the shape of the box shown in Fig. 1. There are clearly two fold crossings of ∂D_i, of opposite sign; there is also one self crossing of ∂D_i under f. Hence $M_f(\partial D_i) \equiv 1 \pmod 2$, as it should be. The reader will find it instructive to draw curves A around p_i of various sorts, and verify that $M_f(A) \equiv 1 \pmod 2$ for each.

In D but outside $D_1, \cdots, D_s$ there are a finite number of fold curves. We may cover each of these by a finite number of adjacent boxes, for each of which the reader will verify at once that $M_f \equiv 0 \pmod 2$. If, for instance, a box is in the shape of a small parallelogram, with one side close to the fold curve, then there will be two fold crossings of the same sign, and also a self crossing of the image; the total contribution is thus 0 (mod 2).

Finally, we may cut the rest of D into small pieces, in each of which f is one-one; clearly M_f is 0 for each. This completes the proof.

30. On the mappings of a projective plane into E

The definition of excellence of a mapping of anys urface into the plane is clear. We prove

THEOREM 30A. *For any excellent mapping f of the projective plane P into the plane E, the number of cusp points is odd.*

Let A be a smooth closed curve in P which reverses orientation in P; let D_1 be a ribbon about A. D_1 is a Möbius strip, whose boundary A_1 is a simple closed curve, and $D_2 = P - D_1$ is a disk, with boundary curve A_1. We may choose A and D_1 so that D_1 contains no cusps of f, $f(A)$ has self crossings only as considered in §26, and so that $f(A_1)$ runs along close to and nearly parallel to $f(A)$. We shall prove

$$M_f(A_1) \equiv 1 \pmod 2;$$

by Theorem 26A, there are an odd number of cusps in D_2, as required.

First we consider the self-crossings of $f(A_1)$. Let $(p_1, q_1), \cdots, (p_s, q_s)$ be the pairs of points of A such that $f(p_i) = f(q_i)$. We may suppose that there are no fold points of f near p_i or q_i. Then each of the two arcs of A_1 near p_i intersect each of the two arcs near q_i under f; thus the contribution to $M_f'(A_1)$ is $4 \equiv 0$ (mod 2). Say there are λ fold points of f on A. The folding of the ribbon D_1 near each such point causes a single self crossing of the boundary A_1 under f. Therefore $M_f'(A_1) \equiv \lambda \pmod 2$.

Now we consider the folds. Cut across the ribbon at an arc B, away from any fold; D_1 is turned into a piece D_1', which we may orient. If $J > 0$ in D_1' at one end, then since A was orientation reversing, $J < 0$ at the other end. Therefore the number λ is odd.

For each piece of a fold cutting A, A_1 has two arcs crossing it; with a fixed orientation of D_2, we shall show that these crossings are of opposite sign. Let the fold cut A_1, then A, then A_1 again, in points p_1, p, p_2 respectively. Let p_1p_1' and p_2p_2' be arcs of A_1 pointing in the same direction. With an orientation of D_1 near p, the Jacobian is of course of the same sign at p_2' as at p_1'. Follow from p_1' along A_1 around to p_2 and to p_2'; crossing over to p_1' again gives an arc which reverses orientation in P. Hence if the chosen orientation of D_1 near p agrees with that of D_2 near p_1', it disagrees with that of D_2 near p_2' (and vice versa). This proves that, in D_2, J has opposite signs at p_1' and at p_2'. Hence, in determining the signs of the crossings of A_1 at p_1 and at p_2, we must direct the arcs in the opposite manner. Since the arcs are nearly parallel, the signs of the crossings are opposite, as stated. Therefore $M''_f(A_1) = 0$. Now Theorem 26A gives $M_f(A_1) \equiv \lambda \equiv 1 \pmod 2$, completing the proof.

THE INSTITUTE FOR ADVANCED STUDY

BIBLIOGRAPHY

[1] E. E. LEVI, *I. problemi dei valori . . .*, Memorie die Mat. e di Fis. d. Soc. Italiana d. Science, Serie Terza, 16 (1910), pp. 88–90.

[2] M. MORSE, *Relations between the critical points of a real function of n independent variables*, Trans. Amer. Math. Soc., 27 (1925), pp. 345–396.

[3] ——, *The critical points of a function of n variables*, Trans. Amer. Math. Soc., 33 (1931), pp. 72–91.

[4] A. W. TUCKER, *Branched and folded coverings*, Bull. Amer. Math. Soc., 42 (1936), pp. 859–862.

[5] H. WHITNEY, *Analytic extensions of differentiable functions defined in closed sets*, Trans. Amer. Math. Soc., 36 (1934), pp. 63–89.

[6] ——, *Differentiable manifolds*, Ann. of Math., 37 (1936), pp. 645–680.

[7] ——, *Differentiability of the remainder term in Taylor's formula*, Duke Math. J., 10 (1943), pp. 153–158. (We note a correction. The lemma on p. 156 is not true. The following theorem should contain the hypothesis that f_n is of class C^p).

[8] ——, *Differentiable even functions*, ibid., pp. 159–160.

[9] ——, *The general type of singularity of a set of $2n - 1$ smooth functions of n variables*, ibid., pp. 161–172.

[10] ——, *The singularities of a smooth n-manifold in $(2n - 1)$-space*, Ann. of Math., 45 (1944), pp. 247–293.

[11] N. Z. WOLFSOHN, *On differentiable maps of Euclidean n-space into Euclidean m-space* Harvard thesis, 1952. See the abstract in Bull. Amer. Math. Soc., 61 (1955), p. 171.

On functions with bounded n^{th} differences,

BY HASSLER WHITNEY,

Institute for Advanced Study, Princeton, N. J.

---

1. THE PRINCIPAL THEOREMS. — Let f be a real valued function with derivatives through the nth order in the closed interval $(0, 1)$, and let the nth derivative $f^{(n)}$ be bounded. Then there is a polynomial P of degree at most $n-1$ such that

$$|f(x) - P(x)| \leqq \frac{1}{n!} \sup |f^{(n)}(y)| \qquad (0 \leqq x \leqq 1); \tag{1.1}$$

we may obtain P for instance from Taylor's formula with remainder at $x = 0$. We could replace $\sup|f^{(n)}(y)|$ by the upper bound of the nth divided differences. It is natural to ask if such a relation holds if we use instead n^{th} differences (1).

We use the definition

$$\Delta_h^n f(x) = \sum_{i=0}^{n} (-1)^{n-i} \binom{n}{i} f(x + ih). \tag{1.2}$$

THEOREM 1. — *For each integer $n \geqq 1$ there is a number K_n with the following property. Let I be any closed interval (which may be infinite in one or both directions). Then for any continuous function f in I*

---

(1) This is a conjecture of H. Burkill, communicated to me by J. C. Burkill. For the case $n = 2$, *see* H. BURKILL, *Proc. London Math. Soc.*, series 3, vol. 2, 1952, p. 157, lemma 5.2.

there is a polynomial P *of degree at most* $n-1$ *such that*

(1.3) $$|f(x)-P(x)|\leq K_n \sup|\Delta_h^n f(y)| \qquad (x \text{ in } I).$$

For a bounded interval I, the smallest possible K_n is clearly independent of the length of I. We may therefore restrict the discussion to three cases

(1.4) $$I_0=(0, 1), \qquad I^*=(0, \infty), \qquad I^{**}=(-\infty, \infty).$$

Let K_n, K_n^*, K_n^{**} denote the smallest possible constants in each case.

THEOREM 2. — *The constants satisfy*

(1.5) $$K_n^{**}\leq K_n^*\leq K_n.$$

We shall show also that

(1.6) $$\frac{1}{2}\leq K_n^*\leq 1.$$

Using the definition

(1.7) $$G_n=\sup_k\binom{n}{k},$$

we shall prove [2]

(1.8) $$K_n^{**}=\frac{1}{2G_{n-1}}=\frac{1}{G_n} \quad \text{if } n \text{ is even},$$

(1.9) $$\frac{1}{2G_{n-1}}\leq K_n^{**}\leq \frac{1}{G_n} \quad \text{if } n \text{ is odd}.$$

In theorem 1, if I is bounded, it is possible to require that P be that polynomial which equals f at the n points which are the end points of a division of I into $n-1$ equal intervals. Of course the constant K'_n in this case may have to be larger than K_n. *See* (2.5).

We shall prove

(1.10) $$K'_n\geq 2K_n^*.$$

Using (1.5) and (1.6) gives

(1.11) $$K_n\geq\frac{1}{2}, \qquad K'_n\geq 1.$$

Most of our knowledge of the particular values of the constants is

---

[2] These upper bounds for the K_n^* and the K_n^{**} were given me by A. Beurling.

contained in the following table. A single entry denotes the correct value; a pair of entries denote known lower and upper bounds.

n........	1.	2.	3.	4.	5.
K_n;.......	$\frac{1}{2}$	$\frac{1}{2}$	$\frac{8}{15}$, $\frac{7}{10}$	$\frac{1}{2}$, 3.2425	$\frac{1}{2}$, 10.4
K'_n........	1	1	$\frac{16}{15}$, $\frac{14}{9}$	1, 3.2425	1, 10.4
K^*_n........	$\frac{1}{2}$	$\frac{1}{2}$	$\frac{8}{15}$, 0.6645	$\frac{1}{2}$, 1	$\frac{1}{2}$, 1
K^{**}_n.......	$\frac{1}{2}$	$\frac{1}{2}$	$\frac{1}{4}$	$\frac{1}{6}$	$\frac{1}{12}$, $\frac{7}{76}$

We shall give an inequality for $(n-1)^{\text{th}}$ differences of bounded functions in I^* or I^{**}. We use the numbers $H_1 = 1$ and

$$(1.12)\qquad H_{2\mu} = \frac{2}{\mu}\binom{2\mu-2}{\mu-1}\binom{2\mu}{2},\qquad H_{2\mu+1} = \frac{2}{\mu}\binom{2\mu-1}{\mu-1}\binom{2\mu+1}{2},$$

$$(1.13)\qquad N^{**}_m = \frac{H_m}{2^m-1},\qquad N^*_m = \frac{2^{m-1}m}{2^m-1}.$$

Theorem 3. — *For bounded functions f in I^* or in I^{**},*

$$(1.14)\qquad |\Delta_h^{n-1} f(x)| \leq N^*_{n-1} \sup |\Delta_h^n f(y)| \quad \text{if } I^* \text{ is used,}$$

$$(1.15)\qquad |\Delta_h^{n-1} f(x)| \leq N^{**}_{n-1} \sup |\Delta_h^n f(y)| \quad \text{if } I^{**} \text{ is used.}$$

2. Discussion, further results. — Theorem 1 would fail if we omitted the hypothesis of continuity; for the discontinuous functions of Hamel ([3]) are not bounded, and yet have the property $\Delta_h^n f(x) = 0$ for $n \geq 2$. It would probably be sufficient to assume that f is measurable. In this connection, *see* (12.1).

Because of the formula (easily proved by induction)

$$(2.1)\qquad \Delta_h^n f(x) = \int_0^h \cdots \int_0^h f^{(n)}(x + t_1 + \ldots + t_n)\, dt_1 \ldots dt_n,$$

we have, in any interval I,

$$(2.2)\qquad \sup |\Delta_h^n f(x)| \leq h^n \sup |f^{(n)}(y)|.$$

Hence, in a small interval, the condition $\sup|\Delta_h^n f(x)| \leq N$ is weaker

([3]) G. Hamel, *Eine basis aller Zahlen*... (*Math. Ann.*, vol. 60, 1905, p. 459-462).

than the condition $\sup|f^{(n)}(y)|\leqq N$ (N fixed). However, in a large interval, the first condition is stronger; this is shown by theorem 1 for infinite intervals, which would fail if $\sup|f^{(n)}(y)|$ were used, as is apparent from the example $f(x)=x^n$.

In theorem 1 for finite intervals, one may of course choose for P the « best approximating polynomial » P_f; it is characterized by the fact that $f-P_f$ takes on alternately its upper and lower bounds, a total of at least $n+1$ times. If I is unbounded, then requiring $f-P$ to be bounded determines P except for the constant term; for the best polynomial P_f, the upper and lower bounds of $f-P_f$ are equal but of opposite sign.

Using the interval I_0, the K'_n mentioned in paragraph 1 are defined by using that polynomial P'_f such that

$$(2.3)\qquad f\left(\frac{i}{n-1}\right)-P'_f\left(\frac{i}{n-1}\right)=0 \qquad (i=0, 1, \ldots, n-1).$$

We shall relate K'_n to K_n. *We may restrict ourselves to functions f such that* $|\Delta_h^n f(x)|\leqq 1$. Using the interpolation formula or a compactness argument, we see that there is a number J_n such that for any polynomial Q of degree at most $n-1$,

$$(2.4)\quad |Q(x)|\leqq J_n \text{ in } I_0 \quad \text{if} \quad \left|Q\left(\frac{i}{n-1}\right)\right|\leqq 1 \quad \text{for } i=0, 1, \ldots, n-1.$$

(For example, $J_3=\frac{5}{4}$.) The definitions of P_f and of P'_f give

$$\left|P_f\left(\frac{i}{n-1}\right)-P'_f\left(\frac{i}{n-1}\right)\right|\leqq K_n \qquad (i=0, 1, \ldots, n-1);$$

hence

$$|P_f(x)-P'_f(x)|\leqq J_nK_n \quad \text{in } I_0.$$

Since P_f is the best polynomial, this gives

$$|f(x)-P'_f(x)|\leqq K_n+J_nK_n.$$

Therefore

$$(2.5)\qquad K_n\leqq K'_n\leqq(1+J_n)K_n.$$

Theorem 3 and the definitions of $\Delta_h^1 f(x)$, K_n^* and K_n^{**} give

$$(2.6)\qquad K_n^*\leqq\frac{1}{2}N_1^*N_2^*\ldots N_{n-1}^*, \qquad K_n^{**}\leqq\frac{1}{2}N_1^{**}N_2^{**}\ldots N_{n-1}^{**};$$

the K_n^* and K_n^{**} are in general considerably smaller than the products shown, though equality holds for $n \leqq 2$.

One may define K_n^{**} (and similarly K_n^*) through the following relation. Considering all bounded continuous functions f such that $|\Delta_h^n f(y)| \leqq 1$,

$$K_n^{**} = \frac{1}{2} \sup |\Delta_h^1 f(x)|. \tag{2.7}$$

This is clear when we note that P_f is constant if f is bounded. In this definition, one could probably omit the hypothesis of continuity; at least, the bounds we give for K_n^{**} (and also for K_n^*) will still apply, as the proofs will show.

We shall see that for any bounded monotone increasing function f in I^{**},

$$\sup |\Delta_h^n f(x)| = G_{n-1}(f(\infty) - f(-\infty)); \tag{2.8}$$

it follows at once that $K_n^{**} \geqq \frac{1}{2 G_{n-1}}$. Also, for any bounded function f in I^* such that

$$(-1)^m \Delta_h^m f(x) \geqq 0 \qquad (m = 1, 2, \ldots, n), \tag{2.9}$$

we will see that

$$0 \leqq (-1)^m \Delta_h^m f(x) \leqq f(0) - f(\infty) \qquad (m = 1, 2, \ldots, n); \tag{2.10}$$

this holds, in particular, for $f(x) = \frac{1}{1+x}$, since $(-1)^m f^{(m)}(x) \geqq 0$. This gives $K_n^* \geqq \frac{1}{2}$.

Instead of using an interval for the domain of definition of f, one may use for instance the set J^* of non-negative integers, or the set J^{**} of all integers. Replacing I^{**} by J^{**} does not change the constants for $n \leqq 4$; *see* (13.1). However, with an obvious notation,

$$K_3(J^*) = \frac{2}{3} > K_3^*. \tag{2.11}$$

To prove the equality, we note first that the proof of theorem 3 below gives (using integral h)

$$|\Delta_h^1 f(x)| \leqq N_1^* N_2^* \sup |\Delta_h^3 f(y)| = \frac{4}{3} \sup |\Delta_h^3 f(y)|; \tag{2.12}$$

hence $K_3(J^*) \leqq \frac{2}{3}$. On the other hand, if f has the values (using integers $p > 0$)

$$(2.13) \quad f(0) = -\frac{2}{3}, \qquad f(3p) = \frac{1}{3}, \qquad f(3p-i) = \frac{2}{3} \qquad (i = 1, 2),$$

we see easily that $\sup |\Delta_h^3 f(x)| = 1$; hence $K_3(J^*) \geqq \frac{2}{3}$.

We obtain the same value of the third constant (omitting the continuity hypothesis) if we replace J^* by the non-negative dyadic rationals. This is seen with the help of the function f defined by

$$(2.14) \quad f(0) = -\frac{2}{3}, \qquad f\left(\frac{3p}{2^k}\right) = \frac{1}{3}, \qquad f\left(\frac{3p-i}{2^k}\right) = \frac{2}{3} \qquad (i = 1, 2).$$

The problem of finding the K_n, K'_n and K^*_n is probably extremely difficult, as should be clear from the proofs we give of bounds for K_3 and K^*_3. The values of the K^{**}_n for odd $n \geqq 5$ and the values of $K_n(J^{**})$ for $n \geqq 5$ should not be so hard to determine.

3. A TYPE OF FORMULA INVOLVING DIFFERENCES. — The nth difference $\Delta_h^n f(x)$ relates the values of f at $n+1$ equally spaced points. We shall show that certain linear combinations of differences relate the values of f at more general sets of $n+1$ points.

Consider the set of integers

$$X = (x_0, x_1, \ldots, x_{n-1}), \qquad 0 = x_0 < x_1 < \ldots < x_{n-1},$$

and let Y be the set of integers between x_0 and x_{n-1} which are not in X. Let s be any element of Y, and let Y_s be the set Y except for s. The number of elements in Y_s is

$$l = x_{n-1} - n.$$

LEMMA 1. — *Given* X *and* s, *there are numbers*

$$a_0, \quad \ldots, \quad a_l, \qquad c_0, \quad \ldots, \quad c_{n-1},$$

with the following properties. Let f be any function defined at the points $y + ih$ (y fixed, $i = 0, 1, \ldots, x_{n-1}$). Then

$$(3.1) \qquad \sum_{i=0}^{l} a_i \Delta_h^n f(y+ih) = f(y+sh) + \sum_{j=0}^{n-1} c_j f(y + x_j h).$$

Moreover, if $Q(x)$ *is the polynomial of degree* $\leqq n-1$ such that

$$Q(y+x_j h)=f(y+x_j h) \qquad (j=0, \ldots, n-1), \tag{3.2}$$

then

$$f(y+sh)=\sum_{i=0}^{l} a_i \Delta_h^n f(y+ih)+Q(y+sh). \tag{3.3}$$

Remark. — It seems probable that all the a_i are of the same sign.

First, define a transformation from any set of numbers $b_0, b_1, \ldots, b_l$ to the set of numbers Z_k by

$$Z_k=\sum_{i=0}^{l}(-1)^{n-k+i}\binom{n}{k-i}b_i. \tag{3.4}$$

We can have $Z_k \neq 0$ at most for $0 \leqq k \leqq l+n=x_{n-1}$, since otherwise all the binomial coefficients vanish.

We now require that

$$Z_k=0 \quad \text{for } k \text{ in } Y_s; \tag{3.5}$$

this is a set of l equations in the $l+1$ unknowns b_i, and hence has a solution $B_0, \ldots, B_l$, not all these being zero.

Using $B_0, \ldots, B_l$, the other Z_i are determined; we show that

$$Z_s \neq 0. \tag{3.6}$$

For suppose $Z_s=0$. Let B_r be the first non-zero B_i; then (3.4) shows that $Z_r=(-1)^n B_r \neq 0$. Hence r is not in Y, but is in X. Let R be the polynomial of degree at most $n-1$ such that

$$R(r)=1, \qquad R(i)=0 \quad \text{for } i \text{ in X} \quad (i \neq r).$$

Then

$$0=\sum_{i=0}^{l} B_i \Delta_1^n R(i)=\sum_{i=0}^{l} B_i \sum_k (-1)^{n-k+i}\binom{n}{k-i}R(k)=\sum_k Z_k R(k)=Z_r \neq 0$$

a contradiction; thus (3.6) is proved.

Set

$$a_i=\frac{B_i}{Z_s} \qquad (i=0, 1, \ldots, l); \tag{3.7}$$

then

$$\sum_{i=0}^{l} a_i \Delta_h^n f(y+ih) = \sum_{i=0}^{l} a_i \sum_k (-1)^{n-k+i} \binom{n}{k-i} f(y+kh) = \sum_k f(y+kh)\frac{Z_k}{Z_s},$$

which gives (3.1), with

$$c_j = \frac{Z_{x_j}}{Z_s} \qquad (j=0, 1, \ldots, n-1). \tag{3.8}$$

Now consider the polynomial Q. Applying (3.1) to it gives

$$Q(y+sh) + \sum_{j=0}^{n-1} c_j Q(y+x_j h) = 0. \tag{3.9}$$

This relation together with (3.1) and (3.2) gives (3.3).

4. A SPECIALIZED FORMULA. — Given the integers $n \geqq 1$ and $\nu \geqq 2$, define the set

$$X = (0, \nu, 2\nu, \ldots, (n-1)\nu), \tag{4.1}$$

and set $s = 1$. Apply lemma 1 to give

$$\sum_{i=0}^{(n-1)\nu-n} a_i \Delta_h^n f(y+ih) = f(y+h) + \sum_{j=0}^{n-1} \gamma_j f(y+j\nu h). \tag{4.2}$$

LEMMA 2. — *Given n and $\varepsilon > 0$, we may choose ν so that*

$$\sigma = |\gamma_1| + |\gamma_2| + \ldots + |\gamma_{n-1}| \leqq \varepsilon. \tag{4.3}$$

Note that we omit $|\gamma_0|$ here. For each set of numbers $u_1, \ldots, u_{n-1}, x$, let R be the polynomial of degree at most $n-1$ such that

$$R(0) = 0, \qquad R(i) = u_i \quad \text{for } i = 1, \ldots, n-1,$$

and set $\varphi(u_1, \ldots, u_{n-1}, x) = R(x)$. Then φ is continuous, and vanishes for $x = 0$; hence we may choose ν so that

$$|\varphi(u_1, \ldots, u_{n-1}, x)| \leqq \varepsilon \quad \text{if } |u_1|, \ldots, |u_{n-1}| \leqq 1 \qquad \left(x \leqq \frac{1}{\nu}\right). \tag{4.4}$$

Now set $u_i = 1$ or $u_i = -1$ according as $\gamma_i \geqq 0$ or $\gamma_i < 0$, and define R as above. Applying (4.2) to R, with $y = 0$, $h = \frac{1}{\nu}$, gives

$$0 = R\left(\frac{1}{\nu}\right) + \sum_{j=1}^{n-1} |\gamma_j|,$$

and (4.3) follows.

5. Preliminary inequalities. — We start the proof of existence of K'_n, defined in paragraph **1**. Set $g(x) = f(x) - P'_f(x)$; then

$$g\left(\frac{i}{n-1}\right) = 0 \qquad (i = 0, 1, \ldots, n-1). \tag{5.1}$$

We may assume that

$$|\Delta_h^n g(y)| \leqq 1; \tag{5.2}$$

we must prove

$$|g(x)| \leqq K'_n. \tag{5.3}$$

For each k, consider the set of numbers in I_0

$$S'_k = \text{all } \frac{i}{2^k(n-1)} \qquad [i = 0, 1, \ldots, 2^k(n-1)],$$

and set $S_0 = S'_0$, $S_k = S'_k - S'_{k-1}$ $(k > 0)$. We shall obtain inequalities on $|g(x)|$ for the points of $S_1, S_2, \ldots$ in turn.

Consider the set of numbers

$$X = (0, 2, 4, \ldots, 2(n-1));$$

the corresponding set Y is $(1, 3, \ldots, 2(n-1)-1)$. For each s in Y, use (3.1) to get

$$g(y + sh) = \sum_{i=0}^{n-2} a_{si} \Delta_h^n g(y + ih) - \sum_{j=0}^{n-1} c_{sj} g(y + 2jh). \tag{5.4}$$

Set

$$a^*_s = |a_{s0}| + \ldots + |a_{s,n-2}|, \qquad c^*_s = |c_{s0}| + \ldots + |c_{s,n-1}|, \tag{5.5}$$

$$a^* = \sup\{a^*_1, a^*_3, \ldots, a^*_{2n-3}\}, \qquad c^* = \sup\{c^*_1, c^*_3, c^*_{2n-3}\}, \tag{5.6}$$

$$\lambda_0 = 0, \qquad \lambda_k = a^*(1 + c^* + \ldots + c^{*k-1}) \qquad (k = 1, 2, \ldots). \tag{5.7}$$

We shall prove

$$|g(x)| \leqq \lambda_k \qquad (x \text{ in } S_k;\ k = 0, 1, 2, \ldots). \tag{5.8}$$

This is true for $k=0$. We shall prove it for k, assuming it holds for smaller values. Take any $x=\frac{p}{2^k(n-1)}$ (p odd) in S_k. We may choose $y=\frac{q}{2^{k-1}(n-1)}$ such that

$$0 \leqq y < x < y+\frac{1}{2^{k-1}} \leqq 1.$$

Then

$$x=y+sh, \qquad h=\frac{1}{2^k(n-1)} \qquad [s \text{ odd}; \ 0<s<2(n-1)].$$

Each point $y+2jh$ is in some S_l with $l<k$; hence, by induction,

$$|g(y+2jh)| \leqq \lambda_{k-1} \qquad (j=0, 1, \ldots, n-1).$$

Because of (5.2), (5.4) gives

$$|g(x)| \leqq \sum_{i=0}^{n-2} |a_{si}| + \lambda_{k-1} \sum_{j=0}^{n-1} |c_{sj}| = a_s^* + \lambda_{k-1} c_s^* \leqq a^* + \lambda_{k-1} c^* = \lambda_k,$$

as required.

6. Proof of theorem 1 for I_0, $n=3$. — The general proof below takes care of this case; we give this case for illustrative purposes.

Note that

$$(6.1) \qquad 3\Delta_h^3 g(y) + \Delta_h^3 g(y+h) = -3g(y) + 8g(y+h) - 6g(y+2h) + g(y+4h).$$

Thus (5.4) holds (taking $s=1$ and setting $a_i=a_{si}, \ldots$), with

$$a_0=\frac{3}{8}, \qquad a_1=\frac{1}{8}, \qquad c_0=-\frac{3}{8}, \qquad c_1=-\frac{3}{4}, \qquad c_2=\frac{1}{8}.$$

With the notations of (4.2), $\nu=2$ and $\sigma=\frac{7}{8}<1$.

We shall prove $|g(x)| \leqq 4$ in S_k, using induction; hence $K_3' \leqq 4$. By (6.1),

$$(6.2) \qquad |g(y+h)| \leqq \frac{1}{2} + \frac{3}{8}|g(y)| + \frac{3}{4}|g(y+2h)| + \frac{1}{8}|g(y+4h)|.$$

Suppose first that $x \leqq \frac{1}{4}$. Then take $y=0$, $h=x$, and use $g(0)=0$

and induction to get

$$|g(x)| \leqq \frac{1}{2} + \left(\frac{3}{4} + \frac{1}{8}\right).4 = 4.$$

If $\frac{1}{4} < x \leqq \frac{1}{2}$, choose y and h so that $y + h = x$, $y + 2h = \frac{1}{2}$; using $g\left(\frac{1}{2}\right) = 0$ and induction gives

$$|g(x)| \leqq \frac{1}{2} + \left(\frac{3}{8} + \frac{1}{8}\right).4 = \frac{5}{2} < 4.$$

The proof is similar for $x > \frac{1}{2}$.

Remark. — This proof of theorem 1 would fail for $n \geqq 4$, since we would find $\sigma \geqq 1$ if $s = 1$.

7. Proof of theorem 1 for I_0. — We continue from paragraph 5. Choose ν by lemma 2 so that $\sigma < 1$. Choose μ and m so that

$$m = 2^{\mu}(n-1) \geqq (n-1)\nu - 1. \tag{7.1}$$

Now by (5.8),

$$|g(x)| \leqq L = \lambda_{\mu} \qquad (x \text{ in } S'_{\mu}). \tag{7.2}$$

Using the a_i and γ_j in (4.2), set

$$\tau = \Sigma |a_i|, \qquad K'_n = \frac{\tau + |\gamma_0| L}{1 - \sigma}. \tag{7.3}$$

We prove $|g(x)| \leqq K'_n$.

Define the sets

$$T'_k = \text{all } \frac{i}{\nu^k m} \qquad (k = 0, 1, \ldots),$$

$$T_0 = T'_0 = S'_{\mu}, \qquad T_k = T'_k - T'_{k-1} \qquad (k \geqq 1).$$

By (7.2), $|g(x)| \leqq K'_n$ if x is in T_0 (we could assume rather than prove $K'_n \geqq L$). Using induction, we shall prove the inequality for x in T_k; it then follows for all x, since g is continuous.

The points of T_0 divide I_0 into m intervals $I^*_0, I^*_1, \ldots, I^*_m$; I^*_p is the union of two intervals

$$I'_p = \left(\frac{p}{m}, \frac{p+1}{m+1}\right), \qquad I''_p = \left(\frac{p+1}{m+1}, \frac{p+1}{m}\right). \tag{7.4}$$

Given x in T_k, $k > 0$, we shall choose y and h so that for some q,

$$y = \frac{q}{m}, \qquad y + h = x,$$

and each $y + j\nu h$ is in T'_{k-1} $(j = 0, 1, \ldots, n-1)$.

Case I: x is in some I'_p. Then set $y = \frac{p}{m}$. Now

$$h = x - y \leqq \frac{p+1}{m+1} - \frac{p}{m} = \frac{m-p}{m(m+1)},$$

and hence, by (7.1),

$$y + (n-1)\nu h \leqq \frac{p}{m} + \frac{m-p}{m} = 1.$$

Thus all the $y + j\nu h$ are in I_0. Also x and y and hence h are integers divided by $\nu^k m$, and therefore each $y + j\nu h$ is in T'_{k-1}.

Case II: x is in some I''_p. Then set $y = \frac{p+1}{m}$. Now

$$-h = y - x \leqq \frac{p+1}{m(m+1)},$$

and we find $y + (n-1)\nu h \geqq 0$. Again the $y + j\nu h$ are in T'_{k-1}.

Using induction, we have

$$|g(y)| \leqq L, \qquad |g(y + j\nu h)| \leqq K'_n \qquad (j = 1, 2, \ldots, n-1).$$

Therefore the relations (4.2) and (7.3) give

$$|g(x)| \leqq \Sigma |a_i| + |\gamma_0| L + \sum_{j=1}^{n-1} |\gamma_j| K'_n = K'_n,$$

which completes the proof of the theorem for I_0 and hence for any bounded I.

8. Proof of theorem 1, I unbounded. — There are essentially two cases to consider, the intervals I^* and I^{**}. We may suppose $n \geqq 2$. For each integer k consider a set

$$X_k = (x_k, x_k + k, x_k + 2k, \ldots, x_k + (n-1)k);$$

for I^* we take $x_k = 0$, and for I^{**} we take $x_k = -\frac{1}{2}(n-1)k$.

For each k, let P_k be the polynomial of degree $\leqq n-1$ such that

(8.1) $$g_k(y)=f(y)-P_k(y)=0 \quad \text{for } y \text{ in } X_k.$$

By theorem 1 for bounded intervals [assuming $|\Delta_h^n f(x)|\leqq 1$],

(8.2) $$|g_k(y)|\leqq K'_n, \qquad x_k\leqq y\leqq x_k+(n-1)k \qquad (k=1, 2, \ldots).$$

In particular, this holds at the points $0, 1, \ldots, n-1$, for k large enough. Hence there is a sequence of numbers $\lambda_1, \lambda_2, \ldots$ such that the limits

$$\lim_{i\to\infty} g_{\lambda_i}(j)=g_j \qquad (j=0, 1, \ldots, n-1)$$

exist. Now $\lim P_{\lambda_i}(y)$ exists at these same points, and hence at all points. Thus we can set

(8.3) $$P(y)=\lim_{i\to\infty} P_{\lambda_i}(y), \qquad g(y)=\lim_{i\to\infty} g_{\lambda_i}(y)=f(y)-P(y).$$

Because of (8.2),

$$|g(y)|=|f(y)-P(y)|\leqq K'_n \quad \text{all } y,$$

completing the proof.

9. Proof of theorem 2 and (1.10). — To prove $K_n^{**}\leqq K_n^*$, we note from the properties of P_f (*see* § 2) that for any $\varepsilon>0$ there is a continuous function f in I^{**} such that

(9.1) $$\sup f(x)\geqq K_n^{**}-\frac{\varepsilon}{2}, \qquad \inf f(x)\leqq -K_n^{**}+\frac{\varepsilon}{2}, \qquad \sup|\Delta_h^n f(x)|\leqq 1.$$

Set $f_\lambda(x)=f(x-\lambda)$ for $x\geqq 0$. We may choose λ so that

$$\sup f_\lambda(x)-\inf f_\lambda(x)\geqq 2K_n^{**}-2\varepsilon, \quad x \text{ in } I^*,$$

which shows that $K_n^*\geqq K_n^{**}-\varepsilon$ and hence $K_n^*\geqq K_n^{**}$.

To prove $K_n^*\leqq K_n$, take any $\varepsilon>0$. There is an η, $0\leqq\eta\leqq\frac{\varepsilon}{3}$, and a continuous function f in I^* such that

$$\sup f(x)=K_n^*-\eta, \qquad \inf f(x)=-K_n^*+\eta, \qquad \sup|\Delta_h^n f(x)|\leqq 1.$$

Set $\varepsilon'=\frac{\eta}{4K_n^*}$. By (4.4) we may choose $\delta>0$ such that for any polynomial Q of degree $\leqq n-1$ which vanishes at $x=0$,

$$|Q(x)|\leqq \varepsilon', \quad 0\leqq x\leqq \delta \quad \text{if} \quad \left|Q\left(\frac{i}{n-1}\right)\right|\leqq 1 \quad (i=1, \ldots, n-1).$$

We may choose N so large that if we define the function $f'(x)=f(Nx)$ in I_0, we have

$$(9.2) \qquad \sup_{0\leqq x\leqq \delta} f'(x) > K_n^* - 2\eta, \qquad \inf_{0\leqq x\leqq \delta} f'(x) < -K_n^* + 2\eta.$$

Let $P_{f'}$ be the best approximating polynomial, and write

$$P_{f'}(x) = A + P'(x), \qquad P'(0)=0.$$

Set $g'(x)=f'(x)-P_{f'}(x)$. Then by definition of $P_{f'}$,

$$\sup|g'(x)|\leqq \sup|f'(x)| < K_n^*,$$
$$|A| = |P_{f'}(0)| \leqq |f'(0)| + |g'(0)| < 2K_n^*,$$
$$\left|P'\left(\frac{i}{n-1}\right)\right| \leqq |A| + \left|f'\left(\frac{i}{n-1}\right)\right| + \left|g'\left(\frac{i}{n-1}\right)\right| < 4K_n^*,$$

and hence

$$|P'(x)| \leqq 4K_n^*\varepsilon' \qquad (0\leqq x\leqq \delta).$$

Therefore, by definition of g',

$$2K_n \geqq \sup g' - \inf g' \geqq \sup_{0\leqq x\leqq \delta}[f'(x)-P'(x)] - \inf_{0\leqq x\leqq \delta}[f'(x)-P'(x)]$$
$$> (K_n^* - 2\eta - 4K_n^*\varepsilon') - (-K_n^* + 2\eta + 4K_n^*\varepsilon') \geqq 2K_n^* - 2\varepsilon,$$

giving the result.

Remark. — Using the function $f(x)=\frac{1}{1+x}-\frac{1}{2}$ (*see* § **2**), construct $g'(x)$ as above. Then, for some small ξ, g' drops sharply from $\frac{1}{2}-\xi$ to $-\frac{1}{2}+\xi$, and then is approximately a polynomial of degree $n-1$, going alternately to $\frac{1}{2}-\xi$ and to $-\frac{1}{2}+\xi$, ending (at $x=1$) at $\pm\left(\frac{1}{2}-\xi\right)$. This function shows directly that $K_n \geqq \frac{1}{2}$.

To prove $K_n' \geqq 2K_n^*$, start with the function f used in proving $K_n^* \leqq K_n$; we may clearly suppose that $f(0)=\inf f(x) = -K_n^* + \eta$.

Define f' again, and the corresponding polynomial P'_f. Write

$$P'_{f'}(x) = B + P''(x), \qquad P''(0) = 0;$$

then $B = P'_{f'}(0) = f'(0) = -K_n^* + \eta$. Since

$$\left|P'_{f'}\left(\frac{i}{n-1}\right)\right| = \left|f'\left(\frac{i}{n-1}\right)\right| < K_n^*, \qquad \left|P''\left(\frac{i}{n-1}\right)\right| < 2K_n^*,$$

we have $|P''(x)| \leqq 2K_n^*\varepsilon'$ for $0 \leqq x \leqq \delta$. Hence, by (9.2),

$$\sup_{0 \leqq x \leqq \delta}[f'(x) - P'_{f'}(x)] \geqq (K_n^* - 2\eta) - (-K_n^* + \eta) - 2K_n^*\varepsilon' > 2K_n^* - 4\eta,$$

which gives the result.

10. Proof of theorem 3. — We first note a formula relating n^{th} differences with different intervals

$$(10.1) \qquad \Delta_{2h}^n f(x) = \sum_{j=0}^{n} \binom{n}{j} \Delta_h^n f(x + jh).$$

This is an immediate consequence of the formula

$$\sum (-1)^i \binom{n}{i} \binom{n}{k-i} = \begin{cases} 0 & (k \text{ odd}), \\ (-1)^p \binom{n}{p} & (k = 2p), \end{cases}$$

which follows by equating coefficients in the expansion of

$$(1 + x)^n (1 - x)^n = (1 - x^2)^n.$$

Since

$$(10.2) \qquad \Delta_h^m f(x + jh) = \Delta_h^m f(x) + \sum_{i=0}^{j-1} \Delta_h^{m+1} f(x + ih),$$

(10.1) gives

$$(10.3) \qquad \Delta_{2h}^m f(x) = 2^m \Delta_h^m f(x) + \sum_{i=0}^{m-1} \sum_{j=i+1}^{m} \binom{m}{j} \Delta_h^{m+1} f(x + ih).$$

Suppose (1.14) were false. Then for some bounded function f in I^* with $|\Delta_h^n f(y)| \leqq 1$, there would be an x and h such that, setting $m = n - 1$,

$$(10.4) \qquad \Delta_h^m f(x) = N_m^* + a \qquad (a > 0).$$

Since

$$(10.5)\qquad \sum_{i=0}^{m-1}\sum_{j=i+1}^{m}\binom{m}{j}=\sum_{j=1}^{m} j\binom{m}{j}=2^{m-1}m,$$

(10.3) gives

$$\Delta_{2h}^{m}f(x)\geqq 2^{m}(N_m^{*}+a)-2^{m-1}m=N_m^{*}+2^{m}a.$$

Repeating the process gives

$$\Delta_{2^p h}^{m}f(x)\geqq N_m^{*}+2^{pm}a \qquad (p=1, 2, \ldots).$$

But this is not possible, since f is bounded.

Now suppose that (1.15) were false. Then there would be a bounded f in I^{**} and an x and h such that (10.4) holds with N_m^{**} in place of N_m^{*}. Write (10.1) in the form

$$\Delta_{2h}^{m}f(x-\mu h)=\sum_{j=0}^{\mu}\binom{m}{\mu-j}\Delta_h^m f(x-jh)+\sum_{k=1}^{m-\mu}\binom{m}{\mu+k}\Delta_h^m f(x+kh).$$

By (10.2),

$$|\Delta_h^m f(x\pm jh)-\Delta_h^m f(x)|\leqq j;$$

hence, using either $m=2\mu$ or $m=2\mu+1$,

$$(10.6)\qquad |\Delta_{2h}^{m}f(x-\mu h)-2^m\Delta_h^m f(x)|\leqq\sum_{j=0}^{\mu} j\binom{m}{\mu-j}+\sum_{k=1}^{m-\mu} k\binom{m}{\mu+k}=H_m.$$

Therefore

$$|\Delta_{2h}^{m}f(x-\mu h)|\geqq 2^m(N_m^{**}+a)-H_m=N_m^{**}+2^m a.$$

This process may be repeated, and we arrive at a contradiction, as before. Thus the theorem is proved.

11. Lower bounds for the constants. — Let f be a bounded monotone increasing function in I^{**}; we shall prove (2.8). First, given x and h, we have $a_i=\Delta_h^1 f(x+ih)\geqq 0$. Hence

$$\Delta_h^n f(x)=\Delta_h^{n-1}\Delta_h^1 f(x)\leqq\binom{n-1}{n-1}a_{n-1}+\binom{n-1}{n-3}a_{n-3}+\ldots$$
$$\leqq G_{n-1}(a_{n-1}+a_{n-3}+\ldots)\leqq G_{n-1}(f(\infty)-f(-\infty)).$$

On the other hand, by proper choice of x and h, we may clearly make a middle term above as near $G_{n-1}(f(\infty)-f(-\infty))$ as we like, proving (2.8). Applying the first inequality we proved to any fixed f gives $K_n^{**} \geqq \frac{1}{2G_{n-1}}$.

Let f be any bounded function in I^* satisfying (2.9); we prove (2.10). This holds for $m=1$; assuming it for numbers $< m$, we prove it for m. Since

$$(-1)^{m-1}\Delta_h^{m-1} f(x)-(-1)^{m-1}\Delta_h^{m-1} f(x+h)=(-1)^m\Delta_h^m f(x)\geqq 0,$$

induction gives

$$0\leqq(-1)^{m-1}\Delta_h^{m-1} f(x+h)\leqq(-1)^{m-1}\Delta_h^{m-1} f(x)\leqq f(0)-f(\infty),$$

and the relation with m follows. In particular,

$$|\Delta_h^n f(x)|\leqq f(0)-f(\infty). \tag{11.1}$$

For $f(x)=\frac{1}{1+x}$, $f(0)-f(\infty)=1$, proving $K_n^*\geqq\frac{1}{2}$.

Note that the proofs show that the *same bounds hold for* $K_n(J^{**})$ *and* $K_n(J^*)$.

Because of (1.5) and (1.10), (1.11) also is proved. Recall the remark in paragraph **9**.

12. UPPER BOUNDS FOR K_m^{**}, K_n^*. — We shall prove the remaining inequalities in (1.6), (1.8) and (1.9). Because of (2.7) (also for K_n^*), it is sufficient to prove the following fact. *Let f be any bounded function in* I^{**} *or* I^* *such that* $|\Delta_h^n f(x)|\leqq 1$; *then*

$$\sup|\Delta_h^1 f(x)|\leqq\frac{2}{G_n} \text{ for } I^{**}, \qquad \sup|\Delta_h^1 f(x)|\leqq 2 \text{ for } I^*. \tag{12.1}$$

The proof we give is due essentially to A. Beurling. Note that the proof of the first relation holds if I^{**} is replaced by any subgroup in which each element is divisible by $n!$; for the second, we may use the part of such a subgroup in I^*.

We first prove the second part of (12.1). Take any x and y, $x<y$. Say $y=x+n!h$. For each integer $j=1, \ldots, n$, write $n!=jj'$.

For $N \geqq n!$, the definition of $\Delta_h^n f(x)$ gives, if $|f(x)| \leqq K$,

$$(-1)^n[f(y)-f(x)] = \frac{1}{N}\sum_{i=1}^{N}[\Delta_{ih}^n f(y) - \Delta_{ih}^n f(x)]$$
$$-\frac{1}{N}\sum_{j=1}^{n}(-1)^{n-j}\binom{n}{j}\sum_{i=1}^{N}[f(y+jih)-f(x+jih)].$$

Now

$$\left|\sum_{i=1}^{N}[f(y+ijh)-f(x+ijh)]\right|$$
$$= \left|\sum_{i=N-j'+1}^{N} f(y+ijh) - \sum_{i=1}^{j'} f(x+ijh)\right| \leqq 2j'K \leqq 2n!\,K,$$

and hence

$$|f(y)-f(x)| \leqq 2 + \frac{1}{N}\sum_{j=1}^{n}\binom{n}{j}.2n!\,K < 2 + \frac{1}{N}.2^{n+1}n!\,K,$$

which is arbitrarily near 2 for N large enough, as required.

The proof with I^{**} is similar. Write $n = 2\mu$ or $n = 2\mu - 1$, and use

$$\frac{1}{N}\sum_{i=1}^{N}[\Delta_{ih}^n f(y-\mu ih) - \Delta_{ih}^n f(x-\mu ih)] = (-1)^{n-\mu}G_n[f(y)-f(x)] + \ldots.$$

13. THE VALUES OF K_1^*, K_2^*, K_1^{**}, K_2^{**}, K_3^{**}, K_4^{**}. — It is clear that $K_1^* = K_1^{**} = \frac{1}{2}$, and (1.8) gives K_n^{**} for n even; we shall find K_2^* and K_3^{**}. We shall also prove that (*see* § **2**).

(13.1) $\qquad K_n(J^*) = K_n^* \quad (n \leqq 2), \qquad K_n(J^{**}) = K_n^{**} \quad (n \leqq 4).$

Recall from (2.11) that $K_3(J^*) > K_3^*$.

By (1.14) and (1.13),

(13.2) $\qquad |\Delta_h^1 f(x)| \leqq N_1^* \sup|\Delta_h^2 f(y)| = \sup|\Delta_h^2 f(y)| \quad \text{in } I^*;$

hence $K_2^* \leqq \frac{1}{2}$, and using (1.6) gives $K_2^* = \frac{1}{2}$. Essentially the same proof gives $K_2(J^*) = \frac{1}{2}$.

To prove $K_3^{**} \leqq \frac{1}{4}$, take any bounded f with $|\Delta_h^3 f(x)| \leqq 1$; by subtracting a constant we may assume that

$$\sup f(x) = a, \qquad \inf f(x) = -a. \tag{13.3}$$

We may choose x and $y = x + h$ so that

$$f(x) = a - \varepsilon, \qquad f(y) = -a + \varepsilon. \tag{13.4}$$

Then

$$\Delta_h^3 f(x - h) = 6(a - \varepsilon) - f(x + h) - f(y + h) \leqq 1,$$

$$6a \leqq 1 + 6\varepsilon + 2a, \qquad a \leqq \frac{1}{4} + \frac{3}{2}\varepsilon, \qquad a \leqq \frac{1}{4}.$$

Note that using $\Delta_h^2 f(x)$ here shows again that

$$K_2^* \leqq \frac{1}{2}, \qquad K_2(J^*) \leqq \frac{1}{2}.$$

The proofs we have used give all of (13.1) except for the fact that $K_4(J^{**}) \leqq \frac{1}{6}$. To show this we use the formula

$$\sum_{k=1}^{p} 4^{p-k} \Delta_{2^{k-1}}^4 f(-2^k) = f(-2^p) - 4^p f(-1) + 2(4^p - 1) f(0) - 4^p f(1) + f(2^p), \tag{13.5}$$

whose proof is immediate. Now take any bounded f with

$$|\Delta_h^4 f(x)| \leqq 1;$$

we suppose (13.3) holds. By a linear transformation we may suppose (13.4) holds, with $x = 0$, $y = 1$. Now (13.5) gives

$$[2(4^p - 1) + 4^p](a - \varepsilon) \leqq \frac{4^p - 1}{3} + (1 + 4^p + 1)a,$$

for all p and ε. Letting $\varepsilon \to 0$ and $p \to \infty$ gives $a \leqq \frac{1}{6}$.

14. Bounds for K_5^{**}, $K_5(J^{**})$. — We prove that the upper bound shown in the table for K_5^{**} holds and that this same bound holds for $K_5(J^{**})$; the same proof will hold for each.

If the bound fails, then there is a function f in I^{**} (or in J^{**}) such that $|\Delta_h^5 f(x)| \leqq 1$, and

$$\sup f(x) = a, \qquad \inf f(x) = -a \qquad \left(a > \frac{7}{76}\right).$$

By a linear transformation we may suppose that

$$f(0) = b, \qquad f(1) = -b, \qquad b > \frac{7}{76}.$$

Set

$$g(x) = \frac{1}{2}[f(x) - f(1-x)];$$

then

$$(14.1) \quad g(0) = b, \qquad g(1) = -b, \qquad g(1-x) = g(x), \qquad |\Delta_h^5 g(x)| \leqq 1.$$

We now apply the formula [using $g(1-x) = g(x)$]

$$(14.2) \qquad 5\Delta_1^5 g(-2) + 2\Delta_2^5 g(-4) = 120g(1) - 30g(2) - 10g(4) + 2g(5) + 2g(6)$$

to give

$$120b \leqq 7 + 44b, \qquad b \leqq \frac{7}{76},$$

a contradiction.

15. The values of K_2 and K'_2. — We must prove $K_2 \leqq \frac{1}{2}$, $K'_2 \leqq 1$. As in H. Burkill [1], suppose $f(0) = f(1) = 0$ [so that $P'_f(x) = 0$],

$$|\Delta_h^2 f(y)| \leqq 1, \qquad \sup|f(y)| = a.$$

Say $f(x) = a$. If $x \leqq \frac{1}{2}$, then

$$a = f(x) = -\frac{1}{2}\Delta_x^2 f(0) + \frac{1}{2}f(2x) \leqq \frac{1}{2} + \frac{1}{2}a,$$

proving $a \leqq 1$. Similarly for $x \geqq \frac{1}{2}$. Hence $K'_2 \leqq 1$.

Next, take any f with $\sup|\Delta_h^2 f(x)| \leqq 1$, and corresponding $g = f - P_f$. Then (if necessary using $-g$ in place of g) there are points x_0, x_1, x_2 such that

$$g(x_0) = -g(x_1) = g(x_2) = a = \sup g(y).$$

If $x_1 - x_0 \leqq x_2 - x_1$, considering $\Delta^2_{x_1-x_0} g(x_0)$ shows at once that $a \leqq \frac{1}{2}$; in the other case, consider $\Delta^2_{x_1-x_2} g(x_2)$. Hence $K_2 \leqq \frac{1}{2}$.

A particular function showing that $K'_2 \geqq 1$ and $K_2 \geqq \frac{1}{2}$ is the following

$$(15.1) \qquad f(x) = \begin{cases} \dfrac{1-\varepsilon}{\varepsilon} x & \text{if } 0 \leqq x \leqq \varepsilon, \\ 1 - x & \text{if } \varepsilon \leqq x \leqq 1. \end{cases}$$

We could change f for $x > \frac{1}{2}$, making $f\left(\frac{1}{2} + y\right) = f\left(\frac{1}{2} - y\right)$. We could also change the original f so that it moves rapidly up to $1 - \varepsilon$ for x near 1.

16. Lower bounds for K_3, K'_3, K^*_3. — We prove $K^*_3 \geqq \frac{8}{15}$; then (1.5) and (1.10) give $K_3 \geqq \frac{8}{15}$, $K'_3 \geqq \frac{16}{15}$.

Define f at certain points as follows

x	0.	2.	3.	4.	6.	8.	12.	16.	24.	32.	48.
$15 f(x)$	-8	4	4	8	4	7	4	6	4	5	4

Let f be linear between any two of these points, and set $f(x) = \frac{4}{15}$ for $x \geqq 48$. It is easily verified that $|\Delta^3_h f(x)| \leqq 1$ for all x and h. Hence

$$K^*_3 \geqq \sup |f(x)| = \frac{8}{15}.$$

17. Some particular formulas. — More detailed study of the values of the K_n for instance for larger n requires combining n^{th} differences in various ways. We give here formulas for the cases $n = 3, 4, 5$, which will be used below. All but (17.4) and (17.5) are of the general form (3.1), perhaps combined with (10.1) or a generalization of this formula.

We tell how to find the a_i and c_j in (3.1). (One naturally prefers to multiply through by some integer to make all the coefficients integral.) To simplify the notation, we take $y = 0$, $h = 1$, and write f_k for $f(k)$.

We first find the polynomial Q of (3.2), and hence the c_j, by (3.9). By the interpolation formula, this gives

$$(17.1)\qquad c_j = -\frac{(s-x_0)\dots(s-x_{j-1})(s-x_{j+1})\dots(s-x_{n-1})}{(x_j-x_0)\dots(x_j-x_{j-1})(x_j-x_{j+1})\dots(x_j-x_{n-1})}.$$

We now substitute into (3.1), to find $a_0, a_1, a_2, \dots$ in turn.

For one example, take $X = (0, 4, 6)$, $s = 2$. We find

$$c_0 = -\frac{1}{3}, \qquad c_1 = -1, \qquad c_2 = \frac{1}{3},$$

and the right hand side of (3.1) is $\frac{1}{3}\Delta_1^3 f_0$. Working out the left hand side gives (10.1) for $n = 3$. We may find $\Delta_k^n f_0$ in terms of the $\Delta_1^n f_i$ in the same manner.

In the applications, it is advantageous to have the sum of the absolute values of the coefficients of the $\Delta_k^n f_i$ as small as possible. We may use (10.1) and the generalization just mentioned to accomplish this.

Formulas with $n = 3$. — Besides (6.1), we have

$$(17.2)\qquad 9\Delta_1^3 f_0 + 3\Delta_1^3 f_1 + \Delta_2^3 f_0 = -10 f_0 + 24 f_1 - 15 f_2 + f_6,$$

$$(17.3)\qquad 18\Delta_1^3 f_0 + 6\Delta_1^3 f_1 + 3\Delta_2^3 f_0 + \Delta_2^3 f = -21 f_0 + 48 f_1 - 28 f_2 + f_8,$$

$$(17.4)\qquad \Delta_1^3 f_1 - 3\Delta_1^3 f_2 - 2\Delta_2^3 f_0 - \Delta_2^3 f_1 = 2 f_0 - 15 f_3 + 16 f_4 - 2 f_6 - f_7,$$

$$(17.5)\qquad 3\Delta_2^3 f_0 + \Delta_3^3 f_0 + \Delta_1^3 f_1 = -4 f_0 - f_1 + 12 f_2 - 8 f_4 + f_9.$$

Formulas with $n = 4$. — Besides (13.5), we have

$$(17.6)\qquad \Delta_1^4 f_0 + 4\Delta_1^4 f_1 + \Delta_1^4 f_2 = f_0 - 9 f_2 + 16 f_3 - 9 f_4 + f_6,$$

$$(17.7)\qquad 5\Delta_1^4 f_0 + 4\Delta_1^4 f_1 + \Delta_1^4 f_2 = 5 f_0 - 16 f_1 + 15 f_2 - 5 f_4 + f_6,$$

$$(17.8)\qquad \begin{aligned} &293.2\,\Delta_1^4 f_0 + 208\,\Delta_1^4 f_1 + 72\,\Delta_1^4 f_4 + 51.2\,\Delta_1^4 f_5 \\ &\quad + 77.8\,\Delta_2^4 f_0 + 4.8\,\Delta_2^4 f_1 + 14\,\Delta_3^4 f_0 \\ &\quad = 385 f_0 - 960 f_1 + 616 f_2 - 55 f_8 + 14 f_{12}, \end{aligned}$$

$$(17.9)\qquad \begin{aligned} &1106\,\Delta_1^4 f_0 + 852\,\Delta_1^4 f_1 - 3\,\Delta_1^4 f_2 - 120\,\Delta_1^4 f_3 - 14\,\Delta_1^4 f_4 \\ &\quad - 7\,\Delta_1^4 f_5 - 55\,\Delta_1^4 f_6 - 2\,\Delta_1^4 f_7 - \Delta_1^4 f_8 \\ &\quad - 27\,\Delta_1^4 f_9 - 55\,\Delta_1^4 f_{10} + 309\,\Delta_2^4 f_0 + 15\,\Delta_2^4 f_5 \\ &\quad + 145\,\Delta_3^4 f_0 + 52\,\Delta_4^4 f_1 + 13\,\Delta_5^4 f_2 \\ &\quad = 1560 f_0 - 3\,520 f_1 + 2\,002 f_2 - 55 f_{14} + 13 f_{22}. \end{aligned}$$

Formulas with $n=5$. — Besides (14.2), we have

$$(17.10)\quad 35\,\Delta_1^5 f_0+47\,\Delta_1^5 f_1+25\,\Delta_1^5 f_2+5\,\Delta_1^5 f_3 = -35 f_0+128 f_1-140 f_2+70 f_4-28 f_6+5 f_8,$$

$$(17.11)\quad 5\,\Delta_1^5 f_0+25\,\Delta_1^5 f_1+15\,\Delta_1^5 f_2+3\,\Delta_1^5 f_3 = -5 f_0+60 f_2-128 f_3+90 f_4-20 f_6+3 f_8,$$

$$(17.12)\quad 282\,\Delta_1^5 f_0+320\,\Delta_1^5 f_1-\Delta_1^5 f_2-165\,\Delta_1^5 f_3+47\,\Delta_1^5 f_5-60\,\Delta_1^5 f_6 - 56\,\Delta_1^5 f_7+103\,\Delta_2^5 f_0+55\,\Delta_2^5 f_1+55\,\Delta_2^5 f_3+11\,\Delta_3^5 f_1 = -385 f_0+1024 f_1-704 f_2+110 f_8-56 f_{12}+11 f_{16},$$

$$(17.13)\quad 3\,760\,\Delta_1^5 f_0+3\,925\,\Delta_1^5 f_1-\Delta_1^5 f_2-1\,655\,\Delta_1^5 f_3 + 5\,\Delta_1^5 f_4+1\,085\,\Delta_1^5 f_5+870\,\Delta_1^5 f_6 + 328\,\Delta_1^5 f_7+1\,740\,\Delta_2^5 f_0+1\,121\,\Delta_2^5 f_2+385\,\Delta_2^5 f_4+77\,\Delta_3^5 f_2 = -5500 f_0+14875 f_1-10472 f_2+1496 f_7-476 f_{12}+77 f_{17}.$$

18. An upper bound for K'_3. — We prove $K'_3 \leq \frac{14}{9}$, and incidentally, for any $g=f-P'_f$,

$$(18.1)\qquad |g(x)| \leq \frac{23}{18}\sup|\Delta_h^3 f(y)|,\qquad \frac{1}{4}\leq x\leq \frac{3}{4}.$$

If $K'_3>\frac{14}{9}$, then there is a function f such that for the corresponding $g=f-P'_f$, $|\Delta_h^3 g(x)|\leq 1$ and

$$(18.2)\qquad g(0)=g\left(\frac{1}{2}\right)=g(1)=0,\qquad \sup|g(x)|=a>\frac{14}{9}.$$

Applying (6.1) gives

$$(18.3)\quad 3\Delta_{\frac14}^3 g(0)+\Delta_{\frac14}^3 g\left(\frac14\right)=8g\left(\frac14\right),\qquad \left|g\left(\frac14\right)\right|\leq\frac12,\qquad \left|g\left(\frac34\right)\right|\leq\frac12.$$

Take any x, $\frac14\leq x\leq\frac34$. Setting $h=\frac12-x$, $y=x-h$, and applying (6.1) gives

$$3\,\Delta_h^3 g(x-h)+\Delta_h^3 g(x)=-3g(x-h)+8g(x)+g\left(\frac12+2h\right);$$

hence we find

$$(18.4)\qquad b=\sup_{\frac14\leq x\leq\frac34}|g(x)|\leq\frac12+\frac12 a.$$

Because of (18.2), $b<a$.

Say $g(x_0) = a$; we may suppose $0 < x_0 < \frac{1}{4}$.

Case I : $\frac{1}{8} \leqq x_0 < \frac{1}{4}$. Apply (17.2) (inserting y and h), with

$$y + h = x, \qquad x + h = \frac{1}{4}.$$

This gives, using (18.3) and (18.4),

$$24a \leqq 13 + 10a + 15 \times \frac{1}{2} + b \leqq 21 + \frac{21}{2}a, \qquad a \leqq \frac{14}{9}.$$

Case II : $0 < x_0 \leqq \frac{1}{8}$. Applying (17.3) (with y and h), with $y = 0$, $h = x$, gives

$$48a \leqq 28 + 28a + a, \qquad a \leqq \frac{28}{19} < \frac{14}{9},$$

again a contradiction. Thus $K'_3 \leqq \frac{14}{9}$. Relation (18.1) follows from this and (18.4).

19. An upper bound for K_3. — We shall prove $K_3 \leqq \frac{7}{10}$. If this is false, then there is a function f satisfying $|\Delta_h^3 f(x)| \leqq 1$, such that if $g = f - P_f$, then

$$\sup g(x) = -\inf g(x) = a > \frac{7}{10}. \tag{19.1}$$

By definition of P_f, g takes on alternately the values a, $-a$, a, $-a$ (or their negatives). We may clearly suppose

$$g(0) = a, \qquad g(p) = -a, \qquad g(q) = a, \qquad g(1) = -a \qquad (p < q). \tag{19.2}$$

We shall call (x, y) a « central interval » if, setting

$$h = y - x, \qquad x' = x - h, \qquad y' = y + h,$$

the points x' and y' are both in I_0. Since $|\Delta_h^3 g(x')| \leqq 1$, (19.1) gives

$$|g(y) - g(x)| \leqq \frac{1}{3} + \frac{2}{3}a \quad \text{for central intervals } (x, y). \tag{19.3}$$

It is not true that $\frac{1}{3} \leqq p$, $q \leqq \frac{2}{3}$; for then (p, q) would be a central interval, and (19.2) would contradict (19.3). Hence we may suppose $p < \frac{1}{3}$.

Suppose $p \leqq \frac{1}{8}$. Then we can apply (17.3) with $h = p$ inserted, giving

$$48a \leqq 28 - 21a + (28 + 1)a, \qquad a \leqq \frac{7}{10},$$

a contradiction. Similarly $q \geqq \frac{7}{8}$ is impossible. Now

$$\frac{1}{8} < p < \frac{1}{3}, \qquad q < \frac{7}{8}. \tag{19.4}$$

To simplify the calculations, we now use the function

$$g'(x) = \frac{7}{10a} g(x), \qquad \sup g' = -\inf g' = a' = \frac{7}{10}. \tag{19.5}$$

The relations above hold with g' and a' in place of g and a.

Suppose $p \leqq \frac{2}{9}$. Applying (17.5) with $h = \frac{p}{2}$ inserted gives

$$8g'(2p) \leqq 5 - 4a' - 12a' + (1+1)a' = -\frac{48}{10}, \qquad g'(2p) \leqq -\frac{3}{5}.$$

Set $p^\star = p$ if $p \geqq \frac{2}{9}$ and $p^\star = 2p$ if $p < \frac{2}{9}$. Then the above relation and (19.4) give

$$\frac{2}{9} \leqq p^\star < \frac{4}{9}, \qquad g'(p^\star) \leqq -\frac{3}{5}. \tag{19.6}$$

Suppose $g'(x) = a'$ for some x, $\frac{4}{9} \leqq x \leqq \frac{2}{3}$. Setting $h = \frac{1}{2}x$ and applying $|\Delta_h^3 g'(0)| \leqq 1$ gives

$$g'(h) \geqq a' - \frac{1}{3} = \frac{11}{30}.$$

Now $(h, p^\star)$ is a central interval, and (19.3) with g' and a' gives

$$|g'(h) - g'(p^\star)| \leqq \frac{1}{3} + \frac{2}{3}a' = \frac{4}{5}, \qquad g'(p^\star) \geqq -\frac{13}{30};$$

but this contradicts (19.6). If $g'(x) = a'$ for some x, $p < x < \frac{4}{9}$, then (x, p^*) is a central interval, and again we are led to a contradiction. Therefore, using (19.4),

$$\frac{2}{3} < q < \frac{7}{8}. \tag{19.7}$$

Define q^* the same way p^* was defined. Then

$$\frac{5}{9} < q^* \leqq \frac{7}{9}, \qquad g'(q^*) \geqq \frac{3}{5}. \tag{19.8}$$

Applying $\Delta_h^3 g'(0) \geqq -1 \left(h = \frac{1}{3}\right)$ gives

$$g'\left(\frac{1}{3}\right) - g'\left(\frac{2}{3}\right) \geqq -\frac{1}{3} + \frac{2}{3}a' = \frac{2}{15}. \tag{19.9}$$

Suppose $g'\left(\frac{1}{3}\right) \geqq \frac{1}{15}$. Note that $\left(p^*, \frac{1}{3}\right)$ is a central interval. Set

$$h = \frac{1}{3} - p^*, \qquad y = p^* - h, \qquad x = \frac{1}{3} + h,$$

and apply $\Delta_h^3 g'(y) \geqq -1$; using (19.6) gives

$$g'(x) \geqq -1 - a' + \frac{9}{5} + \frac{1}{5} = \frac{3}{10}.$$

Now (p^*, x) is also a central interval, and (19.3) gives

$$g'(p^*) \geqq g'(x) - \frac{4}{5} \geqq -\frac{1}{2},$$

contradicting (19.6). Therefore $g'\left(\frac{1}{3}\right) < \frac{1}{15}$. Now (19.9) gives

$$g'\left(\frac{2}{3}\right) < -\frac{1}{15}.$$

Applying the same process as above with q^* in place of p^* again gives a contradiction. Hence $K_3 \leqq \frac{7}{10}$ is proved.

20. An upper bound for K_3^*. — We show that

$$K_3^* \leqq \frac{3\,863}{5\,814} < 0.664431. \tag{20.1}$$

Consider any f in I^* with $|\Delta_h^3 f(x)| \leqq 1$, $\sup f = -\inf f = a$. Take any $\varepsilon > 0$. By a linear transformation, we may suppose

$$f(0) = -a + \varepsilon, \qquad f(3) = a - \varepsilon.$$

We use the notation

$$f(k) = f_k = a - \eta_k.$$

By (17.4),

$$(20.2) \quad 16 f_4 \geqq -7 + 17(a - \varepsilon) + 2(a - \eta_6) - a = 18a - 7 - 2\eta_6 - 17\varepsilon.$$

Now

$$-\Delta_k^3 f_0 = -a + \varepsilon - 3 f_k + 3 f_{2k} - (a - \eta_{3k}) \geqq -1,$$

$$f_{2k} - f_k \geqq \frac{1}{3}(2a - 1 - \varepsilon - \eta_{3k}).$$

Adding these equations for $k = 4, 8, 16, 32$ gives

$$(20.3) \quad f_{64} - f_4 \geqq \frac{4}{3}(2a - 1 - \varepsilon) - \frac{1}{3}\eta', \qquad \eta' = \eta_{12} + \eta_{24} + \eta_{48} + \eta_{96}.$$

By (1.14) and (1.13),

$$-\Delta_3^2 f_0 = -(-a + \varepsilon) + 2(a - \varepsilon) - (a - \eta_6) \leqq \frac{4}{3},$$

$$(20.4) \qquad \eta_6 \leqq \frac{4}{3} - 2a + 3\varepsilon.$$

More generally, for $k \geqq 1$,

$$\eta_{3.2^k} \leqq (2^k - 1)\left(\frac{4}{3} - 2a\right) + (2^{k+1} - 1)\varepsilon,$$

which follows at once from $-\Delta_{3.2^k}^2 f_0 \leqq \frac{4}{3}$ and induction. Hence

$$(20.5) \qquad \eta' \leqq 56\left(\frac{4}{3} - 2a\right) + 116\varepsilon.$$

Combining these equations gives

$$48a \geqq 48 f_{64} \geqq 3\left[18a - 7 - 2\left(\frac{4}{3} - 2a + 3\varepsilon\right) - 17\varepsilon\right]$$
$$+ 64(2a - 1 - \varepsilon) - 16\left[56\left(\frac{4}{3} - 2a\right) + 116\varepsilon\right],$$

$$(1986 - 48)a \leqq 85 + 902 \times \frac{4}{3} + 1989\varepsilon,$$

and since ε is arbitrary, the required inequality follows.

21. UPPER BOUNDS FOR K_4 AND K'_4. — We prove

$$K_4 \leqq K'_4 \leqq A = \frac{1297}{400} = 3.2425. \tag{21.1}$$

Say $|\Delta_h^4 f(x)| \leqq 1$ and

$$f(0) = f\left(\frac{1}{3}\right) = f\left(\frac{2}{3}\right) = f(1) = 0, \qquad f(p) = \sup|f(x)| = a;$$

we need merely prove $a \leqq A$. First apply (17.6) (with h inserted), using $h = \frac{1}{6}$; then apply (17.7) with $h = \frac{1}{6}$ and then with $h = \frac{1}{12}$. This gives

$$\left|f\left(\frac{1}{2}\right)\right| \leqq \frac{3}{8}, \qquad \left|f\left(\frac{1}{6}\right)\right| \leqq \frac{5}{8}, \qquad \left|f\left(\frac{1}{12}\right)\right| \leqq \frac{79}{64}. \tag{21.2}$$

We consider different possibilities for p; we may suppose $p \leqq \frac{1}{2}$.

Suppose $0 \leqq p \leqq \frac{1}{12}$. Apply (17.8), with $h = x$. This gives

$$960a \leqq 721 + (616 + 55 + 14)a, \qquad a \leqq \frac{721}{275} < 3 < A.$$

Suppose $\frac{1}{12} < p \leqq \frac{1}{8}$. We apply (17.9), with the term f_0 at $\frac{1}{12}$ and the term f_1 at p. This gives

$$3\,520a \leqq 2\,776 + 1\,560 \times \frac{79}{64} + 2\,070a, \qquad a \leqq A.$$

Suppose $\frac{1}{8} < p \leqq \frac{1}{6}$. Applying (17.8) with f_1 at p and f_2 at $\frac{1}{6}$ gives

$$960a \leqq 721 + 616 \times \frac{5}{8} + 454a, \qquad a \leqq \frac{1106}{506} < 3 < A.$$

Suppose that $\frac{1}{6} < p \leqq \frac{1}{3}$. Applying (17.7) with f_1 at p and f_2 at $\frac{1}{3}$ gives

$$16a \leqq 10 + 11a, \qquad a \leqq 2.$$

Suppose finally that $\frac{1}{3} < p \leqq \frac{1}{2}$. Applying (17.6) with f_2 at $\frac{1}{3}$ and f_3 at p gives $a \leqq \frac{6}{5}$.

22. UPPER BOUNDS FOR K_5 AND K'_5. — We prove

$$(22.1)\qquad K_5 \leqq K'_5 \leqq B = \frac{6\,262\,837}{602\,624} < 10.392612.$$

Our proof will follow the lines of paragraph **7**. Start as in paragraph **21**, with $f\left(\frac{i}{4}\right) = 0$. Apply (17.10) with f_0 at 0, f_1 at $\frac{1}{8}$, then (17.11) with f_0 at 0, f_3 at $\frac{3}{8}$, then (17.12) with f_0 at 0, f_1 at $\frac{1}{16}$, then (17.11) with f_0 at 0, f_3 at $\frac{3}{16}$, then (17.11) with f_0 at $\frac{1}{2}$, f_3 at $\frac{5}{16}$, and finally (17.11) with f_0 at $\frac{1}{4}$, f_3 at $\frac{7}{16}$, to find the bounds on f as given in the following table :

x	$\frac{1}{8}$	$\frac{3}{8}$	$\frac{1}{16}$	$\frac{3}{16}$	$\frac{5}{16}$	$\frac{7}{16}$
$f(x) \leqq$	$\frac{7}{8}$	$\frac{3}{8}$	$\frac{1771}{1024}$	$\frac{27}{32}$	$\frac{11}{16}$	$\frac{39}{64}$

Of course the same bounds hold with x replaced by $1 - x$.

We now have

$$(22.2)\qquad \left|f\left(\frac{i}{16}\right)\right| \leqq L = \frac{1771}{1024} < B \quad \text{all } i;$$

we prove that $|f(x)| \leqq B$ for any x of the form $\frac{i}{2^k}$ $(k = 5, 6, \ldots)$. With $m = 16$, form the intervals I'_p, I''_p, as in (7.4). If x is in I'_p, we may apply (17.13) with f_0 at $\frac{p}{m}$ and f_1 at x; if x is in I''_p, we may apply (17.13) with f_0 at $\frac{p+1}{m}$ and f_1 at x. In either case, using (22.2) for f_0, we find, using induction,

$$14\,875\,|f(x)| \leqq 14\,952 + 5\,500\,L + 12\,521\,B,$$

which gives $|f(x)| \leqq B$. Since f is continuous, (22.1) is proved.

SINGULARITIES OF MAPPINGS OF EUCLIDEAN SPACES

By Hassler Whitney

1. Introduction

We shall describe here some results and methods pertaining to the following general problem (details will appear elsewhere). Suppose a mapping f_0 of an open set R in n-space E^n into m-space E^m is given (we shall write $f: R^n \to E^m$). How can we alter f_0 slightly, obtaining a mapping f with nicer and simpler properties? By the Weierstrass approximation theorem (generalized), we may require that f be analytic in R; if f_0 was r-smooth (had continuous partial derivatives through the rth order), we may require the partial derivatives of f through the rth order to approximate those of f_0 (we then call f an *r-approximation*). Now take any *regular* point p of f, that is, a point p such that f is of maximum rank $\nu = \inf(n, m)$ at p. (Equivalently, using coordinate systems in E^n and in E^m, the Jacobian matrix of f at p is of rank ν.) Then, by the implicit function theorem, we may choose coordinates so that f has the form

$$(1.1) \qquad y^i = x^i \quad (i = 1, \cdots, \nu), \qquad y^i = 0 \quad (i > \nu, \text{ if } m > n).$$

Hence at any regular point p, f has the structure of the particular mapping (1.1); we shall be satisfied with f here.

Any non-regular point we call a *singular* point of f. In general (unless $m \geqq 2n$; see §4) we cannot avoid the presence of singular points. We would then like to reduce them as much as possible, making them lie in small and simple point sets, and requiring the structure of f to be as simple as possible in the neighborhood of a singular point. We shall show that the set S' of singular points may be made to form a smooth manifold plus boundary. There will be subsets of S', also manifolds with boundaries, consisting of points in the neighborhood of which f is more complicated. This gives a splitting of S' into sets, in each of which f satisfies specified conditions. Thus the singularities of f are divided into various types. We give the geometric basis of this splitting in Part I. Examples of the types of singularities will be found in Part II.

The manner of defining the singularities is as follows. Let L^r be the space of possible values of the differentials of a mapping $f: R^n \to E^m$ through the order r at a point p (see §2). Each L^r contains a certain set of manifolds, say $L^r_{(1)}, L^r_{(2)}, \cdots$, of various dimensions. Given f, the values at p of the differentials of f give a mapping $f^r: R^n \to L^r$, for each r. (We may keep $r \leqq 2n$; see §11.) By a slight change in f, we may require f^r to be "crosswise" to the $L^r_{(i)}$ (see §5); then $f^r(R)$ does not intersect any $L^r_{(j)}$ of small dimension, and it intersects other $L^r_{(i)}$ in as simple a manner as possible. Using a suitable r, we then say f is *locally generic*. The sets in R which map into the $L^r_{(i)}$ are the singular sets of f; at a point of one of these sets, we say f has a *generic singularity*.

Suppose p is a generic singularity; say $q = f^r(p) \in L^r_{(i)}$. Then f^r is crosswise to $L^r_{(i)}$ at q. It follows that for any good $(r + 1)$-approximation f_1 to f, f_1^r will also be crosswise to $L^r_{(i)}$ at some point $q_1 = f_1^r(p_1)$; thus a generic singularity cannot be removed in this manner. (It may be removable under a k-approximation for smaller k; see for instance §22 of [13].)

We say the above f_1 has a singularity at p_1 of the same type that f had at p. A basic question now is, to what extent is f_1 near p_1 like f near p? If the division into singular sets is sufficiently complete, then we would like f_1 to be obtainable from f by "changes of coordinates." Explicitly, there is then a mapping F defined by

$$y^i = \phi^i(x^1, \cdots, x^n) \qquad (i = 1, \cdots, m),$$

such that both f and f_1 may be put into this form by proper choices of coordinate systems. We then choose a particular mapping F, and call it a *normal form* for the type of singularity, and we say that f is *stable* at p.

Our *principal conjecture* is that the division of singularities into types satisfies the above condition. The general program may be described as follows.

(a) Carry out the definition of types of singularities, as proposed below. This will be seen to run into questions about the relation of planes to certain algebraic cones. (The basic theoretical considerations are not difficult, but carrying out the description of the singularities in high dimensional cases seems very complicated.) It is then easy to show that arbitrarily near any f_0 there is a locally generic f.

(b) Show that the division into types is complete, in that any locally generic mapping is stable at each point. This is the most difficult part of the program. (The choice of a normal form for a given type of singularity is relatively easy.)

A further study should include the following:

(c) Find topological properties relating to the singularities, both locally and in the large (with E^n and E^m replaced by smooth manifolds). We shall not discuss this problem here. See [11] and [13]; also Thom, [4], [5] and [6].

The program has been carried through in certain cases, as follows: For $m = 1$, we have a real function f in E^n; the singular points are the critical points of f. The theory of Marston Morse, in [1] and [2], covers this case (see §16). For $m \geqq 2n$, there are no singularities; see §4. For $m = 2n - 1$, we can have singularities at isolated points; see [9] and §20. For $n = m = 2$, S' consists of smooth curves, and there are isolated points of other type on the curves; see [13] and §17.

Suggestions have been made about the possible types of singularities in more general cases. F. Roger [3] described the types we call $S_1^{(k)}$. R. Thom found explicitly the types in low dimensional cases in [4] (the entry $S_2(S_1)$ appears first for $(n, m) = (5, 4)$ (see §25), not $(4, 3)$). But no proofs that the division into types is complete in these cases has been given.

Singularities which can be removed by small deformations (for instance branch points; see §7 of [13]) of course may nevertheless be of importance. We shall not consider these here.

I. Geometric structure of singularities

2. The differentials of a mapping

Let $V(E^n) = V^n$ denote the space of vectors in E^n. The *differential* $df(p)$ of f at p is the linear transformation of V^n into V^m defined by

$$(2.1) \qquad df(p) \cdot v = \lim_{t \to 0+} \frac{1}{t}[f(p + tv) - f(p)] \in V^m, \qquad v \in V^n.$$

The *second differential* $d^2f(p)$ is the bilinear transformation of $V^n \times V^n$ into V^m defined by

$$(2.2) \quad d^2f(p) \cdot (v, w) = \lim_{s,t \to 0+} \frac{1}{st}[f(p + sv + tw) - f(p + sv) - f(p + tw) + f(p)].$$

It is symmetric:

$$(2.3) \qquad d^2f(p) \cdot (v, w) = d^2f(p) \cdot (w, v).$$

Higher differentials are defined similarly, or by induction.

If coordinate systems in the spaces are given, then naming the first r differentials of f is equivalent to naming the partial derivatives of orders up through r. Relation (2.3) corresponds to the symmetry of cross partial derivatives.

Let $L^1 = \mathfrak{L}^1$ denote the space of linear transformations of V^n into V^m; it is a linear space, of dimension nm. More generally, let $\mathfrak{L}^r$ denote the space of multilinear symmetric transformations of the Cartesian product $V^n \times \cdots \times V^n$ (r factors) into V^m. Thus an element of $\mathfrak{L}^r$ is a multilinear symmetric function $F(v_1, \cdots, v_r)$, whose values are vectors of V^m. Set

$$(2.4) \qquad L^r = \mathfrak{L}^1 \oplus \cdots \oplus \mathfrak{L}^r.$$

Now with f given, $d^kf(p)$ is an element of $\mathfrak{L}^k$, for each k. Let $f^r(p)$ denote the set of $d^kf(p)$, $k \leqq r$. Thus

$$(2.5) \qquad d^kf(p) \in \mathfrak{L}^k, \qquad f^r(p) = (df(p), d^2f(p), \cdots, d^rf(p)) \in L^r.$$

3. The structure of L^1

Each point $T \in L^1$ is a transformation of V^n into V^m. The *rank* of T is the dimension of the image space $T(V^n)$. Let L^1_ρ denote the set of points of L^1 of rank ρ. Now

$$(3.1) \qquad L^1 = L^1_0 \cup L^1_1 \cup \cdots \cup L^1_\nu, \quad \nu = \inf(n, m).$$

It is easy to see that each L^1_ρ is a manifold, of dimension

$$(3.2) \qquad \dim(L^1_\rho) = (n + m)\rho - \rho^2 = nm - (n - \rho)(m - \rho).$$

The *codimension* of a manifold in a fixed containing space is the difference of the two dimensions. Hence

$$(3.3) \qquad \operatorname{codim}(L^1_\rho) = (n - \rho)(m - \rho).$$

Clearly any limit points of L^1_ρ not in L^1_ρ lie in the sets $L^1_{\rho'}$, $\rho' > \rho$. We express this by saying that the L^1_ρ form a *manifold collection*.

4. The rank of f

Given $f_0 : R^n \to E^m$, there corresponds a mapping $f_0^1 = df_0 : R^n \to L^1$. Set

$$\rho(n, m) = \sup \{r + 1 : (n - r)(m - r) > n\}; \tag{4.1}$$

this is the largest number $r + 1$ such that the condition shown holds.

Set $\rho' = \rho(n, m) - 1$; then

$$\operatorname{codim} (L^1_{\rho'}) = (n - \rho')(m - \rho') > n.$$

It follows [13, Theorem 11A] that a small deformation of f_0 will give a mapping f such that no point $df(p)$ lies in $L^1_{\rho'}$.

We say *f is of rank ρ at p* if $df(p)$ is of rank ρ. The *rank* of f is the smallest rank at any point. The result above is that *by a small deformation, we may make f of rank $\geqq \rho(n, m)$.*

The *deficiency* of f is (with $\nu = \inf (n, m)$)

$$\operatorname{dfc} (f, p) = \nu - \operatorname{rank} (f, p), \qquad \operatorname{dfc} (f) = \nu - \operatorname{rank} (f). \tag{4.2}$$

Set

$$\delta(n, m) = \nu - \rho(n, m). \tag{4.3}$$

Then *we may make f of deficiency $\leqq \delta(n, m)$.*

Some special cases of this are the following:

(a) If $m \geqq 2n$, then $\rho(n, m) = n$, $\delta(n, m) = 0$; hence we can remove all singularities in this case. Take any smooth manifold M, and set $f_0(p) = q_0 \in E^{2n}$ $(p \in M)$. Applying the result above gives an "immersion" f of M in E^{2n}. (There may be self-intersections of $f(M)$.)

(b) Suppose $m \geqq \frac{3}{2}(n - 1)$. Using $r = n - 2$ in (4.1) shows that $\rho(n, m) \geqq n - 1$. Thus all singular points may be made of deficiency $\leqq 1$.

(c) For $n \geqq 2m - 3$, we find $\rho(n, m) \geqq m - 1$, and $\delta(n, m) \leqq 1$ again.

(d) If $n = m < (k + 1)^2$, we find $\delta(n, m) \leqq k$.

The results given are the best possible, in that there exist mappings such that any sufficiently nearby mapping has the rank $\rho(n, m)$ and deficiency $\delta(n, m)$.

Some values of $\delta(n, m)$ are given in the table. (This was found in large part by Wolfsohn, [14].)

n \ m	1	2	3	4	5	6	7	8	9	10	11	12
1	1	0	0	0	0	0	0	0	0	0	0	0
2	1	1	1	0	0	0	0	0	0	0	0	0
3	1	1	1	1	1	0	0	0	0	0	0	0
4	1	1	1	2	1	1	1	0	0	0	0	0
5	1	1	1	1	2	1	1	1	1	0	0	0
6	1	1	1	1	2	2	2	1	1	1	1	0

Table of $\delta(n, m)$.

5. Crosswise mappings

Consider $F : R^n \to E^N$, and let M be a smooth manifold in E^N. Suppose $F(p) = q \in M$. At q we have the tangent vector space $P(M, q)$ of M, and the image $dF(p)(V^n)$ of V^n under dF at p. We say F is *crosswise* to M if, for all such p, these two vector spaces span V^N:

$$dF(p)(V^n) + P(M, F(p)) = V^N. \tag{5.1}$$

Note that if

$$n + \dim (M) < N, \quad \text{i.e.,} \quad n < \operatorname{codim} (M),$$

then $F(R) \cap M$ must be void.

LEMMA (Thom, [6]). *Let M be a manifold collection in L^r. Then arbitrarily near any $f_0 : R^n \to E^m$ there is an f for which f^r is crosswise to each manifold of M.*

Suppose $f^r : R^n \to L^r$ is crosswise to the smooth manifold M, and $\operatorname{codim}(M) = s \leqq n$ (otherwise $f(R)$ does not intersect M). Say $f^r(p) = q \in M$. Let $\phi_1, \cdots, \phi_s$ be real functions defining M near q; they vanish only in M, and their gradients are independent there. Set $\psi_i(p) = \phi_i(f^r(p))$ in R. Now the functions $\psi_1, \cdots, \psi_s$, near p, vanish only in $S = (f^r)^{-1}(M)$, and their gradients are independent at p, since f^r is crosswise to M; hence S is a manifold in E^n, of codimension s.

6. The singular sets S_k

Let $f : R^n \to E^m$ be locally generic (§1). Then (§5) $f^1 = df$ is crosswise to the L^1_ρ. Set

$$S_k = (f^1)^{-1}(L^1_{\nu-k}), \qquad \nu = \inf (n, m). \tag{6.1}$$

Then (§5) the S_k are smooth manifolds in E^n; they form a manifold collection. Now S_0 is the set of regular points of f, and $S' = S_1 \cup S_2 \cup \cdots$ is the set of singular points; f is of deficiency k in S_k. By (3.3) and §5,

$$\operatorname{codim} (S_{\nu-\rho}) = (n - \rho)(m - \rho), \tag{6.2}$$

provided this number is $\leqq n$; otherwise, $S_{\nu-\rho}$ is void. This gives

$$\operatorname{codim} (S_k) = k(|n - m| + k). \tag{6.3}$$

As special cases (provided the numbers shown are between 0 and n),

$$n \geqq m: \begin{cases} \operatorname{codim} (S_1) = n - m + 1, & \dim (S_1) = m - 1, \\ \operatorname{codim} (S_2) = 2(n - m + 2), & \dim (S_2) = 2m - n - 4, \end{cases}$$

$$n \leqq m: \begin{cases} \operatorname{codim} (S_1) = m - n + 1, & \dim (S_1) = 2n - m - 1, \\ \operatorname{codim} (S_2) = 2(m - n + 2), & \dim (S_2) = 3n - 2n - 4. \end{cases}$$

7. Structure of the S_k

Take any point p where f is of rank ρ; then $p \in S_{\nu-\rho}$, $q = f^1(p) \in L^1_\rho$. It is easy to see that coordinate systems $(x^1, \cdots, x^n)$ and $(y^1, \cdots, y^m)$ may be chosen near p and $f(p)$ respectively so that f has the form

$$(7.1) \qquad y^i = x^i \quad (i \leqq \rho), \qquad y^j = \phi^j(x^1, \cdots, x^n) \quad (j > \rho).$$

Let y^i_k denote $\partial y^i/\partial x^k$. The Jacobian matrix of (7.1) is

$$J = \begin{Vmatrix} I & 0 \\ J'' & J' \end{Vmatrix}, \qquad J' = \begin{Vmatrix} y^{\rho+1}_{\rho+1} \cdots y^{\rho+1}_n \\ \cdots\cdots\cdots\cdots \\ y^m_{\rho+1} \cdots y^m_n \end{Vmatrix};$$

I is the unit matrix of ρ rows and columns, and 0 is a matrix of zeros. Moreover, the elements of J'' and of J' vanish at p.

The y^i_k occurring in J' are a set of $h_\rho = (n - \rho)(m - \rho)$ real functions in E^n near p. Now assume that f is locally generic. Then f^1 is crosswise to L^1_ρ, and as a result, it is easy to see that the gradients of these functions are independent at p. We may choose functions z^j $(h_\rho < j \leqq n)$ which vanish at p and whose gradients are independent of the above gradients; now *these* y^i_k *and* z^j *may be chosen as new coordinates in* E^n *near* p, with origin at p.

Note that

$$(7.2) \qquad \text{rank } (J) = \text{rank } (J') + \rho.$$

Hence the set $\bar{S}_{\nu-(\rho+k)}$ is given (near p) by setting all $(k + 1)$-rowed determinants of J' equal to zero. This is a set of algebraic equations of degree $k + 1$ in h_ρ of the new coordinates. For an example, see §23 below.

8. The null spaces $N(p)$

The *null space* $N(p)$ of f at p is the set of all vectors v mapped into 0 by $df(p)$. Clearly

$$(8.1) \qquad \text{for } p \in S_k, \qquad \dim (N(p)) = \begin{cases} n - m + k & \text{if } n \geqq m, \\ k & \text{if } n \leqq m. \end{cases}$$

For $p \in S_k$, $N(p)$ can be related in various ways to S_k near p, and more generally to the sets $S_{k-1}, \cdots, S_1$ near p. This will be used to split S_k into several sets.

9. The differentials df^r

Given f, we have $f^r(p) \in L^r$, and since $dd^kf = d^{k+1}f$,

$$df^r = d(df, \cdots, d^rf) = (d^2f, \cdots, d^{r+1}f).$$

Thus knowing the value of f^r at p, the possible values for the tangent vectors $df^r(p) \cdot v$ are severely restricted; only the last term $d^{r+1}f(p) \cdot v$ is arbitrary.

There are also restrictions due to the symmetry of higher differentials. Thus, $f^1(p) = df(p) \in L^1$, hence $df^1(p) \cdot v \in L^1$, $[df^1(p) \cdot v] \cdot u \in V^m$, and

$$[df^1(p) \cdot v] \cdot u = d^2f(p) \cdot (u, v) = d^2f(p) \cdot (v, u) = [df^1(p) \cdot u] \cdot v.$$

Suppose we know the point $f^{r+1}(p) \in L^{r+1}$. Then we also know the $d^kf(p)$ $(k \leqq r+1)$, hence $df^r(p)$, and thus the tangent vectors $df^{(r)}(p) \cdot v$ in L^r.

In particular, the statement that f^r is crosswise to a given manifold collection $M \subset L^r$ is a statement about df^r, and thus is equivalent to the statement that $f^{r+1}(R)$ avoids a certain point set $M_1 \subset L^{r+1}$. One might prove Thom's lemma (§5) by showing that any such M_1 is a manifold collection of codimension $>n$. Such a proof seems difficult to carry out.

10. The mapping f^2 into L^2

In the expression (2.4) for L^r, if we drop out the last terms, we obtain any L^k, $k < r$. Hence to each point of L^r corresponds a definite point of each L^k, $k < r$. Let L^r_ρ be the subset of L^r thus corresponding to L^1_ρ in L^1. Clearly

$$\text{codim}\,(L^r_\rho) = \text{codim}\,(L^1_\rho), \tag{10.1}$$

and f^r is transversal to the L^r_ρ if f^1 is crosswise to the L^1_ρ.

Supposing that f^1 is crosswise to the L^1_ρ, we now consider, for each point p, the relation of the null space $N(p)$ to the sets $\bar{S}_k$ which contain p. To give the relation of a vector v to $\bar{S}_k$ $(p \in \bar{S}_k)$ is to give the relation of $df^1(p) \cdot v$ to $L^1_{\nu-k}$; hence a given position of $N(p)$ relative to the $\bar{S}_k$ corresponds to a property of $df^{(1)}(p)$, i.e., a property in L^2 (see §9).

By §§6 and 8,

$$\dim\,(S_1) + \dim\,(N(p)) = n \qquad \text{if} \quad n \geqq m; \tag{10.2}$$

hence we may expect, at most of the points $p \in S_1$, that $N(p)$ and the tangent plane $P(S_1, p)$ have only the zero vector in common. More generally,

$$\dim\,(N(p)) = \text{codim}\,(S_k) + (k-1)(n-m+k) \qquad \text{if} \quad n \geqq m. \tag{10.3}$$

Then

$$h = \dim\,[N(p) \cap P(S_k, p)]$$

might be 0, or >0.

To say that $f^1(p) = q \in L^1_\rho$ and f^1 is crosswise to L^1_ρ at q is to say that $f^2(p) \in L^2_\rho$ but $f^2(p)$ avoids a certain subset of L^2_ρ (of codimension $>n$). Let L^{*2} be the part of L^2 in no such subset; set $L^{*2}_\rho = L^{*2} \cap L^2_\rho$.

The different values of h above correspond to different subsets of $L^{*2}_{\nu-k}$, and give a splitting

$$L^{*2}_{\nu-k} = L^{*2}_{\nu-k,0} \cup L^{*2}_{\nu-k,1} \cup \cdots. \tag{10.4}$$

Set

$$S_{k,h} = (f^2)^{-1}(L^{*2}_{\nu-k,h}); \quad \text{then}\ S_k = S_{k,0} \cup S_{k,1} \cup \cdots. \tag{10.5}$$

Now $S_{k,h}$ is the subset of S_k where, if we consider f in S_k alone, f has deficiency h. The $S_{k,h}$ are called $S_h(S_k)$ in Thom, [4] (provided that $n \geqq m$).

We have similar splittings for $n < m$.

Now consider $p \in S_2$, assuming $n \geqq m$. We must consider the relation of $N(p)$ not only to $P(S_2, p)_r$ but also in relation to $\bar{S}_1$ near p. This is a more complex situation; see §23 below. The possible relationships of $N(p)$ to the structure of $\bar{S}_1$ near p is reflected in a splitting of the $L^{*2}_{\nu-2,h}$ near $f(p)$, and this gives a corresponding splitting of the $S_{2,h}$. We shall not give names to the new sets here. In a similar manner, each $S_{k,h}$ is split into subsets, on considering the possible relationships of $N(p)$ to the $\bar{S}_l$ $(l \leqq k)$ near $p \in S_k$.

Note that S_0 (the set of regular points) is not split. Also, since (using coordinate systems) the possible relationships of the $N(p)$ to the $\bar{S}_k$ are expressible by algebraic equations, and hence the new sets in L^{*2} are algebraic varieties and hence manifold collections, we see that the new sets in E^n are manifold collections. There is an open subset of S' where $N(p)$ is in the most general position; the rest of S' forms a manifold collection of dimension less than that of S_1.

11. Further splitting of the S_k

In L^3, we have the subset L^{*3} corresponding to mappings f such that f^2 is crosswise to the sets in L^{*2}. We have also sets $L^{*3}_\rho \subset L^{*3}$, corresponding to the L^1_ρ (see §10); they also correspond to the L^{*2}_ρ in L^2. But these sets L^{*2}_ρ have been split up; this gives a splitting of the L^{*3}_ρ. We shall split these further.

The S_k in E^n have been split into subsets, forming a manifold collection. At any point p, we have considered the relation of $N(p)$ to the $\bar{S}_k$; we now consider it also in relation to the new sets. The various possible relationships correspond to facts about df^2, and hence to subsets of L^{*3} (see §9); this gives the desired further splitting in L^{*3}. Through $(f^3)^{-1}$, we find a further splitting of the S_k.

As in the last section, the first new manifolds we obtain are of dimension one less than that of the previous new ones at most, and hence of dimension two less than that of S_1 at most.

REMARK. Though we have described the manner of splitting in L^{*2} and in L^{*3} through a discussion of the function f, the definitions of the sets in L^2 and L^3 are clearly intrinsic; they depend on the properties of the $\mathfrak{L}^r$ alone.

We next consider the relation of the $N(p)$ to the new sets, giving a splitting in L^{*4} and hence a new splitting of the S_k, the largest dimension of a new manifold being three less than that of S_1 at most, etc. The process must stop after the new manifolds are of dimension at most zero. Hence the largest f^r and L^r we need use are those for which $r - 1 = \dim (S_1)$. In L^{*r}, we can make f^r crosswise to all manifolds obtained, and this gives us the desired locally generic f; see also §13. The final requirements on f employ $d^{r+1}f$. Thus (see §6) the requirements on f to be locally generic involve derivatives through the order at most

$$\gamma(n, m) = n - |n - m| + 1 = \begin{cases} m + 1 & (n \geqq m), \\ 2n - m + 1 & (n \leqq m). \end{cases} \tag{11.1}$$

12. The singular sets $S_1^{(i)}$

Suppose first that $n \geqq m$. In a neighborhood $U \subset S_1$ of a point p of S_1, we may choose a set $v_1(p'), \cdots, v_{m-1}(p')$ of independent vectors orienting $P(S_1, p')$; choose also $v_m(p'), \cdots, v_n(p')$ orienting $N(p')$ $(p' \in U)$; see (8.1). For most points $p' \in U$, the whole set $v_1(p'), \cdots, v_n(p')$ is independent, and an orientation of E^n is determined. In some parts of U, one orientation of E^n may be determined, and in other parts, the opposite orientation; these parts are divided by the set $S_{1,1} \cup S_{1,2} \cup \cdots$. Hence

$$\dim (S_{1,1}) = \dim (S_1) - 1 = m - 2 \quad (n \geqq m). \tag{12.1}$$

We shall write $S_1^{(2)}$ for the manifold $S_{1,1}$.

For $p \in S_1^{(2)}$, $N(p) \cup P(S_1, p) = Q(p)$ is of dimension 1; it may point from p into $P(S_1, p)$ in either side of $P(S_1^{(2)}, p)$; or at exceptional points, it may lie in $P(S_1^{(2)}, p)$. This exceptional set is a manifold in $S_1^{(2)}$, of dimension $m - 3$; it is the inverse image under f^3 of a certain set in L^3. At these points p, $N(p) \cap P(S_1^{(2)}, p) = Q'(p)$ is of dimension 1 instead of 0; we call this set $S_{1,1,1}$, or $S_1^{(3)}$. At a point of $S_1^{(3)}$, $Q'(p)$ may point into $P(S_1^{(2)}, p)$ in either side of $P(S_1^{(3)}, p)$, or be tangent to the latter; thus we find $S_1^{(4)}$, etc. We have

$$\dim (S_1^{(i)}) = m - i \qquad (1 \leqq i \leqq m \leqq n). \tag{12.2}$$

These singular sets (inverse images of sets in L^1) were found by Roger [3], and appear in Thom [4] with the notation $S_1(S_1)$, $S_1(S_1(S_1))$, etc. Note that the last one corresponds to a set in L^m; this gives actual occurrences of $\gamma(n, m)$ in (11.1), for $n \geqq m$.

This gives the total splitting of S_1, if $n \leqq 4$.

For $m \geqq n$, codim $(S_1) = m - n + 1$ and $\dim (N(p)) = 1 (p \in S_1)$; hence $N(p)$ will lie in $P(S_1, p)$ (for locally generic f) in a manifold $S_{1,1} = S_1^{(2)}$ of codimension $m - n + 1$ in S_1, and hence of dimension $3n - 2m - 2$. For $p \in S_1^{(2)}$, $N(p)$ lies in $P(S_1, p)$; it will lie in $P(S_1^{(2)}, p)$ at each point of a manifold $S_{1,1,1} = S_1^{(3)}$ of codimension $m - n + 1$ in $S_1^{(2)}$, etc. Thus

$$\text{codim } (S_1^{(i)}) = i(m - n + 1) \text{ if present } (1 \leqq i \leqq m \geqq n). \tag{12.3}$$

The definition of the $S_1^{(i)}$ shows that f is locally one-one with nonvanishing Jacobian in each $S_1^{(i)} - S_1^{(i+1)}$; but the image has a cusp manifold in the next set $S_1^{(i+1)}$.

13. On the classification of singularities

As before, if we cut out a certain subset of L^r, $r = \gamma(n, m)$, of codimension $>n$, corresponding to mappings not crosswise to the sets we have found in L^{*r-1}, we have left L^{*r}; any mapping $f_0 : R^n \to E^m$ is arbitrarily near a mapping f with $f^r(p) \in L^{*r}$ $(p \in R)$, and any such f is locally generic. Morevoer, f^r is automatically crosswise to the sets of the splitting of L^{*r}.

The various sets of the splitting in L^{*r} define the different types of singularities for locally generic f.

Note that each set A of the splitting in $L^{*r-1} (r = \gamma(n, m))$ corresponds to a set B in L^r which is not split further; but part of B may be cut out when we reduce L^r to L^{*r}, and what is left may be composed of several connected pieces, which may correspond to different types of singularities. This happens for instance in the case $m = 1$, $n > 1$; see §16.

14. Encasing of singularities

Given $f : R^n \to E^m$, the *s-dimensional encasing* of f is a mapping $F : R^{n+s} \to E^{m+s}$ defined as follows. Write $E^{n+s} = E^n \times E^s$, $E^{m+s} = E^m \times E^s$, and

$$F(p, q) = (f(p), q) \qquad (p \in E^n,\ q \in E^s). \tag{14.1}$$

To each generic singular point of f corresponds a generic singular point of F; the latter is the *s-dimensional encasing* of the former. Examples will be given below.

Clearly

$$\operatorname{rank}(F, (p, q)) = \operatorname{rank}(f, p) + s; \tag{14.2}$$

hence, for each $q \in E^s$,

$$S_k(F) \cap (E^n \times q) = S_k(f) \times q, \qquad \operatorname{codim}(S_k(F)) = \operatorname{codim}(S_k(f)). \tag{14.3}$$

Clearly $N_F(p, q)$ lies in the space V^n of E^n always; hence the relation of null spaces to the S_k is the same for F as for f, and we see step by step that the S_k are split into subsets in the same way for each. Finally, at any p, lying in a certain singular set, f^r (for some r) is crosswise to the corresponding subset in $L^r(n, m)$; hence F^r is crosswise to the corresponding set in $L^r(n + s, m + s)$, showing that F is locally generic.

15. Increasing n

Given $f : R^n \to E^m$, with $n \geqq m$, define $F : R^{n+1} \to E^m$ near a point $p \in S_1$ as follows. Since $\dim [df(p) \cdot V^n] = m - 1$, we may choose a vector w_0 in V^m not in $df(p) \cdot V^n$. (There are essentially two choices.) Set

$$F(p, t) = f(p) + t^2 w_0. \tag{15.1}$$

With a coordinate system in E^n,

$$\frac{\partial F}{\partial x^i} = \frac{\partial f}{\partial x^i} \quad (i = 1, \cdots, n), \qquad \frac{\partial F}{\partial t} = 2tw_0.$$

For p' near p and $t \neq 0$, the last vector is independent of the former; hence rank $(F, (p', t)) = \operatorname{rank}(f, p') + 1 = m = \nu$, and (p', t) is not a singular point of F. For $t = 0$, $N_F(p', 0) = N_f(p') + V^1$; the relation of these spaces to the $S_1^{(i)}$ for F is the same as that of the $N(p')$ to the $S_1^{(i)}$ for f. Thus $S_1^{(i)}$ for f becomes part of $S_1^{(i)}$ for F at $t = 0$.

The above discussion fails for the S_k, $k \geqq 2$. For instance, S_2 may be present for mappings $E^4 \to E^4$, but not for mappings $E^5 \to E^4$ (see §4).

II. Examples of singularities

16. Critical points of real functions

(See M. Morse, [1] and [2]). For $E^n \to E^1$, we let E^1 be the set of real numbers. Here, L^1 is simply the set of real linear functions (i.e., covectors) in V^n; it is of dimension n, and L_0^1 consists of 0 only. Since $\operatorname{codim}(S_1) = \operatorname{codim}(L_0^1) = n$, S_1 consists of isolated points.

With a Cartesian coordinate system in E^n and hence in L^1, the components of $f^1(p) = df(p)$ are

$$(16.1) \qquad f^1(p) = \left(\frac{\partial f}{\partial x^1}, \cdots, \frac{\partial f}{\partial x^n}\right).$$

Now (compare §9)

$$\partial f^1/\partial x^i = (\partial^2 f/\partial x^1 \partial x^i, \cdots, \partial^2 f/\partial x^n \partial x^i),$$

and if f is locally generic, these vectors in L^1 are independent for $i = 1, \cdots, n$. That is, the matrix of second partial derivatives is of rank n.

With proper choice of coordinates, f near p may be written in the form (for some k)

$$(16.2) \qquad y = (x^1)^2 + \cdots + (x^k)^2 - (x^{k+1})^2 - \cdots - (x^n)^2.$$

For $n = 1$, $y = \pm(x^1)^2$. (If we allow coordinate systems reversing orientation, we may write $y = (x^1)^2$.) If we now increase n as in §15, choosing $w_0 \in E^1$ sometimes positive and sometimes negative, we obtain the above critical points.

Note that L_0^{*2} is L_0^2 except for all points where the determinant $|\partial^2 f/\partial x^i \partial x^j| = 0$. The different parts of L_0^{*2} correspond to critical points of different index, i.e., with different numbers of minus signs in (16.2).

17. The case $n = m = 2$

For $E^1 \to E^1$, a typical singularity is given by $y = x^2$. If we encase this (§14), we obtain

$$(17.1) \qquad y^1 = x^1, \qquad y^2 = (x^2)^2;$$

this has a singular set S_1 on the line $x^2 = 0$.

Consider the mapping defined by

$$(17.2) \qquad y^1 = x^1, \qquad y^2 = x^1 x^2 - (x^2)^3.$$

The Jacobian is $J = \partial y^2/\partial x^2 = x^1 - 3(x^2)^2$, which vanishes on the curve S_1, defined by $x^1 = 3(x^2)^2$. With unit vectors e_1, e_2 along the axes, the image of any vector v under df is

$$df(x^1, x^2) \cdot (v^1 e_1 + v^2 e_2) = \left(v^1 \frac{\partial y^1}{\partial x^1} + v^2 \frac{\partial y^1}{\partial x^2}, v^1 \frac{\partial y^2}{\partial x^1} + v^2 \frac{\partial y^2}{\partial x^2}\right).$$

Along S_1, this equals $(v^1, v^1 x^2)$; hence $N(p)$ consists of all vectors $a e_2$ for $p \in S_1$. Except for $p = (0, 0)$, $N(p)$ is not tangent to S_1; hence we are not in $S_1^{(2)}$ here. But we have tangency at $(0, 0)$. (The image under f of S_1 is a curve with a cusp point at the origin.) Also, $N(p)$ is crossing S_1 here (the mapping f^2 is crosswise to the corresponding set), as is easily seen; hence the origin is in $S_1^{(2)}$.

For fuller details, see [13].

18. The case $m = 2$, $n \geq 3$

If we increase n as in §15, we find singular curves S_1, with isolated points $S_1^{(2)}$ on them, as in §17. We find typical mappings, first for a point p of $S_1 - S_1^{(2)}$, with $n = 3$. By (17.1), the images of all vectors at p lie in the y^1-direction; hence we may choose $w_0 = te_2$ in (15.1); t is replaced by x^3. Inserting the sign $\pm$ in (17.1), this gives

$$y^1 = x^1, \quad y^2 = \pm (x^2)^2 \pm (x^3)^2, \quad \text{in } S_1 - S_1^{(2)}. \tag{18.1}$$

Similarly, using (17.2),

$$y^1 = x^1, \quad y^2 = x^1x^2 - (x^2)^3 \pm (x^3)^2 \quad \text{in } S_1^{(2)}. \tag{18.2}$$

Continuing this process gives, for typical singularities in the general case,

$$y^1 = x^1, \quad y^2 = \pm (x^2)^2 \pm \cdots \pm (x^n)^2 \quad \text{in } S_1 - S_1^{(2)}, \tag{18.3}$$

$$y^1 = x^1, \quad y^2 = x^1x^2 - (x^2)^3 \pm (x^3)^2 \pm \cdots \pm (x^n)^2 \quad \text{in } S_1^{(2)}. \tag{18.4}$$

19. The case $n \geqq m = 3$

Here, we have singularities of types S_1, $S_1^{(2)}$, $S_1^{(3)}$. Typical examples are, for $n = 3$,

$$y^1 = x^1, \quad y^2 = x^2, \quad y^3 = (x^3)^2 \quad \text{in } S_1 - S_1^{(2)}, \tag{19.1}$$

$$y^1 = x^1, \quad y^2 = x^2, \quad y^3 = x^2x^3 - (x^3)^3 \quad \text{in } S_1^{(2)} - S_1^{(3)}, \tag{19.2}$$

$$y^1 = x^1, \quad y^2 = x^2, \quad y^3 = x^1x^3 + x^2(x^3)^2 - (x^3)x^4 \quad \text{in } S_1^{(3)}. \tag{19.3}$$

In the last example, the Jacobian is $\partial y^3/\partial x^3$, and S_1 is the surface

$$S_1 : \phi = x^1 + 2x^2x^3 - 4(x^3)^3 = 0.$$

Clearly $N(x^1, x^2, x^3)$ consists of all vectors ae_3, in the x^3-direction, at points of S_1. These lie in S_1 if they are orthogonal to the gradient of ϕ. Thus $S_1^{(2)}$ is the curve given by:

$$S_1^{(2)} : \phi = 0, \qquad \psi = 2x^2 - 12(x^3)^2 = 0.$$

For points of $S_1^{(3)}$, e_3 must be orthogonal to the gradient of ψ also, this gives $x^3 = 0$. Thus $S_1^{(3)}$ contains the origin alone.

For general n, typical singularities are as follows. As before, $y^1 = x^1$ and $y^2 = x^2$; in the above expressions for y^3, we add $\pm(x^4)^2 \pm \cdots \pm (x^n)^2$.

20. The case $m = 2n - 1$

Here, S_1 consists of isolated points. By [9], a typical singularity is given by

$$\begin{cases} y^i = x^i & (i = 1, \cdots, n-1), \\ y^j = x^n x^{j-n+1} & (j = n, n+1, \cdots, 2n-1). \end{cases} \tag{20.1}$$

For instance, for $n = 2$,

$$y^1 = x^1, \quad y^2 = x^1x^2, \quad y^3 = (x^2)^2. \tag{20.2}$$

21. The case $m = 2n - 2$

If we take the singularity above for $E^{n-1} \to E^{2n-3}$, and encase it as in §14, we find a typical singularity in S_1 for $E^n \to E^{2n-2}$. We have extra variables, say x^0 and y^0, and the extra equation $y^0 = x^0$; S_1 is the x^0-axis. For $n \geqq 3$, (12.3) shows that $S_1^{(2)}$ is void; for $n = 2$, we are back to the case $n = m = 2$.

22. Some other cases with $m > n$

The lowest case not considered is the case $(n, m) = (4, 5)$. Here,

$$\dim(S_1) = 2, \qquad \dim(S_1^{(2)}) = 0 \qquad (n = 4, m = 5). \tag{22.1}$$

In general, by (12.3), $S_1^{(2)}$ appears only if $2m \leqq 3n - 2$; $S_1^{(3)}$ appears only if $3m \leqq 4n - 3$ (the first case is $n = 6$, $m = 7$), etc. From the table in §4, we see that the first appearance of S_2 with $m > n$ is also for the case $n = 6$, $m = 7$.

23. The case $n = m = 4$

In this case, S_1 may contain sets $S_1^{(2)}, S_1^{(3)}$ and $S_1^{(4)}$, but no $S_{1,2}$. We may now also have isolated points belonging to S_2. We examine these more closely.

As in §7, we may choose coordinates in the two spaces so that, near the origin p_0, we have $y^1 = x^1$, $y^2 = x^2$. Now the matrix J' in §7 is

$$J' = \begin{Vmatrix} y_3^3 & y_4^3 \\ y_3^4 & y_4^4 \end{Vmatrix} \tag{23.1}$$

We may also (§7) use these y_k^i as new coordinates near p_0:

$$X^1 = y_3^3, \qquad X^2 = y_4^3, \qquad X^3 = y_3^4, \qquad X^4 = y_4^4. \tag{23.2}$$

The parts of $\bar{S}_1$ and of S^2 near p_0 are

$$\bar{S}_1: \begin{vmatrix} X^1 & X^2 \\ X^3 & X^4 \end{vmatrix} = X^1X^4 - X^2X^3 = 0, \tag{23.3}$$

$$S_2: X^1 = X^2 = X^3 = X^4 = 0. \tag{23.4}$$

Thus, in these coordinates, $\bar{S}_1$ is a quadratic cone, with vertex at p_0, p_0 being the single point of S_2. Let T be the tangent cone of $\bar{S}_1$ at p_0; it consists of all vectors v such that $p_0 + v \in \bar{S}_1$ (and hence $p_0 + av \in \bar{S}_1$, all a).

We now ask what relation $N(p_0)$ (of dimension 2) can have to $\bar{S}_1$. If we look for vectors in both $N(p_0)$ and T, we find a quadratic relation; there may be no common vectors $\neq 0$, or there may be common vectors in one or in two distinct directions. There can also be more special relations (for instance, if $N(p_0)$ is the (X^1, X^2)-plane, then $N(p_0) \subset T$).

We give two examples:

$$y^1 = x^1, \qquad y^2 = y^2, \qquad y^3 = x^1x^3 + x^2x^4 + x^3x^4, \tag{23.5}$$

and

$$\begin{cases} \text{(a) } y^4 = x^1x^4 - x^2x^3 + \frac{1}{2}(x^3)^2 - \frac{1}{2}(x^4)^2, \\ \text{(b) } y^4 = x^1x^4 + x^2x^3 - \frac{1}{2}(x^3)^2 - \frac{1}{2}(x^4)^2. \end{cases} \tag{23.6}$$

We find at once that $\bar{S}_1$ is given in the two cases by

$$\text{(a)}\quad (x^1)^2 + (x^2)^2 - (x^3)^2 - (x^4)^2 = 0,$$

$$\text{(b)}\quad (x^1)^2 - (x^2)^2 + (x^3)^2 - (x^4)^2 = 0.$$

In both cases, $N(p_0)$ consists of the vectors in the (x^3, x^4)-plane. In (a), $N(p_0) \cap T$ has no vector $\neq 0$; in (b), it contains those vectors with $v^3 = \pm v^4$. We thus have two distinct types of singularities, which might be called $S_{2,a}$, $S_{2,b}$.

24. The case $n = m = 5$

Here, S_2 consists of curves. We shall illustrate a special singularity at an isolated point of a curve of S_2. Define f by

$$y^i = x^i \ (i = 1, 2, 3), \qquad y^4 = x^2x^4 + (x^4)^2 + x^4x^5,$$

(24.1)

$$y^5 = x^3x^5 + x^4x^5 + \tfrac{1}{2}(x^4)^2 + \tfrac{1}{2}(x^5)^2 + \tfrac{1}{2}x^1(x^4)^2.$$

Then S_2 is the x^1-axis (near the origin p_0), and for $p \in S_2$, $N(p)$ consists of the vectors in the (x^4, x^5)-plane. Note that S_2 is not tangent to $N(p)$; hence f maps S_2 into a smooth curve.

The singular points are given by $y^4_4y^5_5 - y^4_5y^5_4 = 0$, i.e.,

$$(x^2 + 2x^4 + x^5)(x^3 + x^4 + x^5) - x^4(x^4 + x^5 + x^1x^4) = 0.$$

Let $S'(x^1)$ be that part of S' with x^1 fixed; it is a quadratic cone. Let $T(x^1)$ be the corresponding cone of vectors, as in §23, and $N(x^1)$, the null space at $(x^1, 0, \cdots, 0)$. A simple computation shows that $N(x^1)T \cap (x^1)$ has no non-zero vectors if $x^1 < 0$, and has two independent directions if $x^1 > 0$; for $x^1 = 0$, it is the 1-dimensional space of all vectors $a(e_4 - e_5)$. The origin is a new type of singular point.

25. The case $n = 5$, $m = 4$

Here we have no S_2, but there can be points of $S_{1,2}$. Note that $\dim(S_1) = 3$, $\dim(N(p)) = 2 (p \in S_1)$. Consider the mapping f defined by $y^i = x^i$ $(i = 1, 2, 3)$, and

$$y^4 = x^1x^4 + x^2x^5 + x^3x^4x^5 + \tfrac{1}{6}(x^4)^3 + \tfrac{1}{6}(x^5)^3. \tag{25.1}$$

Then S_1 is given by $y^4_4 = y^4_5 = 0$, that is, by

$$x^1 + x^3x^5 + \tfrac{1}{2}(x^4)^2 = x^2 + x^3x^4 + \tfrac{1}{2}(x^5)^2 = 0,$$

and $N(p)$ consists of all vectors in the (x^4, x^5)-plane $(p \in S_1)$. Note that at the origin p_0, the tangent 3-plane to S_1 is the (x^3, x^4, x^5)-plane, which contains $N(p_0)$.

This is a phenomenon which cannot be removed by arbitrarily small alterations of f. To see this, note that the components of the differentials of y^4_4 and y^4_5 are

$$\begin{Vmatrix} dy^4_4 \\ dy^4_5 \end{Vmatrix} = \begin{Vmatrix} y^4_{14} & y^4_{24} & y^4_{34} & y^4_{44} & y^4_{45} \\ y^4_{15} & y^4_{25} & y^4_{35} & y^4_{45} & y^4_{55} \end{Vmatrix}$$

To have the (x^4, x^5)-plane tangent to S_1, we must have

$$dy_4^4(p)\cdot e_i = dy_5^4(p)\cdot e_i = 0, \quad i = 4, 5.$$

These equations give

$$y_{44}^4 = y_{45}^4 = y_{55}^4 = 0,$$

occurring at the origin p_0 in the present example. Note that

$$y_{44}^4 = x^4, y_{45}^4 = x^3, y_{55}^4 = x^5,$$

and

$$J^* = \begin{Vmatrix} dy_{44}^4 \\ dy_{45}^4 \\ dy_{55}^4 \end{Vmatrix} = \begin{Vmatrix} 0 & 0 & 0 & 1 & 0 \\ 0 & 0 & 1 & 0 & 0 \\ 0 & 0 & 0 & 0 & 1 \end{Vmatrix}$$

Letting p move in S_1, this shows that the mapping $(y_{44}^4, y_{45}^4, y_{55}^4)$ is crosswise to $(0, 0, 0)$ in 3-space. Now take any sufficiently good 3-approximation f' to f. It is easy to see that new coordinates $\bar{x}^i$, $\bar{y}^i$ close to the old may be found so that $y^i = \bar{x}^i$ $(i = 1, 2, 3)$. We will have S_1 for f' close to S_1 for f, and J^* for f' close to J^* for f. Hence there will be a solution of $\bar{y}_{44}^4 = \bar{y}_{45}^4 = \bar{y}_{55}^4 = 0$, with crosswiseness, as required. (The crosswiseness can of course be expressed in L^3.) Note that the fact that the second differential is symmetric $(y_{54}^4 = y_{45}^4)$ plays an essential role here.

III. General definitions

26. Local equivalence of mappings

Take $f_i : R_i^n \to E_i^m$ $(i = 1, 2)$, and take $p_i \in R_i$. We say f_2 at p_2 is *equivalent* to f_1 at p_1 if we can write

(26.1) $$f_2 = \psi f_1 \phi \text{ near } p_2,$$

where $\phi : U_2^n \to R_1^n$ and $\psi : U_1^m \to E_2^m$ are smooth homeomorphisms of rank n and m respectively. Let $\mathfrak{A}^s$ denote arithmetic s-space (of all ordered sets of s real numbers). If we take $E_2^n = \mathfrak{A}^n$, $E_2^m = \mathfrak{A}^m$, we have introduced coordinates into E_1^n and E_1^m through ϕ and ψ, and the mapping f_2 becomes

(26.2) $$y^i = \phi^i(x^1, \cdots, x^n) \quad (i = 1, \cdots, m).$$

We noted in §1 that if $f : R^n \to E^m$ is regular at p, then f is equivalent at p to the mapping (1.1). If p is a singular point of f, we would like to choose coordinates in some way to give a particularly simple representation (26.2) for f. Such a choice we call a *normal form* for the singular point. For instance, for $f : E^1 \to E^1$, if f has a minimum at p and $d^2f(p) \neq 0$, an equivalent mapping is given by $y = x^2$. Various normal forms were given in Part II.

27. Local stability

We say $f : R^n \to E^m$ is *stable at* p if there is a neighborhood U of p with the following property. Take any neighborhood U' of p, $U' \subset U$. Then for any sufficiently good s-approximation f' to f in U, for some s, there is a point $p' \in U'$ such that f' at p' is equivalent to f at p.

Clearly the mapping (1.1) is stable at all points (use good 1-approximations). So is the mapping $y = x^2$ at $x = 0$, using 2-approximations; but if we used only 1-approximations, we could obtain a new mapping with y constant in some small interval, and local equivalence would then fail.

One form of the principal conjecture is that any locally generic mapping (defined through the discussion of Part I) is stable at all points. We give a strengthened form in the next section.

28. Topologies in function space.

Let $\mathfrak{S}$ be the set of all pairs (f, R), where $f : R^n \to E^m$ is ∞-smooth. We shall give rough descriptions of two topologies in $\mathfrak{S}$, the *normal topology* and the *stability topology*, turning $\mathfrak{S}$ into topological spaces $\mathfrak{S}_n$ and $\mathfrak{S}_s$ respectively.

In $\mathfrak{S}_n$, we say that (f', R') is near (f, R) if R' is near R (with a simple definition), and for some large r, f' is a good r-approximation to f in $R \cap R'$. In $\mathfrak{S}_s$, we say (f', R') is near (f, R) if we can make (26.1) hold, with the following requirements. There are open sets R_1, R_1' near R, R' respectively, ϕ maps R_1' onto R_1, and ϕ and ψ are good r-approximations to the identity in R_1' and in the domain of ψ respectively, with large r.

There is an identity mapping I of $\mathfrak{S}_n$ into $\mathfrak{S}_s$, which is not continuous at most points (f, R). Let $\mathfrak{S}(U_0)$ denote the set of all (f, U_0). We say f is *strongly stable* at p if there is a neighborhood U of p such that the mapping I, considered in $\mathfrak{S}_n(U)$ alone, is continuous at (f, U). Thus any (f', U) which is sufficiently close to (f, U) in the normal topology is also close in the stability topology.

If I is continuous in $\mathfrak{S}_n(U)$ at (f, U), it is also continuous in $\mathfrak{S}_n(U')$ at (f, U'), for any open ball $U' \subset U$. This may be seen by extending $f' - f$ from U' to U; see [12].

Say f is *locally strongly stable* in R if it is strongly stable at each point of R. If this holds, then we can cover R by open balls $U_1, U_2, \cdots$ such that I is continuous at (f, U_i) in $\mathfrak{S}_n(U_i)$ for each i. It follows that for any sufficiently good r-approximation f' to f in R, for large enough r, we can write $f' = \psi_i f \phi_i$ in each U_i. We cannot expect to have $f' = \psi f \phi$ in R, unless the self-intersections of $f(R)$ have been made generic. For example, if $f : E^1 \to E^1$ is given by $y = (x^2 - 1)^2$, then $df = 0$ at $x = -1$ and at $x = 1$, and $f(-1) = f(1)$; this will not happen for all nearby f'.

29. General problems

For each $R \subset E^n$, we have a subset $\mathfrak{S}(R)$ of $\mathfrak{S}$. Are the locally strongly stable mappings, or the locally stable mappings, in $\mathfrak{S}(R)$, dense in $\mathfrak{S}_n(R)$? Do they form an open set in both cases? If the complete pattern of singularities is as suggested in Part I, these questions are answered in the affirmative.

In §13 we discussed the description of singularities, in terms of the method of splitting of the L^r used; each connected piece $L^r_{(i)}$ corresponds to a singularity type. The test of this is as follows. Let q_1 and q_2 be points of the same $L^r_{(i)}$. Let

$$f_1 : R_1^n \to E^m \quad \text{and} \quad f_2 : R_2^n \to E^m$$

be mappings such that, with γ defined by (11.1),

$$f_1^{\gamma(n,m)}(p_1) = q_1, \qquad f_2^{\gamma(n,m)}(p_2) = q_2.$$

Then ϕ and ψ should exist as in §26, such that $f_2 = \psi f_1 \phi$ near p_2. In other words a single normal form will do for all points of any $L^r_{(i)}$. This has been proved in the following cases: for $m = 1$, in [1] and [2]; for $m = 2n - 1$, in [9]; for the s-dimensional encasing (§14) of the above singularities in [14]; and for $n = m = 2$, in [13].

Say $f : R^n \to E^m$ is *stable* if the following is true. For any sufficiently good approximation f' to f (we would use a $\gamma(n, m)$-approximation, which gets better and better as we approach the boundary of R), we can find ϕ, mapping R^n onto R^n, and ψ, mapping a neighborhood of $f(R^n)$ into E^m, such that $f' = \psi f \phi$ in R. We could require that ϕ and ψ be near the identity; we could also require that ψ be defined in a larger set, perhaps the whole of E^m, We then wish to show that the stable mappings form a dense open set in $\mathfrak{S}(R)$, in a suitable topology. As noted in the last section, we must take care of intersection properties of $f(R)$ to obtain a *generic* mapping (i.e., a stable mapping, assuming the truth of the statement). The study of self-intersections seems less difficult than that of singularities.

Finally, recall that topological questions have been discussed only in certain cases; see §1. (For topological properties related to self-intersections in the case $m = 2n$, see [10].)

THE INSTITUTE FOR ADVANCED STUDY
PRINCETON, NEW JERSEY.

BIBLIOGRAPHY

1. M. MORSE, *Relations between the critical points of a real function of n independent variables*, Trans. Amer. Math. Soc., 27 (1925), pp. 345–396.
2. M. MORSE, *The critical points of a function of n variables*, ibid., 33 (1931), pp. 72–91.
3. F. ROGER, *Sur les variétés critiques* . . . , C. R. Acad. Sci., Paris. 208 (1939), pp. 29–31.
4. R. THOM, *Les singularités des applications différentiables*, Seminaire Bourbaki, Paris, May, 1956.
5. R. THOM, *Les singularités des applications différentiables*, Annales de l'Institut Fourier, VI (1956), pp. 43–87.
6. R. THOM, *Un lemme sur les applications différentiables*, Boletin de la Sociedad Matemática Mexicana, 1, ser. 2 (1956), pp. 59–71.
7. A. W. TUCKER, *Branched and folded coverings*, Bull. Amer. Math. Soc., 42 (1936), pp. 859–862.
8. H. WHITNEY, *Differentiable manifolds*, Ann. of Math. 37 (1936), pp. 645–680.
9. H. WHITNEY, *The general type of singylarity of a set of 2n − 1 smooth functions of n variables*, Duke Math. J., 10 (1943), pp. 161–172.
10. H. WHITNEY, *The self-intersections of a smooth n-manifold in 2n-space*, Ann. of Math., 45 (1944), pp. 220–246.
11. H. WHITNEY, *The singularities of a smooth n-manifold in (2n − 1)-space*, ibid., pp. 247–293.
12. H. WHITNEY, *On the extension of differentiable functions*, Bull. Amer. Math. Soc., 50 (1944), pp. 76–81.
13. H. WHITNEY, *On singularities of mappings of Euclidean spaces, I, Mappings of the plane into the plane*, Ann. of Math., 62 (1955), pp. 374–410.
14. N. Z. WOLFSOHN, *On differentiable maps of Euclidean n-space into Euclidean m-space*, Harvard thesis, 1952. See the abstract in Bull. Amer. Math. Soc., 61 (1955), p. 171.

ON BOUNDED FUNCTIONS WITH BOUNDED nTH DIFFERENCES

HASSLER WHITNEY

We consider real valued functions f defined in a closed interval I (bounded or unbounded), with nth differences

$$\Delta_h^n f(x) = \sum_i (-1)^{n-i} \binom{n}{i} f(x + ih)$$

bounded for some fixed n.

THEOREM. *For each integer $n \geqq 1$ there is a number L_n with the following property. Let f be defined in I, and suppose there is an interval $I' \subset I$ in which f is bounded. Then there is a polynomial P of degree $\leqq n-1$ such that*

$$(1) \qquad |f(x) - P(x)| \leqq L_n \sup |\Delta_h^n f(y)|, \qquad x \in I.$$

Assuming f continuous, this was proved by the author [**4**], and with a different definition of differences, by J. C. Burkill [**1**]. The present theorem, for I unbounded, was proved recently by F. John [**2**], who kindly showed me his manuscript. We show here that the theorem follows easily from the proof of the theorem of [**4**], together with a method of proof [**4**, §12] due to A. Beurling.

It is sufficient to consider the intervals

$$I_0 = (0, 1), \qquad I^* = (0, \infty), \qquad I^{**} = (-\infty, \infty).$$

The smallest constants L_n^*, L_n^{**} in the latter two cases for which (1) holds will be shown to satisfy

$$(2) \qquad L_n^* \leqq 1, \qquad L_n^{**} \leqq 1 \Big/ \sup_k \binom{n}{k}.$$

For further information about the constants, compare [**4**]. The inequality for L_n^{**} is in fact an equality if n is even.

It would be of interest to prove the theorem assuming merely that f is measurable (compare [**3**, Theorem 1]).

We consider first the interval I_0; we may suppose that $|\Delta_h^n f(x)| \leqq 1$. Let T be the set of all numbers of the form $i/2^k(n-1)$, i and k being integers. In [**4**, §4], we may take ν to be a power of 2. Define the quantities in §§4, 5 and 7 of [**4**], in particular, K_n' (not the K_n' of

Received by the editors September 15, 1958.

§1). Choose the polynomial P_f' of degree $\leqq n-1$, equal to f at the points $i/(n-1)$ $(i=0, 1, \cdots, n-1)$, and set $g=f-P_f'$. We show first that

$$(3) \qquad |g(x)| \leqq K_n', \qquad x \in T \cap I_0.$$

Noting that the sets S_k and T_k exactly fill out the set $T\cap I_0$, (3) is proved in [4, §7].

Let U be the open set of points x, $0<x<1$, such that f (and hence g) is bounded in a neighborhood of x. We show next that

$$(4) \qquad |g(x)| \leqq K_n' + 2, \qquad x \in U.$$

Take $x\in U$; say $|g(y)| \leqq K$ in the interval J about x. Take any integer N. We may choose a point $y\in J\cap T$ so close to x that if $y=x+n!h$, then all points $x+ijh$ and $y+ijh$ $(0\leqq i\leqq N, 0\leqq j\leqq n)$ are in J. Since $|\Delta_u^n g(v)| \leqq 1$, we may apply the proof in §12 of [4], giving

$$|g(y) - g(x)| < 2 + 2^{n+1}n!K/N;$$

since $|g(y)| \leqq K_n'$ and N is arbitrary, (4) follows.

To complete the proof of the theorem for I_0 (and show that $L_n \leqq K_n' + 2$), we need merely show that U contains all points interior to I_0. Suppose not. Then U contains an open interval (x_0, x_1), x_0 (or x_1) being interior to I_0, $x_0 \notin U$. We may choose x (near x_0) and h so that

$$|g(x)| \geqq 2^n(K_n' + 2),$$

and the points $x+ih$ $(i=1, 2, \cdots, n)$ are in U. Then by (4),

$$|\Delta_h^n g(x)| \geqq 2^n(K_n' + 2) - \sum_{i=1}^{n} \binom{n}{i} (K_n' + 2) > 1,$$

a contradiction.

The proof in [4, §8], now applies, giving the theorem for unbounded intervals. Applying [4, §12] now gives (2) also.

References

1. J. C. Burkill, *Polynomial approximations to functions with bounded differences*, J. London Math. Soc. vol. 33 (1958) pp. 157–161.

2. F. John, 1958 (unpublished).

3. H. Whitney, *Derivatives, difference quotients and Taylor's formula*, Bull. Amer. Math. Soc. vol. 40 (1934) pp. 89–94.

4. ———, *On functions with bounded nth differences*, J. Math. Pures Appl. (9) vol. 36 (1957) pp. 67–95.

Institute for Advanced Study

Chapter 3

Analytic Spaces

ANNALS OF MATHEMATICS
Vol. 66, No. 3, November, 1957

ELEMENTARY STRUCTURE OF REAL ALGEBRAIC VARIETIES

BY HASSLER WHITNEY

(Received April 4, 1957)

1. Introduction

A real (or complex) algebraic variety V is a point set in real n-space R^n (or complex n-space C^n) which is the set of common zeros of a set of polynomials. The general properties of a real V as a point set have not been the subject of much study recently (but see for instance [2], [3] and [4]); attention has turned more to the complex case, the complex projective case, and especially the abstract algebraic theory. Facts about the real case are sometimes needed in the applications; proofs are commonly very difficult to locate.

Our first object is to give a proof involving no algebraic theory that through a certain splitting process, V may be expressed as a union of "partial algebraic manifolds". (Part of the proof is standard.) To prove the finiteness of the process, the Hilbert basis theorem must be used. The principal concept used is the "rank" of an ideal at a point; it has immediate geometric-analytic meaning. The splitting procedure works equally well in the complex case; in fact, in the case of real or complex analytic varieties.

Each partial manifold M_i may have limit points in later M_j ; this is unavoidable. Also, M_i need not be connected; but it can have at most a finite number of topological components. We prove this by showing that for any subvariety V' of V (V' may be void), $V - V'$ has at most a finite number of topological components. (This conjecture was communicated to me by D. Mostow.) The methods used in the splitting process are used here also, together with some standard algebraic material.

Use is made of the relation of V to the smallest complex variety V^* containing V; the irreducible components of V and of V^* correspond, as does the notion of rank at points of V. The structure of the splitting of Theorem 1 can be seen with the help of the irreducible components of V or V^*. The more obvious splitting, by dimension, is given also; it is not of the same elementary character as the other. Examples are given to show some of the peculiar properties of real algebraic varieties.

We do not enter into the more difficult and important questions about local structure; the possibility[1] of triangulation of V is a theorem in this direction.

2. Definitions

The following definitions hold both for the real and the complex case. The

[1] See [6], Appendix to Chapter IV. Some revision in the proof is necessary. For instance lemma on p. 125 is used in the real domain on p. 128, where it does not apply. Thus a circle in the plane is a real algebraic variety, while its projection into a line is a closed line segment, which is not an algebraic variety.

(polynomial) *ideal* $\mathbf{I}(Q)$ of a point set $Q \subset R^n$ (or $Q \subset C^n$) is the set of polynomials which vanish in Q; if f and g are in $\mathbf{I}(Q)$, so is $f + g$, and so is ϕf for any polynomial ϕ. Any set S of polynomials defines a variety $V = \mathbf{V}(S)$ (§1); S generates an ideal I, consisting of the linear combinations of elements of S with polynomial coefficients, and $I \subset \mathbf{I}(\mathbf{V}(S))$.

The *differential* $df(p)$ of a function f at a point p is a covector, that is, a linear function of vectors v. The definition, geometric or through the use of a coordinate system, is

$$df(p)\cdot v = \lim_{t\to 0+} \frac{1}{t}\,[f(p + tv) - f(p)] = \textstyle\sum_i v_i \frac{\partial f(p)}{\partial x_i}. \tag{2.1}$$

The *rank* $\mathrm{rnk}_p(S)$ of a set S of functions at a point p is the number of independent differentials $df_1(p), \cdots, df_s(p)$, the f_i being in S; if S is an ideal, the set of all covectors $df(p)$ ($f \in S$) forms a vector space, whose dimension is $\mathrm{rnk}(S)$. Clearly $\mathrm{rnk}_p(S)$ is also the number of independent differentials from any set of functions generating S; see (10.6). The *rank* $\mathrm{rnk}_p(Q)$ of a point set Q at $p \in Q$ is $\mathrm{rnk}_p(\mathbf{I}(Q))$; the *rank* $\mathrm{rnk}(Q)$ of Q is the largest value of $\mathrm{rnk}_p(Q)$ for $p \in Q$.

An *algebraic partial manifold* M is a point set, associated with a number ρ, with the following property. Take any $p \in M$. Then there exists a set of polynomials $f_1, \cdots, f_\rho$, of rank ρ at p, and a neighborhood U of p, such that $M \cap U$ is the set of zeros in U of these f_i. The number $n - \rho$ is the *dimension* of the partial manifold. Note that M need not be closed or connected, and that any open subset of M is also a partial algebraic manifold. In the complex case, if M is of (complex) dimension r, it may be considered, topologically, as a real manifold, of real dimension $2r$.

3. The theorems

The first two theorems are stated for the real case, but will be proved first in the complex case.

THEOREM 1. *Let $V \subset R^n$ be a real algebraic variety, and let M_1 be the set of points $p \in V$ where the rank of V is its maximum. Then M_1 is an algebraic partial manifold of dimension $n - \mathrm{rnk}(V)$, and $V_1 = V - M_1$ is void or is a proper algebraic subvariety of V.*

We may now apply this splitting process to V_1 if it is not void, giving $V_1 = V_2 \cup M_2$, etc.

The first part of the following theorem is an immediate consequence of the Hilbert basis theorem.

THEOREM 2. *This process comes to an end after a finite number of steps, giving*

$$V = M_1 \cup M_2 \cup \cdots \cup M_s, \tag{3.1}$$

each M_i being an algebraic partial manifold in V, and the M_i being disjoint. We have

$$s \leqq 2^n - 1. \tag{3.2}$$

Note that each point set

$$V_i = M_{i+1} \cup \cdots \cup M_s \tag{3.3}$$

is an algebraic variety, and hence is a closed set. We say that V is expressed in this manner as a *manifold collection*.

REMARK. Using some algebraic theory, we will give another such splitting in §§9 and 11, with the property that the manifolds are of decreasing dimension; hence there will be at most n terms.

THEOREM 3. *A real algebraic variety has at most a finite number of topological components.*

THEOREM 4. *Let V' be a subvariety of the real variety V; then $V - V'$ has at most a finite number of topological components.*

As a corollary, we have

THEOREM 5. *In the expression* (3.1) (*real case*), *each M_i has but a finite number of topological components.*

4. The variety V_1

Given the functions $f_1, \cdots, f_r$, their differentials are independent at p if and only if some Jacobian determinant (using the natural coordinate system of R^n)

$$\frac{\partial(f_1, \cdots, f_r)}{\partial(x_{\lambda_1}, \cdots, x_{\lambda_r})} = \begin{vmatrix} \partial f_1/\partial x_{\lambda_1} & \cdots & \partial f_1/\partial x_{\lambda_r} \\ \cdots & \cdots & \cdots \\ \partial f_r/\partial x_{\lambda_1} & \cdots & \partial f_r/\partial x_{\lambda_r} \end{vmatrix} \tag{4.1}$$

is $\neq 0$ at p. Set $\rho = \text{rnk}(V)$. By definition of M_1 and V_1, $I = \mathbf{I}(V)$ is of rank ρ at all points of M_1 and of rank $<\rho$ at all points of V_1; that is, all Jacobian determinants of elements $f_1, \cdots, f_\rho$ of I are 0 at all points of V_1, while this is not true for any point of M_1. Let S be this set of Jacobian determinants; these are polynomials. Now V_1 is just the set of points where all polynomials of I and all polynomials of S vanish; hence V_1 is an algebraic variety.

5. A special coordinate system

We prove an elementary lemma.

LEMMA 1. *Let $f_1, \cdots, f_\rho$ have independent differentials at p. Then there is a coordinate system $(x'_1, \cdots, x'_n)$ in a neighborhood U of p such that*

$$x'_i(q) = f_i(q) - f_i(p) \text{ in } U \qquad (i = 1, \cdots, \rho). \tag{5.1}$$

Choose covectors $\xi_{\rho+1}, \cdots, \xi_n$ which, with the $df_i(p)$, form an independent set. Set $x'_i(q) = \xi_i \cdot (q - p)$ for $i > \rho$; then

$$dx'_i(q) = df_i(q) \quad (i \leqq \rho), \qquad dx'_i(q) = \xi_i \quad (i > \rho). \tag{5.2}$$

Hence the Jacobian of the transformation $(x_1, \cdots, x_n) \to (x'_1, \cdots, x'_n)$ is $\neq 0$ at p, and $(x'_1, \cdots, x'_n)$ is a coordinate system.

6. The operations Φ_μ

Take any $p \in M_1$. We may choose polynomials $f_1, \cdots, f_\rho$ in $\mathbf{I}(V)$ with independent differentials at p; we keep these f_i fixed in this section. Given any polynomial g and any $\mu = (\mu_1, \cdots, \mu_{\rho+1})$, define the polynomial

$$\Phi_\mu g = \frac{\partial(f_1, \cdots, f_\rho, g)}{\partial(x_{\mu_1}, \cdots, x_{\mu_\rho}, x_{\mu_{\rho+1}})}. \tag{6.1}$$

Choose a coordinate system $(x'_1, \cdots, x'_n)$ in a neighborhood U of p by Lemma 1. Now $\partial f_i/\partial x'_j = \sum_h (\partial f_i/\partial x_h)(\partial x_h/\partial x'_j)$, and applying the general Lagrange identity to the determinant of such elements gives

$$\frac{\partial(f_1, \cdots, f_\rho, g)}{\partial(x'_1, \cdots, x'_\rho, x'_k)} = \sum_{\mu_1<\cdots<\mu_{\rho+1}} \frac{\partial(f_1, \cdots, f_\rho, g)}{\partial(x_{\mu_1}, \cdots, x_{\mu_\rho}, x_{\mu_{\rho+1}})} \frac{\partial(x_{\mu_1}, \cdots, x_{\mu_\rho}, x_{\mu_{\rho+1}})}{\partial(x'_1, \cdots, x'_\rho, x'_k)}.$$

Because of (5.2), the left hand side is simply $\partial g/\partial x'_k$, if $k > \rho$. Let T^μ_k denote the last term on the right. Then this relation and (6.1) give

$$\frac{\partial g}{\partial x'_k} = \sum_{\mu_1<\cdots<\mu_{\rho+1}} T^\mu_k \Phi_\mu g \text{ in } U \qquad (k > \rho). \tag{6.2}$$

Let M^* be the $(n - \rho)$-dimensional manifold in U defined by the vanishing of the f_i; it is the part of the $(x'_{\rho+1}, \cdots, x'_n)$-coordinate plane in U. Given a function g and a point $p' \in M^*$, $dg(p')\cdot v = 0$ for all vectors v tangent to M^* at p' if and only if $dg(p')$ is dependent on the $df_i(p')$, that is, if and only if all $\Phi_\mu g$ are 0 at p; or again, $\partial g/\partial x'_k = 0$ at p' for $k > \rho$.

We may iterate the operation, forming $\Phi_\lambda\Phi_\mu g$, $\Phi_\sigma\Phi_\lambda\Phi_\mu g$, etc. Let $H(g)$ be the set of all such polynomials, and let $J_U(g)$ be the ideal of analytic functions in U generated by $H(g)$, consisting of all functions $\sum \varphi_i(p)h_i(p)$ with $h_i \in H(g)$ and ϕ_i analytic in U. We show that *each partial derivative of g of any order with respect to the variables $x'_{\rho+1}, \cdots, x'_n$ is an element of $J_U(g)$.*

By (6.2), this holds for first partial derivatives. Using induction, it is sufficient to show that if ϕ is analytic in U and $h = \Phi_\sigma \cdots \Phi_\mu g$, then $\partial(\phi h)/\partial x'_k$ is in $J_U(g)$ for $k > \rho$. Differentiating ϕh and applying (6.2) with h in place of g shows this to be true.

7. The algebraic partial manifold M_1

We shall complete the proof of Theorem 1. Take any $p \in M_1$, and let $f_1, \cdots, f_\rho$ be as in §6. Choose U, the x'_i and M^* again; we may suppose M^* is connected. We need merely show that $M_1 \cap U = M^*$. Since V_1 is closed, we may suppose $V_1 \cap U = 0$; hence $M_1 \cap U = V \cap U \subset M^*$ (since $f_i \in \mathrm{I}(V)$), and there remains to prove $M^* \subset V$.

Take any polynomial $g \in \mathrm{I}(V)$; we must prove that $g = 0$ in M^*. Since $\mathrm{rnk}(V) = \rho$, the differentials $df_1, \cdots, df_\rho, dg$ are dependent throughout V; hence all $\Phi_\mu g$ are 0 throughout V, and $\Phi_\mu g \in \mathrm{I}(V)$. Repeating the argument shows that $H(g) \subset \mathrm{I}(V)$. In particular, all polynomials in $H(g)$ vanish at p. By §6, all partial derivatives of g with respect to $x'_{\rho+1}, \cdots, x'_n$ vanish at p. Now g is analytic in the open connected part of $(x'_{\rho+1}, \cdots, x'_n)$-space which is M^*, and hence $g = 0$ there.

8. Dimension and rank of a complex variety

We give some general facts about a complex algebraic variety V, which will show what the splitting $V = V_1 \cup M_1$ is in the complex case; (3.2) will also follow in this case. The facts will be used to study the real case in §10. Lemmas 2 and 3 will be used in the proofs of the last theorems.

Consider a complex variety $V \subset C^n$, with corresponding ideal of complex polynomials, which we call $I = \mathbf{I}^*(V)$. Suppose that V is irreducible; then I is prime. Adjoint elements $\xi_1, \cdots, \xi_n$ to the field C of complex numbers, with the following algebraic relations: For any polynomial f,

$$(8.1) \qquad f(\xi_1, \cdots, \xi_n) = 0 \quad \text{if and only if} \quad f(x_1, \cdots, x_n) \in I.$$

By definition, the *dimension* of V is the transcendence degree of the set $\xi_1, \cdots, \xi_n$; let this be r. We may suppose $\xi_1, \cdots, \xi_r$ are algebraically independent (after a change of coordinates); then each ξ_i $(i > r)$ satisfies an irreducible equation $P_i = 0$ with coefficients in the field $C(\xi_1, \cdots, \xi_{i-1})$, and these give the full algebraic properties of the ξ_i and hence of I.

LEMMA 2. *A complex irreducible variety V of dimension r is of rank $\leqq n - r$ at all its points; it has points of rank $n - r$. Any proper subvariety of V is of dimension $<r$.*

See [1], Chapter X, §§5 and 14, for the complex projective case; little change is needed in the proof for the present case. The first statement holds essentially because there are only $n - r$ polynomials P_i above.

LEMMA 3. *A 0-dimensional complex irreducible variety consists of a single point.*

See [5], §100. We remark that if $(a_1, \cdots, a_n)$ is the single point, then $\mathbf{I}^*(V)$ is generated by $x_1 - a_1, \cdots, x_n - a_n$.

LEMMA 4. *For any complex varieties V, V',*

$$(8.2) \qquad \text{rnk}_p(V \cup V') = \text{rnk}_p(V) \quad \text{if} \quad p \in V - V'.$$

REMARK. This clearly holds in the real domain also.

Clearly $\leqq$ holds; we prove $\geqq$. Choose $f \in \mathbf{I}^*(V')$ such that $f(p) = 1$. If $\text{rnk}_p(V) = \rho$, choose $f_1, \cdots, f_\rho \in \mathbf{I}^*(V)$ so that $df_1(p), \cdots, df_\rho(p)$ are independent. Set $f_i' = ff_i$; then $f_i' \in \mathbf{I}^*(V \cup V')$, and $df_i'(p) = df_i(p)$, so that the $df_i'(p)$ are independent.

A complex algebraic variety V (also a real one) has a unique minimal expression as a union $V_1' \cup \cdots \cup V_h'$ of irreducible varieties; by definition, $\dim(V) = \sup \dim(V_i')$. If the V_i are all of the same dimension, we say V is of *constant dimension*.

LEMMA 5. *Let V be of constant dimension r. Then in the splitting $V = V_1 \cup M_1$, M_1 is of (complex) dimension r, and* $\dim(V_1) < r$.

See also §9, (b). Write $V = V_1' \cup \cdots \cup V_h'$, as above. For any $p \in V$, say $p \in V_i$; then by Lemma 2, $\text{rnk}_p(V) \leqq \text{rnk}_p(V_i) \leqq n - r$; hence $\text{rnk}(V) \leqq n - r$. Let $V_{i1}' \cup M_{i1}'$ be the splitting of V_i', and set

$$M_i'' = M_{i1}' - \cup_{j \neq i} V_j', \qquad M_1^* = \cup M_i''.$$

Each M_i'' is non void; for if M_1'' is void, for instance, then $V = V_{11}' \cup V_2' \cup \cdots \cup V_h'$, which would clearly contradict uniqueness. By Lemmas 2 and 4, $\text{rnk}_p(V) = n - r$ for $p \in M_1^*$, and hence $M_1^* \subset M_1$. Since M_1 is a manifold, it is easy to see that $M_1 = M_1^*$; also $\dim(M_1) = r$. Now

$$V_1 \subset (\textstyle\bigcup_i V_{i1}') \cup (\bigcup_{i<j} V_i \cap V_j);$$

since V_{i1}' and $V_i' \cap V_j'$ are of dimension $<r$, by Lemma 2, V_1 is of dimension $<r$.

We call the points of M_1 and of V_1 *simple* and *singular* respectively, for varieties V of constant dimension.

9. Two splitting processes, complex case

(a) We show what the splitting $V = V_1 \cup M_1$ of Theorem 1 is in the complex case, in terms of irreducible components of V, and prove (3.2). We may write

$$(9.1) \qquad V = W_1 \cup \cdots \cup W_\alpha, \qquad \dim(W_i) = r_i, \qquad r_1 < \cdots < r_\alpha,$$

the W_i being of constant dimension. The W_i are clearly uniquely determined (in a minimal representation); let us set

$$(9.2) \qquad \tau(V) = 2^{r_1} + \cdots + 2^{r_\alpha}.$$

Now M_1 is the set of simple points of W_1 which are in no other W_j. That is, if $W_1 = W_1' \cup M_1'$ as in Lemma 5, then

$$(9.3) \qquad M_1 = M_1' - W_2 \cup \cdots \cup W_\alpha.$$

For points of $W_2 \cup \cdots \cup W_\alpha$ are of rank $\leqq n - r_2$, and (see the proof of Lemma 5) the only points of W_1 of rank $n - r_1$ in V are those shown.

We prove (3.2) in the complex case. If $\dim(V) = n$, then $V = C^n$, and $V = M_1$. Supposing $\dim(V) < n$, (9.2) shows that

$$(9.4) \qquad \tau(V) \leqq \textstyle\sum_{i=0}^{n-1} 2^i = 2^n - 1.$$

Express W_1' as the union of irreducible varieties; omitting any that lie in any W_j $(j > 1)$ gives a minimal representation

$$(9.5) \qquad V_1 = W_1'' \cup \cdots \cup W_\beta'' \cup W_2 \cup \cdots \cup W_\alpha,$$

with W_i'' of constant dimension r_i'', $r_1'' < \cdots < r_\beta'' < r_1$ (see Lemma 5). Since $\sum 2^{r_i''} < 2^{r_1}$, we find $\tau(V_1) < \tau(V)$. Therefore, with V_i as in (3.3),

$$(9.6) \qquad 2^n - 1 \geqq \tau(V) > \tau(V_1) > \cdots > \tau(V_{s-1}) > 0,$$

and (3.2) follows.

(b) Using the facts of §8, an obvious splitting procedure is as follows. Write V as in (9.1), and write $W_\alpha = W_\alpha' \cup M_\alpha'$ as in Lemma 5. Set

$$(9.7) \qquad M_1^* = M_\alpha' - W_1 \cup \cdots \cup W_{\alpha-1}, \qquad V_1^* = V - M_1^*.$$

Then M_1^* is an algebraic partial manifold of dimension $r_\alpha = \dim(V)$, and V_1^* is an algebraic variety of dimension $<r_\alpha$. We now split V_1^* in the same way, etc.

We thus express V as a manifold collection, using at most $n - 1$ algebraic partial manifolds, of decreasing dimension.

10. Real varieties as parts of complex ones

Let $\mathfrak{R}x$ and $\mathfrak{C}x$ denote the real part and complex conjugate of the complex number x respectively. Now $\mathfrak{C}\mathfrak{C}x = x$, $\mathfrak{C}(x + y) = \mathfrak{C}x + \mathfrak{C}y$, $\mathfrak{C}(xy) = (\mathfrak{C}x)(\mathfrak{C}y)$, and $\mathfrak{C}ix = -i\,\mathfrak{C}x$. For any $p = (x_1, \cdots, x_n) \in C^n$, set $\mathfrak{C}p = (\mathfrak{C}x_1, \cdots, \mathfrak{C}x_n)$. For any polynomial $f = \sum A_{\lambda_1 \cdots \lambda_n} x_1^{\lambda_1} \cdots x_n^{\lambda_n}$,

$$f(\mathfrak{C}(x_1, \cdots, x_n)) = \mathfrak{C} \sum \mathfrak{C}A_{\lambda_1 \cdots \lambda_n} x_1^{\lambda_1} \cdots x_n^{\lambda_n}. \tag{10.1}$$

Define the operation Γ on polynomials f:

$$\Gamma f(p) = \tfrac{1}{2}[f(p) + \mathfrak{C}f(\mathfrak{C}p)]. \tag{10.2}$$

Then (10.1) gives

$$\Gamma \sum A_{\lambda_1 \cdots \lambda_n} x_1^{\lambda_1} \cdots x_n^{\lambda_n} = \sum \mathfrak{R}A_{\lambda_1 \cdots \lambda_n} x_1^{\lambda_1} \cdots x_n^{\lambda_n}; \tag{10.3}$$

either relation shows that Γf is real (i.e. it takes on real values at real points), and $\Gamma f = f$ if and only if f is real. Also,

$$f = \Gamma f - i\Gamma(if). \tag{10.4}$$

For a point set $Q \subset C^n$, let $\mathfrak{C}Q$ denote the set of points $\mathfrak{C}p$, $p \in Q$. Say Q is *real-symmetric* if $\mathfrak{C}Q = Q$.

LEMMA 6. *Given the real variety $V \subset R^n$, there is a unique smallest complex variety $V^* \subset C^n$ containing V; it is defined by the set of polynomials $\mathbf{I}(V)$. Now V^* is real-symmetric, and V is the real part of V^*. If $I = \mathbf{I}(V)$ and $I^* = \mathbf{I}^*(V^*)$ are the associated ideals, then $I \subset I^*$, and I is the set of real polynomials in I^*; it also consists of all Γf with $f \in I^*$, and I^* is the complex ideal generated by I.*

The polynomials $\mathbf{I}(V)$ define a complex variety V^*, whose real part is V; because of (10.1), V^* is real-symmetric. To prove that V^* is the smallest, we show that any complex polynomial f that vanishes in V also vanishes in V^*. Since $f = 0$ in V, Γf and $\Gamma(if)$ are 0 in V and hence in V^*; therefore, by (10.4), $f = 0$ in V^*.

By definition of V^*, $I \subset I^*$; since $V \subset V^*$, I contains all real polynomials in I^*, and hence is just this set of polynomials. For any $f \in I^*$, $\Gamma f \in I \subset I^*$. This set of functions Γf is the set of real polynomials in I^*, and hence is I. Because of (10.4), I generates I^*.

Let $\mathbf{V}^*(V)$ denote the complex variety associated with the real variety V by Lemma 6. Let $\mathfrak{R}(V^*)$ denote the real part of the complex variety V^*.

LEMMA 7. *Let $V^* = \mathbf{V}^*(V)$, and let $V_1^*, \cdots, V_\alpha^*$ be the irreducible components of V^*. Then the $V_i = \mathfrak{R}(V_i^*)$ are the irreducible components of V, and $V_i^* = \mathbf{V}^*(V_i)$.*

Define the V_i^* and V_i as above, and set $V_i'^* = \mathbf{V}^*(V_i)$; then $V_i'^* \subset V_i^*$. Since $V \subset V_1'^* \cup \cdots \cup V_\alpha'^*$ and V^* is minimal, $V^* = V_1'^* \cup \cdots \cup V_\alpha'^*$; hence $V_i'^* = V_i^*$. Suppose $V_{i1} \cup V_{i2}$ were a splitting of V_i into proper subvarieties; then if $V_{ij}^* = \mathbf{V}^*(V_{ij})$, we have $V_i^* = V_{i1}^* \cup V_{i2}^*$ (since V_i^* is minimal), and V_{ij}^* is a proper subvariety of V_i^*, contrary to the irreducibility of V_i^*. Finally, no V_i is contained in the union of the other V_k; for if it were, V_i^*, being minimal, would be contained in the union of the other V_k^*.

Let rnk* denote the rank in the complex case.

LEMMA 8. *With V and V^* as in Lemma* 6,

$$\operatorname{rnk}_p(V) = \operatorname{rnk}_p^*(V^*), \qquad p \in V. \tag{10.5}$$

This is immediate from the fact that I generates I^*, and the relation

$$d\sum \phi_i(p)f_i(p) = \sum \phi_i(p)df_i(p) \quad \text{if all } f_i(p) \text{ are } 0. \tag{10.6}$$

11. The splitting processes, real case

(a) Given the real variety V, set $V^* = \mathbf{V}^*(V)$. Split both V and V^* according to Theorem 1 (for the real and complex cases respectively):

$$V = V_1 \cup M_1, \qquad V^* = V_1^* \cup M_1^*. \tag{11.1}$$

LEMMA 9. *With these notations,*

$$M_1 = \Re(M_1^*), \qquad V_1 = \Re(V_1^*). \tag{11.2}$$

REMARK. $\mathbf{V}^*(V_1)$ need not equal V_1^*; see §12, (f).

First we prove $\operatorname{rnk}(V) = \operatorname{rnk}^*(V^*)$. By Lemma 8, $\leqq$ is true. Suppose $<$ were true; then the points of M_1^* would be of rank (in V^*) greater than $\operatorname{rnk}(V)$, and hence, by (10.5), M_1^* would contain no points of V; but then $V \subset V_1^*$, contrary to the minimal property of V^*. The first relation now follows from (10.5) and the fact that $V = \Re(V^*)$. The second relation now follows also.

We now prove (3.2) in the real case. Define $\tau(V)$ to be $\tau(\mathbf{V}^*(V))$ (see §9). We prove $\tau(V_1) < \tau(V)$; then (9.6) will hold, and (3.2) will follow.

In (9.5) we found V_1^* in the first splitting for $\mathbf{V}^*(V) = V^* = W_1^* \cup \cdots \cup W_\alpha^*$:

$$V_1^* = W_1'^* \cup W_2^* \cup \cdots \cup W_\alpha^*, \qquad W_1'^* = W_1''^* \cup \cdots \cup W_\beta''^*, \tag{11.3}$$

the W_i^* and $W_j''^*$ being of constant dimension as given there, and $W_1^* = W_1'^* \cup M_1'^*$ being the first splitting of W_1^*. By Lemma 9, $V_1 = \Re(V_1^*) \subset V_1^*$; hence

$$\mathbf{V}^*(V_1) \subset V_1^*. \tag{11.4}$$

We show now that

$$W_i^* \subset \mathbf{V}^*(V_1), \qquad i \geqq 2. \tag{11.5}$$

Let $\bigcup_j W_{ij}^*$ be the expression of W_i^* in terms of irreducible components ($i \geqq 2$). By (11.3), $W_{ij}^* \subset V_1^*$; hence, by (11.2), $W_{ij} = \Re(W_{ij}^*) \subset V_1$, and applying Lemma 7 to the irreducible component W_{ij}^* of V^* gives $W_{ij}^* = \mathbf{V}^*(W_{ij}) \subset \mathbf{V}^*(V_1)$, proving (11.5).

These relations give

$$\mathbf{V}^*(V_1) = V^{**} \cup W_2^* \cup \cdots \cup W_\alpha^*, \qquad V^{**} \subset W_1'^*; \tag{11.6}$$

as in §9, $\tau(V_1) < \tau(V)$ follows, and (3.2) is proved.

(b) Given V, the splitting of $V^* = \mathbf{V}^*(V)$ of §9 (b) gives a corresponding splitting of V into at most $n - 1$ real algebraic partial manifolds of decreasing dimension, as we now show. With $W_i^* = \bigcup_j W_{ij}^*$ and $W_{ij} = \mathfrak{R}(W_{ij}^*)$ as above, set

$$W_i = \bigcup_j W_{ij} = \mathfrak{R}(W_i^*);$$

then, as seen above, $W_{ij}^* = \mathbf{V}^*(W_{ij})$, and hence $W_i^* = \mathbf{V}^*(W_i)$.

If $W_\alpha^* = W'^* \cup M'^*$ is the splitting of Theorem 1, Lemma 9 shows that the splitting of W_α is

$$W_\alpha = W' \cup M', \qquad W' = \mathfrak{R}(W'^*), \qquad M' = \mathfrak{R}(M'^*).$$

Set

$$V_1' = W_1 \cup \cdots \cup W_{\alpha-1} \cup W', \qquad M_1' = M' - W_1 \cup \cdots \cup W_{\alpha-1};$$

then $V = V_1' \cup M_1'$, rnk $(W_\alpha) =$ rnk* (W_α^*)(see the proof of Lemma 9), and,

$$\dim(M_1') = \dim(M') = n - \text{rnk}^*(W_\alpha^*) = r_\alpha.$$

Set

$$V_1'^* = \mathbf{V}^*(V_1') = W_1^* \cup \cdots \cup W_{\alpha-1}^* \cup \mathbf{V}^*(W');$$

now $\mathbf{V}^*(W') \subset W'^*$, and hence (applying Lemma 5 to W_α^*) $\dim(V_1'^*) < r_\alpha$. We now repeat the process with V_1' and $V_1'^*$, splitting a manifold of dimension $<r_\alpha$ from V_1', etc.

12. Examples

(a) There may be many different complex varieties containing a given real one. For instance, if p is the origin in the (x, y)-plane, it is a real variety V, with ideal generated by x and y. For each a with $|a| < 1$, $f_a = x^2 + 2axy + y^2$ defines V; it also defines a complex variety V_a^*, and these are distinct. Note that $x + iy$ defines another V'^* with V as real part.

(b) A real algebraic variety V may have several topological components; each of these may or may not be itself a variety; if it is, its rank will in general be greater than its rank as part of V. For instance, let V be defined by $f = x^2 + y^2 - y^3$ in the plane; it consists of a curve C, plus the origin p_0. Since p_0 is a variety and f and hence V is irreducible, C is not a variety; thus any polynomial vanishing on C also vanishes at p_0. Moreover $\partial f/\partial x = \partial f/\partial y = 0$ at p_0, and it follows that V is of rank 0 at p_0. We find therefore $M_1 = C$, $M_2 = V_1 = p_0$, for the splitting.

(c) The phenomenon of (b) does not depend on V being non-connected. Thus, let V be defined by

$$f = (x^2 + y^2)z - y^3$$

in 3-space. It consists of the z-axis Z, plus a surface S; S contains the origin p_0, and the points $z = y^3/(x^2 + y^2)$. Since Z is a variety and V is irreducible, any polynomial vanishing on S also vanishes on Z. V is of rank 1 in $S - p_0$, and of rank 0 in Z.

(d) The variety defined by the equation $x^2 + 2xyz + y^2 = 0$ contains the z-axis and two separate surfaces. Dropping out one of the latter would leave a point set which is locally like an algebraic variety but is not one in the large.

(e) If W is the curve $x^2 - y^3 = 0$ in the plane, then $M_1(W)$ is the curve minus the origin, and $M_2(W)$ is the origin. If X is the cone $z^2 - (x^2 + y^2) = 0$ in R^3, then $M_1(X)$ is the cone minus the vertex and $M_2(X)$ is the vertex. If Y is the surface $x^3 - (y^3 - z^2)^2 = 0$, then $M_1(Y)$ is the part of Y with $x \neq 0$, and we are left with a curve like W; hence $M_2(Y)$ is a curve and $M_3(Y)$ is a point. Now let V be the union in R^3 of a point p_0, and varieties like W, X and Y; then $M_1 = p_0$, M_2 is formed by $M_1(W)$ ($M_2 = M_1(W) - X \cup Y$), and the later M_i are formed by $M_2(W)$, $M_1(X) \cup M_1(Y)$, $M_2(X)$, $M_2(Y)$, and $M_3(Y)$ respectively. Thus $7 = 2^3 - 1$ steps are necessary, showing that (3.2) cannot be improved for $n = 3$. (It clearly cannot be for any n.)

(f) The irreducible polynomial $f = (x^2 + y^2)^2 - z^3$ defines a real variety V and a complex variety V^*; clearly $V^* = \mathbf{V}^*(V)$, and f generates both $\mathbf{I}(V)$ and $\mathbf{I}^*(V^*)$. With the notations of (11.1), V_1 contains (0, 0, 0) only; but V_1^* contains (1, i, 0) for instance, so that $V_1^* \neq \mathbf{V}^*(V_1)$.

13. Proof of Theorem 3

If the theorem is false, we may suppose V is a smallest variety with an infinite number of topological components. For otherwise, we could choose a proper subvariety V_1 with this property, then a proper subvariety V_2 of V_1, etc., contrary to the Hilbert basis theorem.

Consider the splitting $V = V_1 \cup M_1$ of Theorem 1; then V_1 has but a finite number of topological components, and hence M_1 has an infinite number which are closed sets; they are of the same dimension r. We show first that $r > 0$. For suppose $r = 0$. Then the components $p_1, p_2, \cdots$ of M_1 are single points, and V is of rank $n - r = n$ at each; hence, by Lemma 8, the p_i are also topological components of $V^* = \mathbf{V}^*(V)$. Let $V_{\lambda_i}^*$ be an irreducible component of V^* containing p_i; then $\mathrm{rnk}_{p_i}(V_{\lambda_i}^*) \geqq \mathrm{rnk}_{p_i}(V^*) = n$, and by Lemma 2, dim $(V_{\lambda_i}^*) = 0$. By Lemma 3, $V_{\lambda_i}^*$ contains p_i only. But the number of irreducible components of V^* is finite, a contradiction.

Since $r > 0$, the closed topological components $M_1', M_2', \cdots$ of M_1 are not single points, and we may choose a point $q_0 = (a_1, \cdots, a_n)$ which is not equidistant from all points of M_1'; set

$$g = (x_1 - a_1)^2 + \cdots + (x_n - a_n)^2. \tag{13.1}$$

For any polynomials $f_1, \cdots, f_\rho$ $(\rho = n - r)$ and any $\mu = (\mu_1, \cdots, \mu_{\rho+1})$, let $\Phi_\mu(f_1, \cdots, f_\rho)$ denote the right hand side of (6.1). Let V' be the variety defined by the set of all these polynomials with the f_k in $I = \mathbf{I}(V)$, together with the polynomials of I; then $V' \subset V$.

Let p_i be a point of M_i' nearest q_0. Then $dg(p_i)$ vanishes for tangent vectors to M_i' at q_0, hence $df_1, \cdots, df_\rho, dg$ are dependent at p_i if the f_k are in I (see §6), hence the $\Phi_\mu(f_1, \cdots, f_\rho)$ are 0 at p_i, and thus $p_i \in V'$. Therefore V' contains an infinite number of topological components.

Also g is not constant in M_1'; hence (since M_1' is connected) there is a point $p \in M_1'$ such that $dg(p) \cdot v \neq 0$ for some tangent vector v; there are polynomials $f_1, \cdots, f_\rho$ in I with $df_1(p), \cdots, df_\rho(p)$ independent, and now this set, with $dg(p)$, is independent (see §6); hence some $\Phi_\mu(f_1, \cdots, f_\rho)$ is $\neq 0$ at p, and p is not in V'. Therefore V' is a proper subvariety of V, a contradiction, completing the proof.

14. Proof of Theorem 4

As above, if the theorem is not true, we may suppose V is a smallest variety such that for some subvariety V'', $V - V''$ has an infinite number of topological components. Write $V = V_1 \cup M_1$ as in Theorem 1, and set $V' = V'' \cup V_1$. Then $M' = V - V'$ has an infinite number of components; for otherwise, V' would contain an infinite number of the components of $V - V''$, and hence $V' - V''$ would have an infinite number of components, contrary to the minimality of V. Since $M' \subset M_1$, the components of M' are manifolds of some fixed dimension r. (Because of Theorem 3, all but a finite number of these have limit points in V'.)

Set $I = \mathbf{I}(V)$, $I' = \mathbf{I}(V')$. Let $h_1', \cdots, h_\gamma'$ be a basis for I', and set

$$h = (h_1')^2 + \cdots + (h_\gamma')^2; \tag{14.1}$$

then $h = 0$ in V' and $h > 0$ in $R^n - V'$. We show that h is bounded in all but a finite number of the topological components of M'. For if not, then the polynomial $h - 1$ would vanish throughout a proper subset of each of an infinite number of components of M', and hence I, together with $h - 1$, would define a variety with an infinite number of components, contrary to Theorem 3. Let $M_1', M_2', \cdots$ be a set of components of M', in each of which h is bounded.

Choose $q_0 = (a_1, \cdots, a_n)$ not in V, define g by (13.1), and set

$$h_k = g \frac{\partial h}{\partial x_k} - 2(x_k - a_k)h \qquad (k = 1, \cdots, n). \tag{14.2}$$

Then $u = h/g$ is analytic in $R^n - q_0$, and

$$\frac{\partial u}{\partial x_k} = \frac{1}{g^2} h_k \text{ in } R^n - q_0. \tag{14.3}$$

Since h is bounded in M_i' and $h = 0$ in V', there is a compact subset W_i of M_i' where u attains its maximum.

Set

$$\Phi_\mu'(f_1, \cdots, f_\rho) = \begin{vmatrix} \partial f_1/\partial x_{\mu_1} & \cdots & \partial f_1/\partial x_{\mu_{\rho+1}} \\ \cdot & \cdots & \cdot \\ \partial f_\rho/\partial x_{\mu_1} & \cdots & \partial f_\rho/\partial x_{\mu_{\rho+1}} \\ h_{\mu_1} & \cdots & h_{\mu_{\rho+1}} \end{vmatrix}$$

and let $\Psi_\mu(f_1, \cdots, f_\rho)$ be the same, with h_{μ_i} replaced by $\partial u/\partial x_{\mu_i}$; that is, Ψ_μ is given by (6.1), with u in place of g. Then by (14.3),

$$\Phi_\mu'(f_1, \cdots, f_\rho) = g^2 \Psi_\mu(f_1, \cdots, f_\rho) \text{ in } R^n - q_0.$$

Let W be defined by the set of all the polynomials $\Phi'_\mu(f_1, \cdots, f_\rho)$ with the f_j in I, together with I itself. Then, as in the last theorem, $W_i \subset W$ for $i = 1, 2, \cdots$. Also for any point p in any M'_i such that $du(p) \cdot v \neq 0$ for some tangent vector v, we can find $f_1, \cdots, f_\rho$ in I with $\Psi_\mu(f_1, \cdots, f_\rho) \neq 0$ and hence $\Phi'_\mu(f_1, \cdots, f_\rho) \neq 0$ at p, showing that p is not in W.

Take any topological component W'_i of W_i; we show that it is a topological component of W. If W''_i is the topological component of W containing W'_i, we show that u is constant in W''_i; this will prove $W''_i = W'_i$. Let $N_1 \cup \cdots \cup N_t$ be the splitting of W as in (3.1). Take any topological component X of any $X_{ij} = W''_i \cap N_j$; we show that u is locally constant in X. It will follow that u is constant in X; hence u takes on at most a denumerable number of values in W''_i, and W''_i being connected, u is constant in W''_i.

Take any $p \in X$. Since W''_i is a topological component of W and N_j is a manifold, W''_i contains all points of N_j in some neighborhood U_1 of p. Take $U \subset U_1$ so that $N_j \cap U$ is connected; then

$$N_j \cap U = X_{ij} \cap U = X \cap U.$$

By Lemma 1, for some σ, we may suppose U small enough so that there is a coordinate system $(x'_1, \cdots, x'_n)$ in U with $N_j \cap U$ being the part of the $(x'_{\sigma+1}, \cdots, x'_n)$-plane in U; we may also suppose that $N_j \cap U \subset M'_i$. If u is not constant in $N_j \cap U$, then some $\partial u/\partial x'_k$ is $\neq 0$ at some point $p' \in N_j \cap U$, $k > \sigma$. Hence there is a vector v tangent to N_j and therefore tangent to M'_i at p', with $du(p') \cdot v \neq 0$. But as seen above, this means that p' is not in W, a contradiction. We have now proved $W''_i = W'_i$.

We have found a topological component W'_i of W in each M'_i; but this contradicts Theorem 3, and the proof is complete.

THE INSTITUTE FOR ADVANCED STUDY

BIBLIOGRAPHY

[1] W. V. D. HODGE AND D. PEDOE, Methods of algebraic geometry, volume II, University Press, Cambridge, 1952.
[2] S. LEFSCHETZ, L'Analysis situs et la géométrie algébrique, Paris, 1924 and 1950.
[3] J. NASH, *Real algebraic manifolds*, Ann. of Math., 56 (1952), pp. 405–421.
[4] O. A. OLEĬNIK, *Estimates of the Betti numbers of real algebraic hypersurfaces*, Rec. Math. (Mat. Sbornik) N.S., 28 (70), 1951, pp. 635–640.
[5] B. L. VAN DER WAERDEN, Algebra, volume II (third ed.), Springer, Berlin, 1955.
[6] ———, Einführung in die algebraische Geometrie, Springer, Berlin, 1939 (Dover, New York, 1945).

Tirage à part de Commentarii Mathematici Helvetici, vol. 33, fasc. 2, 1959

Quelques propriétés fondamentales des ensembles analytiques-réels

par H. WHITNEY et F. BRUHAT

Ce n'est que depuis peu de temps que les ensembles analytiques dans le domaine réel ont fait l'objet d'études approfondies. La théorie des variétés de STEIN vient d'être transportée au cas des ensembles analytiques-réels «cohérents» par H. CARTAN [6], qui a montré que, dans R^n, un ensemble lieu des zéros d'un faisceau cohérent d'idéaux est «C-analytique», c'est-à-dire est la partie réelle d'un ensemble analytique-complexe, et est globalement définissable par l'annulation d'un nombre fini de fonctions analytiques-réelles. Par ailleurs, F. BRUHAT et H. CARTAN ont étudié dans [2] et [3] le cas général des ensembles analytiques-réels qui ne sont pas C-analytiques et ont montré que, si ces ensembles ont toujours de bonnes propriétés «locales» (i. e. sur un compact), ils peuvent avoir un comportement global très pathologique: en particulier, il n'existe pas toujours de «bonne» décomposition en composantes irréductibles.

Le but de ce travail est d'étudier la structure d'un ensemble C-analytique E (remarquons que localement, tout ensemble analytique-réel est C-analytique). Le premier théorème fondamental (prop. 6) donne une propriété minimale de E dans un voisinage fixe d'un point arbitraire de E. La démonstration repose sur une étude des propriétés de connexion locale de E (prop. 2). Comme conséquence, nous montrons d'une part que l'intersection d'une famille quelconque d'ensembles C-analytiques est C-analytique, d'autre part qu'un ensemble C-analytique E dans R^n est la partie réelle d'un plus petit ensemble analytique-complexe E^* dans un voisinage suffisamment petit de R^n dans C^n, ce qui nous permet de définir la «complexification» $\hat{E}$ de E. Enfin, ceci entraîne l'existence et l'unicité d'une décomposition de E en composantes C-irréductibles, correspondant à la décomposition de E^*.

Ces résultats seront obtenus comme conséquences de l'étude. un peu plus générale, des germes d'ensembles analytiques-complexes au voisinage d'un sous-ensemble analytique-réel arbitraire.

Dans ces considérations, R^n peut être remplacée par n'importe quelle variété analytique-réelle paracompacte Ω: pour cela, nous montrons au n° 1 comment construire une complexification Ω^* de Ω. Par ailleurs, H. GRAUERT vient de montrer que toute variété analytique-réelle peut être plongée analytiquement dans un espace euclidien [7]. Ceci montre en particulier que le résultat d'H. CARTAN rappelé ci-dessus est valable dans Ω.

L'un des auteurs a récemment montré [9] comment une variété algébrique réelle peut être décomposée en réunion de sous-variétés plongées, par des considérations de dimension et de rang. Ceci se généralise au cas C-analytique, grâce aux notions de C-dimension et de C-rang. Quelques exemples illustrant les différences entre le cas complexe, le cas réel et le cas algébrique sont donnés à la fin.

Pour toutes les définitions et résultats relatifs aux sous-ensembles analytiques-complexes et analytiques-réels généraux, nous renverrons à [5], [6] et [8].

1. Complexification d'une variété analytique-réelle

Soit Ω une variété analytique-réelle de dimension n: une *complexification* de Ω est le couple formé d'une variété analytique-complexe de dimension n, Ω^* et d'un isomorphisme φ (analytique-réel) de Ω sur une sous-variété analytique-réelle de Ω^* tels que, pour tout point x de Ω^*, il existe un isomorphisme (analytique-complexe) d'un voisinage ouvert U^* de x sur un ouvert U' de C^n, transformant $\varphi(\Omega) \cap U^*$ en $R^n \cap U'$.

La proposition suivante assure l'existence et en quelque sorte l'unicité d'une telle complexification:

Proposition 1. *Si Ω est paracompacte, elle possède des complexifications. Si (Ω_1^*, φ_1) et (Ω_2^*, φ_2) sont deux complexifications de Ω, il existe un isomorphisme analytique-complexe d'un voisinage ouvert de $\varphi_1(\Omega)$ dans Ω_1^* sur un voisinage ouvert de $\varphi_2(\Omega)$ dans Ω_2^* prolongeant l'isomorphisme $\varphi_2 \circ \varphi_1^{-1}$ de $\varphi_1(\Omega)$ sur $\varphi_2(\Omega)$.*

Démonstration: l'unicité est facile: l'isomorphisme $\varphi_2 \circ \varphi_1^{-1}$ se prolonge en une application analytique ψ_1 d'un voisinage ouvert A_1 de $\varphi_1(\Omega)$ dans Ω_2^*, et on peut supposer que ψ_1 est de rang n en tout point de A_1, donc que $B_2 = \psi_1(A_1)$ est un voisinage ouvert de $\varphi_2(\Omega)$. On construit de même A_2, ψ_2, et B_1. Posons $C_1 = \psi_1^{-1}(A_2 \cap B_2)$; C_1 est un ouvert sur lequel $\psi_2 \circ \psi_1$ est définie. Comme $\psi_2 \circ \psi_1$ est l'identité sur $\varphi_1(\Omega)$, il existe un voisinage ouvert D_1 de $\varphi_1(\Omega)$ sur lequel $\psi_2 \circ \psi_1 = 1$. On construit de même D_2: il est alors immédiat que ψ_1 et ψ_2 sont deux isomorphismes réciproques entre $D_1 \cap \psi_2(D_2)$ et $D_2 \cap \psi_1(D_1)$.

Démontrons maintenant l'existence d'une complexification: d'après les propriétés des variétés paracompactes, on peut trouver trois recouvrements ouverts de type fini de Ω, (V_i'), (U_i') et (T_i'), dépendant du même ensemble d'indices I, tels que V_i' (resp. U_i') soit relativement compact dans U_i' (resp. T_i') et qu'il existe un isomorphisme (pour les structures de variétés analytiques-réelles) φ_i de T_i' sur un ouvert T_i de R^n. Nous poserons:

$$U_i = \varphi_i(U'_i) \qquad V_i = \varphi_i(V'_i) \tag{1}$$

$$U_{i,j} = \varphi_i(U'_i \cap U'_j) \qquad V_{i,j} = \varphi_i(V'_i \cap V'_j) \qquad T_{i,j} = \varphi_i(T'_i \cap T'_j)\,. \tag{2}$$

L'isomorphisme $\varphi_j \circ \varphi_i^{-1}$ de $T_{i,j}$ sur $T_{j,i}$ se prolonge en un isomorphisme analytique-complexe $\psi_{i,j}$ d'un voisinage ouvert $T^*_{i,j}$ de $T_{i,j}$ dans C^n sur un voisinage ouvert $T^*_{j,i}$ de $T_{j,i}$. On peut supposer que $T^*_{i,j}$ est vide si $T_{i,j}$ l'est, que $T^*_{i,j} \cap R^n = T_{i,j}$ et que:

$$\psi_{j,i} = \psi^{-1}_{i,j}\,. \tag{3}$$

Pour tout couple (i, j), choisissons un ouvert $U^*_{i,j}$ de C^n, relativement compact dans $T^*_{i,j}$, tel que $\psi_{i,j}(U^*_{i,j}) = U^*_{j,i}$ et que:

$$U^*_{i,j} \cap R^n = U_{i,j} \qquad \overline{U}^*_{i,j} \cap R^n = \overline{U}_{i,j}\,. \tag{4}$$

Comme $\overline{V}_i \cap \psi_{j,i}(\overline{V}_j \cap \overline{U}_{j,i})$ est un compact contenu dans $U_{i,j}$, on peut choisir un ouvert $W^*_{i,j}$ de C^n relativement compact dans $U^*_{i,j}$, tel que $\psi_{i,j}(W^*_{i,j}) = W^*_{j,i}$ et que:

$$\overline{V}_i \cap \psi_{j,i}(\overline{V}_j \cap \overline{U}_{j,i}) \subset W^*_{i,j}\,. \tag{5}$$

De plus, les complémentaires de $W^*_{i,j}$ dans $\overline{V}_i$ et dans $\psi_{j,i}(\overline{V}_j \cap \overline{U}_{j,i})$ sont des compacts disjoints de C^n, donc sont contenus dans des ouverts disjoints $A^*_{i,j}$ et $B^*_{i,j}$. On a:

$$\overline{V}_i \subset (A^*_{i,j} \cup W^*_{i,j}) \qquad \psi_{j,i}(\overline{V}_j \cap \overline{U}_{j,i}) \subset (W^*_{i,j} \cup B^*_{i,j})\,. \tag{6}$$

Soit A^*_i un ouvert de C^n tel que:

$$A^*_i \cap R^n = V_i \qquad \overline{A}^*_i \cap R^n = \overline{V}_i \tag{7}$$

$$A^*_i \subset (A^*_{i,j} \cup W^*_{i,j}) \tag{8}$$

pour tous les indices j (en nombre fini) tels que $T_{i,j} \neq \varnothing$.

On a:

$$\overline{\psi_{i,j}(A^*_i \cap U^*_{i,j})} \cap R^n \subset \psi_{i,j}(\overline{V}_i \cap \overline{U}_{i,j}) \tag{9}$$

car $\overline{A}^*_i \cap \overline{U}^*_{i,j}$ est un compact contenu dans $T^*_{i,j}$: par suite, $\overline{\psi_{i,j}(A^*_i \cap U^*_{i,j})}$ est contenu dans $\psi_{i,j}(\overline{A}^*_i \cap \overline{U}^*_{i,j})$. Or on a

$$\psi_{i,j}(\overline{A}^*_i \cap \overline{U}^*_{i,j}) \cap R^n = \psi_{i,j}(\overline{A}^*_i \cap \overline{U}^*_{i,j} \cap R^n) = \psi_{i,j}(\overline{V}_i \cap \overline{U}_{i,j})$$

d'après (4) et (7).

Pour tout point x de U_i, choisissons un ouvert $U^*_{i,x}$, contenant x, et satisfaisant aux quatres conditions suivantes:

(10) pour tous les indices j (en nombre fini) tels que x appartienne à $U_{i,j}$, on a $U^*_{i,x} \subset U^*_{i,j}$;

(11) pour tous les indices j (en nombre fini) tels que $x \in \psi_{j,i}(\overline{V}_j \cap \overline{U}_{j,i})$, on a (cf. (6)) $U_{i,x}^* \subset (W_{i,j}^* \cup B_{i,j}^*)$;

(12) pour tout indice j tel que $\varphi_i^{-1}(x) \notin \overline{V}_j'$, on a $U_{i,x}^* \cap \psi_{j,i}(A_i^* \cap U_{j,i}^*) = \varnothing$ (cette condition est trivialement satisfaite si $U_{j,i}^* = \varnothing$; pour les indices j (en nombre fini) tels que $U_{j,i}^* \neq \varnothing$, elle est réalisable parce que, d'après (9), $\varphi_i^{-1}(x) \notin \overline{V}_j'$ entraine $x \notin \overline{\psi_{j,i}(A_j^* \cap U_{j,i}^*)}$);

(13) pour tous les couples (j, k) (en nombre fini) tels que $x \in U_{i,j} \cap U_{i,k}$ (c'est-à-dire tels que $\varphi_i^{-1}(x) \in U_i' \cap U_j' \cap U_k'$), on a

$$U_{i,x}^* \subset \psi_{j,i}(U_{j,i}^* \cap U_{j,k}^*) \cap \psi_{k,i}(U_{k,i}^* \cap U_{k,j}^*)$$

et de plus, on a $\psi_{i,j} = \psi_{k,j} \circ \psi_{i,k}$ sur $U_{i,x}^*$ (cette dernière condition étant réalisable, car on a $\psi_{i,j} = \psi_{k,j} \circ \psi_{i,k}$ sur $U_{i,j} \cap U_{i,k}$).

Soit alors U_i^* la réunion des $U_{i,x}^*$ pour x décrivant U_i et soit V_i^* un voisinage ouvert de V_i dans C^n, contenu dans A_i^* et relativement compact dans U_i^*. D'après (7), on a $V_i^* \cap R^n = V_i$ et $\overline{V}_i^* \cap R^n = \overline{V}_i$. Posons:

$$V_{i,j}^* = V_i^* \cap \psi_{j,i}(V_j^* \cap U_{j,i}^*) \text{ et } V_{i,j,k}^* = V_{i,j}^* \cap V_{i,k}^*. \tag{14}$$

On a $V_{i,j}^* \subset U_{i,j}^*$ et $\psi_{i,j}$ est un isomorphisme de $V_{i,j}^*$ sur $V_{j,i}^*$. D'autre part, un point y de $V_{i,j,k}^*$ est contenu dans un ouvert $U_{i,x}^*$ pour un $x \in U_i$ et cet ouvert $U_{i,x}^*$ rencontre $\psi_{j,i}(V_j^* \cap U_{j,i}^*)$ et $\psi_{k,i}(V_k^* \cap U_{k,i}^*)$ donc a fortiori $\psi_{j,i}(A_j^* \cap U_{j,i}^*)$ et $\psi_{k,i}(A_k^* \cap U_{k,i}^*)$. Ceci entraine (d'après (12) et (2)) que $x \in U_{i,j}^* \cap U_{i,k}^*$, donc que $\psi_{i,k}(y) \in U_{k,j}^* \cap V_k^*$ et que $\psi_{k,j} \circ \psi_{i,k}(y) = \psi_{i,j}(y)$. Par suite le point $z = \psi_{i,j}(y)$ appartient à V_j^*, à $\psi_{i,j}(V_i^* \cap U_{i,j}^*)$ et aussi à $\psi_{k,j}(V_k^* \cap U_{k,j}^*)$, c'est-à-dire finalement appartient à $V_{j,i,k}^*$.

Par suite, on a $\psi_{i,j}(V_{i,j,k}^*) \subset V_{j,i,k}^*$ et de même, $\psi_{j,i}(V_{j,i,k}^*) \subset V_{i,j,k}^*$, ce qui entraine (cf. (3)) que $\psi_{i,j}$ est un isomorphisme de $V_{i,j,k}^*$ sur $V_{j,i,k}^*$, d'inverse $\psi_{j,i}$. D'autre part, on a $\psi_{i,j} = \psi_{k,j} \circ \psi_{i,k}$ sur $V_{i,j,k}^*$.

Considérons alors l'espace *somme* des V_i^* (remarquons que $V_{i,i}^* = V_i^*$) et disons que deux points $x \in V_i^*$ et $y \in V_j^*$ sont *équivalents* si $x \in V_{i,j}^*$, $y \in V_{j,i}^*$ et $y = \psi_{i,j}(x)$: il est immédiat d'après ce qui précède que la relation ainsi définie est bien une relation d'équivalence sur l'espace somme des V_i^*. L'espace topologique quotient, c'est-à-dire l'espace topologique Ω^* obtenu par recollement des V_i^* le long des $V_{i,j}^*$ au moyen des isomorphismes $\psi_{i,j}$ est alors une «variété» analytique-complexe, a priori non nécessairement séparée. De plus, les isomorphismes φ_i définissent un isomorphisme (analytique-réel) φ de Ω sur une sous-variété $\varphi(\Omega)$ de Ω^* et on vérifie facilement que $\varphi(\Omega)$ est fermée et satisfait aux conditions imposées. La démonstration de la proposition 1 sera donc terminée quand nous aurons montré que Ω^* est *séparée*.

Montrons tout d'abord que $\overline{V}^*_{i,j} \subset U^*_{i,j}$ et pour cela, que $V^*_{i,j} \subset W^*_{i,j}$. Il suffit naturellement de faire la démonstration quand $T_{i,j} \neq \varnothing$. Soit $y \in V^*_{i,j}$: comme $V^*_{i,j} \subset V^*_i \subset U^*_i$, il existe un $x \in U_i$ tel que $y \in U^*_{i,x}$. Si $x \notin \psi_{j,i}(\overline{V}_j \cap \overline{U}_{j,i})$, alors $\varphi_i^{-1}(x) \notin \overline{V}'_j$ et par suite (condition (12)) $y \notin \psi_{j,i}(A^*_j \cap U^*_{j,i})$ et a fortiori, $y \notin \psi_{j,i}(V^*_j \cap U^*_{j,i})$, ce qui est contradictoire avec $y \in V^*_{i,j}$ et la définition (14) de $V^*_{i,j}$. On a donc $x \in \psi_{j,i}(\overline{V}_j \cap \overline{U}_{j,i})$, ce qui entraine d'après (11), $y \in (W^*_{i,j} \cup B^*_{i,j})$. Comme, d'après (8), $y \in V^*_i \subset A^*_i \subset (A^*_{i,j} \cup W^*_{i,j})$ et que $A^*_{i,j}$ et $B^*_{i,j}$ sont disjoints, on a bien $y \in W^*_{i,j}$.

Soient alors x' et y' deux points distincts de Ω^* et soient $x \in V^*_i$ et $y \in V^*_j$ tels que x' (resp. y') soit l'image canonique de x (resp. y). Il est clair qu'il suffit de trouver un voisinage A de x dans V^*_i et un voisinage B de y dans V^*_j tels qu'aucun point de A ne soit équivalent à un point de B. Or si cela n'était pas possible, on pourrait trouver deux suites (x_k) et (y_k) de points de C^n, convergeant respectivement vers x et y, avec $x_k \in V^*_{i,j}$, $y_k \in V^*_{j,i}$ et $x_k = \psi_{j,i}(y_k)$. Comme $\overline{V}^*_{i,j} \subset U^*_{i,j}$, on a $x \in U^*_{i,j}$, $y \in U^*_{j,i}$ et $x = \psi_{j,i}(y)$, l'application $\psi_{j,i}$ étant continue dans $U^*_{j,i}$. Par suite, $y \in V^*_j \cap U^*_{j,i}$ et $x \in V^*_i \cap \psi_{j,i}(V^*_j \cap U^*_{j,i}) = V^*_{i,j}$, donc $y \in V^*_{j,i}$ et comme $y = \psi_{i,j}(x)$, les points x et y sont équivalents et on a $x' = y'$ contrairement à l'hypothèse. La démonstration est achevée.

Remarque: On peut choisir Ω^* de telle sorte qu'il existe une involution antiholomorphe $z \to \bar{z}$ de Ω^* sur elle-même, $\varphi(\Omega)$ étant le lieu des points fixes de cette involution. On peut en effet supposer que $\psi_{i,j}(\bar{x}) = \overline{\psi_{i,j}(x)}$ dans $T^*_{i,j}$ ($\bar{x}$ désignant le point de C^n de coordonnées complexes conjuguées de celles de x), et il suffit alors (ce qui est possible) de choisir tous les ensembles envisagés dans C^n invariants par passage aux complexes conjugués.

2. Systèmes de coordonnées et standardisations propres

Soient Ω^* une variété analytique-complexe de dimension n, a un point de Ω^* et $(x_1 \ldots, x_n)$ un système de coordonnées valables dans un voisinage U^* de a et nulles en a. Une *standardisation* au voisinage de a, subordonnée à ce système de coordonnées, est la donnée de n applications F_j (pour $0 \leqslant j \leqslant n-1$) de $(R_+)^j$ dans R_+, où R_+ désigne l'ensemble des nombres réels > 0: l'application F_0 est donc une constante > 0. Un voisinage V^* de a sera dit *standardisé* s'il est contenu dans U^* et y est défini par les inégalités $|x_i| < \varepsilon_i$ (pour $1 \leqslant i \leqslant n$), où les nombres réels strictement positifs ε_i satisfont aux inégalités:

$$\varepsilon_i < F_{n-i}(\varepsilon_{i+1}, \ldots, \varepsilon_n). \tag{15}$$

Soit maintenant Ω une variété analytique-réelle de dimension n, plongée dans une de ses complexifications Ω^*, Ω étant le lieu des points fixes d'une involution antiholomorphe $z \to \bar{z}$ de Ω^*. Soit E^* un sous-ensemble analytique-complexe au voisinage d'un point a de Ω. Un système de coordonnées $\varphi = (x_1, \ldots, x_n)$ dans un voisinage U^* de a, nulles en a, sera dit *p-propre pour E^* au point a* si les p premières coordonnées d'un point quelconque de $E^* \cap U^*$ distinct de a ne sont pas toutes nulles.

On sait (voir p. ex. [8]) que, si E^* est de dimension $\leqslant p$, on peut trouver des systèmes de coordonnées p-propres pour E^* au point a [1]). On peut même se restreindre aux systèmes de coordonnées (que nous appellerons *réels*) pour lesquels toutes les coordonnées x_i sont réelles si et seulement si le point correspondant de U^* est dans Ω.

Dans ce qui suit, nous identifierons C^q (pour $0 \leqslant q \leqslant n$) avec le sous-espace de C^n formé des points dont les $n - q$ dernières coordonnées sont nulles et nous désignerons par π_q la projection $(x_1, \ldots, x_n) \to (x_1, \ldots, x_q)$ de C^n sur C^q.

Soit U^* un voisinage ouvert de a dans Ω^* dont l'image dans C^n par l'application qui a un point fait correspondre ses coordonnées soit un polydisque $| x_i | < \eta_i$. On sait [8] que, si φ est p-propre pour E^* au point a et si les η_i sont suffisamment petits, l'on a les résultats suivants:

a) pour chaque indice $j = 1, \ldots, n - p$, il existe un polynome distingué $Q_j(X; x_1, \ldots, x_p)$ à coefficients analytiques en $(x_1, \ldots, x_p)$ dans $\pi_p(U^*)$, sans facteurs multiples, tel que

$$Q_j(x_{p+j}; x_1, \ldots, x_p) = 0 \quad \text{sur } E^* \cap U^*. \tag{16}$$

On désignera par $\tilde{E}^*$ le sous-ensemble analytique-complexe de U^* défini par les équations (16)[2]).

b) si φ est réel et si E^* est invariant par l'involution $z \to \bar{z}$, on peut choisir les Q_j eux aussi invariants, c'est-à-dire à coefficients réels quand les $x_1, \ldots, x_p$ sont réels.

c) il existe une standardisation subordonnée au système de coordonnées

---

[1]) Cette propriété est prise comme *définition* de la dimension dans [8]. Ici, nous conservons la définition usuelle de la dimension d'un ensemble analytique-complexe (cf. p. ex. [5]): E^* est de dimension $\leqslant p$ si l'ensemble de ses points réguliers est une variété dont toutes les composantes connexes sont de dimension $\leqslant p$.

[2]) Les polynomes Q_j ne sont pas uniques, même en se restreignant aux polynomes de plus petit degré possible, sauf cependant si E^* est purement p-dimensionnel au voisinage de a, cas auquel les Q_j de degré minimum sont uniques ([8], Zusatz II, p. 273). Par suite, la définition de $\tilde{E}^*$ dépend non seulement de E^* et de φ, mais aussi du choix des Q_j. Il en est de même de la notion de standardisation p-propre pour E^* en a et des ensembles $D^*(E^*)$ et $S^*(E^*)$ introduits plus bas. Par abus de langage, nous utiliserons souvent ces notions sans rappeler qu'elles dépendent du choix des polynomes Q_j.

donné, telle que $F_i(\varepsilon_{i+1}, \ldots, \varepsilon_n) < \eta_i$ si $\varepsilon_j < \eta_j$ pour $j > i$ et que, pour tout voisinage standardisé V^* de a, les racines du polynome $Q_j(X; x_1 \ldots, x_p)$ pour $(x_1, \ldots, x_p) \in \pi_p(V^*)$ soient toutes inférieures en module à ε_{p+j}. Une telle standardisation sera dite *p-propre* pour E^* au point a.

d) supposons E^* de dimension p en a et soit E'^* (resp. E''^*) la réunion des composantes irréductibles de dimension p (resp. de dimension $< p$) de $E^* \cap U^*$. Soit D_1^* le sous-ensemble analytique lieu des zéros dans $\pi_p(U^*)$ du produit des discriminants des polynomes Q_j. Posons:

$$D^*(E^*) = D_1^* \cup \pi_p(E''^*) \quad \text{et} \quad S^*(E^*) = E^* \cap \pi_p^{-1}(D^*(E^*)) \cap U^*. \tag{17}$$

$D^*(E^*)$ et $S^*(E^*)$ sont des sous-ensembles analytiques-complexes de dimension $\leqslant p - 1$ dans U^* et $(E^* \cap U^*) - S^*(E^*)$ est *ouvert et fermé* dans $\tilde{E}^* - (\tilde{E}^* \cap \pi_p^{-1}(D^*(E^*)))$.

De plus, si l'on a un système de coordonnées $\varphi = (x_1, \ldots, x_n)$ qui soit p-propre en a relativement à un nombre fini de sous-ensembles analytiques-complexes E_i^* de dimension $\leqslant p - 1$, on peut, par un changement de variables portant uniquement sur les p premières coordonnées, remplacer φ par un système φ' qui soit $(p - 1)$-propre pour chaque E_i^*. En effet, la projection π_p est, d'après b), une application propre[3]) de $F^* = (\bigcup E_i^*) \cap U^*$ sur $\pi_p(F^*)$; par suite (cf. [8bis]) $\pi_p(F^*)$ est un sous-ensemble analytique de dimension $\leqslant p - 1$ dans $\pi_p(U^*)$ et l'on peut trouver dans C^p un système de coordonnées $(x_1', \ldots, x_p')$ qui soit $(p - 1)$-propre pour $\pi_p(F^*)$: il est clair que le système $\varphi' = (x_1', \ldots, x_p', x_{p+1}, \ldots, x_n)$ est $(p - 1)$-propre pour chacun des E_i^*. Si φ est réel, on peut choisir φ' réel.

Soit maintenant E un sous-ensemble analytique-réel dans un voisinage de a dans Ω. On sait (voir [1], [6]) qu'il existe un plus petit germe d'ensemble analytique-complexe (dans Ω^*) en a, dont l'intersection avec Ω coïncide avec le germe de E en a. Par définition, la dimension de E en a est la dimension complexe de ce germe. Pour tout voisinage U^* suffisamment petit de a dans Ω^*, il existe un plus petit ensemble analytique-complexe E^* dans U^*, tel que $E^* \cap \Omega = E \cap U^*$; la dimension de E^* est égale à la dimension de E en a et son germe en a est le complexifié du germe de E en a. Nous appellerons E^* le *complexifié* de E dans U^*: il est clair que $\overline{E}^* = E^*$ (si $\overline{U}^* = U^*$). Nous dirons qu'un système de coordonnées valable dans U^* est *p-propre pour E au point a* s'il l'est pour E^*. Si E est de dimension p (donc aussi E^*), nous poserons:

$$S(E) = S^*(E^*) \cap \Omega .$$

$S(E)$ est un ensemble analytique-réel au voisinage de a dans Ω. Enfin, si le

[3] Rappelons qu'une application est dite *propre* si l'image réciproque d'un compact est compacte.

système de coordonnées est convenablement choisi et le voisinage U^* suffisamment petit, nous définirons par récurrence les ensembles analytiques-réels $S_k(E)$ (avec $S_0(E) = E$ et $S_1(E) = S(E)$) par:

$$S_k(E) = S(S_{k-1}(E)) .$$

Nous aurons besoin par la suite d'utiliser des systèmes de coordonnées qui soient adaptés à l'étude non seulement d'un sous-ensemble analytique (ou même d'une famille finie de sous-ensembles analytiques), mais aussi à l'étude d'un couple (E, F) de deux sous-ensembles analytiques, F étant contenu dans E, et à l'étude des points singuliers de E.

Soit donc (E, F) un couple d'ensembles analytiques-réels dans un voisinage de a dans Ω, F étant contenu dans E. Nous dirons que ce couple est *réduit* si l'on a $\dim F < \dim E$. Dans le cas général, nous appellerons *couple réduit associé* le couple obtenu ainsi: soit U^* un voisinage de a dans Ω^*, suffisamment petit pour que E et F aient des complexifiés E^* et F^* dans U^*. Soit E'^* la réunion des composantes irréductibles de E^* dans U^* qui ne sont pas contenues dans F^*. On remplace le couple (E, F) par le couple (E', F') avec $E' = E'^* \cap \Omega$ et $F' = E' \cap F$ et on vérifie immédiatement que le couple (E', F') est réduit dans le voisinage $U = U^* \cap \Omega$ de a.

Supposons donc le couple (E, F) réduit et posons $p = \dim E$. Nous allons définir par récurrence sur p les notions de *système de coordonnées propre* et de *standardisation propre* pour ce couple. Pour $p = 0$, n'importe quel système de coordonnées nulles en a est propre; supposons donc la notion de système propre définie pour les couples de dimension $< p$. Un système φ sera propre s'il satisfait aux deux conditions suivantes:

(SP 1) *φ est p-propre pour E*, au point a.

Soit alors U^* un voisinage polycylindrique de a dans Ω^*, suffisamment petit pour que E et F aient des complexifiés E^* et F^* dans U^* et que les conditions a) à d) soient satisfaites (pour E^*). Posons:

$$D(E, F) = (D^*(E^*) \cup \pi_p(F^*)) \cap \Omega \tag{18}$$

$$T(E, F) = E \cap U^* \cap \pi_p^{-1}(D(E, F)) . \tag{19}$$

$D(E, F)$ et $T(E, F)$ sont des sous-ensembles analytiques-réels de dimension $\leqslant p - 1$ dans $U = U^* \cap \Omega$ (remarquons que leur définition dépend du choix des polynomes Q_j de a), du moins quand il n'y a pas de choix canonique de ceux-ci, c'est-à-dire quand E^* n'est pas purement p-dimensionnel). Nous pouvons alors poser la deuxième condition:

(SP 2) *φ est propre pour les couples réduits associés aux couples* $(T(E, F), F)$ *et* $(D(E, F), \varnothing)$.

Remarque. Pour $p = n$, la condition (SP 1) est automatiquement satisfaite et (SP 2) signifie que φ est propre pour le couple $(F, \varnothing)$. D'autre part, si φ est propre pour le couple $(E, \varnothing)$, alors les ensembles $S_k(E)$ (ou plus exactement leurs germes en a) sont bien définis au voisinage de a (moyennant toujours des choix successifs pour les polynomes Q) et le système φ est propre pour les couples $(S_k(E), \varnothing)$.

Reste à démontrer l'existence de systèmes propres. Plus précisément, nous allons démontrer par récurrence sur p le Lemme suivant, qui est évident pour $p = 0$:

Lemme 1. *Soit* $((E_i, F_i))_{i \in I}$ *une famille finie de couples réduits au voisinage de* a; *posons* $p = \sup \dim E_i$. *Soit* φ *un système de coordonnées* p*-propre pour chacun des* E_i. *On peut en déduire, par un changement de variables portant uniquement sur les* p *premières coordonnées et en restreignant au besoin le voisinage considéré du point* a, *un système de coordonnées propre pour chacun des couples* (E_i, F_i) *pour* $i \in I$.

Soit en effet J (resp. K) l'ensemble des indices $i \in I$ pour lesquels $\dim E_i = p$ (resp. $\dim E_i < p$): d'après ce que nous avons vu plus haut, on peut, par un changement de variables portant uniquement sur les p premières coordonnées, déduire de φ un système de coordonnées qui soit p-propre pour les $E_j (j \in J)$ et $(p - 1)$-propre pour les $E_k (k \in K)$ ainsi que pour les $D(E_j, F_j)$ (donc les $T(E_j, F_j)$) pour $j \in J$, quitte naturellement à restreindre le voisinage considéré de a. Il suffit alors d'appliquer l'hypothèse de récurrence.

On voit que, pour vérifier qu'un système de coordonnées est propre, on est amené à introduire une famille finie de sous-ensembles analytiques-réels G_s de dimension k_s $(0 \leqslant k_s \leqslant p)$, le système φ étant propre si et seulement si il est k_s-propre pour chacun des G_s. Mais la définition même des G_s dépend de φ et de ses propriétés (ainsi que de choix pour les polynomes Q) et se fait par récurrence descendante sur la dimension. Par exemple, les G_s de dimension $p - 1$ seront définis en supposant φ p-propre pour tous les E_i et seront d'une part les E_k de dimension $p - 1$, d'autre part les $D(E_j, F_j)$ de dimension $p - 1$ et enfin les ensembles de dimension $p - 1$ obtenus en réduisant les couples $(T(E_j, F_j), F_j)$.

Si φ est propre, une standardisation subordonnée à φ sera dite *propre* pour la famille donnée si elle est k_s-propre pour chaque G_s. L'existence de telles standardisations est claire: il suffit de choisir pour chaque indice s une standardisation subordonnée à φ et k_s-propre pour G_s et de prendre la borne inférieure de ces standardisations.

3. Un résultat de connexion locale

Soit Ω une variété analytique-réelle et soient a et b deux points de Ω. Dans la suite de cet article, nous appellerons «arc analytique joignant a à b» une application continue Γ de l'intervalle $[0,1]$ dans Ω telle que $\Gamma(0) = a$, $\Gamma(1) = b$ et telle que Γ soit analytique dans l'intervalle semi-ouvert $[0,1[$. Un tel arc sera dit «contenu» dans un sous-ensemble X de Ω si $\Gamma(t) \in X$ pour $0 \leqslant t < 1$.

Le but de ce numéro est de démontrer le résultat suivant:

Proposition 2[4]). *Considérons une famille finie de couples (E_i, F_i) de sous-ensembles analytiques-réels de Ω, F_i étant contenu dans E_i. Soit a un point de Ω. Il existe un voisinage ouvert V de a dans Ω tel que, pour tout indice i, tout point de $(E_i - F_i) \cap V$ puisse être joint à a par un arc analytique contenu dans $(E_i - F_i) \cap V$.*

Cette proposition étant purement locale, on peut supposer que Ω est un ouvert de R^n et qu'il existe un ouvert Ω^* de C^n, avec $\Omega = \Omega^* \cap R^n$, tel que tous les E_i et les F_i admettent des complexifiés E_i^* et F_i^* dans Ω^*. De plus on peut supposer la famille (E_i, F_i) réduite. Posons $p = \sup \dim E_i$. Nous allons, par récurrence sur le couple (n, p), démontrer le résultat plus précis suivant: pour toute standardisation propre pour la famille (E_i, F_i), on peut prendre pour V l'intersection avec Ω de n'importe quel voisinage standardisé assez petit. Ceci est évident pour $p = 0$, quel que soit n. Supposons donc ce résultat démontré pour les couples (m, q) avec $m < n$ ou avec $m = n$ et $q < p$, et démontrons-le pour le couple (n, p): on peut se borner au cas d'un seul couple (E, F), avec $\dim E = p$ et $\dim F < p$.

1er cas: $p < n$.

Par hypothèse, le système de coordonnées donné est p-propre pour E au point a; soit alors U^* un voisinage polycylindrique de a dans Ω^* assez petit pour que les conditions a) à d) du n° 2 soient satisfaites. Soit V^* un voisinage standardisé de a contenu dans U^*, posons $V = V^* \cap \Omega$ et soit $x \in (E - F) \cap V$. Considérons les ensembles $D(E, F)$ et $T(E, F)$ définis par les formules (18) et (19). Si $x \in T(E, F)$, il suffit d'appliquer l'hypothèse de récurrence, puisque par définition, la standardisation est propre pour le couple $(T(E, F), F)$ et que $\dim T(E, F) \leqslant p - 1$.

Supposons donc $x \notin T(E, F)$ et par suite $\pi_p(x) \notin D(E, F)$. Comme la standardisation donnée est propre pour le couple $(D(E, F), \varnothing)$, sa restric-

[4]) Cette proposition est à rapprocher du Théorème 1 de [2] et le principe de la démonstration en est analogue.

tion à C^p est propre pour le couple $(\pi_p(U), D(E, F))$ (n° 2, Remarque). Puisque $p < n$, on peut appliquer l'hypothèse de récurrence et trouver un arc analytique γ joignant $\pi_p(x)$ à a, contenu dans $\pi_p(V)$ et ne coupant pas $D(E, F)$. Comme $D(E, F)$ contient $D^*(E^*) \cap \Omega$, les discriminants des polynomes Q_j ne s'annulent pas le long de γ, donc leurs racines réelles sont simples, en nombre constant, donc analytiques le long de γ. On en déduit immédiatement l'existence d'un arc analytique unique Γ joignant x à a, contenu dans $\tilde{E}^* \cap V$ et tel que $\pi_p(\Gamma) = \gamma$. Mais comme $x \in E^*$ et que $(E^* \cap U^*) - S^*(E^*)$ est ouvert et fermé dans $\tilde{E}^* - (\tilde{E}^* \cap \pi_p^{-1}(D^*(E^*)))$, cet arc Γ est contenu dans $E^* \cap V = E \cap V$ et comme $\gamma \cap \pi_p(F) = \varnothing$, on a $\Gamma \cap F = \varnothing$ et Γ répond à la question.

2ième cas: $p = n$ (autrement dit, E est une composante connexe de Ω).

Montrons d'abord le résultat auxiliaire suivant: soit G un sous-ensemble analytique-réel au voisinage de a, de dimension $q < n$. Supposons la standardisation donnée propre pour le couple $(G, \varnothing)$ et soit U^* un voisinage standardisé suffisamment petit pour que G ait un complexifié G^* dans U^*, satisfaisant aux conditions a) à d) du n° 2 (en remplaçant E^* par G^* et p par q). Soit x un point de $U = U^* \cap \Omega$ tel que $x \notin G$, que $\pi_k(x) \in \pi_k(G^*)$ pour un entier k avec $q \leqslant k < n$ et que $\pi_q(x) \notin D(G) = D^*(G^*) \cap U$. Dans ces conditions, *il existe un arc analytique Γ joignant x à a dans U et tel que $\pi_k(\Gamma) \subset \pi_k(G^*)$, que $\pi_q(\Gamma) \cap D(G) = \varnothing$ et que $\Gamma \cap G = \varnothing$*. En effet, puisque $\pi_q(x) \notin D(G)$ et que la standardisation donnée est propre par hypothèse pour le couple $(D(G), \varnothing)$ donc pour le couple $(\pi_q(U), D(G))$, il existe par hypothèse de récurrence un arc analytique joignant $\pi_q(x)$ à a dans $\pi_q(U) - D(G)$. Le long d'un tel arc γ, les racines réelles des polynomes Q_j sont simples, en nombre constant, analytiques et tendent vers zéro quand $t \to 1$. Par suite $\pi_k(\tilde{G}^*) \cap \pi_q^{-1}(\gamma) \cap \Omega$ se compose d'un nombre fini d'arcs analytiques disjoints joignant les différents points de $\pi_k(\tilde{G}^*) \cap \pi_q^{-1}(\pi_q(x)) \cap \Omega$ à a. Soit γ' celui de ces arcs qui est d'origine $\pi_k(x)$. On voit de même que $\tilde{G}^* \cap \pi_k^{-1}(\gamma') \cap \Omega$ se compose d'un nombre fini d'arcs analytiques disjoints $\Gamma_1, \ldots, \Gamma_r$ avec $\pi_k(\Gamma_i(t)) = \gamma'(t)$. Si $r = 0$, n'importe quel arc analytique Γ joignant x à a tel que $\pi_k(\Gamma) = \gamma'$ répond à la question. Supposons $r > 0$: pour tout entier $m = k + 1, \ldots, n$, les m-ièmes coordonnées de $\Gamma_i(t)$ et de $\Gamma_j(t)$ sont, pour $i \neq j$, ou bien toujours distinctes ou bien toujours égales. Soient $\lambda_1, \ldots, \lambda_s$ ces m-ièmes coordonnées rangées dans l'ordre croissant; posons $\lambda_0(t) = -\varepsilon_m$ et $\lambda_{s+1}(t) = \varepsilon_m$. Il y a un indice j et un seul $(0 \leqslant j \leqslant s)$ tel que $\lambda_j(0) \leqslant x_m < \lambda_{j+1}(0)$: nous poserons:

$$x_m(t) = \lambda_j(t) + (1 - t)c(\lambda_{j+1}(t) - \lambda_j(t)) \quad \text{pour } j > 0$$

et:

$$x_m(t) = \lambda_1(t) - (1-t)c(\lambda_1(t) - \lambda_0(t)) \qquad \text{pour } j = 0$$

la constante c étant déterminée par la condition $x_m(0) = x_m$ (on a $0 \leqslant c < 1$). Il est alors immédiat que l'arc $t \to y(t)$ défini par $\pi_k(y(t)) = \gamma'(t)$ et $y_m(t) = x_m(t)$ pour $k < m \leqslant n$ répond à la question. Notons que nous avons utilisé la condition c) du n° 2.

Soit alors V^* un voisinage standardisé de a; tous les ensembles $S_k = S_k(F)$ pour $k \geqslant 0$ ont des complexifiés S_k^* dans V^* puisque la standardisation donnée est propre pour tous les S_k (n° 2, Remarque). Comme $S_k = S^*(S_{k-1}^*) \frown \Omega$, on a $S_k^* \subset S^*(S_{k-1}^*)$. Nous poserons $q = \dim F$, et $q_k = \dim S_k$ pour $k \geqslant 1$. On a, d'après (17), pour tout entier r avec $q \leqslant r \leqslant n$:

$$F \frown \pi_r^{-1}(\pi_r(S^*(F^*))) = S(F)\,. \tag{20}$$

Montrons, par récurrence sur k, que:

$$F \frown \pi_q^{-1}(\pi_q(S_k^*)) = S_k\,. \tag{21}$$

C'est évident pour $k = 0$; supposons (21) vraie pour la valeur $k - 1$: on a $F \frown \pi_q^{-1}(\pi_q(S_k^*)) \subset F \frown \pi_q^{-1}(\pi_q(S^*(S_{k-1}^*))) \subset F \frown \pi_q^{-1}(\pi_q(S_{k-1}^*)) = S_{k-1}$, et par suite $F \frown \pi_q^{-1}(\pi_q(S_k^*)) = S_{k-1} \frown \pi_q^{-1}(\pi_q(S_k^*)) \subset S_{k-1} \frown \pi_q^{-1}(\pi_q(S^*(S_{k-1}^*)))$. Or en remplaçant F par S_{k-1} dans (20), on voit que $S_{k-1} \frown \pi_q^{-1}(\pi_q(S^*(S_{k-1}^*))) = S_k$, d'où immédiatement (21).

Soit enfin x un point de $V = V^* \frown \Omega$, n'appartenant pas à F, et soit k le plus grand entier tel que $\pi_q(x) \in \pi_q(S_k^*)$ (remarquons que l'on a $\pi_q(S_0^*) = \pi_q(F^*) \supset \pi_q(V^*)$). Comme $\pi_q(x) \notin \pi_q(S_{k+1}^*)$, on a

$$\pi_{q_k}(x) \notin \pi_{q_k}(S_{k+1}^*) \frown \Omega = D^*(S_k^*) \frown \Omega = D(S_k)$$

et le résultat auxiliaire démontré plus haut s'applique. Soit donc Γ un arc joignant x à a dans V, avec $\pi_q(\Gamma) \subset \pi_q(S_k^*)$ et $\Gamma \frown S_k = \varnothing$. On a bien $\Gamma \frown F = \varnothing$, car si $y \in \Gamma \frown F$, on a $\pi_q(y) \in \pi_q(\Gamma)$ donc $y \in F \frown \pi_q^{-1}(\pi_q(S_k^*)) = S_k$ d'après (21) et $y \in \Gamma \frown S_k$, ce qui est impossible.

La démonstration est achevée.

4. Ensembles analytiques-complexes au voisinage d'un ensemble analytique-réel

Soient Ω^* une variété analytique-complexe et A un sous-ensemble analytique-*réel* de Ω^*: le cas le plus important est celui où A est une variété analytique-réelle dont Ω^* est une complexification. Soit E^* un sous-ensemble analytique-complexe de Ω^*; posons $E = A \frown E^*$ et soit a un point de A.

Proposition 3. *Il existe un voisinage ouvert V de a dans A tel que, quel que soit le voisinage ouvert V^* de V dans Ω^*, toute composante irréductible de $E^* \cap V^*$ dans V^* qui rencontre V, passe par le point a.*

Soit E_p^* la réunion des composantes irréductibles de dimension p de E^* et soit φ un système de coordonnées au voisinage de a qui, pour tout p, soit p-propre pour E_p^*. Soit enfin U^* un voisinage de a dans Ω^* tel que les conditions a) à d) du n° 2 soient satisfaites pour chacun des E_p^*. Nous désignerons par Q_j^p les polynomes de la condition a) associés à E_p^*, et poserons $E_p = E_p^* \cap A$.

Pour tout système $S = (r_1, \ldots, r_{n-p})$ de $n - p$ entiers avec $1 \leqslant r_j \leqslant$ degré de Q_j^p, désignons par $E_{p,S}$ le sous-ensemble analytique-réel de U^* formé des points $(x_1, \ldots, x_n)$ de $E_p \cap U^*$ tels que x_{p+j} soit racine multiple d'ordre $\geqslant r_j$ de $Q_j^p(X; x_1, \ldots, x_p)$. Nous prendrons pour V un voisinage ouvert de a dans A tel que, pour tout p et tout système S, tout point de $E_{p,S} \cap V$ puisse être joint à a par un arc analytique contenu dans $E_{p,S} \cap V$, l'existence d'un tel V résultant de la proposition 2.

Soit donc V^* un voisinage ouvert de V dans Ω^* et soit F^* une composante irréductible de $E^* \cap V^*$: si $\dim F^* = p$, F^* est aussi une composante irréductible de $E_p^* \cap V^*$. Supposons que F^* rencontre V en un point $b = (b_1, \ldots, b_n)$. Soit r_j l'ordre de multiplicité de la racine b_{p+j} de $Q_j^p(X; b_1, \ldots, b_p)$ et soit Γ un arc analytique joignant b à a dans $E_{p,S}$, avec $S = (r_1, \ldots, r_{n-p})$: il est clair qu'il suffit de démontrer que $\Gamma(t) \in F^*$ pour t suffisamment petit, car on aura par prolongement analytique, $\Gamma \subset F^*$ et par suite $a \in F^*$.

Or, pour tout indice $j = 1, \ldots, n-p$, il existe un voisinage convexe B_j de $(b_1, \ldots, b_p)$ dans C^p et un nombre réel $\varepsilon_j > 0$ tels que, pour $(x_1, \ldots, x_p) \in B_j$, le polynome $Q_j^p(X; x_1, \ldots, x_p)$ ait exactement r_j racines satisfaisant à $|X - b_{p+j}| < \varepsilon_j$. Soit B l'intersection des B_j et soit $\hat{B}$ l'ouvert de C^n formé des points $x = (x_1, \ldots, x_n)$ tels que $\pi_p(x) \in B$ et $|x_{p+j} - b_{p+j}| < \varepsilon_j$: on peut supposer (quitte à diminuer les ε_j et les B_j) que $\hat{B} \subset V^* \cap U^*$. Comme F^* est de dimension p et que, pour tout $y \in B$, $F^* \cap \pi_p^{-1}(y)$ se compose d'un nombre fini de points, $\pi_p(F^* \cap \hat{B})$ est un sous-ensemble analytique-complexe de dimension p de B, donc coïncide avec B, qui est connexe.

Soit alors μ un nombre réel > 0 tel que $\Gamma(t) \in \hat{B}$ pour $0 \leqslant t < \mu$. Pour tout t avec $0 \leqslant t < \mu$, il existe un point $z(t) \in F^* \cap \hat{B}$ tel que $\pi_p(\Gamma(t)) = \pi_p(z(t))$. Mais comme la $(p+j)$-ième coordonnée de $\Gamma(t)$ est racine multiple d'ordre $\geqslant r_j$ de $Q_j^p(X; \pi_p(\Gamma(t)))$, elle est la seule racine de ce polynome qui appartienne au cercle $|X - b_{p+j}| < \varepsilon_j$ et par suite, elle est égale à la $(p+j)$-ième coordonnée de $z(t)$, ce qui entraine que $\Gamma(t) = z(t)$ appartient à F^* et achève la démonstration.

Remarquons que si le germe de E^* en a est *irréductible*, il existe une seule composante irréductible de $E^* \cap V^*$ qui rencontre V.

Corollaire. *Soit Ω une variété analytique-réelle et soit f une fonction analytique-réelle sur Ω. Pour tout point a de Ω, il existe un voisinage ouvert V de a tel que, si g est une fonction analytique sur V telle que g/f soit analytique au voisinage de a, alors g/f est analytique dans V.*

En effet, on peut supposer que Ω admet une complexification Ω^* et que f se prolonge en une fonction holomorphe, notée encore f, sur Ω^*. Soit E^* le lieu des zéros de f dans Ω^* et soit V un voisinage de a dans Ω possédant les propriétés exigées dans la proposition 3 (avec $A = \Omega$). La fonction g se prolonge en une fonction holomorphe, notée encore g, sur un voisinage ouvert V^* de V dans Ω^*. L'ensemble des points de V^* au voisinage desquels la fonction méromorphe g/f n'est pas holomorphe est un ensemble analytique-complexe de dimension $n-1$ contenu dans E^* et ne contient aucune des composantes irréductibles de $E^* \cap V^*$ qui passent par a, donc ne rencontre pas V, et g/f est holomorphe au voisinage de tout point de V.

5. Germes d'ensembles analytiques-complexes au voisinage de A

Soient Ω^* une variété analytique-complexe paracompacte et soit A un sous-ensemble analytique-réel de Ω^*. Nous désignerons par $\mathfrak{G}(A)$ l'ensemble des «germes d'ensemble analytique-complexe au voisinage de A», ou plus brièvement des *A-germes:* un tel germe est représenté par un couple formé d'un voisinage ouvert V^* de A dans Ω^* et d'un sous-ensemble analytique-complexe E^* dans V^*, deux tels couples (V_1^*, E_1^*) et (V_2^*, E_2^*) définissant le même A-germe si et seulement si il existe un voisinage V_3^* de A dans Ω^*, tel que $E_1^* \cap V_3^* = E_2^* \cap V_3^*$. Nous dirons que le A-germe Y est de dimension $\leqslant p$ s'il existe un couple (V^*, E^*) définissant Y, avec E^* de dimension (complexe) $\leqslant p$, et que Y est de dimension p s'il est de dimension $\leqslant p$ et n'est pas de dimension $\leqslant p-1$. On désignera par $\mathfrak{G}_p(A)$ l'ensemble des A-germes de dimension p.

Soit $Y \in \mathfrak{G}(A)$ et soit V^* un voisinage de A suffisamment petit pour que la famille $\mathfrak{F}$ des ensembles analytiques-complexes F^* tels que le couple (V^*, F^*) définisse Y ne soit pas vide. Soit Y^* l'intersection des F^* pour $F^* \in \mathfrak{F}$. Les germes des F^* en un point $x \in A$ sont tous égaux, donc sont égaux au germe en x de l'ensemble analytique-complexe Y^*, ce qui montre que le couple (V^*, Y^*) définit aussi Y. Autrement dit, pour tout voisinage V^* suffisamment petit de A, il existe un *plus petit* ensemble analytique-complexe Y^* dans V^* tel que le couple (V^*, Y^*) définisse Y. On a $\dim Y = \dim Y^*$.

Posons en effet $p = \dim Y^*$. Il est clair que $\dim Y \leqslant p$; supposons $\dim Y < p$ et soit (W^*, E^*) un couple définissant Y avec $\dim E^* < p$. En tout point $x \in A$, le germe de Y^* en x est égal au germe de E^* en x, donc est de dimension $< p$. Par suite, le couple (V^*, F^*), où F^* est la réunion des composantes irréductibles de dimension $< p$ de Y^* dans V^*, définit lui aussi Y, ce qui contredit la minimalité de Y^*.

Soient Y_i $(i = 1, 2)$ des éléments de $\mathfrak{G}(A)$ et soient (V_i^*, E_i^*) des couples représentant Y_i. La *réunion* $Y_1 \cup Y_2$ (resp. *l'intersection* $Y_1 \cap Y_2$) est le A-germe défini par le couple $(V_1^* \cap V_2^*, (E_1^* \cup E_2^*) \cap V_1^* \cap V_2^*)$ (respectivement $(V_1^* \cap V_2^*, E_1^* \cap E_2^*)$): il est clair que cette définition ne dépend pas des représentants choisis. On dira que $Y_1 \subset Y_2$ si $Y_2 = Y_1 \cup Y_2$, ou encore si $E_1^* \subset E_2^*$ au voisinage de A. Enfin Y est *irréductible* si on ne peut pas l'écrire comme réunion de deux A-germes Y_1 et Y_2, tous les deux distincts de Y.

Par ailleurs, nous désignerons par $G_p(\Omega^*)$ l'espace des germes analytiques-complexes irréductibles de dimension p aux différents points de Ω^*, muni de sa topologie habituelle (voir [5], [8]): rappelons qu'un système fondamental d'ensembles ouverts $\Gamma_p(U^*; E^*)$ est obtenu comme suit: pour tout couple (U^*, E^*) formé d'un ouvert U^* de Ω^* et d'un sous-ensemble analytique-complexe E^* dans U^*, $\Gamma_p(U^*, E^*)$ est l'ensemble des composantes irréductibles de dimension p des germes de E^* aux différents points de U^*. Soit $G(\Omega^*)$ la somme topologique des $G_p(\Omega^*)$: on sait [5] que $G(\Omega^*)$ est un espace topologique *séparé, localement compact* et *localement connexe.*

Soit π l'application canonique de $G(\Omega^*)$ sur Ω^*, qui à un germe au point a, fait correspondre le point a lui-même. Nous désignerons par $G(A)$ le sous-espace $\pi^{-1}(A)$ de $G(\Omega^*)$: comme π est continue, c'est un sous-espace fermé donc localement compact de $G(\Omega^*)$. On posera $G_p(A) = G(A) \cap G_p(\Omega^*)$: $G(A)$ est la somme topologique des $G_p(A)$.

Soit Y un élément de $\mathfrak{G}(A)$, provenant d'un couple (V^*, E^*): il est immédiat que le sous-ensemble $\Gamma_p(Y) = \Gamma_p(V^*; E^*) \cap \pi^{-1}(A)$ est un sous-ensemble ouvert et fermé de $G_p(A)$, qui ne dépend que de Y et non du couple (V^*, E^*) choisi. Nous poserons $\Gamma(Y) = \cup \Gamma_p(Y)$. Si Y_1 et Y_2 sont deux éléments de $\mathfrak{G}(A)$, $Y_1 = Y_2$ est équivalent à $\Gamma(Y_1) = \Gamma(Y_2)$. D'autre part, si ξ et η sont deux points *distincts* de $\Gamma(Y)$ tels que $\pi(\xi) = \pi(\eta)$, on ne peut pas avoir $\xi \subset \eta$, car deux composantes irréductibles du germe de E^* au point $\pi(\xi)$ ne peuvent pas être contenues l'une dans l'autre, par définition même des composantes irréductibles.

Remarque: En un certain sens, $\mathfrak{G}(A)$ et $G(A)$, avec toutes leurs structures, ne dépendent que de A et non de Ω^*: plus précisément, si l'on a une autre variété analytique-complexe Ω_1^* et un sous-ensemble analytique-réel A_1 de Ω_1^*, tels qu'il existe un isomorphisme φ d'un voisinage de A sur un voisinage de A_1, avec $\varphi(A) = A_1$, il est clair que φ définit un isomorphisme naturel de $\mathfrak{G}(A)$ sur $\mathfrak{G}(A_1)$ et un isomorphisme naturel de $G(A)$ sur $G(A_1)$, compatibles avec les applications Γ et π et avec toutes les opérations définies ci-dessus et ci-dessous. En particulier, si A est une variété analytique-réelle paracompacte Ω, on peut parler des espaces $\mathfrak{G}(\Omega)$ et $G(\Omega)$ sans spécifier quelle complexification on a considérée pour les construire.

Proposition 4. *Pour qu'un sous-ensemble X de $G_p(A)$ soit de la forme $\Gamma(Y)$ pour un $Y \in \mathfrak{G}_p(A)$, il faut et il suffit que X soit ouvert et fermé et que la restriction de π à X soit propre*[3]).

La nécessité de la condition est évidente, compte tenu de ce qui précède et du fait que la restriction de π à $\Gamma_p(V^*, E^*)$ est propre. Réciproquement, soit X un ouvert et fermé de $G_p(A)$ sur lequel π est propre. Soit x un point de A et soient $e_1, \ldots, e_r$ les éléments (en nombre fini) de $X \cap \pi^{-1}(x)$. Puisque X est ouvert, il existe un voisinage ouvert T_x^* de x et des ensembles analytiques-complexes E_i^* de dimension p dans T_x^* tels que E_i^* induise le germe e_i au point x et que $\Gamma_p(T_x^*, E_i^*) \cap \pi^{-1}(A) \subset X$. On peut même trouver un voisinage ouvert U_x^* de x dans Ω^* tel que:

$$X \cap \pi^{-1}(U_x^*) = \Big(\bigcup_i \Gamma_p(T_x^*, E_i^*)\Big) \cap \pi^{-1}(U_x^* \cap A). \tag{22}$$

Sinon, il existerait une suite de points $\xi_q \in X$ telle que ξ_q n'appartienne pas à la réunion des $\Gamma_p(T_x^*, E_i^*)$ et que la suite des $\pi(\xi_q)$ converge vers x. Comme π est propre sur X, on peut supposer que la suite ξ_q converge vers un point $\xi \in X$: on a $\pi(\xi) = x$, donc ξ est l'un des e_i, ce qui entraine que $\xi_q \in \Gamma_p(T_x^*, E_i^*)$ pour q assez grand, contrairement à l'hypothèse.

Comme Ω^* est paracompacte, on peut trouver deux recouvrements (U_i^*) et (V_i^*) de type fini, plus fins que le recouvrement de Ω^* formé des U_x^* et de l'ouvert $\Omega^* - A$ et tels que V_i^* soit relativement compact dans U_i^*. Nous ne considèrerons dans la suite que les ouverts V_i^* et U_i^* tels que $V_i = V_i^* \cap A$ ne soit pas vide: ces ensembles V_i forment un recouvrement ouvert de A et dans chaque U_i^* (rencontrant A, donc contenu dans un ouvert U_x^*), il existe d'après (22) un sous-ensemble analytique-complexe F_i^* de dimension p tel que:

$$X \cap \pi^{-1}(U_i^*) = \Gamma(F_i^*) \cap \pi^{-1}(A). \tag{23}$$

En tout point $x \in A \cap U_i^* \cap U_j^*$, les germes induits par F_i^* et F_j^* sont les mêmes, l'ensemble de leurs composantes irréductibles étant $X \cap \pi^{-1}(x)$. Il existe donc un voisinage $U_{i,j}^* \subset U_i^* \cap U_j^*$ de $A \cap U_i^* \cap U_j^*$ tel que $F_i^* \cap U_{i,j}^* = F_j^* \cap U_{i,j}^*$ (on prend $U_{i,j}^* = U_{j,i}^*$). Soit alors $V_i'^*$ un ouvert de Ω^*, contenu dans V_i^*, et tel que $V_i'^* \cap A = V_i$ et $\overline{V_i'^*} \cap A = \overline{V_i}$ (l'existence d'un tel ouvert est une conséquence immédiate de la régularité de l'espace topologique Ω^*). Pour tout point $x \in V_i$, il existe un voisinage ouvert W_x^* de x dans Ω^* possédant les propriétés suivantes:

$$W_x^* \subset V_i'^* \tag{24}$$

$$\text{si } x \notin \overline{V}_j, \text{ alors } W_x^* \cap V_j'^* = \varnothing \tag{25}$$

(en effet, si $U_i^* \cap U_j^* = \varnothing$, c'est une conséquence de (24) et pour les indices j (en nombre fini) tels que $U_i^* \cap U_j^* \neq \varnothing$, l'existence d'un tel voisinage résulte de ce que $x \notin \overline{V_j'^*}$);

$$\text{si } x \in V_j \text{ (avec } j \neq i\text{), alors } W_x^* \subset U_{i,j}^* \tag{26}$$

(car $U_{i,j}^*$ est un voisinage de $V_i \cap V_j$).

Soit W_i^* la réunion des W_x^* pour x décrivant V_i: on a $W_i^* \subset V_i'^* \subset U_i^*$ et $W_i^* \cap W_j^* \subset U_{i,j}^*$, ce qui entraine l'existence d'un sous-ensemble analytique-complexe de dimension p, F^*, dans $W^* = \bigcup W_i^*$ (qui est un voisinage de A) tel que $F^* \cap W_i^* = F_i^* \cap W_i^*$, donc tel que $X \cap \pi^{-1}(V_i) = \Gamma(F^*) \cap \pi^{-1}(V_i)$, d'où $X = \Gamma(Y)$, Y étant le A-germe défini par F^*.

Corollaire 1. *Soit Y un élément de $\mathfrak{G}(A)$, et soit X un sous-espace de $\Gamma(Y)$. Pour qu'il existe un A-germe $Z \in \mathfrak{G}(A)$ tel que $X = \Gamma(Z)$, il faut et il suffit que X soit ouvert et fermé dans $\Gamma(Y)$.*

La nécessité de la condition est évidente. Réciproquement si X est ouvert et fermé dans $\Gamma(Y)$, il existe pour tout entier p un germe Z_p tel que $\Gamma(Z_p) = X \cap G_p(A)$. Soit Z la réunion des Z_p: on a $\Gamma(Z) \subset \bigcup \Gamma(Z_p)$ et même $\Gamma(Z) = \bigcup \Gamma(Z_p)$ (donc $\Gamma(Z) = X$) car, pour tout point $x \in A$, $\Gamma(Z) \cap \pi^{-1}(x)$ est la réunion des $\Gamma(Z_p) \cap \pi^{-1}(x)$ puisque deux germes distincts au point x appartenant tous les deux à $\Gamma(Y)$, ne peuvent pas être contenus l'un dans l'autre.

Corollaire 2. *Pour qu'un A-germe $Y \in \mathfrak{G}(A)$ soit irréductible, il faut et il suffit que $\Gamma(Y)$ soit connexe.*

En effet, si $Y = Y_1 \cup Y_2$ avec Y_1 et Y_2 distincts de Y, alors $\Gamma(Y)$ est réunion des deux ouverts et fermés distincts de lui-même $\Gamma(Y) \cap \Gamma(Y_1)$ et $\Gamma(Y) \cap \Gamma(Y_2)$. Inversement, si $\Gamma(Y) = X_1 \cup X_2$, X_1 et X_2 étant des ouverts et fermés, distincts de $\Gamma(Y)$, il existe des germes Z_1 et Z_2 distincts de Y tels que $\Gamma(Z_1) = X_1$ et $\Gamma(Z_2) = X_2$ et l'on a $Y = Z_1 \cap Z_2$.

Proposition 5. *Soit Y un A-germe, irréductible de dimension p. Si un A-germe Z est contenu dans Y, ou bien $Z = Y$ ou bien Z est de dimension $< p$.*

En effet, $\Gamma_p(Z)$ est un ouvert et fermé contenu dans $\Gamma(Y)$, qui est connexe: donc ou bien cet ouvert est vide, et Z est de dimension $< p$, ou bien $\Gamma_p(Z) = \Gamma(Y)$ et par suite $Z = Y$ (remarquons que l'on a $\Gamma(Y) = \Gamma_p(Y)$).

Soit U un ouvert de A et soient $\mathfrak{G}(U)$ et $G(U)$ les espaces de germes construits à partir de U considéré comme sous-ensemble analytique-réel d'un voisinage ouvert U^* de U dans Ω^* tel que $U^* \cap A = U$: ces espaces ne dépendent pas du choix de U^*. L'espace $G(U)$ s'identifie à un sous-espace ouvert de $G(A)$ et on a une application canonique de $\mathfrak{G}(A)$ dans $\mathfrak{G}(U)$, qui, à un A-germe défini par un couple (V^*, E^*) fait correspondre le U-germe défini par le couple $(V^* \cap U^*, E^* \cap U^*)$.

Proposition 6. *Soit Y un élément de $\mathfrak{G}(A)$ et soit x un point de $\pi(\Gamma(Y))$. Il existe un voisinage ouvert U de x dans A tel que, si le germe au point x d'un élément Z de $\mathfrak{G}(U)$ contient le germe de Y au point x, alors Z contient l'image canonique de Y dans $\mathfrak{G}(U)$.*

En effet, soit (V^*, E^*) un couple définissant le A-germe Y. D'après la proposition 3, il existe un voisinage ouvert U de x dans A tel que, pour tout voisinage ouvert U^* de U dans Ω^*, toutes les composantes irréductibles de $E^* \cap U^*$ qui rencontrent U, passent par x. Soit alors (U^*, F^*) un couple définissant Z: toutes les composantes irréductibles de $E^* \cap U^*$ qui rencontrent U sont contenues dans F^*, ce qui entraine que le germe de E^* en un point quelconque de U est contenu dans le germe de F^* en ce point, d'où la proposition.

Corollaire. *Soit α un élément de $G(A)$. Il existe un voisinage de α qui est contenu dans tout ensemble ouvert et fermé de $G(A)$ contenant α.*

En effet, soit V un voisinage ouvert de $x = \pi(\alpha)$ dans A tel qu'il existe un élément $Y \in \mathfrak{G}(V)$ avec $\Gamma(Y) \cap \pi^{-1}(x) = \{\alpha\}$. Soit U un voisinage ouvert de x dans V satisfaisant aux conditions exigées dans la proposition 6 (en remplaçant A par V). Pour tout ensemble X ouvert et fermé dans $G(A)$, contenant α, $X \cap \Gamma(Y)$ est de la forme $\Gamma(Z)$ avec $Z \in \mathfrak{G}(V)$ (Corollaire 1 à la proposition 4) et le germe de Z en x contient le germe de Y en x. Par suite, X contient le voisinage fixe $\Gamma(Y) \cap \pi^{-1}(U)$ de α dans $G(A)$.

6. Intersections, réunions et composantes irréductibles des A-germes

Proposition 7. *Soit (Y_i) une famille filtrante décroissante d'éléments de $\mathfrak{G}(A)$. La famille (Y_i) est localement stationnaire, autrement dit, pour tout point $x \in A$, il existe un voisinage U_x de x dans A et un indice i_x tels que les images canoniques de Y_i et de Y_j dans $\mathfrak{G}(U_x)$ coïncident pour $i, j \geqslant i_x$. De plus, il existe un*

A-germe Y et un seul, appelé intersection des A-germes Y_i, tel que le germe de Y au point $x \in A$ soit l'intersection des germes des Y_i en x.

En effet, la famille filtrante décroissante des germes induits par les Y_i au point $x \in A$ est stationnaire: soit i_x un indice tel que les germes au point x de Y_i et Y_j coïncident pour $i, j \geqslant i_x$. D'après la proposition 6, il existe un voisinage ouvert U_x de x dans A tel que, pour $i \geqslant i_x$, l'image canonique de Y_i dans $\mathfrak{G}(U_x)$ contienne donc soit égale à l'image de Y_{i_x}. Soit α_x le germe de Y_i en x, pour $i \geqslant i_x$, et soit X l'ensemble des α_x non vides. On a pour tout $x \in A$, $X \cap \pi^{-1}(U_x) = \Gamma(Y_{i_x}) \cap \pi^{-1}(U_x)$, ce qui entraine immédiatement que X est ouvert et fermé et que π est propre sur X. Par suite, pour tout entier p, $X \cap G_p(A)$ est de la forme $\Gamma(Y_p)$ avec $Y_p \in \mathfrak{G}_p(A)$. Soit Y la réunion des Y_p. Comme $\Gamma(Y_p) \cap \pi^{-1}(U_x) = \Gamma(Y_i) \cap \pi^{-1}(U_x) \cap G_p(A)$ pour $i \geqslant i_x$, et que deux germes au point x distincts appartenant à $\Gamma(Y_i)$ ne sont pas contenus l'un dans l'autre, on a $\Gamma(Y) = \cup\, \Gamma(Y_p)$ et le germe de Y au point x est égal au germe de Y_i en x pour $i \geqslant i_x$, c'est-à-dire à l'intersection des germes des Y_i au point x.

Corollaire 1. *Pour toute famille (Y_i) d'éléments de $\mathfrak{G}(A)$, il existe un A-germe Y et un seul, appelé intersection des Y_i, tel que le germe de Y en un point x de A soit l'intersection des germes des Y_i en x.*

Il suffit d'appliquer la proposition à la famille filtrante décroissante des intersections finies d'éléments Y_i.

Corollaire 2. *Soit B un sous-ensemble de A. Il existe un plus petit A-germe $Y \in \mathfrak{G}(A)$ tel que $\pi(Y) \supset B$.*

Il suffit d'appliquer le Corollaire 1 à la famille des A-germes Y_i tels que $\pi(Y_i) \supset B$.

Nous dirons qu'une famille (Y_i) de A-germes est *localement finie* si, pour tout point $x \in A$, il existe un voisinage U de x dans A tel que $\Gamma(Y_i) \cap \pi^{-1}(U) = \varnothing$ sauf pour un nombre fini d'indices.

Proposition 8. *Soit (Y_i) une famille localement finie d'éléments de $\mathfrak{G}(A)$. Il existe un A-germe Y et un seul, appelé réunion des Y_i, tel que le germe de Y en un point x de A soit la réunion des germes des Y_i en x.*

Considérons la famille filtrante croissante (Z_j) formée des réunions finies de germes Y_i: cette famille est localement stationnaire. Soit X l'ensemble des points $\alpha \in G(A)$ tels que α appartienne à $\Gamma(Z_j)$ pour j assez grand. Pour tout point $x \in A$, il existe un voisinage U_x de x dans A et un indice j_x tels que $X \cap \pi^{-1}(U_x) = \Gamma(Z_j) \cap \pi^{-1}(U_x)$ pour $j \geqslant j_x$. Par suite, X est ouvert et fermé et π est propre sur X. Il existe donc des A-germes Y_p tels que $\Gamma(Y_p) = X \cap G_p(A)$ et on voit comme plus haut que, si Y désigne la réunion des Y_p, on a $\Gamma(Y) = \cup \Gamma(Y_p) = X$, ce qui entraine que le germe de Y en un point x est égal au germe de Z_j pour $j \geqslant j_x$, c'est-à-dire est la réunion des germes des Y_i au point x.

Corollaire 1. *Si un A-germe irréductible Y est réunion d'une famille localement finie de A-germes Y_i, alors $Y = Y_i$ pour au moins un indice i.*

Soit en effet x un point de $\pi(\Gamma(Y))$ et soient $Z_1, \ldots, Z_r$ ceux des Y_i (en nombre fini) tels que $x \in \pi(\Gamma(Y_i))$. Soit Z la réunion des Z_k et soit T la réunion de la famille (localement finie) formée par les autres Y_i. On a $Y = Z \cup T$ et $T \neq Y$ d'où $Y = Z$ et même (par récurrence sur r) $Y = Z_k$ pour au moins un indice k.

Corollaire 2. *Soit $Y \in \mathfrak{G}(A)$. Si Y est irréductible, il existe un couple (V^*, E^*) définissant Y tel que E^* soit irréductible dans V^*.*

Soit en effet (V^*, F^*) un couple définissant Y. Décomposons F^* en composantes irréductibles E_i^* dans V^* et soit Y_i le A-germe défini par E_i^*. La famille (E_i^*), donc la famille (Y_i) sont localement finies et Y est réunion des Y_i. On a donc $Y = Y_i$ pour au moins un indice i, ce qui montre que Y peut être défini par l'un des E_i^*.

Remarque. Les opérations de réunion et d'intersection finies ou non dans $\mathfrak{G}(A)$ se traduisent par les mêmes opérations sur les germes en chaque point: elles possèdent donc les mêmes propriétés de distributivité et d'associativité que les opérations ensemblistes.

Nous pouvons maintenant démontrer l'existence, pour tout A-germe, d'une «bonne» décomposition en composantes irréductibles:

Proposition 9. *Soit Y un élément de $\mathfrak{G}(A)$. Il existe une famille localement finie et une seule de A-germes irréductibles $Y_i \in \mathfrak{G}(A)$ telle que $Y = \cup Y_i$ et $Y_i \not\subset Y_j$ pour $i \neq j$. Les $\Gamma(Y_i)$ sont les composantes connexes de $\Gamma(Y)$ et par suite $\Gamma(Y) = \cup \Gamma(Y_i)$.*

Naturellement, les Y_i seront appelés les composantes irréductibles de Y. L'unicité se démontre par un raisonnement purement latticiel classique, compte tenu du Corollaire 1 à la proposition 8. Pour l'existence, nous aurons besoin du Lemme suivant:

Lemme 2. (H. Cartan) *Soit T un espace topologique. Les trois conditions suivantes sont équivalentes:*

a) *les composantes connexes de T sont ouvertes (et fermées);*

b) *pour tout point* $t \in T$, *il existe un voisinage de t contenu dans tout ensemble ouvert et fermé contenant t*;

c) *pour tout point* $t \in T$, *l'intersection des ensembles ouverts et fermés contenant t est ouverte (et fermée).*

Démonstration du Lemme: a) => b): il suffit de prendre pour voisinage de t la composante connexe de t.

b) => c): si s appartient à l'intersection des ouverts et fermés contenant t, ces derniers sont exactement les ouverts et fermés contenant s et leur intersection est un voisinage de s.

c) => a): l'intersection des ouverts et fermés contenant t est un ensemble ouvert et fermé *connexe* donc est la composante connexe de t.

Ceci étant, comme l'espace $\Gamma(Y)$ satisfait à la condition b) du Lemme 2 (corollaire à la proposition 6), les composantes connexes de $\Gamma(Y)$ sont ouvertes et fermées, donc sont de la forme $\Gamma(Y_i)$ (corollaire 1 à la proposition 4) et les Y_i sont irréductibles (corollaire 2 à la proposition 4). Supposons que Y_i soit contenu dans Y_j et soit $\alpha \in \Gamma(Y_i)$: il existe un $\beta \in \Gamma(Y_j)$ tel que $\alpha \subset \beta$. Mais comme α et β appartiennent à $\Gamma(Y)$, ceci entraine $\alpha = \beta$, donc $\Gamma(Y_i) \cap \Gamma(Y_j) \neq \varnothing$ et par suite $\Gamma(Y_i) = \Gamma(Y_j)$, $Y_i = Y_j$, donc $i = j$. Par suite $Y_i \not\subset Y_j$ pour $i \neq j$.

Pour tout compact K, $\Gamma(Y) \cap \pi^{-1}(K)$ est compact, donc ne rencontre qu'un nombre fini de composantes connexes (ouvertes) de $\Gamma(Y)$, ce qui montre que la famille (Y_i) est localement finie, donc admet une réunion Z. Comme pour tout point $x \in A$, les composantes irréductibles des germes en x des différents Y_i ne sont jamais contenus les unes dans les autres, on a $\Gamma(Z) = \bigcup \Gamma(Y_i) = \Gamma(Y)$, et par suite $Y = Z$, ce qui achève la démonstration.

7. Sous-ensembles C-analytiques

Soit Ω une variété analytique-réelle paracompacte et soit Ω^* une complexification de Ω, que nous supposerons munie d'une involution $z \to \bar{z}$ de lieu de points fixes Ω (n° 1, Remarque). Nous avons vu que les espaces $\mathfrak{G}(\Omega)$ et $G(\Omega)$ associés à Ω plongée dans Ω^* sont en réalité indépendants du choix de Ω^*. Par ailleurs, l'involution $z \to \bar{z}$ de Ω^* donne une involution $\alpha \to \bar{\alpha}$ dans $G(\Omega)$ et une involution $Y \to \bar{Y}$ dans $\mathfrak{G}(\Omega)$. On a $\Gamma(\bar{Y}) = \overline{\Gamma(Y)}$, $\pi(\bar{\alpha}) = \overline{\pi(\alpha)}$ et ces involutions respectent les opérations de réunion et d'intersection.

Définition. *Un sous-ensemble E de Ω est dit C-analytique s'il existe un élément* $Y \in \mathfrak{G}(\Omega)$ *tel que* $E = \pi(\Gamma(Y))$ *(autrement dit E est l'ensemble des points x de Ω tels que le germe de Y au point x ne soit pas vide).*

On peut encore dire que E est C-analytique si et seulement si il existe un voisinage V^* de Ω dans Ω^* et un sous-ensemble analytique-complexe E^* dans V^* tels que $E = E^* \cap \Omega$; on peut supposer (en remplaçant E^* par $E^* \cap \bar{E}^*$) que l'on a $\bar{E}^* = E^*$. Par suite, *un sous-ensemble C-analytique est analytique-réel*, la réciproque étant inexacte[5]).

Les sous-ensembles C-analytiques ont été introduits, dans le cas $\Omega = R^n$, par H. Cartan, qui a démontré la proposition suivante ([6], proposition 15), dont nous redonnerons la démonstration dans le cas général:

Proposition 10. *Soit E un sous-ensemble de Ω. Les trois conditions suivantes sont équivalentes:*

a) *E est C-analytique;*

b) *E est le lieu des zéros d'un faisceau cohérent d'idéaux $\mathfrak{F}$;*

c) *E est le lieu des zéros communs à un nombre fini (qu'on peut prendre égal à 1) de fonctions analytiques-réelles sur Ω.*

a) => b): soit E^* un sous-ensemble analytique-complexe dans un voisinage V^* de Ω, tel que $\bar{E}^* = E^*$ et que $E = E^* \cap \Omega$ et soit, pour $x \in V^*$, $I^*_x(E^*)$ l'idéal des germes de fonctions holomorphes et nulles sur E^* au voisinage de x. On sait que les $I^*_x(E^*)$ forment un faisceau cohérent d'idéaux (voir [4] ou [5]). Pour $x \in \Omega$, soit $I_x(E^*)$ l'idéal des germes de fonctions analytiques-réelles en x, qui se prolongent en des fonctions analytiques-complexes appartenant à $I^*_x(E^*)$: il est immédiat que les $I_x(E^*)$ forment un faisceau cohérent d'idéaux $\mathfrak{J}(E^*)$ qui admet E comme lieu de zéros.

b) => a): On peut en effet ([6], proposition 2 et 5) prolonger le faisceau $\mathfrak{F}$ en un faisceau cohérent (analytique-complexe) $\mathfrak{F}^*$ dans un voisinage V^* de Ω. Le lieu des zéros de $\mathfrak{F}^*$ est un sous-ensemble analytique-complexe E^* tel que $E = E^* \cap \Omega$.

a) => c): On peut supposer d'après [7] que Ω est une sous-variété plongée dans un espace R^N, et que Ω^* est une sous-variété analytique-complexe dans un voisinage B de R^N dans C^N. On peut de plus choisir pour B un domaine d'holomorphie ([6], proposition 1) et supposer qu'il existe un ensemble analytique-complexe E^* dans B avec $\bar{E}^* = E^*$ et $E = E^* \cap \Omega$. Or tout ensemble analytique-complexe dans une variété de Stein, donc dans un domaine d'holomorphie, est définissable par un nombre fini d'équations globales d'où immédiatement c).

c) => a): Si E est défini par un système fini d'équations analytiques $f_i = 0$, les f_i se prolongent simultanément en des fonctions analytiques-complexes sur

[5]) Par exemple, les sous-ensembles analytiques-réels définis aux exemples 1, 2 et 3 de [3] ainsi que les ensembles F_k définis au nº 11, ex. a), ne sont pas C-analytiques.

un voisinage V^* de Ω et le lieu de leurs zéros communs dans V^* est un sous-ensemble analytique-complexe E^* tel que $E = E^* \cap \Omega$.

Soit E un sous-ensemble C-analytique de Ω. D'après le Corollaire 2 à la proposition 7, il existe un *plus petit* germe $Y \in \mathfrak{G}(\Omega)$ tel que $E = \pi(\Gamma(Y))$: ce germe sera appelé *complexifié* de E et noté $\hat{E}$. Comme $\pi(\Gamma(\overline{\hat{E}})) = \pi(\Gamma(\hat{E}))$, on a $\hat{\overline{\hat{E}}} = \hat{E}$. Soit E^* un sous-ensemble analytique-complexe dans un voisinage V^* de Ω, tel que le couple (V^*, E^*) définisse le germe $\hat{E}$. Le faisceau $\mathfrak{J}(E^*)$ sur Ω ne dépend que de $\hat{E}$ et non du choix de E^*: nous le noterons $\mathfrak{J}(\hat{E})$. Soit alors $\mathfrak{F}$ un faisceau cohérent d'idéaux sur Ω, dont le lieu des zéros contienne E: nous dirons que $\mathfrak{F}$ est «nul sur E». Le faisceau $\mathfrak{F}$ se prolonge en un faisceau cohérent $\mathfrak{F}^*$ sur un voisinage de Ω et $\mathfrak{F}^*$ est nul sur E^* au voisinage de Ω. Donc, $\mathfrak{F}^*$ est contenu dans $\mathfrak{J}^*(E^*)$ au voisinage de Ω, et par suite $\mathfrak{F} \subset \mathfrak{J}(\hat{E})$. On voit donc que $\mathfrak{J}(\hat{E})$ est le *plus grand faisceau cohérent* d'idéaux nul sur E. Remarquons qu'en général $\mathfrak{J}(\hat{E})$ n'est pas le faisceau des germes de fonctions analytiques-réelles nulles sur E: il faut et il suffit pour cela que ce dernier faisceau soit cohérent, c'est-à-dire que E soit cohérent au sens de [6], et il existe des ensembles C-analytiques et même algébriques qui ne sont pas cohérents (voir [6], n° 9).

8. Propriétés des ensembles C-analytiques

L'*intersection* E d'une famille quelconque d'ensembles C-analytiques E_i est encore C-analytique. En effet, soit $Y = \bigcap \hat{E}_i$; pour tout point $x \in \Omega$, le germe Y_x de Y en x est l'intersection des germes $(\hat{E}_i)_x$ des $\hat{E}_i$ en x (Corollaire 1 à la proposition 7). Or, dire que $x \in E$ signifie qu'aucun des $(\hat{E}_i)_x$ n'est vide, ce qui équivaut à Y_x non vide ou encore à $x \in \pi(\Gamma(Y))$; on a donc $E = \pi(\Gamma(Y))$ et E est C-analytique. Remarquons que $\hat{E}$ est contenu dans l'intersection des $\hat{E}_i$ mais peut être distinct de cette intersection (voir n° 11, ex. b).

De même, la *réunion* E d'une famille *localement finie* d'ensembles C-analytiques E_i est C-analytique et l'on a:

$$(\bigcup E_i)^\wedge = \bigcup \hat{E}_i . \tag{27}$$

En effet, la famille $(\hat{E}_i)$ est localement finie donc possède une réunion Y. Dire que $x \in \pi(\Gamma(Y))$ signifie que le germe de Y en x n'est pas vide, ou encore que le germe en x de l'un au moins des $\hat{E}_i$ n'est pas vide (proposition 8), c'est-à-dire que $x \in E$. Donc E est C-analytique et $\hat{E} \subset Y$. Mais pour tout indice i, on a $E_i \subset E$, donc $\hat{E}_i \subset \hat{E}$ et par suite $\hat{E} \supset Y$, d'où $\hat{E} = Y$.

Un ensemble C-analytique sera dit *C-irréductible* s'il n'est pas réunion de deux ensembles C-analytiques distincts de lui-même: ceci *n'entraine pas* qu'il soit irréductible en tant qu'ensemble analytique-réel (voir n° 11, ex. a)).

Proposition 11. *Soit E un ensemble C-analytique. Il existe une famille localement finie et une seule d'ensembles C-analytiques C-irréductibles E_i (appelés composantes C-irréductibles de E) telle que E soit la réunion des E_i et que E_i ne soit pas contenu dans E_j pour $i \neq j$. Les $\hat{E}_i$ sont les composantes irréductibles de $\hat{E}$, et E est C-irréductible si et seulement si $\hat{E}$ est irréductible.*

Remarquons tout d'abord que $\hat{E}$ irréductible entraine E C-irréductible, car $E = E_1 \cup E_2$ entraine $\hat{E} = \hat{E}_1 \cup \hat{E}_2$ d'après (27). D'autre part, soient Y_i les composantes irréductibles de $\hat{E}$, et posons $E_i = \pi(\Gamma(Y_i))$. On a $E = \bigcup E_i$. Le complexifié $\hat{E}_i$ de E_i est Y_i: en effet, $\hat{E}_i \subset Y_i$ et par suite le germe $Z = \hat{E}_i \cup (\bigcup_{j \neq i} Y_j)$ est contenu dans $\hat{E}$. Or on a $\pi(\Gamma(Z)) = E$, d'où $Z = \hat{E}$. Comme les $\Gamma(Y_j)$ sont disjoints et que $\Gamma(Z) \supset \Gamma(Y_i)$, on a $\Gamma(\hat{E}_i) \supset \Gamma(Y_i)$ et par suite $\hat{E}_i = Y_i$. Les E_i sont donc C-irréductibles et, comme $Y_i \not\subset Y_j$ pour $i \neq j$, on a $E_i \not\subset E_j$ pour $i \neq j$, d'où l'existence de la décomposition en composantes C-irréductibles. L'unicité est évidente. Enfin, si E est C-irréductible, on a nécessairement $E = E_i$ pour un indice i, d'où $\hat{E} = \hat{E}_i$ et $\hat{E}$ est irréductible.

Proposition 12. *Soit E un ensemble C-analytique. La dimension de E (en tant qu'ensemble analytique-réel) est égale à la dimension du germe complexifié $\hat{E}$.*

On peut supposer que E est C-irréductible. Posons $p = \dim \hat{E}$ et soit (V^*, E^*) un couple définissant $\hat{E}$, E^* étant irréductible de dimension p (Corollaire 2 à la proposition 8). Comme la dimension de E est, par définition, la borne supérieure des dimensions complexes des germes complexifiés des germes de E aux différents points de E, on a $\dim E \leqslant p$. D'autre part, il y a des points de E qui sont des points réguliers de E^*, si non E serait contenu dans le sous-ensemble analytique-complexe des points singuliers de E^* et $\hat{E}$ ne serait pas le plus petit germe contenant E. Or au voisinage d'un tel point, E est une sous-variété de dimension exactement p ([1], p. 121 ou [6], p. 92).

Corollaire. *Soit E un sous-ensemble C-analytique C-irréductible de dimension*

p et soit F un sous-ensemble C-analytique contenu dans E. Ou bien $F = E$, ou bien $\dim F < p$.

Compte tenu de la proposition 12, c'est une traduction de la proposition 5.

Proposition 13. *Soit E un ensemble C-analytique de dimension p et soit $V_p(E)$ l'ensemble des points de E au voisinage desquels E est une sous-variété analytique-réelle de dimension p. Alors, E est réunion de $V_p(E)$ et d'un sous-ensemble C-analytique S de dimension strictement inférieure à p.*

Soit en effet (V^*, E^*) un couple définissant $\hat{E}$, E^* étant de dimension complexe p. Soit $V_p^*(E^*)$ l'ensemble des points de E^* au voisinage desquels E^* est une sous-variété analytique-complexe de dimension p et soit $S^* = E^* - V_p^*(E^*)$. On sait [5] que S^* est un sous-ensemble analytique-complexe de dimension $< p$. Comme $V_p^*(E^*) \cap \Omega \subset V_p(E)$, on a $E = V_p(E) \cup S$ avec $S = S^* \cap \Omega$. Remarquons que l'on n'a pas nécessairement $V_p(E) \cap S = \varnothing$, bien que $V_p^*(E^*) \cap S^* = \varnothing$ (voir n° 11, ex. c)).

Remarque 1. Les énoncés analogues au Corollaire à la proposition 12 et à la proposition 13 sont inexacts pour les sous-ensembles analytiques-réels généraux: pour des contre-exemples, voir [3].

Remarque 2. Nous donnerons au n° 10 un autre procédé de décomposition de E en réunion de variétés.

9. Notion de rang pour les ensembles analytiques-complexes

Soient Ω^* une variété analytique-complexe de dimension n et E^* un sous-ensemble analytique-complexe de Ω^*. Nous désignerons comme au n° 7 par $I_x^*(E^*)$ l'idéal des germes de fonctions holomorphes dans un voisinage de $x \in \Omega^*$ et nulles sur E^* et par $\mathfrak{J}^*(E^*)$ le faisceau cohérent des $I_x^*(E^*)$. Le *rang* $rg_x(E^*)$ au point $x \in E^*$ est par définition le rang de l'idéal $I_x^*(E^*)$, c'est-à-dire le nombre maximum ϱ de fonctions $f \in I_x^*(E^*)$ dont les différentielles df soient indépendantes en x (cf. [9]). En un point x régulier de dimension r de E^*, c'est-à-dire un point au voisinage duquel E^* est une variété de dimension r, on a $rg_x(E^*) = n - r$. Un point régulier de E^* de dimension la plus petite (respectivement la plus grande) possible sera appelé un point régulier minimal (respectivement maximal). Si E^* est de dimension constante, les points réguliers sont à la fois minimaux et maximaux.

Proposition 14. *Si E^* est de rang constant ϱ dans un voisinage d'un point $x \in E^*$, alors x est un point régulier de dimension $n - \varrho$ de E^*.*

Soient en effet $f_1, \ldots, f_\varrho$ des éléments de $I_x^*(E^*)$ dont les différentielles soient indépendantes en x. Les équations $f_1 = \ldots = f_\varrho = 0$ définissent alors dans un voisinage de x une sous-variété analytique-complexe de dimension $n - \varrho$, soit M^*. Le germe de M^* en x est irréductible de dimension $n - \varrho$ et contient le germe de E^* en x. Il est clair qu'il suffit de démontrer que ces deux germes sont égaux. Or s'il n'en était pas ainsi, le germe de E^* en x serait de dimension p strictement inférieure à $n - \varrho$. Or, il y a des points de E^* arbitrairement voisins de x qui sont des points réguliers de dimension p de E^*. En un tel point, le rang de E^* serait $n - p > \varrho$, contrairement à l'hypothèse.

Proposition 15. *Soit M_1^* l'ensemble des points de E^* où le rang de E^* est égal à son maximum ϱ et soit $E_1^* = E^* - M_1^*$. Alors, M_1^* est une variété analytique-complexe qui est l'ensemble des points réguliers minimaux de E^* et E_1^* est un sous-ensemble analytique-complexe.*

Par définition même du rang, M_1^* est ouvert dans E^* et c'est une variété d'après la proposition 14; il est alors clair que M_1^* est l'ensemble des points réguliers minimaux.

Si E^* est irréductible, ou plus généralement de dimension constante (nécessairement égale à $n - \varrho$), alors E_1^* est l'ensemble des points singuliers de E^* donc est analytique. Dans le cas général, soit F^* (respectivement G^*) la réunion des composantes irréductibles de dimension $n - \varrho$ (respectivement de dimension $> n - \varrho$) de E^*: on a $E^* = F^* \cup G^*$. Soit N_1^* (respectivement F_1^*) l'ensemble des points réguliers (respectivement singuliers) de F^*. On a $M_1^* = N_1^* - (N_1^* \cap G^*)$ et par suite $E_1^* = F_1^* \cup G^*$ est analytique.

Corollaire. *Si E^* est irréductible, ou plus généralement de dimension constante, alors le rang de E^* est maximum aux points réguliers et strictement plus petit aux points singuliers.*

Remarque. Pour tout $x \in E_1^*$, choisissons un voisinage U_x^* de x, un système de coordonnées $(x_1, \ldots, x_n)$ dans U_x^*, et des fonctions $g_1, \ldots, g_k$ analytiques dans U_x^* et engendrant $I_y^*(E^*)$ en tout point $y \in U_x^*$. Soit S_x l'ensemble des fonctions g_j et des fonctions $\partial(g_{\nu_1}, \ldots, g_{\nu_\varrho})/\partial(x_{\lambda_1}, \ldots, x_{\lambda_\varrho})$ pour tous les choix possibles des deux systèmes $(\nu_1, \ldots, \nu_\varrho)$ et $(\lambda_1, \ldots, \lambda_\varrho)$ de ϱ indices. Il est facile de montrer d'une part que $E_1^* \cap U^*$ est exactement le lieu des zéros communs aux fonctions de S_x, ce qui prouve à nouveau que E_1^* est analytique, d'autre part que les idéaux engendrés par S_x et S_y en un point $z \in U_x^* \cap U_y^*$ sont les mêmes et forment donc un faisceau cohérent sur Ω^*.

Comme dans [9], § 9, nous avons donc deux méthodes pour décomposer E^*, suivant le rang ou suivant la dimension:

a) Ecrivons $E^* = M_1^* \cup E_1^*$ comme dans la proposition 15, puis $E_1^* = M_2^* \cup E_2^*$, etc. On exprime ainsi E^* comme union de variétés analytiques $M_1^*, \ldots, M_k^*$ et on a $k \leqslant 2^n - 1$ (cf. [9]).

b) Ecrivons $E^* = M_1'^* \cup E_1'^*$, $M_1'^*$ étant l'ensemble des points réguliers maximaux de E^*, puis de même $E_1'^* = M_2'^* \cup E_2'^*$, etc. Alors, les $M_j'^*$ sont des variétés analytiques de dimension décroissante.

10. Notion de rang pour les sous-ensembles C-analytiques

Reprenons les notations du n° 7. Soit E un sous-ensemble C-analytique de Ω, de complexifié $\hat{E}$, et considérons le plus grand faisceau cohérent d'idéaux nul sur E, soit $\mathfrak{J}(\hat{E})$. Par définition le *C-rang* $rg_x^*(E)$ de E au point $x \in E$ est le rang de l'idéal $I_x(\hat{E})$ du faisceau $\mathfrak{J}(\hat{E})$ en x. Si (V^*, E^*) est un couple définissant $\hat{E}$ et si x est un point de E, l'idéal $I_x^*(E^*)$ est engendré par $I_x(\hat{E})$ et par suite, on a $rg_x(E^*) = rg_x^*(E)$.

Remarquons qu'en général, le C-rang de E au point x n'est pas déterminé par la partie de E située dans un voisinage de x (n° 11, ex. e)). D'autre part, si $\Omega = R^n$ et si E est une variété algébrique réelle (qui est évidemment un ensemble C-analytique), le C-rang et le rang défini algébriquement peuvent ne pas être égaux (n° 11, ex. d)).

On peut de manière analogue, définir la *C-dimension* de E en un point x de E comme étant la dimension du germe de $\hat{E}$ en x, ou encore la dimension (complexe) de E^* en x. Cette C-dimension peut être plus grande que la dimension de E en x en tant qu'ensemble analytique-réel (n° 11, ex. e)), mais les bornes supérieures de ces dimensions quand x décrit E sont égales, d'après la proposition 12.

Proposition 16. *Soit M_1 l'ensemble des points de E où le C-rang de E est égal à son maximum ϱ et soit $E_1 = E - M_1$. Alors M_1 est une variété analytique-réelle de dimension $n - \varrho$ et E_1 est C-analytique. De plus, si (V^*, E^*) est un couple définissant $\hat{E}$, E^* étant le plus petit ensemble analytique-complexe dans V^* qui définisse $\hat{E}$, et si $E^* = M_1^* \cup E_1^*$ est la décomposition de E^* introduite à la proposition* 15, *alors on a $M_1 = M^* \cap \Omega$ et $E_1 = E^* \cap \Omega$.*

Cette proposition est une conséquence immédiate de la proposition 15 et de l'égalité $rg_x(E^*) = rg_x^*(E)$ entre rang de E^* et C-rang de E en un même point $x \in E$. En effet, E^* n'a pas de points de rang $> \varrho$, car si non, on aurait $E \subset E_1^*$, contrairement à l'hypothèse de minimalité de E^*. Par suite, on a $M_1 = M_1^* \cap \Omega$ et $E_1 = E_1^* \cap \Omega$, ce qui entraine d'une part que E_1 est C-analytique, d'autre part que M_1 est une variété analytique-réelle de dimension $n - \varrho$ (cf. proposition 12).

Nous obtenons donc, comme dans le cas complexe, deux procédés de décomposition en réunion de variétés, par le C-rang ou par la C-dimension. Ces décompositions s'obtiennent exactement comme dans [9], § 11.

11. Quelques exemples

a) Le sous-ensemble C-analytique E de l'espace R^3 défini par l'équation $x^2 + 2(1 + \sin z)xy + y^2 = 0$ est C-irréductible, mais est réductible en tant qu'ensemble analytique-réel: il est en effet réunion des ensembles analytiques-réels irréductibles F_k, réunion de la droite $x = y = 0$ et du lieu des points de E pour lesquels $2k\pi \leqslant z \leqslant (2k + 1)\pi$ (avec k entier). On notera que la famille (F_k) n'est pas localement finie: le *résidu* de E au sens de [3] est la droite $x = y = 0$.

b) Considérons dans R^3 les deux sous-ensembles C-analytiques E défini par l'équation $x^2 + y^2 + z^2 - 2x = 0$ et F défini par $x = 0$. L'intersection $E \cap F$ est réduite à l'origine et par suite $(E \cap F)^\wedge$ est distinct de $\hat{E} \cap \hat{F}$.

c) (H. Cartan): Soit E le sous-ensemble C-analytique de R^3 défini par l'équation $yz(x^2 - zy^2) + x^4 = 0$. On vérifie aisément que $E - V_2(E)$ se compose de la droite $x = z = 0$ et de la demi-droite $x = y = 0, z \geqslant 0$, donc n'est pas analytique. D'autre part, on montre facilement que le germe de E à l'origine O est irréductible et que O est adhérent à chacune des composantes connexes de $V_2(E)$, ce qui entraine que E est C-irréductible. Enfin, le C-rang de E est 0 sur les droites $x = y = 0$ et $x = z = 0$ et est 1 aux autres points de E, ce qui montre qu'il y a des points de $V_2(E)$ où le C-rang de E est 0.

d) Le sous-ensemble C-analytique E de R^2 défini par l'équation $f = x^2 + y^2 - y^3 = 0$ se compose de l'origine O et d'une courbe S, ne passant pas par O. Bien que E soit algébriquement irréductible, il n'est pas C-irréductible, car S est C-analytique et peut être défini par une équation de la forme $g = y - \psi(x) = 0$, où ψ est analytique. Le rang algébrique de E à l'origine est 0, tandis que le C-rang est 2. Le germe $\hat{E}$ est défini par les deux équations $xg = yg = 0$ et non pas par l'équation $f = 0$.

e) (cf. [3], ex. 2): le sous-ensemble C-analytique E de R^3 défini par l'équation $f = (x^2 + y^2)z - y^3 = 0$ se compose d'une surface S passant par l'origine O et de la droite Z d'équations $x = y = 0$. Il est C-irréductible: comme $Z - \{O\}$ et $S - \{O\}$ sont des variétés connexes, il suffit de montrer que le germe de E à l'origine est irréductible, ou encore que son complexifié l'est, ou encore que f est irréductible dans l'anneau des fonctions holomorphes à l'origine, ce qui est immédiat, f étant un polynome homogène irréductible.

On remarquera que, pour un point $q \in Z - \{O\}$, les ensembles E et Z sont les mêmes dans un voisinage de q, mais $\hat{E} \neq \hat{Z}$ dans tout voisinage de q. L'idéal des fonctions nulles sur E au voisinage de q est engendré par x et y, tandis que l'idéal $I_q(\hat{E})$ est engendré par f.

La décomposition de E (par le rang ou par la dimension) est $E = (S - \{O\}) \cup Z$. Mais l'origine O est singulière par rapport à S et une décomposition plus complète serait $E = (S - \{O\}) \cup (Z - \{O\}) \cup \{O\}$.

f) Appliquons la proposition 16 au sous-ensemble E de R^3 défini par l'équation $(x^2 + y^2)^2 - z^3 = 0$; on voit que $\hat{E}_1$ est strictement plus petit que le germe défini par E_1^*: cf. [9], § 12, (f).

g) Soit E le sous-ensemble C-analytique de R^2 défini par l'équation $y^3 = x^{2k+1}$ avec k entier $\geqslant 2$. Alors E est une courbe différentiable, mais non analytique au voisinage de l'origine O. Il est de rang 0 à l'origine et de rang 1 ailleurs.

Note ajoutée à la correction des épreuves: l'existence de complexifications d'une variété analytique-réelle (prop. 1) vient d'être indépendemment démontrée par H. B. SHUTRICK, *Complex extension*, Quart. J. of Mech. and appl. Math. Series 2, t. 9, (1958), 189–201 et par A. HAEFLIGER, Comment. Math. Helv., 32 (1958), 248–329. Pour le cas compact, voir aussi C. B. MORREY, *The analytic embedding of abstract real-analytic manifolds*, Ann. of Math., *68*, (1958), 159–201.

BIBLIOGRAPHIE

[1] F. BRUHAT, *Sur les représentations induites des groupes de* LIE, Bull. Soc. Math. France, 84 (1956), 97–205; voir § 3, n° 1.

[2] F. BRUHAT et H. CARTAN, *Sur la structure des sous-ensembles analytiques-réels*, C. R. Acad. Sci. Paris, 244 (1957), 988–990.

[3] F. BRUHAT et H. CARTAN, *Sur les composantes irréductibles d'un sous-ensemble analytique-réel*, C. R. Acad. Sci. Paris, 244 (1957), 1123–1126.

[4] H. CARTAN, *Idéaux et modules de fonctions de variables complexes*, Bull. Soc. Math. France, 78 (1950), 29–64.

[5] H. CARTAN, *Séminaires*, E. N. S. 1951/52 et 1953/54.

[6] H. CARTAN, *Variétés analytiques-réelles et variétés analytiques complexes*, Bull. Soc. Math. France, 85 (1957), 77–100.

[7] H. GRAUERT, *On* LEVI's *problem and the imbedding of real analytic mannifolds*, Ann. of Math. 68 (1958), 460–472.

[8] R. REMMERT et K. STEIN, *Über die wesentlichen Singularitäten analytischer Mengen*, Math. Ann., 126 (1953), 263–306.

[8bis] R. REMMERT, *Projektionen analytischer Mengen*, Math. Ann., 130 (1956), 410–441.

[9] H. WHITNEY, *Elementary structure of real algebraic varieties*, Ann. of Math., 66 (1957), 545 to 556.

(Reçu le 6 août 1958)

Local Properties of Analytic Varieties

HASSLER WHITNEY

1. Introduction

Algebraic and analytic varieties have become increasingly important in recent years, both in the complex and the real case. Their local structure has been intensively investigated, by algebraic and by analytic means. Local geometric properties are less well understood. Our principal purpose here is to study properties of tangent vectors and tangent planes in the neighborhood of singular points. We study stratifications of an analytic variety into analytic manifolds; in particular, we may require that a transversal to a stratum is also tranversal to the higher dimensional strata near a given point. A conjecture on possible fiberings of the variety (and of surrounding space) is stated; it is proved at points of strata of codimension 2 in the surrounding space. In the last sections, we see that a variety may have numbers or functions intrinsically attached at points or along strata; also an analytic variety may be locally unlike an algebraic variety.

The original purpose of the present study was to help solidify the theory of singularities of differentiable mappings; this field may be considered as a branch of differential topology. In this theory, a type of singularity is defined by means of a certain algebraic variety lying in a subsidiary space. A mapping has the given singularity at the point p if the corresponding mapping into the subsidiary space maps p into the variety; the mapping is generic at p if the corresponding mapping is transverse to the strata of the variety. It is here that the transversality mentioned above comes into play.

The varieties mentioned above are real algebraic; however, the properties discussed are of an analytic nature, so they might as well be studied in the more general analytic case. The splitting of a real variety into manifolds was studied in an elementary way in [14]; the splitting is in general not a stratification as we define it here; moreover, it has not certain further properties required. In order to carry the theory deeper, it was found most useful to go over to complex varieties, where powerful tools are available. This paper is concerned almost wholly with the complex case; to apply the results to a real variety, one may use the complexification of this variety (see [16]).

In §2 we assemble some properties of analytic varieties which are needed later. At a simple point of a variety there is a natural definition of tangent vector. At a singular point, several different definitions are of importance; we study them in §3. The definitions are given in terms of the injection of the variety in C^n; yet they are intrinsic to the abstract variety. We give the proof of this (essentially known result) in §4. In the algebraic case, it is classical (although apparently only recently proved; see [11]) that the tangent cone is defined by the vanishing of the initial polynomials of the ideal of the variety at the given point. The theorem, in the analytic case, is given in §5; the proof will be completed in a later paper [15]. The connection with the algebraic theory is explained in §6; in particular, we give a purely intrinsic characterization of the tangent cone, which seems not to be in the literature.

To study transversality questions, we must consider not "tangent vectors," but "tangent planes" at a singular point of the variety. In §7 we define such a plane to be the limiting position of a sequence of tangent planes at simple points. We define a stratification of V in §8 to be an expression of V as a union of analytic manifolds, the frontier of each stratum being a union of lower dimensional strata. There is a natural way of defining a primary stratification; this may be used as a starting point for improved stratifications. In particular, we may require that as we approach one stratum M from a higher dimensional stratum M', any limiting position of tangent planes to M' contains the tangent plane to M at the given point. The transversality property mentioned earlier then follows. Moreover, we may require that if we approach a point p of M along a given direction in M', a limiting position of tangent planes to M' contains this direction. Such limiting positions were considered in §7. These properties, in the real case, relate to the paper [13] of R. Thom in this volume. Some of the proofs are deferred to [15].

Let M be a stratum in a stratification of V. It is natural to hope that a neighborhood of any point $p \in M$ in V (in fact, a neighborhood in the containing space) is expressible analytically as a product of an open subset of M with a cross section through p. H. Rossi [10] has shown that this is possible at p if and only if there are holomorphic vector fields tangent to V at all simple points, near p, and forming a basis at p for the tangent space to M at p (see Lemma 9.1). It is not possible in general, however, as we shall see in §13. Instead, we look for a "semi-analytic" fibration of V near p. We conjecture (§9) that every analytic variety has a stratification such that there is a semi-analytic fibration near each point. For a related result, we solve an interpolation problem for a set of analytic functions over a simply connected region of the complex plane, these functions taking on distinct values at each point.

In the Weierstrass Preparation Theorem, there is a hypothesis that the given function does not vanish identically along a certain axis. We elucidate this in §10, showing that the degree of the equivalent pseudopolynomial is a minimum if that axis does not lie in the tangent cone of the corresponding variety, and is larger than the minimum otherwise. Let M be an analytic manifold in V (for instance, a stratum of a stratification), where V is a hypersurface. We show that except in a certain analytic subvariety of M, the tangent cone of V is a continuous function in M. If M is an $(n-2)$-dimensional stratum of the hypersurface $V \subset C^n$, from which we have cut out the above subvariety, then near any $p \in M$ we may express the branches of $V - M$ by holomorphic functions $x_n = \phi(x_1, \ldots, x_{n-1})$. Using this, we show in §12 that a hypersurface may be stratified so that the fibering conjecture holds throughout the $(n-2)$-dimensional strata. An interpolation procedure, described in §11, is used; it gives a semi-analytic fibration, whereas an analytic procedure would not have the required continuity properties.

In §13 we define some varieties for which there is no stratification with analytic fibrations at all points; the varieties are "non-homogeneous" along certain strata. The same type of construction gives (§14) analytic varieties that are not locally algebraic along certain strata. This shows that a certain theorem of N. Levinson [7] to the effect that a fairly general holomorphic function of two variables is, in a new coordinate system, a polynomial, cannot be extended to more variables. In constructing the examples, a basic property of cross ratios is used. In the final section we show that the same properties may be obtained by another method, using fewer sheets than before.

2. Preliminaries

2.1. Let C denote the set of complex numbers, and C^n, complex n-space. There is a natural coordinate system $(x_1, \ldots, x_n)$ in C^n. A (more general) "coordinate system" in the open set $U \subset C^n$ is a one–one holomorphic transformation of an open set $U' \subset C^n$ onto U, with nonvanishing Jacobian (hence the transformation is biholomorphic). If $(x, y, \ldots)$ is a coordinate system in U, then C_x denotes the x-axis, C_{xy} the (x, y)-plane, etc.

2.2. For a point set $Q \subset C^n$, clos $Q = \bar{Q}$ is its closure, and fron $Q = \bar{Q} - Q$ is its frontier. A set $\{Q_1, Q_2, \ldots\}$ of sets is *locally finite* if any compact set has points in but a finite number of the Q_i.

2.3. We use the hermitian metric in C^n. Then for $v = (v_1, \ldots, v_n) \in C^n$,

$|v| = \Sigma_i v_i \bar{v}_i$ ($\bar{v}_i$ = conjugate imaginary of v_i). $|q - p|$ is the distance from p to q. v is orthogonal to w if $\Sigma_i v_i \bar{w}_i = 0$.

2.4. A point set V in the open set $U \subset C^n$ is an *analytic variety* in U provided that V is closed in U, and for each $p \in V$ there is a neighborhood U' of p in C^n and a set $f_1, \ldots, f_s$ of holomorphic functions in U' such that $V \cap U'$ is the set of common zeros of the f_i in U'. Basic properties of analytic varieties may be found in [3], [9] and [6]. For a study of real C-analytic varieties in R^n (R = reals), see [4] and [16].

2.5. The point $p \in V \subset U \subset C^n$ is a *simple* (or *regular* or *ordinary*) point of V if there is a neighborhood U' of p and there are holomorphic functions $f_1, \ldots, f_m$ in U' such that the differentials $df_1(p), \ldots, df_m(p)$ are independent, and $V \cap U'$ is the set of common zeros of the f_i. We may then choose a new coordinate system in a neighborhood U'' of p such that $V \cap U''$ is an open subset of the $(x_1, \ldots, x_{n-m})$-plane. The *dimension* $\dim_p V$ of V at p is then $n - m$. The *tangent space* $T(V, p)$ of V at the simple point p is the set of vectors v expressible as $\lim a_i(q_i - p)$, where $q_i \in V$, $q_i \to p$, and the a_i are complex numbers. With $f_1, \ldots, f_m$ as above, $T(V, p)$ is the set of solutions of $df_i(p) \cdot v = 0$. A point of V which is not simple is *singular*.

2.6. An *analytic manifold* is an analytic variety all of whose points are simple; we require normally that its points all be of the same dimension. (Sometimes it is required to be connected.)

2.7. The *dimension* $\dim V$ of V is the highest dimension of V at simple points. The *dimension at* p of V, $\dim_p V$, is the dimension of $V \cap U'$ for small neighborhoods U' of p. V is of *constant dimension* if $\dim_p V$ is constant; it is of *constant dimension near* p if $V \cap U'$ is of constant dimension for some neighborhood U' of p. V is a *hypersurface* in C^n if it is of constant dimension $n - 1$. An irreducible variety is of constant dimension [9 Satz 5]. If $\dim_p V = d$ then there is an analytic manifold M of dimension $n - d$ containing p such that for some neighborhood U' of p, $V \cap M \cap U' = \{p\}$, and there is no such manifold of higher dimension (see [9]).

2.8. Any analytic variety may be expressed uniquely as the union of a set of irreducible analytic varieties, none of these being contained in the union of the others. This set of sets is locally finite. A proper subvariety of an irreducible variety (in particular, of a connected analytic manifold) is of lower dimension. Given a locally finite set of analytic varieties in an open set, their union is an analytic variety.

2.9. Given $p \in V \subset U \subset C^n$, let $I(V, p)$ denote the ring of germs of holomorphic functions in C^n near p which vanish on V. (In general, we need not distinguish between a function in some neighborhood of p and its germ at p; in §6 we take more care on this point.) The sheaf of all

$I(V, p)$ is *coherent* (see [3, Exposé XVI]); that is, for each $p \in U$ there is a neighborhood U' of p and a set $f_1, \ldots, f_s$ of holomorphic functions in U' which generate $I(V, q)$ for each $q \in U'$. This means that if $g_q \in I(V, q)$, g_q being the germ at q of the holomorphic function g, then there is a neighborhood U'' of q and there are holomorphic functions $\phi_1, \ldots, \phi_s$ in U'' such that $g = \Sigma_i \phi_i g_i$ in U''. (If $q \in U - V$, then $I(V, q)$ contains all holomorphic germs at q.) If V is a hypersurface, $I(V, p)$ is generated by a single function.

2.10. Suppose $V \subset U \subset C^n$. A function f defined in an open subset U^* of V is *holomorphic* in U^* if for each $p \in U^*$ there is a neighborhood U' of p in C^n and a holomorphic function F in U' such that its restriction $F|(V \cap U')$ to $V \cap U'$ equals f there. Let $O(V, p)$ denote the ring of germs of holomorphic functions in V at p. There is a natural isomorphism $O(V, p) \approx O(C^n, p)/I(V, p)$.

2.11. Suppose the space V is given, together with the set of functions on open subsets of V which are "holomorphic"; each $O(V, p)$ must come from an imbedding of a neighborhood of p in V into some C^n. Then we have an *abstract analytic variety*, or, an *analytic space*. See [12] or [6] for details.

2.12. Let V and W be analytic spaces, and let Φ be a mapping of V into W. Then Φ is *holomorphic* if, for each holomorphic function f in some open subset U of W, $\Phi^* f$ is holomorphic in $\Phi^{-1}(U)$. (By definition, $(\Phi^* f)(p) = f(\Phi(p))$.) Suppose Φ is one–one; then Φ is *biholomorphic* provided that both Φ and Φ^{-1} are holomorphic. In particular, if $V \subset U \subset C^n$, then the injection $V \to U$ is biholomorphic.

LEMMA 2.13. *Let Φ be a homeomorphism of V into C^n. Then:*

(a) *For Φ to be holomorphic, it is sufficient that each $\Phi^* x_i$ be holomorphic.*

(b) *For $\psi = \Phi^{-1}$ to be holomorphic, it is sufficient that for each holomorphic function g in an open set U in V, $\psi^* g$ be, near each point of $\Phi(V)$, the restriction to $\Phi(V)$ of a holomorphic function in an open set in C^n.*

(c) *For* (b) *to hold, it is sufficient that the condition hold for holomorphic functions $f_1, \ldots, f_s$, such that each holomorphic function g in V is expressible as $\phi(f_1, \ldots, f_s)$, where ϕ is holomorphic.*

Part (a) above follows from the fact that each holomorphic function in C^n is expressible holomorphically in terms of $x_1, \ldots, x_n$; similarly, (c) follows from (b). Part (b) above follows from the definitions of holomorphic functions in V and of holomorphic mappings.

2.14. We give two examples. (1) Let $V = C_x \cup C_y \cup C_z \subset C^3$, and let W be given by $yx(y - x) = 0$ in C^2. Let Φ be the identity in $C_x \cup C_y$ and let Φ map C_z linearly into $y - x = 0$. Then Φ is holomorphic but not biholomorphic; the function z in V does not go into the restriction to W of a holomorphic function in C^2.

(2) Map $V = C^1 = C_t$ into C^2 by $x = t^2$, $y = t^3$; this is a homeomorphism. By Lemma 2.13, (a), the mapping is holomorphic. It is not biholomorphic, since the function t in the image is not the restriction to the image of V of a holomorphic function (near 0).

LEMMA 2.15. *Let V and W be analytic varieties in $U \subset C^n$. Then $V^* = \operatorname{clos}(V - W)$ and $V^{**} = V^* \cap W$ are analytic varieties, and $\dim_p V^{**} < \dim_p V^*$ for $p \in V^{**}$. Moreover, V^* is the union of the irreducible components of V which have points outside W.*

Express V in terms of its irreducible components:

$$V = (V_1 \cup V_2 \cup \cdots) \cup (X_1 \cup X_2 \cup \cdots), \quad X_i \subset W, \quad V_i \not\subset W;$$

each union is locally finite. Now $V^* = V_1 \cup V_2 \cup \cdots$. For if $p \in V^*$, then $p \in \operatorname{clos}(V_1 \cup V_2 \cup \cdots)$, and hence $p \in \bar{V}_i = V_i$ for some i. Conversely, suppose $p \in V_i$ for some i. If $p \in V_i - W$, then $p \in V^*$. Suppose $p \in V_i \cap W = V_i'$. Since V_i is irreducible, $\dim_p V_i' < \dim_p V_i = \dim V_i$ (see 2.8); hence each neighborhood of p contains points of $V_i - W \subset V^*$, and $p \in V^*$. The rest of the lemma now follows.

LEMMA 2.16. *Let $M_1, M_2, \ldots$ be the connected components of the set of simple points of the analytic variety V. Then the closures of the M_i are the irreducible components of V.*

Each M_i has an open subset contained in some irreducible component V_i of V; by 2.8, $M_i \subset V_i$ and hence $\bar{M}_i \subset V_i$. By [9, Satz 10], V_i contains no point of M_j for any other j. Let W_i be the union of the V_j with $j \neq i$; this is an analytic variety. By the last lemma, $\bar{M}_i = \operatorname{clos}(V_i - W_i)$ is analytic; hence $\bar{M}_i = V_i$.

COROLLARY 2.17. The set of simple points of V is dense in V.

LEMMA 2.18. (a) *The set of singular points of an analytic variety (in C^n) is analytic or is void.* (b) *Let M_0 be the set of simple points of highest dimension of V, and set $V' = V - M_0$; then M_0 is an analytic manifold, and V' is an analytic variety of lower dimension or is void.*

First suppose V is irreducible. By the last lemma, the set M of simple points of V is a connected analytic manifold; it is of dimension $r = \dim V$. Take any $p \in V'$ (if V' is non-void). Let $f_1, \ldots, f_s$ generate $I(V, q)$ for each q in some neighborhood U of p. Let $D_1, \ldots, D_m$ be the set of all determinants of order $n - r$ from the matrix of partial derivatives of the f_i; let W be the analytic variety in U defined by the vanishing of the f_i and the D_j. For $q \in M \cap U$, some set of $n - r$ members of the f_i have independent differentials at q, hence, one of the $D_j(q)$ is $\neq 0$, and $q \notin W$. For $q \in V' \cap U$, all the $D_j(q)$ are 0, and $q \in W$. To show this, suppose some $D_j(q)$ is $\neq 0$. Then the corresponding f_i define a connected analytic manifold M' of dimension r in a neighborhood U' of q and containing q, and $V \cap U' \subset M'$. Since $q \in \bar{M}$, part of $\bar{M}$ lies in M'; hence (see 2.8) $V \cap U' = M'$. Thus q is a simple point of V, a contradiction. We have now proved that $W = V' \cap U$, which shows that V' is analytic. Moreover, $\dim V' < r$, by Lemma 2.15.

In the general case, the lemma follows easily on expressing V in terms of its irreducible components (see Lemma 2.16).

LEMMA 2.19. *Let $W \subset U \subset C^n$ be an analytic variety of dimension $<r$, and let $V \subset U - W$ be an analytic variety with $\dim_p V \geqq r$ for all $p \in V$. Then $\bar{V} \cap U$ is an analytic variety in U.*

For the proof, see [9, Satz 13] or [2, p. 679].

2.20. Let the analytic manifolds M, M' in an open subset of C^n intersect at p (and perhaps at further points). We say M' is *transversal* to M at p if $T(M, p)$ and $T(M', p)$ together span C^n. In particular, we must have $k = \dim M + \dim M' \geqq n$. Then $M \cap M'$, near p, is an analytic manifold of dimension $k - n$.

2.21. A *cone* $Q \subset C^n$ with *vertex* p is a subset of C^n such that for any $q \in Q$ and any complex number a, $p + a(q - p) \in Q$. We call $v = q - p$ a *vector* of Q; the condition is then that av also be a vector of Q. (In the real case, one might require $a > 0$.)

3. Possible definitions of tangent cones

At a singular point of the analytic variety $V \subset C^n$ there are a number of possible definitions of tangency of vectors, several of which are useful for different purposes. Corresponding to each definition we obtain a set of vectors, which in each case is a cone (see 2.21). In later sections, "the tangent cone" will always refer to the cone $C(V, p) = C_3(V, p)$. Each cone is easily seen to be identical with the tangent space $T(V, p)$ at any simple point p.

$C_1(V, p)$: $v \in C_1(V, p)$ if and only if the following is true. There is a neighborhood U of p in C^n and there is a holomorphic vector field $v(q)$ in U such that $v(p) = v$, and $v(q) \in T(V, q)$ at all simple points q of V in U. This criterion appears on Rossi, [10, §3]; compare Lemma 9.1 below.

$C_2(V, p)$: For each sequence $\{q_i\} \to p$ of simple points there is a sequence $\{v_i\} \to v$ of vectors, with $v_i \in T(V, q_i)$. A basic requirement of good stratifications of a variety is expressible in terms of this definition; see §8, (a′). The definition is related to that of C_1; see the proof that $C_2 \subset C_3$ below.

$C_3(V, p) = C(V, p)$: There is a sequence $\{q_i\} \to p$ of points of V and a sequence $\{a_i\}$ of complex numbers such that $a_i(q_i - p) \to v$. In other words, v is in the direction of a limit of secants from p. This is the commonly designated "tangent cone." See also Remark 5.11.

$C_4(V, p)$: There is a sequence $\{q_i\} \to p$ of simple points and a sequence $\{v_i\} \to v$ with $v_i \in T(V, q_i)$.

$C_5(V, p)$: There are sequences $\{q_i\} \to p$ and $\{q_i'\} \to p$ in V and a sequence $\{a_i\}$ such that $a_i(q_i' - q_i) \to v$. This definition seems to be of lesser interest.

$C_6(V, p)$: For every germ $f \in I(V, p)$ we have $df(p) \cdot v = 0$. This is sometimes called the "tangent space" of V at p.

The cones C_1, C_2 and C_6 are vector spaces; the others are not, in general.

If V is a cone with vertex at the origin 0, then $C_3(V, 0) = V$; this is false in general for the other cones.

In the real domain, one might choose somewhat different definitions; for instance, for the curve $y^2 - x^3 = 0$ in C^2, one might wish the cone at 0 to contain the non-negative points of the x-axis only.

We give some examples. Let the varieties $V_1, \ldots, V_4$ be hypersurfaces, defined by the following functions:

In C^2: $\quad f_1 = xy, \quad f_2 = y^2 - x^3;$

In C^3: $\quad f_3 = z^2 - xy, \quad f_4 = y^2 - zx^2.$

The corresponding cones at the origin (except for C_5) are as shown in the table. (Recall that C_x denotes the x-axis, etc.)

	V_1	V_2	V_3	V_4
C_1	$\{0\}$	$\{0\}$	$\{0\}$	$\{0\}$
C_2	$\{0\}$	C_x	$\{0\}$	C_x
C_3	V_1	C_x	V_3	C_{xz}
C_4	V_1	C_x	C^3	C^3
C_6	C^2	C^2	C^3	C^3

For another example concerning C_1, see Remark 13.3.

In common cases, the vector space spanned by C_5 will be C_6; however, this is not always so. Note that if $V \subset C^3$ is the union of the three axes, then $C_5 = C_{xy} \cup C_{xz} \cup C_{yz}$, $C_6 = C^3$. On the other hand, let $W \subset C^3$ be the union of three lines in C_{xy}: $x = 0$, $x - y = 0$, $x + y = 0$, and the curve $y = z - x^2 = 0$; then $C_5 = C_{xy}$ while $C_6 = C^3$.

We now show that the above cones form an increasing sequence:

$$C_1 \subset C_2 \subset C_3 \subset C_4 \subset C_5 \subset C_6. \tag{3.1}$$

$C_1 \subset C_2$: Trivial, since holomorphic functions are continuous.

$C_2 \subset C_3$: Let $v \in C_2$ be given; we may suppose $|v| = 1$. Given $\varepsilon > 0$, choose a neighborhood U of p such that for each simple point $q \in V \cap U$ there is a vector $v' \in T(V, q)$ with $|v' - v| < \varepsilon$. Set $\phi(q) =$ projection of v into $T(V, q)$; then ϕ is holomorphic in the set M of simple points of V in U, and $|\phi(q) - v| < \varepsilon$ in M.

Say U contains the η-neighborhood of p. Choose a point q of M within $\eta\varepsilon$ of p. Considering M as a real manifold and ϕ as a real vector field in M, construct the curve Γ through q so as to have $\phi(q')$ as tangent vector at each point q'. Suppose we can follow Γ till we reach a point q_1 distant η from q. Set $a = 1/\eta$. Now

$$|a(q_1 - p) - v| \leqq |a(q_1 - q) - v| + a|q - p| < 2\varepsilon,$$

and since $q_1 \in V$, this proves $v \in C_3(V, p)$. If Γ runs into a singular point q' of V, then we may start from a point on the opposite side of q' (note that if $C_3(V, q')$ contains a vector u, it contains $-u$ also, by Remark 5.11 below), and start a new arc Γ'; we may in fact let Γ' start from q'. Continue till we find q_1 distant η from q. We shall not go into details here.

$C_3 \subset C_4$: Let V' be the variety of singular points of V. Take any

$$v \in C_3(V, p) = C(V, p).$$

By (3.6) there is an irreducible component V_1 of V with $v \in C(V_1, p)$. Let V_1' be the union of the set of singular points of V_1 and the intersection of V_1 with the other components of V. By Lemmas 5.13 and 2.18,

$$\dim C(V_1', p) = \dim_p V_1' < \dim_p V_1 = \dim C(V_1, p),$$

and $C(V_1, p)$ is of constant dimension. Hence arbitrarily near v we may find a vector v' in $C(V_1, p)$ which is not in $C(V_1', p)$. By Lemma 3.8 there is a 1-dimensional variety $W \subset V_1$ containing p, such that $v' \in C(W, p)$. It is elementary that W contains a sequence $\{q_i\} \to p$ and a sequence $\{v_i\} \to v'$, v_i being tangent to W at q_i. Near p, W contains only simple points of V (except for p); hence $v_i \in T(V, q_i)$ (i large), proving that $v' \in C_4(V, p)$ and hence $v \in C_4(V, p)$.

$C_4 \subset C_5$: Take $v \subset C_4(V, p)$ and corresponding sequences $\{q_i\}$, $\{v_i\}$. We may take q_i'' in V very near q_i and take a_i so that $a_i(q_i'' - q_i)$ is near v_i. Now $a_i(q_i'' - q_i) \to v$, proving $v \subset C_5(V, p)$.

$C_5 \subset C_6$: This is trivial.

LEMMA 3.2. *Suppose $p \in W \subset V$. Then*

$$C_i(W, p) \subset C_i(V, p) \qquad i = 3, 4, 5, 6. \tag{3.3}$$

This is evident for $i = 3, 5, 6$. We shall not go into the proof for $i = 4$. (See [15].) As a consequence, if $p \in V \cap W$,

$$C_i(V \cap W, p) \subset C_i(V, p) \cap C_i(W, p), \qquad i = 3, 4, 5, 6. \tag{3.4}$$

The lemma fails for $i = 1, 2$, as is shown by taking $V = V_1$ as above, $W = C_x$, $p = 0$. For an example with V irreducible, take $V = V_3$ as above, $W = C_x$, $p = 0$. For the variety V of Example 13.2, taking $W = C_t$, (3.3) fails with $i = 1$ at all points of W.

We cannot expect equality in (3.4); this is shown by taking V and W in C^2, defined by $y = 0$ and by $y - x^2 = 0$ respectively.

LEMMA 3.5. *Suppose* $p \in V \cap W$. *Then*

$$C_i(V \cup W, p) = C_i(V, p) \cup C_i(W, p), \qquad i = 3, 4. \tag{3.6}$$

This is clear. For $i = 5, 6$, $\supset$ holds, by Lemma 3.2; however, $=$ does not, as is seen by letting V and W be the irreducible components of the variety $C_x \cup C_y \subset C^2$.

In contrast with this, it is easily seen that

$$C_i(V \cap W, p) \subset C_i(W, p) \qquad (i = 1, 2) \quad \text{if } \dim_p (W \cap V) < \dim_p (W). \tag{3.7}$$

Given a tangent vector v at a simple point V, there is of course an analytic arc through p with v as tangent at p. At singular points, the following is true:

LEMMA 3.8. *Given* $p \in V \subset U \subset C^n$ *and* $v \in C(V, p)$, *there is a 1-dimensional variety* W *with* $p \in W \subset V$ *such that* $v \in C(W, p)$.

The proof is not difficult. It will be given in [15].

PROBLEM 3.9. Given the analytic variety V and $p \in V$, show that there is a neighborhood U of p with the following property. For any $q \in V \cap U$ there is a connected 1-dimensional variety $W \subset V \cap U$ containing p and q, which has no singular points other than possibly at p.

This problem is probably of great difficulty. If we ask only for a real analytic arc from p to q, the problem is much simpler; see [16], Proposition 2, and the reference to Bruhat-Cartan given there.

LEMMA 3.10. *Given* $p \in V \subset C^n$ *and* $\varepsilon > 0$, *there is a neighborhood* U *of* p *such that for each simple point* $q \neq p$ *of* $V \cap U$ *there is a vector* $v \in T(V, q)$ *pointing nearly away from* p:

$$\left| \frac{v}{|v|} - \frac{q - p}{|q - p|} \right| < \varepsilon.$$

This is equally true in the real case. For the proof, see [15]. Note that the size of U cannot be taken uniformly as a function of p; this is shown by several of the varieties defined above.

4. Intrinsic character of tangent cones

Let V be an abstract analytic variety, with the ring $O(V, p)$ of germs of holomorphic functions at p. We give an intrinsic characterization of the tangent cone $C_6(V, p)$ in Lemma 4.1, of importance in studying the effect on cones of holomorphic mappings. (For C_3, see the next section.)

Suppose we have two biholomorphic imbeddings of an abstract variety V_0 into open subsets of complex spaces C^n, C'^m; then there is a biholomorphic homeomorphism ϕ of one imbedded variety onto the other, and this may be extended to a holomorphic mapping Φ of some U^n into C'^m. Now Φ is necessarily biholomorphic over some analytic manifold M containing $V \cap U^n$, and $d\Phi$ carries each tangent cone in C^n onto that in C'^m (Theorem 4.7). Most of the results of this section may be found in Rossi [10, §2].

Recall that a *derivation* δ on a ring O of germs of functions about p is a linear function on O such that

$$\delta(fg) = g(p)\delta(f) + f(p)\delta(g).$$

Let $\Delta(V, p)$ denote the linear space of derivations of $O(V, p)$.

LEMMA 4.1. *Suppose $p \in V \subset U \subset C^n$. Then there is a natural isomorphism θ of $C_6(V, p)$ onto $\Delta(V, p)$, defined as follows: Given $v \in C_6(V, p)$ and $f \in O(V, p)$, choose a holomorphic extension F of f through a neighborhood of p in C^n, and set*

$$[\theta(v)](f) = dF(p) \cdot v. \tag{4.2}$$

First, θ is well defined. For if $F_1|V = F_2|V$, set $G = F_2 - F_1$; then $G \in I(V, p)$, $dG(p) \cdot v = 0$ since $v \in C_6(V, p)$, $dF_1(p) \cdot v = dF_2(p) \cdot v$. We see at once that $\theta(v) \in \Delta(V, p)$.

Clearly θ is linear. To show that it is one–one, suppose $\theta(v) = 0$. Then $dF(p) \cdot v = 0$ for all holomorphic F, and hence $v = 0$.

To show that θ is onto $\Delta(V, p)$, let δ be any derivation in $O(V, p)$. Taking p as origin 0 of coordinates, set

$$\bar{x}_i = x_i|V, \qquad v_i = \delta(\bar{x}_i), \qquad v = (v_i, \ldots, v_n).$$

Take any $f \in O(V, p)$; say $F|V = f$. Expand F at 0:

$$F(x_1, \ldots, x_n) = a + \Sigma b_i x_i + \Sigma R_i(x_1, \ldots, x_n)x_i, \quad R_i(0) = 0. \tag{4.3}$$

Then

$$\delta(f) = \Sigma b_i\delta(\bar{x}_i) = \Sigma v_i\partial F(0)/\partial x_i = dF(0) \cdot v,$$

and hence $\theta(v) = \delta$, if $v \in C_6(V, p)$. This holds; for taking any $F \in I(V, p)$ gives (by the same reasoning) $f = 0$, $dF(p) \cdot v = \delta(0) = 0$.

REMARK 4.4. For $V = C^n$, we have $C^n \approx \Delta(C^n, p)$.

Let ϕ be a holomorphic mapping of V into V'_r, with $\phi(p) = p'$. Then there is a naturally defined linear function

$$\phi_\Delta\colon \Delta(V, p) \to \Delta(V', p'), \tag{4.5}$$

defined as follows: Given $\delta \in \Delta(V, p)$ and $f' \in O(V', p')$, set

$$(\phi_\Delta \delta)(f') = \delta(\phi^* f'). \tag{4.6}$$

ϕ_Δ behaves in the natural manner with regard to mappings.

Part of the next theorem requires the analytic character of the $C_i(V, p)$; see Theorem 5.6 below.

THEOREM 4.7. *Suppose* $p \in V \subset U \subset C^n$, $p' \in V' \subset U' \subset C'^{n'}$, *and* ϕ *is a holomorphic mapping of* V *into* V' *with* $\phi(p) = p'$. *Then there is a holomorphic mapping* Φ *of a neighborhood* U_1 *of* p *in* C^n *into* $C'^{n'}$ *such that* $\Phi|(V \cap U_1) = \phi|(V \cap U_1)$. *For any such* Φ, *we have (letting* θ, θ' *be as in* (4.2))

$$\text{in } C_6(V, p), \qquad d\Phi(p) = \theta'^{-1} \circ \phi_\Delta \circ \theta, \tag{4.8}$$

$$d\Phi(p) \text{ maps } C_i(V, p) \text{ holomorphically into } C_i(V', p') \qquad (i = 3, 4, 5, 6). \tag{4.9}$$

Suppose, furthermore, that ϕ *is a biholomorphic homeomorphism onto* V'. *Then*

$$d\Phi(p)|C_i(V, p) \text{ is one–one biholomorphic onto } C_i(V', p') \qquad (i = 1, \ldots, 6). \tag{4.10}$$

Also, if $n \leqq n'$, *we may make* Φ *a biholomorphism in* U_1. *If* $n > n'$, *we may choose an analytic manifold* $M^{n'} \subset C^n$ *with* $V \cap U_1 \subset M^{n'}$, *and make* Φ *a biholomorphism in* $M^{n'}$.

First, since each coordinate ϕ_i $(i = 1, \ldots, n')$ of ϕ is holomorphic in V, it has a holomorphic extension Φ_i; these functions give Φ.

To prove (4.8), take any $v \in C_6(V, p)$. Say

$$v \xrightarrow{\theta} \delta \xrightarrow{\phi_\Delta} \delta' \xleftarrow{\theta'} v'.$$

Now take any holomorphic function F' in a neighborhood of p', and set

$$F = \Phi^* F', \qquad f' = F'|V', \qquad f = F|V.$$

Then, using (4.6) and (4.2),

$$\begin{aligned} dF'(p') \cdot v' &= \delta'(f') = \delta(f) = dF(p) \cdot v \\ &= d(F' \circ \Phi)(p) \cdot v = dF'(p') \cdot [d\Phi(p) \cdot v]; \end{aligned}$$

hence $v' = d\Phi(p) \cdot v$, and (4.8) follows.

(4.9) for $i = 6$ is an immediate consequence of (4.8). Since $d\Phi(p)$ is linear in $C_6(V, p)$, it is holomorphic in each $C_i(V, p)$. Suppose $v \in C_5(V, p)$. Choose sequences $\{q_j\}$, $\{r_j\} \to p$ in V, and $\{a_j\}$, $a_j(r_j - q_j) \to v$. Given $\varepsilon > 0$,

$$|\Phi(r_j) - \Phi(q_j) - d\Phi(p) \cdot (r_j - q_j)| \leqq \varepsilon |r_j - q_j|$$

for j large enough; hence

$$d\Phi(p) \cdot v = \lim a_j[\Phi(r_j) - \Phi(q_j)] \in C_5(V', p').$$

Replacing q_j by p in the above proof gives (4.9) with $i = 3$.

Now take $v \in C_4(V, p)$; say $\{q_j\} \to p$, q_j simple, $v_j \in T(V, q_j)$, $v_j \to v$. Then (4.9) with $i = 3$ gives

$$v'_j = d\Phi(q_j) \cdot v_j \in C_3(V', q'_j) \subset C_4(V', q'_j);$$

hence there is a simple point q''_j and a vector v''_j such that

$$v''_j \in T(V', q''_j), \quad |q''_j - q'_j| < 1/j, \qquad |v''_j - v'_j| < 1/j.$$

Hence

$$d\Phi(p) \cdot v = \lim v''_j \in C_4(V', p').$$

If ϕ is a biholomorphism, then clearly ϕ_Δ is an isomorphism onto, and (4.10) with $i = 6$ follows from (4.8). By (3.1), $d\Phi(p)$ is one–one in each $C_i(V, p)$; by symmetry, (4.10) for each i will follow from (4.9) for the same i. Thus we now need only prove (4.9) for $i = 1, 2$. For $i = 1$, note that if $v(q)$ is holomorphic in U_1, then $v'(\Phi(q)) = d\Phi(q) \cdot v(q)$ is holomorphic in $\Phi(U_1)$. The proof for $i = 2$ is simple also.

Choose an analytic manifold M as in Lemma 4.12 below. Because of (4.10) and (4.13), Φ is a biholomorphism from $M \cap U_1$ (for U_1 small enough) onto $\Phi(M \cap U_1)$. Set $m = \min(n, n')$. Leaving Φ fixed in $M \cap U_1$, we may clearly alter it in a chosen $M^m \supset M$ so it is biholomorphism there. This completes the proof.

REMARK 4.11. Taking $V = C_x$, $V' = C_x \cup C_y$ in C^2, ϕ = identity in V, shows that (4.9) may fail for $i = 1, 2$.

LEMMA 4.12. *Suppose $p \in V \subset U \subset C^n$. Then there is an analytic manifold M containing the part of V near p, such that*

$$C_6(V, p) = T(M, p). \tag{4.13}$$

For any analytic manifold M' containing V near p, $C_6(V, p) \subset T(M', p)$.

Let $F_1, \ldots, F_s$ be a maximal set of elements of $I(V, p)$ with independent differentials at p; then for some U, $F_1 = \cdots = F_s = 0$ defines an analytic manifold M in U, and $V \cap U \subset M$. If (4.13) is false, there is a vector v in $T(M, p)$ which is not in $C_6(V, p)$. We may then choose $F \in I(V, p)$ with $dF(p) \cdot v \neq 0$. But this shows that $dF(p)$ is independent of the $dF_i(p)$, a contradiction. Thus (4.13) holds. The last part of the lemma is clear.

COROLLARY 4.14. Given $p \in V \subset U < C^n$, there is a coordinate system with origin at p (analytically related to the original coordinates) such that the part of V near p is in the $(x_1, \ldots, x_d)$-plane, where $d = \dim C_6(V, p)$.

5. Algebraic character of tangent cones

For any V and p, the cones C_1, C_2, and C_6 are vector spaces. We shall show that the remaining cones are algebraic varieties; moreover, for $C = C_3$, we give the representation in terms of initial polynomials of functions in $I(V, p)$. First we show that for $i = 4, 5, 6$, the set of cones $C_i(V, p)$ for all $p \in V$ forms an analytic variety.

THEOREM 5.1. *Let V be an analytic variety in the open set $U \subset C^n$. Set*

$$C_i(V) = \{(p, v) \in U \times C^n : p \in V,\ v \in C_i(V, p)\}, \qquad i = 1, \ldots, 6. \tag{5.2}$$

Then for $i = 4, 5, 6$, $C_i(V)$ is an analytic variety in $U \times C^n$. Moreover, $C_4(V)$ is the smallest analytic variety containing all (p, v) with p simple, $v \in T(V, p)$.

First take the case $i = 6$ (Rossi, [10, Lemma 4.1]). Take any $p \in U$; by coherence, there are functions $f_1, \ldots, f_s$, holomorphic in a neighborhood U_1 of p, which generate $I(V, q)$ for each $q \in U_1$. Set

$$A = \{(q, v) : q \in U_1,\ df_i(q) \cdot v = 0\ (i = 1, \ldots, s)\};$$

then A is analytic in $U_1 \times C^n$. So is $(V \cap U_1) \times C^n$. Also

$$C_6(V) \cap (U_1 \times C^n) = [(V \cap U_1) \times C^n] \cap A,$$

showing that $C_6(V)$ is analytic.

For the case $i = 4$, we need merely note that, if V' is the set of singular points of V, then

$$C_4(V) = \operatorname{clos}[C_6(V) - V' \times C^n] \cap (U \times C^n); \tag{5.3}$$

now apply Lemma 2.15. The last statement of the theorem is clear.

Now consider the case $i = 5$. Using the coordinates in C^n, define the holomorphic functions in $C^n \times C^n \times C^n$

$$\alpha_{jk}(p, q, v) = \begin{vmatrix} q_j - p_j & q_k - p_k \\ v_j & v_k \end{vmatrix}; \tag{5.4}$$

these all vanish if and only if $q = p$ or v is a multiple of $q - p$. Set

$$B = \{(p, q, v) : p, q \in U;\ \alpha_{jk}(p, q, v) = 0 \quad (\text{all } j, k)\}.$$

This is an analytic variety, and hence so is

$$B' = B \cap (V \times V \times C^n).$$

Set $D = \{(p, p) : p \in V\}$ and

$$B'' = \operatorname{clos}[B' - D \times C^n] \cap (U \times U \times C^n)$$

this also is analytic, as is

$$C_5' = B'' \cap (D \times C^n).$$

Now clearly

$$(p, v) \in C_5(V) \cap (U \times C^n) \text{ if and only if } (p, p, v) \in C_5',$$

showing that $C_5(V)$ is analytic. (If C_5' is defined near (p_0, p_0, v_0) by functions f_i, then the functions $f_i'(p, v) = f_i(p, p, v)$ define $C_5(V)$ near (p_0, v_0).)

REMARK 5.5. The sets $C_i(V)$ $(i = 1, 2, 3)$ are not closed, and hence are not analytic, in general. (See for instance, Example 7.8.)

THEOREM 5.6. *Given $p \in V \subset U \subset C^n$, each cone $C_i(V, p)$ is an algebraic variety.*

By a theorem of Chow (see for instance [9, Satz 14]), it is sufficient to show that the cone is analytic. This is trivial for $i = 1, 2, 6$, since the cone is then a vector space. For $i = 4, 5$, it follows from the last theorem. We must still consider the case $i = 3$. Set

$$E_p = \{(q, v) \colon q \in U,\ \alpha_{jk}(p, q, v) = 0 \text{ (all } j, k)\};$$

this is analytic in $U \times C^n$. So is $E_p' = E_p \cap (V \times C^n)$, and so is

$$C_p' = \operatorname{clos}[E_p' - \{p\} \times C^n] \cap (U \times C^n).$$

Noting that $v \in C(V, p)$ if and only if $(p, v) \in C_p'$, the result follows. (It follows also from Theorem 5.8 below.)

We shall now find polynomials defining $C(V, p)$. Suppose f is a holomorphic function in an open set U containing $p = (a_1, \ldots, a_n)$. Expanding about p and collecting terms of the same degree gives

$$f = f^{[m]} + f^{[m+1]} + \cdots, \qquad f^{[m]} \not\equiv 0, \tag{5.7}$$

where each $f^{[k]}$ is a homogeneous polynomial of degree k in the $x_i - a_i$. We then say that f is of *order* m at p, and call $f^{[m]}$ the *initial polynomial* of f at p. The number m is characterized as the order to which f first fails to vanish at p, and $f^{[m]}$ is the function defining $C(V_f, p)$, where V_f is defined by f (see Lemma 10.6).

THEOREM 5.8. *Given $p \in V \subset U \subset C^n$, $C(V, p)$ is defined by the vanishing of all initial polynomials at p of functions in $I(V, p)$.*

One part of the theorem is easy to prove: Taking p as origin of coordinates, let f be any function in $I(V, p)$; we shall show that for each $v \in C(V, p)$, $f^{[m]}(v) = 0$ if $\operatorname{ord} f = m$. For any complex number λ, we have

$$f(\lambda v) = \lambda^m[f^{[m]}(v) + \lambda R(v, \lambda)], \tag{5.9}$$

where R is holomorphic. Take $\{q_i\} \to 0$ in V and $\{a_i\}$ so that $a_i q_i \to v$. Set $\lambda_i = 1/a_i$ (we may suppose $v \neq 0$). Now

$$0 = f(q_i) = \lambda_i^m [f^{[m]}(a_i q_i) + \lambda_i R(a_i q_i, \lambda_i)].$$

Since $a_i q_i \to v$, $\lambda_i \to 0$ and R is bounded, we must have $f^{[m]}(v) = 0$.

The other half of the theorem is more difficult to prove; see [15].

REMARK 5.10. In general, $I(C(V, p), p)$ is not generated by the initial parts of a given set of generators of $I(V, p)$. For an example, let $V^1 \subset C^3$ be defined by the vanishing of

$$f = xy - y^3, \qquad g = xz - z^3;$$

then f and g generate $I(V, 0)$, but $f^{[2]}$ and $g^{[2]}$ do not generate $I(C(V, 0), 0)$. The latter ideal is generated by $f^{[2]} = xy$, $g^{[2]} = xz$, and

$$(zf - yg)^{[4]} = yz(z^2 - y^2).$$

REMARK 5.11. In the definition of $C = C_3$, we could require that the a_i be real and positive; for Theorem 5.8 may be proved with this definition, and it then follows that $av \in C$ if and only if $v \in C$ (a complex).

How near is V to $C(V, p)$ near p? We give an answer in terms of the Hausdorff distance between two sets: Let $S_\rho(p) = \{q: |q - p| = \rho\}$ be sphere of radius ρ about p. Let $U_\rho(Q)$ denote the ρ-neighborhood of the point set Q, i.e., the set of points, each within ρ of some point of Q. Set

$$\text{dist}\,(P, Q) = \inf\{\rho: P \subset U_\rho(Q) \text{ and } Q \subset U_\rho(P)\}.$$

Then

$$\rho^{-1}\,\text{dist}\,[V \cap S_\rho(p), C(V, p) \cap S_\rho(p)] \to 0 \text{ as } \rho \to 0. \tag{5.12}$$

To show this, call the two sets V_ρ, C_ρ. That $V_\rho \subset V_{\rho\varepsilon}(C_\rho)$ for small ρ follows at once from the definition of $C(V, p)$. With the help of Lemma 3.8 it is easy to see that $C_\rho \subset U_{\rho\varepsilon}(V_\rho)$ for small ρ. (Use a finite set of vectors in $(C(V, p)$.)

LEMMA 5.13. *For $p \in V \subset U \subset C^n$,*

$$\dim C(V, p) = \dim_p V.$$

If V is of constant dimension near p, so is $C(V, p)$.

The truth of the equality follows from the proof concerning $C(V, p)$ in Theorem 5.6. The last statement will be proved in [15]; it makes use of (5.12).

REMARK 5.14. $C(V, p)$ may be reducible even if V is irreducible, as is shown by the variety $z^3 - xy = 0$, at the origin.

6. Intrinsic study of $C(V, p)$

We wish to give an intrinsic characterization of the tangent cone of V at p as a set of derivations on $O(V, p)$. We need some definitions.

Let $\mathfrak{m}$ be the maximal ideal of $O(V, p)$; it consists of those elements of $O(V, p)$ which vanish at p. The kth power $\mathfrak{m}^k$ of $\mathfrak{m}$ consists of those elements expressible as a product of k elements of $\mathfrak{m}$, together with finite sums of such elements.

Example 6.1. Let $y^2 - x^3$ define V in C^2, and set $\bar{x} = x|V$, $\bar{y} = y|V$. Then the element $\bar{y}^2$ of $\mathfrak{m}^2$ equals $\bar{x}^3$ and hence lies also in $\mathfrak{m}^3$.

Let $C_\Delta(V, p)$ denote the set of those derivations δ of $O(V, p)$ with the following property: given any s, k, and elements $u_{ij} (i = 1, \ldots, s; j = 1, \ldots, k)$ of $\mathfrak{m}$,

$$\text{(6.2)} \qquad \text{if } \sum_i u_{i1} \cdots u_{ik} \in \mathfrak{m}^{k+1} \text{ then } \sum_i \delta(u_{i1}) \cdots \delta(u_{ik}) = 0.$$

THEOREM 6.3. *Suppose $p \in V \subset U \subset C^n$. Then the isomorphism θ of Lemma* 4.1, *restricted to $C(V, p)$, is one-one onto $C_\Delta(V, p)$.*

First, take any $\delta \in C_\Delta(V, p)$. Say $\theta(v) = \delta$; then $v_i = \delta(\bar{x}_i)$. Take any $f = f^{[\mu]} + f' \in I(V, p)$ (f' of order $>\mu$). Now $f^{[\mu]}|V = -f'|V$ is of order $>\mu$, and hence is in $\mathfrak{m}^{\mu+1}$; applying (6.2) with $k = \mu$ to the expression of $f^{[\mu]}$ as a polynomial, we find $f^{[\mu]}(v) = 0$. By Theorem 5.8, $v \in C(V, p)$.

Conversely, take any $v \in C(V, p)$ and corresponding δ. Suppose

$$\sum_{i=1}^{s} u_{i1} \cdots u_{i\mu} = \sum_{j=1}^{t} w_{j1} \cdots w_{j\mu'}, \quad \mu' = \mu + 1,$$

with $u_{ik}, w_{ik} \in \mathfrak{m}$. Expand each u_{ij}:

$$u_{ij} = \sum_h a_{ijh}\bar{x}_h + g_{ij}(\bar{x}_1, \ldots, \bar{x}_n), \text{ ord } g_{ij} > 1,$$

and expand each w_{ij} similarly. Now if

$$F(x_1, \ldots, x_n) = \sum_{i=1}^{s} \Big(\sum_{h=1}^{n} a_{i1h}x_h\Big) \cdots \Big(\sum_{h=1}^{n} a_{i\mu h}x_h\Big),$$

we have

$$\sum_i u_{i1} \cdots u_{i\mu} - \sum_j w_{j1} \cdots w_{j\mu'} = (F + G)|V, \quad \text{ord } G > \mu.$$

Since $F + G = 0$ in V and $v \in C(V, p)$, $F(v) = 0$. Using $v_i = \delta(\bar{x}_i)$ and $\delta(u_{ij}) = \Sigma_h a_{ijh}\delta(\bar{x}_h)$ gives $\Sigma_i \delta(u_{i1}) \cdots \delta(u_{i\mu}) = 0$, proving $\delta \in C_\Delta(V, p)$.

We next connect $C_\Delta(V, p)$ with the graded ring G of $\mathfrak{m}$; see Samuel [11, pp. 68–69] and Zariski-Samuel [17, pp. 248–250].

Set $\mathfrak{m}^0 = O(V, p)$,

$$L_i = \mathfrak{m}^i/\mathfrak{m}^{i+1}, \quad G = \sum_{i=0}^{\infty} L_i; \tag{6.4}$$

the elements of G are the finite sums $\alpha_0 + \cdots + \alpha_k$, $\alpha_i \in L_i$. The multiplication in $\mathfrak{m}$ defines a multiplication in G, making it into a graded ring.

LEMMA 6.5. *Let γ be a linear function on $L_1 = \mathfrak{m}/\mathfrak{m}^2$. Then there is an extension Γ of γ over G which is a ring homomorphism into C if and only if the following condition is satisfied: for any s, μ, and set $\{\alpha_{ij}\}(i = 1, \ldots, s; j = 1, \ldots, \mu)$ of elements of L_1,*

$$\text{if } \sum_i \alpha_{i1} \cdots \alpha_{i\mu} = 0, \text{ then } \sum_i \gamma(\alpha_{i1}) \cdots \gamma(\alpha_{i\mu}) = 0. \tag{6.6}$$

If Γ exists, it must be given as follows: in L_0, Γ is the natural isomorphism onto C. For any element β of any L_μ ($\mu > 0$), choose a representation $\beta = \sum_i \alpha_{i1} \cdots \alpha_{i\mu}$; set $\Gamma(\beta) = \sum_i \gamma(\alpha_{i1}) \cdots \gamma(\alpha_{i\mu})$. That the result is independent of the representation follows at once from (6.6). Clearly Γ has the required properties.

Any derivation δ on $O(V, p)$ is defined in $\mathfrak{m}$ and vanishes in $\mathfrak{m}^2$; hence it gives rise to a linear function $\bar{\delta}$ on L_1.

THEOREM 6.7. *A derivation δ on $O(V, p)$ is in $C_\Delta(V, p)$ if and only if the corresponding linear function $\bar{\delta}$ on L_1 can be extended to be a ring homomorphism of the graded ring G into C.*

This is an immediate consequence of the definition of C_Δ and of the last lemma.

We have been working in V directly. Let us now return to the situation $p \in V \subset U \subset C^n$. Choose coordinates as in Corollary 4.14; the part of V near p lies in $(x_1, \ldots, x_d)$-space, and $d = \dim C_6(V, p)$. Let J be the injection of V in C^n. For a function f in V, let $\pi_p f$ denote the germ of f at p. For a germ u lying in $\mathfrak{m}^\mu$, let $\rho_\mu u$ denote the corresponding element of $L_\mu = \mathfrak{m}^\mu/\mathfrak{m}^{\mu+1}$. Set

$$\bar{x}_i = x_i | V = J^* x_i, \quad \pi_p \bar{x}_i = u_i, \quad \rho_1 u_i = \alpha_i.$$

It is easily seen that $\alpha_1, \ldots, \alpha_d$ form a base in L_1; hence

(6.8) $C_6(V, p)$ or $\Delta(V, p)$ may be considered as the dual of L_1.

Consider the polynomial ring $C[x_1, \ldots, x_d]$. We shall define a homomorphism ϕ of this ring into the graded ring G as follows: for any homogeneous polynomial f of degree μ, set

$$\phi(f) = \rho_\mu \pi_p J^* f = f(\alpha_1, \ldots, \alpha_d) \in L_\mu. \tag{6.9}$$

For any polynomial $f = f^{[0]} + \cdots + f^{[k]}$, set $\phi(f) = \Sigma_i \phi(f^{[i]})$. Let K_ϕ denote the kernel of ϕ.

LEMMA 6.10. *With the above notations, the following are equivalent for homogeneous polynomials f:*

(a) $f \in K_\phi$.

(b) $f(\alpha_1, \ldots, \alpha_d) = 0$.

(c) *f is the initial polynomial of a function F in $I(V, p)$.*

This is easy to verify; it is practically implicit in results above.

As a consequence, we have the following characterization of $C(V, p)$:

THEOREM 6.11. *Suppose $p \in V \subset U \subset C^n$, and coordinates are chosen as above. Then with the above notations, $v \in C(V, p)$ if and only if for all homogeneous polynomials f in K_ϕ we have $f(v) = 0$.*

REMARK 6.12. One may give an intrinsic definition of an analytic manifold as follows: it is an analytic variety M such that for each $p \in M$, $O(M, p)$ is a regular ring, that is, is isomorphic with $C[x_1, \ldots, x_d]$ for some d. If we have $M \subset C^n$, then with coordinates as above, M near p is an open subset of $(x_1, \ldots, x_d)$-space, $K_\phi = 0$, and $O(M, p)$ is regular; conversely, if $O(M, p)$ is regular, say isomorphic with $C[x_1, \ldots, x_d]$, then considering $x_1, \ldots, x_d$ as coordinates near p imbeds M near p biholomorphically onto an open subset of C^d. Note that p is a simple point of the analytic variety V if and only if $C(V, p) = C_6(V, p)$.

Example 6.13. For $V: y^2 - x^3 = 0$ in C^2, 0 is a singular point. In V add the new holomorphic function t, with $t^2 = x$, $t^3 = y$; V becomes an analytic manifold, holomorphically but not biholomorphically imbedded in C^2.

7. Tangent spaces

So far we have considered vectors at a singular point p of an analytic variety V, relating them to vectors at nearby simple points. We now take up similar questions, studying tangent planes $T(V, q)$ (q simple) as elements of a Grassmann space, instead of the individual vectors in these spaces.

Let $P^{n-1,r-1}$ be the space of projective alternating contravariant $(r-1)$-vectors in projective $(n-1)$-space P^{n-1}. An element α of $P^{n-1,r-1}$ has components $\alpha_{\lambda_1 \cdots \lambda_r}$, each λ_i running from 1 to n. Certain $(r-1)$-vectors (the simple ones) are the $(r-1)$-directions of (unoriented) r-planes in C^n; these form the Grassmann space $G^{n-1,r-1}$.

Suppose we have $V \subset U \subset C^n$, *V being of constant dimension r.* For each simple point q of V, let $T_0(V, q)$ denote the $(r-1)$-direction of $T(V, q)$. Consider the set of all pairs $(q, T_0(V, q))$ in $U \times G^{n-1,r-1}$. Let $\tau(V)$ denote the closure of this set in $U \times G^{n-1,r-1}$; we call it the *tangent space* of V.

(Using individual vectors in place of tangent planes gives $C_4(V)$, except that the affine case was used previously; see Theorem 5.1.) If V is an analytic manifold, the set of pairs shown above is already closed.

Set $\tau(V, p) = \{\alpha: (p, \alpha) \in \tau(V)\}$; we may call this the *tangent space set* of V at p. For q simple, $\tau(V, p)$ has the single element $T_0(V, p)$. For an example, consider the variety V_4 of §3, defined by $y^2 - zx^2 = 0$. For $p = (0, 0, z)$ with $z \neq 0$, $\tau(V, p)$ has two elements, corresponding to the tangent planes of the two sheets of V there. At 0, $\tau(V, 0)$ is the set of 1-directions of all planes through the x-axis, as we see by considering the parabolas in which the planes $x = x_0$ cut V.

THEOREM 7.1. *Let V be an analytic variety of constant dimension r in the open set $U \subset C^n$. Then $\tau(V)$ is an analytic variety in $U \times G^{n-1,r-1}$, and for each $p \in V$, $\tau(V, p)$ is an algebraic variety in $G^{n-1,r-1}$.*

The last statement is a consequence of the first and of the Theorem of Chow. To prove the first, take any $p \in V$. There are functions $f_1, \ldots, f_s$, holomorphic in a neighborhood U_1 of p, which generate $I(V, q)$ for each $q \in U_1$. We shall sketch the proof that $\tau(V) \cap (U_1 \times G^{n-1,r-1})$ is analytic. For full details, see [15].

Given $q \in U_1$, $\lambda = (\lambda_1, \ldots, \lambda_{n-r})$, $\nu = (\nu_1, \ldots, \nu_{n-r})$, $(\lambda_i \leqq s, \nu_i \leqq n)$, let $D_{\lambda\nu}(q)$ be the determinant with elements $\partial f_{\lambda_i}(q)/\partial x_{\nu_j}$. Given any $\mu = (\mu_1, \ldots, \mu_{n-r+1})$ $(\mu_i \leqq n)$ and λ, define the vector function $v_{\lambda\mu}(q)$ by means of its components: letting $\mu^{(i)}$ denote μ with the ith component omitted, set

$$v_{\lambda\mu}^{\mu_i}(q) = (-1)^{i-1} D_{\lambda\mu^{(i)}}(q), \text{ other } v_{\lambda\mu}^{j}(q) = 0. \tag{7.2}$$

Take any simple point q of $V \cap U_1$, and any λ and μ. If the $df_{\lambda_i}(q)$ are dependent, then $v_{\lambda\mu}(q) = 0$. Otherwise, the f_{λ_i} define V near q, and $df_{\lambda_i}(q) \cdot v_{\lambda\mu}(q) = 0$ for each i, proving $v_{\lambda\mu}(q) \in T(V, q)$. It is easy to see in fact that the set of all $v_{\lambda\mu}(q)$ span $T(V, q)$.

Given $\alpha \in G^{n-1,r-1}$, let us write down the condition that α be $T_0(V, q)$ at the simple point q. Geometrically, the exterior product of any vector of $T(V, q)$ with α must be 0. It is sufficient to use a spanning set of vectors. Analytically, this is expressed as follows: for any λ, μ and $\sigma = (\sigma_1, \ldots, \sigma_{r+1})$, letting $\hat{\imath}$ denote the omission of the corresponding symbol,

$$\sum_{i=1}^{r+1} (-1)^{i-1} v_{\lambda\mu}^{\sigma_i}(q) \alpha^{\sigma_1 \cdots \hat{\imath} \cdots \sigma_{r+1}} = 0. \tag{7.3}$$

Let W_0 consist of all pairs $(q, \alpha) \in U_0 \times G^{n-1,r-1}$ such that $q \in V$ and

(7.3) holds for all λ, μ and σ; then W_0 is analytic. Let V' be the singular variety of V. Clearly

$$(7.4)\quad \tau(V) \cap (U_1 \times G^{n-1,r-1}) = \operatorname{clos}(W_0 - V' \times G^{n-1,r-1}) \cap (U_1 \times G^{n-1,r-1}),$$

and just as in §5, this shows that $\tau(V)$ is analytic.

REMARK 7.5. The natural projection of $\tau(V)$ onto V is holomorphic, and is one–one except over V'. Thus $\tau(V)$ is a modification, or "blowing-up," of V (see [5]).

We shall need to study not only what limits of tangent planes we obtain on approaching a singular point of V, but also what limits are obtained on the approach through a given direction. This will be carried out along the points of an analytic manifold M lying in V (in practice, lying in the singular set of V).

A tangent cone relative to M will be useful; we define it as follows:

Let $C(V, M, p)$ $(p \in M)$ denote the set of all vectors v with the following property: there are sequences $\{p_i\} \to p$ in M, $\{q_i\} \to p$ in V, and $\{a_i\}$, such that $a_i(q_i - p_i) \to v$. Clearly

$$(7.6)\qquad C(V, p) \subset C(V, M, p) \subset C_5(V, p),$$

$$(7.7)\qquad C(V, \{p\}, p) = C(V, p), \quad C(V, V, p) = C_5(V, p).$$

Example 7.8. Note that $y^2 = x^2(1 - x)$ defines a curve with a crossing point at the origin, and no other singularity. Now in (t, x, y)-space, form such a curve through each point $\rho_t = (t, 0, 0)$, contracting it by the factor t^2; this gives the variety defined by $t^2y^2 - t^2x^2 + x^3 = 0$. For later purposes, we wish to get rid of the singular line that coincides with the y-axis. To this end, we add the term $-y^4$, giving the variety V defined by

$$f(t, x, y) = t^2y^2 - t^2x^2 + x^3 - y^4.$$

It is easy to see that the singular variety (defined by $f = df = 0$) of V consists of the t-axis C_t only. For $t \neq 0$, $C(V, \rho_t)$ is the pair of planes $y = \pm x$, while $C(V, 0)$ is the plane $x = 0$. Also

$$C(V, C_t, \rho_t) = C(V, \rho_t) \ (t \neq 0), \quad C(V, C_t, 0) = C^3.$$

Note that the sequence $a_i(q_i - p_i)$ with

$$p_i = \rho_{1/i}, \quad q_i = (1/i, 1/i^2, 0), \quad a_i = i^2,$$

defines the vector $e_x = (0, 1, 0)$ in $C(V, C_t, 0)$, and that $T(V, q_i)$ is nearly orthogonal to e_x for i large. We are interested in the set of points of M where there is such an occurrence.

We shall form a modification of V in which the points of M are blown up. Choose a holomorphic function $N(p)$ in M which, at each p, is an analytic plane of maximum dimension transverse to M at p; supposing that M is small enough, this function will exist. Let U_0 be a neighborhood of M such that the connected components of the $N(p)$ in U_0 and containing p fill out U_0 in a one–one holomorphic way; then setting $\gamma(q) = p$ for q in the component of $N(p)$ gives a holomorphic retraction γ of U_0 onto M. Let η denote the natural projection of $C^n - \{0\}$ onto P^{n-1}. Let V_0 denote the set of simple points of V. Consider the set of pairs

$$(q, \eta(q - \gamma(q))), \quad q \in V_0 \cap U_0 - M;$$

let $V^* \subset U_0 \times P^n$ denote the closure of the set of these pairs. We see easily that V^* is analytic (compare the proof of Theorem 5.1, $i = 5$). If π is the natural projection onto $V \cap U_0$, π is one–one in $V_0 \cap U_0 - M$, and the points over $p \in M$ correspond to the vectors in

$$C_N(V, M, p) = C(V, M, p) \cap N(p).$$

Now consider the set of triples

$$\phi(q) = (q, \eta(q - \gamma(q))), T_0(V, q)), q \in V_0 \cap U_0 - M;$$

set

$$\tau^*(V, M) = \operatorname{clos}\{\phi(q)\} \cap (M \times P^{n-1} \times G^{n-1,r-1}). \tag{7.9}$$

(The points away from M do not interest us.) By the usual methods one may show that this set is analytic. Set

$$\tau^*(V, M, p, \beta) = \{\alpha: (p, \beta, \alpha) \in \tau^*(V, M)\}, p \in M, \beta \in P^{n-1}; \tag{7.10}$$

this is the set of limits of r-directions $T_0(V, q)$ as $q \to p$ in such a manner that the direction from $\gamma(q)$ to q approaches β. (It is void if β is not parallel to $N(p)$.)

Recall that for the V_4 considered in §3, setting $M = C_z$, some elements of $\tau(V, 0)$ do not contain M; on the other hand, for $p \in M$, each α in $\tau^*(V, C_z, p, \beta)$ contains β. In contrast with this, take the variety V of Example 7.8, and set $M = C_t$. For $p \in M$, each element of $\tau(V, p)$ contains the 0-direction of C_t; on the other hand, $\tau^*(V, C_t, 0, \eta(e_x))$ contains the 1-direction of C_{ty}, which does not contain $\eta(e_x)$. In the next section we look further into these phenomena.

REMARK 7.11. Changes of coordinates cause linear transformations of differentials, and hence do not alter the structure of the tangent spaces.

8. Stratifications

If V is an analytic variety of dimension r, then $V = V_0 \cup V'$, where V_0 is the set of simple points of dimension r of V and hence is an analytic manifold, and V' is an analytic variety of lower dimension (Lemma 2.18). We may therefore split V' similarly, and continue until we have expressed V as a union of analytic manifolds. This suggests the following definition:

A *stratification* of the analytic variety V is an expression of V as the disjoint union of a locally finite set of analytic manifolds, each of constant dimension, called the *strata*, such that the frontier of each stratum is the union of a set of lower dimensional strata. (We may require the strata to be connected.)

Example 8.1. Let W be the variety V_3 of §3 (defined by $z^2 - xy = 0$), together with the z-axis. Then with the splitting $W = W_0 \cup W'$ as above, $W' = C_z$, an analytic manifold. Note that fron $(W_0) = \{0\} \neq C_z$; the splitting is not a stratification.

Lemma 8.2. *The closure of each stratum of a stratification is an analytic variety whose dimension is that of the stratum. The frontier of a stratum is an analytic variety of lower dimension.*

We prove this by induction on the dimension of the stratum. Given the stratum M, of dimension d, note that fron (M) is a closed set; hence we may write

$$\text{fron}\,(M) = \bar{M} - M = M_1 \cup M_2 \cup \cdots = \bar{M}_1 \cup \bar{M}_2 \cup \cdots = A,$$

where $\dim M_i < d$. By induction, each $\bar{M}_i$, and hence A, is an analytic variety; $\dim A < d$. That $\bar{M}$ is analytic now follows from Lemma 2.19.

Take any analytic variety $V \subset U \subset C^n$. We prove the *existence* of a stratification by defining the *primary stratification*: The highest dimensional strata are the connected components of the set V_0; the complement V' is analytic. Now suppose the strata of dimension $> d$ have been defined, the closure of each is analytic, and the complement in V of the union of these strata is an analytic variety V^* of dimension d. Let $M_1^*, M_2^*, \ldots$ be the (locally finite) set of connected components of the set M^* of simple points of V^* of dimension d; by Lemma 2.16, each set $\bar{M}_i^*$ is analytic.

Consider any M_i^*, and any previously defined stratum M_j. Set

$$A_j = \bar{M}_j \cap V^*, \quad A_{ij} = A_j \cap \bar{M}_i^*;$$

these are analytic. The set of sets A_{ij} is locally finite; hence the union B_i of those A_{ij} of dimension $<d$ is analytic. Set

$$M_i = M_i^* - B_i;$$

this is a connected analytic manifold. The M_i form new strata; the $\bar{M}_i = \bar{M}_i^*$ are analytic. The complement in V of the union of all the strata so far is $(V^* - M^*) \cup (B_1 \cup B_2 \cup \cdots)$, which is analytic. Hence we may continue defining the strata.

We must show that the frontier K of any stratum M_j is a union of strata of lower dimension. It is clearly contained in such a union; hence we need merely show that if K contains a point p of a stratum M_i, then $M_i \subset K$. With the notations as above, $p \notin B_i$; hence (since $p \in A_{ij}$) $\dim A_{ij} = d$. Therefore, by 2.8, $M_i^* \subset \bar{M}_j$ and hence $M_i \subset K$. Thus we have a stratification.

In this connection, see [15, Theorem 18.11].

REMARK 8.3. Applying the definition of a stratification directly in the real case leads to difficulties. For instance, take V_4 as in §3, in the real domain. The highest dimensional stratum M^2 should consist naturally of the simple points of V_4; now fron (M^2) consists of half the z-axis, and this is not an analytic variety. For C-analytic sets V (see [16]), one may take the complexification V^*, stratify it, and let the strata of V be the real parts of the strata of V^*, in spite of the failure of certain properties.

We now look into the question of improving a given stratification.

Example 8.4. The primary stratification of V_4 is $V_4 = M^2 \cup M^1$, where $M^1 = C_z$. As noted at the end of the last section, the point 0 of C_z has an unpleasant property; we may remove it by taking 0 out of M^1 and making it into a new stratum M^0. A suggestion for what should be considered a "complete stratification" will be given in the next section.

We shall say that a stratification is *regular with respect to tangent planes* if the following is true: suppose $p \in M_i \subset \bar{M}_j$. Then (recalling that $\bar{M}_j$ is an analytic variety of constant dimension)

(a) Each plane T corresponding to an element of $\tau(\bar{M}_j, p)$ contains $T(M_i, p)$.

(b) Each plane T corresponding to an element of $\tau(\bar{M}_j, M_i, p, \beta)$ contains β, for any 1-direction β.

In simpler language, the following must hold: with the retraction γ of a neighborhood of M_i onto M_i as in §7, take any simple point q of $\bar{M}_j$ sufficiently near p; then $T(\bar{M}_j, q)$ must nearly contain both $T(M_i, p)$ and the vector $q - \gamma(q)$. Note that (a) may be phrased as follows:

$$\text{(a}'\text{)} \qquad T(M_i, p) \subset C_2(\bar{M}_j, p),$$

Some examples of failure of these properties were given at the end of §7.

THEOREM 8.5. *Any analytic variety $V \subset U \subset C^n$ has a stratification that is regular with respect to tangent planes.*

We shall give the method of proof only; see [15] for details.

Starting with the primary stratification, examine first each stratum M of next highest dimension relative to the strata of highest dimension. The points of M of bad behavior form an analytic subvariety S; we remove these from M (they will go into lower dimensional strata). With the original highest dimensional strata and the new strata of next highest dimension, reform a stratification. Now examine the strata of third highest dimension, and remove the points of bad behavior relative to higher dimensional strata. Continuing in this manner gives the required stratification.

At any given stage, we are examining a stratum M relative to a higher dimensional stratum M', such that $M \subset \bar{M}'$. Let S_a be the subset of M where (a) fails for this pair. We find S_a as follows: for each $p \in M$ let $G(p) \subset G^{n-1,d-1}$ ($d = \dim M'$) be the set of $(d-1)$-directions whose planes contain $T(M, p)$; we say p is "good" if $\tau(M', p) \subset G(p)$, and p is "bad" otherwise. The sets

$$W_0 = \{(p, \alpha): p \in M,\ \alpha \in G(p)\},$$
$$W_1 = \{(p, \alpha): p \in M,\ \alpha \in \tau(\bar{M}', p)\}$$

are analytic varieties in $M \times G^{n-1,d-1}$ (the first is a manifold). By Lemma 2.15, clos $(W_1 - W_0)$ is an analytic variety; it projects onto an analytic variety $S_a \subset M$, clearly containing all bad points. We must show that $\dim S_a < \dim M$. If not, then a neighborhood of some $q \in W_1$ projects onto a neighborhood of $p \in M$, with maximum rank. Let Γ be an arc in M through p in the direction of a vector v not in the plane corresponding to some element of $\tau(\bar{M}', p)$. We may take a section in $U \times G^{n-1,r-1}$ (U being a neighborhood of p in C^n) to replace $\bar{M}'$ by a 2-dimensional variety containing Γ, with tangent 2-planes not containing v; this leads at once to a contradiction (as suggested to me by R. Thom).

By a similar process, find a bad set S_b, analytic, of dimension $< \dim M$. Now $S = S_a \cup S_b$ is the set we remove from M.

COROLLARY 8.6. $V \subset U \subset C^n$ may be stratified so that the following is true: given p in the stratum M of dimension d and given a transversal plane P^k to M through p (then $k \geqq n - d$), there is a neighborhood U' of p in C^n such that P^k is transversal to $M' \cap U'$ for each stratum M'.

This holds for the stratification of the last theorem; only (a) is used.

Consider now a *real C-analytic variety* $V \subset U \subset R^n$. Take the complexification V^* (see [16]) and stratify it by the last theorem. Let the strata of V be the real parts of the strata of V^*. Then properties corresponding to (a) and (b) hold for the real strata. One may show therefore that the stratification is of the type required by R. Thom in his paper [13] in this volume.

9. The fibering conjecture

Suppose we have a stratification of the variety $V \subset U \subset C^n$ that we consider to be "good." Then it is reasonable to suppose that a neighborhood of any point p_0 of V is expressible as a product, as follows. Suppose p_0 is in the stratum M. Let N be the analytic plane orthogonal to M at p_0. Then a neighborhood U_0 of p_0 is expressible as the disjoint union of *fibers* $F(q)$, as follows. Set $M_0 = M \cap U_0$, $N_0 = N \cap U_0$. Then $F(p_0) = M_0$, and for each $q \in N_0$, $F(q)$ is a manifold biholomorphically homeomorphic with M_0; the fibration is *consistent* with the stratification, in that each fiber is either disjoint from or lies in each stratum. Moreover the fibration is expressible through a function $\phi(p, q)$ $(p \in M_0,\ q \in N_0)$ such that

$$\phi(p, p_0) = p\ (p \in M_0), \quad \phi(p_0, q) = q\ (q \in N_0),$$
$$F(q) = \{q' = \phi(p, q) : p \in M_0\};$$

$\phi(p, q)$ is analytic in p and continuous in q. We call this a *semi-analytic fibration* of a neighborhood of p_0 in C^n. It is *analytic* if ϕ is analytic in both variables.

In general there is no analytic fibration: see Remark 13.3 below.

Lemma 9.1. *Suppose $p \in M$, M being a stratum of dimension d. Then there is an analytic fibration of a neighborhood of p in C^n consistent with the fibration if and only if there are holomorphic vector fields $v_1(q), \ldots, v_d(q)$ in a neighborhood of p in C^n, consistent with the fibration, such that $v_1(p), \ldots, v_d(p)$ span $T(M, p)$.*

This is essentially contained in Rossi, [10, Corrolary 3.4].

In the particular case that $d = 1$, the essential condition is simply that $T(M, p) \subset C_1(V, p)$.

Conjecture 9.2. Any analytic variety $V \subset U \subset C^n$ has a stratification such that each point has a neighborhood with a semi-analytic fibration.

We shall prove a special case of the conjecture in §12.

A stratification satisfying the conjecture (possibly with further conditions on the functions ϕ) would probably be sufficient for all needs. For instance, condition (a) of the last section follows at once. To see this, note that any fiber $F(q)$ with $q \subset M_j$ sufficiently near p_0 is near $F(p_0) = M_0$. Supposing coordinates chosen so that M_0 is part of the $(x_1, \ldots, x_d)$-plane $(d = \dim M)$, $F(q)$ is expressed by holomorphic functions $x_i = f_i(x_1, \ldots, x_d)$, $i = 1, \ldots, n - d$. These functions are small throughout M_0; hence their partial derivatives are small in a smaller neighborhood of p_0. Since $F(q) \subset M_j$ if $q \in M_j$, this clearly implies the condition (a).

To obtain condition (b) in a simple manner, one should probably require more than just continuity of ϕ in the second variable.

In the case of a *real analytic variety*, one might look for real analytic fibers. This is a much simpler problem. Probably by continuing the method begun in §12 one may carry out a proof, with relatively mild conditions on the stratification.

PROBLEM 9.3. With a good stratification, may one fiber a complete neighborhood of any stratum?

PROBLEM 9.4. Given $p \in V$, can one express a neighborhood of p as the union of a set of 1-dimensional analytic varieties, each without singularities except at p, and disjoint except at p?

This would be a generalization of Problem 3.8.

We give one rather trivial case in which a complete region may be fibered:

LEMMA 9.5. *Let U be a simply connected region of the plane C, and let $f_1, \ldots, f_s$ be holomorphic functions in U such that*

$$f_i(x) \neq f_j(x) \quad \text{if } i \neq j,\ x \in U. \tag{9.6}$$

Then there is a consistent analytic fibration of $U \times C$, defined by means of a holomorphic function ϕ.

Say $0 \in U$. We wish to define $\phi(x, y)$ $(x \in U, y \in C)$ such that

$$\phi(0, y) = y, \quad \phi(x, f_i(0)) = f_i(x); \tag{9.7}$$

we require also that for each fixed $x \in U$, ϕ is a homeomorphism of C with itself. Then the fibers

$$F(y) = \{(x, y'): x \in U,\ y' = \phi(x, y)\} \qquad (y \in C)$$

have the required properties.

We defines the fibers as the solutions of a differential equation in $U \times C$. We need merely define $\partial\phi/\partial x$ at each point. We do this by means of the usual interpolation formula:

$$\frac{\partial\phi(x, y)}{\partial x} = \sum_{i=1}^{s} \frac{[y - f_1(x)] \cdots \hat{\imath} \cdots [y - f_s(x)]}{[f_i(x) - f_1(x)] \cdots \hat{\imath} \cdots [f_i(x) - f_s(x)]} \frac{df_i(x)}{dx}. \tag{9.8}$$

The given fibers are solutions; analyticity is clear.

10. Tangent cones to hypersurfaces

Given the holomorphic function f defined near $p_0 \in C^n$, let $f^{[*]}(p_0; p)$ denote the initial polynomial of f, when expanded about p_0 (see §5). Note that

$$f^{[*]}(p_0; p) = f(p_0) \quad \text{if } f(p_0) \neq 0. \tag{10.1}$$

Clearly

$$(fg)^{[*]}(p_0; p) = f^{[*]}(p_0; p) g^{[*]}(p_0; p). \tag{10.2}$$

We first relate the tangent cones of a hypersurface to the Weierstrass Preparation Theorem. This theorem reads as follows: Let f be analytic in a neighborhood of 0 in C^n, and suppose $f \not\equiv 0$ on the x_n-axis. Then there is a pseudopolynomial ω:

$$(10.3) \qquad \omega(x_1, \ldots, x_n) = x_n^k + \sum_{i=1}^{k} A_i(x_1, \ldots, x_{n-1}) x_n^{k-i},$$

where the A_i are holomorphic and vanish at the origin in C^{n-1}, and there is a holomorphic function Φ near 0, such that

$$(10.4) \qquad f(x) = \Phi(x)\omega(x), \quad \Phi(0) \neq 0.$$

Hence

$$(10.5) \qquad f^{[*]}(0; x) = \Phi(0)\omega^{[*]}(0; x).$$

Lemma 10.6. *Let V be a hypersurface in C^n, and suppose $0 \in V$. Let f be a holomorphic function near 0 generating $I(V, 0)$; see 2.9. Define ω as above. Then $v \in C(V, 0)$ if and only if $f^{[*]}(0; v) = 0$, or, $\omega^{[*]}(0; v) = 0$.*

Moreover:

(a) *If the unit vector e_n in the x_n-direction is not in $C(V, 0)$, then the degree of ω equals the order of f at 0, and each A_i is of order at least i.*

(b) *Otherwise, the degree of ω is greater than the order of f at 0, and some A_i is of order $<i$.*

First, take any vector v satisfying $f^{[*]}(0, v) = 0$. For any $g \in I(V, p)$, we may write $g = \phi f$ for some holomorphic ϕ; by (10.2), $g^{[*]}(0, v) = 0$ also. Now Theorem 5.8 and (10.5) show that $C(V, p)$ is defined both by $f^{[*]}$ and by $\omega^{[*]}$.

Say the order of f and hence of ω at 0 is m. In Case (a), $e_n \notin C(V, 0)$, we have $\omega^{[*]}(0; e_n) = \omega^{[m]}(e_n) \neq 0$; since $\omega(\lambda e_n) = \lambda^k$, this shows that $m = k$. No A_i is of order $<i$, for otherwise ω would be of order $<k = m$. The reasoning in Case (b) is similar.

Lemma 10.7. *Let M be an analytic manifold in an open subset of the analytic hypersurface V, and let f generate $I(V, p)$ for each $p \in M$. Let m be the lowest order of f at points of M. Then:* (a) *the set S_c of points of M where the order of f is $>m$ is an analytic subvariety of M.* (b) *$C(V, p)$ is continuous in $M - S_c$.*

Take any $p_0 \in M$. Choose new axes $(t_1, \ldots, t_d, x_1, \ldots, x_{n-d})$ about p_0 (which does not affect the order of f) so that the part of M near p_0 is an open subset of the d-plane C_t; take the origin at p_0. Since f is of order at least m at each $(t, 0)$, we may write

$$(10.8) \qquad f(t, x) = \sum_{|\mu|=m} a_\mu(t) x_1^{\mu_1} \cdots x_{n-d}^{\mu_{n-d}} + R_t(x),$$

where the a_μ and R are holomorphic, and R_t is of order $>m$ for each t. (For $\mu = (\mu_1, \ldots, \mu_{n-d})$, we write $|\mu| = \mu_1 + \cdots + \mu_{n-d}$.) Clearly f is of order $>m$ at $p_t = (t, 0)$ if and only if all the $a_\mu(t)$ are 0; this defines the analytic variety S_c in M near p_0.

For $p_t \in M - S_c$, set

$$(10.9) \qquad g_t(x) = f^{[*]}(p_t; t', x) = \sum_{|\mu|=m} a_\mu(t) x_1^{\mu_1} \cdots x_{n-d}^{\mu_{n-d}}$$

(which is independent of t'); the elements of $C(V, p_t)$ are the solutions of $g_t(x) = 0$, and these are continuous functions of the $a_\mu(t)$ (where the latter do not all vanish). This proves (b).

Let us set (with coordinates as above)

$$\sigma_x = (0, x), \quad \sigma_x' = (0, x/|x|) \text{ if } x \neq 0.$$

We strengthen (b) above as follows:

LEMMA 10.10. *With the hypotheses of the last lemma and with the above notations, suppose* $p_{t_0} \in M - S_c$ *and* $\varepsilon > 0$. *Then there is a* $\delta > 0$ *such that if*

$$(t, x) \in V, \quad |t - t_0| < \delta, \quad |t' - t_0| < \delta, \quad 0 < |x| < \delta,$$

we have

$$(10.11) \qquad \operatorname{dist}(\sigma_x', C(V, p_{t'})) < \varepsilon.$$

Because of (b) above, we may take $t' = t_0$. By compactness, we may find $\gamma > 0$ so that if $|x| = 1$ and $|g_{t_0}(x)| < \gamma$ then $\operatorname{dist}(\sigma_x, C(V, p_{t_0})) < \varepsilon$. As in (5.9), we may write

$$f(t, \lambda x) = \lambda^m[g_t(x) + \lambda R'(t, x, \lambda)],$$

where R' is holomorphic. Say $|R'| < N$ if $|x| = 1$, $|t| < \delta_0$, $|\lambda| < \delta_0$. Choose $\delta < \min(\delta_0, \gamma/2N)$ so that

$$\text{if } |t - t_0| < \delta \text{ then } |a_\mu(t) - a_\mu(t_0)| < \lambda/2(n - d + 1)^m, \text{ all } \mu.$$

Now if (t, x) satisfies the required conditions, then $f(t, x) = 0$, and hence, setting $y = x/|x|$, $x = \lambda y$,

$$-g_{t_0}(y) = \Sigma[a_\mu(t) - a_\mu(t_0)] y_1^{\mu_1} \cdots y_{n-d}^{\mu_{n-d}} + \lambda R'(t, y, \lambda).$$

Hence $|g_{t_0}(y)| < \gamma$ and (10.11) with $t' = t_0$ follows.

We now give a particular result, which will be used in §12.

Let V be a hypersurface in an open subset of C^n; choose a stratification. If M_1 is a stratum of dimension $n - 2$, remove the analytic subsets S_a and S_b as in the proof of Theorem 8.5. Remove also the subset S_c defined in Lemma 10.7. This leaves an analytic manifold M. We may thus form a stratification of V such that no $(n - 2)$-dimensional stratum has points of type S_a, S_b or S_c.

REMARK 10.12. It seems quite possible that (in the present case) $S_c \subset (S_a \cup S_b)$.

LEMMA 10.13. *Let M be an $(n-2)$-dimensional stratum of the hypersurface V as above. Take any $p_0 \in M$ and any $v_0 \notin C(V, p_0)$. Then there is a neighborhood U of p_0 in C^n such that if q is any point of $(V - M) \cap U$, then q is simple and $v_0 \notin T(V, q)$.*

We may choose coordinates $(t_1, \ldots, t_{n-2}, x, y)$ with p_0 as origin so that for some U_0 about p_0, $M \cap U_0$ is an open subset of C_t, and so that v_0 is the unit vector e_y (see Remark 7.11). Let $e_1, \ldots, e_{n-2}$ be the unit vectors in C_t. Let C' be the set of unit vectors in $C(V, 0) \cap C_{xy}$. Since C' is compact we may find $\varepsilon > 0$ such that if

$$|w^{(i)} - e_i| \leqq \varepsilon \ (i \leqq n-2), \ \mathrm{dist}(w^{(n-1)}, C') \leqq 2\varepsilon,$$

then $w^{(1)}, \ldots, w^{(n-1)}, e_y$ are independent.

By the choice of stratification, there is a $\delta > 0$ with the following property. Take any $q = (t, x, y) \in V - M$ within δ of 0. Set $\gamma(q) = (t, 0, 0)$ (see §7). Now $q - \gamma(q) = (0, x, y) = v$ say. Set $v' = v/|v|$. Because of the stratification, q must be simple. Now:

(a′) Each e_i (which is in $T(M, 0)$) is within ε of a vector $w^{(i)}$ of $T(V, q)$.

(b′) v' is within ε of a vector $w^{(n-1)}$ of $T(V, q)$.

(c′) v' is within ε of a vector w of $C(V, 0)$ (see Lemma 10.10).

Since (see (10.9)) $C(V, 0)$ is the set of solutions of an equation $\sum a_i x^i y^{m-i} = 0$, we may take $w \in C_{xy}$; we may clearly also take $|w| = 1$. Thus $w \in C'$.

By the choice of ε, the vectors $w^{(1)}, \ldots, w^{(n-1)}, e_y$ are independent. Since all but the last are in $T(V, q)$, a vector space of dimension $n - 1$, $e_y \notin T(V, q)$. This proves the lemma.

COROLLARY 10.14. With the conditions of the last lemma and the coordinates chosen there, a neighborhood of any point q of $V - M$ in V with q sufficiently near p_0 may be represented in the form $y = f(t_1, \ldots, t_{n-2}, x)$, f being holomorphic.

11. An interpolation formula

We describe a method of interpolation between functions that will be used in the proof of the fibering theorem in the next section.

Let T be a set, and let Z be a finite dimensional affine metric space. Let functions $\phi_1, \ldots, \phi_m : T \to Z$ be given and have distinct values at each t. Let t_0 be a fixed point of T. Set

$$a_i = \phi_i(t_0), \quad \Delta_i(t) = \phi_i(t) - a_i \qquad (i = 1, \ldots, m).$$

Suppose that

$$\gamma = \sup \frac{|\Delta_k(t) - \Delta_j(t)|}{|a_k - a_j|} \tag{11.1}$$

is finite. We shall then define a function $\psi\colon T \times Z \to Z$, with the following properties:

(a) For all $z, z' \in Z$ and $t \in T$,

$$\psi(t_0, z) = z \qquad (z \in Z) \tag{11.2}$$

$$\psi(t, a_i) = \phi_i(t) \qquad (i = 1, \ldots, m), \tag{11.3}$$

$$|[\psi(t, z') - \psi(t, z)] - [z' - z]| \leqq 4(m-1)\gamma|z' - z|. \tag{11.4}$$

(b_0) If T is a topological space and the ϕ_i are continuous, then ψ is continuous.

(b_1) If T is an affine space and the ϕ_i are differentiable to a certain order, or are real analytic, then so is $\psi(t, z)$, except perhaps at $z = a_i$.

(b_2) If T is an open subset of C^k, $Z = C^\ell$, and the ϕ_i are holomorphic, then so is each $\psi_z'(t) = \psi(t, z)$.

(c) If $\gamma < 1/4(m-1)$, then for each t, $\psi_t(z) = \psi(t, z)$ is a homeomorphism of Z onto Z.

REMARK 11.5. Let $F(z) = \psi_z'(T)$ be the set of all $\psi(t, z)$ for $t \in T$. Then with the condition of (c), these form a fibering of $T \times Z$. With the conditions of (b_2) also, the fibering is semi-analytic (§9).

We first define weighting functions σ_i as follows. Set

$$\sigma_k(a_k) = 1, \quad \sigma_j(a_k) = 0 \quad (j \neq k).$$

If $z \neq a_k$ for all k, set

$$\rho_i(z) = |z - a_i|, \quad \mu_i(z) = 1/\rho_i(z), \quad \sigma_i(z) = \mu_i(z)/\Sigma_j\mu_j(z).$$

We let the change in $\psi(t, z)$ from its initial value at t_0 be a weighted average of the changes Δ_i in the ϕ_i:

$$\psi(t, z) = z + \sum_{i=1}^{m} \sigma_i(z)\Delta_i(t). \tag{11.6}$$

Then (11.2) and (11.3) hold.

We need an inequality on the rate of change of the σ_i. Take any $z \in Z$ and any unit vector v; consider the functions ρ_i, μ_i, σ_i at $z + \lambda v$ as functions of the real variable λ. Since $|v| = 1$, we have

$$|\partial\rho_i/\partial\lambda| \leqq 1.$$

Also

$$\frac{\partial\mu_i}{\partial\lambda} = -\frac{1}{\rho_i^2}\frac{\partial\rho_i}{\partial\lambda}, \quad \left|\frac{\partial\mu_i}{\partial\lambda}\right| \leqq \mu_i^2.$$

Moreover,

$$\frac{\partial \sigma_i}{\partial \lambda} = \frac{1}{\Sigma \mu_k} \frac{\partial \mu_i}{\partial \lambda} - \frac{\mu_i}{(\Sigma \mu_k)^2} \sum \frac{\partial \mu_k}{\partial \lambda},$$

and hence

$$\left| \rho_i \frac{\partial \sigma_i}{\partial \lambda} \right| \leqq \frac{\mu_i}{\Sigma \mu_k} + \frac{\Sigma \mu_k^2}{(\Sigma \mu_k)^2} \leqq 2. \tag{11.7}$$

We now prove (11.4). Given z, z', t, we may write

$$z' = z + av, \quad |v| = 1, \quad a \geqq 0.$$

Set

$$\vartheta(\lambda) = \psi(t, z + \lambda v).$$

Now

$$[\psi(t, z') - \psi(t, z)] - [z' - z] = \theta(a) - \theta(0) - av = \int_0^a \left[\frac{\partial \theta}{\partial \lambda} - v \right] d\lambda.$$

If we prove

$$\left| \frac{\partial \theta}{\partial \lambda} - v \right| \leqq 4(m - 1)\gamma, \tag{11.8}$$

then (11.4) will follow, since $a = |z' - z|$.

Take any fixed λ, $0 \leqq \lambda \leqq a$. Choose s so that

$$\rho_s(z + \lambda v) \leqq \rho_i(z + \lambda v), \quad \text{all } i. \tag{11.9}$$

Set $z_\lambda = z + \lambda v$. Since $\Sigma \sigma_i = 1$, (11.6) gives

$$\theta(\lambda) = z + \lambda v + \Delta_s(t) + \sum_{i \neq s} \sigma_i(z_\lambda)[\Delta_i(t) - \Delta_s(t)],$$

$$\frac{\partial \theta}{\partial \lambda} - v = \sum_{i \neq s} \frac{\partial \sigma_i(z_\lambda)}{\partial \lambda} [\Delta_i(t) - \Delta_s(t)].$$

By (11.1) and (11.9),

$$|\Delta_i(t) - \Delta_s(t)| \leqq \gamma |a_i - a_s| \leqq \gamma[\rho_i(z_\lambda) + \rho_s(z_\lambda)] \leqq 2\gamma \rho_i(z_\lambda);$$

now (11.8) follows with the help of (11.7). Thus (11.4) is proved.

Properties (b) are clear. Under the condition of (c), it is clear from (11.4) that ψ is one–one; we must still show that, for t fixed, $\psi_t(z) = \psi(t, z)$ maps Z onto Z. Set $N = \Sigma |\Delta_i(t)| + 1$. Take any $z_0 \in Z$, and let S be the sphere about z_0 of radius N. Now

$$|\psi_t(z) - z| = |\Sigma \sigma_i(z) \Delta_i(t)| < N.$$

Therefore no line segment from $\psi_t(z)$ to z ($z \in S$) touches z_0. We may deform ψ_t along these line segments into the identity, showing that the degree of ψ_t in S about z_0 is 1. By a standard theorem of algebraic topology, ψ_t maps some point of Z inside S onto z_0. This completes the proof.

12. A fibering theorem

Recalling the definition of §9, we prove the following theorem:

THEOREM 12.1. *An analytic hypersurface V in an open subset of C^n may be stratified so that a neighborhood of each point of each stratum of dimension $n-2$ has a semi-analytic fibration.*

We use the stratification employed in Lemma 10.13. Take $p_0 \in M$ as there, and choose coordinates about p_0 as in the proof of the lemma. Then in a neighborhood U_0 of p_0, the points of $V - M$ may be represented locally in the form $y = f(t, x)$, where $t = (t_1, \ldots, t_{n-2})$ (see Corollary 10.14). If we take a fixed $x_0 \neq 0$, the points above (t, x_0) in U_0 are all simple, and are given by holomorphic functions $\phi_1(t, x_0), \ldots, \phi_m(t, x_0)$ (for some m), these m roots being distinct for each t.

In order to apply §11, we need the following lemma.

LEMMA 12.2. *With the above notations, let $\varepsilon > 0$ be given. Then there is a $\delta > 0$ with the following property: take any x with $0 < |x| < \delta$. Then the functions ϕ_i are defined above for $|t| < \delta$. Take any j and k, and set*

$$\alpha(t, x) = \phi_k(t, x) - \phi_j(t, x).$$

Then

$$\frac{|\alpha(t, x) - \alpha(0, x)|}{|\alpha(0, x)|} < \varepsilon \quad \text{if } |t| < \delta,\ 0 < |x| < \delta. \tag{12.3}$$

If we keep t fixed and let x make m circuits about 0, α comes back to its original value; hence we may define a holomorphic function β near 0 such that

$$\beta(t, \xi) = \alpha(t, \xi^m).$$

Since the ϕ_i are distinct for $x \neq 0$, we have $\beta \neq 0$ for $\xi \neq 0$. Hence if we let ξ make a circuit about 0, $\beta(t, \xi)$ will make a number h of circuits about 0, where h is independent of the choice of circuit and of t. Therefore, as a function of ξ, β is of order h at $\xi = 0$ for each t, and we may write

$$\beta(t, \xi) = \xi^h[a(t) + \xi R(t, \xi)], \quad a(t) \neq 0.$$

Take $\delta_0 > 0$ so that $|a(t) - a(0)| < |a(0)|\varepsilon/3$ if $|t| < \delta_0$. Take δ_1 and N so that $|R(t, \xi)| < N$, $2\delta_1 N < |a(0)|\varepsilon/3$ if $|t| < \delta_0$, $|\xi| < \delta_1$. For such (t, ξ),

$$\frac{|\beta(t, \xi) - \beta(0, \xi)|}{|\beta(0, \xi)|} \leq \frac{|a(t) - a(0)| + |\xi||R(t, \xi) - R(0, \xi)|}{|a(0)| - |\xi||R(0, \xi)|}$$

which is $<\varepsilon$ (if $\varepsilon < 1$). Set $\delta = \min(\delta_0, \delta_1^m)$; then (12.3) follows.

Set $\varepsilon = 1/4m$, and choose δ accordingly. Let U be the set of all (t, x, y) with $|t|$, $|x| < \delta$; we shall fiber U. Let U_x be the part of U with this x; we fiber each U_x separately.

First take $x \neq 0$. Set $t_0 = 0$. Then the conditions of §11 are satisfied, including (b_2) and (c); (11.1) holds, with $\gamma \leqq \varepsilon$, by (12.3). We may therefore form a semi-analytic fibering of U_x by the formula (11.6); see Remark 11.5. It is clear that the resulting fibering of the part of U over all $x \neq 0$ is continuous.

For $x = 0$, we have at the start the single fiber $M: y = 0$; let $y =$ constant give the remaining fibers. We must show that the fibering is continuous in a neighborhood of $x = 0$. Because of the continuity of the ϕ_i, we may, given $\varepsilon > 0$, find $\eta > 0$ so that

$$|\phi_k(t, x)| < \varepsilon \quad \text{if} \quad |t| < \delta,\ |x| < \eta.$$

Using (11.6) and $\Sigma\sigma_i(y) = 1$ gives (for each x)

$$|\psi_x(t, y) - y| \leqq \Sigma\sigma_i(y)|\Delta_i(t)| < 2\varepsilon, \quad |x| < \eta.$$

Thus the fibers defined by the $\psi_x(t, y)$ converge to the fibers $\psi_0(t, y) = y$. This completes the proof.

REMARK 12.4. If we used Lemma 9.5 in place of (11.6) for the fibering of the U_x ($x \neq 0$), an analytic instead of semi-analytic fibration would result. However, we could not then in general fiber U_0 to obtain a continuous fibering. The impossibility in general is shown by Example 13.2 below. In the present case, letting $V \subset C^3$ be defined by

$$y(y - x)(y - (2 + t)x) = 0,$$

fibering by means of Lemma 9.5 gives fibers which do not converge as $x \to 0$.

13. Examples of non-homogeneity

First we show that with certain analytic varieties there may be a complex number intrinsically attached.

Example 13.1. Let $V \subset C^2$ be defined by the vanishing of

$$f = xy(y - x)(y - ax).$$

V consists of four intersecting analytic lines. Taking them in the order of the factors of f, they have a cross ratio, which is $\sigma_a = a$. Taking them without regard to order, we have a set σ_a^* of six cross ratios; these are distinct if $|a| > 2$, and $|a|$ is then the largest.

Now consider V as an abstract analytic variety. Take any neighborhood of the intersection point, and any imbedding of it into an open set in C'^2.

By Theorem 4.7 there is a biholomorphism Φ of a neighborhood of 0 in C^2 into C'^2 which extends the biholomorphism between the varieties; moreover, $d\Phi(0)$ is linear in $C_6(V, 0) = C^2$, and is one–one between the tangent cones. Hence the four tangent lines of the newly imbedded variety have the same set of cross ratios as before.

Thus the set σ_a^* is an invariant of V itself. In particular, taking a near 3, *the number a is an invariant.*

Next we define a variety with a naturally defined function along the singular locus.

Example 13.2. Let $V \subset U \subset C^3$ be defined by the vanishing of

$$f(t, x, y) = xy(y - x)(y - (3 + t)x);$$

require $|t| < 1$ in U. Near 0, the singular locus V' is the t-axis C_t. Consider any point $p_t = (t, 0, 0)$ of C_t. The tangent cone $C(V, p_t)$ consists of four analytic planes, intersecting in C_t. These four planes have a set of cross ratios, which equals that of the four lines in which $C(V, p_t)$ intersects any transverse plane, for instance, one orthogonal to C_t. One of these cross ratios is $\sigma_t = 3 + t$. Again by Theorem 4.7, this number is intrinsically associated with the point p_t, for t small. Thus, considering V as an abstract analytic variety, the function σ_t is invariantly defined on V'.

Because of this, we may say that V is *non-homogeneous along* V'. This means the following: take two distinct points p_1, p_2 of V'. Then it is not true that there are neighborhoods U_1, U_2 of these points in V and there is a biholomorphism Φ of U_1 onto U_2.

REMARK 13.3. For the above V, there is a semi-analytic fibration of a neighborhood U_1 of 0, by Theorem 12.1. There is no analytic fibration. For if there were, then we could form a biholomorphic transformation in U_1 simply by changing t along the fibers, contrary to the non-homogeneity proved above. (In particular, $C_1(V, p) = \{0\}$, $p \in C_t$; compare Rossi, [10, Theorem 3.2].) Note that with the chosen semi-analytic fibration, the fibering of each individual sheet of V is analytic.

Example 13.4. Taking f as in Example 13.2, define V by the vanishing of

$$g(t, x, y, z) = z^2 - [f(t, x, y)]^2.$$

(We could use $g = z^2 - tf^2$, making V irreducible; it would still be reducible at other points of the singular variety V'.) Now V consists of two sheets, intersecting along the singular locus V', defined by $z = f = 0$; thus V' is biholomorphic with the variety of Example 13.2. The present variety has the following property, not shared by the previous one: If W is an analytic variety and Φ is a holomorphic homeomorphism onto a neighborhood of 0 in V, then W consists similarly of two intersecting sheets, and is therefore similarly non-homogeneous along the intersection of the four sheets of the singular variety W'. (Φ need not be biholomorphic.)

The property obtained above may hold for a normal variety, though it takes more theory to prove it:

Example 13.5. With f as in Example 13.2 again, define V by the vanishing of

$$g(t, x, y, z) = z^5 - f(t, x, y).$$

V is clearly irreducible at all points. The singular variety V' of V is again C_t (near 0). Since V' is of codimension 2 in V, it follows that V is normal (see for instance [1]). Hence (see [6, Satz 22]) any continuous function in V near 0 which is holomorphic at all simple points is holomorphic in V near 0. Moreover [6, Satz 23] any holomorphic homeomorphism of an analytic variety onto a neighborhood of 0 in V is biholomorphic.

By Lemma 10.6, $C(V, p)$ is defined by the vanishing of the initial polynomial of g, expanded about p. For $p = p_{t_0} = (t_0, 0, 0)$, this polynomial is $-f(t_0, x, y)$, which is independent of t. The cone consists of four planes through C_t, with a largest cross ratio $3 + t_0$; again we have non-homogeneity along V'.

I am indebted to D. B. Mumford for pointing out the properties of this example to me.

14. Varieties not locally algebraic

We say the analytic variety V is *locally algebraic* at the point p if there is an algebraic variety W with a point q such that there is a biholomorphic homeomorphism of some neighborhood of q in W onto a neighborhood of p in V.

Example 14.1. Define V in a neighborhood of 0 in C^3 by the vanishing of

$$f(t, x, y) = xy(y - x)(y - (3 + t)x)(y - \gamma(t)x),$$

where γ is a transcendental function, and $\gamma(0) = 4$. V has five sheets, intersecting along C_t. The largest cross ratio of the first four sheets at $p_t = (t, 0, 0)$ is $3 + t$; the largest cross ratio of the first three and the last sheet is $\gamma(t)$. As in the last section, each cross ratio is intrinsically related to the variety at the given point. Hence the set of pairs $(3 + t, \gamma(t))$, and hence the function γ, is intrinsic to the variety. We show that the variety is not locally algebraic at any point of C_t.

Suppose then that $W \subset C^n$ is an algebraic variety, and ϕ is a biholomorphic homeomorphism of a neighborhood $W \cap U_0$ of 0 in W onto a neighborhood of 0 in V. By Theorem 4. dim 7, $C_6(W, 0) = \dim C_6(V, 0) = 3$; hence, by Lemma 4.12, there is an analytic manifold M of dimension 3 about $0 \subset C^n$ containing W near 0. Since $\dim W \cap U_0 = \dim V = 2$, we may suppose $\dim W = 2$. Let P be the tangent plane to M at 0. We may choose an analytic line L through 0 which avoids $W - \{0\}$, so that if π is the projection along L of C^n onto some $C^{n-1} \supset P$, then for some U_1 about 0, $\pi(W) \cap U_1 = \pi(W \cap U_0) \cap U_1$. It is a standard fact that $\pi(W)$ is an algebraic variety. Moreover, π is a biholomorphism in $M \cap U_1$; hence we may use $\pi(W) \subset C^{n-1}$ in place of $W \subset C^n$. Continuing, we may suppose $W \subset C^3$.

Now W is an algebraic hypersurface in C^3; hence it is defined by a polynomial g. Let W' be the singular variety of W. With ϕ and U_0 as above, ϕ extends to a biholomorphism Φ of some $U \subset U_1$ about $p_0 \in W'$ with $\phi(p_0) = 0$. Now Φ^{-1} carries each sheet in V into an analytic sheet in W. Hence g may be factored analytically, and hence algebraically, into five factors $g_1, \ldots, g_5$, which may clearly be supposed irreducible. Now g_i generates $I(W_i, p)$ for the corresponding sheet; hence g generates $I(W, p)$, for each $p \in W' \cap U$. By Lemma 10.6, for each such p, the vectors $v \in C(W, p)$ are the solutions of $g^{[*]}(p; v) = 0$. The solutions are algebraic in terms of the coefficients, which in turn are algebraic in terms of the value of a parameter t defining $p \in W'$; hence both cross ratios under discussion are algebraic functions of t, and the pair of them determines an algebraic function. But this function is γ, and we have a contradiction.

Example 14.2. With f as above, set $g = z^6 - f$. Then (compare Example 13.5) V, defined by $g = 0$, is normal, and not locally algebraic at any singular point near 0.

REMARK 14.3. Any analytic curve is locally algebraic. N. Levinson [7] has proved the following stronger result: Let F be any holomorphic function of the form

$$F(x, y) = y^k \Phi(x, y) \omega(x, y), \quad \Phi(0, 0) \neq 0,$$

$$\omega(x, y) = y^m + \sum_{i=1}^{m} a_i(x) y^{m-1}, \quad a_i(0) = 0 \qquad (i = 1, \ldots, m),$$

the discriminant of ω not being identically 0. Then there is an analytic change of coordinates under which F becomes a polynomial.

For a function $F(x_1, \ldots, x_n, y)$ of more variables, of a similar form, Levinson [8] has shown that it may be transformed into a polynomial in y with coefficients analytic in the x_i. However, it cannot in general (say with $n = 2$) be transformed into a polynomial. For if it could, then it would define an algebraic variety, biholomorphic to the variety defined by the given function, which may not be possible. Thus the functions of Examples 14.1 and 14.2 cannot be transformed through analytic changes of coordinates into polynomials.

15. Some further examples

First we study certain 1-dimensional varieties with just two sheets; they turn out to be biholomorphic with one of the varieties V_μ^* of (15.2). For certain 1-dimensional varieties with three sheets, there is an intrinsic number attached. (In Example 13.1 we needed four sheets.) We then show that the phenomena of Examples 13.2 and 14.1, and hence of non-homogeneity and non-algebraic character, are obtainable through varieties with fewer sheets than used there.

15.1. We suppose given an analytic variety V expressible as $V_1 \cup V_2$, each V_i being an analytic 1-dimensional manifold (compare Remark 6.12), these manifolds intersecting in a single point p. We shall show that V is (near p) biholomorphically homeomorphic with one of the varieties

$$(15.2) \qquad V_\mu^*: y(y - x^\mu) = 0 \text{ in } C^2 \ (\mu \text{ an integer } \geqq 1).$$

Note that $V_1^*: y(y - x) = 0$ is biholomorphic with $V_1': yx = 0$.

To start, we may suppose V imbedded in C^n. Since V_1 is an analytic manifold, we may take new axes near p, with origin at p, so that V_1 is on the x_1-axis.

We wish the function $x_1|V_2$ to generate $O_2 = O(V_2, 0)$. If this is not so, then some $x_i|V_2$ generates O_2. Set $x_1' = x_1 + x_i$. With the coordinates $(x_1', x_2, \ldots, x_n)$, $x_1'|V_2$ generates O_2. Take this coordinate system, dropping primes. V_1 (near 0) is still on the x_1-axis.

Now V_2 near 0 is given by equations $x_k = \phi_k(x_1)$ $(k \geqq 2)$. Let μ be the lowest order occurring among the ϕ_k; we may suppose it occurs for $k = 2$. Then the equations for V_2 have the form

$$x_2 = x_1^\mu(a + x_1 R(x_1)), \quad a \neq 0,$$
$$x_k = x_1^\mu R_k(x_1), \qquad k > 2.$$

Set

$$a' = 1/a, \quad S(x_1) = \frac{-a' R(x_1)}{a + x_1 R(x_1)} \text{ near } x_1 = 0,$$
$$x_2' = [a' + x_1 S(x_1)]x_2.$$

With the new coordinates $(x_1, x_2', \ldots, x_n)$, V_1 is again the x_1-axis, and (as we see by insertion) $x_2' = x_1^\mu$ on V_2. Drop primes again; let Φ denote the resulting biholomorphic imbedding of V (near p) into C^n.

Taking the functions $x_1 = x$ and $x_2 = y$ only in V gives a holomorphic mapping Ψ of V into C^2, which is clearly a homeomorphism. We show that Ψ is biholomorphic. By Lemma 2.13, it is sufficient to show that the functions $x_3, \ldots, x_n$ in $\Phi(V)$ are expressible in terms of x_1 and x_2. The relations $x_k = x_1^\mu R_k(x_1) (k > 2)$ hold in V_2 in the new coordinates; hence

$x_k = x_2 R_k(x_1)$ for $k > 2$ in both V_1 and V_2, as required. Thus $\Psi(V) = V_\mu$ near 0, completing the proof.

We shall say that the two sheets of V_μ^* have *μ-contact* at p.

15.3. We now study a variety $V = V_1 \cup V_2 \cup V_3$, each V_i being a 1-dimensional analytic manifold, with p in each, with no other intersections, and each pair of sheets having 2-contact; moreover, we assume $\dim C_6(V, p) = 2$. There is then a biholomorphic imbedding (near p) into C^2; by the considerations above, we may find an imbedding so that two of the sheets go into the variety $y(y - x^2) = 0$. Because of the invariance of μ-contact, we see that the imbedded V is given by the vanishing of

$$f(x, y) = y(y - x^2)(y - x^2\phi(x)), \quad \phi(0) = c \neq 0, 1.$$

We shall show that *c is an invariant of V* (taking the sheets in a fixed order).

Consider any other imbedding into C^2 such that the first two sheets go into $y = 0$ and $y - x^2 = 0$. By Theorem 4.7 the biholomorphism from one imbedding into the other is the restriction of a biholomorphism from C^2 into C^2. Taking coordinates (x, y) and (u, v), write this as

$$u = ax + a'y + \cdots, \quad v = bx + b'y + b''x^2 + \cdots.$$

Since V_1 is carried into itself, we must have $v = 0$ if $y = 0$; hence $b = b'' = 0$. If $y = x^2$, we must have $v = u^2$; hence (with $y = x^2$)

$$u = ax + a'x^2 + \cdots, \qquad u^2 = a^2x^2 + \cdots,$$
$$v = b'x^2 + \cdots, \qquad v - u^2 = (b' - a^2)x^2 + \cdots;$$

hence $b' = a^2$. Now on the image V_3' of V_3: $y = cx^2 + \cdots$ we have

$$u = ax + \cdots, \qquad u^2 = a^2x^2 + \cdots, \qquad v = a^2cx^2 + \cdots;$$

hence V_3' is given by $v = cu^2 + \cdots$, with the same c.

Remark 15.4. We may find c intrinsically as follows: choose generators f, g of $O(V, p)$ (for instance, the functions x, y above) such that $f|V_1$ generates $O(V_1, p)$ (and hence each $O(V_i, p)$), and $g|V_1 = 0$. For any $\varepsilon \neq 0$, let $p_{i\varepsilon}$ be the point of V_i where f has the value ε. Then

$$c = \lim_{\varepsilon \to 0} g(p_{3\varepsilon})/g(p_{2\varepsilon}).$$

Example 15.5. Let $V \subset C^3$ near 0 be defined by the vanishing of

$$f(t, x, y) = y(y - x^2)(y - (2 + t)x^2);$$

then at $p_t = (t, 0, 0)$, the number $2 + t$ is intrinsically associated. Therefore the variety is non-homogeneous along C_t.

Example 15.6. Let $V \subset C^3$ near 0 be defined by the vanishing of

$$f(t, x, y) = y(y - x^2)(y - (2 + t)x^2)(y - \gamma(t)x^2),$$

where γ is transcendental and $\gamma(0) = 3$. Then the pair $(2 + t, \gamma(t))$ for each t and hence the function γ is intrinsically associated with V. For an algebraic variety, γ is algebraic: hence V is not locally algebraic along C_t.

INSTITUTE FOR ADVANCED STUDY

REFERENCES

[1] S. ABHYANKAR, Concepts of order and rank on a complex space, and a condition for normality, *Math. Annalen, 141* (1960), p. 171–192.

[2] E. BISHOP, Partially analytic spaces, *Am. Journal of Math., 83* (1961), p. 669–692.

[3] H. CARTAN, Séminaire 1951–52, École Normal Sup., Paris.

[4] ———, Variétés analytiques-réelles et variétés analytiques-complexes, *Bull. Soc. Math. France, 85* (1957), p. 77–100.

[5] H. GRAUERT and R. REMMERT, Zur Theorie der Modifikationen. I. Stetige und eigentliche Modifikationen komplexer Räume., *Math. Annalen, 129* (1955), p. 274–296.

[6] ———, Komplexe Räume, *Math. Annalen, 136* (1958), p. 245–318.

[7] N. LEVINSON, A polynomial canonical form for certain analytic functions of two variables at a critical point, *Bull. Am. Math. Soc., 66* (1960), p. 366–368.

[8] ———, Transformation of an analytic function of several complex variables to a canonical form, *Duke Math. Journal, 28* (1961), p. 345–354.

[9] R. REMMERT and K. STEIN, Über die wesentlichen Singularitäten analytischer Mengen, *Math. Annalen, 126* (1953), p. 263–306.

[10] H. ROSSI, Vector fields on analytic spaces, *Annals of Math., 78* (1963), p. 455–467.

[11] P. SAMUEL, Méthodes d'algèbre abstraite en géométrie algébrique, *Ergebnisse der Math.*, (N. F.) Heft 4, Springer, Berlin, 1955.

[12] J.-P. SERRE, Géométrie algébrique et géométrie analytique, *Annales de l'Institut Fourier, VI* (1955), p. 1–42.

[13] R. THOM, Sur l'homologie des variétés algébriques réelles, this volume.

[14] H. WHITNEY, Elementary structure of real algebraic varieties, *Annals of Math., 66* (1957), p. 545–556.

[15] ———, Tangents to an analytic variety (to appear in *Annals of Math.*).

[16] ———, and F. BRUHAT, Quelques propriétés fondamentales des ensembles analytiques-réels, *Comm. Math. Helvetici, 33* (1959), p. 132–160.

[17] O. ZARISKI and P. SAMUEL, *Commutative Algebra*, vol. II. Princeton, N.J.: D. van Nostrand Co., 1960.

Tangents to an Analytic Variety

By HASSLER WHITNEY

1. Introduction

The purpose of this paper is to study the structure of the sets of tangent vectors and tangent planes to a complex analytic variety, particularly in the neighborhood of singular points of the variety. We prove the existence of a stratification of the variety which has nice properties relative to tangent planes. Corresponding properties of real analytic varieties (which are the real parts of of complex ones) may be found by considering the corresponding complex analytic variety.

The principal theorems were stated in another paper, [9]; we refer to that paper for further results and discussion. The present paper is, however, self-contained.

Part I gives preliminary material. A natural metric in projective space P^{n-1} will be used. The local study of a variety is based on a lemma of Remmert and Stein, given here in expanded form. Properties of the part of a variety lying outside another variety are needed. Certain modifications of a variety (known as Hopf's σ-process, monoidal transformations) are important in imbedding a tangent cone in a larger variety which reflects the given variety. Elementary facts about the intersections of varieties are given. An important tool in studying a variety near a manifold is to cut down the dimension of the variety to one more than that of the manifold; two lemmas of this nature are presented.

Part II is devoted to the study of tangent cones. Through the use of the modifications mentioned above, it is shown that each tangent cone has a dimension equal to that of the variety at the given point. Moreover, holomorphic mappings of varieties cause corresponding mappings of tangent cones. Conditions for the image of the cone to be all of the cone of the image are given. The classical theorem on the expression of the tangent cone in terms of the initial polynomials of the ideals of the variety at the point is given; its proof does not seem to be easily available elsewhere. A tangent vector is defined by means of sequences of points; we show that it may also be defined by means of of differentiable arcs from the given point.

In §§12–14 we make a study of a variety V near an imbedded manifold M. The tangent cones along the manifold need not form a continuous family. We construct a variety $K_M(V)$ about V, whose sections through M are cones containing the projected tangent cone at each point of M; we may do this, keeping the dimension of $K_M(V)$ equal to that of V. If $\dim M = m$, we mentioned that the dimension of V could be cut down to $m+1$ near a point p of M; here we cut V down further, forming a real differentiable manifold of real dimension $2m+1$ stretching from an open subset of M into V; we call this a *wing*. Wings will be used later in the paper (and elsewhere). Assuming $\dim V = m+1$ at each point of M, we show that at most points of M, the sheets of V attach in a smooth manner, taking into account the tangents at points of M.

Tangent planes are taken up in Part III. Of first importance is to form a variety from all tangent planes and their limits. The essential tool is the existence of a set of holomorphic vector fields defining all tangent planes at all simple points near p; this follows from the coherence of the sheaf of ideals. We form also a *directed* variety, relative to a subvariety. A *stratification* of a variety V is a partition into manifolds with certain basic properties. If the strata fit together in a natural manner with regard to tangent planes, the stratification is *regular*. A basic theorem is the existence of a regular stratification. We mention an application in the final section which has equal meaning in the real case: Near enough any point of a variety there is a tangent vector pointing nearly towards the given point.

Let C denote the set of complex numbers, and C^n, complex n-space. We speak of either *points* or *vectors* in C^n. If C' is a subspace (through the origin O), $p + C'$ denotes $\{q = p + v: v \in C'\}$; this is an *analytic plane* or *plane* for short. For any plane T, a *vector in* T means a vector of the form $q - p$, where $p, q \in T$. We write: point plus vector equals point, point minus point equals vector. A *line* is a plane of dimension 1.

Let P^{n-1} denote the projective space of C^n; it is in natural correspondence with the set of lines through O. For any line L, let $\kappa(L)$ denote the corresponding element of P^{n-1}; we call it the *direction* or 1-*direction* of L. For $v \in L$, $v \neq 0$, write $\kappa(v) = \kappa(L)$. An m-plane T defines an element $\kappa(T)$, the *direction* or *m-direction* of T, which is an element of the projective Grassman space $G = G^{n-1,m-1}$. For any $\alpha \in G$, let $T(\alpha)$ denote the corresponding plane through O. Writing $\alpha \subset \beta$ means $T(\alpha) \subset T(\beta)$; $\kappa(v) \subset \alpha$ means $v \in T(\alpha)$, etc.

Let C_x and $C_{x_1 x_2}$ denote the (complex) x-axis and (x_1, x_2)-coordinate plane, etc. By e_{x_i} or e_i we mean the unit vector in the corresponding direction.

For any subset A of an open set H, let H-clos A denote the part of the closure of A in H. The H-frontier of A, H-fron A, is H-clos $A - A$. Given $p \in C^n$ and $\delta > 0$, set $U_\delta(p) = \{q: |q - p| < \delta\}$; $\bar{U}_\delta(p)$ is the closure of this set. Set $U_\delta(A) = \{q: |q - p| < \delta$ for some $p \in A\}$. Write also $\partial U =$ fron U.

A set of sets in H is locally finite in H if any compact subset of H intersects but a finite number of the sets.

PART I. PRELIMINARIES

2. Metrics in P^{n-1} and in $G^{n-1,m-1}$

We recall some facts about the hermitian metric in C^n: We have $u \cdot v = \sum u_i \bar{v}_i$, $|v|^2 = v \cdot v = \sum v_i \bar{v}_i$. ($\bar{a}$ is the complex conjugate of a.) The orthogonal complement of a plane T is $T^\perp = \{v: u \cdot v = 0, \text{ all } u \in T\}$. We have $\dim T^\perp = n - \dim T$, $T^{\perp\perp} = T$. Let $\pi, \pi^\perp$ denote projections onto the corresponding planes. Then $v = \pi v + \pi^\perp v$, and $|v|^2 = v \cdot v$ gives

$$|v|^2 = |\pi v|^2 + |\pi^\perp v|^2 = |\pi v|^2 + |v - \pi v|^2 . \tag{2.1}$$

Hence the distance from v to T is

$$\operatorname{dist}(v, T) = \inf\{|v - u|: u \in T\} = |v - \pi v| . \tag{2.2}$$

We define the sine of the positive angle θ between u and v by

$$\sin\theta = |u, v|^* = |u - \pi_v u|/|u| , \qquad \text{where } u \neq 0, v \neq 0 , \tag{2.3}$$

and π_v is the projection onto the line $T(v)$ of v. Clearly

$$|au, v|^* = |u, bv|^* = |u, v|^* \qquad (a, b \neq 0) ; \tag{2.4}$$

hence we may define

$$|\kappa(u), \kappa(v)|^* = |u, v|^* \qquad (u, v \neq 0) . \tag{2.5}$$

To show that this is a metric in P^{n-1}, we prove symmetry and the triangle inequality:

$$|u, v|^* = |v, u|^* , \qquad |u, w|^* \leqq |u, v|^* + |v, w|^* . \tag{2.6}$$

We may suppose u, v, w are unit vectors. Say $\pi_v u = av$, $\pi_u v = bu$. Now

$$0 = v \cdot (u - \pi_v u) = v \cdot u - \bar{a}|v|^2 = v \cdot u - \bar{a} ;$$

hence $a = u \cdot v$. Similarly $b = v \cdot u$. Hence, by (2.1), $|u - \pi_v u|^2 = 1 - |a|^2 = 1 - |b|^2 = |v - \pi_u v|^2$, and symmetry follows.

Next, say $\pi_v u = av$; then $|a| \leqq 1$. Since $\pi_w u$ is the nearest point of $T(w)$ to u,

$$|u - \pi_w u| \leqq |u - \pi_v u| + |\pi_v u - \pi_w \pi_v u| \leqq |u, v|^* + |a||v - \pi_w v| ,$$

giving the inequality.

Since $\pi_v u = (u \cdot v)v/|v|^2$, (2.3) gives

$$|u, v|^* = \frac{||v|^2 u - (u \cdot v)v|}{|v|^2 |u|} . \tag{2.7}$$

If $u, v, u + v$ are $\neq 0$, then

$$(2.8) \qquad |u, v|^* = \frac{|u - v|}{|u|}|u - v, v|^* = \frac{|u - v|}{|v|}|u, u - v|^* .$$

This follows directly on applying (2.7) to $|u - v, v|^*$ and by symmetry.

As a consequence of (2.8),

$$(2.9) \qquad |u, v|^* \leqq \frac{|u - v|}{|u|}, \qquad |u, v|^* \leqq \frac{|u - v|}{|v|} .$$

A formula analogous to the expansion of $\sin(\theta - \varphi)$ is

$$(2.10) \qquad |v + au, v + bu|^* = \frac{|b - a||u||v|}{|v + au||v + bu|}|v, u|^* .$$

For (2.8) gives

$$|v + au, v + bu|^* = \frac{|b - a||u|}{|v + bu|}|v + au, (a - b)u|^* ,$$

and we now apply (2.8) to $|v + au, (a - b)u|^* = |v + au, au|^*$.

Let T_1 and T_2 be orthogonal planes, and let π_i be the projection onto T_i. Then for $v \in T_1 + T_2$, we set

$$(2.11) \qquad |v, T_2|^* = \inf_{u \in T_2}|v, u|^* = \frac{|\pi_2 v - v|}{|v|} = \frac{|\pi_1 v|}{|v|} .$$

Let T_1, T_2 and T_3 be mutually orthogonal planes; take π_i as above, and let π_{ij} be the projection onto $T_i + T_j$. Now for $v \in T_1 + T_2 + T_3$,

$$(2.12) \qquad |v, T_3|^* \geqq |\pi_{13}v, T_3|^* \geqq |v, T_2 + T_3|^* \qquad \text{if } \pi_{13}v \neq 0 .$$

This is geometrically clear, and easy to prove analytically.

If π_T is the projection onto T, we mention that

$$(2.13) \qquad |\pi_T u, \pi_T v|^* \leqq \frac{|u|}{|\pi_T u|}|u, v|^* \qquad \text{if } \pi_T u, \pi_T v \neq 0 .$$

We now define a metric in G. For m-planes K, L, set

$$(2.14) \qquad |K, L|_m^* = \sup_{u \in K} \inf_{v \in L}|u, v|^* = \sup_{u \in K}\frac{|u - \pi_L u|}{|u|} ,$$

and set $|\kappa(K), \kappa(L)|_m^* = |K, L|_m^*$.

To prove symmetry, it is sufficient to show that $|K, L|^* \leqq |L, K|^*$. This is fairly straightforward, and we shall not go into it. The proof of the triangle inequality is simple.

3. Analytic varieties

A *local analytic variety* V in C^n is a set (possibly void) such that, for each $p \in V$, there is a neighborhood U of p and a set of holomorphic functions in U whose set of common zeros is $V \cap U$. There is then a finite set of such functions in U. An *analytic variety* in the open set $H \subset C^n$ is a local analytic variety lying in H and forming a closed subset of H. A *variety* will always mean an analytic variety. We often write $V \subset H \subset C^n$ to describe this. A local variety is clearly a variety in some open set. A *subvariety* of the variety V in H is a subset of V which is itself a variety in H. We speak of a variety *near* p if for some U about p, $V \cap U$ is a variety in U.

The intersection of any set of varieties in H is a variety in H, as is the union of a locally finite set of varieties in H. For general properties of varieties, see for instance [1a], [2], [4], [4a] and [7].

The point $p \in V$ is a *simple* (or *regular* or *ordinary*) point of V if U and the functions $f_1, \cdots, f_s$ defining V in U may be taken so that $df_1(p), \cdots, df_s(p)$ are independent. (Then V is of dimension $n - s$ at p.) Other points of V are *singular*.

There are two definitions of the dimension of V at p:

(3.1) $\dim_p V =$ the highest dimension of V at simple points near p.

(3.2) $\dim_p V =$ the lowest codimension of planes P through p such that p is an isolated point $V \cap P$.

The equivalence of the two definitions is proved in [7]. Note that, by (3.1), the set $\{p \in V: \dim_p V \leqq s\}$ is open in V, for each s.

Define $\dim V = \sup\{\dim_p V: p \in V\}$. Say V is of *constant dimension* r, $\dim V \equiv r$, if $\dim_p V = r$, all $p \in V$. (This does not require V to be non-void.) Note that $\dim_p V = 0$ if and only if p is an isolated point of V. We write $\dim V < 0$ if and only if V is void: $V = \varnothing$. A *hypersurface* in C^n is a local variety of constant dimension $n - 1$. Because of the Weierstrass preparation theorem, a variety defined by $f = 0$ (where $f \not\equiv 0$) is a hypersurface.

An *analytic manifold* is a local variety of constant dimension, all of whose points are simple. (It need not be connected, and it need not be a variety in H.) In this connection, see § 18. We generally omit *analytic*.

For any manifold M in H, set $\bar{M} = H$-clos M. Note that fron $M = \bar{M} - M$ is closed in H; for $\bar{M}$ is closed in H, and M is clearly open in $\bar{M}$.

For any variety V, define

(3.3) $\qquad V_{sp} = \{p \in V:\ p \text{ simple}, \dim_p V = \dim V\}, \quad V_{sg} = V - V_{sp}.$

Now V_{sp} is a manifold; V_{sg} is a variety (see for instance [9, Lemma 2.18]) which consists of the singular points of V and the irreducible components of dimension less than $\dim V$.

Let $I(V, p)$ denote the ring of germs of holomorphic functions in C^n about p which vanish on the germ of V at p; this is an ideal in the ring of holomorphic germs about p. The sheaf of these ideals is coherent (see [2]; the meaning will be described in § 15).

A *pseudopolynomial* ω in y over $C_{x_1\cdots x_N}$ is a function of the form

$$\omega(x_1, \cdots, x_N, y) = y^h + \sum_{k=1}^{h} A_k(x_1, \cdots, x_N)y^{h-k} ,$$

where the A_k are holomorphic in some open set H. It is *distinguished* if $(0, \cdots, 0) \in H$ and $A_k(0, \cdots, 0) = 0$ for each k. A *polycylindrical neighborhood* of a point $(a_1, \cdots, a_n)$ is a set expressible in the form $|x_i - a_i| < \varepsilon_i (i = 1, \cdots, n)$.

The following lemma gives some basic structural properties of varieties. It is contained essentially in [7].

LEMMA 3.4. *Let V be a variety containing the origin O in $C^n = C^N \times C^M$, and suppose*

(3.5) *O is an isolated point of $V \cap C^M$, hence* $\dim_o V = r \leqq N$.

Let π denote the projection onto C^N. Then there is a neighborhood U_0 of O with the following properties: $\dim(V \cap U_0) = r$. *Take any polycylindrical neighborhood $U = U^N \times U^M \subset U_0$ of O such that $V \cap (U^N \times \partial U^M)$ is void. Then:*

(a) *$W = \pi(V \cap U)$ is a variety of dimension r in U^N, and (for $q \in W$)*

$$\dim_q W = \sup_{p \in \pi^{-1}q} \dim_p (V \cap U) , \quad \dim_o W = r .$$

(b) *There are distinguished pseudopolynomials*

$$\omega_j(x_1, \cdots, x_N, x_j) \qquad (j = N + 1, \cdots, n) \tag{3.6}$$

with the following properties. The discriminant D_j of ω_j is $\not\equiv 0$. The set of common zeros of the ω_j in $U \cap \pi^{-1}(W)$ forms a variety V' in U, an "imbedding variety" of $V \cap U$; V' is of dimension r, and $V \cap U \subset V'$.

(c) *If* $\dim(V \cap U) \equiv r$, *then* $\dim W \equiv r$ *and* $\dim V' \equiv r$, *and $V \cap U$ is the union of certain irreducible components of V'.*

(d) *If $V \cap U$ is irreducible, then the ω_j are irreducible.*

(e) *Suppose $r = N$. Then $W = U^N$. Also, let W_1 denote the union of the varieties of zeros of the D_j in U^N, and set $V_1 = V \cap U \cap \pi^{-1}(W_1)$; then V_1 is a variety in U, and* $\dim W_1 < N$, $\dim V_1 < N$.

(f) *Suppose* $\dim(V \cap U) \equiv N$. *Then, using* (e), *each point $p \in V \cap U - V_1$ is in a neighborhood U_p such that $V \cap U_p = V' \cap U_p$ and $\pi | (V \cap U_p)$ is an analytical equivalence (see* § 4). *Also, if π_j is the projection onto $C_{x_1\cdots x_N x_j}$, then $\pi_j(V \cap U)$ is the variety in $\pi_j(U)$ defined by $\omega_j = 0$, and ω_j generates $I(\pi_j(V \cap U), p)$ (over $C_{x_1\cdots x_N x_j}$), $p \in \pi_j(U)$. Moreover, the roots of ω_j in x_j lie in $U \cap C_{x_j}$ for $(x_1, \cdots, x_N)$ in $\pi(U)$.*

REMARK 3.7. Suppose V is a hypersurface, and we apply the lemma with $M = 1$. Then π_n is the identity, and (f) shows that ω_n generates $I(V, p)$.

To prove the lemma, take U_0 as in [7, Satz 1], small enough so that

$\dim(V \cap U_0) = r$. That theorem gives the analyticity of W. Suppose $\dim_q W = s$. Then [7, proof of (b), p. 269] shows that there is a simple point p' of $V \cap U$ mapping into a simple point q' of W near q, with $\dim_{p'} V \geqq s$; hence we find $p \in \pi^{-1}q$ with $\dim_p (V \cap U) \geqq s$. On the other hand, there is a plane P in C^N of dimension $N - s$ such that q is isolated in $W \cap P$; set $P' = \pi^{-1}(P)$. Now if $p \in V \cap U$, $\pi(p) = q$, then p is isolated in $V \cap P'$ (this follows on using (b)), and $\dim P' = N - s + M = n - s$, showing that $\dim_p V \leqq s$. Thus (a) is proved.

The main part of (b) is in [7, Satz 1]; it is clear that $\dim V' = \dim W$. The statements in (c) follow from what has already been proved. For the proof of (d) see [7, Bemerkung, p. 271]. The statements in (e) follow from [7, Zusatz I and Satz 1], and a simple consideration of dimension. As to (f), see [7, Zusatz I and Zusatz II]. The statement about ideals in (f) may be proved by taking any holomorphic function f in $\pi_j(U)$, vanishing in $\pi_j(V \cap U)$, and noting that $g = f/\omega_j$, defined first in $\pi_j(U) - \pi_j(V \cap U)$, may be extended over $\pi_j(U)$ so as to be holomorphic.

Suppose V and W are varieties in the open set $H \subset C^n$. Then $V - W$ is clearly a variety in $H - W$; it is easy to see (using (3.1)) that

$$\dim V = \sup \{\dim (V \cap W),\ \dim (V - W)\}\,. \tag{3.8}$$

LEMMA 3.9. *Let V and W be varieties in the open set H. Then $V_0 =$ H-clos$(V - W)$ and $V_1 = V_0 \cap W$ are varieties in H. V_0 is the union of the irreducible components of V which have points in $H - W$. We have*

$$\dim V_0 = \dim (V - W) \leqq \dim V\,, \quad \dim_p V_1 < \dim_p V_0 \quad \textit{for } p \in V_1\,. \tag{3.10}$$

See [9, Lemma 2.15]. Part of this lemma is strengthened by the following theorem of Remmert and Stein:

LEMMA 3.11. *Suppose $W \subset H \subset C^n$ and $V \subset H - W$ are varieties. Suppose $\dim W < r$ and $\dim_p V \geqq r$ for all $p \in V$. Then H-clos V is a variety in H.*

See [7, Satz 13] or [1, p. 679].

The following lemmas will be used in § 18.

LEMMA 3.12. *Suppose $V \subset H \subset C^n$. Let $M_1, M_2, \cdots$ be the connected pieces of V_{sp}. Then, setting $\bar{M}_i = H$-clos M_i, the $\bar{M}_i$ are the irreducible components of highest dimension of V, and $\bar{M}_i - M_i = \bar{M}_i \cap V_{sg}$.*

The proof is simple; see [9, Lemma 2.16].

LEMMA 3.13. *Let $M \subset H$ be a manifold of dimension m. If $\bar{M} = H$-clos M is a variety in H, then $\dim \bar{M} \equiv m$; if also $\bar{M} - M$ is a variety in H, then $\dim(\bar{M} - M) < m$. If $\bar{M} - M$ is a variety in H and $\dim(\bar{M} - M) < m$, then*

$\bar{M}$ is a variety in H.

Take any simple point p of $\bar{M}$; choose a neighborhood U of p so that $M' = \bar{M} \cap U$ is a manifold, of dimension k say. There is a point $q \in M \cap U$; now for a small enough U' about q, $\bar{M} \cap U' = M \cap U'$, showing that $k = m$; hence $\dim \bar{M} \equiv m$. Next, set $W = \bar{M} - M$. Then

$$[H\text{-clos}\,(\bar{M} - W)] \cap W = \bar{M} \cap (\bar{M} - M) = \bar{M} - M\,;$$

by Lemma 3.9, this is of dimension $<m$. The last statement follows from Lemma 3.11, since M is clearly a variety in $H - (\bar{M} - M)$.

In the next lemma, note that the intersection of two manifolds is a local variety.

LEMMA 3.14. *Let $M \subset H$ and $M' \subset H$ be manifolds of dimension m, suppose* $\dim (M \cap M') < m$, *and suppose $V = H$-clos M and $V' = H$-clos M' are varieties. Set $X = V \cap V'$. Then* $\dim X < m$.

First suppose $M \cap M' = \varnothing$. Let $V_1, V_2, \cdots$ and $V'_1, V'_2, \cdots$ be the irreducible components of V and of V' respectively; by Lemma 3.13, these are of dimension m. Set $X_{ij} = V_i \cap V'_j$; it is sufficient to show that $\dim X_{ij} < m$. If this is not so, then since V_i and V'_j are irreducible, $V_i = X_{ij} = V'_j$. Set $V_i^* = \bigcup_{k \neq i} V_k$. Since $V_i \not\subset V_i^*$, we may find $p \in M - V_i^*$. Choose a neighborhood U of p so that

$$M \cap U = V \cap U = V_i \cap U = V'_j \cap U\,.$$

Set $V'^*_j = \bigcup_{k \neq j} V'_k$. Since $\dim (V'_j \cap V'^*_j) < m$, we may find $p' \in M' \cap U - V'^*_j$. Now $p' \in V'_j \cup U \subset M$, so that $M \cap M' \neq \varnothing$, a contradiction.

In the general case, set $W = M \cap M'$; then $\dim W < m$. Set $M'' = M' - M$. Since $\dim W < \dim M'$, clearly $M' \subset H$-clos M''; hence H-clos $M'' = V'$. Applying the case already proved to M and M'' gives $\dim X < m$.

4. Holomorphic mappings

Let $V \subset H \subset C^n$ be a variety. A *holomorphic function* in V is a complex valued function such that for each $p \in V$ there is a neighborhood U of p and a holomorphic function g in U such that $g\,|\,(V \cap U) = f\,|\,(V \cap U)$. (In some papers a function is called holomorphic in V if it is continuous in V and holomorphic at all simple points; the two definitions agree for normal varieties.)

Given $V \subset H \subset C^n$, a mapping f: $V \to C^m$ is *holomorphic* if each of the m components of f is holomorphic. Then the same local extension property of f as above holds. Suppose $V \subset H \subset C^n$, $W \subset H' \subset C^m$, and f: $V \to W$ is holomorphic. Then for any holomorphic function φ in W, $\varphi \circ f$ is holomorphic in V. Suppose, further, that f is one-one onto W. Then f is *biholomorphic* if

f^{-1} is holomorphic. We then call f a *biholomorphic homeomorphism*, or, an *analytical equivalence*.

LEMMA 4.1. *Suppose $V \subset H \subset C^n$, $W \subset H' \subset C^m$, and $f: V \to W$ is an analytical equivalence of V with W. Then for each $p \in V$, we may find neighborhoods U of p and U' of $f(p)$ and manifolds $M \subset U$, $M' \subset U'$ of the same dimension, such that $V \cap U \subset M$, $W \cap U' \subset M'$, and there is an analytical equivalence F of M with M' which is an extension of $f|(V \cap U)$.*

See for instance [9, Theorem 4.7].

An analytical equivalence F between manifolds M, M' has non-vanishing jacobian; that is, $dF(p)$ sends independent vectors at p in M into independent vectors at $F(p)$ in M'.

If M is a manifold of dimension m and $p \in M$, a *coordinate system in M about p* is an analytical equivalence of a neighborhood of the origin in C^m with a neighborhood of p in M. A special case is: $M = C^m$. The coordinate system then gives a new metric in C^m and alters orthogonality, but preserves topology and vectorial properties. This is clear from the fact that, given the holomorphic function F in $H \subset C^n$, $p_0 \in H$ and $\varepsilon > 0$, there is a $\delta > 0$ such that, if p and q are within δ of p_0, then

$$F(q) = F(p) + dF(p) \cdot (q - p) + R\,, \qquad |R| \leqq \varepsilon |q - p|\,. \tag{4.2}$$

Using the F of Lemma 4.1, and coordinate systems about p in M and about $f(p)$ in M', we see that *an analytical equivalence between varieties preserves vectorial properties at each point.*

In any particular case, the above will have meaning if we can apply (4.2). For instance, a tangent vector v to a variety V at a singular point p is defined as the limit of a sequence of vectors $v_i = a_i(q_i - p)$; hence an analytical equivalence between two varieties extends to an analytical equivalence between the cones at corresponding points. (This fact is contained in [9, Theorem 4.7].)

EXAMPLES. (a) Define $f: C_t \to C_{xy}$ by $f(t) = (t^2, t^3)$; f is one-one and holomorphic. Since $df(0) \cdot v = 0$, all v, f is not biholomorphic.

(b) Define $f: (C_x \cup C_y) \to C_{xy}$ by $f(x, 0) = (x, 0)$, $f(0, y) = (y, y^2)$. Then $df(0) \cdot (e_y - e_x) = 0$.

(c) See [9, 2.14, (a)].

Suppose $V \subset H \subset C^n$, $W \subset H' \subset C^m$ (for instance, $W = C^m$), and $f: V \to W$ is holomorphic. Then for any simple point p of V, the rank of f at p (i.e., of $df(p)$) is defined, considering the image in C^m. The *rank* $\operatorname{rnk} f$ of f is the maximum of the rank of f at simple points of V. This definition agrees essentially with that of Remmert in [6].

LEMMA 4.3. *With V, W, f as above, if V_1 is a subvariety of V then* $\operatorname{rnk} f \mid V_1 \leqq \operatorname{rnk} f$; *if* $V_1 \subset H\text{-clos}(V - V_1)$, *then* $\operatorname{rnk} f \mid (V - V_1) = \operatorname{rnk} f$.

(This is an easy consequence of Lemma 13.5 below.)

The mapping $f: V \to H' \subset C^m$ is *proper into* H' if, for any compact set $Q \subset H'$, $f^{-1}(Q)$ is compact. We shall need the following theorem of Remmert; see [6] or [8]:

LEMMA 4.4. *Suppose $V \subset H \subset C^n$ is a variety, and $f: V \to H'$ is holomorphic and proper into H'. Then $f(V)$ is a variety in H', and* $\dim f(V) = \operatorname{rnk} f \leqq \dim V$. *If* $\dim_p V$ *and* $\operatorname{rnk}_p f$ *are constant, then* $\dim f(V)$ *is constant.*

The following lemmas are useful. The first is elementary in character; the second may be proved easily with the help of the vector fields of § 15.

LEMMA 4.5. *Let f be a smooth mapping of the manifold M of dimension m into the manifold M', and suppose f is of rank ρ at all points. Then:*

(a) *For each $q \in f(M)$, $f^{-1}(q)$ is a manifold of dimension $m - \rho$.*

(b) *Each $p \in M$ is in a neighborhood $U \subset M$ such that $f(U)$ is a manifold of dimension ρ.*

The above holds in both the complex and real cases; *smooth* means *continuously differentiable*. (The manifold need not be analytic here.)

LEMMA 4.6. *Let $f: V \to W$ be holomorphic, and let V_1 be V_{s0} together with the points of V_{sp} where f is of rank less than the maximum. Then V_1 is a subvariety of V.*

5. Modifications of C^n

We shall *blow up* C^n at the point $p \in C^n$ by applying Hopf's *σ-process*, as follows. First, define the relation dep for sets of vectors:

(5.1) $\operatorname{dep}(v_1, \cdots, v_s)$ means that $v_1, \cdots, v_s$ are dependent.

In the s-fold cartesian product of C^n, this relation is described by the vanishing of certain determinants, and therefore defines a variety. Since replacing v_i by $a_i v_i$ does not alter the relation if $a_i \neq 0$, we may define $\operatorname{dep}(\kappa(v_1), \cdots, \kappa(v_s)) = \operatorname{dep}(v_1, \cdots, v_s)$; more generally, $\operatorname{dep}(u, \kappa(v)) = \operatorname{dep}(u, v)$, etc.

Now, given $p \in C^n$, define

$$C^{*n}(p) = \{(q, \nu) \in C^n \times P^{n-1}: \operatorname{dep}(q - p, \nu)\}\ ; \tag{5.2}$$

this is a variety in $C^n \times P^{n-1}$. Note that $p \times P^{n-1} \subset C^{*n}$. We have

$$C^{*n}(p) - p \times P^{n-1} = \{(q, \kappa(q - p)): q \neq p\}\ , \tag{5.3}$$

$$C^{*n}(p) = \operatorname{clos}\left(C^{n*}(p) - p \times P^{n-1}\right)\ . \tag{5.4}$$

Let π^* denote the projection of $C^n \times P^{n-1}$ onto C^n. Then $\pi^* \mid C^{*n}(p)$ is holomorphic, $\pi^*(p \times P^{n-1}) = \{p\}$, and π^* is one-one in $C^{*n}(p) - p \times P^{n-1}$; thus $C^{*n}(p)$ is a modification of C^n (see [3]).

LEMMA 5.5. *$C^{*n}(p)$ is a manifold of dimension n, and $p \times P^{n-1}$ is a submanifold of dimension $n-1$.*

We may see this by choosing a coordinate system near (p, ν) for any ν, as follows. Say $\nu = \kappa(v)$; we may suppose v_1 is $\neq 0$, and hence that $v = (1, v_2, \cdots, v_n)$. Now take any $(u_1, u_2, \cdots, u_n)$ near $(0, v_2, \cdots, v_n)$, and set

$$\varphi(u_1, u_2, \cdots, u_n) = \bigl(p + u_1(1, u_2, \cdots, u_n),\quad \kappa(1, u_2, \cdots, u_n)\bigr) ; \tag{5.6}$$

then

$$\varphi(0, v_2, \cdots, v_n) = (p, \nu) , \qquad \varphi(0, u_2, \cdots, u_n) \in p \times P^{n-1} .$$

Generalizing the above, we now blow up C^n along $C^m = C_{x_1 \cdots x_m}$, where $m < n$. Let P' be the projective space of $C_{x_{m+1} \cdots x_n}$. Let π denote the projection of C^n onto C^m, and π^*, the projection of $C^n \times P'$ onto C^n. Define

$$C^{*n}(C^m) = C_m^{*n} = \{(q, \nu) \in C^n \times P' : \operatorname{dep}\bigl(q - \pi(q), \nu\bigr)\} ; \tag{5.7}$$

this is a variety in $C^n \times P'$. It is a modification of C^n, and $C^m \times P' \subset C_m^{*n}$, $\pi^*(C^m \times P') = C^m$, while π^* is one-one elsewhere in C_m^{*n}. We have

$$C_m^{*n} - C^m \times P' = \{\bigl(q, \kappa\bigl(q - \pi(q)\bigr)\bigr) : q \in C^n - C^m\} , \tag{5.8}$$

and the closure of this set is C_m^{*n}. Just as before, we have

LEMMA 5.9. *C_m^{*n} is a manifold of dimension n, with $C^m \times P'$ as submanifold of dimension $n-1$. There is a natural expression $C_m^{*n} = C^{*,n-m}(O) \times C^m$.*

For more information on modifications, see [3] and [5].

6. Intersections of varieties

We shall need the following classical lemma:

LEMMA 6.1. *Let V and W be varieties in $H \subset C^n$. Then*

$$\dim_p (V \cap W) \geqq \dim_p V + \dim_p W - n , \qquad p \in V \cap W . \tag{6.2}$$

Hence $\operatorname{codim}_p (V \cap W) \leqq \operatorname{codim}_p V + \operatorname{codim}_p W$. Since the proof is not easily available, we give it here. Say $\dim_p V = r$, $\dim_p W = s$, $V_1 = V \cap W$. Suppose first that W is a plane. If (6.2) is false, then $k = \dim_p V_1 < r + s - n$. By (3.1), since $V_1 \subset W$, we may interpret this dimension in W; by (3.2), there is a plane P in W through p, of dimension $s - k$, such that p is isolated in $V_1 \cap P$. Now p is isolated in $V \cap P$, and hence $\dim_p V \leqq n - (s - k) < r$, a contradiction.

Now consider the general case. Set

$$X = V \times W \subset H \times H \subset C^n \times C^n .$$

X is a variety, as is easily seen, and using (3.1) shows that $\dim_x X = \dim_p V + \dim_p W$, where $x = (p, p)$. The diagonal $\Delta = \{(q, q): q \in C^n\}$ is a plane of dimension n; by the case already proved,

$$\dim_x (X \cap \Delta) \geqq \dim_x X + \dim_x \Delta - 2n = \dim_p V + \dim_p W - n .$$

Now $X \cap \Delta = \{(p, p): p \in V \cap W\}$ is analytically equivalent to $V \cap W$, and (6.2) follows from (3.1).

REMARK 6.3. The theorem of course remains true if C^n is replaced by a manifold of dimension n. It fails in more general cases, however, as we show by two examples.

EXAMPLE 6.4. Let Z consist of two intersecting 2-planes in C^3. Let V be one of the planes, and let W be a line in the other; let V and W intersect at p. Then $\operatorname{codim}(V, Z) = 0$, $\operatorname{codim}(W, Z) = 1$, $\operatorname{codim}(V \cap W, Z) = 2$.

EXAMPLE 6.5. Define Z, V and W in C^4 by $x_1x_4 - x_2x_3 = 0$, $x_3 = x_4 = 0$, $x_1 = x_2 = 0$ respectively. Then V, W, $V \cap W$ are of codimension 1, 1, 3 in Z respectively. Note that Z is irreducible, and since $d(x_1x_4 - x_2x_3)$ vanishes at O only, O is the only singular point of Z; furthermore, V and W are manifolds intersecting transversally.

We shall now compare the dimension of a variety with the dimensions of plane sections of the variety. We give first a special case.

LEMMA 6.6. *Let U be a polycylindrical neighborhood of O, let $V \subset U \subset C^n$ be a variety, write $C^n = C^1 \times C^{n-1}$, and suppose*

$$\dim [V \cap (p + C^{n-1})] \geqq s , \qquad \text{all } p \in C^1 \cap U .$$

Then $\dim V \geqq s + 1$.

Suppose $\dim V \leqq s$. Let $V_1, V_2, \cdots$ be the irreducible components of V. For each i, set $V_i(q) = V_i \cap (q + C^{n-1})$, and let Q_i be the set of points $q \in C^1$ such that $\dim V_i(q) \geqq s$. Suppose $q \in Q_i$. Then since V_i is irreducible, $V_i(q) = V_i$; hence each Q_i contains at most one point. Hence the hypothesis cannot hold.

LEMMA 6.7. *Suppose $V \subset U \subset C^n = C^m \times C^{n-m}$ (U polycylindrical), and*

$$\dim [V \cap (p + C^{n-m})] \geqq s , \qquad \text{all } p \in C^m .$$

Then $\dim V \geqq s + m$.

We use induction on m; it is proved for $m = 1$. For each $q \in C_{x_1} \cap U$, set

$$C(q) = q + C_{x_2 \cdots x_n} , \qquad C'(q) = C^m \cap C(q) , \qquad V_q = V \cap C(q) .$$

For each $p \in C'(q) \cap U$, $V_q \cap (p + C^{n-m}) = V \cap (p + C^{n-m})$ is of dimension $\geqq s$, and by induction, $\dim V_q \geqq s + m - 1$. Now the previous lemma gives $\dim V \geqq$

$s + m$.

LEMMA 6.8. *Suppose $V \subset U \subset C^n = C^m \times C^{n-m}$ (U polycylindrical), Q is a closed subset of $C^m \cap U$, and*

$$\dim [V \cap (q + C^{n-m})] \geqq s\,, \qquad \text{all } q \in Q\,.$$

Then if $\dim V < s + m$, *Q contains no inner points; if* $\dim V \leqq s + h (h \leqq m)$ *and Q is a variety, then* $\dim Q \leqq h$.

First, if Q contains inner points, then the last lemma applies, giving $\dim V \geqq s + m$. Next, if Q is a variety and $\dim Q = k > h$, then there is a manifold $M \subset Q$ of dimension k; with new coordinates $(x_1, \cdots, x_k)$ in M, the union of the $q + C^{n-m}$ $(q \in M)$ forms a manifold of the form $C^k \times C^{n-m}$, to which we may apply the last lemma, giving $\dim V \geqq \dim [V \cap (C^k \times C^{n-m})] \geqq s + k > s + h$, a contradiction.

7. Cutting down dimensions

Suppose V is a variety and M is a manifold in V, with $\dim V > \dim M + 1$. We wish to replace V by a subvariety W of dimension k, $\dim M < k < \dim V$, so that $W \cap M = V \cap M$. Also, if V' is a subvariety of V (for instance, V_{sg}), we wish to cut down the dimension of V' by the same amount. Lemma 7.8 shows how to do this in the neighborhood of a point.

The proof is carried out by intersecting V with a plane (in a suitable coordinate system). For some purposes, the approximate direction of this plane must be chosen in advance. Fuller information is given by the first lemma.

For planes of the same dimension intersecting in a given point, we may define *nearness* through the metric in (2.14). If M is a manifold $T(M, p)$ denotes the tangent plane to M at p. Nearness of manifolds of a fixed dimension near a point may be defined through nearness of their tangent plane.

LEMMA 7.1. *Let V and V′ be varieties in $H \subset C^n$, and M a manifold in H, such that*

$$V' \subset V\,, \qquad \dim V \equiv r\,, \qquad \dim (V' - M) \leqq s\,.$$

Let M_0 be a manifold in H such that $M \cap M_0$ is a manifold, and suppose

$$p_0 \in M \cap M_0 \cap V\,, \qquad \dim M_0 = n - h\,, \qquad M_0 \not\subset M\,, \tag{7.2}$$

$$T(M, p_0) \cap T(M_0, p_0) = T(M \cap M_0, p_0)\,, \tag{7.3}$$

$$0 \leqq h \leqq r - r'\,, \qquad \text{where } r' = \dim (V \cap M \cap M_0)\,. \tag{7.4}$$

Then there is a neighborhood U of p_0 and a manifold $M_1 \subset U$ arbitrarily close to $M_0 \cap U$, such that

$$M_1 \cap M \cap U = M_0 \cap M \cap U\,, \tag{7.5}$$

(7.6) $$\dim (V \cap U \cap M_1) \equiv r - h ,$$

(7.7) $$\dim (V' \cap U \cap M_1 - M) \leqq s - h .$$

We prove this by induction on h; it is trivial if $h = 0$.

Suppose first that $h = 1$. Because of (7.3) we may choose new coordinates in a neighborhood U of p_0 so that M and M_0 become parts of planes there, and

$$M \cap M_0 \cap U = C_{x_1 \cdots x_k} \cap U , \qquad M_0 \cap U = C_{x_1 \cdots x_{n-1}} \cap U , \qquad k < n - 1 .$$

For any $\lambda = (\lambda_{k+1}, \cdots, \lambda_{n-1})$, set

$$\varphi_\lambda(x_1, \cdots, x_n) = x_n - (\lambda_{k+1}x_{k+1} + \cdots + \lambda_{n-1}x_{n-1}) .$$

Now $\varphi_\lambda = 0$ defines a plane $M(\lambda) \subset U$, and $M(0) = M_0 \cap U$.

Since V has a finite number of irreducible components near p_0, we may choose U so that $V \cap U$ has a finite number of irreducible components with points not in M, say $V_1, \cdots, V_a$. We may suppose $V_1', \cdots, V_b'$ defined similarly for V'. Choose $p_i \in V_i \cap U - M$, $p_j' \in V_j' \cap U - M$.

Say $p_1 = (x_1, \cdots, x_n)$. Then $x_i \neq 0$ for some $i > k$, and we may choose an arbitrarily small λ' such that $\varphi_{\lambda'}(p_1) \neq 0$. We may find λ'' arbitarily near λ' such that $\varphi_{\lambda''}(p_1) \neq 0$, $\varphi_{\lambda''}(p_2) \neq 0$. Continuing gives an arbitarily small λ such that $\varphi_\lambda(p_i) \neq 0$, $\varphi_\lambda(p_j') \neq 0$ for each i, j. Set $M_1 = M(\lambda)$. Then M_1 is near M_0, and (7.5) holds. Take any i. Since $V_i \not\subset M_1$ and V_i is irreducible, $\dim(V_i \cap M_1) < r$. By Lemma 6.1, the dimension is $\geqq r - 1$ at each point; hence the dimension is $\equiv r - 1$. For any other component X of $V \cap U$, $X \subset M$, and

$$X \cap M_1 \subset V \cap M \cap M_0 , \qquad \dim (X \cap M_1) \leqq r' \leqq r - 1 ;$$

since $\dim V \equiv r$, $\dim (X \cap M_1) \geqq r - 1$. Thus we see that (7.6) holds. Part of the above discussion proves (7.7).

Now suppose $h > 1$. Choose coordinates as before, so that

$$M \cap M_0 \cap U_0 = C_{x_1 \cdots x_k} \cap U_0 , \qquad M_0 \cap U_0 = C_{x_1 \cdots x_{n-h}} , \qquad k < n - h .$$

Set $M_0^{n-1} = C_{x_1 \cdots x_{n-1}}$; then (7.3) holds for M and M_0^{n-1}. Say $\dim(V \cap M \cap M_0^{n-1}) = r''$. Applying Lemma 6.1 to the pair $V \cap M \cap M_0^{n-1}$, M_0 in $M_0^{n-1} \cap U_0$ shows that $r' \geqq r'' - h + 1$; since $h \leqq r - r'$, this gives $1 \leqq r - r''$. We may therefore apply the proof above, with M_0 replaced by M_0^{n-1} and h replaced by 1, giving U_1 and M_1^{n-1} near M_0^{n-1} such that $M_1^{n-1} \cap M \cap U_1 = M_0^{n-1} \cap M \cap U_1$ and

$$\dim (V \cap U_1 \cap M_1^{n-1}) \equiv r - 1 , \qquad \dim (V' \cap U_1 \cap M_1^{n-1} - M) \leqq s - 1 .$$

We may choose a plane $M_2 \subset M_1^{n-1}$ near M_0 so that $M_2 \cap M = M_0 \cap M$. Now the conditions of the lemma are satisfied, with the replacements

C^n	V	V'	M	M_0	r	s	h
M_1^{n-1}	$V \cap M_1^{n-1}$	$V' \cap M_1^{n-1}$	$M \cap M_1^{n-1}$	M_2	$r - 1$	$s - 1$	$h - 1$.

Hence, by induction, we may find $U \subset U_1$ and $M_1 \subset M_1^{n-1}$ near M_2 (and hence near M_0) so that the conclusions of the lemma hold.

LEMMA 7.8. *Suppose* $p_0 \in M \subset V \subset H \subset C^n$, $V' \subset V$, $\dim V \equiv r$, $\dim(V' - M) \leqq s$, $\dim M = m$, *and* $0 \leqq h \leqq r - m$. *Then there is a variety* W *in a neighborhood* U *of* p_0 *such that* $M \cap U \subset W \subset V$, *and*

$$\dim W \equiv r - h , \qquad \dim (W \cap V' - M) \leqq s - h . \tag{7.9}$$

If $h < n - m$, we may choose a manifold $M_0 \supset M \cap U$ with $\dim M_0 = n - h$, and coordinates in U as in the last lemma, for some U. The conditions of that lemma are satisfied; finding $M_1 \subset U$ and setting $W = V \cap M_1$ gives the required facts. If $h = n - m$, then $r = n$, and we may set $W = M$.

PART II. TANGENT CONES

8. Elementary properties

A *cone* in C^n is a set K of vectors such that if $v \in K$ and $a \in C$, then $av \in K$. All cones we use will be closed sets. A *cone from* p is a set of points of the form $p + v$, where v is any vector of some cone. Given a cone K, the corresponding *projective cone* K^* is the set $\kappa(K) = \{\kappa(v): v \in K\}$; this is a subset of P^{n-1}. Thus κ gives a one-one correspondence between sets in P^{n-1} and cones in C^n; the null set in P^{n-1} corresponds to the trivial cone $\{O\}$.

The *projective distance* from the vector $v \neq 0$ to the cone K, or from the direction of v to K^*, is

$$| v, K |^* = | \kappa(v), \kappa(K) |^* = \inf \{ | v, u |^* : u \in K\} ; \tag{8.1}$$

$| v, K^* |^*$ and $| \kappa(v), K |^*$ mean the same quantity. This extends the definition of $| v, T |^*$ for planes T in § 2.

Given the variety $V \subset H \subset C^n$ and $p \in V$, the *tangent cone* $C(V, p)$ of V at p is the set of vectors v with the following property: There are sequences $\{p_i\}$ of points of V with $p_i \to p$ and $\{a_i\}$ of complex numbers such that $a_i(p - p_i) \to v$. This is clearly a closed cone. Note that this is a cone from O. The cone $p + C(V, p)$ from p will be used in § 9.

REMARK 8.2. It will not alter the definition if we require that the a_i be real and positive; in fact, there is an arc in V defining any $v \in C(V, p)$, as will be seen in Theorem 11.8.

The following properties are clear:

$$C(W, p) \subset C(V, p) \qquad \text{if } p \in W \subset V , \tag{8.3}$$

$$C(V \cup W, p) = C(V, p) \cup C(W, p) \qquad \text{if } p \in V \cap W , \tag{8.4}$$

$$C(V \cap W, p) \subset C(V, p) \cap C(W, p) \qquad \text{if } p \in V \cap W . \tag{8.5}$$

We cannot replace "$\subset$" by "$=$" in (8.5), as is shown by considering V and W in C^2 defined by $y = 0$ and $y - x^2 = 0$ respectively.

The *projective tangent cone* of V at p is $C^*(V, p) = \kappa(C(V, p))$.

Given $p \in V \subset H$, we blow up V through tangents at p, forming V_p^*, as follows: Recalling the definition of $C^{*n}(p)$ in § 5, set

$$(8.6) \qquad V_p^{**} = C^{*n}(p) \cap (V \times P^{n-1}) ;$$

this is a subvariety of $C^{*n}(p)$. Note that, by (5.3),

$$(8.7) \qquad V_p^{**} - p \times P^{n-1} = \{(q, \kappa(q - p)) : q \neq p, q \in V\} .$$

Set

$$(8.8) \qquad V_p^* = (H \times P^{n-1})\text{-clos}\,(V_p^{**} - p \times P^{n-1}) .$$

By Lemma 3.9, this is a variety in $H \times P^{n-1}$. Let π^* be the projection of $C^n \times P^{n-1}$ onto C^n. Then $\pi^* \mid V_p^*$ is holomorphic, and by (8.7), it is one-one except over p. Thus V_p^* is a modification of V over p.

We now show that the part of V_p^* over p is the projective tangent cone:

$$(8.9) \qquad V_p^* \cap (p \times P^{n-1}) = p \times C^*(V, p) .$$

First, take any $\nu \in C^*(V, p)$; say $\nu = \kappa(v)$. Then there are sequences $\{p_i\} \to p$ in V, $\{a_i\}$, such that $p_i \neq p$, $a_i(p_i - p) \to v$. Now $(p_i, \kappa(p_i - p)) = p_i^* \in V_p^{**}$, by (8.7), and $p_i^* \to (p, \nu)$, proving that $(p, \nu) \in V_p^*$. The converse is similar.

THEOREM 8.10. *$C^*(V, p)$ is an algebraic variety in P^{n-1}.*

Since V_p^* is a variety, so is $C^*(V, p)$; by the Theorem of Chow, it is algebraic. A proof, finding an explicit representation, will be given in § 10.

LEMMA 8.11. *If $p \in V$ and $\dim_p V = r \geqq 1$, then*

$$(8.12) \qquad \dim C(V, p) = \dim_p V = \dim C^*(V, p) + 1 .$$

If V is of constant dimension near p, then $C(V, p)$ and $C^*(V, p)$ are of constant dimension.

For some U about p, $\dim(V \cap U) = r$; hence suppose $\dim V = r$. By (8.7), $\dim(V_p^{**} - p \times P^{n-1}) = r$; hence, by Lemma 3.9, $\dim C^*(V, p) < r$. Next, there is a sequence $p_i \to p$ of simple points of V where V is of dimension r; since P^{n-1} is compact, there are simple points of V_p^* arbitrarily near some $(p, \nu) = p^*$ where V_p^* is of dimension r, and $\dim_{p^*} V_p^* = r$. By Lemma 5.5 and Lemma 6.1 with Remark 6.3, $\dim C^*(V, p) \geqq r - 1$, and (8.12) follows.

If V is of constant dimension near p, then the reasoning above may be applied to any $(p, \nu) \in V_p^*$, and the rest of the lemma follows.

REMARK 8.13. It is clear that, if K is a cone from p and K is a variety, then

$$\dim K = \dim_p K . \tag{8.14}$$

LEMMA 8.15. *Given $p \in V$ and $\varepsilon > 0$, there is a neighborhood U of p such that*

$$|q - p, C(V, p)|^* < \varepsilon \qquad \text{if } q \in V \cap U, q \neq p . \tag{8.16}$$

This is evident from the compactness of P^{n-1}. The lemma will be extended in Lemma 11.11 below.

LEMMA 8.17. *If V is irreducible at p, then $C^*(V, p)$ is connected.*

If not, then (since the projective cone is closed) we may write

$$C^*(V, p) = A \cup B , \qquad |A, B|^* = 2\varepsilon > 0 .$$

Choose U about p by Lemma 8.15; then $V \cap U - \{p\}$ falls into two disjoint closed sets Q_A, Q_B; each is a variety in $U - \{p\}$, and each is of dimension $\equiv r \geqq 1$. By Lemma 3.11, $Q_A \cup \{p\}$ and $Q_B \cup \{p\}$ are distinct varieties in U, whose union is $V \cap U$, contrary to the irreducibility of V at p.

EXAMPLE 8.18. Let $V \subset C^3$ be defined by $z^3 - xy = 0$; this is irreducible at O. By Theorem 10.3 below, $C(V, O)$ is defined by $xy = 0$; thus $C(V, O) = C_{yz} \cup C_{xz}$, and though $C^*(V, O)$ is connected, it is reducible.

REMARK 8.19. The set of all pairs (p, v) such that $p \in V$ and $v \in C(V, p)$ does not form a variety in $H \times C^n$; see for instance Example 12.18. We obtain a variety $C_4(V)$ on forming the closure of this set; see § 22.

9. Images of tangent cones under mappings

The first lemma below shows that holomorphic mappings carry tangent cones into tangent cones of any variety containing the image. Conditions sufficient for the mapping of the tangent cone to be onto are given in the next two lemmas. A condition about a plane intersecting a tangent cone at its vertex only is used; the final lemma shows the existence of such planes.

If $f: V \to C^m$ is holomorphic and $p \in V$, we may define $df(p) \cdot v$ for $v \in C(V, p)$ by means of an extension of f through a neighborhood of p; the result is clearly independent of the extension chosen.

LEMMA 9.1. *Given $p \in V$ and a holomorphic mapping $f: V \to W$,*

$$df(p) \cdot C(V, p) \subset C(W, q) , \qquad q = f(p) . \tag{9.2}$$

Take any $v \in C(V, p)$; say $a_i(p_i - p) \to v$, where $p_i \in V$ and $p_i \to p$. We may suppose $|v| = 1$ and $|a_i| = 1/|p_i - p|$. Then, as in (4.2),

$$a_i[f(p_i) - f(p)] = df(p) \cdot a_i(p_i - p) + a_i R_i ,$$

where $a_i R_i \to 0$; hence $a_i[f(p_i) - f(p)] \to df(p) \cdot v$, which is therefore a vector in $C(W, q)$.

LEMMA 9.3. *Suppose* $p \in V$, $f\colon V \to C^m$ *is holomorphic, and for some* $\eta > 0$,

$$|f(p') - f(p)| \geqq \eta |p' - p| , \qquad p' \in V . \tag{9.4}$$

Then $W = f(V)$ *is a variety near* $q = f(p)$, *and*

$$df(p) \cdot C(V, p) = C(W, q) . \tag{9.5}$$

Take $\delta > 0$ so that $\bar{U}_\delta(p) \subset H$ (where $V \subset H$), and set

$$V_1 = V \cap f^{-1}(U_{\eta\delta}(q)) .$$

Then V_1 is open in V and $V_1 \subset U_\delta(p)$. For each $p' \in V_1$, choose a neighborhood $U_{p'} \subset U_\delta(p)$ of p' so that $U_{p'} \cap V = U_{p'} \cap V_1$; let U_1 be the union of these neighborhoods. Now V_1 is a variety in U_1, and $V_1 = V \cap U_1$.

We show next that $f | V_1$ is proper into $U' = U_{\eta\delta}(q)$. Take any compact set $Q \subset U'$; then $f^{-1}(Q) \cap V$ is a closed subset of the compact set $V \cap \bar{U}_\delta(p)$ and hence is compact; but $f^{-1}(Q) \cap V \subset V_1$, and thus $(f | V_1)^{-1}(Q)$ is compact. We may now apply Lemma 4.4, showing that $f(V) \cap U' = f(V_1)$ is a variety in U'.

Take any $w \in C(W, q)$; we show that for some $v \in C(V, p)$, $df(p) \cdot v = w$; because of Lemma 9.1, this will complete the proof.

If $w = 0$, set $v = 0$. Now suppose $w \neq 0$. Say

$$b_i(q_i - q) \to w , \qquad q_i \in W \cap U' , \qquad q_i \neq q, \qquad q_i \to q .$$

Say $f(p_i) = q_i$, $p_i \in V$. Using a subsequence if necessary, we may suppose that $\kappa(p_i - p) \to \nu \in P^{n-1}$. Because of (9.4), $p_i \to p$; hence if $a_i = 1/|p_i - p|$, $a_i(p_i - p) \to u \in C(V, p)$. Now (as in Lemma 9.1)

$$df(p) \cdot u = \lim a_i(q_i - q) = w_1$$

for some w_1. Also, by (9.4),

$$| a_i(q_i - q) | = |f(p_i) - f(p)|/|p_i - p| \geqq \eta ;$$

hence $w_1 \neq 0$. Since $b_i(q_i - q) \to w$, $w = cw_1$ for some $c \neq 0$. Now if $v = cu$, $df(p) \cdot v = w$, as required.

Define the *tangent cone from* p by

$$C_p(V) = p + C(V, p) . \tag{9.6}$$

LEMMA 9.7. *Suppose* $p \in V$, T_1 *is a plane through* p, T_2 *is the orthogonal plane through* p, π_i *is the projection onto* T_i, *and*

$$C_p(V) \cap T_1 = \{p\} . \tag{9.8}$$

Then for some neighborhood U *of* p, $W = \pi_2(V \cap U)$ *is a variety near* $\pi_2(p) = p$, *and*

$$d\pi_2(p)\cdot C(V,p) = C(W,p)\,. \tag{9.9}$$

Since $C^*(V, p)$ is a closed set, there is an $\eta > 0$ such that

$$|v, T_1|^* \geqq 2\eta \qquad \text{if } v \in C(V,p),\ v \neq 0\,.$$

Hence, by Lemma 8.15, we may take a neighborhood U of p so small that

$$|q - p, T_1|^* \geqq \eta \qquad \text{if } q \in V \cap U,\ q \neq p\,.$$

Since $|q - p, T_1|^* = |q - \pi_1 q|/|q - p| = |\pi_2 q - p|/|q - p|$, we have

$$|\pi_2 q - p| \geqq \eta |q - p| \qquad \text{if } q \in V \cap U,\ q \neq p\,.$$

Thus the conditions of Lemma 9.3 hold for $V \cap U$, and (9.5) gives (9.9).

EXAMPLE 9.10. Let $V \subset C_{xy}$ be defined by $y^2 - x^3 = 0$, and let π be the projection onto C_y. Then $C(V, O) = \{ae_x\}$, $\pi(V) = C_y$, but $d\pi(O)\cdot e_x = 0$.

LEMMA 9.11. *Suppose* $p \in V \subset H \subset C^n$, $\dim_p V = r$. *Then there is an* $(n - r)$*-plane* K *through* p *such that*

$$C_p(V) \cap K = \{p\}\,. \tag{9.12}$$

In fact, if J *is an* h*-plane through* p *which satisfies* (9.12), *with* $0 \leqq h < n - r$, *and* $K_0 \supset J$ *is an* $(n - r)$*-plane, we may choose* $K \supset J$ *arbitrarily near* K_0 *so that* (9.12) *holds.*

We prove the second part; then the first part follows on taking $J = \{p\}$, or on using Lemma 8.11. We use induction on n. We may suppose $r < n$.

If $\dim J \geqq 1$, let $L \subset J$ be a line through p; otherwise, since $\dim C(V, p) = r < n$, we may choose L arbitarily near K_0 so that $C(V, p) \cap L = \{p\}$. Let π be the projection onto the plane T orthogonal to L through p. By Lemma 9.7, we may choose a neighborhood U of p so that $W = \pi(V \cap U)$ is a variety near p, and (9.9) holds with π. We may also apply Lemmas 3.4 and 8.11, giving $\dim C(W, p) = \dim W = r$.

Set $K_0' = K_0 \cap T$, $J' = J \cap T$; now $K_0' \supset J'$, and $\dim K_0' = n - r - 1$. Suppose $w \in C(W, p)$, $p + w \in J'$. Then for some $v \in C(V, p)$, $d\pi(p)\cdot v = w$; if $w \neq 0$, then $J' \neq \{p\}$, $L \subset J$, and $p + v \in J$, $v \neq 0$, contradicting the hypothesis about J; thus $w = 0$, and the hypotheses of the lemma are satisfied for W, K_0' and J' at p in T. Hence, by induction, we may choose K' arbitrarily near K_0' so that

$$J' \subset K' \subset T\,, \qquad C_p(W) \cap K' = \{p\}\,.$$

Now K' and L span an $(n - r)$-plane K, which is near K_0. If (9.12) is false, say $v \in C(V, p)$, $p + v \in K$, $v \neq 0$. By the choice of L, $p + v \notin L$; hence

$$w = \pi v \in C(W, p)\,, \qquad p + w \in K', \qquad w \neq 0\,.$$

But this is a contradiction, and the lemma is proved.

10. Algebraic expression of tangent cones

We show here that $C(V, p)$ is defined by the vanishing of the initial polynomials of the functions in the ideal $I(V, p)$. The proof is given first for a hypersurface, then for the general case. (The author is indebted to the referee for suggesting simplifications in the proof.)

Let f be a holomorphic function in a neighborhood of p. Then expanding about p gives

$$(10.1) \qquad f(p + v) = f_p^{[0]} + f_p^{[1]}(v) + f_p^{[2]}(v) + \cdots, \qquad f_p^{[0]} = f(p) ,$$

where $f_p^{[m]}(v)$ is a homogeneous polynomial of degree m in the components $v_1, \cdots, v_n$ of v. If m is the smallest number with $f_p^{[m]} \not\equiv 0$, we say f is of *order* m at p, and we call $f_p^{[m]} = f_p^{[*]}$ the *initial polynomial* of f at p. Note that

$$(10.2) \qquad (fg)_p^{[*]} = f_p^{[*]} g_p^{[*]} , \qquad (fg)_p^{[*]} = g(p) f_p^{[*]} \qquad \text{if } g(p) \neq 0 .$$

THEOREM 10.3. *Let $V \subset H \subset C^n$ be a hypersurface, defined near p by the vanishing of the holomorphic function f. Then $C(V, p)$ is defined by the vanishing of $f_p^{[*]}$.*

REMARK 10.4. By Remark 3.7, V is defined near p by the vanishing of a single function. In the theorem, f need not generate $I(V, p)$.

To prove the theorem, say f is of order m at p; then we may write

$$(10.5) \qquad f(p + \lambda v) = \lambda^m [f_p^{[m]}(v) + \lambda R(p, v, \lambda)] , \qquad p + v \in U, |\lambda| \leqq 1 ,$$

if f is defined in U; R is holomorphic. We show that $f_p^{[m]} = 0$ defines the cone $C(V, p)$.

First, take any $v \in C(V, p)$. There are sequences $\{p_i\} \to p$ in V and $\{a_i\}$, $a_i \neq 0$, $a_i \to \infty$, such that $a_i(p_i - p) \to v$. Set $\lambda_i = 1/a_i$; replacing λ and v in (10.5) by λ_i and $a_i(p_i - p)$ respectively and letting $i \to \infty$ shows at once that $f_p^{[m]}(v) = 0$.

Now take any $v \notin C(V, p)$. Let $F(u, \lambda) = f(p + \lambda u)$. The definition of $C(V, p)$ shows that, for some neighborhood U of $(v, 0)$ in $C^n \times C$,

$$F(u, \lambda) \neq 0 \qquad \text{if } (u, \lambda) \in U, \lambda \neq 0 .$$

Since $F(u, 0) = 0$ for all u, applying the Weierstrass preparation theorem to F shows that for some integer m' and holomorphic function G,

$$f(p + \lambda u) = \lambda^{m'} G(u, \lambda) , \qquad G(v, 0) \neq 0 .$$

Since $f_p^{[m]} \not\equiv 0$, (10.5) shows that $m' = m$, and hence also that $f_p^{[m]}(v) = G(v, 0) \neq 0$, completing the proof.

We turn now to the general case.

THEOREM 10.6. *Given $p \in V$, $C(V, p)$ is the set of solutions of $f_p^{[*]}(v) = 0$,* *for all f whose germs at p are in $I(V, p)$.*

Take any $v \in C(V, p)$. For any f with germ in $I(V, p)$, defining $V_f \supset V$ near p, we have $v \in C(V_f, p)$, and the last theorem shows that $f_p^{[*]}(v) = 0$.

Now take any $v \notin C(V, p)$; we shall find a function f with germ in $I(V, p)$ such that $f_p^{[*]}(v) \neq 0$. We may suppose $|v| = 1$.

First suppose that V is irreducible at p. Say $\dim V = r$. By Lemma 9.11 there is a plane C' through p, of dimension $n - r$, such that (see (9.6))

$$(10.7) \qquad v \in C', \qquad C_p(V) \cap C' = \{p\} .$$

Take new axes so that $v = e_n$ $C' = C_{x_{r+1}\cdots x_n}$. By (10.7) and Lemma 8.15, there is a neighborhood U_0 of p and an $\alpha > 0$ such that

$$(10.8) \qquad |y - p, C'|^* > \alpha \qquad \text{if } y \in V \cap U_0, y \neq p .$$

We may therefore apply Lemma 3.4 (particularly (b) and (f)), giving a polycylindrical $U_1 \subset U_0$ and a holomorphic function

$$f(x_1, \cdots, x_n) = \omega_n(x_1, \cdots, x_r, x_n)$$

in U_1, such that:

(a) $f = 0$ in $V \cap U_1$; hence f has germ in $I(V, p)$.

(b) If π' is the projection onto $C_{x_1\ldots x_r x_n}$, then $\pi'(V \cap U_1)$ is the set of zeros of f in $\pi'(U_1)$.

Say $f = 0$ defines the variety Z in U_1; then $V \cap U_1 \subset Z$. Take any $x \in Z$. By (b), we may choose $y \in V \cap U_1$ such that $\pi' y = \pi' x$. Set $v_x = x - p$, $v_y = y - p$. By (2.12) and (10.8),

$$|v_x, C_{x_n}|^* \geqq |\pi' v_x, C_{x_n}|^* = |\pi' v_y, C_{x_n}|^* \geqq |v_y, C'|^* > \alpha .$$

(If $\pi' v_x = 0$, then $|v_x, C_{x_n}|^* = 1$.) It follows that $v = e_n \notin C(Z, p)$. By Theorem 10.3, $C(Z, p)$ is defined by $f_p^{[*]} = 0$; hence $f_p^{[*]}(v) \neq 0$, as required.

In the general case, we may write V as the union of irreducible components $V_1, \cdots, V_s$ in some neighborhood U of p. Given $v \notin C(V, p)$, we may find f_i with germ in $I(V_i, p)$ such that $f_{i,p}^{[*]}(v) \neq 0$. Set $f = f_1 \cdots f_s$; then f has germ in $I(V, p)$ and $f_p^{[*]}(v) = f_{1,p}^{[*]}(v) \cdots f_{s,p}^{[*]}(v) \neq 0$, completing the proof.

REMARK 10.9. A set $\{f_i\}$ generating $I(V, p)$ may be insufficient to generate $C(V, p)$ as in the theorem; see [9, Remark 5.10].

11. One-dimensional sections

Given $p \in V$ and $v \in C(V, p)$, $v \neq 0$, we show that there is a one-dimensional subvariety W of V with $v \in C(W, p)$, and an arc $\Gamma \subset W$ with v as tangent vector. First we note:

LEMMA 11.1. *If* $\dim V \equiv 1$ *and V is irreducible at p, then $C^*(V, p)$ contains a single element.*

By Lemma 8.11, $\dim C^*(V, p) \equiv 0$; now apply Lemma 8.17.

LEMMA 11.2. *Given $p \in V$ and $\nu \in C^*(V, p)$, there is a neighborhood U of p and an irreducible variety $W \subset V \cap U$ such that $\dim W \equiv 1$ and $C^*(W, p) = \{\nu\}$.*

Because of (8.4), we may clearly suppose that V is irreducible; say $\dim V \equiv r$. Blow up V through tangents at p, forming V_p^*, as in (8.8). By (8.8), (8.7) and Lemma 3.9, $\dim V_p^* \equiv r$. Set $p^* = (p, \nu)$; by (8.9), $p^* \in V_p^*$. Set $X^* = V_p^* \cap (p \times P^{n-1})$, and apply Lemma 7.8 with the subsitutions

$$\begin{array}{cccccc} p_0 & V & V' & M & s & h \\ p^* & V_p^* & X^* & \{p^*\} & r-1 & r-1\,. \end{array}$$

This gives a variety W^* near p^* such that

$$p^* \in W^* \subset V_p^*\,, \qquad \dim W^* \equiv 1\,, \qquad \dim\big(W^* \cap X^* - \{p^*\}\big) \leqq 0\,.$$

Now p^* is isolated in $W^* \cap (p \times P^{n-1})$; hence we may apply Lemma 3.4 to the projection π^* of $C^n \times P^{n-1}$ onto C^n. Setting $W_0 = \pi^* W^*$ near p^*, W_0 is a variety near p, and $W_0 \subset V$, $\dim W_0 \equiv 1$.

Since $\dim W^* \equiv 1$, we may choose $p_i^* \in W^*$ near p^*, $p_i^* \notin X^*$. Now $p_i^* = (p_i, \nu_i) \in V_p^* \subset C^{*n}(p)$; by (5.3), $\nu_i = \kappa(p_i - p)$. Now $p_i \in W_0$, showing that $\nu \in C^*(W_0, p)$. Let W be an irreducible component of W_0 (near p) such that $C^*(W, p)$ contains ν; then $C^*(W, p) = \{\nu\}$.

EXAMPLE 11.3. Define $V \subset C^2$ by $y(y - x^2) = 0$, and set $\nu = \kappa(e_x)$; then for W we may take either irrreducible component of V.

A branch of a real analytic curve has a one-sided continuous derivative at a singular point. The following lemma expresses part of this; see also Lemma 13.10. (Second derivatives need not be bounded, as is shown by the variety $y^2 - x^3 = 0$ in C_{xy}.)

LEMMA 11.4. *Let $f(x, y) \not\equiv 0$ be holomorphic near O, and let φ be a continuous complex valued function of the real variable λ such that*

$$f(\lambda, \varphi(\lambda)) = 0\,, \qquad \lambda \in I_\alpha = \{\lambda: 0 \leqq \lambda \leqq \alpha\}\,. \tag{11.5}$$

Suppose

$$\lim_{\lambda \to 0} \varphi(\lambda)/\lambda = 0\,. \tag{11.6}$$

Then, if $\varphi' = d\varphi/d\lambda$, φ' is continuous in I_α, and hence

$$\lim_{\lambda \to 0} \varphi'(\lambda) = \varphi'(0) = 0\,. \tag{11.7}$$

Splitting f into its irreducible factors near O, it is clear that we may suppose f irreducible at O. Now if $D(x)$ is the discriminant of f, then $D \not\equiv 0$; hence $f(x, y) = 0$ determines y as a multi-valued function of x near O. For some smallest h, y is a single valued function of ξ, where $\xi^h = x$; that is, there is a holomorphic function $\psi(\xi)$ for $|\xi| < \beta$ such that

$$\psi(0) = 0\,; \qquad f(\xi^h, \psi(\xi)) = 0\,, \qquad |\xi| < \beta\,.$$

With a fixed $\gamma \in I_\beta$ such that $\gamma^h \in I_\alpha$, since $f(\gamma^h, \psi(\gamma)) = 0$ and $f(\gamma^h, \varphi(\gamma^h)) = 0$, we may suppose ψ chosen so that $\psi(\gamma) = \varphi(\gamma^h)$. Now by continuity (if γ is small enough),

$$\psi(\lambda) = \varphi(\lambda^h) \qquad \text{if } \lambda \in I_\gamma\,.$$

Keeping $\lambda \in I_\gamma$, (11.6) gives $\psi(\lambda)/\lambda^h \to 0$ as $\lambda \to 0$; hence we may write

$$\psi(\xi) = \xi^h\theta(\xi)\,, \qquad \theta(\xi) \text{ holomorphic for } |\xi| < \gamma,\ \theta(0) = 0\,.$$

Differentiating $\varphi(\lambda^h) = \lambda^h\theta(\lambda)$ and dividing by $h\lambda^{h-1}$ gives

$$\varphi'(\lambda^h) = \theta(\lambda) + \lambda\theta'(\lambda)/h \qquad (\lambda > 0)\,.$$

Letting $\lambda \to 0$ gives (11.7), and the conclusion follows.

We now show that tangent vectors to V may be defined by differentiable arcs:

THEOREM 11.8. *Given $p \in V$ and $v \in C(V, p)$, $v \neq 0$, there is an arc $\Gamma \subset V$, defined by a function $\Phi(\lambda)$ $(0 \leqq \lambda \leqq \lambda_1)$, such that $\Phi(0) = p$ and*

$$\text{(11.9)} \qquad \lim_{\lambda\to 0} \Phi'(\lambda) = \lim_{\lambda\to 0} \frac{\Phi(\lambda) - p}{\lambda} = v\,.$$

Choose the 1-dimensional variety $W \subset V$ by Lemma 11.2, and choose axes so that v is in the x_1-direction. By Lemma 3.4, there are pseudopolynomials $\omega_j(x_1, x_j)(j = 2, \cdots, n)$ in a polycylindrical neighborhood U of $p = O$, which vanish over W, and whose discriminants are not $\equiv 0$ and hence may be taken to be $\neq 0$ in $C_{x_1} \cap U - \{O\}$. Choose $\lambda_1 > 0$ so that $q = O + \lambda_1 e_1 \in U$. Since $\pi(W \cap U) = C_{x_1} \cap U$ (π = projection onto C_{x_1}), we may choose $q' \in W \cap U$ with $\pi q' = q$. Set $q_\lambda = O + \lambda e_1$; as $\lambda \to 0$, we may follow the solutions of the equations $\omega_j = 0$, defining the arc Γ in W from q' to O; say Γ is defined by $x_j = \varphi_j(\lambda)$ $(j = 2, \cdots, n)$. Since $C^*(W, p)$ contains $\kappa(v)$ only, we have $\varphi_j(\lambda)/\lambda \to 0$, all j. Applying Lemma 11.4 to $f = \omega_j$ for each j gives (11.9).

REMARK 11.10. By Theorem 11.8, we may define $C(V, p)$ as the set of tangent vectors to differentiable arcs in V from p. In particular, letting $\{p_i\}$ be a sequence along such an arc approaching p and setting $a_i = |v|/|p_i - p|$, we have $a_i(p_i - p) \to v$, with the a_i real and positive. (That there is such a sequence follows also from the proof of Theorems 10.3 and 10.6.)

We end this section by showing that V and $p + C(V, p)$ are near each other near p in a certain strict sense. Define the ρ-sphere $S_\rho(p) = \partial U_\rho(p) = \{q: |q - p| = \rho\}$ about p. Let dist (A, B) denote the Hausdorff distance between the sets A, B (see the explanation below).

LEMMA 11.11. *Given $p \in V$, consider the intersections with $S_\rho(p)$:*

$$V'_\rho = V \cap S_\rho(p)\,, \qquad C'_\rho = [p + C(V, p)] \cap S_\rho(p)\;;$$

then

$$\rho^{-1}\,\mathrm{dist}\,(V'_\rho, C'_\rho) \to 0 \qquad \textit{as } \rho \to 0\,. \tag{11.12}$$

Explicitly, for each $\varepsilon > 0$, there is a $\rho_0 > 0$ such that

$$V'_\rho \subset U_{\varepsilon\rho}(C'_\rho), \qquad C'_\rho \subset U_{\varepsilon\rho}(V'_\rho)\,, \qquad \text{if } \rho \leqq \rho_0\,. \tag{11.13}$$

The first part is a direct consequence of the compactness of P^{n-1} and the definition of $C(V, p)$. To prove the second part, set $\delta = \varepsilon/2$, and choose the unit vectors $v_1, \cdots, v_s$ in $C(V, p)$ so that, if $q_i = p + v_i$, then $C'_1 \subset U_\delta(\{q_1, \cdots, q_s\})$. Let Γ_i be an arc in V from p with the direction of v_i at p (Theorem 11.8). Say Γ_i and the v_i-ray from p intersect $S_\rho(p)$ at $p_{i\rho}$ and $q_{i\rho}$ respectively. Now for some $\rho_0 > 0$, $|p_{i\rho} - q_{i\rho}| < \rho\delta$ (all i) if $\rho \leqq \rho_0$, and the second inclusion follows at once.

12. Tangents at points of a submanifold

Supposing V contains the manifold M, we shall find (near $p_0 \in M$) a variety $K_M(V)$ whose cross section at $p \in M$ is the *M-tangent cone* of V at p; this cross section contains the projection of $C(V, p)$ onto the plane orthogonal to M at p.

A manifold, near a point, may be taken as part of a coordinate plane. Hence, throughout this section, we assume the following:

$$M = C_{x_1\cdots x_m} \cap H\,, \qquad M \subset V \subset H \subset C^n\,, \qquad \dim_p V > m \qquad \text{if } p \in M\,. \tag{12.1}$$

Set $C^m = C_{x_1\cdots x_m}$, $C' = C_{x_{m+1}\cdots x_n}$, and let π and π' be the projections onto these spaces. Define the *M-tangent cone* of V at $p \in M$:

$$C_M(V, p) = \{v\text{: for some sequences } \{q_i\}, \{a_i\}, \text{ we have } q_i \in V,\ q_i \to p,\ a_i(q_i - \pi q_i) \to v\}\,. \tag{12.2}$$

(This is closely related to the cone $C(V, M, p)$ in [9, § 7].) Set

$$K_M(V, p) = p + C_M(V, p)\,, \tag{12.3}$$

$$K_M(V) = \textstyle\bigcup_{p \in M} K_M(V, p)\,. \tag{12.4}$$

We call $K_M(V)$ the *M-tangent set* of V.

We shall find further expressions for $K_M(V)$. First, in the space $H \times C'$, set

$$\begin{cases} K'_M(V) = \{(q, v)\colon q \in V,\ v \in C',\ \mathrm{dep}\,(q - \pi q, v)\}\,, \\ K''_M(V) = (H \times C')\text{-clos}\,[K'_M(V) - M \times C']\,, \\ K'''_M(V) = [K''_M(V) \cap (M \times C')]\,. \end{cases} \tag{12.5}$$

Clearly $K'_M(V) - M \times C'$ is a variety, of dimension $r + 1$, where $r = \dim V$.

Hence, by Lemma 3.9, $K''_M(V)$ and $K'''_M(V)$ are varieties, and

(12.6) $$\dim K''_M(V) = r + 1\,, \qquad \dim K'''_M(V) \leqq r\,.$$

Noting that $\varphi(p, v) = p + v$ defines an analytical equivalence between $M \times C'$ and $\pi^{-1}M$, we show that

(12.7) $\varphi \mid K'''_M(V)$ is an analytical equivalence onto $K_M(V)$;

hence $K_M(V)$ is a variety. First, by (12.1), $M \times 0 \subset K'''_M(V)$; it corresponds to M in $K_M(V)$. Now, given $p \in M$, $q \in p + C'$, $q \neq p$, the condition that q be in $\varphi(K'''_M(V))$ is that $(p, q - p) = (p, v)$ be in $K'''_M(V)$, or, that there exists a sequence $\{(q_i, v_i)\} \in K'_M(V) - M \times C'$ such that $(q_i, v_i) \to (p, v)$, that is, for some $\{a_i\}$, $v_i = a_i(q_i - \pi q_i)$ and the conditions in (12.2) hold; thus, finally, that $v = q - p$ be in $C_M(V, p)$, or, that q be in $K_M(V, p)$. Thus (12.7) is proved.

Next, form the modification $C^{*n}(M)$ of C^n over M as in (5.7). Let P' be the projective space of C', and set

(12.8) $$\begin{cases} K'^*_M(V) = C^{*n}(M) \cap (V \times P')\,, \\ K''^*_M(V) = (\pi^{-1}M \times P')\text{-clos}\,[K'^*_M(V) - M \times P']\,, \\ K^*_M(V) = K''^*_M(V) \cap (M \times P')\,. \end{cases}$$

Note that, by (5.8),

(12.9) $$K'^*_M(V) - M \times P' = \{(q, \kappa(q - \pi q)) \colon q \in V \cap \pi^{-1}(M) - M\}\,;$$

thus this variety is of dimension r. By Lemma 3.9, $K''^*_M(V)$ and $K^*_M(V)$ are varieties.

Take any $p \in M$. Choose a neighborhood U of p so that $\dim (V \cap U) = \dim_p V$. Applying Lemma 3.9 in U gives, using (12.9) in U,

$$\dim K^*_M(V \cap U) < \dim\,[K'^*_M(V \cap U) - M \times P'] = \dim_p V\,.$$

Moreover, for some ν, setting $p^* = (p, \nu)$, we have

$$\dim_{p^*} K''^*_M(V) = \dim_p V$$

(compare the proof of Lemma 8.11). Since (Lemma 5.9) $M \times P'$ is a manifold of codimension 1 in $C^{*n}(M)$, Lemma 6.1 with Remark 6.3 gives

(12.10) $$\sup_{\nu \in P'} \dim_{(p,\nu)} K^*_M(V) = \dim_p V - 1\,,$$

and hence also

(12.11) $$\dim K^*_M(V) = \sup_{p \in M} \dim_p V - 1\,.$$

Comparing (12.5) with (12.8) and (12.9) and using (12.7), we see that:

(12.12) Given $p \in M$, $v \in C'$, $v \neq 0$, $p + v \in K_M(V, p)$ if and only if $(p, \kappa(v)) \in K^*_M(V)$.

We now prove the following properties of $K_M(V)$:

(12.13) $K_M(V)$ is a variety in $\pi^{-1}(M)$,

(12.14) $\dim_p K_M(V) = \dim_p V$, $p \in M$,

(12.15) $p + \pi' C(V, p) \subset K_M(V, p)$, $p \in M$,

(12.16) $\pi' C(V, p) \subset C_M(V, p)$, $p \in M$.

The first property we know, and the next is an easy consequence of (12.12) and (12.10). To prove (12.16), take any $v \in C(V, p)$; we may suppose $\pi' v \neq 0$. Then there are sequences $\{q_i\}$, $\{a_i\}$ such that $q_i \in V - M$, $q_i \to p$, $a_i(q_i - p) \to v$; now $a_i(q_i - \pi q_i) \to \pi' v$, and $\pi' v \in C_M(V, p)$. (12.15) is another way of stating (12.16).

EXAMPLE 12.17. Let $V \subset C_{xy}$ be defined by $y(x - ay) = 0$; set $M = C_x$. Then $K_M(V) = C_x \cup C_y$. We obtain the same $K_M(V')$ from V': $y(y - x^2) = 0$.

EXAMPLE 12.18. Let $V \subset C_{txy}$ be defined by $t^2y^2 - t^2x^2 + x^3 - y^4 = 0$ (see [9, Example 7.8]); set $M = C_t$. Set $\rho_t = (t, 0, 0)$. Now if $t \neq 0$, then $C(V, \rho_t)$ is the pair of planes $y = \pm x$, and

$$K_M(V, \rho_t) = C(V, \rho_t) \cap (\rho_t + C_{xy}) ;$$

for $t = 0$,

$$C(V, O) = C_{ty} , \qquad K_M(V, O) = C_M(V, O) = C_{xy} .$$

Note that $\pi' C(V, O) = C_y \neq K_M(V, O)$.

EXAMPLE 12.19. Define $V \subset C^4$ by $f = x_1x_4 - x_2x_3 = 0$ (see Example 6.5); set $M = C_{x_1x_2}$. Now if $a \in M$, $a \neq O$, then f is of order 1 at a, and

$$f_a^{[*]}(v) = a_1v_4 - a_2v_3 , \qquad C(V, a) = \{v \colon \operatorname{dep}((a_1, a_2), (v_3, v_4))\} ,$$
$$C_M(V, a) = C(V, a) \cap C_{x_3x_4} ,$$

while f is of order 2 at O, and

$$f_O^{[*]}(v) = v_1v_4 - v_2v_3 , \qquad C(V, O) = \{v \colon \operatorname{dep}((v_1, v_2), (v_3, v_4))\} ,$$
$$C_M(V, O) = C_{x_3x_4} .$$

We now define certain subvarieties of M, *assuming* (12.1) holds and $\dim_p V \equiv r$, $p \in M$ (recall Remark 8.13):

(12.20) $K_i = \{p \in M \colon \dim_p K_M(V, p) \geqq n - m - i\}$, $i = 0, \cdots, n - r$.

LEMMA 12.21. *K_i is a subvariety of M, and*

(12.22) $\dim K_i \leqq r - n + m + i$.

Since $r > m$, we have $i < n - m$.

First we find another expression for K_i. Take any plane T in C' through O, of dimension $i + 1$, and set

$$T^* = \{\kappa(v) \colon v \in T\} , \qquad X_i^*(T) = K_M^*(V) \cap (M \times T^*) ,$$
$$X_i(T) = \pi^* X_i^*(T) , \qquad X_i = \textstyle\bigcap_T X_i(T) ,$$

where π^* is the projection of $C^n \times P^{n-1}$ onto C^n. Since $K_M^*(V)$ is a variety in $M \times P'$, so is $X_i^*(T)$. Since π^* is proper into M, Lemma 4.4 shows that $X_i(T)$ is a variety in M; hence so is X_i. We now show that $K_i = X_i$, proving that K_i is a variety in M.

First, take any $p \in X_i$. Take any T as above. Then $p \in X_i(T)$, and hence for some $\nu \in P'$, $(p, \nu) \in X_i^*(T)$; by (12.12), if $\nu = \kappa(v)$,

$$p + v \in K_M(V, p) , \qquad v \in T .$$

This holds also for $av (a \in C)$; thus p is not isolated in $K_M(V, p) \cap (p + T)$. Since this is true for all such T, we have $\dim_p K_M(V, p) > n - m - i - 1$, and hence $p \in K_i$.

Conversely, if $p \in K_i$, then for any T as above, p is not isolated in $K_M(V, p) \cap (p + T)$, and this set contains some $v \neq 0$, hence $(p, \kappa(v)) \in K_M^*(V) \cap (M \times T^*)$, and $\pi^*(p, \kappa(v)) = p \in X_i(T)$; therefore $p \in X_i$.

We now apply Lemma 6.8 to $K_M(V)$, with $Q = K_i$; since $K_M(V) \cap (p + C') = K_M(V, p)$ and (12.14) holds, we may use $s = n - m - i$, $s + h = r$. This gives (12.22).

COROLLARY 12.23. *Suppose* $\dim_p V = r$, *all* $p \in M$. *Then there is a closed subset* Q *of* M *without inner points, such that*

$$\dim K_M(V, p) = \dim_p K_M(V, p) = r - m , \qquad p \in M - Q . \tag{12.24}$$

Set $Q = K_{n-r-1}$; by (12.22), $\dim Q \leqq m - 1$, and by (12.20), $\dim_p K_M(V, p) \leqq r - m$, $p \in M - Q$. On the other hand, for $p \in M - Q$, Lemma 6.1 applied to $K_M(V)$ and $p + C'$, together with (12.14), gives

$$\dim_p K_M(V, p) \geqq \dim_p K_M(V) + (n - m) - n = r - m .$$

REMARK 12.25. In the proof of the corollary, Lemma 12.21 is not needed. For $Q = K_{n-r-1}$ is closed; Lemma 6.8 shows that Q contains no inner points.

13. Wings

We shall show that, if M is a manifold in the variety V and $\dim V \equiv r > m = \dim M$, then near any point of M there is a *wing* B spreading from M into V; we may avoid a lower dimensional subvariety V' of V (for instance, V_{sg}). B will be a real differentiable manifold of real dimension $2m + 1$, with $2m$-dimensional sections which are (complex) m-manifolds.

A wing will be the image of a set $Z^\sim$ defined as follows: Let t denote $(t_1, \cdots, t_m)$, each t_k a complex number; λ will denote a real number. Let J_α, $J_\alpha^\sim$ be the intervals $0 < \lambda < \alpha$, $0 \leqq \lambda < \alpha$ respectively. For some fixed numbers $\alpha_1, \cdots, \alpha_m, \alpha$, set

$$(13.1)\qquad \begin{cases} Z = \{(t, \lambda): |t_k| < \alpha_k \text{ (all } k),\ \lambda \in J_\alpha\}\,, \\ Z_\lambda = \{(t, \lambda): |t_k| < \alpha \text{ (all } k)\}\,, \end{cases} \qquad Z^{\sim} = Z \cup Z_0\,.$$

Let $\partial/\partial\lambda$ and $\partial/\partial t_k$ denote partial derivatives in the real and complex senses respectively.

Suppose $F: Z^{\sim} \to C^n$ is holomorphic in t and differentiable in λ. Write $t_k = t'_k + it''_k$, t'_k and t''_k being real. Then $\partial F(p)/\partial t_k \neq 0$ implies that the vectors $\partial F(p)/\partial t'_k$, $\partial F(p)/\partial t''_k$ are independent in the real sense. The independence of $\partial F/\partial t_1, \cdots, \partial F/\partial t_m$ in the complex sense at p implies the independence of the corresponding $2m$ vectors in the real sense at p. Condition (c) below means that the corresponding $2m + 1$ vectors are independent in the real sense.

A *wing* in C^n *from* B_0 *into* Q is a set B, together with a homeomorphism F of some $Z^{\sim}$ onto B, with the following properties:

(a) For each $\lambda \in J_\alpha^{\sim}$, $F_\lambda(t) = F(t, \lambda)$ is biholomorphic, and each $\partial F_\lambda/\partial t_k$ is continuous in $Z^{\sim}$.

(b) $\partial F/\partial\lambda$ is continuous in Z.

(c) $\partial F/\partial t_1, \cdots, \partial F/\partial t_m, \partial F/\partial\lambda$ are independent in Z.

(d) $F(Z_0) = B_0$, $F(Z) \subset Q$.

We say the wing is *smooth* if (b) and (c) hold in $Z^{\sim}$.

We note the following property of wings:

(e) Set $B_\lambda = F(Z_\lambda)$. Suppose $q_i \in B_{\lambda_i}$, $q_i \to p \in B_0$. Then $T(B_{\lambda_i}, q_i) \to T(B_0, p)$.

Since each F_λ is an analytical equivalence, the $T(B_{\lambda_i}, q_i)$ and $T(B_0, p)$ are defined. Now (e) follows from (a).

REMARK 13.2. If the conditions (a) through (d) hold, omitting the continuity of the $\partial F_\lambda/\partial t_k$ at $\lambda = 0$, then this holds also.

To show this, set $G_\lambda(t) = F_\lambda(t) - F_0(t)$; since $\partial F_0(t)/\partial t_k$ is continuous, it is sufficient to show that given k, $t^{(0)} \in Z_0$ and $\varepsilon > 0$, there is a $\delta > 0$ such that

$$(13.3)\qquad |\partial G_\lambda(t)/\partial t_k| < \varepsilon \qquad \text{if } |t - t^{(0)}| < \delta,\ |\lambda| < \delta\,.$$

Take $\rho > 0$ so that $U_{3\rho}(t^{(0)}) \cap C^m \subset Z_0$; since F is continuous, we may choose δ, $0 < \delta \leqq \rho$, so that

$$|G_\lambda(t)| < \rho\varepsilon \qquad \text{if } |t - t^{(0)}| < 2\rho,\ \lambda < \delta\,.$$

Now with (t, λ) as in (13.3), set $\varphi(z) = G_\lambda(t + ze_k)$; applying the Schwarz lemma over the circle of radius ρ about 0 gives $|d\varphi(0)/dz| < \varepsilon$, and (13.3) follows.

REMARK 13.4. Because of (c), $B - B_0$ is a real differentiable manifold of real dimension $2m + 1$. If the wing is smooth, (c) shows that B_0 lies smoothly on the boundary of B; hence the differentiable manifold may be extended across B_0. We shall not need this fact.

LEMMA 13.5. *Suppose* $p_0 \in M \subset V \subset H \subset C^n$, *where* M *is a manifold,* V *is a variety, and* $\dim M = m$, $\dim V \equiv r > m$. *Let* V' *be a subvariety of* V *with* $\dim V' < r$. *Then there is an open subset* M' *of* M *arbitrarily near* p_0, *and a wing* B *from* M' *into* $V - V'$. *We may require* B *to be smooth.*

We first prove the lemma without the final statement; this will suffice for most applications. The proof of the full lemma will require part of the results of the next section.

First we apply Lemma 7.8 with $h = r - m - 1$ to give a neighborhood U_0 of p_0 and a variety $W \subset U_0$, such that

$$M \cap U_0 \subset W \subset V , \qquad \dim W \equiv m + 1 , \qquad \dim (W \cap V' - M) \leqq m .$$

Now (Lemma 3.9) $\dim [(U_0\text{-clos}(W \cap V' - M)) \cap M] < m$, so that we may find $p_1 \in M \cap U_0$ and a neighborhood U_1 of p_1 such that

$$W \cap U_1 - M \subset V - V' . \tag{13.6}$$

Choose coordinates in a neighborhood $U_2 \subset U_1$ of p_1 so that $M \cap U_2 = C^m \cap U_2$. By Lemma 9.11, we may find an $(n - m - 1)$-plane C'' through p_1 such that p_1 is isolated in $C_p(W) \cap C''$. Since $M \cap U_2 \subset W \cap U_2$, C'' has no vector $\neq 0$ in common with C^m; hence we may suppose the last coordinates chosen so that $C'' = C_{x_{m+2} \cdots x_n}$. Let π be the projection onto $C^{m+1} = C_{x_1 \cdots x_{m+1}}$.

Applying Lemma 3.4, there is a polycylindrical neighborhood $U_3 \subset U_2$ of p_1 such that $\pi(W \cap U_3) = C^{m+1} \cap U_3$, and writing $\tilde{x} = (x_1, \cdots, x_m)$, there are pseudopolynomials

$$\omega_j(\tilde{x}, x_{m+1}, x_j) , \qquad j = m + 2, \cdots, n , \tag{13.7}$$

with the following properties: The discriminant $D_j(\tilde{x}, x_{m+1})$ of ω_j is $\not\equiv 0$ (all j); and if $C_j = C_{x_1 \cdots x_{m+1} x_j}$ and π_j is the projection onto C_j, then

$$\pi_j(W \cap U_3) = \{p \in \pi_j(U_3) \colon \omega_j(p) = 0\} . \tag{13.8}$$

Let X_j be the variety $D_j = 0$ in C^{m+1}; set $X = X_{m+2} \cup \cdots \cup X_n$. Set $Y = [U_3\text{-clos}(X - M)] \cap M$; then we may choose $p_2 \in M \cap U_3 - Y$ and a polycylindrical neighborhood $U_4 \subset U_3$ of p_2 such that $U_4 \cap (X - M) = \emptyset$. Thus for all j,

$$D_j(\tilde{x}, x_{m+1}) \neq 0 \qquad \text{if } (\tilde{x}, x_{m+1}) \in U_4,\ x_{m+1} \neq 0 . \tag{13.9}$$

Consider any solution $(x_{m+2}, \cdots, x_n)$ of the set of equations $\omega_j(\tilde{x}, 0, x_j) = 0$. Since the degree of each ω_j in x_j is constant, the set of such solutions forms a continuous function of $\tilde{x}$. Moreover, since the ω_j are holomorphic, the multiplicity of the root $(0, \cdots, 0)$ has a minimum in $M \cap U_4$, which is attained over a dense open subset. Choose p_3 in this subset; say the neighborhood $U_5^m \subset M$ of p_3 lies in the subset. Then for some U_5^{n-m-1} about $(0, \cdots, 0)$, if $(\tilde{x}, 0, z)$ $(\tilde{x} \in U_5^m, z \in \bar{U}_5^{n-m-1})$ is a root of the equations $\omega_j = 0$, then $z = (0, \cdots, 0)$.

Hence for some $\delta > 0$, if $U_\delta^1 = \{y: |y| < \delta\}$, there is no root $(\tilde{x}, y, z)$ of the equations $\omega_j = 0$ such that $\tilde{x} \in U_\delta^m$, $y \in U_\delta^1$, $z \in \partial U_\delta^{n-m-1}$. Set $U_\delta = U_\delta^m \times U_\delta^1 \times U_\delta^{n-m-1}$.

Say $U_\delta = \{x: |x_i| < a_i, i = 1, \cdots, n\}$. Define $Z, Z_0, Z^\sim$ by (13.1), with $\alpha = \alpha_{m+1} = \delta$. Since $M \subset W$ and $\dim W \equiv m + 1$, there is a point $x^{(0)} \in W \cap U_\delta - M$; now $x_{m+1}^{(0)} \neq 0$, and $\omega_j(x^{(0)}) = 0$ (all j). We may clearly suppose $x_{m+1}^{(0)}$ real and positive. Because of (13.9), and since Z is simply connected, we may continue this root of the set $\{\omega_j = 0\}$ over Z uniquely, giving a continuous function $z = F(\tilde{x}, \lambda)$. Since $F(\tilde{x}, 0) \subset M$, $F(Z_0)$ is the open subset $M \cap U_\delta$ of M. By (13.6), $F(Z) \subset V - V'$.

Since the ω_j are holomorphic and (13.9) holds, $F_\lambda(t)$ is holomorphic for $\lambda > 0$; it is trivially if $\lambda = 0$. Also $\partial F_\lambda/\partial t_k$ is continuous for $\lambda > 0$. Since $F(t, \lambda)$ is real analytic in λ, (b) holds. Since $\pi F(t, \lambda) = (t, \lambda)$, (c) is true. Now Remark 13.2 completes the proof of property (a), and the lemma is proved, except for the last statement.

To complete the proof, we extend Lemma 11.4:

LEMMA 13.10. *With $t = (t_1, \cdots, t_m)$, let $f(t, x, y)$ be holomorphic in the polycylindrical neighborhood $U \subset C^{m+2}$ of O: $|t_i| < \alpha_i$ (all i), $|x| < \alpha$, $|y| < \alpha'$. Suppose*

$$f(t, x, y) \neq 0 \qquad \text{if } |y| = \alpha', \text{ or if } x = 0, y \neq 0 .$$

Let $D(t, x)$ be the discriminant of f, and suppose $D \neq 0$ if $x \neq 0$. Let $\varphi(t, \lambda)$ be a continuous complex valued function in $Z^\sim$ (see (13.1)), and suppose

$$(13.11) \qquad f(t, \lambda, \varphi(t, \lambda)) \equiv 0 , \qquad \lim_{\lambda \to 0} \frac{\varphi(t, \lambda)}{\lambda} = 0 \qquad \text{for } |t_i| < \alpha_i .$$

Then $\varphi'(t, \lambda) = \partial\varphi/\partial\lambda$ is continuous in $Z^\sim$ and vanishes in Z_0.

Since $D \neq 0$ if $x \neq 0$ and $|y| \neq \alpha'$ if $f = 0$, $f = 0$ determines y as a holomorphic function of (t, x), near any solution (t^0, x^0, y^0) with $x^0 \neq 0$, continuable in $|t_i| < \alpha_i$, $|x| < \alpha$. Hence for some integer h, y is determined as a single valued holomorphic function $\psi(t, \xi)$, where $\xi^h = x$, $|\xi| < \alpha^{1/h}$; that is,

$$f(t, \xi^h, \psi(t, \xi)) \equiv 0 , \qquad |\psi(t, \xi)| < \alpha' .$$

As in Lemma 11.4, we may choose ψ so that $\psi(t, \lambda) = \varphi(t, \lambda^h)$ for any real λ, $\lambda^h \in J_\alpha^\sim$. By (13.11), $\psi(t, \lambda)/\lambda^h \to 0$ as $\lambda \to 0$; hence we may write

$$\psi(t, \xi) = \xi^h\theta(t, \xi) , \qquad \theta \text{ holomorphic, } \theta(t, 0) \equiv 0 .$$

Now $\varphi(t, \lambda^h) = \lambda^h\theta(t, \lambda)$. Differentiating gives

$$\varphi'(t, \lambda^h) = \theta(t, \lambda) + \frac{\lambda}{h}\frac{\partial}{\partial\lambda}\theta(t, \lambda) .$$

Letting $\lambda \to 0$ gives the lemma.

Returning to Lemma 13.5, we shall use the more explicit facts about a neighborhood of a point of M given in Theorem 14.4 (for W in place of V); we shall not use the last part of (14.3).

Say $B \subset W_i$, an irreducible component of $W \cap U_0$. With $v^i(\tilde{x})$ as in (14.1), make the following transformation of coordinates: $x'_i = x_i$ for $i \leqq m+1$, and

$$x'_j = x_j - v^i_j(x_1, \cdots, x_m)x_{m+1}, \qquad j = m+2, \cdots, n.$$

In the new coordinates, dropping primes, we have (for this i)

$$v^i(\tilde{x}) = e_{m+1}, \qquad \tilde{x} \in M \cap U_0.$$

Now the wing is defined as in the proof above, by

$$F(\tilde{x}, \lambda) = x = (\tilde{x}, \lambda, \varphi(\tilde{x}, \lambda)),$$

where $\lambda = x_{m+1} \in J^{\sim}_a$ and φ has components $\varphi_{m+2}, \cdots, \varphi_n$; it satisfies $\omega_j(\tilde{x}, \lambda, \varphi_j) \equiv 0$, all j. Write, for $\lambda > 0$,

$$a = 1/\lambda, \qquad \Delta = x - \pi_{m+1}x = (0, \cdots, 0, 0, \varphi).$$

Take any ε, $0 < \varepsilon < 1/2$. By the first part of (d) of § 14, we may find $\delta > 0$ such that if $|x - \tilde{x}| < \delta$, then

$$|x - \tilde{x}, v^i(\tilde{x})|^* < \varepsilon.$$

Since $x - \tilde{x} = \lambda e_{m+1} + \Delta = \lambda(e_{m+1} + a\Delta)$ and $e_{m+1} \cdot \Delta = 0$, (2.3) and (2.1) give

$$(|x - \tilde{x}, v^i(\tilde{x})|^*)^2 = |a\Delta|^2/(1 + |a\Delta|^2) < \varepsilon^2,$$

and $|a\Delta| < 2\varepsilon$. Taking the jth component gives $\varphi_j(\tilde{x}, \lambda)/\lambda \to 0$ as $\lambda \to 0$ (all j). The conditions of Lemma 13.10 for ω_j and φ_j are satisfied, and hence $\partial\varphi_j/\partial\lambda$ is continuous in $Z^{\sim}$ and vanishes in Z_0. Thus (b) is proved in $Z^{\sim}$. (c) clearly holds in $Z^{\sim}$, and the proof of Lemma 13.5 is complete.

14. The neighborhood of a submanifold of codimension 1

If $\dim V \equiv m + 1$ and $M \subset V$, $\dim M = m$, we show that, except in a nowhere dense closed subset Q of M, the sheets of V *attach smoothly* to M. It is a reasonable conjecture that Q may be intrinsically defined as a proper subvariety of M; we shall not look into this here.

With V and M as above, we say V *attaches smoothly to* M *near* $p_0 \in M$ if there is a coordinate system in a neighborhood U_0 of p_0 so that $M \cap U_0 = C^m \cap U_0$, and the following are true:

(a) $V \cap U_0$ falls into irreducible components $V_1, \cdots, V_s$.

(b) If $p \in V \cap U_0 - M$, then for some neighborhood U of p, $V \cap U$ is given by equations of the form

$$x_j = \varphi_j(x_1, \cdots, x_{m+1}) , \qquad j = m + 2, \cdots, n ,$$

the φ_j being holomorphic.

(c) Write $\tilde{x} = (x_1, \cdots, x_m)$; let π' be the projection onto $C' = C_{x_{m+1}\cdots x_n}$. There is a set of holomorphic vector fields (not necessarily distinct)

(14.1) $$v^i(\tilde{x}) = (0, \cdots, 0, 1, v^i_{m+2}(\tilde{x}), \cdots, v^i_n(\tilde{x})) , \qquad i = 1, \cdots, s ,$$

such that

(14.2) $$C(V_i, \tilde{x}) = \{v: \pi' v = a v^i(\tilde{x}) \text{ for some } a\} , \qquad \tilde{x} \in M \cap U_0 .$$

(d) For each $\varepsilon > 0$, there is a $\delta > 0$ such that if $\tilde{x} \in M \cap U_0$, $x \in V_i \cap (\tilde{x} + C')$, $0 < |x - \tilde{x}| < \delta$, and $T(x)$ is the $(m + 1)$-plane containing C^m and x, then

(14.3) $$|x - \tilde{x}, v^i(\tilde{x})|^* < \varepsilon , \qquad |T(x), T(V, x)|^*_m < \varepsilon .$$

THEOREM 14.4. *Suppose* $M \subset V \subset H \subset C^n$, $\dim M = m$, $\dim V \equiv m + 1$. *Then there is a closed subset* Q *of* M *without inner points such that, for each* $p \in M - Q$, V *attaches smoothly to* M *near* p.

For each $p \in M$ and $\varepsilon > 0$ we shall find $p' \in M$ such that $|p' - p| < \varepsilon$ and V attaches smoothly to M near p' and hence near all points of a neighborhood $A(p, \varepsilon) \subset M$ of p'. Let A be the union of the sets $A(p, \varepsilon)$ for all pairs (p, ε), and set $Q = M - A$; this is the required set.

Given $p_0 \in M$ and $\varepsilon_0 > 0$, choose a coordinate system in a neighborhood $U_1 \subset U_{\varepsilon_0}(p_0)$ of p_0 so that $M \cap U_1 = C^m \cap U_1$. By Corollary 12.23, there is a proper closed subset Q of $M_1 \cap U_1$ such that, choosing $p_1 \in M_1 \cap U_1 - Q$, we have

$$\dim K_M(V, p_1) = 1 .$$

Hence we may choose the last coordinates so that if $C'' = C_{x_{m+2}\cdots x_n}$, then p_1 is isolated in $K_M(V, p_1) \cap (p_1 + C)''$. Since $K_M(V)$ is closed and each $K_M(V, p)$ is a cone, we see easily that we may choose U_2 about p_1 and find $\zeta > 0$ such that

(14.5) $$|K_M(V, p), p + C''|^* \geqq \zeta , \qquad p \in M \cap U_2 ;$$

in fact (with π as below) by (12.2) and (12.3),

(14.5′) $$|q - \pi q, C''|^* \geqq \zeta , \qquad q \in V \cap U_2 - M .$$

Let π, π', π'' denote the projections onto C^m, C', C'' respectively. By (12.15) and (14.5), writing $v = (v_1, \cdots, v_n)$,

(14.6) $$\text{if } v \in C(V, p), \pi' v \neq 0, p \in M \cap U_2, \text{ then } v_{m+1} \neq 0.$$

Set

$$C^{m+1} = C_{x_1\cdots x_{m+1}} , \qquad C_j = C_{x_1\cdots x_{m+1}x_j} , \qquad C'_j = C_{x_{m+1}x_j} ,$$

and let π^{m+1}, π_j, π'_j be the corresponding projections.

Because of (14.5′), we may apply Lemma 3.4 to the projection π^{m+1}. Hence there are pseudopolynomials $\omega_j(\tilde{x}, x_{m+1}, x_j)$ in U_2 (taking U_2 small enough) as in (13.7), such that the discriminant $D_j(\tilde{x}, x_{m+1})$ of ω_j is $\not\equiv 0$, and so that

$$\pi_j(V \cap U_2) = \{p \in \pi_j(U_2): \omega_j(p) = 0\} . \tag{14.7}$$

Let $X_j \subset C^{m+1} \cap U_2$ be the variety $D_j = 0$, and set

$$W_j = [U_2\text{-clos}\,(X_j - M)] \cap M . \tag{14.8}$$

Say ω_j is of degree h_j. Take any $\tilde{x} \in M \cap U_2$. By (14.5′), $e_j \notin C(\pi_j(V \cap U_2), \tilde{x})$; hence, by (14.7) and Theorem 10.3, if $A_j = \omega_j^{[m]}$ is the initial polynomial of ω_j at $\tilde{x}$, $A_j(e_j) \neq 0$. Because of (14.5′) and the last statement in Lemma 3.4, $\omega_j(\tilde{x}, 0, x_j) = x_j^{h_j}$; hence $m = h_j$. Expanding, we may therefore write $\omega_j = A_j + R_j$, where

$$A_j(\tilde{x}, x_{m+1}, x_j) = \sum_{k=0}^{h_j} a_{jk}(\tilde{x}) x_{m+1}^k x_j^{h_j - k} ; \tag{14.9}$$

the a_{jk} and R_j are holomorphic, $a_{j0}(\tilde{x}) \equiv 1$, and R_j is of order $> h_j$ in (x_{m+1}, x_j) for each $\tilde{x}$. For each j, there is a proper subvariety W'_j of $M \cap U_2$ and there is a number $b_j \leqq h_j$ such that, for any $\tilde{x}^{(0)} \in M \cap U_2 - W'_j$ and all $\tilde{x}$ sufficiently near $\tilde{x}^{(0)}$, the solutions of $A_j(\tilde{x}, x_{m+1}, x_j) = 0$ fall into b_j distinct lines, whose directions

$$\nu_{j1}(\tilde{x}), \cdots, \nu_{jb_j}(\tilde{x}) \in P'_j$$

(P'_j = projective space of C'_j) are holomorphic functions of $\tilde{x}$. (The $\nu_{jk}(\tilde{x})$ may interchange if $\tilde{x}$ runs around W'_j in $M \cap U_2$.) Let W be the union of the W_j and the W'_j; this is a proper subvariety of $M \cap U_2$.

Now take any $p_2 = \tilde{x}^{(2)} \in M \cap U_2 - W$. For some $\eta > 0$ we have

$$|\nu_{jk}(\tilde{x}^{(2)}), \nu_{jl}(\tilde{x}^{(2)})|^* \geqq 4\eta , \qquad \text{all } k \neq l, \text{ all } j . \tag{14.10}$$

Using (14.8), we may take a polycylindrical neighborhood U_0 about p_2 so that

$$U_0 \cap M \cap W = \emptyset , \qquad (U_0 - M) \cap X_j = \emptyset , \qquad \text{all } j , \tag{14.11}$$

$$|\nu_{jk}(\tilde{x}), \nu_{jk}(\tilde{x}^{(2)})|^* < \eta \qquad \text{if } \tilde{x} \in M \cap U_0, \text{ all } j, k . \tag{14.12}$$

Moreover, for any $t \in C$, $|t| \leqq 1$, we may write

$$\omega_j(\tilde{x}, tx_{m+1}, tx_j) = t^{h_j}[A_j(\tilde{x}, x_{m+1}, x_j) + tR'_j(\tilde{x}, x_{m+1}, x_j, t)] ,$$

where R'_j is holomorphic. Since the a_{jk} are continuous, R'_j is bounded, and P'_j is compact, it is easily seen (compare the proof of Lemma 10.10 of [9]) that for some $\lambda_0 > 0$, if

$$\omega_j(\tilde{x}, x_{m+1}, x_j) = 0 , \qquad \tilde{x} \in M \cap U_0, \qquad |(x_{m+1}, x_j)| < \lambda_0 ,$$

then there is a pair $(x'_{m+1}, x'_j) \neq (0, 0)$ such that

$$A_j(\tilde{x}, x'_{m+1}, x'_j) = 0 , \qquad |(x_{m+1}, x_j), (x'_{m+1}, x'_j)|^* < \eta . \tag{14.13}$$

We may now require of U_0 that for any $x = (x_1, \cdots, x_n) \in U_0$, we have $|(x_{m+1}, \cdots, x_n)| < \lambda_0$, and (a) holds.

Consider any V_i. For each j, take any

$$x = (\tilde{x}, x_{m+1}, x_j) = \tilde{x} + v \in \pi_j(V_i) \cap U_0 - M , \qquad v \neq 0 ; \tag{14.14}$$

by (14.7) and (14.13) with (14.10) and (14.12), there is a unique k such that $|v, \nu_{jk}(\tilde{x})|^* < \eta$. Since $V_i - M$ is connected, k is the same for all $\tilde{x}$. Write therefore $k = \gamma(i, j)$, and set $\nu_j^i(\tilde{x}) = \nu_{j,\gamma(i,j)}(\tilde{x})$. Now

(14.15) if $\tilde{x} \in M \cap U_0$ and (14.14) holds, then $|v, \nu_j^i(\tilde{x})|^* < \eta$.

Note that (14.13) and hence (14.15) could have been proved with η replaced by any $\rho > 0$, provided that λ_0 is taken small enough. Hence

(14.16) for each $\rho > 0$ there is a $\delta > 0$ such that if $\tilde{x} \in M \cap U_0$, (14.14) holds and $|v| < \delta$, then $|v, \nu_j^i(\tilde{x})|^* < \rho$.

Next we show that for $\tilde{x} \in M \cap U_0$,

(14.17) if $v \in C(V_i, \tilde{x})$, $\pi' v \neq 0$, then $\kappa\pi_j' v = \nu_j^i(\tilde{x})$ (all j).

By (14.6), $v_{m+1} \neq 0$; hence $\pi_j' v \neq 0$. We may find sequences $y^k \in V_i$, $y_{m+1}^k \neq 0$, $y^k \to \tilde{x}$, $a_k(y^k - \tilde{x}) \to v$. Now $u^k = a_k(y^k - \pi y^k) \to \pi' v$. Applying (14.16) to

$$(\pi y^k, y_{m+1}^k, y_j^k) = \pi y^k + a_k^{-1}\pi_j' u^k \in \pi_j(V_i \cap U_0) - M$$

gives $|\pi_j' u^k, \nu_j^i(\pi y^k)|^* \to 0$. Since $\pi_j'\pi' v = \pi_j' v$, (14.17) follows.

Since $\dim C(V_i, \tilde{x}) = \dim V_i = m + 1$, there is a vector $v \in C(V_i, \tilde{x})$ with $\pi' v \neq 0$; using this v with (14.17) and (14.6) shows that we may write

$$\nu_j^i(\tilde{x}) = \kappa\big(1, v_j^i(\tilde{x})\big) , \qquad \text{all } i, j . \tag{14.18}$$

Now define $v^i(\tilde{x})$ by (14.1); then v^i is holomorphic, and

$$\nu_j^i(\tilde{x}) = \kappa\pi_j' v^i(\tilde{x}) , \qquad \text{all } i, j, \tilde{x} \in M \cap U_0 . \tag{14.19}$$

We now prove (14.2) and hence (c). First, take any $v \in C(V_i, \tilde{x})$. We may suppose $\pi' v \neq 0$; then (14.17) and (14.18) show that $\kappa\pi' v = \kappa v^i(\tilde{x})$, and $\pi' v = a v^i(\tilde{x})$ for some a.

Conversely, we must show that if $u \in C^m$, then $v^i(\tilde{x}) + u \in C(V_i, \tilde{x})$. Choose real numbers $\lambda_1, \lambda_2, \cdots \to 0$. We may define y^k and find z^k so that

$$y^k = \tilde{x} + \lambda_k(e_{m+1} + u) \in C^{m+1} \cap U_0 , \qquad z^k \in V_i \cap U_0, \qquad \pi^{m+1} z^k = y^k .$$

Now $(\pi y^k, \lambda_k, z_j^k) = \pi_j z^k \in \pi_j(V_i)$; hence, setting $a_k = 1/\lambda_k$, we may apply (14.16), giving

$$|(1, a_k z_j^k), \nu_j^i(\pi y^k)|^* \longrightarrow 0 .$$

Hence $a_k z_j^k \to v_j^i(\tilde{x})$ as $k \to \infty$, and

$$a_k(z^k - \tilde{x}) = (u_1, \cdots, u_m, 1, a_{m+1}z_{m+1}^k, \cdots, a_n z_n^k) \longrightarrow u + v^i(\tilde{x}) .$$

Since $z^k \in V_i$, the statement is proved.

Next we note that, if $\tilde{x} \in M \cap U_0$, (14.2) and (12.16) give

$$v^i(\tilde{x}) \in C(V, \tilde{x}) \cap C' \subset \pi' C(V, \tilde{x}) \subset C_M(V, \tilde{x}) ;$$

Now (14.1), (12.3) and (14.5) show that

$$(14.20) \qquad |v^i(\tilde{x}), C''|^* = 1/|v^i(\tilde{x})| \geqq \zeta , \qquad |v^i(\tilde{x})| \leqq 1/\zeta .$$

Note that, by (14.5′),

$$(14.21) \qquad \text{if } \tilde{x} \in M \cap U_0,\ v \in C',\ y = \tilde{x} + v \in V_i \cap U_0,\ y \neq \tilde{x}, \text{ then } y_{m+1} \neq 0.$$

To prove (b), take any $y \in V \cap U_0 - M$; say $y \in V_i$. Set $\tilde{x} = \pi y$. By (14.21), $y_{m+1} \neq 0$. Now (14.11) shows that $\pi^{m+1} y \notin X_j$ (all j), and the existence of the φ_j follows.

We now prove the first part of (d). Given ε, $0 < \varepsilon < 1/2$, set $\rho = \varepsilon\zeta/n$, and choose $\delta > 0$ so that (14.16) holds for each i and j. Now suppose $\tilde{x} \in M \cap U_0$, $y \in V_i$, $v = y - \tilde{x} \in C'$, $0 < |v| < \delta$. By (14.21), $y_{m+1} \neq 0$, and we may set $a = 1/y_{m+1}$. Since

$$(\tilde{x}, y_{m+1}, y_j) = \tilde{x} + \pi_j v \in \pi_j V_i , \qquad 0 < |\pi_j v| < \delta ,$$

(14.16) gives

$$|\pi_j v, \nu^i_j(\tilde{x})|^* < \rho .$$

We now apply (2.10), with v and u replaced by $(1, 0)$ and $(0, 1)$ in C'_j respectively; this gives

$$\rho > |(1, ay_j), (1, v^i_j(\tilde{x}))|^* = \frac{|ay_j - v^i_j(\tilde{x})|}{|(1, ay_j)|\,|(1, v^i_j(\tilde{x}))|} ,$$

$$|ay_j - v^i_j(\tilde{x})| < |av|\,|v^i(\tilde{x})|\,\rho , \qquad j = m+2, \cdots, n .$$

Now (2.9) gives, with (14.20),

$$|v, v^i(\tilde{x})|^* = |av, v_i(\tilde{x})|^* \leqq \frac{|av - v^i(\tilde{x})|}{|av|}$$

$$\leqq \sum_{j=m+2}^{n} \frac{|ay_j - v^i_j(\tilde{x})|}{|av|} < n\,|v^i(\tilde{x})|\,\rho \leqq \varepsilon .$$

As to the last part of (d), we note first that it follows (possibly if Q is enlarged) from Theorem 19.2 below. It may be proved here as follows: Given $x^{(0)} \in V \cap U_0 - M$, we have $a = x^{(0)}_{m+1} \neq 0$; now $\omega_j(\tilde{x}, a, x_j) = 0$ (all j) defines a manifold $B(a)$ near $M \cap U_0$. Property (e) of § 13 shows that $T(B(a), x)$ is near C^m. That $T(V, x)$ nearly contains the vector $x - \tilde{x}$ may be proved as the continuity of φ' was proved in Lemma 13.10. From these facts, the rest of (d) follows.

PART III. TANGENT PLANES, STRATIFICATIONS

15. A special set of vector fields

Through the use of coherence, we shall find a set of vector fields near a point of a variety which determine the tangent plane at each simple point.

LEMMA 15.1. *Let V be an analytic variety of constant dimension r, and suppose $p \in V$. Then there is a neighborhood U of p and a set $v^{(1)}, \cdots, v^{(h)}$ of holomorphic vector fields in U with the following properties:*

(a) *$v^{(k)}(q) = 0$ for each k if $q \in V_{sg}$.*

(b) *The $v^{(k)}(q)$ are in $T(V, q)$ and span $T(V, q)$ if $q \in V_{sp}$.*

Due to the coherence of the set of ideals $I(V, q)$ (see §3) there is a set of holomorphic functions $f_1, \cdots, f_s$ in a neighborhood U of p which vanish in $V \cap U$, such that the following is true: For each $q \in U$ and for each $g \in I(V, q)$ there is a neighborhood U' of q and a set $\varphi_1, \cdots, \varphi_s$ of holomorphic functions in U' such that $g = \varphi_1 f_1 + \cdots + \varphi_s f_s$ in U'.

Given any increasing sequences

$$\lambda = (\lambda_1, \cdots, \lambda_{n-r}) \subset (1, \cdots, s) , \qquad \nu = (\nu_1, \cdots, \nu_{n-r}) \subset (1, \cdots, n) ,$$

let $D_{\lambda\nu}(q)$ be the determinant with the elements $\partial f_{\lambda_i}(q)/\partial x_{\nu_j}$. For any increasing sequence $\mu = (\mu_1, \cdots, \mu_{n-r+1}) \subset (1, \cdots, n)$, let $\mu(i)$ denote this sequence with μ_i omitted. Now for any λ and μ, we define the holomorphic vector field $v_{\lambda\mu}$ in U by means of its components (written as superscripts here):

$$v_{\lambda\mu}^{\mu_i}(q) = (-1)^{i-1} D_{\lambda\mu(i)}(q) , \qquad v_{\lambda\mu}^{j}(q) = 0 \qquad \text{if } j \notin \mu . \tag{15.2}$$

(a) Suppose $q \in V_{sg} \cap U$. Then the $df_i(q)$ have fewer than $n-r$ independent elements, hence all the $D_{\lambda\nu}(q)$ are 0, and the $v_{\lambda\mu}(q)$ are 0.

(b_1) Given $q \in V_{sp} \cap U$, we show that each $v_{\lambda\mu}(q)$ is in $T(V, q)$. This is so if $v_{\lambda\mu}(q) = 0$; suppose not. Then some $D_{\lambda\nu}(q)$ is $\neq 0$, and hence $df_{\lambda_1}(q), \cdots, df_{\lambda_{n-r}}(q)$ are independent; therefore $df_{\lambda_k}(q) \cdot v = 0$ (all k) implies $v \in T(V, q)$. By (15.2),

$$df_{\lambda_k}(q) \cdot v_{\lambda\mu}(q) = \sum_{i=1}^{n-r+1} (-1)^{i-1} D_{\lambda\mu(i)}(q) \partial f_{\lambda_k}(q)/\partial x_{\mu_i} ,$$

and the right hand side is the expansion of a vanishing determinant. Thus (b_1) is true.

(b_2) The $v_{\lambda\mu}(q)$ span $T(V, q)$ if $q \in V_{sp}$. For there are holomorphic functions $g_1, \cdots, g_{n-r}$ near q and vanishing on V, with the $dg_i(q)$ independent. We may write $g_i = \sum_j \varphi_{ij} f_j$ near q; then $dg_i(q) = \sum_j \varphi_{ij}(q) df_j(q)$, showing that for some λ, $df_{\lambda_1}(q), \cdots, df_{\lambda_{n-r}}(q)$ are independent. Therefore, for some ν, $D_{\lambda\nu}(q)$ is $\neq 0$. Let ν' be the subset $(\nu'_1, \cdots, \nu'_r)$ of $(1, \cdots, n)$ complementary to ν, and form μ^k by inserting ν'_k into ν at the proper point. Consider the vectors $v_k^* = v_{\lambda\mu^k} (k = 1, \cdots, r)$; we need merely show that these are independent. In fact, let Δ be the determinant formed from the v_k^* with the components ν'. Since $v_{\lambda\mu^k}^{\nu'_j} \neq 0$ implies $\nu'_j \in \mu^k$ and hence $j = k$, Δ has non-zero elements only on the main diagonal. Each diagonal element is $\pm D_{\lambda\nu}(q)$; hence $\Delta \neq 0$, completing the proof.

16. Spaces of tangent planes

The tangent plane $T(V, p)$ is defined at any simple point p of V. If p is a singular point, we shall not define such a plane there; instead, *assuming V is of constant dimension*, we shall consider all possible limits of tangent planes $T(V, q_i)$, q_i simple, $q_i \to p$. The set of all planes thus considered at all points of V form the *tangent plane space* of V.

If $q_i \to p$, the q_i being simple, we are interested also in the directions $\kappa(q_i - p)$ along which we approach p, for p in a submanifold M of V. Taking the directions into account, we form a space to be studied in the next section.

The concepts are independent of the imbedding of V in C^n; see § 4.

In this section, $V \subset H \subset C^n$ is always a variety of constant dimension r. If $p \in V$ is simple, $T^*(V, p) \in G = G^{n-1,r-1}$ is the $(r-1)$-direction of the tangent plane $T(V, p)$.

Define

$$\tau_0^*(V) = \{(q, T^*(V, q)) \colon q \in V_{sp}\} ; \tag{16.1}$$

this is a variety in $(H - V_{sg}) \times G$. Set

$$\tau^*(V) = (H \times G)\text{-clos}\, \tau_0^*(V) ; \tag{16.2}$$

this is the *tangent plane space* of V. We show below that it is a variety.

For each $p \in V$, let $\tau^*(V, p)$ be the elements of G over p in $\tau^*(V)$; thus

$$p \times \tau^*(V, p) = \tau^*(V) \cap (p \times G) . \tag{16.3}$$

For $p \in V_{sp}$, $\tau^*(V, p)$ contains the single element $T^*(V, p)$. If $p \in V_{sg}$, then $\tau^*(V, p)$ may contain more than one element; it contains at least one, since $V_{sg} \subset \text{clos}\, V_{sp}$ and G is compact.

Let $\tau(V, p)$ be the set of r-planes whose $(r-1)$-directions are in $\tau^*(V, p)$. For $p \in V_{sp}$, $\tau(V, p) = \{T(V, p)\}$.

THEOREM 16.4. *$\tau^*(V)$ is a variety in $H \times G$; $\tau^*(V, p)$ is a variety (in fact, an algebraic variety) in G.*

We prove the first statement; the second one then follows. (That $\tau^*(V, p)$ is algebraic follows from the Theorem of Chow; see for instance [7, Satz 14]. We do not need this fact here.)

Given $\alpha \in G$, let $\alpha^{\sigma_1 \cdots \sigma_r}$ $(1 \leq \sigma_1 < \cdots < \sigma_r \leq n)$ denote its (homogeneous) components. Now the vector v lies in the plane of α if and only if the exterior product of v with α vanishes. Analytically, this is expressed as follows:

(16.5) $$(v \vee \alpha)^{\sigma_1 \cdots \sigma_{r+1}} = \sum_{k=1}^{r+1} (-1)^{k-1} v^{\sigma_k} \alpha^{\sigma_1 \cdots \hat{k} \cdots \sigma_{r+1}} = 0 ,$$

where $\hat{k}$ denotes the omission of the corresponding symbol.

Now take any $p_0 \in V$. Find the neighborhood $U \subset H$ of p_0 and the holomorphic vector fields $v^{(1)}(p), \cdots, v^{(h)}(p)$ in U by Lemma 15.1. Define

(16.6) $$\tau_U^{**}(V) = \{(p, \alpha): p \in V \cap U, v^{(i)}(p) \vee \alpha = 0 \text{ (all } i)\} .$$

This is a variety in $U \times G$. For any $p \in V_{sp}$, $(p, \alpha) \in \tau_U^{**}(V)$ if and only if all the $v^{(i)}(p)$ lie in the plane of α; since the $v^{(i)}(p)$ span $T(V, p)$, this requires that $\alpha = T^*(V, p)$. Comparing with (16.1) shows that

$$\tau_U^{**}(V) \cap (p \times G) = \tau_0^*(V) \cap (p \times G) , \qquad p \in V_{sp} \cap U .$$

Hence also

(16.7) $$\tau^*(V) \cap (U \times G) = (U \times G)\text{-clos}\,[\tau_U^{**}(V) - V_{sg} \times G] ,$$

which, with Lemma 3.9, shows that $\tau^*(V)$ is a variety. The lemma also shows that

(16.8) $$\dim \tau^*(V) \equiv r , \qquad \dim \tau^*(V) \cap (V_{sg} \times G) \leqq r - 1 .$$

If π^* is the projection of $C^n \times G$ onto C^n, then π^* is holomorphic, and is one-one in $\tau^*(V) \cap (V_{sp} \times G)$. Thus *$\tau^*(V)$ is a modification of V.*

17. Directed tangent plane spaces

We assume $W \subset V \subset H \subset C^n$, $\dim V \equiv r$, $\dim W < r$. Define $G = G^{n-1,r-1}$ as before, and set $Q = P^{n-1} \times G$. Set

(17.1) $$\tau_{W,0}^{\#}(V) = \{(p, q, \kappa(q - p), T^*(V, q)): p \in W, q \in V_{sp}, q \neq p\} ;$$

this is a subset of $H \times H \times Q$. Set

(17.2) $$\tau_W^{\#}(V) = (H \times H \times Q)\text{-clos}\, \tau_{W,0}^{\#}(V) .$$

We call this the *W-directed tangent plane space* of V.

Let $\Delta = \{(p, p): p \in C^n\}$ be the diagonal in $C^n \times C^n$. Let φ be the natural analytical equivalence of $\Delta \times Q$ with $C^n \times Q$: $\varphi(p, p, \nu, \alpha) = (p, \nu, \alpha)$. Set

(17.3) $$\tau^{\#}(V, W) = \varphi[\tau_W^{\#}(V) \cap (\Delta \times Q)] ;$$

this is the *directed tangent plane space* of V over W.

Note that $(p, \nu, \alpha) \in \tau^{\#}(V, W)$ if and only if there are sequences $\{p_i\} \in W$, $\{q_i\} \subset V_{sp}$, $q_i \neq p_i$, such that

(a) $p_i \to p , \quad q_i \to p , \quad \kappa(q_i - p_i) \to \nu , \quad T^*(V, q_i) \to \alpha .$

If $\tau^{\#}(V, W, p, \nu)$ is the corresponding subset of G with given p, ν, then

(17.4) $$p \times \nu \times \tau^{\#}(V, W, p, \nu) = \tau^{\#}(V, W) \cap (p \times \nu \times G) .$$

THEOREM 17.5. *$\tau^{\sharp}_{W}(V)$ is a variety in $H \times H \times Q$; $\tau^{\sharp}(V, W)$ is a variety in $H \times Q$; $\tau^{\sharp}(V, W, p, \nu)$ is a variety (hence an algebraic variety) in G.*

As in Theorem 16.4, we need only prove the first statement.

Take any $p_0 \in W$, $q_0 \in V$. Find neighborhoods U of p_0 and U' of q_0 and the vector fields $v^{(i)}(q)$ in U' relative to V, as in § 15. Let

$$\tau^{\sharp\sharp} = \{(p, q, \nu, \alpha): p \in W \cap U, q \in V \cap U', \operatorname{dep}(q - p, \nu), v^{(i)}(q) \vee \alpha = 0 (\text{all } i)\} ;$$

this is a variety in $U \times U' \times Q$. It is clear that

$$\tau^{\sharp}_{W,0}(V) \cap (U \times U' \times Q) = \tau^{\sharp\sharp} - [(W \times V_{sg}) \cup \Delta] \times Q .$$

The required fact now follows from (17.2) and Lemma 3.9.

We shall consider a slightly simpler space near a point of W_{sp}.

Suppose $p_0 \in W_{sp}$. Let N be the orthogonal vector plane (or a transverse plane) to $T(W, p_0)$. Since $p + N$ is transverse to $p + T(W, p)$ for all p near p_0, we may choose an open subset N' of N about p_0 and an open subset M of W about p_0 so that if $U = \{q \in p + N' : p \in M\}$, then

$$U \subset H , \qquad W \cap U = M ,$$

and M is a manifold, say of dimension m. Moreover, setting $\pi q = p$ if $q \in p + N'$, $p \in M$, π is a holomorphic mapping of U onto M.

We define certain subsets of $U \times Q'$, where P' is the projective space of N, and $Q' = P' \times G$:

$$\sigma^{\sharp}_{M,U,0}(V) = \{(q, \kappa(q - \pi q), T^*(V, q)): q \in V_{sp} \cap U - M\} , \tag{17.6}$$

$$\sigma^{\sharp}_{M,U}(V) = (U \times Q')\text{-clos}\, \sigma^{\sharp}_{M,U,0}(V) , \tag{17.7}$$

$$\sigma^{\sharp}(V, M) = \sigma^{\sharp}_{M,U}(V) \cap (M \times Q') . \tag{17.8}$$

To show that these are varieties, consider the variety

$$\tau' = \tau^{\sharp}_{W}(V) \cap (U \times U \times Q') , \tag{17.9}$$

in $U \times U \times Q'$. Take any $(p, q, \nu, \alpha) \in \tau'$. Then there are sequences $\{p_i\} \subset M$, $\{q_i\} \subset V_{sp} \cap U$, $q_i \neq p_i$, such that

$$p_i \to p , \qquad q_i \to q , \qquad \kappa(q_i - p_i) \to \nu , \qquad T^*(V, q_i) \to \alpha .$$

Since $p, q \in U$ and $\nu \in P'$, this implies that $p = \pi q$. Now (generalizing φ) consider the projection

$$\pi_1 : U \times (U \times Q') \to U \times Q' , \qquad \pi_1(p, q, \nu, \alpha) = (q, \nu, \alpha) .$$

If $(p, q, \nu, \alpha) \in \tau'$, then for no other p' is $(p', q, \nu, \alpha) \in \tau'$; hence we may apply Lemma 3.4 near each point, showing that $\pi_1\tau'$ is a variety in $U \times Q'$. (Or we could apply Lemma 4.4.)

Now $(q, \nu, \alpha) \in \pi_1\tau'$ $(q \in V \cap U, \nu \in P')$ if and only if there is a sequence $\{q_i\} \in V_{sp} \cap U - M$ such that

$$q_i \to q \ , \qquad \kappa(q_i - \pi q_i) \to \nu \ , \qquad T^*(V, q_i) \to \alpha \ .$$

For if $(q, \nu, \alpha) \in \pi_1\tau'$, then $(\pi q, q, \nu, \alpha) \in \tau'$, and there are corresponding sequences $\{p_i\}$, $\{q_i\}$; since $\nu \in P'$, we may suppose $q_i \notin M$; now $\{q_i\}$ has the properties shown. The converse is clear, on setting $p_i = \pi q_i$. But clearly this condition is precisely the condition that (q, ν, α) be in $\sigma^{\sharp}_{M,U}(V)$. Therefore

$$\sigma^{\sharp}_{M,U}(V) = \pi_1\tau' \tag{17.10}$$

is a variety in $U \times Q'$, and hence so is $\sigma^{\sharp}(V, M)$.

It is evident that

$$\sigma^{\sharp}(V, M) \subset \tau^{\sharp}(V, W) \ . \tag{17.11}$$

We shall derive a converse relation. For any $p \in M$ and vector v, write

$$v = \pi_p v + \pi'_p v \ , \qquad \pi_p v \in T(M, p) \ , \ \pi'_p v \in N \ .$$

If P_p is the projective space of $T(M, p)$, let π'_p also denote the projection of $P - P_p$ onto P': $\pi'_p\kappa(v) = \kappa(\pi'_p v)$ if $\pi'_p v \neq 0$. Let π'^* denote the corresponding projection in product spaces:

$$\pi'^*(p, \nu, \alpha) = (p, \pi'_p\nu, \alpha) \ , \qquad p \in M, \ \nu \notin P_p \ . \tag{17.12}$$

Define the following subvariety of $U \times Q$:

$$\tau_0(M) = \{(p, \nu, \alpha) \colon p \in M, \ \nu \in P_p, \ \alpha \in G\} \ . \tag{17.13}$$

We now show that

$$\pi'^*[\tau^{\sharp}(V, W) \cap (U \times Q) - \tau_0(M)] = \sigma^{\sharp}(V, M) \ . \tag{17.14}$$

Take any $(p, \nu, \alpha) \in \tau^{\sharp}(V, W)$, $p \in M$, $\nu \notin P_p$. There are sequences $\{p_i\} \subset M$, $\{q_i\} \subset V_{sp} \cap U - M$ such that (a) holds. Also, if $\nu = \kappa(v)$, we may write $a_i(q_i - p_i) \to v$ and hence $a_i(q_i - \pi q_i) \to \pi'_p v$. Clearly

$$\pi'^*(p, \nu, \alpha) = \bigl(p, \kappa(\pi'_p v), \alpha\bigr) \in \sigma^{\sharp}(V, M) \ ,$$

proving part of (17.14). The other part is clear from (17.11).

18. Stratifications

Given the variety V, we may split it: $V = V_{sp} \cup V_{sg}$. We may now split V_{sg} into $(V_{sg})_{sp}$ and $(V_{sg})_{sg}$, etc. Thus, setting $M_1 = V_{sp}$, $M_2 = (V_{sg})_{sp}$, $M_3 = ((V_{sg})_{sg})_{sp}$ etc. gives

$$V = M_1 \cup M_2 \cup \cdots \ , \qquad \dim V = \dim M_1 > \dim M_2 > \cdots \ ; \tag{18.1}$$

each M_i is a manifold. We have thus constructed the *partition of V by dimension*. If we partition each of the above M_i into its connected pieces, we form the *full partition of V by dimension.*

Let M be a manifold in the open set H, of dimension m. We say M is *H-strict* if the H-closure $\bar{M} = H\text{-clos}\, M$ and the H-frontier $\bar{M} - M$ are varieties in H; then by Lemma 3.13, $\dim \bar{M} \equiv m$ and $\dim(\bar{M} - M) < m$. A *strict partition* of a variety is a partition into strict manifolds, the partition being locally finite.

In this section, $\bar{A}_i$ will denote H-clos A_i, etc.

LEMMA 18.2. *The partition and the full partition of V by dimension are both strict.*

Let $M_{11}, M_{12}, \cdots$ be the connected pieces of $V_{sp} = M_1$; by Lemma 3.12, the $\bar{M}_{1i}$ are the irreducible components of highest dimension of V, and $\bar{M}_{1i} - M_{1i} = \bar{M}_{1i} \cap V_{sg}$; thus the M_{1i} are strict. Also

$$\bar{M}_1 = H\text{-clos}\,(M_{11} \cup M_{12} \cup \cdots) = \bar{M}_{11} \cup \bar{M}_{12} \cup \cdots ,$$

and $\bar{M}_1 - M_1 = \bar{M}_1 \cap V_{sg}$; thus M_1 is strict. Now apply the same reasoning to V_{sg} and M_2, etc.

We say that a set of manifolds in H has the *frontier property* (in H) if, for any two of the manifolds, say M, M', with $M \neq M'$,

$$\text{if } M' \cap \bar{M} \neq \emptyset, \text{ then } M' \subset \bar{M} \text{ and } \dim M' < \dim M. \tag{18.3}$$

We say a partition into manifolds has the frontier property if the set of manifolds has.

A *stratification* of $V \subset H \subset C^n$ is a strict partition of V which has the frontier property. The manifolds are the *strata* of the partition.

LEMMA 18.4. *A locally finite partition of V with the frontier property is a stratification. Moreover, for each stratum M, $\bar{M} - M$ is the union of the other closed strata in $\bar{M}$.*

We prove that each manifold of the partition is strict, by induction on the dimension of the manifold. Given M, of dimension m, let $M_1, M_2 \cdots$ be the other manifolds with points in $\bar{M}$; then each lies in $\bar{M} - M$. Since $\bar{M} - M$ is closed in H (§ 3),

$$\bar{M} - M = M_1 \cup M_2 \cup \cdots = \bar{M}_1 \cup \bar{M}_2 \cup \cdots = X .$$

Since the union is locally finite in H, and the $\bar{M}_i$ are varieties, by induction, X is a variety in H. By the frontier property, $\dim X < m$. Now Lemma 3.11 shows that $\bar{M}$ is a variety in H, and the lemma is proved.

EXAMPLE 18.5. Set $V = C_x \cup C_y \subset C^2$, and $M_1 = C_x$, $M_2 = C_y - \{O\}$; this this is a strict partition but not a stratification.

EXAMPLE 18.6. Let $f = z^2 - xy = 0$ define $W \subset C^3$, and set $V = W \cap C_z$. Now $V = V_{sp} \cup V_{sg} = (V - C_z) \cup C_z$ is not a stratification.

EXAMPLE 18.7. Let $f = y^2 - tx^2 = 0$ define $V \subset C_{txy}$; then $V_{sg} = C_t$, and $V = V_{sp} \cap V_{sg}$ is a stratification. Note that for simple points $p \in V$ near $(t, 0, 0)$ with $t \neq 0$, $T(V, p)$ nearly contains C_t; but for $p = (0, x, 0)$ with $x \neq 0$, $T(V, p)$ is orthogonal to C_t. The singular set S_a for the pair (V_{sp}, V_{sg}) (see § 20) is $\{O\}$. A preferable stratification is $V = V_{sp} \cup (C_t - \{O\}) \cup \{O\}$.

EXAMPLE 18.8. Let $f = t^2y^2 - t^2x^2 + x^3 - y^4 = 0$ define $V \subset C_{txy}$ (see Example 12.18). Then the partition by dimension: $V = (V - C_t) \cup C_t$ is a stratification. Note that $p_t = (t, t^2, 0) \in V_{sp}$ if $t \neq 0$; moreover, $T(V, p_t)$ is near C_{ty} for small t. Now $T(V, p_t)$ nearly contains C_t; on the other hand, it is nearly orthogonal to the vector $v_t = p_t - \pi_t p_t = (0, t^2, 0)$ (π_t = projection onto C_t). The singular set S_b for the pair of manifolds is $\{O\}$. A preferable stratification is to cut out $\{O\}$ from $V_{sg} = C_t$.

REMARK 18.9. Let $\mathfrak{S}$ be a strict partition of V. Then in forming a strict refinement, no connected piece M' of a manifold M of $\mathfrak{S}$ can be cut apart; at most, a proper subvariety is taken out of M' and partitioned into manifolds. In contrast with this, in the real case, a manifold may be cut into several manifolds.

REMARK 18.10. The intersection of two strict partitions need not be a strict partition. For instance, let M_1 and M_2 be the manifolds in C^3 given by $z = 0$ and $z - y^2 + x^3 = 0$ respectively. Then if $V = C^3$, and we partition V into $(V - M_1) \cup M_1$ and into $(V - M_2) \cup M_2$, the intersection contains the variety $M_1 \cap M_2$ which is not a manifold.

THEOREM 18.11. *Any strict partition of a variety V has a refinement which is a stratification into connected strata.*

To aid the proof, we introduce a further definition: We say a set $\mathfrak{S}$ of manifolds has the frontier property over a set $\mathfrak{S}'$ of manifolds if, for any distinct elements $M \in \mathfrak{S}$ and $M' \in \mathfrak{S}'$, (18.3) holds.

Let the strict partition $\mathfrak{S}$ of V be given. Set $\mathfrak{S}_{r+1} = \mathfrak{S}$, where $r = \dim V$. We shall find a succession of strict refinements $\mathfrak{S}_r, \mathfrak{S}_{r-1}, \cdots, \mathfrak{S}_0$, with the following properties. Let $\mathfrak{S}'_m$ and $\mathfrak{S}''_m$ denote the sets of manifolds of $\mathfrak{S}_m$ of dimensions $\geqq m$ and of dimensions $> m$ respectively. Then:

(a) $\mathfrak{S}''_m = \mathfrak{S}'_{m+1}$.

(b) The set $\mathfrak{S}_m$ has the frontier property over the set $\mathfrak{S}'_m$.

(c) The manifolds of $\mathfrak{S}'_m$ are connected.

Then clearly $\mathfrak{S}_0$ is the required refinement of $\mathfrak{S}$.

Having found $\mathfrak{S}_{m+1}$, we construct $\mathfrak{S}_m$ as follows: Let $M_1^h, M_2^h, \cdots$ be the manifolds of $\mathfrak{S}_{m+1}$ of dimension h. Let $M_{i1}, M_{i2}, \cdots$ be the connected pieces of $(\bar{M}_i^m)_{sp}$; by Lemma 3.13, $\dim \bar{M}_i^m \equiv m$, and by Lemma 3.12, the M_{ij} are strict, they are disjoint (for fixed i), and the $\bar{M}_{ij}$ are the irreducible components of $\bar{M}_i^m$. Consider any M_{ij}. For each M_k^h, $h > m$, set

$$(18.12)\qquad \begin{aligned} X_{ijk}^h &= \emptyset && \text{if } M_{ij} \subset \bar{M}_k^h\,, \\ X_{ijk}^h &= \bar{M}_{ij} \cap \bar{M}_k^h && \text{otherwise.} \end{aligned}$$

Set

$$(18.13)\qquad X_{ijk} = \bar{M}_{ij} \cap \bar{M}_k^m \; (k \neq i)\,, \qquad X_{ijk}^h = \bar{M}_{ij} \cap \bar{M}_k^h \; (h < m)\,.$$

Let X_{ij} be the union of all the X_{ijk}^h and the X_{ijk}, and set

$$(18.14)\qquad M'_{ij} = M_{ij} - X_{ij}\,.$$

We prove the following facts:

$$(18.15)\qquad \dim X_{ij} < m\,,$$

$$(18.16)\qquad \bar{M}_k^{h'} \cap M_l^h = \emptyset \qquad \text{if } h > m,\ h' \leqq m\,,$$

$$(18.17)\qquad \bar{M}'_{ij} = H\text{-clos}\, M'_{ij} = \bar{M}_{ij}\,, \qquad M'_{ij} \subset M_i^m\,.$$

First, if $h < m$, then $\dim X_{ijk}^h < m$. If $h > m$ and $X_{ijk}^h \neq \emptyset$, then $M_{ij} \not\subset \bar{M}_k^h$, and X_{ijk}^h is a proper subvariety of the irreducible variety $\bar{M}_{ij}$ and hence is of dimension $<m$. Finally, if $\dim X_{ijk} = m$, then by Lemma 3.14, $\dim (M_{ij} \cap M_k^m) = m$; hence this set contains a manifold M of dimension m. Now $M \subset \bar{M}_i^m$, hence (by Lemma 3.13) M contains points of M_i^m, and $M_i^m \cap M_k^m \neq \emptyset$, a contradiction. Thus (18.15) is proved.

Next, if (18.16) is false, then (b) for $\mathfrak{S}_{m+1}$ (with $M = M_k^{h'}$, $M' = M_l^h$) gives $h < h'$, a contradiction. Next, since $\dim X_{ij} < m$, $M_{ij} \subset H\text{-clos}\,(M_{ij} - X_{ij})$, and the first part of (18.17) follows. To prove the second part, take any $p \in M'_{ij}$. By (18.16), p is in no M_k^h with $h > m$. Since $p \notin X_{ij}$, p is in no M_k^h with $h < m$ and in no M_k^m with $k \neq i$; hence $p \in M_i^m$.

We show next that

$$(18.18)\qquad \bar{M}'_{ij} - M'_{ij} = \bar{M}_{ij} - M'_{ij} = X_{ij} \cup [\bar{M}_{ij} \cap (\bar{M}_i^m)_{sg}]\,.$$

The first relation follows from (18.17). Write the next as $A = B$. Since $X_{ij} \cap M'_{ij} = \emptyset$ and $(\bar{M}_i^m)_{sg} \cap M_{ij} = \emptyset$, we have $B \subset A$. Also, using Lemma 3.12,

$$\bar{M}_{ij} - M_{ij} = \bar{M}_{ij} \cap (\bar{M}_i^m)_{sg}\,, \qquad M_{ij} - M'_{ij} \subset X_{ij}\,,$$

giving $A \subset B$.

Let each M_k^h with $h \neq m$ be a manifold of $\mathfrak{S}_m$; now (a) holds. Let the M'_{ij} be the manifolds of $\mathfrak{S}_m$ of dimension m; by (18.17) and (18.18), these manifolds are strict. We must still partition the disjoint sets

$$A_i = M_i^m - \textstyle\bigcup_j M'_{ij}\,.$$

Set

$$A_{ij} = A_i \cap M_{ij} ;$$

since $A_i \subset M_i^m \subset (\bar{M}_i^m)_{sp} = \bigcup_j M_{ij}$, $A_i = \bigcup_j A_{ij}$. Also these sets are disjoint, since the M_{ij} (i fixed) are. Thus we need merely partition each A_{ij}.

Since $A_{ij} = M_i^m \cap M_{ij} - \bigcup_k M'_{ik} = M_i^m \cap (M_{ij} - M'_{ij})$,

$$A_{ij} = X_{ij} \cap M_{ij} \cap M_i^m . \tag{18.19}$$

Note next that

$$\begin{aligned} F_{ij} = X_{ij} - A_{ij} &= X_{ij} - M_{ij} \cap M_i^m \\ &= (X_{ij} - M_{ij}) \cup (X_{ij} - M_i^m) \\ &= [X_{ij} \cap (\bar{M}_{ij} - M_{ij})] \cup [X_{ij} \cap (\bar{M}_i^m - M_i^m)] , \end{aligned}$$

is a variety, since the M_{ij} and M_i^m are strict.

Let $N_1 \cup N_2 \cup \cdots$ be the full partition of X_{ij} by dimension; it is strict, by Lemma 18.2, and the $\bar{N}_i$ are irreducible. By (18.15), $\dim N_k < m$. Set $N'_k = N_k - F_{ij}$; then, omitting void sets,

$$A_{ij} = \bigcup_k N'_k ,$$

is the desired partition of A_{ij}. It is a partition, since $X_{ij} - F_{ij} = A_{ij}$.

Since N_k is strict, $\bar{N}_k$ and $Z_k = \bar{N}_k - N_k$ are varieties. Also, if N'_k is non-void, then $\dim (F_{ij} \cap \bar{N}_k) < \dim N_k$, and hence clearly $N_k \subset \bar{N}'_k$; thus $\bar{N}'_k = \bar{N}_k$ is a variety. Also

$$\begin{aligned} N'_k &= (\bar{N}_k - Z_k) - F_{ij} = \bar{N}_k - Z_k \cup F_{ij} , \\ \bar{N}'_k - N'_k &= \bar{N}_k - (\bar{N}_k - Z_k \cup F_{ij}) = \bar{N}_k \cap (Z_k \cup F_{ij}) , \end{aligned}$$

which is also a variety; thus the N'_k are strict.

The set of manifolds is clearly locally finite; hence we have formed a strict refinement $\mathfrak{S}_m$ of $\mathfrak{S}_{m+1}$. We must still prove (b).

Take any $M \in \mathfrak{S}_m$, $M' \in \mathfrak{S}'_m$, $M \neq M'$, and suppose $M' \cap \bar{M} \neq \emptyset$; we must show that $M' \subset \bar{M}$ and $\dim M' < \dim M$. If $\dim M' = h > m$, then $M' = M_k^h$, and by (18.16), we clearly have $\dim M > m$, so that $M \in \mathfrak{S}_{m+1}$, $M' \in \mathfrak{S}'_{m+1}$, and we may apply (b) for $\mathfrak{S}_{m+1}$. We now suppose $\dim M' = m$; say $M' = M'_{ij}$.

If $h<m$, then since $\bar{M}_k^h \cap \bar{M}_{ij} = X_{ijk}^h$, $\bar{M}_k^h \cap M'_{ij} = \emptyset$; thus $M \neq M_k^h (h<m)$. Similarly $M \not\subset M_k^m$, $k \neq i$. Also $\bar{M}_i^m - M'_{ij}$ is closed in H, and hence $M \not\subset M_i^m$. Therefore $M = M_k^h$ for some $h > m$ and k; we have $\dim M' < \dim M$. If $M' \not\subset \bar{M}$, then $M_{ij} \not\subset \bar{M}_k^h$, hence, by (18.12), $X_{ijk}^h = \bar{M}_{ij} \cap \bar{M}_k^h$, and by (18.14), $M'_{ij} \cap \bar{M}_k^h = \emptyset$, a contradiction; thus $M' \subset \bar{M}$, and the theorem is proved.

REMARK 18.20. If we start with a strict partition $\mathfrak{S}$ in which V_{sp} is an element of $\mathfrak{S}$ or a union of elements of $\mathfrak{S}$, then we do not need to partition V_{sp}

further. Any point $p \in V$ may be made into one of the elements of the final stratification.

This is clear from the proof.

19. Regular stratifications

Suppose M is a manifold and V is a variety, $\dim V \equiv r$. We say V is *regular over* M if the following conditions are satisfied:

(a) For each $p \in M$ and r-plane T, if there is a sequence $\{q_i\} \subset V_{sp}$ such that

$$q_i \to p\,, \qquad T(V, q_i) \to T\,,$$

then $T(M, p) \subset T$.

(b) For each $p \in M$, vector v and r-plane T, if there are sequences $\{p_i\} \subset M$, $\{q_i\} \subset V_{sp}$ and $\{a_i\}$ such that

$$p_i \to p\,, \qquad q_i \to p\,, \qquad a_i(q_i - p_i) \to v\,, \qquad T(V, q_i) \to T\,,$$

then $v \in T$.

Note that these hold trivially if $M \cap V = \emptyset$.

We say V is a-*regular* (b-*regular*) *over* M if the first (second) condition holds.

The conditions may clearly be phrased as follows:

(a′) If $p \in M$, $T \in \tau(V, p)$, then $T(M, p) \subset T$.

(b′) If $p \in M$, $v \in C^n$, $\kappa(T) \in \tau^*(V, M, p, \kappa(v))$, then $v \in T$.

EXAMPLE 19.1. Let $f = z^2 - xy = 0$ define $V \subset C^3$; set $M = C_x$. Setting $v = (-1, 1, 0)$, $T = C_{yz}$, $p_i = (1/i, 0, 0)$, $q_i = (0, 1/i, 0)$, $a_i = i$, shows that (b) fails at O. If $M = \{O\}$, then (b) holds.

In Example 18.7, setting $M = C_t$, (a) fails at O.

In Example 18.8, setting $M = C_t$, (b) fails at O; for setting $q_i = (t_i, t_i^2, 0)$, $p_i = (t_i, 0, 0)$ $(t_i \to 0)$ shows that $\kappa(C_{ty}) \in \tau^*(V, M, O, \kappa(e_x))$; but $e_x \notin C_{ty}$. Note that $C(V, O) = C_{ty}$; if we required $v \in C(V, p)$ in (b), then (b) would hold here.

We say a stratification of a variety V is *regular* if, for any two strata M, M', $\bar{M}'$ is regular over M. Recall that the closure of each stratum is a variety; that M and $\bar{M}'$ are disjoint if $\dim M \geqq \dim M'$, and that if $M \cap \bar{M}' \neq \emptyset$, then $M \subset \bar{M}'$.

THEOREM 19.2. *Any variety V has a regular stratification. Any stratification has a regular refinement.*

The proof is based on the following lemma, whose proof will be given in the next two sections:

LEMMA 19.3. *Let W and V be varieties, and suppose* $\dim W \equiv m$,

$\dim V \equiv r > m$. *Then there is a subvariety S of W of dimension $< m$ such that V is regular over $W_{sp} - S$.*

Assuming the lemma, the theorem is proved as follows. By Lemma 18.2 and Theorem 18.11, V has a stratification; hence we need only prove the second part. Let the stratification $\mathfrak{S}$ be given; we shall find a regular refinement $\mathfrak{S}_0$. We shall in fact find stratifications $\mathfrak{S}_r = \mathfrak{S}, \mathfrak{S}_{r-1}, \cdots, \mathfrak{S}_0$ such that each is a refinement of the preceding, the strata of any $\mathfrak{S}_m$ of dimension $>m$ are those of $\mathfrak{S}_{m+1}$ (if $m < r$), and so that for any two strata M^m, M^h $(h > m)$ of $\mathfrak{S}_m$ of the dimensions shown, $\bar{M}^h$ is regular over M^m. Then $\mathfrak{S}_0$ is a regular refinement of $\mathfrak{S}$.

Given $\mathfrak{S}_{m+1}(m < r)$, we find $\mathfrak{S}_m$ as follows: For each $h \geqq m$, let $M_1^h, M_2^h, \cdots$ be the strata of $\mathfrak{S}_{m+1}$ of dimension h. Set

$$W^h = \bar{M}_1^h \cup \bar{M}_2^h \cup \cdots .$$

Apply Lemma 19.3 to the pair W^m, $W^h (h > m)$ to give the subvariety $S^{(h)}$ of W^m of dimension $<m$; set $S = S^{(m+1)} \cup \cdots \cup S^{(r)}$. Now $\dim S < m$. Let the strata of $\mathfrak{S}_m$ of dimension $>m$ be those of $\mathfrak{S}_{m+1}$. The strata of dimension m are the connected pieces of the sets $M_i^m - S$. The strata of $\mathfrak{S}_m$ of lower dimension are formed just as the corresponding strata in the proof of Theorem 18.11 were formed.

Take any stratum M_k^m, $M_i^h (h > m)$ of $\mathfrak{S}_m$. From the construction, W^h is regular over M_k^m. Since $(W^h)_{sp}$ is dense in $(\bar{M}_i^h)_{sp}$, it follows easily that $\bar{M}_i^h$ is regular over M_k^m, as required.

20. The singular set S_a

In this and the following section we have

$$W \subset V \subset H \subset C^n , \qquad \dim W \equiv m , \qquad \dim V \equiv r > m .$$

Here we shall find a subvariety S_a of W of dimension $<m$ such that V is a-regular over $W_{sp} - S_a$. In the next section we find S_b for b-regularity. Then setting $S = S_a \cup S_b$, Lemma 19.3 and hence Theorem 19.2 will follow.

We shall use notations

$$G = G^{n-1,r-1} , \qquad G_m = G^{n-1,m-1} , \qquad G' = G_m \times G .$$

$T(\alpha)$ is the plane corresponding to $\alpha \in G$; similarly for $\alpha' \in G_m$. Writing $\alpha' \subset \alpha$ means $T(\alpha') \subset T(\alpha)$. We shall use the set

(20.1) $$R^* = \{(p, \alpha', \alpha): p \in C^n, \alpha' \in G_m, \alpha \in G, \alpha' \subset \alpha\} ;$$

this is a variety in $C^n \times G'$.

Form the products (see § 16) $\tau^*(W) \times G$, $\tau^*(V) \times G_m$, as follows:

$$\tau_1^*(W) = \{(p, \alpha', \alpha): p \in W, \alpha' \in \tau^*(W, p), \alpha \in G\} ,$$
$$\tau_1^*(V) = \{(p, \alpha', \alpha): p \in V, \alpha' \in G_m, \alpha \in \tau^*(V, p)\} ;$$

these are varieties in $H \times G'$. Set

$$\tau_1^* = \tau_1^*(W) \cap \tau_1^*(V) = \{(p, \alpha', \alpha): p \in W, \alpha' \in \tau^*(W, p), \alpha \in \tau^*(V, p)\} ; \tag{20.2}$$

this again is a variety in $H \times G'$. With the projection π_0^* of $C^n \times G'$ onto C^n, set

$$S_a^* = (H \times G')\text{-clos}\,(\tau_1^* - R^*) , \qquad S_a = \pi_0^* S_a^* ; \tag{20.3}$$

this is the required subvariety of W. (Since G' is compact, π_0^* is proper into H, and Lemma 4.4 applies.)

To show that V is a-regular over $M = W_{sp} - S_a$, take any $p \in M$ and r-plane $T \in \tau(V, p)$; we prove that $T(M, p) \subset T$. Since $M \subset W_{sp}$, $T^*(M, p) = \alpha'$ is the sole element of $\tau^*(W, p)$. Also $\alpha = \kappa(T) \in \tau^*(V, p)$. Hence $p^* = (p, \alpha', \alpha) \in \tau_1^*$. Suppose $\alpha' \not\subset \alpha$; then $p^* \notin R^*$, hence $p^* \in S_a^*$, and $p = \pi_0^* p^* \in S_a$, a contradiction. Therefore $\alpha' \subset \alpha$, as required.

We must show still that $\dim S_a < m$. Suppose not; then $\dim S_a = m$. By Lemma 4.4 there is a simple point p_0^* of S_a^* at which $\pi_0^* \mid S_a^*$ is of rank m. By the second part of Lemma 4.3, with V and V_1 replaced by S_a^* and $S_a^* \cap R^*$ respectively, we may suppose p_0^* is a simple point of $S_a^* - R^* = \tau_1^* - R^*$. Since the image of $\pi^* \mid S_a^*$ near p_0^* contains a manifold of dimension m, and such a manifold must contain points of W_{sp}, we may suppose also that $\pi_0^*(p_0^*)$ is in the m-manifold $M = W_{sp}$. Say $p_0^* = (p_0, \alpha_0', \alpha_0)$.

Consider the projections

$$\pi^*: C^n \times G_m \times G \to C^n \times G , \qquad \pi': C^n \times G \to C^n .$$

Set $p_0' = (p_0, \alpha_0) = \pi^* p_0^*$. Choose neighborhoods U_0 of p_0, $U_0'^*$ of α' and $U_0''^*$ of α such that

$$W \cap U_0 \subset M , \tag{20.4}$$

$$\text{if } p \in W \cap U_0, \text{ then } T^*(M, p) \subset U_0'^*. \tag{20.5}$$

Set $U_0^* = U_0 \times U_0'^* \times U_0''^*$. Using (20.2), (20.4) and (20.5) gives

$$\tau_1^* \cap U_0^* = \{(p, T^*(M, p), \alpha): p \in M \cap U_0, \alpha \in \tau^*(V, p) \cap U_0''^*\} .$$

Set

$$\tau' = \tau^*(V) \cap (W \times G') = \{(p, \alpha): p \in W, \alpha \in \tau^*(V, p)\} . \tag{20.6}$$

Comparing these expressions shows that, if $U_0' = U_0 \times U_0''^*$, then

$$\pi^* \mid (\tau_1^* \cap U_0^*) \text{ is an analytical equivalence onto } \tau' \cap U_0' . \tag{20.7}$$

Now p_0' is a simple point of τ'.

Since $p_0^* \notin R^*$, $T(M, p_0) \not\subset T(\alpha_0)$, and we may choose v_0 so that

(20.8) $$v_0 \in T(M, p_0)\ , \qquad v_0 \notin T(\alpha_0)\ .$$

Since π_0^* is of rank m in τ_1^* at p_0^*, (20.7) shows that π' is of rank m in τ' at p_0'. Hence we may choose $v_0' \in T(\tau', p_0')$ such that $d\pi'(p_0') \cdot v_0' = v_0$. Let N' be the orthogonal plane to τ' at p_0' in $C^n \times G$, and let N_1' be the plane spanned by N' and v_0'. Set

$$L' = \tau' \cap N_1'\ .$$

Now L' is a 1-manifold in τ' near p_0', $v_0' \in T(L', p_0')$, and π' is of rank 1 in L' at p_0'; hence we may take a neighborhood U':

$$U' = U \times U^*\ , \quad U \subset U_0\ , \quad U^* \subset U_0''^*\ ,$$

of p_0' so that $\pi' \mid (L' \cap U')$ is an analytical equivalence onto a 1-manifold L in M, and, by (20.8), so that

(20.9) if $p \in L$, $v \in T(L, p)$, $v \neq 0$, $\alpha \in U^*$, then $v \notin T(\alpha)$.

Set

$$\tau'' = \tau^*(V) \cap (V_{sg} \times G)\ .$$

By (16.8), $\dim \tau^*(V) \equiv r$, $\dim \tau'' < r$. Now

$$p_0' \in L' \cap U' \subset \tau^*(V) \cap U' \subset U'\ ,$$

and we may apply Lemma 13.5: There is an open subset L_1' of $L' \cap U'$ and a wing B' from L_1' into $(\tau^*(V) - \tau'') \cap U'$; let B' be defined by F' over $Z^\sim$. For $\lambda > 0$, $B_\lambda' \subset V_{sp} \times G$ (see § 13, (e)); hence $\pi' \mid B_\lambda'$ is an analytical equivalence for $\lambda > 0$. Recall that $\pi' \mid L_1'$ is also an analytical equivalence. Hence if we set $F = \pi' \circ F'$, all the properties of a wing are satisfied for $B = F(Z^\sim)$ except possibly for (a) at $\lambda = 0$. By Remark 13.2, (a) holds also; hence *$B = F(Z^\sim)$ is a wing from $L_1 = \pi'(L_1')$ into V_{sp}*. (The projection property in § 13 by means of which (c) was proved clearly holds here also; actually, we do not need this property here.)

Set $B_\lambda = F(Z_\lambda)$. Choose $q_i \in B_{\lambda_i}$ so that $q_i \to p \in B_0 = L_1$. By the property (e) of wings,

(20.10) $$T(B_{\lambda_i}, q_i) \to T(B_0, p) = T(L, p)\ .$$

Say $q_i = F(t^{(i)}, \lambda_i)$. Since $q_i \in V_{sp}$,

$$q_i' = F'(t^{(i)}, \lambda_i) = \big(q_i, T^*(V, q_i)\big) \in B'\ .$$

If $p = F(t^{(0)}, 0)$, $F'(t^{(0)}, 0) = (p, \alpha)$, then

(20.11) $$q_i' \to p' = (p, \alpha) \in L_1'\ , \qquad T^*(V, q_i) \to \alpha \in U^*\ .$$

Choose $v \neq 0$ in $T(L, p)$. By (20.10) we may choose $v_i \in T(B_{\lambda_i}, q_i)$ so that

$v_i \to v$. Since $B_{\lambda_i} \subset V_{sp}$, $v_i \in T(V, q_i)$. Now (20.11) shows that $v \in T(\alpha)$. But this contradicts (20.9). Therefore $\dim S_a < m$.

21. The singular set S_b

We assume $W \subset V$, $\dim W \equiv m$, $\dim V \equiv r > m$ again. With Q as in § 17, set

$$(21.1) \qquad R^\sharp = \{(p, \nu, \alpha) \in C^n \times Q : \nu \subset \alpha\} ;$$

this is a variety in $C^n \times Q$. Letting $\pi^\sharp$ be the projection of $C^n \times Q$ onto C^n, set

$$(21.2) \qquad S_b^\sharp = (H \times Q)\text{-clos}\,[\tau^\sharp(V, W) - R^\sharp] , \qquad S_b = \pi^\sharp S_b^\sharp .$$

Now S_b is a subvariety of W.

We show that V is b-regular over $W_{sp} - S_b$. Take any

$$p^\sharp = (p, \kappa(v), \kappa(T)) \in \tau^\sharp(V, W) , \qquad p \in W_{sp} ;$$

suppose $v \notin T$. Then $p^\sharp \in \tau^\sharp(V, M) - R^\sharp$, and $p \in S_b$, proving the statement.

We shall prove still that

$$(21.3) \qquad \dim S_b < m ;$$

then, setting $S = S_a \cup S_b$, $\dim S < m$, and V is both a-regular and b-regular over $W_{sp} - S$, completing the proof of Lemma 19.3 and hence of Theorem 19.2.

Suppose (21.3) is false. Then since $\dim S_a < m$, S_b contains some connected piece M_0 of $W_{sp} - S_a$. Take $N, M \subset M_0$, U, $\sigma^\sharp(V, M)$ and π'_p as in § 17. First we show that, for $p \in M$,

$$(21.4) \qquad \text{if } (p, \nu, \alpha) \in \tau^\sharp(V, W) - R^\sharp, \text{ then } (p, \pi'_p\nu, \alpha) \in \sigma^\sharp(V, M) - R^\sharp.$$

Say $\nu = \kappa(v)$, $\alpha = \kappa(T)$; then $v \notin T$. Since $p \notin S_a$, $T(M, p) \subset T$; hence $v \notin T(M, p)$. Since $\pi_p v \in T(M, p) \subset T$, $\pi'_p v \notin T$. Also, by (17.14),

$$\pi'^*(p, \nu, \alpha) = (p, \kappa(\pi'_p v), \alpha) \in \sigma^\sharp(V, M) ,$$

proving the statement.

Since Q is compact, $\pi^\sharp \mid (S_b^\sharp \cap (U \times Q))$ is proper into U. By Lemma 4.4 there is a simple point p_0 of the above set at which $\pi^\sharp$ is of rank m. By Lemma 4.3, we may suppose p_0 is a simple point of $\tau^\sharp(V, W) - R^\sharp$. By (21.4) and (17.12), over M,

$$[\pi^\sharp \mid (\sigma^\sharp(V, M) - R^\sharp)] \circ [\pi'^* \mid (\tau^\sharp(V, W) - R^\sharp)] = \pi^\sharp \mid (\tau^\sharp(V, W) - R^\sharp) ;$$

hence there is a point

$$(21.5) \qquad p_0^\sharp = (p_0, \nu_0, \alpha_0) \in \sigma^\sharp(V, M) - R^\sharp ,$$

which is simple in $\sigma^\sharp(V, M)$, such that

$$(21.6) \qquad \pi^\sharp \mid \sigma^\sharp(V, M) \text{ is of rank } m \text{ at } p_0^\sharp.$$

We introduce some shorthand notations (with Q' as in § 17):

$$V^{\sharp} = \sigma^{\sharp}_{M,U}(V) , \qquad V'^{\sharp} = V^{\sharp} \cap (V_{sg} \times Q') , \tag{21.7}$$

$$M^{\sharp} = M \times Q' , \qquad W^{\sharp} = \sigma^{\sharp}(V, M) = V^{\sharp} \cap M^{\sharp} . \tag{21.8}$$

Say $W^{\sharp}$ is of dimension $m + s$ at p_0. Consider the part $W^{\sharp}(p) \subset Q'$ of $W^{\sharp}$ over $p \in M$:

$$W^{\sharp} \cap (p \times Q') = p \times W^{\sharp}(p) , \qquad p \in M ; \tag{21.9}$$

by Lemma 4.5, if

$$U_0^{\sharp} = X_0 \times Y_0 \times Z_0 ,$$

is a neighborhood of $p_0^{\sharp}$ such that $W^{\sharp} \cap U_0^{\sharp} \subset (W^{\sharp} - R^{\sharp}) \cap (W^{\sharp})_{sp}$ and $\pi^{\sharp}$ is of rank m there, then

$$W'(p) = W^{\sharp}(p) \cap (Y_0 \times Z_0) \tag{21.10}$$

is a manifold of dimension s in Q'.

Let K_0 be the orthogonal complement in Q' of the tangent plane of $W'(p_0)$ at (ν_0, α_0); we may suppose $U_0^{\sharp}$ so small that: K_0 is transverse to $T(W'(p), (\nu, \alpha))$ *if* $(p, \nu, \alpha) \in U_0^{\sharp}$, and

$$K_0 \cap W'(p_0) = \{(\nu_0, \alpha_0)\} . \tag{21.11}$$

Set

$$m' = n - r + m + s + 1; \qquad \text{then } m' > m.$$

Now (21.7) and (21.8) with (17.6) through (17.8) and Lemma 3.9 show that

$$m + s = \dim W^{\sharp} < \dim V^{\sharp} \equiv r ;$$

hence $m' \leqq n$. Therefore we may choose a manifold M' in some neighborhood $X_0' \subset X_0$ of p_0 such that

$$M \cap X_0' \subset M' , \qquad \dim M' = m' > m .$$

Set $K_0^{\sharp} = M' \times K_0$. Then adding two codimensions shows that

$$\operatorname{codim}(K_0^{\sharp}, C^n \times Q') = n - m' + s = r - m - 1 .$$

Using Lemma 3.9 again shows that $\dim V'^{\sharp} \leqq r - 1$. Note that

$$M^{\sharp} \cap K_0^{\sharp} = M \times K_0 \qquad \text{near } p_0^{\sharp} ,$$

$$\dim (V^{\sharp} \cap M^{\sharp} \cap K_0^{\sharp}) = \dim \big(W^{\sharp} \cap (M \times K_0)\big) = m .$$

Now we see that all the conditions of Lemma 7.1 are satisfied, with the following substitutions:

$$\begin{array}{cccccccc} C^n & V & V' & r & s & M & M_0 & h \\ C^n \times Q' & V^{\sharp} & V'^{\sharp} & r & r-1 & M^{\sharp} & K_0^{\sharp} & r - m - 1 . \end{array}$$

Hence we can find a neighborhood $U_1^{\sharp} = X_1 \times Y_1 \times Z_1 \subset U_0^{\sharp}$ of $p_0^{\sharp}$ $(X_1 \subset X_0')$ and a manifold $K_1^* \subset U_1^*$ arbitrarily close to $K_0^* \cap U_1^*$ such that

$$K_1^* \cap M^* \cap U_1^* = K_0^* \cap M^* \cap U_1^* , \tag{21.12}$$

$$\dim (V_1^*) \equiv m + 1 , \qquad \dim (V_1'^* - M^*) \leqq m \tag{21.13}$$

where we have set

$$V_1^* = V^* \cap U_1^* \cap K_1^* , \qquad V_1'^* = V'^* \cap U_1^* \cap K_1^* . \tag{21.14}$$

Because of (21.5) and (21.1) we may find $\eta > 0$, and may take U_1^* so small that

$$\text{if } \nu \in Y_1, \alpha \in Z_1, | \nu, \nu' |^* < \eta, \text{ then } \nu' \not\subset \alpha. \tag{21.15}$$

Using formulas numbered 14, 12, 8, 9, 10 and 11, with $K_0^* = M' \times K_0$, gives

$$\begin{aligned}(p_0 \times Q') \cap V_1^* &= (p_0 \times Q') \cap M^* \cap V^* \cap U_1^* \cap K_1^* \\ &= (p_0 \times Q') \cap W^* \cap U_1^* \cap K_0^* \\ &= (p_0 \times W^*(p_0)) \cap (X_1 \times Y_1 \times Z_1) \cap (M' \times K_0) \\ &= p_0 \times [K_0 \cap W'(p_0) \cap (Y_1 \times Z_1)] = \{p_0^*\} .\end{aligned}$$

Therefore we may apply Lemma 3.4 to π^*, with both V_1^* and $V_1'^*$ in U_1^*. (If $p_0^* \notin V_1'^*$, we may suppose $V_1'^* \cap U_1^* = \emptyset$.) Set

$$V_1 = \pi^*(V_1^*) , \qquad V_1' = \pi^*(V_1'^*) ; \tag{21.16}$$

these are then varieties in X_1. From (c) and (a) of the lemma, we have

$$\dim V_1 \equiv m + 1 , \qquad \dim (V_1' - M) \leqq m . \tag{21.17}$$

Recall from § 17 that $N'(p_0)$ is an open subset of the orthogonal complement of $T(M, p_0)$. We may apply Lemma 7.1 with the substitutions

$$\begin{array}{cccccccc} H & V & V' & r & s & M & M_0 & h \\ X_1 & V_1 & V_1' & m+1 & m & M \cap X_1 & N'(p_0) & m . \end{array}$$

There is therefore a neighborhood $X_2 \subset X_1$ of p_0 and a manifold (actually, a plane) $N_1 \subset X_2$ arbitrarily close to $N'(p_0) \cap X_2$, such that

$$N_1 \cap M \cap X_2 = \{p_0\} , \qquad \dim V_2 \equiv 1 , \qquad \dim (V_2' - M) \leqq 0 , \tag{21.18}$$

where we have set

$$V_2 = V_1 \cap X_2 \cap N_1 , \qquad V_2' = V_1' \cap X_2 \cap N_1 . \tag{21.19}$$

We may clearly take N_1 so close to $N'(p_0)$ that (with π as in § 17)

$$| q - p_0, q - \pi q |^* < \eta/3 \qquad \text{if } q \in N_1 - M . \tag{21.20}$$

By (21.18) and Lemma 8.11, $\dim C(V_2, p_0) = 1$; hence we may choose $v_0 \in C(V_2, p_0)$ with $| v_0 | = 1$. By Theorem 11.8 there is an arc $\Gamma \subset V_2$, defined by a function $\Phi(\lambda) (0 \leqq \lambda \leqq \lambda_1)$ with $\Phi(0) = p_0$, such that

$$\lim_{\lambda\to 0}\Phi'(\lambda) = \lim_{\lambda\to 0}\frac{\Phi(\lambda) - p_0}{\lambda} = v_0 .$$

Using the last relation in (21.18), we may choose $\lambda > 0$ so that, if we set $q = \Phi(\lambda)$, $v = \Phi'(\lambda)$, we have

(21.21) $$q \notin V_2' , \qquad |v, v_0|^* < \eta/3 , \qquad |q - p_0, v_0|^* < \eta/3 .$$

Since $q \in N_1$, we may apply (21.20); using (21.21) gives

(21.22) $$|v, q - \pi q|^* < \eta .$$

Since $q \in V_2 - V_2'$, (21.19) gives $q \in V_1 - V_1'$; hence (21.16) gives $q \in \pi^\sharp(V_1^\sharp) - \pi^\sharp(V_1'^\sharp)$; and now (21.14) and (21.7) show that

(21.23) $$q \in \pi^\sharp(V^\sharp \cap U_1^\sharp \cap K_1^\sharp) - \pi^\sharp(V^\sharp \cap (V_{sg} \times Q') \cap U_1^\sharp \cap K_1^\sharp) ;$$

this shows that $q \in V - V_{sg}$, and q is a simple point of V. Hence also, by (21.7), (17.7) and (17.6),

$$(\pi^\sharp)^{-1}(q) \cap V^\sharp = \{(q, \kappa(q - \pi q), T^*(V, q))\} = \{q^\sharp\} .$$

By (21.23), $q^\sharp \in U_1^\sharp$; hence

$$\kappa(q - \pi q) \in Y_1 , \qquad T^*(V, q) \in Z_1 .$$

Using (21.22) and (21.15) shows that $v \notin T(V, q)$. But v is tangent to $\Gamma \subset V_2 \subset V$ and hence $v \in T(V, q)$, a contradiction. This completes the proof of Theorem 19.2.

22. Some applications

First we show that sufficiently near any point p_0 of a variety V, any tangent plane $T(V, p)$ nearly contains the vector $p - p_0$.

THEOREM 22.1. *Given $p_0 \in V$ and $\varepsilon > 0$, there is a $\delta > 0$ such that, if $p \in V_{sp}$ and $|p - p_0| < \delta$, then there is a vector $v \in T(V, p)$ with $|p - p_0, v|^* < \varepsilon$.*

REMARK 22.2. Given a stratification of V, it follows that for any point p near p_0, in some element M of the stratification, there is a vector $v \in T(M, p)$ near $p - p_0$.

To prove the theorem, form a stratification of V; by Remark 18.20, we may take $M = V_{sp}$ as one of the strata, and $N = \{p_0\}$ as another. (We may suppose $p_0 \in V_{sg}$.) We may suppose $N \subset \bar{M}$. Applying Lemma 19.3 to the pair N, $\bar{M}$ shows that $\bar{M}$ is regular over N; hence if $(p_0, \kappa(v), \kappa(T)) \in \tau^\sharp(\bar{M}, N)$ then $v \in T$.

Since Q is compact, it is easy to see that given $\varepsilon > 0$ we may find $\delta > 0$ such that if $p \in M$ and $|p - p_0| < \delta$, then for some ν and α,

$$(p_0, \nu, \alpha) \in \tau^\sharp(\bar{M}, N) , \qquad |p - p_0, \nu|^* < \varepsilon/2 , \qquad |T(\bar{M}, p), \alpha|_m^* < \varepsilon/2 .$$

With this δ, take any such p. Choose $\nu = \kappa(v)$ and $\alpha = \kappa(T)$ with the above properties. Since $\bar{M}$ is regular over N, $v \in T$. Therefore (by definition of distance in G) we may choose $v' \in T(V, p)$ so that $|v, v'|^* < \varepsilon/2$. Now $|p - p_0, v'|^* < \varepsilon$, proving the theorem.

For the second application, we recall the definition of the cone $C_4(V, p)$ in [9, § 3]; *we assume* $\dim V \equiv r$:

$C_4(V, p)$ is the set of all vectors v such that, for some sequences $\{q_i\}$, $\{v_i\}$, we have

$$q_i \in V_{sp}\,, \qquad q_i \to p\,, \qquad v_i \in T(V, q_i)\,, \qquad v_i \to v\,.$$

This is the part over p of a variety $C_4(V)$ formed as follows:

$$C_{4,0}(V) = \{(q, v) \colon q \in V_{sp},\, v \in T(V, q)\}\,, \tag{22.3}$$

$$C_4(V) = (H \times C^n)\text{-clos}\, C_{4,0}(V)\, : \tag{22.4}$$

clearly we have

$$p \times C_4(V, p) = C_4(V) \cap (p \times C^n)\,. \tag{22.5}$$

We prove, assuming $\dim V \equiv r$, $\dim W \equiv s$,

$$\text{if } p \in W \subset V, \text{ then } C_4(W, p) \subset C_4(V, p). \tag{22.6}$$

Take any $v \in C_4(W, p)$; say $q_i \in W_{sp}$, $q_i \to p$, $v_i \in T(W, q_i)$, $v_i \to v$. Since some subsequence of $\{q_i\}$ lies in an irreducible component of W, we may suppose that W is irreducible.

Form a regular stratification of V (Theorem 19.2). Let N be a smallest stratum such that $W \subset \bar{N}$. By the frontier property (§ 18) $X = \bar{N} - N$ is a union of strata N_i', in fact of the $\bar{N}_i'$, of dimension $< m = \dim N$. Since $W \not\subset \bar{N}_i'$ and W is irreducible, $\dim (W \cap \bar{N}_i') < s$; hence $\dim (W \cap X) < s$. We may therefore choose $q_i' \in W_{sp} \cap N$ arbitrarily near q_i, and choose $v_i' \in T(W, q_i')$ near v_i. We now have

$$q_i' \in N\,, \qquad q_i' \to p\,, \qquad v_i' \in T(N, q_i')\,, \qquad v_i' \to v\,.$$

If $m = r$, we are through. Suppose not. Then since $\dim V \equiv r$, there is a stratum M of dimension r such that $N \subset \bar{M}$. Since $\bar{M}$ is a-regular over N, we may find $p_i \in M$ arbitrarily near q_i' so that $T(M, p_i)$ almost contains $T(N, q_i')$; hence we may find $w_i \in T(M, p_i)$ near v_i'. Now the pair of sequences $\{p_i\}$, $\{w_i\}$ shows that $v \in C_4(V, p)$, and (22.6) is proved.

INSTITUTE FOR ADVANCED STUDY

REFERENCES

1a. S. S. ABHYANKAR, Local Analytic Geometry, Academic Press, New York, 1964.
1. E. BISHOP, *Partially analytic spaces*, Amer. J. Math., 83 (1961), 669-692.
2. H. CARTAN, Séminaire 1951-52, Ecole Normal Sup., Paris.
3. H. GRAUERT and R. REMMERT, *Zur Theorie der Modifikationen*: I. *Stetige und eigentliche Modifikationen komplexer Räume.*, Math. Ann., 129 (1955), 274-296.

4. ———, *Komplexe Räume*, Math. Ann., 136 (1958), 245–318.

4a. M. HERVE, Several Complex Variables, Local Theory, Oxford University Press, London, 1963.

5. E. KREYSZIG, *Stetige Modifikationen komplexer Mannigfaltigkeiten*, Math. Ann., 128 (1955), 479-492.

6. R. REMMERT, *Holomorphe und meromorphe Abbildungen komplexer Räume*, Math. Ann., 133 (1957), 328–370.

7. ——— and K. STEIN, *Über die wesentlichen Singularitäten analytischer Mengen*, Math. Ann., 126 (1953), 263-306.

8. H. ROSSI, *Analytic spaces with compact subvarieties*, Math. Ann., 146 (1962), 129–145.

9. H. WHITNEY, "Local properties of analytic varieties", Differential and Combinatorial Topology, Princeton University Press, 1965.

(Received April 10, 1964)
(Revised September 22, 1964)

PERMISSIONS
Volume I

Birkhäuser Boston thanks the original publishers of the papers of Hassler Whitney for granting permission to reprint the papers in this collection.

[3] Reprinted from *Annals of Mathematics* **32**, ©1931 by Princeton University Press.

[5] Reprinted from *AMS Transactions* **34**, ©1932 by the American Mathematical Society.

[6] Reprinted from *American Journal of Mathematics* **54**, ©1932 by the Johns Hopkins University Press.

[10] Reprinted from *Annals of Mathematics* **33**, ©1932 by Princeton University Press.

[12] Reprinted from *American Journal of Mathematics* **55**, ©1933 by the Johns Hopkins University Press.

[13] Reprinted from *American Journal of Mathematics* **55**, ©1933 by the Johns Hopkins University Press.

[14] Reprinted from *American Journal of Mathematics* **55**, ©1933 by the Johns Hopkins University Press.

[17] Reprinted from *Fundamenta Mathematicae* **21**, ©1933 by Fundamenta Mathematicae.

[18] Reprinted from *AMS Transactions* **36**, ©1934 by the American Mathematical Society.

[19] Reprinted from *AMS Bulletin* **40**, ©1934 by the American Mathematical Society.

[20] Reprinted from *AMS Transactions* **36**, ©1934 by the American Mathematical Society.

[21] Reprinted from *Annals of Mathematics* **35**, ©1934 by Princeton University Press.

[22] Reprinted from *Annals of Mathematics* **35**, ©1934 by Princeton University Press.

[23] Reprinted from *American Journal of Mathematics* **57**, ©1935 by the Johns Hopkins University Press.